Cooperative Communications and Networking

Presenting the fundamental principles of cooperative communications and networking, this book treats the concepts of space, time, frequency diversity, and MIMO, with a holistic approach to principal topics where significant improvements can be obtained.

Beginning with background and MIMO systems, Part I includes a review of basic principles of wireless communications, space–time diversity and coding, and broadband space–time–frequency diversity and coding. Part II then goes on to present topics on physical layer cooperative communications, such as relay channels and protocols, performance bounds, optimum power control, multi-node cooperation, distributed space–time and space–frequency coding, relay selection, differential cooperative transmission, and energy efficiency. Finally, Part III focuses on cooperative networking including cooperative and content–aware multiple access, distributed routing, source–channel coding, source–channel diversity, coverage expansion, broadband cooperative communications, and network lifetime maximization.

With end-of-chapter review questions included, this text will appeal to graduate students of electrical engineering and is an ideal textbook for advanced courses on wireless communications. It will also be of great interest to practitioners in the wireless communications industry.

Presentation slides for each chapter and instructor-only solutions are available at www.cambridge.org/9780521895132

K. J. Ray Liu is Professor in the Electrical and Computer Engineering Department, and Distinguished Scholar-Teacher, at the University of Maryland, College Park. Dr. Liu has received numerous honours and awards including best paper awards from IEEE Signal Processing Society, IEEE Vehicular Technology Society, and EURASIP, the IEEE Signal Processing Society Distinguished Lecturer, and National Science Foundation Young Investigator Award.

Ahmed K. Sadek is Senior Systems Engineer with Corporate Research and Development, Qualcomm Incorporated. He received his Ph.D. in Electrical Engineering from the University of Maryland, College Park, in 2007. His research interests include communication theory and networking, information theory and signal processing, with current focus on cognitive radios, spectrum sharing, cooperative communications, and interface management.

Weifeng Su is Assistant Professor at the Department of Electrical Engineering, State University of New York (SUNY) at Buffalo. He received his Ph.D. in Applied Mathematics from Nankai University, China in 1999, followed by his Ph.D. in Electrical Engineering from the University of Delaware, Newark in 2002. His research interests span a broad range of areas from signal processing to wireless communications and networking, and he won the Invention of the Year Award from the University of Maryland in 2005.

Andres Kwasinski is with Texas Instruments Inc., Communication Infrastructure Group. After receiving his Ph.D. in Electrical and Computer Engineering from the University of Maryland, College Park in 2004, he became Faculty Research Associate in the University's Department of Electrical and Computer Engineering. His research interests are in the areas of multimedia wireless communications, cross layer designs, digital signal processing, and speech and video processing.

Cooperative Communications and Networking

K. J. RAY LIU
University of Maryland, College Park

AHMED K. SADEK
Qualcomm, San Diego, California

WEIFENG SU
State University of New York (SUNY) at Buffalo

ANDRES KWASINSKI
Texas Instruments, Germantown, Maryland

CAMBRIDGE
UNIVERSITY PRESS

CAMBRIDGE UNIVERSITY PRESS
Cambridge, New York, Melbourne, Madrid, Cape Town, Singapore, São Paulo, Delhi

Cambridge University Press
The Edinburgh Building, Cambridge CB2 8RU, UK

Published in the United States of America by Cambridge University Press, New York

www.cambridge.org
Information on this title: www.cambridge.org/9780521895132

First published 2009

Printed in the United Kingdom at the University Press, Cambridge

A catalog record for this publication is available from the British Library

Library of Congress Cataloging in Publication data
Cooperative communications and networking / K. J. Ray Liu ... [et al.].
p.   cm.
Includes bibliographical references and index.
ISBN 978-0-521-89513-2
1. Wireless communication systems.   2. Internetworking (Telecommunication)   3. Radio relay systems.
4. MIMO systems.   I. Liu, K. J. Ray, 1961–   II. Title.
TK5103.2.C663   2009
621.384–dc22
2008042422

ISBN 978-0-521-89513-2 hardback

Contents

Preface

Wireless communications technologies have seen a remarkably fast evolution in the past two decades. Each new generation of wireless devices has brought notable improvements in terms of communication reliability, data rates, device sizes, battery life, and network connectivity. In addition, the increase homogenization of traffic transports using Internet Protocols is translating into network topologies that are less and less centralized. In recent years, ad-hoc and sensor networks have emerged with many new applications, where a source has to rely on the assistance from other nodes to forward or relay information to a desired destination.

Such a need of cooperation among nodes or users has inspired new thinking and ideas for the design of communications and networking systems by asking whether cooperation can be used to improve system performance. Certainly it means we have to answer what and how performance can be improved by cooperative communications and networking. As a result, a new communication paradigm arose, which had an impact far beyond its original applications to ad-hoc and sensor networks.

First of all, why are cooperative communications in wireless networks possible? Note that the wireless channel is broadcast by nature. Even directional transmission is in fact a kind of broadcast with fewer recipients limited to a certain region. This implies that many nodes or users can "hear" and receive transmissions from a source and can help relay information if needed. The broadcast nature, long considered as a significant waste of energy causing interference to others, is now regarded as a potential resource for possible assistance. For instance, it is well known that the wireless channel is quite bursty, i.e., when a channel is in a severe fading state, it is likely to stay in the state for a while. Therefore, when a source cannot reach its destination due to severe fading, it will not be of much help to keep trying by leveraging repeating-transmission protocols such as ARQ. If a third party that receives the information from the source could help via a channel that is independent from the source–destination link, the chances for a successful transmission would be better, thus improving the overall performance.

Then how to develop cooperative schemes to improve performance? The key lies in the recent advances in MIMO (multiple-input multiple-output) communication technologies. In the soon-to-be-deployed fourth-generation (4G) wireless networks, very high data rates can only be expected for full-rank MIMO users. More specifically, full-rank MIMO users must be equipped multiple transceiver antennas. In practice, most users either do not have multiple antennas installed on small-size devices, or the propagation environment cannot support MIMO requirements. To overcome the limitations of achieving MIMO gains in future wireless networks, one must think of new techniques beyond traditional point-to-point communications.

A wireless network system is traditionally viewed as a set of nodes trying to communicate with each other. However, from another point of view, because of the broadcast nature of wireless channels, one may think of those nodes as a set of antennas distributed in the wireless system. Adopting this point of view, nodes in the network may cooperate together for distributed transmission and processing of information. A cooperating node can act as a relay node for a source node. As such, cooperative communications can generate independent MIMO-like channel links between a source and a destination via the introduction of relay channels.

Indeed, cooperative communications can be thought of as a generalized MIMO concept with different reliabilities in antenna array elements. It is a new paradigm that draws from the ideas of using the broadcast nature of the wireless channels to make communicating nodes help each other, of implementing the communication process in a distribution fashion, and of gaining the same advantages as those found in MIMO systems. Such a new viewpoint has brought various new communication techniques that improve communication capacity, speed, and performance; reduce battery consumption and extend network lifetime; increase the throughput and stability region for multiple access schemes; expand the transmission coverage area; and provide cooperation tradeoff beyond source–channel coding for multimedia communications.

The main goals of this textbook are to introduce the concepts of space, time, frequency diversity, and MIMO techniques that form the foundation of cooperative communications, to present the basic principles of cooperative communications and networking, and to cover a broad range of fundamental topics where significant improvements can be obtained by use of cooperative communications. The book includes three main parts:

- **Part I: Background and MIMO systems**   In this part, the focus is on building the foundation of MIMO concepts that will be used extensively in cooperative communications and networking. Chapter 1 reviews of fundamental material on wireless communications to be used in the rest of the book. Chapter 2 introduces the concept of space–time diversity and the development of space–time coding, including cyclic codes, orthogonal codes, unitary codes, and diagonal codes. The last chapter in this part, Chapter 3, concerns the maximum achievable space–time–frequency diversity available in broadband wireless communications and the design of broadband space–frequency and space–time–frequency codes.
- **Part II: Cooperative communications**   This part considers mostly the physical layer issues of cooperative communications to illustrate the differences and improvements under the cooperative paradigm. Chapter 4 introduces the concepts of relay channels and various relay protocols and schemes. A hierarchical scheme that can achieve linear capacity scaling is also considered to give the fundamental reason

for the adoption of cooperation. Chapter 5 studies the basic issues of cooperation in the physical layer with a single relay, including symbol error rate analysis for decode-and-forward and amply-and-forward protocols, performance upper bounds, and optimum power control. Chapter 6 analyses multi-node scenarios. Chapter 7 presents distributed space–time and space–frequency coding, a concept similar to the conventional space–time and space–frequency coding but different in that it is now in a distributed setting where assumptions and conditions vary significantly. Chapter 8 concerns the issue of minimizing the inherent bandwidth loss of cooperative communications by considering when to cooperate and whom to cooperate with. The main issue is on devising a scheme for relay selection and maximizing the code rate for cooperative communications while maintaining significant performance improvement. Chapter 9 develops differential schemes for cooperative communications to reduce transceiver complexity. Finally, Chapter 10 studies the issues of energy efficiency in cooperative communications by taking into account the practical transmission, processing, and receiving power consumption and illustrates the trade-off between the gains in the transmit power and the losses due to the receive and processing powers when applying cooperation.

- **Part III: Cooperative networking**   This part presents impacts of cooperative communications beyond physical layer, including MAC, networking, and application layers. Chapter 11 considers the effect of cooperation on the capacity and stability region improvement for multiple access. Chapter 12 studies how special properties in speech content can be leveraged to efficiently assign resources for cooperation and further improve the network performance. Chapter 13 discusses cooperative routing with cooperation as an option. Chapter 14 develops the concept of source–channel–cooperation to consider the tradeoff of source coding, channel coding, and diversity for multimedia content. Chapter 15 focuses on studying how source coding diversity and channel coding diversity interact with cooperative diversity, and the system behavior is characterized and compared in terms of the asymptotic performance of the distortion exponent. Chapter 16 presents the coverage area expansion with the help of cooperation. Chapter 17 considers the various effects of cooperation on OFDM broadband wireless communications. Finally, Chapter 18 discusses network lifetime maximization via the leverage of cooperation.

This textbook primarily targets courses in the general field of cooperative communications and networking where readers have a basic background in digital communications and wireless networking. An instructor could select Chapters 1, 2, 4, 5, 6, 7.1, 8, 10, 11, 13, 14, and 16 to form the core of the material, making use of the other chapters depending on the focus of the course.

It can also be used for courses on wireless communications that partially cover the basic concepts of MIMO and/or cooperative communications which can be considered as generalized MIMO scenarios. A possible syllabus may include selective chapters from Parts I and II. If it is a course on wireless networking, then material can be drawn from Chapter 4 and the chapters in Part III.

This book comes with presentation slides for each chapter to aid instructors with the preparation of classes. A solution manual is also available to instructors upon request. Both can be obtained from the publisher via the proper channels.

This book could not have been made possible without the contributions of the following people: Amr El-Sherif, T. Kee Himsoon, Ahmed Ibrahim, Zoltan Safar, Karim Seddik, and W. Pam Siriwongpairat. We also would like to thank them for their technical assistance during the preparation of this book.

Part I

Background and MIMO systems

1 Introduction

Wireless communications have seen a remarkably fast technological evolution. Although separated by only a few years, each new generation of wireless devices has brought significant improvements in terms of link communication speed, device size, battery life, applications, etc. In recent years the technological evolution has reached a point where researchers have begun to develop wireless network architectures that depart from the traditional idea of communicating on an individual point-to-point basis with a central controlling base station. Such is the case with ad-hoc and wireless sensor networks, where the traditional hierarchy of a network has been relaxed to allow any node to help forward information from other nodes, thus establishing communication paths that involve multiple wireless hops. One of the most appealing ideas within these new research paths is the implicit recognition that, contrary to being a point-to-point link, the wireless channel is broadcast by nature. This implies that any wireless transmission from an end-user, rather than being considered as interference, can be received and processed at other nodes for a performance gain. This recognition facilitates the development of new concepts on distributed communications and networking via cooperation.

The technological progress seen with wireless communications follows that of many underlying technologies such as integrated circuits, energy storage, antennas, etc. Digital signal processing is one of these underlying technologies contributing to the progress of wireless communications. Perhaps one of the most important contributions to the progress in recent years has been the advent of MIMO (multiple-input multiple-output) technologies. In a very general way, MIMO technologies improve the received signal quality and increase the data communication speed by using digital signal processing techniques to shape and combine the transmitted signals from multiple wireless paths created by the use of multiple receive and transmit antennas.

Cooperative communications is a new paradigm that draws from the ideas of using the broadcast nature of the wireless channel to make communicating nodes help each other, of implementing the communication process in a distribution fashion and of gaining the same advantages as those found in MIMO systems. The end result is a set of new tools that improve communication capacity, speed, and performance; reduce battery consumption and extend network lifetime; increase the throughput and stability region for multiple access schemes; expand the transmission coverage area; and provide cooperation tradeoff beyond source–channel coding for multimedia communications.

In this chapter we begin with the study of basic communication systems and concepts that are highly related to user cooperation, by reviewing a number of concepts that will be useful throughout this book. The chapter starts with a brief description of the relevant characteristics of wireless channels. It then follows by discussing orthogonal frequency division multiplexing followed by the different concepts of channel capacity. After this, we describe the basic ideas and concepts of MIMO systems. The chapter concludes by describing the new paradigm of user cooperative communications.

1.1   Wireless channels

Communication through a wireless channel is a challenging task because the medium introduces much impairment to the signal. Wireless transmitted signals are affected by effects such as noise, attenuation, distortion and interference. It is then useful to briefly summarize the main impairments that affect the signals.

1.1.1   Additive white Gaussian noise

Some impairments are additive in nature, meaning that they affect the transmitted signal by adding noise. Additive white Gaussian noise (AWGN) and interference of different nature and origin are good examples of additive impairments. The additive white Gaussian channel is perhaps the simplest of all channels to model. The relation between the output $y(t)$ and the input $x(t)$ signal is given by

$$y(t) = x(t)/\sqrt{\Gamma} + n(t), \tag{1.1}$$

where Γ is the loss in power of the transmitted signal $x(t)$ and $n(t)$ is noise. The additive noise $n(t)$ is a random process with each realization modeled as a random variable with a Gaussian distribution. This noise term is generally used to model background noise in the channel as well as noise introduced at the receiver front end. Also, the additive Gaussian term is frequently used to model some types of inter-user interference although, in general, these processes do not strictly follow a Gaussian distribution.

1.1.2   Large-scale propagation effects

The *path loss* is an important effect that contributes to signal impairment by reducing its power. The path loss is the attenuation suffered by a signal as it propagates from the transmitter to the receiver. The path loss is measured as the value in decibels (dB) of the ratio between the transmitted and received signal power. The value of the path loss is highly dependent on many factors related to the entire transmission setup. In general, the path loss is characterized by a function of the form

$$\Gamma_{\text{dB}} = 10\nu \log(d/d_0) + c, \tag{1.2}$$

where Γ_{dB} is the path loss Γ measured in dB, d is the distance between transmitter and receiver, ν is the path exponent, c is a constant, and d_0 is the distance to a power measurement reference point (sometimes embedded within the constant c). In many practical scenarios this expression is not an exact characterization of the path loss, but is still used as a sufficiently good and simple approximation. The path loss exponent v characterizes the rate of decay of the signal power with the distance, taking values in the range of 2 (corresponding to signal propagation in free space) to 6. Typical values for the path loss exponent are 4 for an urban macro cell environment and 3 for urban micro cell. The constant c includes parameter related to the physical setup of the transmission such as signal wavelength, antennas height, etc.

Equation (1.2) shows the relation between the path loss and the distance between the transmit and the receive antenna. In practice, the path losses of two receive antennas situated at the same distance from the transmit antenna are not the same. This is, in part, because the transmitted signal is obstructed by different objects as it travels to the receive antennas. Consequently, this type of impairment has been named *shadow loss* or *shadow fading*. Since the nature and location of the obstructions causing shadow loss cannot be known in advance, the path loss introduced by this effect is a random variable. Denoting by S the value of the shadow loss, this effect can be added to (1.2) by writing

$$\Gamma_{\mathrm{dB}} = 10v \log(d/d_0) + S + c. \tag{1.3}$$

It has been found through experimental measurements that S when measured in dB can be characterized as a zero-mean Gaussian distributed random variable with standard deviation σ (also measured in dB). Because of this, the shadow loss value is a random value that follows a log-normal distribution and its effect is frequently referred as *log-normal fading*.

1.1.3 Small-scale propagation effects

From the explanation of path loss and shadow fading it should be clear that the reason why they are classified as large-scale propagation effects is because their effects are noticeable over relatively long distances. There are other effects that are noticeable at distances in the order of the signal wavelength; thus being classified as small-scale propagation effects. We now review the main concepts associated with these propagation effects.

In wireless communications, a single transmitted signal encounters random reflectors, scatterers, and attenuators during propagation, resulting in multiple copies of the signal arriving at the receiver after each has traveled through different paths. Such a channel where a transmitted signal arrives at the receiver with multiple copies is known as a multipath channel. Several factors influence the behavior of a multipath channel. One is the already mentioned random presence of reflectors, scatterers and attenuators. In addition, the speed of the mobile terminal, the speed of surrounding objects and the transmission bandwidth of the signal are other factors determining the behavior of the channel. Furthermore, due to the presence of motion at the transmitter, receiver, or surrounding objects, the multipath channel changes over time. The multiple copies of the transmitted signal, each having a different amplitude, phase, and delay, are added at the receiver creating either constructive or destructive interference with each other. This results in a received signal whose shape changes over time. Therefore, if we denote the transmitted signal by $x(t)$ and the received signal by $y(t)$, we can write their relation as

$$y(t) = \sum_{i=1}^{L} h_i(t)x(t - \tau_i(t)), \qquad (1.4)$$

where $h_i(t)$ is the attenuation of the i-th path at time t, $\tau_i(t)$ is the corresponding path delay, and L is the number of resolvable paths at the receiver. This relation implicitly assumes that the channel is linear, for which $y(t)$ is equal to the convolution of $x(t)$ and the channel response at time t to an impulse sent at time τ, $h(t, \tau)$. From (1.4), this impulse response can be written as

$$h(t, \tau) = \sum_{i=1}^{L} h_i(t)\delta(t - \tau_i(t)), \qquad (1.5)$$

Furthermore, if it is safe to assume that the channel does not change over time, the received signal can be simplified as

$$y(t) = \sum_{i=1}^{L} h_i x(t - \tau_i),$$

and the channel impulse response as

$$h(t) = \sum_{i=1}^{L} h_i \delta(t - \tau_i). \qquad (1.6)$$

In many situations it is convenient to consider the discrete-time baseband-equivalent model of the channel, for which the input–output relation derived from (1.4) for sample m can be written as

$$y[m] = \sum_{k=l}^{L} h_k[m]x[m - k], \qquad (1.7)$$

where $h_k[m]$ represents the channel coefficients. In this relation it is implicit that there is a sampling operation at the receiver and that all signals are considered as in the baseband equivalent model. The conversion to a discrete-time model combines all the paths with arrival time within one sampling period into a single channel response coefficient $h_l[m]$. Also, note that the model in (1.7) is nothing more than a time-varying FIR digital filter. In fact, it is quite common to call the channel model based on the impulse response as the tapped-delay model. Since the nature of each path, its length, and the presence of reflectors, scatterers, and attenuators are all random, the channel coefficients h_k of a time-invariant channel are random variables (and note that the redundant time index needs not be specified). If, in addition, the channel changes randomly over time, then the channel coefficients $h_k[m]$ are random processes. Such an effect needs to be taken into consideration with functions that depend on the coefficients, since now they become random functions.

1.1.4 Power delay profile

The function determined by the average power associated with each path is called the *power delay profile* of the multipath channel. Figure 1.1 shows the power delay profile for a typical wireless channel slightly modified from the ITU reference channel model called "Vehicular B" [87]. Several parameters are derived from the power delay profile or its spectral response (Fourier transform of the power delay profile), which are used to both characterize and classify different multipath channels:

- The *channel delay spread* is the time difference between the arrival of the first measured path and the last. If the duration of the symbols used for signaling over the channel exceeds the delay spread, then the symbols will suffer from inter-symbol interference. Note that, in principle, there may be several signals arriving through very attenuated paths, which may not be measured due to sensitivity of the receiver. This makes the concept of delay spread tied to the sensitivity of the receiver.
- The *coherence bandwidth* is the range of frequencies over which the amplitude of two spectral components of the channel response are correlated. The coherence bandwidth provides a measurement of the range of frequencies over which the channel shows a flat frequency response, in the sense that all the spectral components have approximately the same amplitude and a linear change of phase. This means that if the transmitted signal bandwidth is less than the channel coherence bandwidth, then all the spectral components of the signal will be affected by the same attenuation and by a linear change of phase. In this case, the channel is said to be a *flat fading channel*. In another way, since the signal sees a channel with flat frequency response, the channel is often called a *narrowband channel*. If on the contrary, the transmitted signal

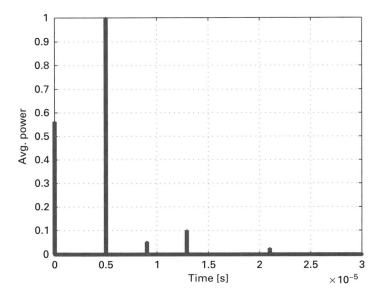

Fig. 1.1   The power delay profile of a typical wireless channel.

bandwidth is more than the channel coherence bandwidth, then the spectral components of the signal will be affected by different attenuations. In this case, the channel is said to be a *frequency selective channel* or a *broadband channel*.

Example 1.1   There are a large number of different channel models that have been used over time for evaluation of communications systems. The large number is due to the different settings found in the plethora of communication systems already in the market or under development. In Tables 1.1 through 1.4 we summarize the parameters of the power delay profile for some of the channels defined in the ITU recommendation M.1225, which is intended for a system operating at a carrier frequency of 2 GHz. In the ITU recommendation, several channel models are discussed so as to account for typically large variability of wireless channels. In this example, Tables 1.1 and 1.2 show the parameters for channels corresponding to a pedestrian setting. As its names indicates, this environment is designed to model pedestrian users, either outside on a street or inside a residence, with small cells, low transmit power and outside base stations with low antenna heights. Tables 1.3 and 1.4 show the parameters for channels corresponding to a vehicular setting. In contrast with the pedestrian environment, the vehicular case models larger cell sizes and transmit power. Also to account for the large variability of wireless channels, two types of channel models are specified for both the pedestrian and vehicular cases. The two types of channels are called "type A" and "type B", where the channel type A is defined as that of a low delay spread case that occurs frequently and channel type B is defined as that of the median delay spread case.

Table 1.1 ITU-R M.1225 Pedestrian A channel parameters.

Tap	Relative delay [ns]	Average power [dB]
1	0	0
2	110	−9.7
3	190	−19.2
4	410	−22.8

Table 1.2 ITU-R M.1225 Pedestrian B channel parameters.

Tap	Relative delay [ns]	Average power [dB]
1	0	0
2	200	−0.9
3	800	−4.9
4	1200	−8.0
4	2300	−7.8
4	3700	−23.9

Table 1.3 ITU-R M.1225 Vehicular A channel parameters.

Tap	Relative delay [ns]	Average power [dB]
1	0	0
2	310	−1.0
3	710	−9.0
4	1090	−10.0
4	1730	−15.0
4	2510	−20.0

Table 1.4 ITU-R M.1225 Vehicular B channel parameters.

Tap	Relative delay [ns]	Average power [dB]
1	0	−2.5
2	300	0
3	8900	−12.8
4	12900	−10.0
4	17100	−25.2
4	20000	−16.0

Figures 1.2 and 1.3 show in the time and frequency domain, respectively, the impulse response in Tables 1.1 through 1.4. The figures illustrate the typical variability of channel models, both in terms of delay spread and coherence bandwidth. Also note how, within the same type A or type B channels, the vehicular channels exhibit a larger delay spread. ▲

Whether a particular channel will appear as flat fading or frequency selective depends, of course, on the channel delay spread, but it also depends on the characteristics of the signal being sent through the channel. Figure 1.4 shows a section of the spectral response of the channel with power delay profile shown in Figure 1.1. We can see that if the transmitted signal has a bandwidth larger than a few tens of kilohertz, then the channel will affect differently those spectral components of the transmitted signal that are sufficiently apart.

This can be seen in Figure 1.5, which shows the time and frequency domain input and output signals to the channel in Figures 1.1 and 1.2. In Figure 1.5, the input signal is a raised cosine pulse with roll off factor 0.25 and symbol period 0.05 µs. For this pulse, the bandwidth is approximately 2 MHz. This makes the channel behave like a frequency selective channel. As can be seen in the frequency domain representation of the output pulse in Figure 1.5, the typical result of the frequency selectivity is that there are large differences in how each spatial component is affected. In the time domain, it can be

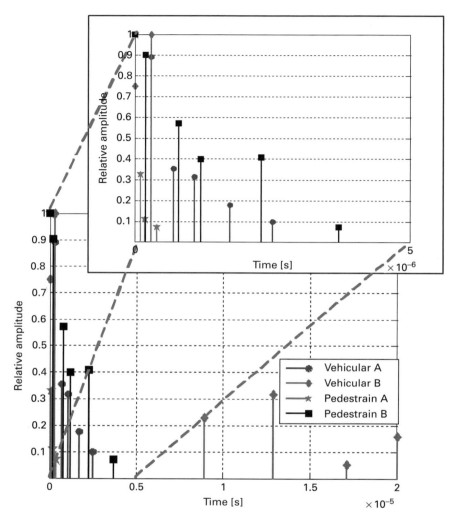

Fig. 1.2   The amplitude of the different paths for the channels in Tables 1.1 through 1.4. The amplitudes
of each path are shown relative to the value of the path with larger gain.

seen that the single pulse at the input of the channel appears repeated at the output with
different delays corresponding to each path.

Such a phenomenon can also be seen in detail in Figure 1.6, which shows the output
pulse and each of the pulses arriving through a different path, with their corresponding
delay. Since the delay associated with some path is larger than the symbol period, the
multipath, frequency selective channel is suffering from intersymbol interference (ISI).
The fact that a time domain phenomenon such as instances of a signal arriving with
different delays, translate into a frequency domain effect, such as frequency selectivity,
can be understood in the following way. When the signals with different delays from
the multipath get superimposed at the receive antenna, the different delay translates

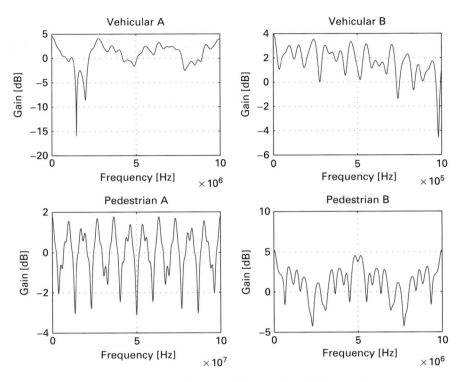

Fig. 1.3 Spectral response of the channels with power delay profile in Tables 1.1 through 1.4.

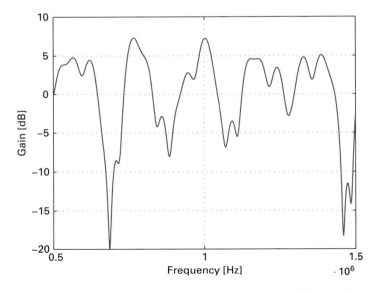

Fig. 1.4 Section of the spectral response of the channel with power delay profile shown in Figure 1.1.

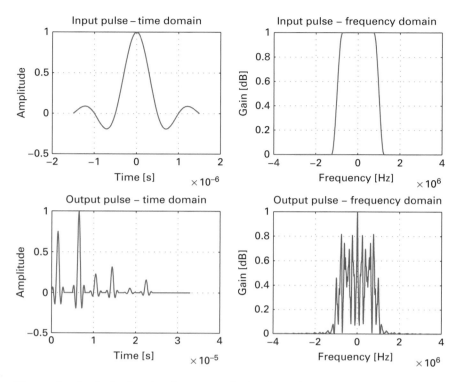

Fig. 1.5   The input and output pulses to a frequency selective channel.

into different phases. Depending on the phase difference between the spectral components, their superposition may result into destructive or constructive interference. Even
more, because the relation between phase and path delay for each spectral component of
the arriving signal varies with the frequency of the spectral component, the signal will
undergo destructive or constructive interference of different magnitude for each spectral
component, resulting in the frequency response of the channel not appearing of constant
amplitude.

Figures 1.7 and 1.8 show the time and frequency domains input and output signals to
the channel in Figures 1.1 and 1.4 when the input pulse have a transmission period long
enough that the channel behaves as non frequency selective. In this case, the input pulse
has a bandwidth of approximately 2 KHz, for which the frequency response of the channel appears roughly flat. Consequently, the transmitted pulse suffers little alterations in
both time and frequency domains. Also, note that now with the longer duration of the
pulse, the delays associated with different channel paths can be practically neglected
and there is no ISI.

In addition to power delay profile and channel delay spread, there are other parameters related to time-varying characteristics of the wireless channel. As we have said,
the motions of the transmitter, the receiver or the reflectors along the signal propagation
path creates a change of the channel transfer characteristics over time. Such motions also
introduce frequency shifts due to the Doppler shift effect. To characterize the channel in

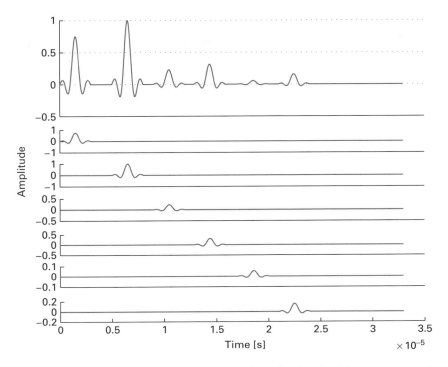

Fig. 1.6 The pulse at the output of a frequency selective channel and each of the component pulses due to multipath.

terms of Doppler shift it is necessary to look at the variation of the channel power profile over time. In other words, instead of considering the statistics of the channel between two frequencies at a fixed time instant, we now look at the same frequency component as it changes over time. The parameters usually considered in these cases are:

- For a time-invariant channel, the coefficient of the channel impulse response corresponding to one path is a random variable (generally a complex-valued Gaussian random variable). When the channel changes over time, the coefficient becomes a random process, with each realization at different time instants being a random variable. These random variables may or may not be correlated. The *channel coherence time* is the time difference that makes the correlation between two realizations of the channel impulse response be approximately zero. The Fourier transform of the correlation function between the realizations of a channel coefficient is known as the channel *Doppler power spectrum*, or simply the Doppler spectrum. The Doppler spectrum characterizes in a statistical sense how the channel response widens an input signal spectrum due to Doppler shift, i.e., if a single tone of frequency f_c is sent through a channel with a Doppler shift f_d, the Doppler spectrum will have components in the range from $f_c - f_d$ to $f_c + f_d$.

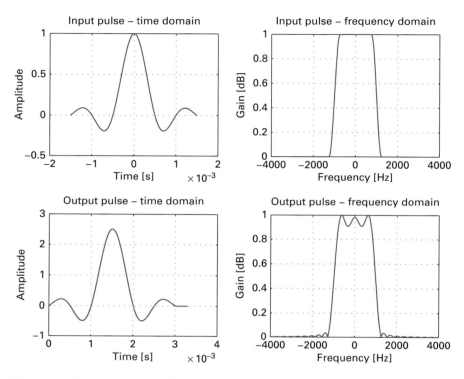

Fig. 1.7   The input and output pulses to a frequency non-selective channel.

- The *Doppler spread* is defined as the range of frequencies over which the Doppler power spectrum is nonzero. The Doppler spread is the inverse of the channel coherence time and, as such, provides information on how fast the channel changes over time. Here, again, the notion on how fast the channel is changing depends also on the input signal. If the channel coherence time is larger than the transmitted signal symbol period; or equivalently, if the Doppler spread is smaller than the signal bandwidth, the channel will be changing over a period of time longer than the input symbol duration. In this case, the channel is said to have *slow fading*. If the converse applies, the channel is said to have *fast fading*.

1.1.5   Uniform scattering environment models

As previously mentioned, the channel coefficients are complex-valued random variables or processes. This raises the important question of what are the statistical properties of the coefficients and what kind of mathematical model can characterize this behavior. One of the most common models for the random channel coefficients is based on an environment known as the "uniform scattering environment." Since this model was introduced by R. H. Clarke and later developed by W. C. Jakes, the model is also known as Clarke's model or Jakes' model. In the model, it is assumed that a waveform arrives at a receiver after being scattered on a very large number of scatterers. These scatterers

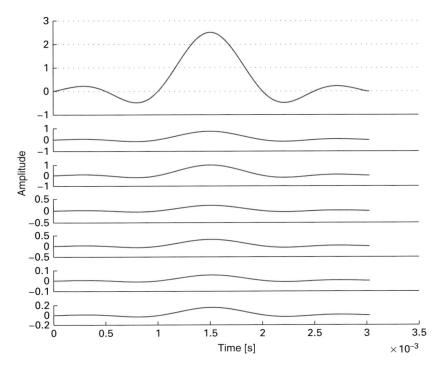

Fig. 1.8 The pulse at the output of a frequency non-selective channel and each of the component pulses due to multipath.

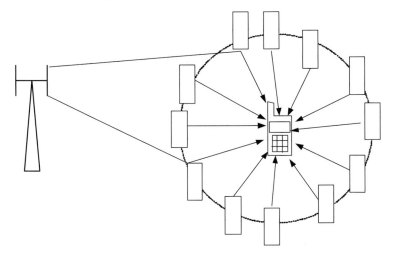

Fig. 1.9 The uniform scattering environment.

are assumed to be randomly located on a circle centered on the receiver (see Figure 1.9). In the environment it is assumed that there is no line-of-sight (LOS) signal with a power notably larger than the rest. Consequently, the received waveform is made of the super-position of many waveforms arriving from the scatterers at an angle that is uniformly distributed between 0 and 2π. For the purpose of this study, let us first introduce the complex baseband representation of the transmitted bandpass signals $s(t)$,

$$s(t) = \Re\left\{x(t)e^{j2\pi f_c t}\right\},$$

where f_c is the carrier frequency. Expanding this expression, we get

$$s(t) = \Re\{x(t)\}\cos(2\pi f_c t) + \Im\{x(t)\}\sin(2\pi f_c t)$$
$$= s_I(t)\cos(2\pi f_c t) + s_Q(t)\sin(2\pi f_c t),$$

where $s_I(t)$ and $s_Q(t)$ are the in-phase and quadrature components of $s(t)$, respectively. When this signal is transmitted through a channel with baseband impulse response $h(t)$, the resulting received signal is

$$y(t) = \Re\left\{\left(x(t) * h(t)\right)e^{j2\pi f_c t}\right\}.$$

If the channel has L paths, with path n having amplitude $h_n(t)$, an associated delay $\tau_n(t)$, and a Doppler phase shift φ_n (which accounts for the Doppler shift due to the motion of the receiver of each received wave), the received signal can be written as

$$y(t) = \Re\left\{\sum_{n=1}^{L} h_n(t)x\left(t - \tau_n(t)\right)e^{j\,[2\pi f_c(t-\tau_n(t))+\varphi_n]}\right\}.$$

If, for the purpose of this characterization, we assume that the transmitted signal is a single tone with the same frequency as the carrier frequency, the received signal becomes

$$y(t) = \Re\left\{\sum_{n=1}^{L}\left[h_n(t)e^{-j(2\pi f_c\tau_n(t)-\varphi_n)}\right]e^{j2\pi f_c t}\right\}$$
$$= y_I(t)\cos(2\pi f_c t) + y_Q(t)\sin(2\pi f_c t), \tag{1.8}$$

where

$$y_I(t) = \sum_{n=1}^{L} h_n(t)\cos(2\pi f_c\tau_n(t) - \varphi_n), \tag{1.9}$$

$$y_Q(t) = \sum_{n=1}^{L} h_n(t)\sin(2\pi f_c\tau_n(t) - \varphi_n). \tag{1.10}$$

This result shows that both the in-phase and the quadrature components of the received signal are actually composed of the superposition of multiple copies of the signal arriving with a change of amplitude and phase as determined by the characteristics of each of the channel paths.

Next, as part of the settings associated with the uniform scattering environment, we assume that all the received signals arrive with the same amplitude. This is a reasonable assumption because, given the geometry of the uniform scattering environment, in the absence of a direct LOS path, each signal arriving at the receiver would experience similar attenuations. Furthermore, in the uniform scattering environment, it is reasonable to assume that the number of paths L is very large. Therefore, resorting to the Central Limit Theorem, it follows that each coefficient can be modeled as a circularly symmetric complex Gaussian random variable with zero mean (i.e., as a random variable made of two quadrature components, with each component being a zero mean Gaussian random variable with the same variance as the other component σ^2). We denote this observation as $y_I \sim \mathcal{N}(0, \sigma^2)$, $y_Q \sim \mathcal{N}(0, \sigma^2)$.

To better understand the channel behavior, it is important to find the statistics (in terms of probability density function (pdf)) of the envelope and phase of the channel coefficients. To get this, it is necessary to consider the transformation of those random variables representing a channel coefficient in Cartesian coordinates into those representing the coefficient in polar coordinates. This means that, if we write the coefficient as $h = h_I + jh_Q$ (h_I and h_Q represent the in-phase and the quadrature phase components, respectively), we want to find the pdf of the random variables r and θ, obtained through the transformations

$$r = \sqrt{h_I^2 + h_Q^2},$$
$$\theta = \arctan(h_Q/h_I), \tag{1.11}$$

which represent a channel coefficient as $h = r\, e^{j\theta}$. Equivalently, we may consider the inverse transform

$$h_I = r \cos \theta,$$
$$h_Q = r \sin \theta. \tag{1.12}$$

For the general case of transforming random variables $V = t_1(X, Y)$ and $W = t_2(X, Y)$, the transformation of the joint pdf $f_{X,Y}$ of the random variables X and Y into the joint pdf $f_{V,W}$ of the random variables V and W, is given by the expression [112]

$$f_{V,W}(v, w) = \frac{f_{X,Y}\big(s_1(v, w), s_2(v, w)\big)}{|\mathcal{J}(x, y)|},$$

$$\mathcal{J}(x, y) = \det \begin{bmatrix} \frac{\partial v}{\partial x} & \frac{\partial v}{\partial y} \\ \frac{\partial w}{\partial x} & \frac{\partial w}{\partial y} \end{bmatrix},$$

where $\mathcal{J}(x, y)$ is the Jacobian of the transformation and where $x = s_1(v, w)$ and $y = s_2(v, w)$ are the inverse transformations of t_1 and t_2, respectively. Applying this relation to the transformation (1.12) results in the Jacobian $\mathcal{J}(r, \theta) = r/\sigma^2$, which leads to the joint pdf

$$f(r, \theta) = \frac{r}{2\pi\sigma^2} e^{-r^2/(2\sigma^2)}, \quad r \geq 0, 0 \leq \theta \leq 2\pi,$$

where we have used for easier readability a slightly modified notation for the pdf. From the joint pdf it is possible to find the marginal pdfs

$$f(r) = \int_0^{2\pi} f(r,\theta)d\theta = \frac{r}{\sigma^2}e^{-r^2/(2\sigma^2)}, \quad r \geq 0,$$

$$f(\theta) = \int_0^{\infty} f(r,\theta)dr = \frac{1}{2\pi}, \quad 0 \leq \theta \leq 2\pi.$$

This result shows that the magnitude of the channel coefficients is a random variable with a Rayleigh distribution and the phase is also a random variable with a uniform distribution in the range $[0, 2\pi]$. Because the magnitude of the channel coefficients follow a Rayleigh distribution, this model is frequently called a *Rayleigh fading* model.

Also, for the case of two nonnegative random variables related by the transformation $Y = X^2$, using similar random variables transformation techniques yields the relation between pdfs

$$f_Y(y) = \frac{f_X(x)}{dy/dx}.$$

With this relation it can be shown that a random variable X that is defined as the squared magnitude of a Rayleigh-distributed channel coefficient ($X = |h|^2$) follows an exponential distribution, with pdf

$$f_X(x) = \frac{1}{\sigma^2}e^{-x/\sigma^2}, \quad x \geq 0. \tag{1.13}$$

In addition, the sum of the squared magnitude of channel coefficient, $\sum_i |h_i|^2$, where each is the sum of two real i.i.d. Gaussian random variables representing the in-phase and quadrature components (i.e., $h_i = h_{I_i} + jh_{Q_i}$), results in a Chi-square random variable with $2L$ (L being the number of channel coefficients in the sum) degrees of freedom. The pdf of this distribution is

$$f(x) = \frac{x^{L-1}}{(L-1)!}e^{-x}, \quad x \geq 0. \tag{1.14}$$

To consider the statistics of the received signal and the channel as they change over time, the two most important results are the time correlation and its Fourier transform, the power spectral density (PSD). Using as a starting point (1.8), (1.9), and (1.10), the time correlation of the received signal is

$$\begin{aligned}C_y(\tau) &= E[y(t)y(t+\tau)] \\ &= C_{y_I}(\tau)\cos(2\pi f_c\tau) + C_{y_I,y_Q}(\tau)\sin(2\pi f_c\tau), \tag{1.15}\end{aligned}$$

where $E[\cdot]$ is the expectation operator. The autocorrelation of y_I, $C_{y_I}(\tau)$ equals

$$C_{y_I}(\tau) = E[y_I(t)y_I(t+\tau)].$$

Considering the expression for $y_I(t)$ in (1.9), the magnitude of the Doppler phase shift, φ_n, depends on the velocity of the receiver relative to the scatterer from where the wave comes from. This relative velocity is equal to $v\cos\alpha_k$, where v is the absolute velocity of the receiver. If we denote the angle associated with the k-th wave as α_k, the Doppler shift for waveform k equals

$$\varphi_k = 2\pi \frac{v}{\lambda} t \cos \alpha_k = 2\pi f_D t \cos \alpha_k,$$

where λ is the wavelength and $f_D = v/\lambda$ is the Doppler frequency. Now we can write (1.9) as

$$y_I(t) = \sum_{n=1}^{L} h_n(t) \cos\left(2\pi f_c \tau_n(t) - 2\pi \frac{v}{\lambda} t \cos \alpha_n\right). \tag{1.16}$$

In the uniform scattering environment the phase $2\pi f_c \tau_n(t)$ changes more rapidly than the Doppler phase shift. In addition, since the distance from the scatterers to the mobile is much larger than the signal wavelength, it is possible to assume that the angle associated with the k-th wave, α_k, is a random variable uniformly distributed in $[0, 2\pi]$ and independent of the angle associated with other paths. Under these conditions, the autocorrelation of y_I, $C_{y_I}(\tau)$ equals

$$C_{y_I}(\tau) = \frac{P_r}{2\pi} \int_0^{2\pi} \cos(\pi v \tau \cos \alpha / \lambda) d\alpha$$
$$= P_r J_0(2\pi f_c v \tau / c),$$

where c is the speed of light, P_r is the received power, and $J_0(\cdot)$ is the Bessel function of the first kind and zeroth order, defined as

$$J_0(x) = \frac{1}{\pi} \int_0^{\pi} e^{-jx \cos \theta} d\theta.$$

Using similar reasoning, the cross-correlation between the in-phase and quadrature components of the received signal, $C_{y_I, y_Q}(\tau)$, can be found to equal zero. Therefore, using (1.15), the time correlation of the received signal is

$$C_y(\tau) = P_r J_0(2\pi f_c v \tau / c) \cos(2\pi f_c \tau). \tag{1.17}$$

In this result, the $\cos(2\pi f_c \tau)$ component indicates the correlation of the received signal to a complete period shift due to being a single tone of frequency f_c. Taking the Fourier transform of (1.17) yields the power spectral density of the received signal

$$S_y(f) = \begin{cases} \dfrac{P_r}{4\pi f_D} \dfrac{1}{\sqrt{1 - \left(\frac{|f - f_c|}{f_D}\right)^2}} & \text{if } |f - f_c| \le f_D \\ 0 & \text{else.} \end{cases} \tag{1.18}$$

Note here that the frequency shift $f - f_c$ is a consequence of the $\cos(2\pi f_c \tau)$ component in (1.17) and the frequency shift property of Fourier transforms. Since the input signal is a single tone, ignoring the frequency shift in the power spectral density of the channel effects, which consequently has a time correlation equal to

$$C_h(\tau) = P_r J_0(2\pi f_c v \tau / c). \tag{1.19}$$

Example 1.2   In this example we show typical cases of the cross-correlation and the power spectral density functions we just have discussed. We assume a system setup where $f_c = 1\,\mathrm{GHz}$, $v = 30\,\mathrm{km/h}$, and $P_r = 1$. Figure 1.10 shows the power spectral density as obtained from (1.18). For the settings in this example, $f_D \approx 27\,\mathrm{Hz}$, so in the figure the rapid increase of PSD at frequencies near $f = f_c \pm f_D$ can be observed. Since the uniform scattering environment is nothing more than a model to represent physical channel behaviors, the PSD in (1.18) could become infinite at frequencies $f = f_c \pm f_D$. This cannot happen in practice, but (1.18) tells us that the PSD will be maximum at these frequencies. Next, Figure 1.11 shows the time correlation as given by (1.19). Note that there are values of τ for which the correlation is 0, which means that the channel will decorrelate signals arriving with these delays.   ▲

1.1.6   Other channel coefficients models

The Rayleigh fading model is not the only model for the channel coefficients. In fact, when we derived the Rayleigh fading for the uniform scattering environment from (1.8)–(1.10), we assumed that all the received signals arrive with the same amplitude due to the absence of a direct LOS path and the symmetric geometry of the environment. When there is one line-of-sight path, then it can no longer be assumed that both the in-phase and quadrature components can be approximated as a zero mean Gaussian random variable. Now, the two components are Gaussian random variables but one has

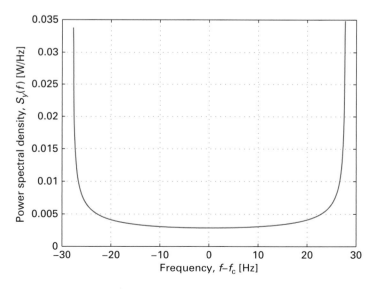

Fig. 1.10   Power spectral density of a received tone of frequency $f_c = 1\,\mathrm{GHz}$ for a mobile in a uniform scattering environment moving at $v = 30\,\mathrm{km/h}$.

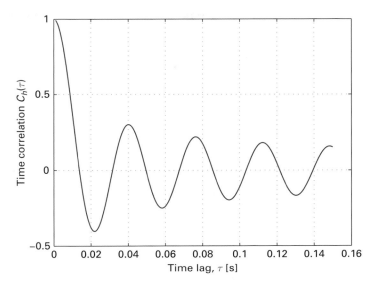

Fig. 1.11 Time correlation for a signal in a uniform scattering environment over a channel with central frequency $f_c = 1\,\text{GHz}$ and for a mobile speed of $v = 30\,\text{km/h}$.

mean A, which is the peak amplitude of the signal from the line-of-sight path, and the other still has zero mean. In this case, we can still use the useful transformations (1.11), for which the cumulative distribution function (CDF) of r is

$$
\begin{aligned}
F_r(z) &= \Pr[r \leq z] \\
&= \frac{1}{2\pi\sigma^2} \iint_{\sqrt{h_I^2 + h_Q^2} \leq z} \exp\left[-\frac{1}{2}\left(\left[\frac{h_I - A}{\sigma}\right]^2 + \left[\frac{h_Q}{\sigma}\right]^2 \right) \right] dx\, dy \\
&= \frac{e^{-\frac{1}{2}\left(\frac{A}{\sigma}\right)^2}}{2\pi\sigma^2} \int_0^z e^{-\frac{1}{2}\left(\frac{r}{\sigma}\right)^2} \left(\int_0^{2\pi} e^{rA} \cos\frac{\theta}{\sigma^2}\, d\theta \right) u(z) r\, dr,
\end{aligned}
\tag{1.20}
$$

where $u(z)$ is the unit step function so that (1.20) is valid only for $z \geq 0$. Next, the integral in θ in (1.20) can be written in terms of the modified Bessel function of the first kind and zeroth order, defined as

$$
I_0(x) = J_0(\mathrm{j}x) = \frac{1}{2\pi} \int_0^{2\pi} e^{x\cos\theta}\, d\theta,
\tag{1.21}
$$

to get

$$
F_r(z) = \frac{e^{-\frac{1}{2}\left(\frac{A}{\sigma}\right)^2}}{\sigma^2} \int_0^z r I_0\left(\frac{rA}{\sigma^2}\right) e^{-\frac{1}{2}\left(\frac{z-A}{\sigma}\right)^2} u(z)\, dr.
\tag{1.22}
$$

From this result, the pdf is obtained by differentiating with respect to z, resulting in

$$
f_r(z) = \frac{z}{\sigma^2} e^{-\left(\frac{z^2}{2\sigma^2} + K\right)} I_0\left(\frac{2Kx}{A}\right), \quad z \geq 0.
\tag{1.23}
$$

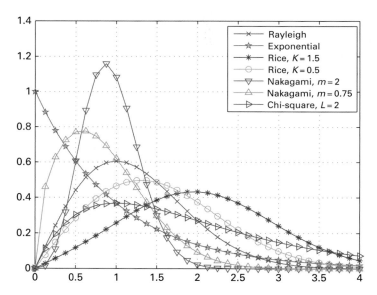

Fig. 1.12   Different probability density functions used to model random fading.

This is the pdf of a Rician distribution. In (1.23), K is a parameter of the Ricean distribution defined as $K = A^2/(2\sigma^2)$. Note that, when $K = 0$, the Rician pdf becomes equal to the Rayleigh pdf, which is consistent with the fact that $K = 0$ means there is no LOS path.

In some cases, it is convenient to model the channel by taking samples of channel realizations and then matching them to a mathematical model. For this it is useful to have a probability density function that can be easily matched to the data samples. This function is provided by the Nakagami fading distribution, which is given by

$$f(x) = \frac{2m^m x^{2m-1}}{\Gamma(m)\sigma^{2m}} e^{-mx^2/\sigma^2}, \quad m \geq 1/2, \tag{1.24}$$

where $\Gamma(\cdot)$ is the Gamma function and m is a parameter used to adjust the pdf of the Nakagami distribution to the data samples. For example, if $m = 1$, then the Nakagami distribution becomes equal to the Rayleigh distribution. One advantage of the Nakagami distribution is that it matches empirical data better than other distributions. In fact, the Nakagami distribution was originally proposed due to this reason.

Figure 1.12 shows the different probability density functions that were discussed in this section.

1.2   Characterizing performance through channel capacity

In this book we will be studying different communication schemes. One important way of characterizing their achievable performance is through the use of information theory concepts, most notably through the use of concepts such as mutual information and the characterization of performance limits through system capacity. At its core, information theory deals with the information provided by the outcome of a random variable. The information provided by the outcome x of a discrete random variable X is defined as

$$I_X(x) = \log \frac{1}{\Pr[X = x]} = -\log \Pr[X = x], \qquad (1.25)$$

where $\Pr[X = x]$ is the probability of the outcome $X = x$ and the logarithm can be, in principle, of any base but is most frequently taken with base 2, followed by base e in some fewer cases. Intuitively, the rarer an event is, the more information it provides. Since the communication process is inherently a process relating more than one random variable (e.g. the input and output of a channel, an uncompressed and a compressed representation of a signal, etc.), it is also important to define a magnitude relating the information shared by two random variables. This magnitude is the *mutual information*, which for two discrete random variables X and Y is defined as

$$I(X; Y) = \sum_{x \in X} \sum_{y \in Y} \Pr[X = x, Y = y] \log \frac{\Pr[X = x, Y = y]}{\Pr[X = x]\Pr[Y = y]},$$

where $\Pr[X = x, Y = y]$ is the joint probability mass function and $\Pr[X = x]$ and $\Pr[Y = y]$ are marginal probability mass functions. Following Bayes theorem ($\Pr[X = x, Y = y] = \Pr[X = x | Y = y]\Pr[Y = y]$, with $\Pr[X = x | Y = y]$ being the conditional probability mass function of X given that $Y = y$), the mutual information can also be written as

$$I(X; Y) = \sum_{x \in X} \sum_{y \in Y} \Pr[X = x, Y = y] \log \frac{\Pr[X = x | Y = y]}{\Pr[X = x]}.$$

Furthermore, we can write

$$I(X; Y) = -\sum_{x \in X} \log \Pr[X = x] \sum_{y \in Y} \Pr[X = x, Y = y]$$

$$+ \sum_{x \in X} \sum_{y \in Y} \Pr[X = x, Y = y] \log \Pr[X = x | Y = y]$$

$$= -\sum_{x \in X} \Pr[X = x] \log \Pr[X = x]$$

$$+ \sum_{x \in X} \sum_{y \in Y} \Pr[X = x, Y = y] \log \Pr[X = x | Y = y]. \qquad (1.26)$$

The first term in this result is called the *entropy* of the random variable X,

$$H(X) = -\sum_{x \in X} \Pr[X = x] \log \Pr[X = x],$$

and the second term can be written in terms of the *conditional entropy* of X,

$$H(X|Y) = -\sum_{x \in X} \Pr[X = x, Y = y] \log \Pr[X = x | Y = y].$$

Considering (1.25), the entropy of the random variable can also be read as the mean value of the information provided by all its outcomes. Likewise, the conditional entropy can be regarded as the mean value of the information provided by all the outcomes of a random variable (X) given than the outcome of a second random variable (Y) is known, or how much uncertainty about a random variable (X) remains after knowing the outcome of a second random variable (Y). Therefore, the mutual information as in (1.26) can now be written as

$$I(X; Y) = H(X) - H(X|Y),$$

and intuitively be interpreted as the mean amount of uncertainty about one random variable (X) that is resolved after learning about the outcome of another random variable (Y), or the average amount of information shared by the two random variables. Finally, we note here that although the concepts we have been introducing were tailored to discrete random variables, the same concepts apply to continuous random variables with the only differences that the sums are replaced by integrals and the probability mass functions by probability density functions.

In information theory, one of the main measures of performance is system capacity. Nevertheless, due to the fact that the calculation of capacity always involves a number of assumptions and simplifications, the measurement of capacity does not come in a "one size fits all" solution. In particular, the notion of capacity is influenced by how much the channel changes over the duration of a coding interval and the properties of the random process associated with the channel fluctuations.

When the random variations of the channel are a stationary and ergodic process it is possible to consider the traditional notion of capacity as introduced by Claude Shannon [181]. In this case, coding is assumed to be done using arbitrary long blocks. Also, the random process driving the channel changes needs to be stationary and ergodic. Because of this, this notion of capacity is known as ergodic capacity or Shannon capacity. The capacity of an AWGN channel with fast flat fading, when only the receiver has knowledge of the channel state,

$$C = \mathrm{E}\left[\log\left(1 + \frac{|h|^2 P}{N_0}\right)\right], \tag{1.27}$$

where $\mathrm{E}[\cdot]$ is the expectation operator (operating on the random channel attenuation), P is the power of the transmitted signal (assumed i.i.d. Gaussian, with zero mean so as to achieve capacity), N_0 is the variance of the background noise, and $|h|^2$ is the envelope of the channel attenuation.

Although the notion of Shannon capacity is quite useful, there are also many design settings where the assumptions of using arbitrary long codes or that the channel is a stationary and ergodic random process do not hold. In these cases, Shannon capacity may not yield useful results. For example, in the case of a non-ergodic slowly fading channel following a Rayleigh distribution, the Shannon capacity is arbitrary small or zero. This is because the result is affected by those realizations of the channel corresponding to deep fades. Nevertheless, an arbitrary small capacity is not the true depiction of many realizations of the fading process (which is confirmed by the many communications taking place every day under these conditions!). Therefore, for these cases it is more appropriate to consider the notion of outage capacity. This notion is tied to the concept of an outage event.

There are many ways of defining an outage event but, from an information theory point of view, an outage event is defined as the set of channel realizations that cannot support reliable transmission at a rate R. In other words, the outage event is the set of channel realizations with an associated capacity less than a transmit rate R. Considering now that the setup that led to (1.27) corresponds to a non-ergodic channel, the outage condition for a realization of the fading can be written as

$$\log \left(1 + \frac{|h|^2 P}{N_0} \right) < R. \tag{1.28}$$

From here, the outage probability is calculated as the one associated with the outage event,

$$P_{out} = Pr\left[\log \left(1 + \frac{|h|^2 P}{N_0} \right) < R \right], \tag{1.29}$$

where $Pr[\cdot]$ is the probability operator (once again on the random channel attenuation).

Once we have introduced the concepts of outage event and outage probability, the Pr_{out} outage capacity, C_{out}, is defied as the information rate that can be reliably communicated with a probability $1 - Pr_{out}$, that is

$$Pr\left[C \leq C_{out} \right] = Pr_{out}, \tag{1.30}$$

where C is the Shannon capacity associated with the channel.

We finally note here that, as mentioned above, the outage probability may be defined differently from (1.29). Another way of defining the outage probability is by considering the event that the received signal to noise ratio is below a threshold. This definition can be related to (1.29) by simple algebraic operations that expose the received signal-to-noise ratio as the random variable, i.e., for the signal-to-noise ratio (SNR) γ and SNR threshold γ_T,

$$P_{out} = Pr\left[\gamma < 2^R - 1 \right] = Pr[\gamma < \gamma_T].$$

1.3   Orthogonal frequency division multiplexing (OFDM)

In Section 1.1.3 we discussed that when the signal bandwidth is much larger than the channel coherence bandwidth, the channel is frequency selective. We also explained that these channels present such impairments as intersymbol interference, which deform the shape of the transmitted pulse, risking the introduction of detection errors at the receiver. This impairment can be addressed with different techniques. One of these techniques is multicarrier modulation. In multicarrier modulation, the high bandwidth signal to be transmitted is divided over multiple mutually orthogonal signals of a bandwidth small enough such that the channel appears to be non-frequency selective. Different multicarrier modulation techniques may differ based on the choice of orthogonal signals. Among the many possible multicarrier modulation techniques, orthogonal frequency division multiplexing (OFDM) is the one that has gained more acceptance as the modulation technique for high-speed wireless networks and 4G mobile broadband standards. In OFDM, the orthogonal signals used for multicarrier modulation are truncated complex exponentials. Assume an OFDM transmitter where the high rate serial input stream is split into N parallel substreams. Assume also that, at some instant of time, the sequence $\{d_k\}_{k=0}^{N-1}$ represents the N complex symbols that are input to the OFDM modulator for transmission as a single OFDM symbol of duration T_s. This translates in practice into an operation where the input stream to the OFDM modulator is divided and organized into blocks of N symbols, which are modulated into a single OFDM symbol. The resulting OFDM modulated symbol is, for $0 \leq t \leq T_s$,

$$s(t) = \sum_{k=0}^{N-1} d_k \phi_k(t) = \sum_{k=0}^{N-1} d_k e^{j2\pi f_k t}, \tag{1.31}$$

where $f_k = f_0 + k\Delta f$ and $\Delta f = 1/T_s$. In (1.31) the signals $\phi_k(t)$, which are defined as

$$\phi_k(t) = \begin{cases} e^{j2\pi f_k t} & \text{if } 0 \leq t \leq T_s \\ 0 & \text{else,} \end{cases} \tag{1.32}$$

form an orthonormal set that are used as the carrier signal of each subcarrier in this multicarrier modulation technique. Because these signals are truncated complex exponential, in frequency domain they are of the form $\sin(x)/x$. These signals are shown in Figure 1.13, which also illustrates how OFDM splits a carrier with large bandwidth into multiple orthogonal subcarriers of much smaller bandwidth.

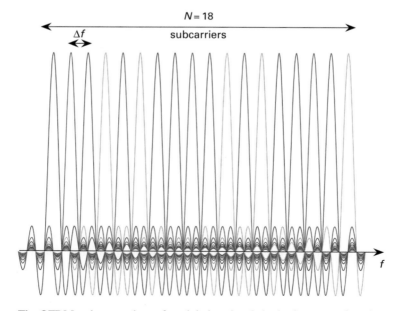

Fig. 1.13   The OFDM orthonormal set of modulation signals in the frequency domain.

Assume, next, that the OFDM symbol in (1.31) is sampled with a period $T_{sa} = T_s/N$. Then, we can write the resulting sampled signal $s[n]$ as

$$s[n] = s(nT_{sa}) = \sum_{k=0}^{N-1} d_k e^{j2\pi f_k nT_s/N}, \quad 0 \le n \le N - 1.$$

If we assume, without loss of generality, that $f_0 = 0$ we get $f_k = k\Delta f = k/T_s$, leading to

$$s[n] = \sum_{k=0}^{N-1} d_k e^{j2\pi nk/N}. \tag{1.33}$$

This result can be read as $s[n]$ being the inverse Fourier transform of d_k, which is a simple way of generating an OFDM symbol and one of its main advantages.

In practice, the OFDM symbol as defined in (1.33) is extended with the addition of a cyclic prefix. To understand the construction of the prefix, assume a multipath channel with L taps defined through the coefficients $h[0], h[1], \ldots, h[L-1]$. With this in mind, the original channel input sequence $s[0], s[1], \ldots, s[N-L], s[N-L+1], \ldots, s[N-1]$, becomes $s[N-L+1], \ldots, s[N-1], s[0], s[1], \ldots s[N-L], s[N-L+1], \ldots, s[N-1]$ after adding the cyclic prefix. Note that the prefix is built by just copying the last $L-1$ elements of the original channel input sequence at the beginning of it. This operation does not affect the OFDM signal or its properties, such as the orthogonality between the multicarrier modulated signals, because it is simply a reaffirmation of the periodicity of the OFDM symbol (period equal to N), as follows from (1.33). Also note that, following the assumption of a multipath channel with delay spread L, the samples corresponding to the cyclic prefix will be affected by intersymbol interference from the previous OFDM symbol. To combat this interference, the prefix can be eliminated at the receiver without any loss of information in the original sequence and without intersymbol interference affecting the original sequence. Next, let us illustrate the effect of adding the cyclic prefix on transforming a frequency selective fading channel into a set of parallel flat fading channels.

Let us call the channel input sequence, after adding the cyclic prefix, as $\mathbf{x}$ where

$$\mathbf{x} = [s[N - L + 1], \ldots, s[N - 1], s[0], s[1], \ldots, s[N - L] ,$$
$$s[N - L + 1], \ldots, s[N - 1]] . \tag{1.34}$$

The output of the channel can be written as

$$y[n] = \sum_{l=0}^{L-1} h[l]x[n - l] + v[n], \quad n = 1, 2, \ldots, N + L - 1, \tag{1.35}$$

where $v[n]$ is additive white Gaussian noise.

The multipath channel affects the first $L-1$ symbols and therefore the receiver ignores these symbol. The received sequence is then given by

$$\mathbf{y} = [y[L], y[L + 1], \ldots, y[N + L - 1]] . \tag{1.36}$$

Equivalently, one can write the received signal in terms of the original channel input as

$$y[n] = \sum_{l=0}^{L-1} h[l]s\left[(n - l - L)\mathrm{mod}N\right] + v[n]. \tag{1.37}$$

This can be also written in terms of cyclic convolution as follows:

$$\mathbf{y} = \mathbf{h} \otimes \mathbf{s} + \mathbf{v}, \tag{1.38}$$

where $\otimes$ denotes cyclic convolution. At the receiver, after taking the discrete Fourier transform (DFT) of the received signal and after removing the cyclic prefix, we get

$$Y_n = H_n S_n + V_n, \tag{1.39}$$

where Y_n, H_n, S_n, and V_n are the n-th point of the N-point DFT of the received signal, channel response, channel input, and noise vector, respectively.

From (1.39), at the receiver side the frequency selective fading channel has been transformed to a set of parallel flat fading channels. Therefore, one can see the benefit of OFDM and how it reduces the complexities associated with time equalization.

Finally, Figure 1.14 summarizes the operations involved in the OFDM communication link by showing a block diagram of a transmitter and a receiver. It is worth highlighting here that an OFDM symbol is made of a block of N input symbols. At the OFDM transmitter the N input symbols are converted into an OFDM symbol with

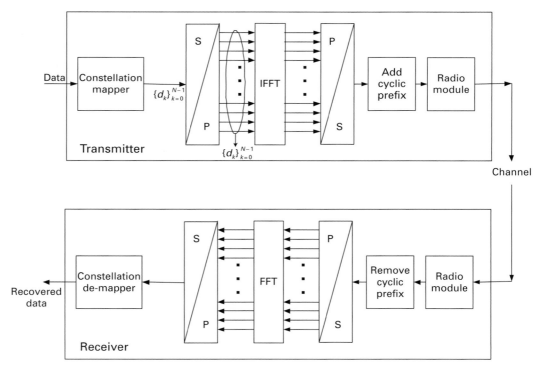

Fig. 1.14   Block diagram of an OFDM transmitter and a receiver.

N subcarriers. If we now consider the successive transmission of several OFDM symbols, the data organization on the channel can be conceptually pictured as a grid in a *frequency × time* plane with a width of N subcarriers (in the frequency dimension) and a depth equal to the number of transmitted OFDM symbols (in the time dimension).

1.4 Diversity in wireless channels

As we have explained, fading wireless channels present the challenge of being changing over time. In communication systems designed around a single signal path between source and destination, a crippling fade on this path is a likely event that needs to be addressed with such techniques as increasing the error correcting capability of the channel coding block, reducing the transmission rate, using more elaborate detectors, etc. Nevertheless, these solutions may still fall short for many practical channel realizations.

Viewing the problem of communication through a fading channel with a different perspective, the overall reliability of the link can be significantly improved by providing more than one signal path between source and destination, each exhibiting a fading process as much independent from the others as possible. In this way, the chance that there is at least one sufficiently strong path is improved. Those techniques that aim at providing multiple, ideally independent, signal paths are collectively known as *diversity techniques*. In its simplest form, akin to repetition coding (where signal redundancy is achieved by simply repeating the signal symbols multiple times), the multiple paths may carry multiple distorted copies of the original message. Nevertheless, better performance may be achieved by applying some kind of coding across the signals sent over the multiple paths and by combining in a constructive way the signals received through the multiple paths.

Also important is the processing performed at the receiver, where the signals arriving through the multiple paths are constructively combined. The goal in combining is to process the multiple received signals so as to obtain a resulting signal of better quality or with better probability of successful reception than each of the received ones. The nature of the processing that is applied to each signal during combining is a function of the particular design goals. If the goal is to linearly combine the signals so that the signal-to-noise ratio (SNR) is maximized at the resulting signal, then the resulting mechanism is called a *maximal ratio combiner* (MRC). Suppose that at the input of the MRC there are L signal samples, $y_0, y_1, \ldots, y_{L-1}$, that are to be combined into a signal sample y_M. Each of the received signals correspond to a unit-energy transmited signal that have been received through the corresponding L different paths characterized as $h_0 e^{j\phi_0}, h_1 e^{j\phi_1}, \ldots, h_{L-1} e^{j\phi_{L-1}}$. Since the MRC is a linear combiner, the input and output are related through the relation

$$y_M = \sum_{k=0}^{L-1} c_k e^{-j\phi_k} y_k, \tag{1.40}$$

where c_k are the coefficients of the MRC combiner and the complex exponential is used for equalizing the phases of each term (cophasing). Assuming that the signals to be combined are equally affected by noise at the receiver with power density N_0, the SNR at the output of the MRC, γ_M is

$$\gamma_M = \frac{\left(\displaystyle\sum_{k=0}^{L-1} c_k h_k\right)^2}{N_0 \displaystyle\sum_{k=0}^{L-1} c_k^2}, \qquad (1.41)$$

because the noise is also processed as part of the received signal samples. The MRC coefficients that maximize (1.41) also maximize its numerator. Then, the maximizing coefficients can be found by using the Cauchy–Schwarz inequality,

$$\left(\sum_{k=0}^{L-1} c_k h_k\right)^2 \leq \left(\sum_{k=0}^{L-1} c_k^2\right)\left(\sum_{k=0}^{L-1} h_k^2\right). \qquad (1.42)$$

The SNR in (1.41) is maximized when (1.42) is an equality. This is achieved by letting

$$c_k = \frac{h_k}{\sqrt{N_0}}.$$

The resulting maximized SNR at the output of the MRC is

$$\gamma_M = \frac{\left(\displaystyle\sum_{k=0}^{L-1} h_k^2\right)}{N_0}. \qquad (1.43)$$

Intuitively, the MRC combines multiple signals by first cophasing them, followed by weighting each sample proportionally to the corresponding path SNR and finally adding them. The resulting signal at the output of the MRC will have an SNR equal to the sum of the SNRs corresponding to each path.

As mentioned earlier, the MRC is not the only known combiner. Other cases are the *selection combiner*, where the output is the input with best SNR, and the *threshold combiner*, which sequentially scans the received signals and outputs the first one with SNR exceeding a threshold.

For any diversity technique, the performance improvement is manifested by the communication error probability decreasing at a much larger rate at a high channel signal-to-noise ratio (SNR) than systems with less or no diversity. When using log–log scales, this rate of decrease in the communication error probability becomes the slope of the line representing the communication error probability at high SNR and is known as the *diversity gain*. Strictly speaking, the diversity gain is defined as [239]

$$m = -\lim_{\gamma \to \infty} \frac{\log P_{SER}}{\log \gamma}, \qquad (1.44)$$

where γ is the SNR and P_{SER} is the probability of symbol error (a function of the SNR). This definition establishes an implicit behavior at high SNR for the probability of symbol error as being a linear function of the SNR when seen in a plot with log–log scales. Then it can be seen that, as previously stated, in these conditions the diversity gain is the slope of the linear relation. It is better to have as large a diversity gain as possible, since it means that the probability of symbol error is reduced at a faster rate.

For MIMO systems, as will be shown in Chapters 2 and 3, it is possible to achieve a diversity gain equal to the product between the number of transmit and receive antennas. Also, it is important to note that, depending on the particular diversity scheme and the system setup, other measures of probability of error can be used. For example, the outage probability is used in some cases, instead of the probability of symbol error.

There are many different forms of diversity in addition to the spatial diversity mentioned above. For example, in time diversity, multiple (possibly coded) copies of a symbol are sent at different time instants, and in frequency diversity, multiple (possibly coded) copies of a symbol are sent through channels of different carrier frequency. Furthermore, multiple diversity techniques can be combined to provide even greater performance improvement. Next, a succinct introduction to time, frequency, and antenna diversity systems is provided. Some of the following chapters will consider techniques that are derived from or that combine these forms of diversity.

1.4.1   Time diversity

It is quite common to find communication scenarios where the channel coherence time equals or exceeds several symbol transmission periods. This implies that two symbols transmitted with a separation in time longer than the coherence time will experience channel realizations that are highly uncorrelated and can be used to obtain diversity. The simplest way to achieve this is to form the two symbols by using a repetition coding scheme. Also, to guarantee that the repeated symbols will be transmitted over uncorrelated channel realization, an appropriate interleaver is applied to the stream of symbols to be transmitted.

At the receiver, the copies of the symbol will have to be combined together. As explained above, if the transmission of each symbol can be represented by an input–output expression of the form

$$y_i = h_i x + n_i, \tag{1.45}$$

where x is the unit-energy transmitted symbol, y_i is the symbol received over path i, n_i is the background noise modeled as a circularly symmetric Gaussian random variable with zero mean and variance N_0, and h_i is the channel realization over path i, assumed to be following a Rayleigh fading, optimal combining that maximizes the received SNR is achieved with a maximal ratio combiner. Recall that the SNR at the output of an MRC equals the sum of the SNRs of the branches at the input of the MRC. From the explanation above on Rayleigh fading, the MRC output SNR will have a Chi-squared distribution as in (1.14).

If, for example, we assume BPSK modulation, and M copies of the transmitted symbol being combined with path channel gains, $h_1, h_2, \ldots, h_M$, the error probability that can be obtained using MRC combiner is

$$\mathcal{Q}\left(\sqrt{\frac{2}{N_0}\sum_{n=1}^{M}|h_n|^2}\right).$$

From here, the average probability of symbol error is [45]

$$P_{SER} = \int_0^\infty \mathcal{Q}(\sqrt{2\gamma})f(\gamma)\mathrm{d}\gamma \tag{1.46}$$

$$= \left(\frac{1-\sqrt{\frac{\bar{\gamma}}{1+\bar{\gamma}}}}{2}\right)^M \sum_{m=0}^{M-1}\binom{M+m-1}{m}\left(\frac{1+\sqrt{\frac{\bar{\gamma}}{1+\bar{\gamma}}}}{2}\right)^m, \tag{1.47}$$

where $\gamma = (1/N_0)\sum_{n=1}^{M}|h_n|^2$ is the SNR at the output of the MRC, $f(\gamma)$ is the probability density function of the Chi-squared distribution, as in (1.14), and $\bar{\gamma}$ is the mean SNR at the output of the MRC.

Example 1.3   Although it is possible to derive useful results from (1.47), in the study of systems with some form of diversity (and, in fact, in many other communication systems also), it is sometimes better to work with upper bounds derived from the Chernoff bound. The Chernoff bound is a useful inequality in applications of probability and random processes to signal processing problems that establishes an upper bound on the tail probability of a random variable X,

$$\Pr[X \geq a] \leq \min_t\{\mathrm{e}^{-at}\mathrm{E}[\mathrm{e}^{tX}]\}, \tag{1.48}$$

where a is a constant. When X is a standard Gaussian random variable $\mathcal{N}(0,1)$ (zero mean, unit variance), the Chernoff bound becomes

$$\Pr[X \geq a] \leq \min_t\{\mathrm{e}^{(-at+t^2/2)}\} = \mathrm{e}^{-a^2/2}, \tag{1.49}$$

where the minimization is done by simply equating the derivative of $\mathrm{e}^{(-at+t^2/2)}$ to zero. When applying the Chernoff bound to the $\mathcal{Q}$ function in (1.46), we obtain the upper bound on average probability [45]

$$P_{SER} \leq \prod_{m=1}^{M} \frac{1}{1+\frac{\bar{\gamma}_i}{2}}. \tag{1.50}$$

At large SNR, this upper bound becomes tighter, leading to the approximation

$$P_{SER} \approx \left(\frac{\bar{\gamma}}{2}\right)^{-M}. \tag{1.51}$$

▲

The result (1.51) is revealing in presenting the physical meaning of diversity gain. If now, we look at this expression using log–log scales we get

$$\log(P_{SER}) \approx -M\log(\bar{\gamma}), \tag{1.52}$$

which is the equation of a linear function. Furthermore, notice that by applying the definition of diversity gain (1.44), it can be seen not only that the diversity gain equals $-M$ (the number of repetitions of the symbol), but also that the diversity gain represents the slope of the approximately linear relation (1.52). Since the diversity gain equals the number of repetitions of the symbol, we may say that the time-diversity system with repetition coding achieves full diversity gain. Nevertheless, the use of repetition coding sacrifices the total bit rate. This drawback can be addressed through other coding schemes as will be seen in later chapters.

1.4.2   Frequency diversity

Analogous to time diversity, in those wideband systems where the available bandwidth exceeds the channel coherence bandwidth, it is possible to realize diversity by using channels that are a partition of the available bandwidth and that are separated by more than the channel coherence bandwidth.

Realizing frequency diversity as a partition of the whole system bandwidth into channels with smaller bandwidth and independent frequency response is perhaps the most intuitively natural approach. This approach is applicable in multicarrier systems, where transmission is implemented by dividing the wideband channel into non-overlapping narrowband subchannels. The symbol used for transmission in each subchannel has a transmission period long enough for the subchannel to appear as a flat fading channel. Different subchannels are used together to achieve frequency diversity by ensuring that each is separated in the frequency domain from the rest of the subchannels in the transmission by more than the coherence bandwidth. In this way, the fading processes among the subchannels will show a small cross-correlation. As will be seen in later chapters, an example of these systems are those using orthogonal frequency division multiplexing (OFDM).

Although not as intuitively natural, frequency diversity can also be achieved through processing based on a time-domain phenomenon. Recall that the frequency response of multipath channels is not of constant amplitude and linear phase because each spectral component of the signal undergoes destructive or constructive interference of different magnitude depending on the delay of each path and the frequency of the spectral component. These multipath channels provide diversity through each of the copies of the signal arriving through each path. Because of this, the overall channel appears as frequency selective (see Figures 1.5 and 1.6). It is then possible to achieve diversity of an order equal to the number of independent paths.

1.4.3   MIMO systems

By using multiple antennas at the transmitter side and/or the receiver side as shown in Figure 1.15, we may exploit diversity in the spatial domain which is called spatial diversity (also called antenna diversity). This configuration of deploying multiple antennas is

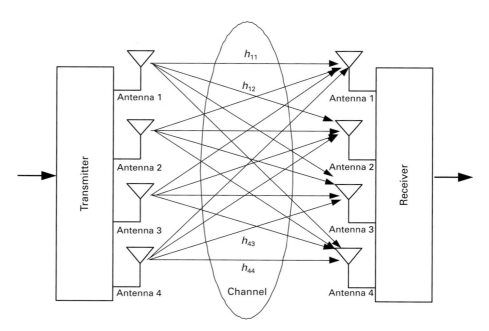

Fig. 1.15   A four-transmit, four-receive MIMO system.

often referred as multiple-input-single-output (MISO) systems if only a single antenna is deployed at the receiver side, single-input-multiple-output (SIMO) systems if a single transmit antenna is used, or, in general, multiple-input-multiple-output (MIMO) systems with multiple transmit antennas and multiple receive antennas. With more than one transmit/receive antenna, different channels are established between each pair of transmit and receive antennas. With such a configuration, the transmitted information can go through different channels to arrive at the receiver side. As long as one of the channels is strong enough, the receiver should be able to recover the transmitted information. If we assume that different channels are independent or correlated with a low correlation, then the chance that all channel links fail is low. The greater the number of antenna pairs, the greater the redundancy (diversity) of the received signals, i.e., the higher the reliability of the transceiver detection. The assumption of low-correlated or independent channel links can be achieved by appropriate separation of the antennas at both the transceiver sides. The necessary antenna separation at each side depends on the scattering in the neighborhood of the antenna and on the signal carrier frequency. For a mobile, the typical separation is between half to one carrier wavelength, for base stations the necessary separation is in the order of tens of wavelengths, which can be easily satisfied.

1.4.3.1   Two motivated examples

Note that, in a MISO system, the signal present at the receive antenna is the combination of signals from all transmit antennas after having traveled through the different fading channels established from each transmit antenna to the receive antenna. The redundancy is also termed as transmit diversity, which depends on the number of transmit antennas. While in a SIMO system, a transmitted signal goes through different channels and is received at each receive antenna. Signals from each receive antenna are combined and jointly detected at the receiver side. The corresponding redundancy is often called receive diversity, which is related to the number of receive antennas. In general, in a MIMO system, both transmit and receive diversities are realized. As we will discuss in Chapter 2, the overall signal redundancy, or diversity order, is in this case the product of the numbers of transmit and receive antennas. The MIMO configuration can be exploited through different designs that differ, among other factors, in the antenna configuration at the receiver and the transmitter, as well as the particular form of performance improvement that it is intended to obtain. For better understanding the spatial diversity, we consider the following two motivated examples.

Example 1.4 Consider a system that implements receive diversity by using one antenna at the transmitter and M_r antennas at the receiver (i.e., a SIMO system). There are M_r paths between transmitter and receiver. The signal arriving from all paths need to be combined at the receiver. If the additive background noise is complex-valued, zero-mean circularly symmetric Gaussian, and independent from each path, then the optimal combiner is the maximal ratio combiner (MRC) (a fact that can be accepted after realizing the similarity of the combining problem here and the one in diversity from repetition coding over time, with the diversity branches now over space instead of over time). Assuming BPSK modulation, and conditioning on the M_r channel gains of the paths, $h_1, h_2, \ldots, h_{M_r}$, the error probability that can be obtained using MRC combiner is $Q\left(\sqrt{2\gamma \sum_{j=1}^{M_r} |h_j|^2}\right)$. This error probability can be written in the following form, which provides more insight into the achievable gains,

$$P_e = Q\left(\sqrt{2(M_r\gamma)\left(\frac{\sum_{j=1}^{M_r} |h_j|^2}{M_r}\right)}\right).$$

In this expression, the factor $M_r\gamma$ shows that the use of the MRC results in a linear increase of SNR with the number of paths. This gain is called the array gain. Also, the factor $(1/M_r)\sum_{j=1}^{M_r} |h_j|^2$ shows an averaging effect on the paths gain where the paths in deeper fade are compensated by those in a good condition, thus resulting in a lower probability that the overall link attenuation will be too large. This effect is the materialization of the diversity gain in this scheme. To see this more clearly, consider that each h_j is the sum of two i.i.d. real Gaussian random variables, and recall from Section 1.1.3 that $\sum_{j=1}^{M_r} |h_j|^2$ follows a Chi-square distribution with $2M_r$ degrees of freedom (see (1.14)). Integrating over this distribution to obtain the average probability of symbol error results in

$$P_{SER} = \left(\frac{1 - \sqrt{\frac{\gamma}{1+\gamma}}}{2}\right)^{M_r} \sum_{m=0}^{M_r-1} \binom{M_r + m - 1}{m} \left(\frac{1 + \sqrt{\frac{\gamma}{1+\gamma}}}{2}\right)^m. \qquad (1.53)$$

Unsurprisingly, this expression is very similar to (1.47). This is due to the analogy between achieving diversity by transmitting at different time instants and by transmitting through different paths. The essence of this similarity resides in the common use of an MRC combiner, which results in a similar expression for the probability of error. Furthermore, we can now draw on this similarly and apply the Chernoff bound to derive expressions similar to (1.50) and (1.51) to get

$$P_{SER} \approx \alpha_{M_r} \left(\frac{\beta_{M_r} \gamma}{2} \right)^{-M_r},$$

which shows the diversity order M_r, since the SNR decays as γ^{-M_r}.   ▲

Example 1.5 (Alamouti scheme) Consider now a system that implements transmit diversity, with two antennas at the transmitter and one antenna at the receiver. Transmission is done by sending two symbols, s_1 and s_2, over the duration of two symbols periods. We assume a flat fading channel, i.e., the channel does not change during the two symbol periods. Instead of sending the symbol s_1 during the first symbol period and the symbol s_2 during the second period, the Alamouti scheme treats the two symbols as a block and applies a form of precoding to the symbols. Specifically, during the first transmission period, s_1 is sent from antenna 1 and s_2 is sent from antenna 2; and during the second transmission period, $-s_2^*$ is sent from antenna 1 and s_1^* from antenna 2 (here s_1^* denotes the complex conjugate of s_1).

At the receiver, we denote by y_1 and y_2 the signals received during the first and second symbol periods, respectively. The two received signals are processed as a vector given by

$$\begin{bmatrix} y_1 \\ y_2 \end{bmatrix} = \begin{bmatrix} s_1 & s_2 \\ -s_2^* & s_1^* \end{bmatrix} \begin{bmatrix} h_1 \\ h_2 \end{bmatrix} + \begin{bmatrix} z_1 \\ z_2 \end{bmatrix}$$

The above transceiver signals can be rearranged as

$$\begin{bmatrix} y_1 \\ y_2^* \end{bmatrix} = \begin{bmatrix} h_1 & h_2 \\ h_2^* & -h_1^* \end{bmatrix} \begin{bmatrix} s_1 \\ s_2 \end{bmatrix} + \begin{bmatrix} z_1 \\ z_2^* \end{bmatrix}.$$

Denote $\mathbf{y} = [y_1 \ y_2^*]^T$ and

$$\mathbf{H} = \begin{bmatrix} h_1 & h_2 \\ h_2^* & -h_1^* \end{bmatrix}.$$

Then the receiver processes the received vector $\mathbf{y}$ by taking $\tilde{\mathbf{y}} = \mathbf{H}^H \mathbf{y}$, resulting in

$$\tilde{\mathbf{y}} = \begin{bmatrix} |h_1|^2 + |h_2|^2 & 0 \\ 0 & |h_1^2| + |h_2^2| \end{bmatrix} \begin{bmatrix} s_1 \\ s_2 \end{bmatrix} + \mathbf{H}^H \begin{bmatrix} z_1 \\ z_2^* \end{bmatrix}.$$

Since the processed noise vector is still complex Gaussian with zero mean, but now with a covariance matrix equal to the diagonal matrix diag($|h_1^2| + |h_2^2|$, $|h_1^2| + |h_2^2|$) times the received noise power, the received SNR for a symbol detected with $\mathbf{z}$ is

$$\gamma = \frac{|h_1^2| + |h_2^2|}{2} \gamma_S,$$

where γ_S is the SNR of a symbol transmitted without the Alamouti scheme and the factor 2 is due to the fact that each s_i is transmitted at half γ_S. We finally note here that this result shows in a fashion similar to Example 1.4 that the Alamouti scheme can achieve a diversity order 2. ▲

1.4.3.2   MIMO capacity

In this subsection, we address advantages of MIMO systems from an information theory aspect by reviewing a fundamental capacity result indicating that the capacity of a MIMO system increases at least linearly with the minimum number of transmit or receive antennas.

We consider a MIMO system with M_t transmit antennas and M_r receive antennas. The channel coefficient between transmit antenna i, $1 \leq i \leq M_t$, and receive antenna j, $1 \leq j \leq M_r$, is denoted by $h_{i,j}$. These coefficients are modeled as independent circularly symmetric complex Gaussian random variables with zero mean and variance one. The MIMO transceiver can be modeled as

$$Y = \sqrt{\frac{\rho}{M_t}} \, XH + Z, \tag{1.54}$$

where $X = [x_1 \ x_2 \ \dots \ x_{M_t}]$ is a signal vector transmitted by the M_t antennas, $Y = [y_1 \ y_2 \ \dots \ y_{M_r}]$ is a signal vector received by the M_r receive antennas, and $Z = [z_1 \ z_2 \ \dots \ z_{M_r}]$ is a noise vector whose elements are modeled as independent circularly symmetric complex Gaussian random variables with zero mean and variance one. The channel coefficient matrix $H = \{h_{i,j} : 1 \leq i \leq M_t, 1 \leq j \leq M_r\}$ is assumed to be known at the receiver side, but unknown at the transmitter side. The signal vector is assumed to satisfy the energy constraint $E||X||_F^2 = M_t$, where $||X||_F$ is the Frobenius norm of X, defined as

$$||X||_F^2 = \sum_{i=1}^{M_t} |x_i|^2.$$

In (1.54), the factor $\sqrt{\rho/M_t}$ ensures that ρ is the average signal to noise ratio (SNR) at each receive antenna, and it is independent of the number of transmit antennas.

If the input signal X is a circularly symmetric complex Gaussian random vector with zero mean and variance $E\{X^H X\} = Q$, then the output signal is also a circularly symmetric complex Gaussian random vector with zero mean and variance

$$E\{Y^H Y\} = \frac{\rho}{M_t} H^H Q H + I_{M_r},$$

in which I_{M_r} is an identity matrix of size M_r by M_r. So, for any given channel H, the mutual information between the input X and the output Y is

$$I(X; Y | H) = \log_2 \det \left(I_{M_r} + \frac{\rho}{M_t} H^H Q H \right). \tag{1.55}$$

Since $E||X||_F^2 = M_t$, trace$(Q) =$ trace $\left(E\{X^H X\}\right) = M_t$. Next, we would like to maximize the mutual information $I(X; Y|H)$ as in (1.55) over the choice of nonnegative Q with the constraint trace$(Q) = M_t$. Since we can write Q as $Q = U^H Q_0 U$, where U is unitary and Q_0 is diagonal, so

$$\log_2 \det \left(I_{M_r} + \frac{\rho}{M_t} H^H Q H \right) = \log_2 \det \left(I_{M_r} + \frac{\rho}{M_t} \tilde{H}^H Q_0 \tilde{H} \right), \qquad (1.56)$$

in which $\tilde{H} = UH$. Note that $\tilde{H}$ has the same distribution as that of H since U is unitary. So we just need to maximize

$$\log_2 \det \left(I_{M_r} + \frac{\rho}{M_t} \tilde{H}^H Q_0 \tilde{H} \right)$$

over the nonnegative diagonal matrix Q_0 with the constraint trace$(Q_0) = M_t$. It has been shown in [215] that

$$\log_2 \det \left(I_{M_r} + \frac{\rho}{M_t} \tilde{H}^H Q_0 \tilde{H} \right)$$

is maximized when Q_0 has equal diagonal elements, i.e., $Q_0 = I_{M_t}$. Intuitively, it is easy to understand that if the transmitter has no prior information about the channel, each transmit antenna should be treated equally and allocated the same weight, i.e., the variance of the input signal vector over M_t transmit antennas should be an identity matrix. Therefore, the maximal mutual information is

$$\log_2 \det \left(I_{M_r} + \frac{\rho}{M_t} \tilde{H}^H \tilde{H} \right).$$

Finally, we assume that the channel is memoryless, i.e., the channel H changes independently from each use of the channel to another. Thus the average capacity of the MIMO system is given by

$$C = E_H \left\{ \log_2 \det \left(I_{M_r} + \frac{\rho}{M_t} H^H H \right) \right\}, \qquad (1.57)$$

in which the expectation is taken over the fading channel H. We can see that when $M_t = M_r = 1$, the above result reduces to the capacity in (1.27) for a conventional single-input-single-output (SISO) system.

In the following, we further interpret the capacity result in (1.57) for SIMO, MISO, and MIMO systems, respectively. For a SIMO system, i.e., $M_t = 1$ and $M_r > 1$, the channel H is a vector as $H = [h_{1,1} \ h_{1,2} \ldots h_{1,M_r}]$. Since

$$\det \left(I_{M_r} + \rho H^H H \right) = \det \left(I_{M_t} + \rho H H^H \right)$$

$$= 1 + \rho \sum_{j=1}^{M_r} |h_{1,j}|^2,$$

in which the first equality follows from the determinant identity $\det(I_m + AB) = \det(I_n + BA)$ for any matrices A and B of sizes $m \times n$ and $n \times m$ respectively. Therefore, the capacity of the SIMO system is

$$C = E_H \left\{ \log_2 \left(1 + \rho \sum_{j=1}^{M_r} |h_{1,j}|^2 \right) \right\}, \tag{1.58}$$

in which $\sum_{j=1}^{M_r} |h_{1,j}|^2$ is a Chi-square random variable with $2M_r$ degrees of freedom, compared with the SISO case where $C = E_h \left\{ \log_2 \left(1 + \rho |h|^2 \right) \right\}$ and $|h|^2$ is a Chi-square random variable with two degrees of freedom.

For a MISO system, i.e., $M_r = 1$ and $M_t > 1$, the channel H is a column vector as $H = [h_{1,1} \ h_{1,2} \ldots h_{M_t,1}]^T$. In this case,

$$\det \left(I_{M_r} + \frac{\rho}{M_t} H^H H \right) = 1 + \frac{\rho}{M_t} \sum_{i=1}^{M_t} |h_{i,1}|^2.$$

So the corresponding capacity can be specified as

$$C = E_H \left\{ \log_2 \left(1 + \rho \frac{\sum_{i=1}^{M_t} |h_{i,1}|^2}{M_t} \right) \right\}. \tag{1.59}$$

We can see that when M_t is large, $\left(\sum_{i=1}^{M_t} |h_{i,1}|^2 \right) / M_t \approx E\{|h|^2\}$, in which $|h|^2$ is a Chi-square random variable with two degrees of freedom. Thus, the capacity in (1.59) is almost the same as that of the SISO system.

For a MIMO system, without loss of generality, we assume $M_t = M_r > 1$. Note that

$$H^H H = \begin{bmatrix} \sum_{i=1}^{M_t} |h_{i,1}|^2 & \sum_{i=1}^{M_t} h_{i,1}^* h_{i,2} & \cdots & \sum_{i=1}^{M_t} h_{i,1}^* h_{i,M_r} \\ h_{i,2}^* h_{i,1} & \sum_{i=1}^{M_t} |h_{i,2}|^2 & \cdots & \sum_{i=1}^{M_t} h_{i,2}^* h_{i,M_r} \\ \vdots & \vdots & \ddots & \vdots \\ h_{i,M_r}^* h_{i,1} & \sum_{i=1}^{M_t} h_{i,M_r}^* h_{i,2} & \cdots & \sum_{i=1}^{M_t} |h_{i,M_r}|^2 \end{bmatrix}.$$

Since $h_{i,j}$ are i.i.d. complex Gaussian random variables with zero mean and variance one, so for large M_t, $\left(\sum_{i=1}^{M_t} |h_{i,j}|^2 \right) / M_t \rightarrow 1$ for each $1 \le j \le M_r$ and $\left(\sum_{i=1}^{M_t} h_{i,j_1}^* h_{i,j_2} \right) / M_t \rightarrow 0$ for any $1 \le j_1 \ne j_2 \le M_r$. Thus, for large M_t, $(1/M_t) H^H H \rightarrow I_{M_r}$. Therefore, for large M_t, the capacity in (1.57) is

$$\begin{aligned} C &\rightarrow \log_2 \det \left(I_{M_r} + \rho I_{M_r} \right) \\ &= M_r \log_2(1 + \rho), \end{aligned} \tag{1.60}$$

which increases linearly with the number of receive antennas. Note that the above discussion is also true for any $M_t \ge M_r$. Based on the above discussion for SIMO, MISO, and MIMO systems, we can see that the capacity of a multiple-antenna system increases at least linearly with the minimum number of transmit or receive antennas.

1.4.3.3 Diversity–multiplexing tradeoff

In MIMO systems, the multiple paths created between any pair of transmit-receive antennas can be used to obtain diversity gain. On the other hand, these paths can also be used to transmit independent messages from each transmit antenna, in which case it is possible to achieve an increase in transmit bit rate given by a *multiplexing gain*.

At the receiver, the MIMO configuration allows for separation of each data stream. It is readily apparent that there should be a tradeoff between diversity and multiplexing gains because the later is achieved at the expense of signal paths that otherwise could be used to increase the former.

The diversity–multiplexing tradeoff is specified through the choice of an achievable combination of diversity and multiplexing gains, or, in other words, by specifying the achievable diversity gain as a function of the multiplexing gain. Let $d^*(r)$ be the diversity–multiplexing tradeoff curve for the slow fading channel. A point on this curve, a diversity gain $d^*(r)$, is achieved at multiplexing gain r if

$$d^*(r) = -\lim_{\gamma \to \infty} \frac{\log P_{\text{out}}(r \log \gamma)}{\log \gamma}. \tag{1.61}$$

This definition means that when communicating at a rate $R = r \log \gamma$, the achievable diversity gain is that for which the outage probability decays as $P_{\text{out}}(r \log \gamma) \approx \gamma^{-d^*(r)}$ at arbitrary large SNR. Also, this formulation can be extended to any type of fading channel, beyond the slow fading channel, by replacing the outage probability with the probability of error, i.e.,

$$d^*(r) = -\lim_{\gamma \to \infty} \frac{\log P_{\text{e}}(r \log \gamma)}{\log \gamma}. \tag{1.62}$$

Example 1.6   Consider a system that transmits a single symbol using QAM modulation at SNR γ over a channel with unit-average power Rayleigh fading and complex-valued, circularly symmetric additive Gaussian background noise with zero mean and unit power. At the receiver, the detection performance is driven by the minimum distance between constellation points $d_{\min}$, which at high SNR is approximately given by [219]

$$d_{\min} \approx \frac{\sqrt{\gamma}}{2^{R/2}}.$$

This leads to an error probability at high SNR approximately equal to

$$P_{\text{e}} \approx \frac{1}{d_{\min}^2} = \frac{2^R}{\gamma},$$

which results in the diversity–multiplexing tradeoff

$$d(r) = 1 - r,$$

for r between 0 and 1.   ▲

1.5   Cooperation diversity

The proliferation of wireless communication applications in the last few years is unprecedented. Voice communication is no longer the only application people need.

High data rate applications, wireless broadband Internet, gaming, and many other applications have emerged recently. Most future wireless systems such as ultra mobile broadband (UMB), Long Term Evolution (LTE), and IEEE 802.16e (WiMAX) promise very high data rates per user over high bandwidth channels (5, 10, and 20 MHz). For example, in the fourth generation wireless networks to be deployed in the next couple of years, namely, mobile broadband wireless access (MBWA) or IEEE 802.20, peak date rates of 260 Mbps can be achieved on the downlink, and 60 Mbps on the uplink [80]. These data rates can, however, only be achieved for full-rank MIMO users. More specifically, full-rank MIMO users must have multiple antennas at the mobile terminal, and these antennas must see independent channel fades to the multiple antennas located at the base station. In practice, not all users can guarantee such high rates because they either do not have multiple antennas installed on their small-size devices, or the propagation environment cannot support MIMO because, for example, there is not enough scattering. In the later case, even if the user has multiple antennas installed, full-rank MIMO is not achieved because the paths between several antenna elements are highly correlated.

To overcome the above limitations of achieving MIMO gains in future wireless networks, we must think of new techniques beyond traditional point-to-point communications. The traditional view of a wireless system is that it is a set of nodes trying to communicate with each other. From another point of view, however, because of the broadcast nature of the wireless channel, we can think of those nodes as a set of antennas distributed in the wireless system. Adopting this point of view, nodes in the network can cooperate together for distributed transmission and processing of information. The cooperating node acts as a relay node for the source node.

Cooperative communications is a new communication paradigm which generates independent paths between the user and the base station by introducing a relay channel. The relay channel can be thought of as an auxiliary channel to the direct channel

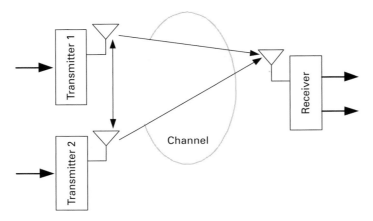

Fig. 1.16   Two transmitters associated in a user-cooperative configuration.

between the source and destination. Since the relay node is usually several wavelengths distant from the source, the relay channel is guaranteed to fade independently from the direct channel, which introduces a full-rank MIMO channel between the source and the destination. In the cooperative communications setup, there are a-priori few constraints to different nodes receiving useful energy that has been emitted by another transmitting node. The new paradigm in user cooperation is that, by implementing the appropriate signal processing algorithms at the nodes, multiple terminals can process the transmissions overheard from other nodes and be made to collaborate by relaying information for each other (Figure 1.16). The relayed information is subsequently combined at a destination node so as to create spatial diversity. This creates a network that can be regarded as a system implementing a distributed multiple antenna where collaborating nodes create diverse signal paths for each other.

Hence, cooperative communications is a new paradigm shift for the fourth generation wireless system that will guarantee high data rates to all users in the network, and we anticipate that it will be the key technology aspect in fifth generation wireless networks.

In terms of research "ascendance," cooperative communications can be seen as related to research in relay channel and MIMO systems. The concept of user cooperation itself was introduced in two-part series of papers [179, 180]. In these works, Sendonaris *et al.* proposed a two user cooperation system, in which pairs of terminals in the wireless network are coupled to help each other forming a distributed two-antenna system. Parts II and III of this book will study in detail the design and analysis of cooperative communications, so we defer further explanation until then.

1.6   Bibliographical notes

Because this is an introductory chapter, we have not covered in detail any of the topics studied here. Nevertheless, there are plenty of excellent textbooks and research papers that complement our presentation. On the topic of wireless channels, the reader can find more information in the books by Proakis [146], Rappaport [150], Tse and Viswanath [219], and Goldsmith [45]. For a practical presentation, the description of channel models used in different standards such as [87] are always a good reference. The books by Tse and Viswanath [219] and by Goldsmith [45] were the main sources for most of the topics covered in this introduction and are where further explanations can be found. The book by Cover and Thomas [26] presents very good in-depth study of many topics in information theory. Also, in-depth study of the concept of outage capacity can be found in [141]. Extra coverage on the topic of diversity can be found in books such as [91], tutorial papers such as [144], or books covering communications over fading channels [189]. Finally, the topic of diversity–multiplexing tradeoff has been studied in [239] and related papers.

2 Space–time diversity and coding

The idea of using multiple transmit and receive antennas in wireless communication systems has attracted considerable attention with the aim of increasing data transmission rate and system capacity. A key issue is how to develop proper transmission techniques to exploit all of the diversities available in the space, time, and frequency domains. In the case of narrow-band wireless communications, the channel fading is frequency non-selective (flat) and diversities are available only in the space and time domains. The modulation and coding approach that is developed for this scenario is termed space–time (ST) coding, exploiting available spatial and temporal diversity.

In this chapter, we first describe the MIMO communication system architecture with frequency-non-selective fading channels, which are often termed as narrow-band wireless channels, and discuss design criteria in achieving the full space–time diversity. Then, we introduce several well-known ST coding techniques that can be guaranteed to achieve full space–time diversity.

2.1 System model and performance criteria

Assume that the MIMO systems have M_t transmit and M_r receive antennas. Channel state information (CSI) is assumed to be known at the receiver, but not at the transmitter. In narrowband transmission scenario, the fading channel is frequency-non-selective or flat, and is assumed to be quasi-static, i.e., the channel stays constant during one codeword transmission and it may change independently from one codeword transmission to another. In this case, diversity is available only in the space and time domains.

An ST-coded MIMO system is shown in Figure 2.1. The ST encoder divides input data stream into b bit long blocks and, for each block, selects one ST codeword from the codeword set of size $L = 2^b$. The selected codeword is then transmitted through the channel over the M_t transmit antennas and T time slots. Each codeword can be represented as a $T \times M_t$ matrix

$$C = \begin{bmatrix} c_1^1 & c_1^2 & \cdots & c_1^{M_t} \\ c_2^1 & c_2^2 & \cdots & c_2^{M_t} \\ \vdots & \vdots & \ddots & \vdots \\ c_T^1 & c_T^2 & \cdots & c_T^{M_t} \end{bmatrix} \triangleq \{c_t^i : i = 1, 2, \ldots, M_t\}, \qquad (2.1)$$

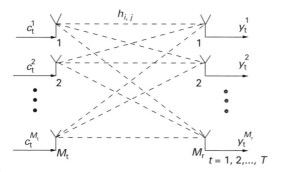

$t = 1, 2, \ldots, T$

Fig. 2.1   MIMO communication system with M_t transmit and M_r receive antennas

where c_t^i denotes the channel symbol transmitted by transmit antenna i, $i = 1, 2, \ldots, M_t$, at discrete time t, $t = 1, 2, \ldots, T$. The codewords are assumed to satisfy the energy constraint $E\|C\|_F^2 = M_t T$, where E stands for the expectation and $\|C\|_F$ is the Frobenius norm of C, defined as

$$\|C\|_F^2 = \mathrm{tr}(C^H C) = \mathrm{tr}(CC^H) = \sum_{t=1}^{T} \sum_{i=1}^{M_t} |c_t^i|^2.$$

The channel coefficient between transmit antenna i and receive antenna j is denoted by $h_{i,j}$. These coefficients are modeled as zero-mean, complex Gaussian random variables with unit variance. The received signal y_t^j at receive antenna j at time t can be expressed as

$$y_t^j = \sqrt{\frac{\rho}{M_t}} \sum_{i=1}^{M_t} c_t^i h_{i,j} + z_t^j, \quad t = 1, 2, \ldots, T, \tag{2.2}$$

where z_t^j is the complex Gaussian noise component at receive antenna j at time t with zero mean and unit variance. The factor $\sqrt{\rho/M_t}$ in (2.2) ensures that ρ is the average signal-to-noise ratio (SNR) at each receive antenna, and it is independent of the number of transmit antennas. The received signal (2.2) can be rewritten in a more compact form as

$$Y = \sqrt{\frac{\rho}{M_t}} CH + Z, \tag{2.3}$$

where $Y = \{y_t^j : 1 \le t \le T, 1 \le j \le M_r\}$ is the received signal matrix of size $T \times M_r$, $H = \{h_{i,j} : 1 \le i \le M_t, 1 \le j \le M_r\}$ is the channel coefficient matrix of size $M_t \times M_r$, $Z = \{z_t^j : 1 \le t \le T, 1 \le j \le M_r\}$ is the noise matrix of size $T \times M_r$, and C is the space–time codeword, as defined in (2.1).

Assume that the perfect channel information is available at the receiver, then the maximum-likelihood (ML) decoding of the transmitted matrix is

$$\hat{C} = \arg \min_{C} \|Y - \sqrt{\frac{\rho}{M_t}} CH\|_F^2.$$

Suppose that codeword C is transmitted and the receiver erroneously in favor of codeword $\tilde{C}$. Since the noise term is a zero-mean Gaussian random variable, the pairwise error probability (PEP) for a fixed channel realization can be determined as (leave the proof as an exercise)

$$P(C \to \tilde{C}|H) = Q\left(\sqrt{\frac{\rho}{2M_t}}||(C-\tilde{C})H||_F\right)$$

$$= \frac{1}{\pi}\int_0^{\pi/2}\exp\left(-\frac{\rho}{4M_t\sin^2\theta}||(C-\tilde{C})H||_F^2\right)d\theta, \qquad (2.4)$$

where $Q(x) = 1/\sqrt{2\pi}\int_x^\infty \exp\left(-t^2/2\right)dt$ is the Gaussian error function, and the second equality comes from the Craig expression

$$Q(x) = \frac{1}{\pi}\int_0^{\pi/2}\exp\left(-\frac{x^2}{2\sin^2\theta}\right)d\theta.$$

Averaging over the Rayleigh fading channel H, the PEP can be determined as follows (leave the proof as an exercise):

$$P(C \to \tilde{C}) = \frac{1}{\pi}\int_0^{\pi/2}\prod_{i=1}^{\gamma}\left(1+\frac{\rho\lambda_i}{4M_t\sin^2\theta}\right)^{-M_r}d\theta, \qquad (2.5)$$

where $\gamma = \mathrm{rank}(C-\tilde{C})$, and $\lambda_1, \lambda_2, \ldots, \lambda_\gamma$ are the non-zero eigenvalues of $(C-\tilde{C})(C-\tilde{C})^H$. The superscript H stands for the complex conjugate and transpose of a matrix. By taking $\theta = \pi/2$ in (2.5), we have the well known upper bound

$$P(C \to \tilde{C}) \leq \frac{1}{2}\prod_{i=1}^{\gamma}\left(1+\frac{\rho\lambda_i}{4M_t}\right)^{-M_r} \qquad (2.6)$$

$$\leq \frac{1}{2}\left(\frac{\rho}{4M_t}\right)^{-\gamma M_r}\left(\prod_{i=1}^{\gamma}\lambda_i\right)^{-M_r}. \qquad (2.7)$$

On the other hand, for high enough SNR, it is easy to see that the exact PEP in (2.5) can be upper bounded as

$$P(C \to \tilde{C}) \leq \frac{1}{\pi}\int_0^{\pi/2}\prod_{i=1}^{\gamma}\left(\frac{\rho\lambda_i}{4M_t\sin^2\theta}\right)^{-M_r}d\theta$$

$$= \left(\frac{2\gamma M_r - 1}{\gamma M_r - 1}\right)\left(\frac{\rho}{M_t}\right)^{-\gamma M_r}\left(\prod_{i=1}^{\gamma}\lambda_i\right)^{-M_r}, \qquad (2.8)$$

where the equality comes from

$$\frac{1}{\pi}\int_0^{\pi/2}(\sin\theta)^{2\gamma M_r}d\theta = 2^{2\gamma M_r}\left(\begin{array}{c}2\gamma M_r - 1 \\ \gamma M_r - 1\end{array}\right).$$

Let us compare the three upper bounds in (2.6–2.8) with the exact PEP (2.5) in Figure 2.2. We can see that, at high SNR, the upper bound (2.8) is much tighter than

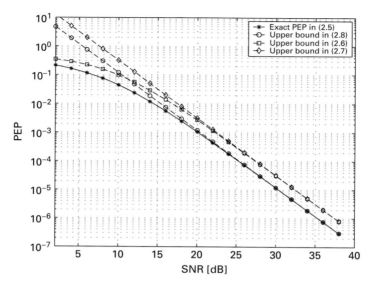

Fig. 2.2   Comparison between the exact PEP and the three upper bounds. Assume that there are $M_t = 2$
transmit and $M_r = 1$ receive antennas, $\gamma = 2$ and $\lambda_1 = \lambda_2 = 1$.

that in (2.7). Note that they share the same term on the product of the nonzero eigenvalues and the order of SNR, and the difference between them is a constant. We observe
that the term $(\rho/M_t)^{-\gamma M_r}$ in the upper bounds is a dominant term when the SNR ρ
is high, thus for a given SNR, the rank γ should be maximized in order to minimize
the PEP error rate. Two ST code design criteria can be developed based on the upper
bound (2.7):

- *Rank criterion* or *diversity criterion*: The minimum rank of the code difference matrix
  $C - \tilde{C}$ overall distinct codewords C and $\tilde{C}$ should be as large as possible. If the matrix
  $C - \tilde{C}$ is always of full rank for a specific ST code, we say that this ST code achieves
  full diversity.
- *Product criterion*: The minimum value of the product $\prod_{i=1}^{\gamma} \lambda_i$ over all distinct codewords C and $\tilde{C}$, which is often termed as coding gain, should be as large as possible.
  This quantity is referred to as the coding advantage achieved by the ST code.

The diversity criterion is the more important of the two since it determines the slope
of the performance curve. In order to achieve the maximum diversity, the difference
matrix $C - \tilde{C}$ has to be full rank for any pair of distinct codewords C and $\tilde{C}$. The
product criterion is of secondary importance and should be optimized if the full diversity
is achieved. If $(C - \tilde{C})(C - \tilde{C})^{\mathrm{H}}$ is of full rank, then the product $\lambda_1 \lambda_2 \ldots \lambda_n$ is equal
to the determinant of $(C - \tilde{C})(C - \tilde{C})^{\mathrm{H}}$. In this case, which implies $M_t \geq T$, a helpful
quantity ζ called *diversity product* is given by

$$\zeta = \frac{1}{2\sqrt{M_t}} \min_{C \neq \tilde{C}} \left| \det\left[(C - \tilde{C})(C - \tilde{C})^{\mathrm{H}} \right] \right|^{1/(2T)}, \qquad (2.9)$$

which is a normalized coding gain. The factor $1/(2\sqrt{M_t})$ guarantees that $0 \leq \zeta \leq 1$. When all codewords are square matrices, i.e., $T = M_t$, the diversity product can be simplified as

$$\zeta = \frac{1}{2\sqrt{M_t}} \min_{C \neq \tilde{C}} \left| \det(C - \tilde{C}) \right|^{1/M_t}, \qquad (2.10)$$

which is often used as a benchmark in designing ST codes.

2.2 Space–time coding

In the previous section, we analyzed the performance of the ST-coded MIMO systems and obtained ST code design criteria in achieving the spatial and temporal diversity. In this section, we introduce several well-known ST coding techniques that can guarantee to achieve the full space–time diversity.

2.2.1 Cyclic and unitary ST codes

A simple and effective coding scheme achieving full space–time diversity is the cyclic ST coding approach which follows the following code structure:

$$C_l = \sqrt{M_t}\, \mathrm{diag}\{e^{ju_1\theta_l},\ e^{ju_2\theta_l},\ \ldots,\ e^{ju_{M_t}\theta_l}\}, \qquad l = 0, 1, \ldots, L-1, \qquad (2.11)$$

where $\mathrm{diag}\{e^{ju_1\theta_l},\ e^{ju_2\theta_l},\ \ldots,\ e^{ju_{M_t}\theta_l}\}$ is a diagonal matrix with diagonal entries $e^{ju_1\theta_l},\ e^{ju_2\theta_l},\ \ldots,$ and $e^{ju_{M_t}\theta_l}$, in which $\theta_l = (l/L)2\pi$ and $u_1, u_2, \ldots, u_{M_t} \in \{0, 1, \ldots, L-1\}$ are some integers to be optimized. We observe that if we denote

$$V_1 = \mathrm{diag}\{e^{ju_1\frac{2\pi}{L}},\ e^{ju_2\frac{2\pi}{L}},\ \ldots,\ e^{ju_{M_t}\frac{2\pi}{L}}\}, \qquad (2.12)$$

then $C_l = \sqrt{M_t}\, V_1^l$ for $l = 0, 1, \ldots, L-1$, and $V_1^L = V_1^0$, which has a cyclic structure, hence the term *cyclic ST codes*.

In the following, we discuss how the cyclic structure can guarantee the full diversity and we will also optimize the parameters $u_1, u_2, \ldots, u_{M_t}$. For any two distinct codewords C_l and $C_{l'}$, $l \neq l'$, we have

$$\begin{aligned}
C_l - C_{l'} &= \sqrt{M_t}\, \mathrm{diag}\{e^{ju_1\theta_l} - e^{ju_1\theta_{l'}},\ e^{ju_2\theta_l} - e^{ju_2\theta_{l'}},\ \ldots,\ e^{ju_{M_t}\theta_l} - e^{ju_{M_t}\theta_{l'}}\} \\
&= \sqrt{M_t}\, \mathrm{diag}\{e^{ju_1\theta_{l'}},\ e^{ju_2\theta_{l'}},\ \ldots,\ e^{ju_{M_t}\theta_{l'}}\} \\
&\quad \times \mathrm{diag}\{e^{ju_1\theta_{l-l'}} - 1,\ e^{ju_2\theta_{l-l'}} - 1,\ \ldots,\ e^{ju_{M_t}\theta_{l-l'}} - 1\} \\
&= C_{l'}\, \mathrm{diag}\{e^{ju_1\theta_{l-l'}} - 1,\ e^{ju_2\theta_{l-l'}} - 1,\ \ldots,\ e^{ju_{M_t}\theta_{l-l'}} - 1\},
\end{aligned}$$

in which $\theta_{l-l'} = \frac{l-l'}{L}2\pi$. So the determinant of the difference matrix is

$$|\det(C_l - C_{l'})| = |\det(C_{l'})| \cdot \left|\det(\mathrm{diag}\{e^{ju_1\theta_{l-l'}} - 1,\right.$$

$$\left. e^{ju_2\theta_{l-l'}} - 1, \ldots, e^{ju_{M_t}\theta_{l-l'}} - 1\})\right|$$

$$= M_t^{\frac{M_t}{2}} \prod_{i=1}^{M_t} \left|e^{ju_i\theta_{l-l'}} - 1\right|$$

$$= M_t^{\frac{M_t}{2}} \prod_{i=1}^{M_t} \left|2\sin\frac{u_i(l-l')\pi}{L}\right|. \tag{2.13}$$

From (2.13), we can see that if $\sin\frac{u_i(l-l')}{L} \neq 0$ for $u_1, u_2, \ldots, u_{M_t}$ and for any $l \neq l'$, then the cyclic code achieves the full diversity according to the rank criterion discussed in the previous section.

From (2.13), we can calculate the diversity product of the cyclic code as follows:

$$\zeta = \frac{1}{2\sqrt{M_t}} \min_{l\neq l'} |\det(C_l - C_{l'})|^{1/M_t}$$

$$= \min_{l\neq l'} \left|\prod_{i=1}^{M_t} \sin\frac{u_i(l-l')\pi}{L}\right|^{1/M_t}$$

$$= \min_{1\leq l\leq L-1} \left|\prod_{i=1}^{M_t} \sin\frac{u_i l\pi}{L}\right|^{1/M_t}. \tag{2.14}$$

Thus, the parameters $u_i \in \{0, 1, \ldots, L-1\}$ should be chosen such that the diversity product ζ is maximized. For small L and M_t, exhaustive computer search can be performed to find the optimum parameters $u_1, u_2, \ldots, u_{M_t} \in \{0, 1, \ldots, L-1\}$.

Example 2.1   For $M_t = 2$ and $L = 4$ (i.e., $R = 1\,\text{bit/s/Hz}$), by exhaustive computer search, the optimum parameters in this case are $[u_1\ u_2] = [1\ 1]$. The corresponding cyclic ST codes are given by:

$$C_0 = \sqrt{2}\begin{bmatrix} 1 & 0 \\ 0 & 1 \end{bmatrix}, \qquad C_1 = \sqrt{2}\begin{bmatrix} j & 0 \\ 0 & j \end{bmatrix},$$

$$C_2 = \sqrt{2}\begin{bmatrix} -1 & 0 \\ 0 & -1 \end{bmatrix}, \qquad C_3 = \sqrt{2}\begin{bmatrix} -j & 0 \\ 0 & -j \end{bmatrix}.$$

The diversity product $\zeta = \min_{1\leq l\leq 3} \left|\prod_{i=1}^{2} \sin\frac{u_i l\pi}{4}\right|^{1/2} = \frac{\sqrt{2}}{2}$.   ▲

Example 2.2   For $M_t = 2$ and $L = 16$ (i.e., $R = 2\,\text{bits/s/Hz}$), by exhaustive computer search, the optimum parameters in this case are $[u_1\ u_2] = [1\ 7]$. The corresponding cyclic ST codes are given by:

$$C_l = \sqrt{2} \begin{bmatrix} e^{j\theta_l} & 0 \\ 0 & e^{j\theta_l} \end{bmatrix}, \quad \theta_l = \frac{l\pi}{8}, \quad l = 0, 1, \ldots, 15.$$

▲

Cyclic ST codes are special unitary codes in which all off-diagonal entries are zero. In general, unitary matrices with nonzero diagonal entries can also be used to achieve the full space–time diversity and provide a larger diversity product. Unitary ST codes with codewords $C_0, C_1, \ldots, C_{L-1}$ satisfy

$$\left(\frac{1}{\sqrt{M_t}} C_l \right)^{H} \left(\frac{1}{\sqrt{M_t}} C_l \right) = I_{M_t \times M_t}, \quad \text{or} \quad C_l^{H} C_l = M_t \, I_{M_t \times M_t}, \tag{2.15}$$

for $l = 0, 1, \ldots, L - 1$. Unitary ST codes can be designed through Fourier transforms or by using unitary matrices with some special structures. For example, the following unitary ST code has a larger diversity product than the cyclic code shown in Example 2.1.

Example 2.3   For $M_t = 2$ and $L = 4$ (i.e., $R = 1\,\mathrm{bit/s/Hz}$), there are four unitary matrices with size 2×2 that can guarantee the full space–time diversity:

$$C_0 = \sqrt{\frac{2}{3}} \begin{bmatrix} j & 1-j \\ -1-j & -j \end{bmatrix}, \qquad C_1 = \sqrt{\frac{2}{3}} \begin{bmatrix} -j & -1-j \\ 1-j & j \end{bmatrix},$$

$$C_2 = \sqrt{\frac{2}{3}} \begin{bmatrix} -j & 1+j \\ -1+j & j \end{bmatrix}, \qquad C_3 = \sqrt{\frac{2}{3}} \begin{bmatrix} j & -1+j \\ 1+j & -j \end{bmatrix}.$$

The diversity product of this unitary ST code is $\zeta = \sqrt{\frac{2}{3}} \approx 0.8165$.   ▲

2.2.2   ST codes from orthogonal designs

ST codes from orthogonal designs have received considerable attention in MIMO wireless communications. Such codes can guarantee to achieve the full space–time diversity and also provide simple fast ML decoding algorithms. Some codes have been adopted in the WLAN standard IEEE 802.11n.

A motivated example: Alamouti scheme
Alamouti proposed in 1998 [5] a simple scheme for MIMO systems with two transmit antennas as follows:

$$G_2(x_1, x_2) = \begin{bmatrix} x_1 & x_2 \\ -x_2^* & x_1^* \end{bmatrix}, \tag{2.16}$$

in which x_1 and x_2 are arbitrary complex symbols and an energy constraint is $E\|G_2\|_{\mathrm{F}}^2 = 4$. The proposed scheme in (2.16) has an interesting property that for arbitrary complex x_1 and x_2, the columns of G_2 are orthogonal to each other, i.e.,

$$G^H G = (|x_1|^2 + |x_2|^2) I_2. \tag{2.17}$$

Due to the special structure, the Alamouti scheme can guarantee to achieve the full space–time diversity. In fact, for any two distinct codewords $G_2(x_1, x_2)$ and $G_2(\tilde{x}_1, \tilde{x}_2)$ with $(x_1, x_2) \neq (\tilde{x}_1, \tilde{x}_2)$, the difference matrix is

$$\Delta G \triangleq G_2(x_1, x_2) - G_2(\tilde{x}_1, \tilde{x}_2) = \begin{bmatrix} x_1 - \tilde{x}_1 & x_2 - \tilde{x}_2 \\ -(x_2 - \tilde{x}_2)^* & (x_1 - \tilde{x}_1)^* \end{bmatrix}.$$

Since

$$(\Delta G)^H (\Delta G) = (|x_1 - \tilde{x}_1|^2 + |x_2 - \tilde{x}_2|^2) I_2,$$

so

$$\det \left[(\Delta G)^H (\Delta G) \right] = (|x_1 - \tilde{x}_1|^2 + |x_2 - \tilde{x}_2|^2)^2,$$

which means

$$\det(\Delta G) = |x_1 - \tilde{x}_1|^2 + |x_2 - \tilde{x}_2|^2. \tag{2.18}$$

We observe that if $(x_1, x_2) \neq (\tilde{x}_1, \tilde{x}_2)$, $\det(\Delta G) \neq 0$, which implies that the Alamouti scheme can achieve full diversity. Based on (2.18), we can easily calculate the diversity product of the Alamouti scheme as follows:

$$
\begin{aligned}
\zeta &= \frac{1}{2\sqrt{2}} \min_{(x_1, x_2) \neq (\tilde{x}_1, \tilde{x}_2)} |\det(\Delta G)|^{1/2} \\
&= \frac{1}{2\sqrt{2}} \min_{(x_1, x_2) \neq (\tilde{x}_1, \tilde{x}_2)} \left(|x_1 - \tilde{x}_1|^2 + |x_2 - \tilde{x}_2|^2 \right) \\
&= \frac{1}{2\sqrt{2}} \min_{x_1 \neq \tilde{x}_1} |x_1 - \tilde{x}_1|.
\end{aligned} \tag{2.19}
$$

Example 2.4 If x_1 and x_2 are chosen from QPSK symbols $\{\pm 1, \pm j\}$, in this case the spectrum efficiency of the Alamouti scheme is 2 bits/s/Hz. The corresponding diversity product is $\zeta = 1/2$.   ▲

In the following, we show that the Alamouti scheme has a simple fast ML decoding algorithm. Assuming that the MIMO fading channel is quasi-static, the MIMO transceiver signal model in (2.3) can be specified as

$$Y = \sqrt{\frac{\rho}{2}} \, G_2(x_1, x_2) H + Z, \tag{2.20}$$

where Y, H, and Z have a size of $2 \times M_{\rm r}$. The ML decoding of x_1 and x_2 at the receiver is

$$
\begin{aligned}
(\hat{x}_1, \hat{x}_2) &= \arg \min_{x_1, x_2} \| Y - \sqrt{\frac{\rho}{2}} \, G_2(x_1, x_2) H \|_{\rm F}^2 \\
&= \arg \min_{x_1, x_2} \left(\| Y \|_{\rm F}^2 - \sqrt{2\rho} \, \mathrm{Re} \left\{ \mathrm{tr} \left(Y^H G_2 H \right) \right\} + \frac{\rho}{2} \, \mathrm{tr} \left(H^H G_2^H G_2 H \right) \right) \\
&= \arg \min_{x_1, x_2} \left(-\sqrt{2\rho} \, \mathrm{Re} \{ \mathrm{tr} \left(G_2 H Y^H \right) \} + \frac{\rho}{2} \left(|x_1|^2 + |x_2|^2 \right) \| H \|_{\rm F}^2 \right).
\end{aligned}
$$

Note that the size of HY^H is 2×2. Denote

$$HY^H \triangleq \begin{bmatrix} a & b \\ c & d \end{bmatrix},$$

then

$$\mathrm{tr}(G_2 HY^H) = (ax_1 + dx_1^*) + (cx_2 + bx_2^*).$$

So, the ML decoding of x_1 and x_2 can be further specified as

$$(\hat{x}_1, \hat{x}_2) = \arg \min_{x_1, x_2} \left(-\sqrt{2\rho} \, \mathrm{Re}\{(ax_1 + dx_1^*) + (cx_2 + bx_2^*)\} \right.$$

$$\left. + \frac{\rho}{2} (|x_1|^2 + |x_2|^2)||H||_F^2 \right)$$

$$= \arg \min_{x_1, x_2} f_1(x_1) + f_2(x_2)$$

$$= \left(\arg \min_{x_1} f_1(x_1), \ \arg \min_{x_2} f_2(x_2) \right), \qquad (2.21)$$

in which

$$f_1(x_1) = -\sqrt{2\rho} \, \mathrm{Re}(ax_1 + dx_1^*) + \frac{\rho}{2} |x_1|^2 ||H||_F^2,$$

$$f_2(x_2) = -\sqrt{2\rho} \, \mathrm{Re}(cx_2 + bx_2^*) + \frac{\rho}{2} |x_2|^2 ||H||_F^2.$$

We can see that x_1 and x_2 can be decoded separately, not jointly. Suppose that $x_1, x_2 \in \mathcal{A}$, then the decoding complexity of the above algorithm is $2|\mathcal{A}|$, while the complexity of the original ML decoding is $|\mathcal{A}|^2$. For example, if $x_1, x_2 \in 16 - \mathrm{QAM}$, then $2|\mathcal{A}| = 32$, while $|\mathcal{A}|^2 = 256$.

2.2.2.1   ST codes from orthogonal designs

The Alamouti scheme was proposed only for MIMO systems with $M_t = 2$ transmit antennas. A natural question is: for MIMO systems with a higher number of transmit antennas ($M_t \geq 3$), do there exist similar structures like G_2 that can guarantee the full diversity and have a fast ML decoding algorithm? The answer is positive and they are related to the orthogonal designs.

An *orthogonal design* in variables $x_1, x_2, \ldots, x_k$ is a $p \times n$ matrix G such that:

(i) the entries of G are complex linear combinations of $x_1, x_2, \ldots, x_k$ and their complex conjugates $x_1^*, x_2^*, \cdots, x_k^*$;
(ii) the columns of G are orthogonal to each other, i.e.,

$$G^H G = (|x_1|^2 + |x_2|^2 + \cdots + |x_k|^2) I_n. \qquad (2.22)$$

Each of such orthogonal designs is ready to be used to form a ST code that can guarantee to achieve the full diversity and have a simple fast ML decoding algorithm, in which n is related to the number of transmit antennas, M_t, and p is related to the time delay of each code, T. The *rate* of an orthogonal design G is defined as $R = k/p$, which means that the resulting ST code with block length p carries k information symbols.

The first ST code from orthogonal design was proposed by Alamouti for systems with $M_t = 2$ transmit antennas, which is specified in (2.16). Clearly, the rate of $G_2(x_1, x_2)$ is 1. For three and four transmit antennas, ST codes from orthogonal designs with rate $R = 3/4$ are given by

$$G_3(x_1, x_2, x_3) = \begin{bmatrix} x_1 & x_2 & x_3 \\ -x_2^* & x_1^* & 0 \\ -x_3^* & 0 & x_1^* \\ 0 & -x_3^* & x_2^* \end{bmatrix}, \tag{2.23}$$

$$G_4(x_1, x_2, x_3) = \begin{bmatrix} x_1 & x_2 & x_3 & 0 \\ -x_2^* & x_1^* & 0 & x_3 \\ -x_3^* & 0 & x_1^* & -x_2 \\ 0 & -x_3^* & x_2^* & x_1 \end{bmatrix}. \tag{2.24}$$

In fact, G_3 is obtained by taking the first three columns of G_4.

For $M_t = 2^k$ ($k = 1, 2, 3, \ldots$), a recursive expression for orthogonal designs can be given as follows. Let

$$G_1(x_1) = x_1 I_1,$$

and

$$G_{2^k}(x_1, \ldots, x_{k+1}) = \begin{bmatrix} G_{2^{k-1}}(x_1, \ldots, x_k) & x_{k+1} I_{2^{k-1}} \\ -x_{k+1}^* I_{2^{k-1}} & G_{2^{k-1}}^{H}(x_1, \ldots, x_k) \end{bmatrix}, \tag{2.25}$$

for $k = 1, 2, 3, \ldots$. Then, $G_{2^k}(x_1, x_2, \ldots, x_{k+1})$ is an orthogonal design of size $2^k \times 2^k$ with complex variables $x_1, x_2, \ldots, x_{k+1}$. The rate of G_{2^k} is $(k + 1)/2^k$, which is the maximum rate for orthogonal designs of square size. If the number of transmit antennas is not a power of two, a ST code can be obtained by deleting some columns from a larger ST code with a number of transmit antennas that is a power of two.

Example 2.5   According to the recursive method, ST codes from orthogonal designs are illustrated for $M_t = 2, 4, 8$ as follows:

$$G_2(x_1, x_2) = \begin{bmatrix} G_1(x_1) & x_2 \\ -x_2^* & G_1^{H}(x_1) \end{bmatrix} = \left[\begin{array}{c|c} x_1 & x_2 \\ \hline -x_2^* & x_1^* \end{array} \right];$$

$$G_4(x_1, x_2, x_3) = \begin{bmatrix} G_2(x_1, x_2) & x_3 I_2 \\ -x_3^* I_2 & G_2^{H}(x_1, x_2) \end{bmatrix} = \left[\begin{array}{cc|cc} x_1 & x_2 & x_3 & 0 \\ -x_2^* & x_1^* & 0 & x_3 \\ \hline -x_3^* & 0 & x_1^* & -x_2 \\ 0 & -x_3^* & x_2^* & x_1 \end{array} \right];$$

$$G_8(x_1, x_2, x_3, x_4) = \begin{bmatrix} G_4(x_1, x_2, x_3) & x_4 I_4 \\ -x_4^* I_4 & G_4^H(x_1, x_2, x_3) \end{bmatrix}$$

$$= \left[\begin{array}{cccc|cccc} x_1 & x_2 & x_3 & 0 & x_4 & 0 & 0 & 0 \\ -x_2^* & x_1^* & 0 & x_3 & 0 & x_4 & 0 & 0 \\ -x_3^* & 0 & x_1^* & -x_2 & 0 & 0 & x_4 & 0 \\ 0 & -x_3^* & x_2^* & x_1 & 0 & 0 & 0 & x_4 \\ \hline -x_4^* & 0 & 0 & 0 & x_1^* & -x_2 & -x_3 & 0 \\ 0 & -x_4^* & 0 & 0 & x_2^* & x_1 & 0 & -x_3 \\ 0 & 0 & -x_4^* & 0 & x_3^* & 0 & x_1 & x_2 \\ 0 & 0 & 0 & -x_4^* & 0 & x_3^* & -x_2^* & x_1^* \end{array}\right].$$

▲

For ST codes from orthogonal designs with non-square size ($p \neq n$), a systematic design method can provide higher rates which are $(n_0 + 1)/(2n_0)$ if the number of transmit antennas is $n = 2n_0$ or $n = 2n_0 - 1$. For example, for $n = 4$ transmitter antennas, an orthogonal ST code with non-square size is given by

$$G_6 = \begin{bmatrix} s_1 & s_2 & s_3 & 0 \\ -s_2^* & s_1^* & 0 & s_4^* \\ -s_3^* & 0 & s_1^* & s_5^* \\ 0 & -s_3^* & s_2^* & s_6^* \\ 0 & -s_4 & -s_5 & s_1 \\ s_4 & 0 & -s_6 & s_2 \\ s_5 & s_6 & 0 & s_3 \\ -s_6^* & s_5^* & -s_4^* & 0 \end{bmatrix}.$$

2.2.2.2 ST Codes from quasi-orthogonal designs

ST codes from orthogonal designs have advantages of achieving full diversity and have fast ML decoding algorithms. However, the maximum rate of an orthogonal design is only 3/4 for three and four transmit antennas, and it is difficult to construct orthogonal designs with rates higher than 1/2 for more than four transmit antennas. To improve the symbol transmission rate, one natural way is to relax the requirement of the orthogonality, i.e., to consider ST codes from quasi-orthogonal designs. With the quasi-orthogonal structure, the ML decoding at the receiver can be done by searching pairs of symbols, similar to the codes from orthogonal designs where the ML decoding can be done by searching single symbols.

For four transmit antennas, a quasi-orthogonal ST code with a symbol transmission rate of one can be constructed from the Alamouti scheme as follows:

$$C = \begin{bmatrix} A & B \\ -B & A \end{bmatrix} = \left[\begin{array}{cc|cc} x_1 & x_2 & x_3 & x_4 \\ -x_2^* & x_1^* & -x_4^* & x_3^* \\ \hline -x_3^* & -x_4^* & x_1^* & x_2^* \\ x_4 & -x_3 & -x_2 & x_1 \end{array}\right], \tag{2.26}$$

where

$$A = \begin{bmatrix} x_1 & x_2 \\ -x_2^* & x_1^* \end{bmatrix}, \quad B = \begin{bmatrix} x_3 & x_4 \\ -x_4^* & x_3^* \end{bmatrix}, \quad (2.27)$$

and $\overline{A}$ and $\overline{B}$ are the complex conjugates of A and B, respectively. One can check that

$$C^H C = \begin{bmatrix} a & 0 & 0 & b \\ 0 & a & -b & 0 \\ 0 & -b & a & 0 \\ b & 0 & 0 & a \end{bmatrix}, \quad (2.28)$$

where $a = |x_1|^2 + |x_2|^2 + |x_3|^2 + |x_4|^2$, and $b = x_1 x_4^* + x_4 x_1^* - x_2 x_3^* - x_3 x_2^*$. We can see that the ML decision metric of this code can be written as the sum of two terms $f_1(s_1, s_4) + f_2(s_2, s_3)$, where f_1 depends only on s_1 and s_4, and f_2 depends only on s_2 and s_3. Thus, the minimization can be done separately on these two terms, i.e., symbol pairs (s_1, s_4) and (s_2, s_3) can be decoded separately, which leads to a fast ML decoding. However, according to (2.28), the minimum rank of the difference matrix between two distinct codewords is 2, which means that the code (2.26) does not have the full diversity.

A similar quasi-orthogonal STBC for four transmit antennas has the following structure:

$$C = \begin{bmatrix} A & B \\ B & A \end{bmatrix} = \begin{bmatrix} x_1 & x_2 & x_3 & x_4 \\ -x_2^* & x_1^* & -x_4^* & x_3^* \\ x_3 & x_4 & x_1 & x_2 \\ -x_4^* & x_3^* & -x_2^* & x_1^* \end{bmatrix}, \quad (2.29)$$

where A and B are the same as those in (2.27). Similarly,

$$C^H C = \begin{bmatrix} a & 0 & b & 0 \\ 0 & a & 0 & b \\ b & 0 & a & 0 \\ 0 & b & 0 & a \end{bmatrix},$$

where $a = |x_1|^2 + |x_2|^2 + |x_3|^2 + |x_4|^2$, and $b = x_1 x_3^* + x_3 x_1^* - x_2 x_4^* - x_4 x_2^*$. The behaviors of (2.29) are similar to those of (2.26).

The performance of the quasi-orthogonal ST codes in (2.26) and (2.29) is better than that of the codes from orthogonal designs at low SNR due to the higher rate, but worse at high SNR since it does not guarantee full diversity. In fact, in both (2.26) and (2.29), the symbols are chosen from the same signal constellation arbitrarily and the resulting ST codes cannot guarantee the full diversity. A method of optimum signal rotation can be developed for the quasi-orthogonal ST codes to achieve full diversity.

The main idea of the optimum signal rotation is to choose the signal constellations properly to ensure that the resulting codes achieve the full diversity. In the following, we focus on the code structure in (2.29). The discussion with the code structure in (2.26) is similar. Assume that $G_p(x_1, x_2, \ldots, x_k)$ is a $p \times p$ orthogonal design in complex variables $x_1, x_2, \ldots, x_k$. A quasi-orthogonal design $Q_{2p}(x_1, x_2, \ldots, x_{2k})$ of size $2p \times 2p$ in complex variables $x_1, x_2, \ldots, x_{2k}$ is defined as

$$Q_{2p} = \begin{bmatrix} A & B \\ B & A \end{bmatrix},$$ (2.30)

where $A = G_p(x_1, x_2, \ldots, x_k)$ and $B = G_p(x_{k+1}, x_{k+2}, \ldots, x_{2k})$. Since both A and B are orthogonal designs, we have

$$Q_{2p}^{\mathrm{H}} Q_{2p} = \begin{bmatrix} A^{\mathrm{H}}A + B^{\mathrm{H}}B & A^{\mathrm{H}}B + B^{\mathrm{H}}A \\ B^{\mathrm{H}}A + A^{\mathrm{H}}B & B^{\mathrm{H}}B + A^{\mathrm{H}}A \end{bmatrix} = \begin{bmatrix} aI_p & bI_p \\ bI_p & aI_p \end{bmatrix},$$ (2.31)

where $a = \sum_{i=1}^{2k} |x_i|^2$, and $b = \sum_{i=1}^{k} (x_i x_{k+i}^* + x_{k+i} x_i^*)$.

For each i, $1 \le i \le k$, let $\mathcal{A}_i$ denote a signal constellation with average energy 1, and $\mathcal{A}_{k+i}$ denote the signal constellation generated by rotating $\mathcal{A}_i$ with an angle of ϕ_i, i.e., $\mathcal{A}_{k+i} = \{e^{j\phi_i} s : s \in \mathcal{A}_i\}$, denoted as $e^{j\phi_i} \mathcal{A}_i$. Furthermore, let $|\mathcal{A}_i|$ denote the number of elements in $\mathcal{A}_i$. For any information sequence of $\log_2(\prod_{i=1}^{2k} |\mathcal{A}_i|)$ bits, it chooses $2k$ symbols $s_1 \in \mathcal{A}_1$, $s_2 \in \mathcal{A}_2$, $\ldots$, $s_{2k} \in \mathcal{A}_{2k}$. In $Q_{2p}(x_1, x_2, \ldots, x_{2k})$, if we replace $x_1, x_2, \ldots, x_{2k}$ by $s_1, s_2, \ldots, s_{2k}$, respectively, we have a new matrix $Q_{2p}(s_1, s_2, \ldots, s_{2k})$. Consequently, an ST code from a quasi-orthogonal design can be formed as

$$C = \sqrt{\gamma} Q_{2p}(s_1, s_2, \ldots, s_{2k}).$$ (2.32)

The factor $\sqrt{\gamma}$ ensures that the transmitted signal in (2.32) obeys the energy constraint. According to (2.31), the ML decoding of (2.32) can be done separately on each pair of symbols s_i and s_{k+i}.

For any pair of distinct transmitted signals $C = \sqrt{\gamma} Q_{2p}(s_1, s_2, \ldots, s_{2k})$ and $\tilde{C} = \sqrt{\gamma} Q_{2p}(\tilde{s}_1, \tilde{s}_2, \ldots, \tilde{s}_{2k})$, the difference matrix is $\sqrt{\gamma} Q_{2p}(s_1 - \tilde{s}_1, s_2 - \tilde{s}_2, \ldots, s_{2k} - \tilde{s}_{2k})$, denoted as ΔC for simplicity. Then, from (2.31), we have

$$(\Delta C)^{\mathrm{H}}(\Delta C) = \gamma \begin{bmatrix} (\Delta a)I_p & (\Delta b)I_p \\ (\Delta b)I_p & (\Delta a)I_p \end{bmatrix},$$ (2.33)

where $\Delta a = \sum_{i=1}^{2k} |s_i - \tilde{s}_i|^2$, and

$$\Delta b = \sum_{i=1}^{k} \left[(s_i - \tilde{s}_i)(s_{k+i} - \tilde{s}_{k+i})^* + (s_{k+i} - \tilde{s}_{k+i})(s_i - \tilde{s}_i)^* \right].$$

The determinant of (2.33) can be calculated as

$$\det \left[(\Delta C)^{\mathrm{H}}(\Delta C) \right] = \gamma^{2p} \left[(\Delta a)^2 - (\Delta b)^2 \right]^p.$$ (2.34)

Note that for both codes in (2.26) and (2.29), $\phi_i = 0$ for all i, therefore the determinant in (2.34) can be zero, for example when $s_i - \tilde{s}_i = s_{k+i} - \tilde{s}_{k+i}$, which means that the space–time signals do not have the full diversity. Now we can properly choose the rotation angle ϕ_i to ensure that the determinant in (2.34) is nonzero.

Example 2.6 If all of $\mathcal{A}_i$, $1 \le i \le k$, are BPSK, i.e., $\mathcal{A}_i = \{1, -1\}$, then we choose the rotation angle as $\pi/2$, i.e., $\phi_i = \pi/2$. The resulting signal constellation $\mathcal{A}_{k+i} = e^{j\pi/2} \mathcal{A}_i = \{j, -j\}$. Therefore, for any two symbols s_i and $\tilde{s}_i$ in $\mathcal{A}_i$, the difference $s_i - \tilde{s}_i$

belongs to set $\{0, 2, -2\}$; and for any two symbols s_{k+i} and $\tilde{s}_{k+i}$ in $\mathcal{A}_{k+i}$, the difference $s_{k+i} - \tilde{s}_{k+i}$ belongs to the set $\{0, 2j, -2j\}$. It is easy to check that Δb in (2.33) is zero. Thus, the determinant of $(\Delta C)^H (\Delta C)$ is nonzero, which means that the space–time signals achieve the full diversity.   ▲

According to (2.34), the diversity product of the quasi-orthogonal ST codes with signal rotation can be calculated as

$$
\begin{aligned}
\zeta &= \frac{1}{2\sqrt{2p}} \min_{\Delta C \neq 0} \left| \det \left[(\Delta C)^H (\Delta C) \right] \right|^{1/(4p)} \\
&= \frac{1}{2} \sqrt{\frac{\gamma}{2p}} \min_{1 \leq i \leq k} \min_{\substack{u, \tilde{u} \in \mathcal{A}_i;\ v, \tilde{v} \in \mathcal{A}_{k+i} \\ (u, v) \neq (\tilde{u}, \tilde{v})}} \left| (u - \tilde{u})^2 - (v - \tilde{v})^2 \right|^{1/2}.
\end{aligned}
$$

$$(2.35)$$

For convenience, let us define the *minimum ζ-distance* between two signal constellations $\mathcal{A}$ and $\mathcal{B}$ as follows:

$$
d_{\min, \zeta}(\mathcal{A}, \mathcal{B}) \triangleq \min_{(s_1, s_2) \neq (\tilde{s}_1, \tilde{s}_2)} \left| (s_1 - \tilde{s}_1)^2 - (s_2 - \tilde{s}_2)^2 \right|^{1/2}, \tag{2.36}
$$

where $s_1, \tilde{s}_1$ and $s_2, \tilde{s}_2$ are understood as $s_1, \tilde{s}_1 \in \mathcal{A}$, and $s_2, \tilde{s}_2 \in \mathcal{B}$. Obviously, we have

$$
d_{\min, \zeta}(\mathcal{A}, \mathcal{B}) \leq \min \{ d_{\min}(\mathcal{A}), d_{\min}(\mathcal{B}) \}, \tag{2.37}
$$

where $d_{\min}(\mathcal{A})$ and $d_{\min}(\mathcal{B})$ are the minimum Euclidean distances of the signal constellations $\mathcal{A}$ and $\mathcal{B}$, respectively. Thus, we have

$$
d_{\min, \zeta}(\mathcal{A}, e^{j\phi}\mathcal{A}) \leq d_{\min}(\mathcal{A}). \tag{2.38}
$$

We now go back to (2.35). The diversity product can be rewritten as

$$
\begin{aligned}
\zeta &= \frac{1}{2} \sqrt{\frac{\gamma}{2p}} \min_{1 \leq i \leq k} d_{\min, \zeta}(\mathcal{A}_i, \mathcal{A}_{k+i}) \\
&\leq \frac{1}{2} \sqrt{\frac{\gamma}{2p}} \min_{1 \leq i \leq k} d_{\min}(\mathcal{A}_i).
\end{aligned} \tag{2.39}
$$

We observe that the diversity product is determined by the minimum ζ-distance of each pair of signal constellations $\mathcal{A}_i$ and $\mathcal{A}_{k+i}$, and the minimum ζ-distance is upper bounded by the minimum Euclidean distance of each signal constellation.

In the example we discussed before, where the signal constellations $\mathcal{A}_i$ are BPSK and the rotation angle is chosen as $\pi/2$, the minimum ζ-distance between $\mathcal{A}_i$ and $e^{j\pi/2}\mathcal{A}_i$ is equal to the minimum Euclidean distance of $\mathcal{A}_i$. However, for a general signal constellation, $d_{\min}(\mathcal{A})$ is not always achieved by $d_{\min, \zeta}(\mathcal{A}, e^{j\phi}\mathcal{A})$. We will show later that if $\mathcal{A}$ is 8-PSK, then the minimum ζ-distance $d_{\min, \zeta}(\mathcal{A}, e^{j\phi}\mathcal{A})$ cannot be greater than $(2 \sin \pi/8)^{1/2} d_{\min}(\mathcal{A})$, which is strictly less than $d_{\min}(\mathcal{A})$.

In the following, we determine an optimum rotation angle ϕ for a given signal constellation $\mathcal{A}$ such that the minimum ζ-distance between $\mathcal{A}$ and $e^{j\phi}\mathcal{A}$ is maximized. First, let us derive a necessary condition for the minimum ζ-distance to reach the minimum

Euclidean distance. Specifically, for a fixed signal constellation $\mathcal{A}$, assume s_b and s_e are two signals in $\mathcal{A}$ such that the Euclidean distance between s_b and s_e is the minimum Euclidean distance of $\mathcal{A}$, i.e., $|s_b - s_e| = d_{\min}$. From (2.36), we have

$$d_{\min,\zeta}(\mathcal{A}, e^{j\phi}\mathcal{A}) \le \left| (s_b - s_e)^2 - (e^{j\phi}s_b - e^{j\phi}s_e)^2 \right|^{1/2}$$
$$= |1 - e^{j2\phi}|^{1/2} \cdot |s_b - s_e|$$
$$= |2\sin\phi|^{1/2} d_{\min}.$$

Thus, we can conclude that the minimum ζ-distance between $\mathcal{A}$ and $e^{j\phi}\mathcal{A}$ cannot be greater than $|2\sin\phi|^{1/2}d_{\min}$, i.e.,

$$d_{\min,\zeta}(\mathcal{A}, e^{j\phi}\mathcal{A}) \le |2\sin\phi|^{1/2}d_{\min}. \tag{2.40}$$

A necessary condition for the minimum ζ-distance to reach the minimum Euclidean distance is $|2\sin\phi|^{1/2} \ge 1$, i.e., $|\sin\phi| \ge 1/2$ or $\pi/6 \le |\phi| \le 5\pi/6$. If the signal constellation $\mathcal{A}$ is r-PSK, then the effective rotation angle is in the interval $[-\pi/r, \pi/r]$ from the symmetry of signals in r-PSK. According to (2.40), we have $d_{\min,\zeta}(\mathcal{A}, e^{j\phi}\mathcal{A}) \le |2\sin(\pi/r)|^{1/2}d_{\min}$. Thus, for r-PSK with $r > 6$, the minimum ζ-distance is strictly less than $d_{\min}$. Note that for different signal constellations, the corresponding optimum rotation angles may be different.

Example 2.7 Assume that $\mathcal{A}$ is a signal constellation drawn from a square lattice (grid), where the side length of the squares is equal to $d_{\min}(\mathcal{A})$. Then, the minimum ζ-distance between $\mathcal{A}$ and $e^{j\pi/4}\mathcal{A}$ is $d_{\min}(\mathcal{A})$, i.e.,

$$d_{\min,\zeta}(\mathcal{A}, e^{j\pi/4}\mathcal{A}) = d_{\min}(\mathcal{A}).$$

The signal constellation $\mathcal{A}$ can be of any arbitrary subset of the square lattice. The only requirement is that the side length of the squares in the lattice is equal to $d_{\min}(\mathcal{A})$. Thus, for the commonly used QAM[1] constellations, the optimum rotation angle is $\phi = \pi/4$.

▲

Example 2.8 Assume that $\mathcal{A}$ is a signal constellation drawn from an equilaterally triangular lattice, where the side length of the equilateral triangles is equal to $d_{\min}(\mathcal{A})$. Then, the minimum ζ-distance between $\mathcal{A}$ and $e^{j\pi/6}\mathcal{A}$ is $d_{\min}(\mathcal{A})$, i.e.,

$$d_{\min,\zeta}(\mathcal{A}, e^{j\pi/6}\mathcal{A}) = d_{\min}(\mathcal{A}).$$

The signal constellation $\mathcal{A}$ could be of any subset of the lattice of equilateral triangles with the requirement that the side length of the equilateral triangles is equal to $d_{\min}(\mathcal{A})$. For such a signal constellation $\mathcal{A}$, $\phi = \pi/6$ is an optimum rotation angle.

▲

---

[1] Strictly, QAM stands for a constellation drawn from a square lattice, where the side length of the squares in the lattice is equal to the minimum Euclidean distance of the constellation.

2.2.3   Diagonal algebraic ST codes

Another approach to design full-diversity ST codes is to apply transforms (such as Hadamard transforms and Vandermonde transforms) over input data symbols and feed the output to different transmit antennas at different times [28, 122]. Specifically, for K information symbols, $s_1 \, s_2 \, \cdots \, s_K$ (from arbitrary signal constellations such as PSK, QAM, and so on), let

$$[x_1 \, x_2 \, \cdots \, x_K] = [s_1 \, s_2 \, \cdots \, s_K] \cdot V(\theta_1, \theta_2, \ldots, \theta_K), \qquad (2.41)$$

where $V(\theta_1, \theta_2, \ldots, \theta_K)$ is a Vandermonde matrix with variables $\theta_1, \theta_2, \ldots, \theta_K$, which is a $K \times K$ matrix:

$$V(\theta_1, \theta_2, \ldots, \theta_K) \triangleq \begin{bmatrix} 1 & 1 & \cdots & 1 \\ \theta_1 & \theta_2 & \cdots & \theta_K \\ \vdots & \vdots & \ddots & \vdots \\ \theta_1^{K-1} & \theta_2^{K-1} & \cdots & \theta_K^{K-1} \end{bmatrix}. \qquad (2.42)$$

The variables $\theta_1, \theta_2, \ldots, \theta_K$ should be optimized for different signal constellations. With $[x_1 \, x_2 \, \cdots \, x_K]$, a diagonal ST codeword can be formed as follows:

$$C = \mathrm{diag}\{x_1, \, x_2, \, \ldots, \, x_K\}, \qquad (2.43)$$

which can be sent out by K transmit antennas.

The ST code in (2.43) satisfies the energy constraint of $E||C||_F^2 = K^2$, which can be verified as follows:

$$E||C||_F^2 = \sum_{i=1}^{K} E|x_i|^2 = \sum_{i=1}^{K} E \left| [s_1 \, s_2 \, \cdots \, s_K] \cdot \begin{bmatrix} 1 \\ \theta_i \\ \vdots \\ \theta_i^{K-1} \end{bmatrix} \right|^2$$

$$= \sum_{i=1}^{K} E \left| \sum_{k=1}^{K} s_k \theta_i^{k-1} \right|^2$$

$$= \sum_{i=1}^{K} \sum_{k=1}^{K} E|s_k|^2 = K^2,$$

in which we use the assumptions that s_k are independent of each other, $E s_k = 0$, $E|s_k|^2 = 1$, and $|\theta_i| = 1$.

The full diversity of the diagonal code in (2.43) can be guaranteed by the Vandermonde matrix with algebraic numbers $\theta_1, \theta_2, \ldots, \theta_K$. For any two distinct codewords $C = \mathrm{diag}\{x_1, \, x_2, \, \ldots, \, x_K\}$ (with input information symbols $s_1 \, s_2 \, \cdots \, s_K$) and $\tilde{C} = \mathrm{diag}\{\tilde{x}_1, \, \tilde{x}_2, \, \ldots, \, \tilde{x}_K\}$ (with input information symbols $\tilde{s}_1 \, \tilde{s}_2 \, \cdots \, \tilde{s}_K$), the determinant of the difference matrix can be calculated as follows:

$$\det(C - \tilde{C}) = \prod_{i=1}^{K} |x_i - \tilde{x}_i|$$

$$= \prod_{i=1}^{K} \left| \sum_{k=1}^{K} (s_k - \tilde{s}_k)\theta_i^{k-1} \right|.$$

For a given signal constellation, the minimum value of the determinant can be maximized by properly selecting the variables $\theta_1, \theta_2, \ldots, \theta_K$. We summarize in the following some of the best-known results from [44, 14]:

(i) If $K = 2^s$ ($s \geq 1$), the optimum transform $\mathcal{M}_K$ for a signal constellation Ω from $\mathbf{Z}[\mathbf{j}] \triangleq \{a + b\mathbf{j} : \text{both } a \text{ and } b \text{ are integers}, \mathbf{j} = \sqrt{-1}\}$ is given by

$$\mathcal{M}_K = \frac{1}{\sqrt{K}} V(\theta_1, \theta_2, \ldots, \theta_K), \tag{2.44}$$

where $\theta_1, \theta_2, \ldots, \theta_K$ are the roots of the polynomial $\theta^K - \mathbf{j}$ over field $\mathbf{Q}[\mathbf{j}] \triangleq \{c + d\mathbf{j} : \text{both } c \text{ and } d \text{ are rational numbers}\}$, and they can be determined as

$$\theta_k = e^{\mathbf{j}\frac{4k-3}{2K}\pi}, \quad k = 1, 2, \ldots, K. \tag{2.45}$$

(ii) If $K = 3 \cdot 2^s$ ($s \geq 0$), the optimum transform $\mathcal{M}_K$ for a signal constellation Ω from $\mathbf{Z}[\omega] \triangleq \{a + b\omega : \text{both } a \text{ and } b \text{ are integers}, \omega = (-1 + \mathbf{j}\sqrt{3})/2\}$ is given by

$$\mathcal{M}_K = \frac{1}{\sqrt{K}} V(\theta_1, \theta_2, \ldots, \theta_K), \tag{2.46}$$

where $\theta_1, \theta_2, \ldots, \theta_K$ are the roots of the polynomial $\theta^K + \omega$ over field $\mathbf{Q}[\omega] \triangleq \{c + d\omega : \text{both } c \text{ and } d \text{ are rational numbers}\}$, and they can be specified as

$$\theta_k = e^{\mathbf{j}\frac{6k-1}{3K}\pi}, \quad k = 1, 2, \ldots, K. \tag{2.47}$$

The signal constellations Ω from $\mathbf{Z}[\mathbf{j}]$ such as QAM and PAM constellations are of practical interest.

Example 2.9 When $K = 2$, for two 4-QAM symbols s_1 and s_2, let $[x_1 \ x_2] = [s_1 \ s_2] \cdot V(\theta_1, \theta_2)$, where the optimum variables $\theta_1 = e^{\mathbf{j}\frac{\pi}{4}}$ and $\theta_2 = e^{\mathbf{j}\frac{5\pi}{4}}$. Thus, the corresponding 2×2 diagonal algebraic ST code is formed as:

$$C = \begin{bmatrix} s_1 + s_2 e^{\mathbf{j}\frac{\pi}{4}} & 0 \\ 0 & s_1 - s_2 e^{\mathbf{j}\frac{\pi}{4}} \end{bmatrix}.$$

▲

2.3   Chapter summary and bibliographical notes

In this chapter, we reviewed ST code design criteria for MIMO communications systems for narrowband wireless communications. We derived ST code pair-wise error probability and developed several error bounds. ST code design criteria were then developed based on a tight error probability bound. We reviewed several well-known ST block codes that can achieve full spatial and temporal diversity, including ST cyclic codes, ST unitary codes, ST codes from orthogonal and quasi-orthogonal designs, and diagonal algebraic codes.

We provide some bibliographical notes as follows. Both [214] and [40] discovered in the late 1990s that MIMO communication systems deploying multiple transmit and receive antennas have a huge capacity gain compared to that of conventional single antenna communication systems. Since then, a large number of works have been carried out to exploit MIMO capacity and performance diversity. The fundamental performance criteria of ST codes were derived in [52, 212]. Two ST code design criteria based on the upper bound (2.7) were developed in [52, 212]. The quantity of diversity product was given in [70, 184]. To achieve the maximum diversity (or the maximum degrees of freedom available in the multiple antenna systems), the difference matrix between any two distinct codewords should be of full rank. Some popular ST codes achieving full space–time diversity include cyclic ST codes in [70, 77], orthogonal ST codes in [5, 213], quasi-orthogonal ST codes in [90, 216, 202], and diagonal algebraic ST codes in [28]. In addition, diversity analysis of space–time modulation for time-correlated Rayleigh fading channels can be found in [201].

The cyclic ST coding approach was developed specifically in [70, 77] in 2000, independently. Unitary ST codes have been designed through Fourier transforms [71, 72] or by using unitary matrices with some special structures [71, 72, 184, 63]. The first ST code from orthogonal design was proposed by Alamouti in [5] for systems with $M_t = 2$ transmit antennas. For three and four transmit antennas, ST codes from orthogonal designs with rate $R = 3/4$ were presented in [213, 43, 217]. For $M_t = 2^k$ ($k = 1, 2, 3, \ldots$), a recursive expression for orthogonal designs was given in [200]. Designs for space–time trellis codes were presented in [164, 165]. For ST code from orthogonal designs with non-square size ($p \neq n$), a general design with rate $1/2$ for any number of transmit antennas was shown in [213]. Later, a systematic design method was presented in [203] that provide higher rates.

For 4 transmit antennas, the quasi-orthogonal ST code with symbol transmission rate 1 was proposed in [90, 216] but without achieving full diversity. Later, a method of rotating signal constellations was proposed in [202] for the quasi-orthogonal ST codes to achieve full diversity, in which optimum rotation angles were also developed. Diagonal algebraic ST codes were proposed in [28] and related code transforms were also developed in [44, 14, 122].

Exercises

2.1 Show the derivation in (2.4), i.e., to show that the instantaneous PEP between two distinct codewords C and $\tilde{C}$ can be determined as

$$P(C \to \tilde{C}|H) = Q\left(\sqrt{\frac{\rho}{2M_t}} \, ||(C - \tilde{C})H||_F\right)$$

$$= \frac{1}{\pi} \int_0^{\pi/2} \exp\left(-\frac{\rho}{4M_t \sin^2 \theta} ||(C - \tilde{C})H||_F^2\right) d\theta,$$

where $Q(x) = (1/\sqrt{2\pi}) \int_x^\infty \exp\left(-t^2/2\right) dt$ is the Gaussian error function. Moreover, show that by averaging over the Rayleigh fading channel H, the average PEP can be determined as

$$P(C \to \tilde{C}) = \frac{1}{\pi} \int_0^{\pi/2} \prod_{i=1}^{\gamma} \left(1 + \frac{\rho \lambda_i}{4M_t \sin^2 \theta}\right)^{-M_r} d\theta,$$

where $\gamma = \text{rank}(C - \tilde{C})$, and $\lambda_1, \lambda_2, \ldots, \lambda_\gamma$ are the nonzero eigenvalues of $(C - \tilde{C})(C - \tilde{C})^H$.

2.2 A MIMO system with $M_t = 2$ transmit antennas uses the following set of ST signals:

$$C_0 = \sqrt{\frac{2}{3}} \begin{bmatrix} j & 1-j \\ -1-j & -j \end{bmatrix}, \quad C_1 = \sqrt{\frac{2}{3}} \begin{bmatrix} -j & -1-j \\ 1-j & j \end{bmatrix},$$

$$C_2 = \sqrt{\frac{2}{3}} \begin{bmatrix} -j & 1+j \\ -1+j & j \end{bmatrix}, \quad C_3 = \sqrt{\frac{2}{3}} \begin{bmatrix} j & -1+j \\ 1+j & -j \end{bmatrix}.$$

Show that this set of ST signals can guarantee the full diversity, and determine the diversity product of the ST signals.

2.3 An orthogonal ST code for MIMO systems with $M_t = 4$ transmit antennas is given by

$$G_4 = \begin{bmatrix} x_1 & x_2 & x_3 & 0 \\ -x_2^* & x_1^* & 0 & x_3 \\ -x_3^* & 0 & x_1^* & -x_2 \\ 0 & -x_3^* & x_2^* & x_1 \end{bmatrix},$$

where x_1, x_2, and x_3 can be chosen from arbitrary signal constellations.

(a) Show that

$$G_4^H G_4 = (|x_1|^2 + |x_2|^2 + |x_3|^2) I_4.$$

(b) If x_1, x_2 and x_3 are chosen independently from QPSK signals $\{1, -1, j, -j\}$, what is the spectral efficiency (bits per time slot) of the ST code?

(c) Determine the normalized coding gain of the orthogonal ST code when x_1, x_2, and x_3 are chosen independently from QPSK signals $\{1, -1, j, -j\}$.

(d) Show that the orthogonal ST code has fast maximum-likelihood (ML) decoding at the receiver, i.e., x_1, x_2, and x_3 can be decoded separately, not jointly.

2.4   For $k = 1, 2, 3, \ldots$, let

$$G_{2^k}(x_1, \ldots, x_{k+1}) = \begin{bmatrix} G_{2^{k-1}}(x_1, \ldots, x_k) & x_{k+1} I_{2^{k-1}} \\ -x_{k+1}^* I_{2^{k-1}} & G_{2^{k-1}}^H(x_1, \ldots, x_k) \end{bmatrix},$$

and $G_1(x_1) = x_1 I_1$. Show that

$$G_{2^k}^H G_{2^k} = G_{2^k} G_{2^k}^H = (|x_1|^2 + |x_2|^2 + \cdots + |x_{k+1}|^2) I_{2^k},$$

i.e., $G_{2^k}(x_1, x_2, \ldots, x_{k+1})$ is an orthogonal design of size $2^k \times 2^k$ with complex variables $x_1, x_2, \ldots, x_{k+1}$.

2.5   A quasi-orthogonal ST code for MIMO systems with $M_t = 4$ transmit antennas is given by

$$C = \begin{bmatrix} x_1 & x_2 & x_3 & x_4 \\ -x_2^* & x_1^* & -x_4^* & x_3^* \\ x_3 & x_4 & x_1 & x_2 \\ -x_4^* & x_3^* & -x_2^* & x_1^* \end{bmatrix},$$

where x_1, x_2, x_3, and x_4 can be chosen from arbitrary signal constellations.

(a) Show that

$$C^H C = \begin{bmatrix} a & 0 & b & 0 \\ 0 & a & 0 & b \\ b & 0 & a & 0 \\ 0 & b & 0 & a \end{bmatrix},$$

where $a = |x_1|^2 + |x_2|^2 + |x_3|^2 + |x_4|^2$, and $b = x_1 x_3^* + x_3 x_1^* - x_2 x_4^* - x_4 x_2^*$.

(b) Show that the quasi-orthogonal ST code has a fast ML decoding that the symbol pairs (x_1, x_3) and (x_2, x_4) can be decoded independently.

(c) Show that if x_1, x_2, x_3, and x_4 are chosen from a same signal constellation, the code achieves only a diversity order of 2.

(d) If x_1 and x_2 are chosen from a QAM signal constellation $\mathcal{A}$, and x_3 and x_4 are chosen from a rotated signal constellation $e^{j\phi} \mathcal{A}$ with a rotation angle $\phi \in (0, \pi/2)$, show that the resulting quasi-orthogonal ST code guarantees a full diversity order of 4 and the optimum rotation angle is $\phi = \pi/4$ in this case.

2.6   A diagonal algebraic ST code for $M_t = 2$ transmit antennas is given by

$$C = \begin{bmatrix} s_1 + s_2 e^{j\frac{\pi}{4}} & 0 \\ 0 & s_1 - s_2 e^{j\frac{\pi}{4}} \end{bmatrix},$$

in which s_1 and s_2 are chosen from 4-QAM constellation. Show that the code achieves a full diversity, and determine the corresponding diversity product.

2.7   (Simulation project)   Consider a MIMO system with $M_t = 2$ transmit and $M_r = 1$ receive antennas. It uses a set of two ST signals:

$$C_0 = \begin{bmatrix} 1 & 1 \\ -1 & 1 \end{bmatrix}, \quad C_1 = \begin{bmatrix} -1 & -1 \\ 1 & -1 \end{bmatrix}.$$

The transceiver signal can be modeled as:

$$\begin{bmatrix} y_1 \\ y_2 \end{bmatrix} = \sqrt{\frac{\rho}{M_t}} \, C_i \begin{bmatrix} h_1 \\ h_2 \end{bmatrix} + \begin{bmatrix} \eta_1 \\ \eta_2 \end{bmatrix}, \tag{E2.1}$$

where y_1 and y_2 are received signals at time slots 1 and 2, respectively, and h_1 and h_2 are channel coefficients from the two transmit antennas to the receive antenna which are modeled as independent complex Gaussian random variables with zero mean and variance one, i.e., $CN(0, 1)$. The noise η_1 and η_2 are also modeled as $CN(0, 1)$.

(a) Determine the average signal-to-noise ratio (SNR) at the receiver.

(b) For fixed channel realizations h_1 and h_2, show that the instantaneous pairwise error probability between C_0 and C_1 is

$$Pr\{C_0 \to C_1 \,|\, h_1, h_2\} = Q\left(\sqrt{2\rho(|h_1|^2 + |h_2|^2)}\right), \tag{E2.2}$$

where $Q(x) = (1/\sqrt{2\pi}) \int_x^\infty e^{-t^2/2} dt$.

(c) Determine a PEP between C_0 and C_1 by averaging the instantaneous PEP over the Rayleigh fading channels h_1 and h_2.

(d) Simulate the system by Matlab and plot the symbol error rate curve. In this case, the symbol error rate is the same as the PEP between C_0 and C_1. In the same figure, plot the theoretical PEP result from (c) and compare it with the simulation curve.

2.8 (Simulation project) Design cyclic ST code for MIMO systems with two transmit antennas ($M_t = 2$) by exhaustive computer searching:

$$C_l = \sqrt{2} \begin{pmatrix} e^{ju_1\theta_l} & 0 \\ 0 & e^{ju_2\theta_l} \end{pmatrix},$$

where $\theta_l = \frac{l}{L} 2\pi$, $l = 0, 1, \ldots, L - 1$.

(a) When size $L = 8$, search optimal parameters $u_1, u_2 \in \{0, 1, \ldots, L - 1\}$ such that the coding gain is maximized. Determine the normalized coding gain accordingly.

(b) Can we get larger coding gain if we allow non-integer parameters u_1 and u_2? To answer this question, please determine parameters u_1 and u_2 by exhaustive searching in the interval $[0, \ L - 1]$ (you may use searching step 0.1 or smaller).

(c) Repeat questions (a) and (b) for size $L = 32$.

3 Space–time–frequency diversity and coding

In broadband wireless communications, the channel exhibits frequency selectivity (delay spread), resulting in inter-symbol interference (ISI) that can cause serious performance degradation. A mature technique to mitigate the frequency selectivity is to use orthogonal frequency division multiplexing (OFDM), which eliminates the need for high complexity equalization and offers high spectral efficiency. In order to combine the advantages of both the MIMO systems and the OFDM, space–frequency (SF)-coded MIMO-OFDM systems, where two-dimensional coding is applied to distribute channel symbols across space (transmit antennas) and frequency (OFDM tones) within one OFDM block, can be developed to exploit the available spatial and frequency diversity.

If longer decoding delay and higher decoding complexity are allowable, one may consider coding over several OFDM block periods, resulting in space–time–frequency codes to exploit all of the spatial, temporal, and frequency diversity.

The chapter is organized as follows. First, we focus on SF-coded MIMO-OFDM systems in broadband scenario and introduce two systematic approaches to design SF codes to achieve full spatial and frequency diversity within each OFDM block. Then, we consider STF coding for MIMO-OFDM systems, where the coding is applied across multiple OFDM blocks to exploit the spatial, temporal, and frequency diversity available in broadband MIMO wireless communications.

3.1 Space–frequency diversity and coding

In this section, we focus on SF-coded MIMO-OFDM systems to achieve the spatial and frequency diversity in broadband wireless communications. First, we specify an SF-coded MIMO-OFDM system model and discuss design criteria for achieving the full spatial and frequency diversity. Then, we review two systematic approaches to design SF codes to achieve the full spatial and frequency diversity within each OFDM block.

3.1.1 MIMO-OFDM system model

An SF-coded MIMO-OFDM system with M_t transmit antennas, M_r receive antennas, and N subcarriers is shown in Figure 3.1. Suppose that the frequency selective fading channels between each pair of transmit and receive antennas have L independent delay paths and the same power delay profile. The MIMO channel is assumed to be constant

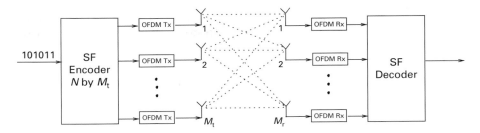

Fig. 3.1 An SF-coded MIMO-OFDM system with M_t transmit and M_r receive antennas.

over each OFDM block period. The channel impulse response from transmit antenna i to receive antenna j can be modeled as

$$h_{i,j}(\tau) = \sum_{l=0}^{L-1} \alpha_{i,j}(l)\delta(\tau - \tau_l), \tag{3.1}$$

where τ_l is the delay of the l-th path, and $\alpha_{i,j}(l)$ is the complex amplitude of the l-th path between transmit antenna i and receive antenna j. The $\alpha_{i,j}(l)$'s are modeled as zero-mean, complex Gaussian random variables with variances $E|\alpha_{i,j}(l)|^2 = \delta_l^2$, where E stands for the expectation. Note that the time delay τ_l and the variance δ_l^2 are assumed to be the same for each transmit–receive link. The powers of the L paths are normalized such that $\sum_{l=0}^{L-1} \delta_l^2 = 1$. From (3.1), the frequency response of the channel is given by

$$H_{i,j}(f) = \sum_{l=0}^{L-1} \alpha_{i,j}(l)e^{-j2\pi f\tau_l}, \quad j = \sqrt{-1}. \tag{3.2}$$

Assume that the MIMO channel is spatially uncorrelated, i.e. the channel taps $\alpha_{i,j}(l)$ are independent for different indices (i, j).

The input bit stream (uncoded or coming from a channel encoder) is divided into b bit long segments, forming 2^b-ary source symbols. These source symbols are parsed into blocks and mapped onto an SF codeword to be transmitted over the M_t transmit antennas. Each SF codeword can be expressed as an $N \times M_t$ matrix

$$C = \begin{bmatrix} c_1(0) & c_2(0) & \cdots & c_{M_t}(0) \\ c_1(1) & c_2(1) & \cdots & c_{M_t}(1) \\ \vdots & \vdots & \ddots & \vdots \\ c_1(N-1) & c_2(N-1) & \cdots & c_{M_t}(N-1) \end{bmatrix}, \tag{3.3}$$

where $c_i(n)$ denotes the channel symbol transmitted over the n-th subcarrier by transmit antenna i, and N is the number of subcarriers. The SF code is assumed to satisfy the energy constraint $E||C||_F^2 = NM_t$, where $||C||_F$ is the Frobenius norm of C. The OFDM transmitter applies an N-point IFFT to each column of the matrix C. After appending a cyclic prefix, the OFDM symbol corresponding to the i-th ($i = 1, 2, \ldots, M_t$) column of C is transmitted by transmit antenna i. Note that all of the M_t OFDM symbols are transmitted simultaneously from different transmit antennas.

At the receiver, after matched filtering, removing the cyclic prefix, and applying FFT, the received signal at the n-th subcarrier at receive antenna j is given by

$$y_j(n) = \sqrt{\frac{\rho}{M_t}} \sum_{i=1}^{M_t} c_i(n) H_{i,j}(n) + z_j(n), \qquad (3.4)$$

where

$$H_{i,j}(n) = \sum_{l=0}^{L-1} \alpha_{i,j}(l) e^{-j2\pi n \Delta f \tau_l} \qquad (3.5)$$

is the channel frequency response at the n-th subcarrier between transmit antenna i and receive antenna j, $\Delta f = 1/T$ is the subcarrier separation in the frequency domain, and T is the OFDM symbol period. We assume that the channel state information $H_{i,j}(n)$ is known at the receiver, but not at the transmitter. In (3.4), $z_j(n)$ denotes the additive complex Gaussian noise with zero mean and unit variance at the n-th subcarrier at receive antenna j. The noise samples $z_j(n)$ are assumed to be uncorrelated for different j's and n's. The factor $\sqrt{\rho/M_t}$ in (3.4) ensures that ρ is the average signal to noise ratio (SNR) at each receive antenna, independently of the number of transmit antennas.

3.1.1.1   General performance criteria

The received signal (3.4) can be rewritten in vector form as

$$\mathbf{Y} = \sqrt{\frac{\rho}{M_t}} \mathbf{DH} + \mathbf{Z}, \qquad (3.6)$$

where $\mathbf{D}$ is an $NM_r \times NM_tM_r$ matrix constructed from the SF codeword C in (3.3) as follows:

$$\mathbf{D} = \begin{bmatrix} D_1 & D_2 & \cdots & D_M & 0 & 0 & \cdots & 0 & \cdots & 0 & 0 & \cdots & 0 \\ 0 & 0 & \cdots & 0 & D_1 & D_2 & \cdots & D_M & \cdots & 0 & 0 & \cdots & 0 \\ & & \vdots & & & & \vdots & & \ddots & & \vdots & & \\ 0 & 0 & \cdots & 0 & 0 & 0 & \cdots & 0 & \cdots & D_1 & D_2 & \cdots & D_M \end{bmatrix}, \qquad (3.7)$$

in which

$$D_i = \text{diag}\{c_i(0),\ c_i(1),\ \ldots,\ c_i(N-1)\}, \quad i = 1, 2, \ldots, M_t. \qquad (3.8)$$

Each D_i in (3.8) is related to the i-th column of the SF codeword C. The channel vector $\mathbf{H}$ of size $NM_tM_r \times 1$ is formatted as

$$\mathbf{H} = [H_{1,1}^{T} \cdots H_{M_t,1}^{T}\ H_{1,2}^{T} \cdots H_{M_t,2}^{T} \cdots H_{1,M_r}^{T} \cdots H_{M_t,M_r}^{T}]^{T}, \qquad (3.9)$$

where

$$H_{i,j} = [H_{i,j}(0)\ H_{i,j}(1) \cdots H_{i,j}(N-1)]^{T}. \qquad (3.10)$$

The received signal vector $\mathbf{Y}$ of size $NM_r \times 1$ is given by

$$\mathbf{Y} = \begin{bmatrix} y_1(0) \cdots y_1(N-1)\ y_2(0) \cdots y_2(N-1) \cdots y_{M_r}(0) \cdots y_{M_r}(N-1) \end{bmatrix}^{T}, \qquad (3.11)$$

and the noise vector $\mathbf{Z}$ has the same form as $\mathbf{Y}$, i.e.,

$$\mathbf{Z} = \begin{bmatrix} z_1(0) & \cdots & z_1(N-1) & z_2(0) & \cdots & z_2(N-1) & \cdots & z_{M_r}(0) & \cdots & z_{M_r}(N-1) \end{bmatrix}^{\mathrm{T}}. \tag{3.12}$$

Suppose that $\mathbf{D}$ and $\tilde{\mathbf{D}}$ are two different matrices related to two different SF codewords C and $\tilde{C}$, respectively. Then, the pairwise error probability between $\mathbf{D}$ and $\tilde{\mathbf{D}}$ can be upper bounded as (leave proof as an exercise)

$$P(\mathbf{D} \to \tilde{\mathbf{D}}) \leq \binom{2r-1}{r} \left(\prod_{i=1}^{r} \gamma_i \right)^{-1} \left(\frac{\rho}{M_t} \right)^{-r}, \tag{3.13}$$

where r is the rank of $(\mathbf{D} - \tilde{\mathbf{D}})\mathbf{R}(\mathbf{D} - \tilde{\mathbf{D}})^{\mathrm{H}}$, $\gamma_1, \gamma_2, \ldots, \gamma_r$ are the nonzero eigenvalues of $(\mathbf{D} - \tilde{\mathbf{D}})\mathbf{R}(\mathbf{D} - \tilde{\mathbf{D}})^{\mathrm{H}}$, and $\mathbf{R} = E\{\mathbf{H}\mathbf{H}^{\mathrm{H}}\}$ is the correlation matrix of $\mathbf{H}$. The superscript H stands for the complex conjugate and transpose of a matrix.

Based on the upper bound on the pairwise error probability in (3.13), two general SF code performance criteria can be proposed as follows:

- *Diversity (rank) criterion:* The minimum rank of $(\mathbf{D} - \tilde{\mathbf{D}})\mathbf{R}(\mathbf{D} - \tilde{\mathbf{D}})^{\mathrm{H}}$ over all pairs of different codewords C and $\tilde{C}$ should be as large as possible.
- *Product criterion:* The minimum value of the product $\prod_{i=1}^{r} \gamma_i$ over all pairs of different codewords C and $\tilde{C}$ should be maximized.

However, it is hard to design SF codes directly based on the discussion on $(\mathbf{D} - \tilde{\mathbf{D}})\mathbf{R}(\mathbf{D} - \tilde{\mathbf{D}})^{\mathrm{H}}$, which is related to an $NM_tM_r \times NM_tM_r$ correlation matrix $\mathbf{R}$.

3.1.1.2 Performance criteria

In case of spatially uncorrelated MIMO channels, i.e., the channel taps $\alpha_{i,j}(l)$ are independent for different transmit antenna i and receive antenna j, the correlation matrix $\mathbf{R}$ of size $NM_tM_r \times NM_tM_r$ becomes

$$\mathbf{R} = E\{\mathbf{H}\mathbf{H}^{\mathrm{H}}\}$$
$$= \mathrm{diag}\Big(R_{1,1}, \ldots, R_{M_t,1}, R_{1,2}, \ldots, R_{M_t,2},$$
$$\ldots, R_{1,M_r}, \ldots, R_{M_t,M_r} \Big), \tag{3.14}$$

where

$$R_{i,j} = E\left\{ H_{i,j} H_{i,j}^{\mathrm{H}} \right\} \tag{3.15}$$

is the correlation matrix of the channel frequency response from transmit antenna i to receive antenna j. Using the notation $w = \mathrm{e}^{-\mathrm{j}2\pi\Delta f}$, from (3.5) and (3.10), we have

$$H_{i,j} = \left[\sum_{l=0}^{L-1} \alpha_{i,j}(l) \quad \sum_{l=0}^{L-1} \alpha_{i,j}(l) w^{\tau_l} \quad \cdots \quad \sum_{l=0}^{L-1} \alpha_{i,j}(l) w^{(N-1)\tau_l} \right]^{\mathrm{T}}$$
$$= W \cdot A_{i,j}, \tag{3.16}$$

where

$$
W = \begin{bmatrix} 1 & 1 & \cdots & 1 \\ w^{\tau_0} & w^{\tau_1} & \cdots & w^{\tau_{L-1}} \\ \vdots & \vdots & \ddots & \vdots \\ w^{(N-1)\tau_0} & w^{(N-1)\tau_1} & \cdots & w^{(N-1)\tau_{L-1}} \end{bmatrix}_{N \times L},
$$

and

$$
A_{i,j} = \begin{bmatrix} \alpha_{i,j}(0) & \alpha_{i,j}(1) & \cdots & \alpha_{i,j}(L-1) \end{bmatrix}^{\mathrm{T}}.
$$

Note that, in general, W is not a unitary matrix. If all of the L delay paths fall at the sampling instances of the receiver, then W is part of the DFT-matrix, which is unitary. Substituting (3.16) into (3.15), $R_{i,j}$ in (3.15) can be expressed as

$$
\begin{aligned}
R_{i,j} &= E\left\{ W A_{i,j} A_{i,j}^{\mathrm{H}} W^{\mathrm{H}} \right\} \\
&= W E\left\{ A_{i,j} A_{i,j}^{\mathrm{H}} \right\} W^{\mathrm{H}} \\
&= W \operatorname{diag}(\delta_0^2, \delta_1^2, \ldots, \delta_{L-1}^2) W^{\mathrm{H}} \triangleq R.
\end{aligned} \tag{3.17}
$$

The third equality follows from the assumption that the path gains $\alpha_{i,j}(l)$ are independent for different paths and different pairs of transmit and receive antennas. Note that the correlation matrix R is independent of the transmit and receive antenna indices i and j. From (3.14) and (3.17), we obtain

$$
\mathbf{R} = I_{M_{\mathrm{t}} M_{\mathrm{r}}} \otimes R, \tag{3.18}
$$

where $I_{M_{\mathrm{t}} M_{\mathrm{r}}}$ is the identity matrix of size $M_{\mathrm{t}} M_{\mathrm{r}} \times M_{\mathrm{t}} M_{\mathrm{r}}$, and $\otimes$ denotes the tensor product which is defined as follows: for any two matrices $A = \{a_{i,j}\}$ and $B = \{b_{i,j}\}$ of size $m \times n$,

$$
A \otimes B \triangleq \begin{bmatrix} a_{1,1}B & \cdots & a_{1,n}B \\ \cdots & \cdots & \cdots \\ a_{m,1}B & \cdots & a_{m,n}B \end{bmatrix}. \tag{3.19}
$$

Therefore, combining (3.3), (3.7), (3.8), and (3.18), the expression for $(\mathbf{D} - \tilde{\mathbf{D}})\mathbf{R} (\mathbf{D} - \tilde{\mathbf{D}})^{\mathrm{H}}$ in (3.13) can be rewritten as

$$
\begin{aligned}
(\mathbf{D} - \tilde{\mathbf{D}})\mathbf{R}(\mathbf{D} - \tilde{\mathbf{D}})^{\mathrm{H}} &= I_{M_r} \otimes \left[\sum_{i=1}^{M_t} (D_i - \tilde{D}_i) R (D_i - \tilde{D}_i)^{\mathrm{H}} \right] \\
&= I_{M_r} \otimes \left\{ \left[(C - \tilde{C})(C - \tilde{C})^{\mathrm{H}} \right] \circ R \right\},
\end{aligned} \tag{3.20}
$$

where $\circ$ denotes the Hadamard product, which is defined as follows: for any two matrices $A = \{a_{i,j}\}$ and $B = \{b_{i,j}\}$ of size $m \times n$,

$$
A \circ B \triangleq \begin{bmatrix} a_{1,1}b_{1,1} & \cdots & a_{1,n}b_{1,n} \\ \cdots & \cdots & \cdots \\ a_{m,1}b_{m,1} & \cdots & a_{m,n}b_{m,n} \end{bmatrix}. \tag{3.21}
$$

Denoting a matrix Δ as

$$\Delta = (C - \tilde{C})(C - \tilde{C})^{\mathrm{H}}, \tag{3.22}$$

and substituting (3.20) into (3.13), the pairwise error probability between C and $\tilde{C}$ can be upper bounded as

$$P(C \to \tilde{C}) \leq \binom{2K M_{\mathrm{r}} - 1}{K M_{\mathrm{r}}} \left(\prod_{i=1}^{K} \lambda_i \right)^{-M_{\mathrm{r}}} \left(\frac{\rho}{M_{\mathrm{t}}} \right)^{-K M_{\mathrm{r}}}, \tag{3.23}$$

where K is the rank of $\Delta \circ R$, and $\lambda_1, \lambda_2, \ldots, \lambda_K$ are the nonzero eigenvalues of $\Delta \circ R$. As a consequence, we can formulate the performance criteria as follows:

- *Diversity (rank) criterion:* The minimum rank of $\Delta \circ R$ over all pairs of distinct signals C and $\tilde{C}$ should be as large as possible.
- *Product criterion:* The minimum value of the product $\prod_{i=1}^{K} \lambda_i$ over all pairs of distinct signals C and $\tilde{C}$ should also be maximized.

According to a rank inequality on Hadamard products ([74], p.307), we have the relationship

$$\mathrm{rank}(\Delta \circ R) \leq \mathrm{rank}(\Delta)\mathrm{rank}(R). \tag{3.24}$$

Since the rank of Δ is at most M_{t}, the rank of R is at most L, and the rank of $\Delta \circ R$ is at most N, so

$$\mathrm{rank}(\Delta \circ R) \leq \min\{L M_t, N\}. \tag{3.25}$$

Thus, the maximum achievable diversity is at most $\min\{L M_t M_{\mathrm{r}}, N M_{\mathrm{r}}\}$. When N is large, the full diversity is $L M_t M_{\mathrm{r}}$, which is the product of the number of multiple delays and the number of transmit and receive antennas.

3.1.2 Full-diversity SF code design via mapping

In this subsection, we present an interesting approach to systematically design full-diversity SF codes from ST codes. It shows that using a simple repetition mapping, full-diversity SF codes can be constructed from *any* ST (block or trellis) code designed for quasi-static flat Rayleigh fading channels.

3.1.2.1 SF code design via mapping

In the sequel, the SF encoder will consist of an ST encoder and a mapping $\mathcal{M}_l$, as shown in Figure 3.2. For each $1 \times M_{\mathrm{t}}$ output vector $[g_1 \ g_2 \ \cdots \ g_{M_{\mathrm{t}}}]$ from the ST encoder and a fixed number l ($1 \leq l \leq L$), the mapping $\mathcal{M}_l$ is defined as

$$\mathcal{M}_l : [g_1 \ g_2 \ \cdots \ g_{M_{\mathrm{t}}}] \to \mathbf{1}_{l \times 1}[g_1 \ g_2 \ \cdots \ g_{M_{\mathrm{t}}}], \tag{3.26}$$

where $\mathbf{1}_{l \times 1}$ is an all one matrix of size $l \times 1$. The resulting $l \times M_{\mathrm{t}}$ matrix is actually a repetition of the vector $[g_1 \ g_2 \ \cdots \ g_{M_{\mathrm{t}}}]$ l times. Suppose that $l M_{\mathrm{t}}$ is not greater than the number of OFDM subcarriers, N, and k is the largest integer such that $k l M_{\mathrm{t}} \leq N$.

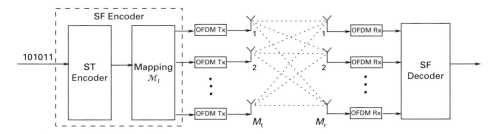

Fig. 3.2   An SF encoder consisting of an ST encoder and a mapping.

Denote the output code matrix of the ST encoder by G. Then, the SF code C of size $N \times M_t$ is constructed as

$$C = \begin{bmatrix} \mathcal{M}_l(G) \\ \mathbf{0}_{(N-klM_t) \times M_t} \end{bmatrix}, \tag{3.27}$$

where

$$\mathcal{M}_l(G) = \begin{bmatrix} I_{kM_t} \otimes \mathbf{1}_{l \times 1} \end{bmatrix} G. \tag{3.28}$$

In fact, the SF code C is obtained by repeating each row of G l times and adding some zeros. The zero padding used here ensures that the space–frequency code C has size $N \times M_t$. Typically, the size of the zero padding is small, and it can be used to drive the trellis encoder to the zero state.

Example 3.1   If G is the Alamouti code and $l = 2$, then an SF code of size 4×2 can be obtained by repeating each row of the Alamouti code twice as follows:

$$C = \begin{bmatrix} x_1 & x_2 \\ x_1 & x_2 \\ -x_2^* & x_1^* \\ -x_2^* & x_1^* \end{bmatrix}. \tag{3.29}$$

▲

The following theorem states that if the employed ST code G has full diversity for quasi-static flat fading channels, the space–frequency code constructed by (3.27) will achieve a diversity of at least lM_tM_r.

THEOREM 3.1.1   *Suppose that an MIMO-OFDM system equipped with M_t transmit and M_r receive antennas has N subcarriers, and the frequency selective channel has L independent paths, in which the maximum path delay is less than one OFDM block period. If an ST (block or trellis) code designed for M_t transmit antennas achieves full diversity for quasi-static flat fading channels, then the SF code obtained from this ST code via the mapping $\mathcal{M}_l$ ($1 \leq l \leq L$) defined in (3.28) will achieve a diversity order of at least $\min\{lM_tM_r, NM_r\}$.*

Proof   Since in typical MIMO-OFDM systems the number of subcarriers, N, is greater than LM_t, we provide the proof for the $lM_t \leq N$ case for a given mapping $\mathcal{M}_l$,

$1 \le l \le L$. If $lM_t > N$, the proof is similar to the one described below and is omitted for brevity. Assume that k is the largest integer such that $klM_t \le N$.

For two distinct SF codewords C and $\tilde{C}$ of size $N \times M_t$, there are two corresponding distinct ST codewords G and $\tilde{G}$ of size $kM_t \times M_t$ such that

$$C - \tilde{C} = \begin{bmatrix} \mathcal{M}_l(G - \tilde{G}) \\ \mathbf{0}_{(N-klM_t) \times M_t} \end{bmatrix}, \tag{3.30}$$

in which

$$\mathcal{M}_l(G - \tilde{G}) = [I_{kM_t} \otimes \mathbf{1}_{l \times 1}](G - \tilde{G}). \tag{3.31}$$

Since the ST code achieves full diversity for quasi-static flat fading channels, $G - \tilde{G}$ is of full rank for two distinct G and $\tilde{G}$, i.e., the rank of $G - \tilde{G}$ is M_t.

Based on the SF code performance criteria, the objective of the proof is to show that the matrix $\Delta \circ R$ has a rank of at least lM_t. From (3.22) and (3.30), we have

$$\Delta = \begin{bmatrix} \mathcal{M}_l(G - \tilde{G})[\mathcal{M}_l(G - \tilde{G})]^H & \mathbf{0}_{klM_t \times (N-klM_t)} \\ \mathbf{0}_{(N-klM_t) \times klM_t} & \mathbf{0}_{(N-klM_t) \times (N-klM_t)} \end{bmatrix}. \tag{3.32}$$

Thus, in $\Delta \circ R$, all entries are zero except a $klM_t \times klM_t$ submatrix. Denote this $klM_t \times klM_t$ submatrix as $(\Delta \circ R)_{klM_t \times klM_t}$.

On the other hand, from (3.17), it can be verified that the entries of the correlation matrix $R \overset{\triangle}{=} \{r_{i,j}\}_{1 \le i,j \le N}$ can be expressed as

$$r_{i,j} = \sum_{s=0}^{L-1} \delta_s^2 w^{(i-j)\tau_s}, \quad 1 \le i, j \le N.$$

Therefore, from (3.30), (3.31), and (3.32), we obtain

$$(\Delta \circ R)_{klM_t \times klM_t} = \left\{ \mathcal{M}_l(G - \tilde{G})[\mathcal{M}_l(G - \tilde{G})]^H \right\} \circ P$$

$$= \left\{ [I_{kM_t} \otimes \mathbf{1}_{l \times 1}](G - \tilde{G})(G - \tilde{G})^H [I_{kM_t} \otimes \mathbf{1}_{l \times 1}]^H \right\} \circ P$$

$$= \left\{ \left[(G - \tilde{G})(G - \tilde{G})^H \right] \otimes \mathbf{1}_{l \times l} \right\} \circ P, \tag{3.33}$$

where $P = \{p_{i,j}\}_{1 \le i,j \le klM_t}$ is a $klM_t \times klM_t$ matrix with entries

$$p_{i,j} = \sum_{s=0}^{L-1} \delta_s^2 w^{(i-j)\tau_s}, \quad 1 \le i, j \le klM_t. \tag{3.34}$$

The last equality in (3.33) follows from the identities $[I_{kM_t} \otimes \mathbf{1}_{l \times 1}]^H = I_{kM_t} \otimes \mathbf{1}_{1 \times l}$ and $(A_1 \otimes B_1)(A_2 \otimes B_2)(A_3 \otimes B_3) = (A_1 A_2 A_3) \otimes (B_1 B_2 B_3)$ ([74], p.251).

We further partition the $klM_t \times klM_t$ matrix P into $l \times l$ submatrices as follows:

$$P = \begin{bmatrix} P_{1,1} & P_{1,2} & \cdots & P_{1,kM_t} \\ P_{2,1} & P_{2,2} & \cdots & P_{2,kM_t} \\ \vdots & \vdots & \ddots & \vdots \\ P_{kM_t,1} & P_{kM_t,2} & \cdots & P_{kM_t,kM_t} \end{bmatrix}, \tag{3.35}$$

where each submatrix $P_{m,n}$, $1 \le m, n \le kM_t$, is of size $l \times l$. Denoting the entries of $P_{m,n}$ as $p_{m,n}(i, j)$, $1 \le i, j \le l$, we obtain

$$p_{m,n}(i, j) = \sum_{s=0}^{L-1} \delta_s^2 w^{[(m-n)l+(i-j)]\tau_s}, \quad 1 \le i, j \le l. \tag{3.36}$$

As a consequence, each submatrix $P_{m,n}$, $1 \le m, n \le kM_t$, can be expressed as

$$P_{m,n} = W_m \mathrm{diag}\{\delta_0^2, \delta_1^2, \ldots, \delta_{L-1}^2\} W_n^{\mathrm{H}}, \tag{3.37}$$

where

$$W_m = \begin{bmatrix} w^{(m-1)l\tau_0} & w^{(m-1)l\tau_1} & \cdots & w^{(m-1)l\tau_{L-1}} \\ w^{[(m-1)l+1]\tau_0} & w^{[(m-1)l+1]\tau_1} & \cdots & w^{[(m-1)l+1]\tau_{L-1}} \\ \vdots & \vdots & \ddots & \vdots \\ w^{[(m-1)l+(l-1)]\tau_0} & w^{[(m-1)l+(l-1)]\tau_1} & \cdots & w^{[(m-1)l+(l-1)]\tau_{L-1}} \end{bmatrix}, \tag{3.38}$$

for $m = 1, 2, \ldots, kM_t$. In (3.38), W_m can be further decomposed as

$$W_m = W_1 \mathrm{diag}\{w^{(m-1)l\tau_0}, w^{(m-1)l\tau_1}, \ldots, w^{(m-1)l\tau_{L-1}}\} \tag{3.39}$$

for $1 \le m \le kM_t$, where

$$W_1 = \begin{bmatrix} 1 & 1 & \cdots & 1 \\ w^{\tau_0} & w^{\tau_1} & \cdots & w^{\tau_{L-1}} \\ \vdots & \vdots & \ddots & \vdots \\ w^{(l-1)\tau_0} & w^{(l-1)\tau_1} & \cdots & w^{(l-1)\tau_{L-1}} \end{bmatrix}. \tag{3.40}$$

Let us denote the matrix consisting of the first l columns of W_1 by W_0. We observe that W_0 is an $l \times l$ Vandermonde matrix in l variables $w^{\tau_0}, w^{\tau_1}, \ldots, w^{\tau_{l-1}}$ ([74], p.400). The determinant of W_0 can be calculated as follows:

$$\det(W_0) = \prod_{0 \le i < j \le l-1} (w^{\tau_j} - w^{\tau_i})$$

$$= \prod_{0 \le i < j \le l-1} \left[e^{-j2\pi \Delta f(\tau_j - \tau_i)} - 1 \right] e^{-j2\pi \Delta f \tau_i}.$$

Since Δf is the inverse of the OFDM block period T and the maximum path delay is less than T, we have $\Delta f(\tau_j - \tau_i) < 1$ for any $0 \le i < j \le l - 1$. Thus, W_0 is of full rank, and so is W_1. It follows that for any $m = 1, 2, \ldots, kM_t$, the rank of W_m is l.

We now go back to (3.33) to investigate the rank of $\Delta \circ R$. For convenience, we use the notation

$$(G - \tilde{G})(G - \tilde{G})^{\mathrm{H}} \stackrel{\triangle}{=} \begin{bmatrix} a_{1,1} & a_{1,2} & \cdots & a_{1,kM_t} \\ a_{2,1} & a_{2,2} & \cdots & a_{2,kM_t} \\ \vdots & \vdots & \ddots & \vdots \\ a_{kM_t,1} & a_{kM_t,2} & \cdots & a_{kM_t,kM_t} \end{bmatrix},$$

and $\Lambda = \text{diag}\{\delta_0^2, \delta_1^2, \ldots, \delta_{L-1}^2\}$. Then, substituting (3.35) and (3.37) into (3.33), we obtain

$$(\Delta \circ R)_{klM_t \times klM_t}$$

$$= \begin{bmatrix} a_{1,1}P_{1,1} & a_{1,2}P_{1,2} & \cdots & a_{1,kM_t}P_{1,kM_t} \\ a_{2,1}P_{2,1} & a_{2,2}P_{2,2} & \cdots & a_{2,kM_t}P_{2,kM_t} \\ \vdots & \vdots & \ddots & \vdots \\ a_{kM_t,1}P_{kM_t,1} & a_{kM_t,2}P_{kM_t,2} & \cdots & a_{kM_t,kM_t}P_{kM_t,kM_t} \end{bmatrix}$$

$$= \begin{bmatrix} W_1 a_{1,1}\Lambda W_1^H & W_1 a_{1,2}\Lambda W_2^H & \cdots & W_1 a_{1,kM_t}\Lambda W_{kM_t}^H \\ W_2 a_{2,1}\Lambda W_1^H & W_2 a_{2,2}\Lambda W_2^H & \cdots & W_2 a_{2,kM_t}\Lambda W_{kM_t}^H \\ \vdots & \vdots & \ddots & \vdots \\ W_{kM_t} a_{kM_t,1}\Lambda W_1^H & W_{kM_t} a_{kM_t,2}\Lambda W_2^H & \cdots & W_{kM_t} a_{kM_t,kM_t}\Lambda W_{kM_t}^H \end{bmatrix}$$

$$= Q\left\{\left[(G - \tilde{G})(G - \tilde{G})^H\right] \otimes \Lambda\right\} Q^H, \tag{3.41}$$

where

$$Q = \text{diag}\left\{W_1, \; W_2, \; \ldots, \; W_{kM_t}\right\}.$$

Since the rank of $G - \tilde{G}$ is M_t, there are M_t linearly independent rows in $G - \tilde{G}$. Suppose that the f_i-th, $1 \le f_1 < f_2 < \cdots < f_{M_t} \le kM_t$, rows of $G - \tilde{G}$ are linearly independent of each other. Then, the matrix

$$A \triangleq \begin{bmatrix} a_{f_1,f_1} & a_{f_1,f_2} & \cdots & a_{f_1,f_{M_t}} \\ a_{f_2,f_1} & a_{f_2,f_2} & \cdots & a_{f_2,f_{M_t}} \\ \vdots & \vdots & \ddots & \vdots \\ a_{f_{M_t},f_1} & a_{f_{M_t},f_2} & \cdots & a_{f_{M_t},f_{M_t}} \end{bmatrix}$$

is a submatrix of $(G - \tilde{G})(G - \tilde{G})^H$, and the rank of A is M_t. Using the notation

$$Q_0 = \text{diag}\{W_{f_1}, \; W_{f_2}, \; \ldots, \; W_{f_{M_t}}\},$$

from (3.41), we can see that $Q_0 \{A \otimes \Lambda\} Q_0^H$ is an $lM_t \times lM_t$ submatrix of $(\Delta \circ R)_{klM_t \times klM_t}$. Therefore, to show that the rank of $\Delta \circ R$ is at least lM_t, it is sufficient to show that the submatrix $Q_0 \{A \otimes \Lambda\} Q_0^H$ has rank lM_t.

Since the rank of A is M_t and the rank of Λ is L, according to a rank equality on tensor products ([74], p.246), we have

$$\text{rank}(A \otimes \Lambda) = \text{rank}(A)\text{rank}(\Lambda) = M_t L,$$

so the matrix $A \otimes \Lambda$ is of full rank. Recall that for any $m = 1, 2, \ldots, kM_t$, the rank of W_m is l, so the rank of Q_0 is lM_t. Therefore, the rank of $Q_0 \{A \otimes \Lambda\} Q_0^H$ is lM_t. This proves the theorem. ∎

In addition, from the proof of Theorem 3.1.1, we can see that the SF code obtained from a space–time block code of square size via the mapping M_l $(1 \le l \le L)$ will achieve a diversity of lM_tM_r exactly. Since the maximum achievable diversity is upper bounded by $\min\{LM_tM_r, NM_r\}$, we arrive at the following result.

Corollary 3.1.1 *Under the assumptions of Theorem 3.1.1, the SF code obtained from a full diversity ST code via the mapping* M_L *defined in (3.28) achieves the maximum achievable diversity* $\min\{LM_tM_r, NM_r\}$.

The symbol rate of the resulting SF codes obtained via the mapping M_l (3.28) is $1/l$ times that of the corresponding ST codes. For example, for a system with two transmit antennas, eight subcarriers and a two-ray delay profile, the symbol rate of the full-diversity SF codes discussed here is $1/2$. In certain practical situations, this effect can be compensated by expanding the constellation size, maintaining the same spectral efficiency. Furthermore, from a system performance point of view, there is a tradeoff between the diversity order and the coding rate. Theorem 3.1.1 offers a flexible choice on the diversity order.

3.1.2.2   Coding advantage

In the following, we characterize the coding advantage of the resulting SF codes in terms of the coding advantage of the underlying ST codes by defining and evaluating the diversity product for SF codes.

We recall from (2.9) that the diversity product or the normalized coding advantage of a full-diversity ST code for quasi-static flat fading channels is

$$\zeta_{ST} = \frac{1}{2\sqrt{M_t}} \min_{G \neq \tilde{G}} \left| \prod_{i=1}^{M_t} \beta_i \right|^{\frac{1}{2M_t}}, \tag{3.42}$$

where $\beta_1, \beta_2, \ldots, \beta_{M_t}$ are the nonzero eigenvalues of $(G - \tilde{G})(G - \tilde{G})^{\mathrm{H}}$ for any pair of distinct ST codewords G and $\tilde{G}$. Similarly, the diversity product of a full-diversity SF code can be defined as

$$\zeta_{SF, R} = \frac{1}{2\sqrt{M_t}} \min_{C \neq \tilde{C}} \left| \prod_{i=1}^{LM_t} \lambda_i \right|^{\frac{1}{2LM_t}}, \tag{3.43}$$

where $\lambda_1, \lambda_2, \ldots, \lambda_{LM_t}$ are the nonzero eigenvalues of $\Delta \circ R$ for any pair of distinct space–frequency codewords C and $\tilde{C}$. In the rest of this section, without loss of generality, we assume that the number of subcarriers, N, is not less than LM_t, i.e., $LM_t \leq N$.

The relationship between the diversity products of the full-diversity SF codes obtained via the repetition mapping and the underlying ST codes is characterized by the following theorem.

THEOREM 3.1.2   *The diversity product of the full-diversity SF code in Corollary 3.1.1 is bounded by that of the corresponding ST code as follows:*

$$\sqrt{\eta_L}\, \Phi\, \zeta_{ST} \leq \zeta_{SF, R} \leq \sqrt{\eta_1}\, \Phi\, \zeta_{ST}, \tag{3.44}$$

where $\Phi = \left(\prod_{l=0}^{L-1} \delta_l\right)^{1/L}$, *and* η_1 *and* η_L *are the largest and smallest eigenvalues, respectively, of the matrix* H *defined as*

$$H = \begin{bmatrix} H(0) & H(1)^* & \cdots & H(L-1)^* \\ H(1) & H(0) & \cdots & H(L-2)^* \\ \vdots & \vdots & \ddots & \vdots \\ H(L-1) & H(L-2) & \cdots & H(0) \end{bmatrix}_{L \times L}, \qquad (3.45)$$

and the entries of H *are given by*

$$H(n) = \sum_{l=0}^{L-1} e^{-j2\pi n \Delta f \tau_l}, \quad n = 0, 1, \ldots, L-1.$$

Proof In the following, we use the notation developed in the proof of Theorem 3.1.1 by replacing the repetition factor l with L, since the full diversity is achieved in Corollary 3.1.1 by using the mapping M_L. For any $n \times n$ nonnegative definite matrix A, we denote its eigenvalues in a non-increasing order as: $\text{eig}_1(A) \geq \text{eig}_2(A) \geq \cdots \geq \text{eig}_n(A)$.

For two distinct SF codewords C and $\tilde{C}$, there are two corresponding ST codewords G and $\tilde{G}$ such that the relationship of $C - \tilde{C}$ and $G - \tilde{G}$ in (3.30) and (3.31) holds. According to (3.25) and Corollary 3.1.1, the rank of $\Delta \circ R$ is exactly LM_t. It means that $\Delta \circ R$ has totally LM_t nonzero eigenvalues, which are the same as the nonzero eigenvalues of $(\Delta \circ R)_{kLM_t \times kLM_t}$. Thus,

$$\zeta_{\text{SF}, R} = \frac{1}{2\sqrt{M_t}} \min_{C \neq \tilde{C}} \left| \prod_{i=1}^{LM_t} \text{eig}_i \left((\Delta \circ R)_{kLM_t \times kLM_t}\right) \right|^{\frac{1}{2LM_t}}$$

$$= \frac{1}{2\sqrt{M_t}} \min_{G \neq \tilde{G}} \left| \prod_{i=1}^{LM_t} \text{eig}_i \left(Q \left\{\left[(G - \tilde{G})(G - \tilde{G})^{\text{H}}\right] \otimes \Lambda\right\} Q^{\text{H}}\right) \right|^{\frac{1}{2LM_t}}$$

$$= \frac{1}{2\sqrt{M_t}} \min_{G \neq \tilde{G}} \left| \prod_{i=1}^{LM_t} \theta_i \, \text{eig}_i \left(\left[(G - \tilde{G})(G - \tilde{G})^{\text{H}}\right] \otimes \Lambda\right) \right|^{\frac{1}{2LM_t}}, \qquad (3.46)$$

where $\text{eig}_{kLM_t}(QQ^{\text{H}}) \leq \theta_i \leq \text{eig}_1(QQ^{\text{H}})$ for $i = 1, 2, \ldots, LM_t$. In (3.46), the second equality follows from (3.41), and the last equality follows by Ostrowski's theorem ([73], p.224). Since $Q = \text{diag}\{W_1, W_2, \ldots, W_{kM_t}\}$, we have

$$QQ^{\text{H}} = \text{diag}\left\{W_1 W_1^{\text{H}}, \; W_2 W_2^{\text{H}}, \; \ldots, \; W_{kM_t} W_{kM_t}^{\text{H}}\right\}.$$

As a requirement of Ostrowski's theorem, the matrix Q should be nonsingular, which is guaranteed by the fact that each matrix W_m is of full rank for any $m = 1, 2, \ldots, kM_t$. Furthermore, from (3.39), we know that for any $1 \leq m \leq kM_t$,

$$W_m W_m^{\text{H}} = W_1 DD^{\text{H}} W_1^{\text{H}} = W_1 W_1^{\text{H}},$$

where $D = \text{diag}\{w^{(m-1)L\tau_0}, w^{(m-1)L\tau_1}, \ldots, w^{(m-1)L\tau_{L-1}}\}$. From (3.40), it is easy to verify that $W_1 W_1^{\text{H}}$ is the matrix H defined in (3.45). Thus, $QQ^{\text{H}} = I_{kM_t} \otimes H$. Therefore, we can conclude that $\text{eig}_L(H) \leq \theta_i \leq \text{eig}_1(H)$ for any $i = 1, 2, \ldots, LM_t$.

Since the set of LM_t nonzero eigenvalues of $\left[(G - \tilde{G})(G - \tilde{G})^{\text{H}} \right] \otimes \Lambda$ can be expressed as ([73], p.246)

$$\left\{ \text{eig}_i \left((G - \tilde{G})(G - \tilde{G})^{\text{H}} \right) \cdot \text{eig}_j(\Lambda) : 1 \leq i \leq M_t, 1 \leq j \leq L \right\}, \qquad (3.47)$$

substituting (3.47) into (3.46), we arrive at

$$\zeta_{\text{SF}, R} = \frac{1}{2\sqrt{M_t}} \min_{G \neq \tilde{G}} \left(\prod_{i=1}^{LM_t} \theta_i \right)^{\frac{1}{2LM_t}}$$

$$\left| \prod_{i=1}^{M_t} \text{eig}_i \left((G - \tilde{G})(G - \tilde{G})^{\text{H}} \right) \right|^{\frac{1}{2M_t}} \left(\prod_{j=1}^{L} \text{eig}_j(\Lambda) \right)^{\frac{1}{2L}}$$

$$= \left(\prod_{i=1}^{LM_t} \theta_i \right)^{\frac{1}{2LM_t}} \left(\prod_{j=1}^{L} \delta_j^2 \right)^{\frac{1}{2L}} \zeta_{\text{ST}}. \qquad (3.48)$$

Since $\eta_L \leq \theta_i \leq \eta_1$ for any $i = 1, 2, \ldots, LM_t$, we have the inequalities in (3.44). ∎

From Theorem 3.1.2, we can see that the larger the coding advantage of the ST code, the larger the coding advantage of the resulting SF code, suggesting that to maximize the performance of the SF codes, we should look for the best-known ST codes existing in the literature. Moreover, the coding advantage of the SF code depends on the power delay profile:

- First, it depends on the power distribution through the square root of the geometric average of path powers, i.e., $\Phi = \left(\prod_{l=0}^{L-1} \delta_l \right)^{1/L}$. Since the sum of the powers of the paths is unity, this implies that the best performance is expected in the case of uniform power distribution (i.e., $\delta_l^2 = 1/L$).
- Second, the entries of the matrix H defined in (3.45) are functions of the path delays, so the coding advantage also depends on the delay distribution of the paths.

For example, in case of a two-ray delay profile (i.e., $L = 2$), the matrix H in (3.45) has two eigenvalues:

$$\eta_1 = 2 + 2|\cos \pi (\tau_1 - \tau_0)/T|,$$
$$\eta_2 = 2 - 2|\cos \pi (\tau_1 - \tau_0)/T|.$$

Typically, the ratio of $(\tau_1 - \tau_0)/T$ is less than $1/2$, so $\cos \pi (\tau_1 - \tau_0)/T$ is nonnegative. Thus, the smaller the separation of the two rays, the smaller the eigenvalue η_2. If the two rays are very close compared to the duration of one OFDM symbol, T, the lower bound in (3.44) approaches zero. Simulation results seem to suggest that the behavior of the coding advantage is close to the lower bound.

3.1.2.3 Code design examples and performance comparisons

In the following, we give some SF code design examples based on the introduced approach and compare their performance and complexity tradeoff. The SF block codes were obtained from orthogonal ST block codes for two and four transmit antennas, respectively. For two transmit antennas, the used 2×2 orthogonal ST block code was Alamouti's structure, given by

$$G_2 = \begin{bmatrix} x_1 & x_2 \\ -x_2^* & x_1^* \end{bmatrix}. \tag{3.49}$$

The SF block code for four transmit antennas was obtained from the 4×4 orthogonal design

$$G_4 = \begin{bmatrix} x_1 & x_2 & x_3 & 0 \\ -x_2^* & x_1^* & 0 & x_3 \\ -x_3^* & 0 & x_1^* & -x_2 \\ 0 & -x_3^* & x_2^* & x_1 \end{bmatrix}. \tag{3.50}$$

In both cases, the x_i's were taken from BPSK or QPSK constellations. Note that the 2×2 orthogonal design could carry one channel symbol per subcarrier, whereas the 4×4 block code had a symbol rate of only $3/4$.

Code performances with two-ray delay profiles

First, let us assume a simple two-ray, equal-power delay profile, with a delay of $\tau\,\mu s$ between the two rays. Let us consider two cases: (i) $\tau = 5\,\mu s$ and (ii) $\tau = 20\,\mu s$. The simulated communication system has $N = 128$ subcarriers, and the total bandwidth is $BW = 1\,\mathrm{MHz}$. Thus, the OFDM block duration is $T = 128\,\mu s$ without the cyclic prefix. We set the length of the cyclic prefix to $20\,\mu s$ for all cases. The MIMO-OFDM systems have one receive antenna.

Figure 3.3 depicts the performance of the SF block codes obtained from the two-antenna orthogonal design. We used BPSK modulation for the non-repeated case and QPSK for the repeated case. Therefore, both systems have a spectral efficiency of $\frac{128 \times 1\,\mathrm{bits}}{(128\,\mu s + 20\,\mu s) \times 1\,\mathrm{MHz}} = 0.86\,\mathrm{bits/s/Hz}$. The figure shows that in case of $\tau = 20\,\mu s$, the performance curve of the full-diversity SF code has a steeper slope than that of the code without repetition. We can observe a performance improvement of about $4\,\mathrm{dB}$ at a BER of 10^{-4}. The performance of the full-diversity SF code degraded significantly from the $\tau = 20\,\mu s$ case to the $\tau = 5\,\mu s$ case, while the performance of the SF code using ST code without repetition was almost the same for the two delay profiles.

This observation is consistent with the theoretical result that the coding advantage depends on the delay distribution of the multiple paths. It also indicates that using ST codes directly as SF codes can exploit only the spatial diversity, and cannot exploit the frequency diversity.

Figure 3.4 shows the performance of the SF block codes obtained from the orthogonal ST code for four transmit antennas. The full-diversity SF code with repetition used QPSK modulation, and the non-repeated code used BPSK modulation. Thus, the spectral efficiency of both codes was $\frac{128 \times 3/4\,\mathrm{bits}}{(128\,\mu s + 20\,\mu s) \times 1\,\mathrm{MHz}} = 0.65\,\mathrm{bits/s/Hz}$. The tendencies

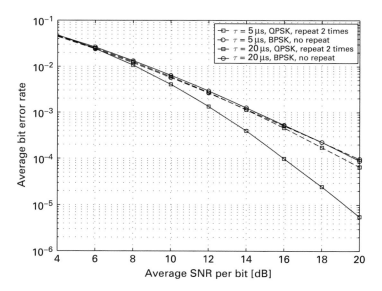

Fig. 3.3   Performance of two-antenna SF block codes with the two-ray channel model.

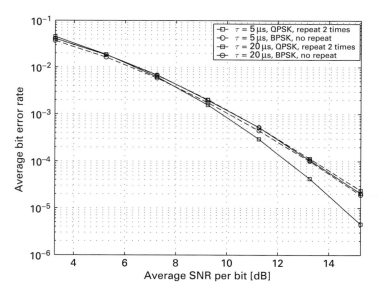

Fig. 3.4   Performance of four-antenna SF block codes with the two-ray channel model.

observed in Figure 3.4 are similar to those observed in Figure 3.3. In case of $\tau = 20\,\mu s$, the full-diversity SF code has a steeper performance curve than the SF code using ST without repetition, and it has an improvement of about 1 dB at a BER of 10^{-4}. In case of $\tau = 5\,\mu s$, the performance of the proposed SF code is a little worse than that of the SF code using ST code without repetition. The worse performance is due to the smaller coding advantage of the proposed SF code since, in order to keep the same spectral efficiency of the two schemes, we used QPSK modulation for the proposed SF code and BPSK modulation for the non-repeated code.

Code performances with the COST207 six-ray delay profile

Let us consider a more realistic channel model, the COST207 typical urban (TU) six-ray channel model [196]. The power delay profile of the channel is shown in Figure 3.5. Let us consider an MIMO-OFDM system having $N = 128$ subcarriers with two different bandwidths: (i) $BW = 1$ MHz (denoted by dashed lines) and (ii) $BW = 4$ MHz (denoted by solid lines). The cyclic prefix was $20\,\mu$s long for both cases.

The performance of the SF block codes from the 2×2 orthogonal design with and without repetition are shown in Figure 3.6 for two transmit and one receive antennas. The repeated code (using QPSK modulation) and the non-repeated code (using BPSK modulation) had the same spectral efficiency of 0.86 bits/s/Hz for the 1 MHz system and 0.22 bits/s/Hz for the 4 MHz system, respectively. We can see from the figure that in case of $BW = 4$ MHz, the code with repetition has a steeper performance curve than the code without repetition. There is a performance improvement of about 2 dB at a BER of 10^{-4}. In case of $BW = 1$ MHz, the maximum delay of the TU profile ($5.0\,\mu$s) is "short" compared to the "long" duration of the OFDM block ($128\,\mu$s), which means that there is little frequency diversity available in the fading channel. From the figure, we observe that the performance of the repeated code is worse than the non-repeated code, due to the smaller coding advantage, which is a result of the larger constellation size.

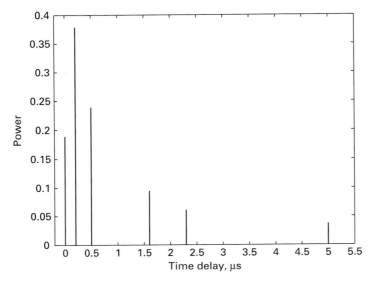

Fig. 3.5   The COST207 typical urban (TU) six-ray power delay profile.

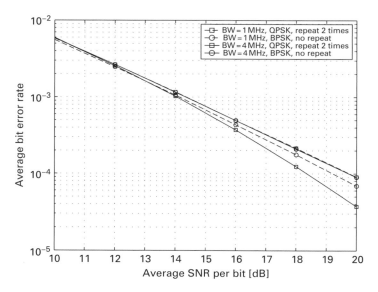

Fig. 3.6   Performance of two-antenna SF block codes with the COST six-ray channel model.

Observations and discussion

Based on the above design examples and their performance results, we can make some observations. It is apparent that by repeating each row of the space–time code matrix, we could construct codes whose error performance curve is steeper than that of the codes without repetition, i.e., the obtained codes have higher diversity order. However, the actual performance of the code depends heavily on the underlying channel model. In all cases, both the absolute performance and the performance improvement obtained by repetition are considerably better in case of the longer delay profile (i.e., $\tau = 20\,\mu s$), and the performance of the obtained full-diversity SF codes degrade significantly in case of the delay profile with $\tau = 5\,\mu s$.

These phenomena can be explained as follows. The delay distribution of the channel has a significant effect on the SF code performance. If the delays of the paths are large with respect to one OFDM block period, there will be fast variations in the spectrum of the channel impulse response, so the probability of simultaneous deep fades in adjacent subchannels will be smaller. This observation is in accordance with Theorem 3.1.2. As discussed in Section 3.1.2.2, we should expect better BER performance when transmitting data over channels with larger path delays. On the other hand, if the two delay paths are very close, the channel will cause performance degradation.

3.1.3   Full-rate full-diversity SF code design

In this subsection, we describe a systematic method to obtain full-rate SF codes achieving full diversity. Specifically, the design approach provides a class of SF codes that can achieve a diversity order of $\Gamma M_t M_r$ for any fixed integer Γ ($1 \leq \Gamma \leq L$).

3.1.3.1 Code structure

The basic idea of the full-rate full-diversity SF codes is similar to the diagonal algebraic code design. Specifically, let us consider a coding strategy where each SF codeword C is a concatenation of some matrices G_p:

$$C = \begin{bmatrix} G_1^T & G_2^T & \cdots & G_P^T & \mathbf{0}_{N-P\Gamma M_t}^T \end{bmatrix}^T, \tag{3.51}$$

where $P = \lfloor N/(\Gamma M_t) \rfloor$, and each matrix G_p, $p = 1, 2, \ldots, P$, is of size $\Gamma M_t \times M_t$. The zero padding in (3.51) is used if the number of subcarriers N is not an integer multiple of ΓM_t. Each matrix G_p ($1 \le p \le P$) has the same structure given by

$$G = \sqrt{M_t}\, \mathrm{diag}\left(X_1, X_2, \ldots, X_{M_t}\right), \tag{3.52}$$

where $\mathrm{diag}(X_1, X_2, \ldots, X_{M_t})$ is a block diagonal matrix, $X_i = [x_{(i-1)\Gamma+1} x_{(i-1)\Gamma+2} \cdots x_{i\Gamma}]^T$, $i = 1, 2, \ldots, M_t$, and x_k, $k = 1, 2, \ldots, \Gamma M_t$, are complex symbols and will be specified later. The energy constraint is

$$E\left(\sum_{k=1}^{\Gamma M_t} |x_k|^2\right) = \Gamma M_t.$$

For a fixed p, the symbols in G_p are designed jointly, but the design of G_{p_1} and G_{p_2}, $p_1 \ne p_2$, is independent of each other. The symbol rate of the code is $P\Gamma M_t/N$, ignoring the cyclic prefix. If N is a multiple of ΓM_t, the symbol rate is 1. If not, the rate is less than 1, but since usually N is much greater than ΓM_t, the symbol rate is very close to 1.

Now we derive sufficient conditions for the SF codes described above to achieve a diversity order of $\Gamma M_t M_r$. Suppose that C and $\tilde{C}$ are two distinct SF codewords which are constructed from $G_1, G_2, \ldots, G_P$ and $\tilde{G}_1, \tilde{G}_2, \ldots, \tilde{G}_P$, respectively. We would like to determine the rank of $\Delta \circ R$, where Δ is defined in (3.22) and R is the correlation matrix defined in (3.17). For two distinct codewords C and $\tilde{C}$, there exists at least one index p_0 ($1 \le p_0 \le P$) such that $G_{p_0} \ne \tilde{G}_{p_0}$. We may further assume that $G_p = \tilde{G}_p$ for any $p \ne p_0$ since the rank of $\Delta \circ R$ does not decrease if $G_p = \tilde{G}_p$ for some $p \ne p_0$ ([74], Corollary 3.1.3, p.149).

From (3.17), we know that the correlation matrix $R \overset{\triangle}{=} \{r_{i,j}\}_{1 \le i,j \le N}$ is a Toeplitz matrix. The entries of R are given by

$$r_{i,j} = \sum_{l=0}^{L-1} \delta_l^2 w^{(i-j)\tau_l}, \quad 1 \le i, j \le N. \tag{3.53}$$

Under the assumption that $G_p = \tilde{G}_p$ for any $p \ne p_0$, we observe that the nonzero eigenvalues of $\Delta \circ R$ are the same as those of $[(G_{p_0} - \tilde{G}_{p_0})(G_{p_0} - \tilde{G}_{p_0})^H] \circ Q$, where $Q = \{q_{i,j}\}_{1 \le i,j \le \Gamma M_t}$ is also a Toeplitz matrix whose entries are

$$q_{i,j} = \sum_{l=0}^{L-1} \delta_l^2 w^{(i-j)\tau_l}, \quad 1 \le i, j \le \Gamma M_t. \tag{3.54}$$

Note that Q is independent of the index p_0, i.e., it is independent of the position of $G_{p_0} - \tilde{G}_{p_0}$ in $C - \tilde{C}$. Suppose that G_{p_0} and $\tilde{G}_{p_0}$ have symbols $\mathbf{X} = [x_1 \ x_2 \ \cdots \ x_{\Gamma M_t}]$ and $\tilde{\mathbf{X}} = [\tilde{x}_1 \ \tilde{x}_2 \ \cdots \ \tilde{x}_{\Gamma M_t}]$, respectively. Then, the difference matrix between G_{p_0} and $\tilde{G}_{p_0}$ is

$$\begin{aligned} G_{p_0} - \tilde{G}_{p_0} &= \sqrt{M_t}\,\mathrm{diag}(X_1 - \tilde{X}_1,\ X_2 - \tilde{X}_2, \ldots, X_{M_t} - \tilde{X}_{M_t}) \\ &= \sqrt{M_t}\,\mathrm{diag}(\mathbf{X} - \tilde{\mathbf{X}})\left(I_{M_t} \otimes \mathbf{1}_{\Gamma \times 1}\right), \end{aligned} \tag{3.55}$$

where $\mathrm{diag}(\mathbf{X} - \tilde{\mathbf{X}}) \overset{\triangle}{=} \mathrm{diag}(x_1 - \tilde{x}_1, x_2 - \tilde{x}_2, \ldots, x_{\Gamma M_t} - \tilde{x}_{\Gamma M_t})$, I_{M_t} is the identity matrix of size $M_t \times M_t$, $\mathbf{1}_{\Gamma \times 1}$ is an all one matrix of size $\Gamma \times 1$, and $\otimes$ stands for the tensor product. Thus, we have

$$\begin{aligned} &\left[(G_{p_0} - \tilde{G}_{p_0})(G_{p_0} - \tilde{G}_{p_0})^{\mathrm{H}} \right] \circ Q \\ &= M_t \left[\mathrm{diag}(\mathbf{X} - \tilde{\mathbf{X}})\left(I_{M_t} \otimes \mathbf{1}_{\Gamma \times 1}\right)\left(I_{M_t} \otimes \mathbf{1}_{\Gamma \times 1}\right)^{\mathrm{H}} \mathrm{diag}(\mathbf{X} - \tilde{\mathbf{X}})^{\mathrm{H}} \right] \circ Q \\ &= M_t \left[\mathrm{diag}(\mathbf{X} - \tilde{\mathbf{X}})\left(I_{M_t} \otimes \mathbf{1}_{\Gamma \times \Gamma}\right) \mathrm{diag}(\mathbf{X} - \tilde{\mathbf{X}})^{\mathrm{H}} \right] \circ Q \\ &= M_t\,\mathrm{diag}(\mathbf{X} - \tilde{\mathbf{X}})\left[\left(I_{M_t} \otimes \mathbf{1}_{\Gamma \times \Gamma}\right) \circ Q \right] \mathrm{diag}(\mathbf{X} - \tilde{\mathbf{X}})^{\mathrm{H}}. \end{aligned} \tag{3.56}$$

In the above derivation, the second equality follows from the identities

$$[I_{M_t} \otimes \mathbf{1}_{\Gamma \times 1}]^{\mathrm{H}} = I_{M_t} \otimes \mathbf{1}_{1 \times \Gamma}$$

and

$$(A_1 \otimes B_1)(A_2 \otimes B_2)(A_3 \otimes B_3) = (A_1 A_2 A_3) \otimes (B_1 B_2 B_3),$$

and the last equality follows from a property of the Hadamard product ([74], p.304). If all of the eigenvalues of $[(G_{p_0} - \tilde{G}_{p_0})(G_{p_0} - \tilde{G}_{p_0})^{\mathrm{H}}] \circ Q$ are nonzero, the product of the eigenvalues is

$$\begin{aligned} &\det\left(\left[(G_{p_0} - \tilde{G}_{p_0})(G_{p_0} - \tilde{G}_{p_0})^{\mathrm{H}} \right] \circ Q \right) \\ &= M_t^{\Gamma M_t} \prod_{k=1}^{\Gamma M_t} |x_k - \tilde{x}_k|^2 \cdot \det\left(\left(I_{M_t} \otimes \mathbf{1}_{\Gamma \times \Gamma}\right) \circ Q \right) \\ &= M_t^{\Gamma M_t} \prod_{k=1}^{\Gamma M_t} |x_k - \tilde{x}_k|^2 \cdot \left(\det(Q_0)\right)^{M_t}, \end{aligned} \tag{3.57}$$

where $Q_0 = \{q_{i,j}\}_{1 \le i,j \le \Gamma}$ and $q_{i,j}$ is specified in (3.54). Q_0 can also be expressed as

$$Q_0 = W_0 \mathrm{diag}(\delta_0^2, \delta_1^2, \ldots, \delta_{L-1}^2) W_0^{\mathrm{H}}, \tag{3.58}$$

where

$$
W_0 = \begin{bmatrix}
1 & 1 & \cdots & 1 \\
w^{\tau_0} & w^{\tau_1} & \cdots & w^{\tau_{L-1}} \\
\vdots & \vdots & \ddots & \vdots \\
w^{(\Gamma-1)\tau_0} & w^{(\Gamma-1)\tau_1} & \cdots & w^{(\Gamma-1)\tau_{L-1}}
\end{bmatrix}_{\Gamma \times L}.
$$

Clearly, with $\tau_0 < \tau_1 < \cdots < \tau_{L-1}$, Q_0 is non-singular. Therefore, from (3.57) we observe that if $\prod_{k=1}^{\Gamma M_t} |x_k - \tilde{x}_k| \neq 0$, the determinant of $[(G_{p_0} - \tilde{G}_{p_0})(G_{p_0} - \tilde{G}_{p_0})^{\mathrm{H}}] \circ Q$ is nonzero. This implies that the SF code achieves a diversity order of $\Gamma M_t M_r$.

The assumption that $G_p = \tilde{G}_p$ for any $p \neq p_0$ is also sufficient to calculate the diversity product. If the rank of $\Delta \circ R$ is ΓM_t and $G_p \neq \tilde{G}_p$ for some $p \neq p_0$, the product of the nonzero eigenvalues of $\Delta \circ R$ cannot be less than that with the assumption that $G_p = \tilde{G}_p$ for any $p \neq p_0$ ([74], Corollary 3.1.3, p.149). Specifically, the diversity product can be calculated as

$$
\zeta = \frac{1}{2\sqrt{M_t}} \min_{G_{p_0} \neq \tilde{G}_{p_0}} \left| \det\left(\left[(G_{p_0} - \tilde{G}_{p_0})(G_{p_0} - \tilde{G}_{p_0})^{\mathrm{H}} \right] \circ Q \right) \right|^{\frac{1}{2\Gamma M_t}}
$$

$$
= \frac{1}{2} \min_{\mathbf{X} \neq \tilde{\mathbf{X}}} \left(\prod_{k=1}^{\Gamma M_t} |x_k - \tilde{x}_k| \right)^{\frac{1}{\Gamma M_t}} |\det(Q_0)|^{\frac{1}{2\Gamma}}
$$

$$
= \zeta_{\mathrm{in}} \cdot |\det(Q_0)|^{\frac{1}{2\Gamma}}, \tag{3.59}
$$

and

$$
\zeta_{\mathrm{in}} = \frac{1}{2} \min_{\mathbf{X} \neq \tilde{\mathbf{X}}} \left(\prod_{k=1}^{\Gamma M_t} |x_k - \tilde{x}_k| \right)^{\frac{1}{\Gamma M_t}} \tag{3.60}
$$

is termed as the "intrinsic" diversity product of the SF code. The "intrinsic" diversity product ζ_{in} does not depend on the power delay profile of the channel. Thus, we have the following theorem.

THEOREM 3.1.3 *For any SF code constructed by (3.51) and (3.52), if $\prod_{k=1}^{\Gamma M_t} |x_k - \tilde{x}_k| \neq 0$ for any pair of distinct sets of symbols $\mathbf{X} = [x_1\ x_2\ \cdots\ x_{\Gamma M_t}]$ and $\tilde{\mathbf{X}} = [\tilde{x}_1\ \tilde{x}_2\ \cdots\ \tilde{x}_{\Gamma M_t}]$, the SF code achieves a diversity order of $\Gamma M_t M_r$, and the diversity product is*

$$
\zeta = \zeta_{\mathrm{in}} |\det(Q_0)|^{\frac{1}{2\Gamma}}, \tag{3.61}
$$

where Q_0 is defined in (3.58), and ζ_{in} is the "intrinsic" diversity product defined in (3.60).

From Theorem 3.1.4, we observe that $|\det(Q_0)|$ depends only on the power delay profile of the channel, and the "intrinsic" diversity product ζ_{in} depends only on $\min_{\mathbf{X} \neq \tilde{\mathbf{X}}} (\prod_{k=1}^{\Gamma M_t} |x_k - \tilde{x}_k|)^{1/(\Gamma M_t)}$, which is called the *minimum product distance* of the set of symbols $\mathbf{X} = [x_1\ x_2\ \cdots\ x_{\Gamma M_t}]$. Therefore, given the code structure (3.52), it is desirable to design the set of symbols $\mathbf{X}$ such that the minimum product distance is as large as possible.

3.1.3.2   Maximizing the "intrinsic" diversity product

The problem of maximizing the minimum product distance of a set of signal points has arisen previously as the problem of constructing signal constellations for Rayleigh fading channels. In this subsection, we will discuss two approaches to design the set of variables $\mathbf{X} = [x_1 \ x_2 \ \cdots \ x_{\Gamma M_t}]$. For simplicity, we will use the notation $K = \Gamma M_t$.

One approach to designing the signal points $\mathbf{X} = [x_1 \ x_2 \ \cdots \ x_K]$ is to apply a transform over a K-dimensional signal set. Specifically, assume that Ω is a set of signal points (a constellation such as QAM, PAM, and so on). For any signal vector $S = [s_1 \ s_2 \ \cdots \ s_K] \in \Omega^K$, let

$$\mathbf{X} = S\mathcal{M}_K, \tag{3.62}$$

where $\mathcal{M}_K$ is a $K \times K$ matrix. For a given signal constellation Ω, the transform $\mathcal{M}_K$ should be optimized such that the minimum product distance of the set of $\mathbf{X}$ vectors is as large as possible. Both Hadamard transforms and Vandermonde matrices have been proposed for constructing $\mathcal{M}_K$ [44, 14]. The results have been used recently to design space–time block codes with full diversity. Note that the transforms $\mathcal{M}_K$ based on Vandermonde matrices result in larger minimum product distance than those based on Hadamard transforms. Some best-known transforms $\mathcal{M}_K$ based on Vandermonde matrices can be found in the previous section for diagonal algebraic ST codes.

The other approach to designing the signal set $\mathbf{X}$ is to exploit the structure of the diagonal space–time block codes. Suppose that the spectral efficiency of the SF code is r bits/s/Hz. We may consider designing the set of $L_0 = 2^{rK}$ variables directly under the energy constraint $E\|\mathbf{X}\|_{\mathrm{F}}^2 = K$. We can take advantage of the diagonal space–time block codes which are given by:

$$C_l = \mathrm{diag}(e^{ju_1\theta_l}, \ e^{ju_2\theta_l}, \ \ldots, \ e^{ju_K\theta_l}), \quad l = 0, 1, \ldots, L_0 - 1, \tag{3.63}$$

where $\theta_l = (l/L_0)2\pi$, $0 \le l \le L_0 - 1$, and $u_1, u_2, \ldots, u_K \in \{0, 1, \ldots, L_0 - 1\}$. The parameters $u_1, u_2, \ldots, u_K$ need to be optimized such that the metric

$$\min_{l \ne l'} \prod_{k=1}^{K} \left| e^{ju_k\theta_l} - e^{ju_k\theta_{l'}} \right| = \min_{1 \le l \le L_0 - 1} \prod_{k=1}^{K} \left| 2 \sin\left(\frac{u_k l}{L_0}\pi\right) \right| \tag{3.64}$$

is maximized. Then, we can design a set of variables $\mathbf{X} = [x_1 \ x_2 \ \cdots \ x_K]$ as follows. For any $l = 0, 1, \ldots, L_0 - 1$, let

$$x_k = e^{ju_k\theta_l}, \quad k = 1, 2, \ldots, K. \tag{3.65}$$

As a consequence, the minimum product distance of the set of the resulting signal vectors $\mathbf{X}$ is determined by the metric in (3.64). The optimum parameters $\mathbf{u} = [u_1 \ u_2 \ \cdots \ u_K]$ can be obtained via computer search. With exhaustive search [70], one can find

$$K = 4, \quad L_0 = 16, \quad \mathbf{u} = [1 \ 3 \ 5 \ 7];$$
$$K = 4, \quad L_0 = 256, \quad \mathbf{u} = [1 \ 25 \ 97 \ 107];$$
$$K = 6, \quad L_0 = 64, \quad \mathbf{u} = [1 \ 9 \ 15 \ 17 \ 23 \ 25];$$
$$K = 6, \quad L_0 = 1024, \quad \mathbf{u} = [1 \ 55 \ 149 \ 327 \ 395 \ 417]; \quad \ldots.$$

3.1.3.3 Maximizing the coding advantage by permutations

In the previous subsection, we obtained a class of SF codes with full rate and full diversity assuming that the transmitter has no *a priori* knowledge about the channel. In this case, the performance of the SF codes can be improved by random interleaving, as it can reduce the correlation between adjacent subcarriers. However, if the power delay profile of the channel is available at the transmitter side, further improvement can be achieved by developing a permutation (or interleaving) method that explicitly takes the power delay profile into account. This possibility will be explored in this subsection.

Suppose that the path delays $\tau_0, \tau_1, \ldots, \tau_{L-1}$ and powers $\delta_0^2, \delta_1^2, \ldots, \delta_{L-1}^2$ are available at the transmitter. Our objective is to develop an optimum permutation (or interleaving) method for the SF codes defined by (3.51) and (3.52) such that the resulting coding advantage is maximized. By permuting the rows of an SF codeword C, we obtain an interleaved codeword $\sigma(C)$. We know that for two distinct SF codewords C and $\tilde{C}$ constructed from $G_1, G_2, \ldots, G_P$ and $\tilde{G}_1, \tilde{G}_2, \ldots, \tilde{G}_P$, respectively, there exists at least one index p_0 ($1 \leq p_0 \leq P$) such that $G_{p_0} \neq \tilde{G}_{p_0}$. In order to determine the minimum rank of $[\sigma(C - \tilde{C})\sigma(C - \tilde{C})^H] \circ R$, we may further assume that $G_p = \tilde{G}_p$ for any $p \neq p_0$ for the same reason as stated in the previous section.

Suppose that G_{p_0} and $\tilde{G}_{p_0}$ consist of symbols $\mathbf{X} = [x_1 \ x_2 \ \cdots \ x_{\Gamma M_t}]$ and $\tilde{\mathbf{X}} = [\tilde{x}_1 \ \tilde{x}_2 \ \cdots \ \tilde{x}_{\Gamma M_t}]$, respectively, with $x_k \neq \tilde{x}_k$ for all $1 \leq k \leq \Gamma M_t$. For simplicity, we use the notation $\Delta x_k = x_k - \tilde{x}_k$ for $k = 1, 2, \ldots, \Gamma M_t$. After row permutation, we assume that the k-th ($1 \leq k \leq \Gamma M_t$) row of $G_{p_0} - \tilde{G}_{p_0}$ is located at the n_k-th ($0 \leq n_k \leq N-1$) row of $\sigma(C - \tilde{C})$, i.e., the k-th row of G_{p_0} will be transmitted at the n_k-th subcarrier. Then, all the $(n_{(m-1)\Gamma+i}, n_{(m-1)\Gamma+j})$-th, $1 \leq i, j \leq \Gamma$ and $1 \leq m \leq M_t$, entries of $\sigma(C - \tilde{C})\sigma(C - \tilde{C})^H$ are nonzero, and the other entries are zero. Thus, all the entries of $[\sigma(C - \tilde{C})\sigma(C - \tilde{C})^H] \circ R$ are zeros except the $(n_{(m-1)\Gamma+i}, n_{(m-1)\Gamma+j})$-th entries for $1 \leq i, j \leq \Gamma$ and $1 \leq m \leq M_t$. For convenience, we define the matrices A_m, $m = 1, 2, \ldots, M_t$, such that the (i, j)-th ($1 \leq i, j \leq \Gamma$) entry of A_m is the $(n_{(m-1)\Gamma+i}, n_{(m-1)\Gamma+j})$-th entry of $[\sigma(C - \tilde{C})\sigma(C - \tilde{C})^H] \circ R$. Since the correlation matrix R is a Toeplitz matrix (see (3.53)), the (i, j)-th, $1 \leq i, j \leq \Gamma$, entry of A_m can be expressed as

$$A_m(i, j) = M_t \Delta x_q \Delta x_r^* \sum_{l=0}^{L-1} \delta_l^2 \omega^{(n_q - n_r)\tau_l}, \qquad (3.66)$$

where $q = (m - 1)\Gamma + i$ and $r = (m - 1)\Gamma + j$. Note that the nonzero eigenvalues of $[\sigma(C - \tilde{C})\sigma(C - \tilde{C})^H] \circ R$ are determined by the matrices $A_m, m = 1, 2, \ldots, M_t$. It can be shown that the product of the nonzero eigenvalues of $[\sigma(C - \tilde{C})\sigma(C - \tilde{C})^H] \circ R$, $\lambda_1, \lambda_2, \ldots, \lambda_{\Gamma M_t}$, can be calculated as (leave proof as an exercise)

$$\prod_{k=1}^{\Gamma M_t} \lambda_k = \prod_{m=1}^{M_t} |\det(A_m)|. \qquad (3.67)$$

From (3.66), for each $m = 1, 2, \ldots, M_t$, the $\Gamma \times \Gamma$ matrix A_m can be decomposed as follows:

$$A_m = D_m W_m \Lambda W_m^H D_m^H, \qquad (3.68)$$

where

$$\Lambda = \text{diag}(\delta_0^2, \ \delta_1^2, \ \ldots, \ \delta_{L-1}^2),$$

$$D_m = \sqrt{M_t} \, \text{diag}(\Delta x_{(m-1)\Gamma+1}, \ \Delta x_{(m-1)\Gamma+2}, \ \ldots, \ \Delta x_{m\Gamma}),$$

$$W_m = \begin{bmatrix} w^{n_{(m-1)\Gamma+1}\tau_0} & w^{n_{(m-1)\Gamma+1}\tau_1} & \cdots & w^{n_{(m-1)\Gamma+1}\tau_{L-1}} \\ w^{n_{(m-1)\Gamma+2}\tau_0} & w^{n_{(m-1)\Gamma+2}\tau_1} & \cdots & w^{n_{(m-1)\Gamma+2}\tau_{L-1}} \\ \vdots & \vdots & \ddots & \vdots \\ w^{n_{m\Gamma}\tau_0} & w^{n_{m\Gamma}\tau_1} & \cdots & w^{n_{m\Gamma}\tau_{L-1}} \end{bmatrix}. \quad (3.69)$$

As a consequence, the determinant of A_m is given by

$$\det(A_m) = M_t^{\Gamma} \prod_{i=1}^{\Gamma} \left| \Delta x_{(m-1)\Gamma+i} \right|^2 \det(W_m \Lambda W_m^{\text{H}}). \quad (3.70)$$

Substituting (3.70) into (3.67), the expression for the product of the nonzero eigenvalues of $[\sigma(C - \tilde{C})\sigma(C - \tilde{C})^{\text{H}}] \circ R$ takes the form

$$\prod_{k=1}^{\Gamma M_t} \lambda_k = M_t^{\Gamma M_t} \prod_{k=1}^{\Gamma M_t} |\Delta x_k|^2 \prod_{m=1}^{M_t} \left| \det(W_m \Lambda W_m^{\text{H}}) \right|. \quad (3.71)$$

Therefore, the diversity product of the permuted SF code can be calculated as

$$\zeta = \frac{1}{2} \min_{\mathbf{x} \neq \tilde{\mathbf{x}}} \left(\prod_{k=1}^{\Gamma M_t} |\Delta x_k| \right)^{\frac{1}{\Gamma M_t}} \left(\prod_{m=1}^{M_t} \left| \det(W_m \Lambda W_m^{\text{H}}) \right| \right)^{\frac{1}{2\Gamma M_t}} = \zeta_{\text{in}} \cdot \zeta_{\text{ex}}, \quad (3.72)$$

where ζ_{in} is the "intrinsic" diversity product defined in (3.60), and ζ_{ex}, the "extrinsic" diversity product, is defined by

$$\zeta_{\text{ex}} = \left(\prod_{m=1}^{M_t} \left| \det(W_m \Lambda W_m^{\text{H}}) \right| \right)^{\frac{1}{2\Gamma M_t}}. \quad (3.73)$$

The "extrinsic" diversity product ζ_{ex} depends only on the permutation and the power delay profile of the channel. The permutation does not effect the "intrinsic" diversity product ζ_{in}.

From (3.69), for each $m = 1, 2, \ldots, M_t$, W_m can be written as

$$W_m = V_m \cdot \text{diag}(w^{n_{(m-1)\Gamma+1}\tau_0}, \ w^{n_{(m-1)\Gamma+1}\tau_1}, \ \ldots, \ w^{n_{(m-1)\Gamma+1}\tau_{L-1}}), \quad (3.74)$$

where

$$V_m = \begin{bmatrix} 1 & \cdots & 1 \\ w^{[n_{(m-1)\Gamma+2}-n_{(m-1)\Gamma+1}]\tau_0} & \cdots & w^{[n_{(m-1)\Gamma+2}-n_{(m-1)\Gamma+1}]\tau_{L-1}} \\ \vdots & \ddots & \vdots \\ w^{[n_{m\Gamma}-n_{(m-1)\Gamma+1}]\tau_0} & \cdots & w^{[n_{m\Gamma}-n_{(m-1)\Gamma+1}]\tau_{L-1}} \end{bmatrix}. \quad (3.75)$$

Thus, $\det(W_m \Lambda W_m^{\text{H}}) = \det(V_m \Lambda V_m^{\text{H}})$. We observe that the determinant of $W_m \Lambda W_m^{\text{H}}$ depends only on the relative positions of the permuted rows with respect to the position $n_{(m-1)\Gamma+1}$, not on their absolute positions.

Let us summarize the above discussion in the following result and obtain upper bounds on the "extrinsic" diversity product ζ_{ex} for arbitrary permutations.

THEOREM 3.1.4 *For any subcarrier permutation, the diversity product of the resulting SF code is*

$$\zeta = \zeta_{in} \cdot \zeta_{ex}, \qquad (3.76)$$

where ζ_{in} and ζ_{ex} are the "intrinsic" and "extrinsic" diversity products defined in (3.60) and (3.73), respectively. Moreover, the "extrinsic" diversity product ζ_{ex} is upper bounded as:

(i) *$\zeta_{ex} \leq 1$; and, more precisely,*

(ii) *if we sort the power profile $\delta_0, \delta_1, \ldots, \delta_{L-1}$ in a non-increasing order as: $\delta_{l_1} \geq \delta_{l_2} \geq \cdots \geq \delta_{l_L}$, then*

$$\zeta_{ex} \leq \left(\prod_{i=1}^{\Gamma} \delta_{l_i} \right)^{\frac{1}{\Gamma}} \left| \prod_{m=1}^{M_t} \det(V_m V_m^H) \right|^{\frac{1}{2\Gamma M_t}}, \qquad (3.77)$$

where equality holds when $\Gamma = L$. As a consequence,

$$\zeta_{ex} \leq \sqrt{L} \left(\prod_{i=1}^{\Gamma} \delta_{l_i} \right)^{\frac{1}{\Gamma}}. \qquad (3.78)$$

We leave the proof of Theorem 3.1.4 as an exercise. We observe from Theorem 3.1.4 (ii) that the "extrinsic" diversity product ζ_{ex} depends on the power delay profile in two ways. First, it depends on the power distribution through the square root of the geometric average of the largest Γ path powers, i.e., $(\prod_{i=1}^{\Gamma} \delta_{l_i})^{1/\Gamma}$. In case of $\Gamma = L$, the best performance is expected if the power distribution is uniform (i.e., $\delta_l^2 = 1/L$) since the sum of the path powers is unity. Second, the "extrinsic" diversity product ζ_{ex} also depends on the delay distribution and the applied subcarrier permutation. On the other hand, the "intrinsic" diversity product, ζ_{in}, is not affected by the power delay profile or the permutation method. It only depends on the signal constellation and the SF code design via the achieved minimum product distance.

3.1.3.4 Maximizing the "extrinsic" diversity product

By carefully choosing the applied permutation method, the overall performance of the SF code can be improved by increasing the value of the "extrinsic" diversity product ζ_{ex}. Toward this end, we consider a specific permutation strategy.

We decompose any integer n ($0 \leq n \leq N - 1$) as

$$n = e_1 \Gamma + e_0, \qquad (3.79)$$

where $0 \leq e_0 \leq \Gamma - 1$, $e_1 = \lfloor n/\Gamma \rfloor$, and $\lfloor x \rfloor$ denotes the largest integer not greater than x. For a fixed integer μ ($\mu \geq 1$), we further decompose e_1 in (3.79) as

$$e_1 = v_1 \mu + v_0, \qquad (3.80)$$

where $0 \leq v_0 \leq \mu - 1$ and $v_1 = \lfloor e_1/\mu \rfloor$.

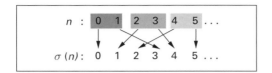

Fig. 3.7   An illustration of the permutation with $\Gamma = 2$ and separation factor $\mu = 3$.

We permute the rows of the $N \times M_t$ SF codeword constructed from (3.51) and (3.52) in such a way that the n-th ($0 \leq n \leq N - 1$) row of C is moved to the $\sigma(n)$-th row, so that

$$\sigma(n) = v_1 \mu \Gamma + e_0 \mu + v_0, \tag{3.81}$$

where e_0, v_0, v_1 come from (3.79) and (3.80). We call the integer μ as the *separation factor*. The separation factor μ should be chosen such that $\sigma(n) \leq N$ for any $0 \leq n \leq N - 1$, or, equivalently, $\mu \leq \lfloor N/\Gamma \rfloor$. Moreover, in order to guarantee that the mapping (3.81) is one-to-one over the set $\{0, 1, \ldots, N - 1\}$ (i.e., it defines a permutation), μ must be a factor of N. The role of the permutation specified in (3.81) is to separate two neighboring rows of C by μ subcarriers. An example of this permutation method is depicted in Figure 3.7.

The following result characterizes the extrinsic diversity product of the SF code that is permuted with the above described method. A proof of the result can be found in the exercise.

THEOREM 3.1.5   *For the permutation specified in (3.81) with a separation factor μ, the "extrinsic" diversity product of the permuted SF code is*

$$\zeta_{\text{ex}} = |\det(V_0 \Lambda V_0^{\text{H}})|^{\frac{1}{2\Gamma}}, \tag{3.82}$$

where

$$V_0 = \begin{bmatrix} 1 & 1 & \cdots & 1 \\ w^{\mu \tau_0} & w^{\mu \tau_1} & \cdots & w^{\mu \tau_{L-1}} \\ w^{2\mu \tau_0} & w^{2\mu \tau_1} & \cdots & w^{2\mu \tau_{L-1}} \\ \vdots & \vdots & \ddots & \vdots \\ w^{(\Gamma-1)\mu \tau_0} & w^{(\Gamma-1)\mu \tau_1} & \cdots & w^{(\Gamma-1)\mu \tau_{L-1}} \end{bmatrix}_{\Gamma \times L}. \tag{3.83}$$

Moreover, if $\Gamma = L$, the "extrinsic" diversity product ζ_{ex} can be calculated as

$$\zeta_{\text{ex}} = \left(\prod_{l=0}^{L-1} \delta_l \right)^{\frac{1}{L}} \left(\prod_{0 \leq l_1 < l_2 \leq L-1} \left| 2 \sin \left(\frac{\mu(\tau_{l_2} - \tau_{l_1})\pi}{T} \right) \right| \right)^{\frac{1}{L}}. \tag{3.84}$$

The permutation (3.81) is determined by the separation factor μ. Our objective is to find a separation factor μ_{op} that maximizes the "extrinsic" diversity product ζ_{ex}:

$$\mu_{\text{op}} = \arg \max_{1 \leq \mu \leq \lfloor N/\Gamma \rfloor} |\det(V_0 \Lambda V_0^{\text{H}})|. \tag{3.85}$$

If $\Gamma = L$, the optimum separation factor μ_{op} can be expressed as

$$\mu_{\text{op}} = \arg \max_{1 \leq \mu \leq \lfloor N/\Gamma \rfloor} \prod_{0 \leq l_1 < l_2 \leq L-1} \left| \sin\left(\frac{\mu(\tau_{l_2} - \tau_{l_1})\pi}{T} \right) \right|, \tag{3.86}$$

which is independent of the path powers. The optimum separation factor can be easily found via low complexity computer search. However, in some cases, closed-form solutions can also be obtained.

Example 3.2   If $\Gamma = L = 2$, the "extrinsic" diversity product ζ_{ex} is

$$\zeta_{\text{ex}} = \sqrt{\delta_0 \delta_1} \left| 2 \sin\left(\frac{\mu(\tau_1 - \tau_0)\pi}{T} \right) \right|^{\frac{1}{2}}. \tag{3.87}$$

Suppose that the system has $N = 128$ subcarriers, and the total bandwidth is BW $= 1$ MHz. Then, the OFDM block duration is $T = 128\,\mu s$ without the cyclic prefix. If $\tau_1 - \tau_0 = 5\,\mu s$, then $\mu_{\text{op}} = 64$ and $\zeta_{\text{ex}} = \sqrt{2\delta_0\delta_1}$. If $\tau_1 - \tau_0 = 20\,\mu s$, then $\mu_{\text{op}} = 16$ and $\zeta_{\text{ex}} = \sqrt{2\delta_0\delta_1}$. In general, if $\tau_1 - \tau_0 = 2^a b\,\mu s$, where a is a nonnegative integer and b is an odd integer, $\mu_{\text{op}} = 128/2^{a+1}$. In all of these cases, the "extrinsic" diversity product is $\zeta_{\text{ex}} = \sqrt{2\delta_0\delta_1}$, which achieves the upper bound (3.78) of Theorem 3.1.4.   ▲

Example 3.3   Assume that $\tau_l - \tau_0 = l N_0 T / N$, $l = 1, 2, \ldots, L-1$, and N is an integer multiple of $L N_0$, where N_0 is a constant and not necessarily an integer. If $\Gamma = L$ or $\delta_0^2 = \delta_1^2 = \cdots = \delta_{L-1}^2 = 1/L$, the optimum separation factor is

$$\mu_{\text{op}} = \frac{N}{L N_0}, \tag{3.88}$$

and the corresponding "extrinsic" diversity product is $\zeta_{\text{ex}} = \sqrt{L} (\prod_{l=0}^{L-1} \delta_l)^{1/L}$ (see Appendix D for the proof). In particular, in case of $\delta_0^2 = \delta_1^2 = \cdots = \delta_{L-1}^2 = 1/L$, $\zeta_{\text{ex}} = 1$. In both cases, the "extrinsic" diversity products achieve the upper bounds of Theorem 3.1.4. Note that if $\tau_l = lT/N$ for $l = 0, 1, \ldots, L-1$, $\Gamma = L$ and N is an integer multiple of L.   ▲

We now determine the optimum separation factors for two commonly used multipath fading models. The COST 207 six-ray power delay profiles for *typical urban* (TU) and *hilly terrain* (HT) environments are described in Tables 3.1 and 3.2, respectively. We consider two different bandwidths: (a) BW $= 1$ MHz, and (b) BW $= 4$ MHz. Suppose that the OFDM has $N = 128$ subcarriers. The plots of the "extrinsic" diversity product ζ_{ex} as the function of the separation factor μ for the TU and HT channel models are shown in Figures 3.8 and 3.9, respectively. In each figure, the curves of the "extrinsic" diversity product are depicted for different Γ ($2 \leq \Gamma \leq L$) values. Note that for a fixed Γ, $\Gamma = 2, 3, \ldots, L$, the separation factor μ cannot be greater than $\lfloor N/\Gamma \rfloor$.

Table 3.1 Typical urban (TU) six-ray power delay profile.

Delay profile (μs):	0.0	0.2	0.5	1.6	2.3	5.0
Power profile:	0.189	0.379	0.239	0.095	0.061	0.037

Table 3.2 Hilly terrain (HT) six-ray power delay profile.

Delay profile (μs):	0.0	0.1	0.3	0.5	15.0	17.2
Power profile:	0.413	0.293	0.145	0.074	0.066	0.008

Let us focus on the case where $\Gamma = 2$. For the TU channel model with BW $= 1$ MHz (Figure 3.8(a)), the maximum "extrinsic" diversity product is $\zeta_{\mathrm{ex}} = 0.8963$. The corresponding separation factor is $\mu_{\mathrm{op}} = 40$. However, to ensure one-to-one mapping, we choose $\mu = 64$, which results in an "extrinsic" diversity product $\zeta_{\mathrm{ex}} = 0.8606$. For the TU channel model with BW $= 4$ MHz (Figure 3.8(b)), the maximum "extrinsic" diversity product is $\zeta_{\mathrm{ex}} = 0.9998$, which approaches the upper bound 1 stated in Theorem 3.1.4. The corresponding separation factor is $\mu_{\mathrm{op}} = 51$. Similarly, we choose $\mu = 64$ to generate a permutation. The resulting "extrinsic" diversity product is $\zeta_{\mathrm{ex}} = 0.9751$, which is a slight performance loss compared to the maximum value $\zeta_{\mathrm{ex}} = 0.9998$. Finally, in case of the six-ray HT channel model with BW $= 1$ MHz (Figure 3.9(a)), the maximum "extrinsic" diversity product is $\zeta_{\mathrm{ex}} = 0.8078$. The corresponding separation factor is $\mu_{\mathrm{op}} = 64$, which is desirable. For the HT channel model with BW $= 4$ MHz (Figure 3.9(b)), the maximum "extrinsic" diversity product is $\zeta_{\mathrm{ex}} = 0.9505$, and the corresponding separation factor is $\mu_{\mathrm{op}} = 46$. To ensure one-to-one mapping, we choose $\mu = 64$, which results in an "extrinsic" diversity product $\zeta_{\mathrm{ex}} = 0.9114$.

3.1.3.5 Code design examples and performance comparisons

In the following, we provide some full-rate full-diversity SF code design examples and show their performances. Assume that the MIMO-OFDM system has $M_{\mathrm{t}} = 2$ transmit antennas, $M_{\mathrm{r}} = 1$ receive antenna and $N = 128$ subcarriers. The simulated full-rate full-diversity SF codes are constructed according to (3.51) and (3.52) with $\Gamma = 2$, yielding the code block structure

$$G = \sqrt{2} \begin{bmatrix} x_1 & 0 \\ x_2 & 0 \\ 0 & x_3 \\ 0 & x_4 \end{bmatrix}. \tag{3.89}$$

The symbols x_1, x_2, x_3, x_4 were obtained as

$$[x_1 \ x_2 \ x_3 \ x_4] = [s_1 \ s_2 \ s_3 \ s_4] \cdot \frac{1}{2} V(\theta, -\theta, j\theta, -j\theta), \tag{3.90}$$

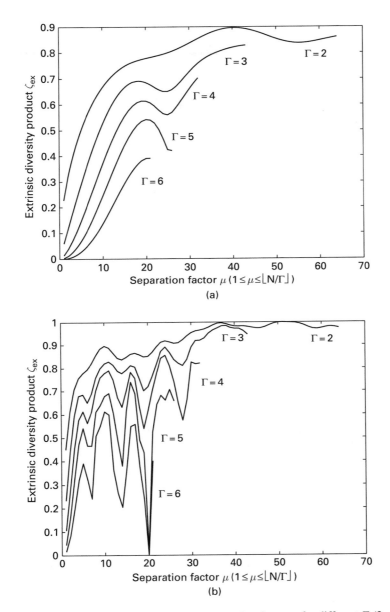

Fig. 3.8 Extrinsic diversity product ζ_{ex} versus separation factor μ for different Γ ($2 \leq \Gamma \leq 6$), TU channel model. (a) BW = 1 MHz, (b) BW = 4 MHz.

where s_1, s_2, s_3, s_4 were chosen from BPSK constellation ($s_i \in \{1, -1\}$) or QPSK constellation ($s_i \in \{\pm 1, \pm j\}$), $V(\cdot)$ is the Vandermonde matrix, and $\theta = e^{j\pi/8}$. This code targets a frequency diversity order of $\Gamma = 2$, thus it achieves full diversity only if the number of delay paths is $L \leq 2$.

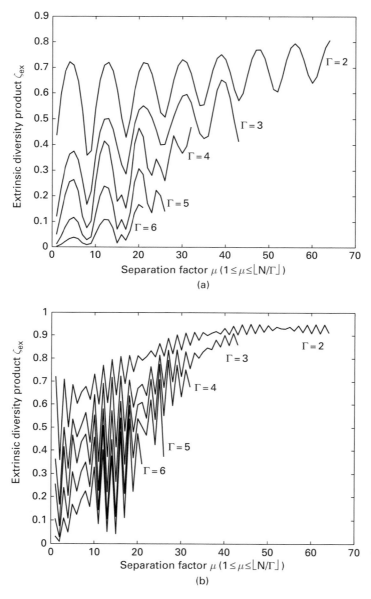

Fig. 3.9   Extrinsic diversity product ζ_{ex} versus separation factor μ for different Γ ($2 \leq \Gamma \leq 6$), HT channel model. (a) BW = 1 MHz, (b) BW = 4 MHz.

Let us compare the performances of the considered SF codes with three permutation schemes: no permutation, random permutation, and the optimum permutation. The random permutation was generated by the Takeshita–Constello method [210], which is given by

$$\sigma(n) = \left(\frac{n(n+1)}{2} \right) \bmod N, \quad n = 0, 1, \cdots, N - 1. \tag{3.91}$$

Code performances with different permutation schemes

Let us compare the performance of the full-rate full-diversity SF codes using different permutation schemes. For example, let us focus on the code (3.89) with the channel symbols s_1, s_2, s_3, s_4 chosen from BPSK constellation. The symbol rate of this code is 1, and its spectral efficiency is 1 bit/s/Hz, ignoring the cyclic prefix.

First, we consider a simple two-ray, equal-power delay profile, with a delay $\tau\,\mu$s between the two rays. We consider two cases: (a) $\tau = 5\,\mu$s, and (b) $\tau = 20\,\mu$s with OFDM bandwidth BW = 1 MHz. From the BER curves, shown in Figures 3.10(a) and (b), we observe that the performance of the proposed SF code with the random permutation is better than that without permutation. In case of $\tau = 5\mu$s, the performance improvement is more significant. With the optimum permutation, the performance is further improved. In case of $\tau = 5\mu$s, there is a 3 dB gain between the optimum permutation ($\mu_{op} = 64$) and the random permutation at a BER of 10^{-5}. In case of $\tau = 20\mu$s, the performance improvement of the optimum permutation ($\mu_{op} = 16$) over the random permutation is about 2 dB at a BER of 10^{-5}. If no permutation is used, the performance of the code in the $\tau = 5\,\mu$s case ($\zeta_{ex} = 0.3499$) is worse than that in the $\tau = 20\,\mu$s case ($\zeta_{ex} = 0.6866$). However, if we apply the optimum permutation, the performance of the SF code in both the $\tau = 5\,\mu$s case ($\mu_{op} = 64$, $\zeta_{ex} = 1$) and the $\tau = 20\,\mu$s case ($\mu_{op} = 16$, $\zeta_{ex} = 1$) is approximately the same. This confirms that by careful interleaver design, the performance of the SF codes can be significantly improved. Moreover, the consistency between the theoretical diversity product values and the simulation results suggests that the "extrinsic" diversity product ζ_{ex} is a good indicator of the code performance.

Let us simulate the code (3.89) with a more practical TU channel model. We consider two situations: (a) BW = 1 MHz, and (b) BW = 4 MHz. Figure 3.11 provides the performance results of the code with different permutations for the TU channel model. From Figures 3.11(a) and (b), we observe that in both cases, the code with random permutation has a significant improvement over the non-permuted code. Using the proposed permutation with a separation factor $\mu = 64$, there is an additional gain of 1.5 dB and 1 dB at a BER of 10^{-5} in case of BW = 1 MHz and BW = 4 MHz, respectively.

Comparisons of the full-rate full-diversity SF codes with other SF codes

Let us compare the performance of the full-rate full-diversity SF codes with that of the full-diversity SF codes obtained via mapping. We consider the full-rate full-diversity SF code (3.89) with symbols s_1, s_2, s_3, s_4 chosen from QPSK constellation. The symbol rate of the code is 1, and the spectral efficiency is 2 bits/s/Hz, ignoring the cyclic prefix. The full-diversity SF code is a repetition of the Alamouti scheme two times as follows:

$$G = \begin{bmatrix} x_1 & x_2 \\ x_1 & x_2 \\ -x_2^* & x_1^* \\ -x_2^* & x_1^* \end{bmatrix}, \tag{3.92}$$

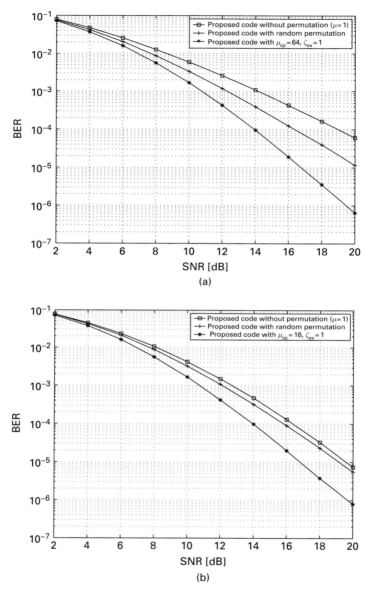

Fig. 3.10   Performance of the proposed SF code with different permutations, two-ray channel model. (a) Two rays at 0 and 5 μs, (b) Two rays at 0 and 20 μs.

where the channel symbols x_1 and x_2 were chosen from 16-QAM in order to maintain the same spectral efficiency.

First, we consider the two-ray, equal-power profile, with (a) $\tau = 5\,\mu$s, and (b) $\tau = 20\,\mu$s. The total bandwidth was BW $= 1\,$MHz. From the BER curve of the $\tau = 5\,\mu$s case, depicted in Figure 3.12(a), we observe that without permutation, the

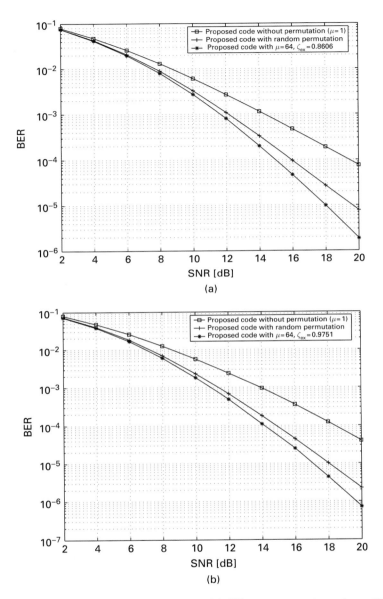

Fig. 3.11   Performance of the proposed SF code with different permutations, six-ray TU channel model.
(a) BW = 1 MHz, (b) BW = 4 MHz.

full-rate SF code outperforms the SF code from orthogonal design by about 3 dB at a
BER of 10^{-4}. With the random permutation (3.91), the full-rate code outperforms the
code from orthogonal design by about 2 dB at a BER of 10^{-4}. With the optimum permutation ($\mu_{op} = 64$), the full-rate code has an additional gain of 3 dB at a BER of
10^{-4}. Compared to the code from orthogonal design with the random permutation, the
full-rate code with the optimum permutation has a total gain of 5 dB at a BER of 10^{-4}.

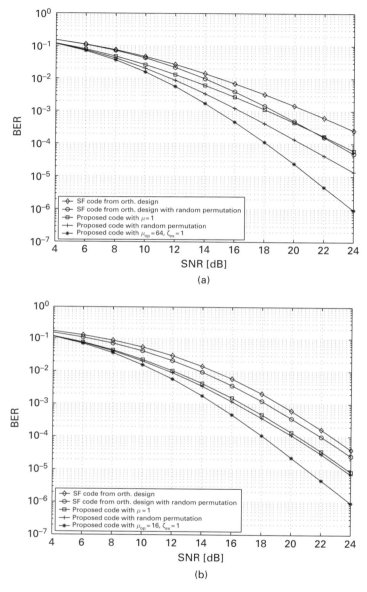

Fig. 3.12   Comparison of the proposed SF code and the code from orthogonal design, two-ray channel model. (a) Two rays at 0 and 5, (b) two rays at 0 and 20 μs.

Figure 3.12(b) shows the performance of the SF codes in the $\tau = 20\,\mu\text{s}$ case. It can be seen that without permutation, the full-rate code outperforms the code (3.92) by about $2\,\text{dB}$ at a BER of 10^{-4}. With the random permutation (3.91), the performance of the full-rate code is better than that of the code (3.92) by about $2\,\text{dB}$ at a BER of 10^{-4}. With the optimum permutation ($\mu_{\text{op}} = 16$), an additional improvement of $2\,\text{dB}$ at a BER of 10^{-4} is achieved by the full-rate code.

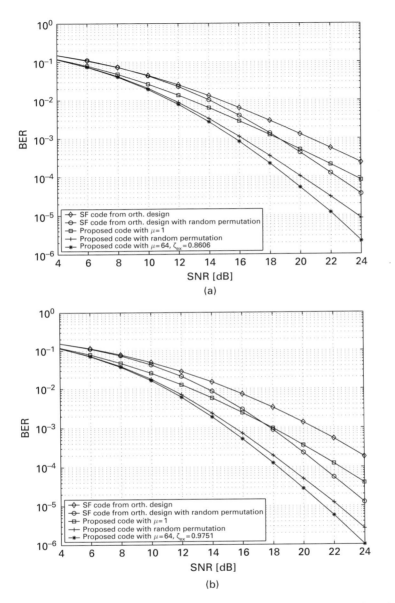

Fig. 3.13   Comparison of the proposed SF code and the code from orthogonal design, six-ray TU channel
model. (a) BW = 1 MHz, (b) BW = 4 MHz.

Let us compare the performances of the two SF codes with a more practical TU
channel model. We consider two situations: (a) BW = 1 MHz, and (b) BW = 4 MHz.
Figure 3.13 depicts the simulation results for the TU channel model. In case of BW =
1 MHz, from Figure 3.13(a), we can see that without permutation, the full-rate SF code
outperforms the SF code (3.92) by about 2 dB at a BER of 10^{-4}. With the random permutation (3.91), the performance of the full-rate code is better than that of the code from

orthogonal design by about 2.5 dB at a BER of 10^{-4}. With the permutation ($\mu = 64$), an additional improvement of 1 dB at a BER of 10^{-4} is achieved by the full-rate SF code. In case of BW = 4 MHz, from Figure 3.13(b), we observe that without permutation, the performance of the full-rate code is better than that of the code from orthogonal design by about 3 dB at a BER of 10^{-4}. With the random permutation, the full-rate SF code outperforms the SF code (3.92) by about 2 dB at a BER of 10^{-4}. With the permutation ($\mu = 64$), there is an additional gain of about 1 dB at a BER of 10^{-4}. Compared to the SF code from orthogonal design with the random permutation, the full-rate SF code with the optimum permutation has a total gain of 3 dB at a BER of 10^{-4}.

3.2 Space–time–frequency diversity and coding

In this section, we consider transmission techniques to exploit all of the spatial, temporal, and frequency diversity available in broadband wireless communications. First, we briefly review a STF-coded MIMO-OFDM system model and determine the maximum achievable diversity in this case. Then, we review two systematic approaches to design STF codes to achieve the maximum achievable diversity in MIMO-OFDM systems.

3.2.1 STF-coded MIMO-OFDM system model

A STF-coded MIMO-OFDM system with M_t transmit antennas, M_r receive antennas and N subcarriers is shown in Figure 3.14. Suppose that the frequency selective fading channels between each pair of transmit and receive antennas have L independent delay paths and the same power delay profile. The MIMO channel is assumed to be constant over each OFDM block period, but it may vary from one OFDM block to another. At the k-th OFDM block, the channel impulse response from transmit antenna i to receive antenna j at time τ can be modeled as

$$h_{i,j}^k(\tau) = \sum_{l=0}^{L-1} \alpha_{i,j}^k(l)\delta(\tau - \tau_l), \qquad (3.93)$$

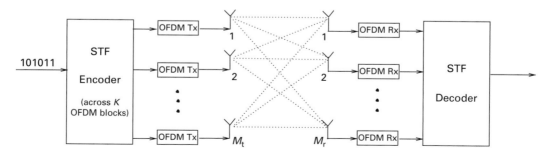

Fig. 3.14 STF-coded MIMO-OFDM system with M_t transmit and M_r receive antennas.

where τ_l is the delay and $\alpha^k_{i,j}(l)$ is the complex amplitude of the l-th path between transmit antenna i and receive antenna j. The $\alpha^k_{i,j}(l)$'s are modeled as zero-mean, complex Gaussian random variables with variances $E|\alpha^k_{i,j}(l)|^2 = \delta^2_l$, where E stands for the expectation. The powers of the L paths are normalized such that $\sum^{L-1}_{l=0} \delta^2_l = 1$. Assume that the MIMO channel is spatially uncorrelated, so the channel coefficients $\alpha^k_{i,j}(l)$ are independent for different indices (i, j). From (3.93), the frequency response of the channel is given by

$$H^k_{i,j}(f) = \sum^{L-1}_{l=0} \alpha^k_{i,j}(l) e^{-j2\pi f \tau_l}, \qquad (3.94)$$

where $j = \sqrt{-1}$.

Let us consider STF coding across M_t transmit antennas, N OFDM subcarriers' and K consecutive OFDM blocks. Each STF codeword can be expressed as a $KN \times M_t$ matrix

$$C = [C^T_1 \ C^T_2 \ \cdots \ C^T_K]^T, \qquad (3.95)$$

where the channel symbol matrix C_k is given by

$$C_k = \begin{bmatrix} c^k_1(0) & c^k_2(0) & \cdots & c^k_{M_t}(0) \\ c^k_1(1) & c^k_2(1) & \cdots & c^k_{M_t}(1) \\ \vdots & \vdots & \ddots & \vdots \\ c^k_1(N-1) & c^k_2(N-1) & \cdots & c^k_{M_t}(N-1) \end{bmatrix}, \qquad (3.96)$$

and $c^k_i(n)$ is the channel symbol transmitted over the n-th subcarrier by transmit antenna i in the k-th OFDM block. The STF code is assumed to satisfy the energy constraint $E||C||^2_F = KNM_t$, where $||C||_F$ is the Frobenius norm of C. During the k-th OFDM block period, the transmitter applies an N-point IFFT to each column of the matrix C_k. After appending a cyclic prefix, the OFDM symbol corresponding to the i-th ($i = 1, 2, \ldots, M_t$) column of C_k is transmitted by transmit antenna i.

At the receiver, after matched filtering, removing the cyclic prefix, and applying FFT, the received signal at the n-th subcarrier at receive antenna j in the k-th OFDM block is given by

$$y^k_j(n) = \sqrt{\frac{\rho}{M_t}} \sum^{M_t}_{i=1} c^k_i(n) H^k_{i,j}(n) + z^k_j(n), \qquad (3.97)$$

where

$$H^k_{i,j}(n) = \sum^{L-1}_{l=0} \alpha^k_{i,j}(l) e^{-j2\pi n \Delta f \tau_l} \qquad (3.98)$$

is the channel frequency response at the n-th subcarrier between transmit antenna i and receive antenna j, $\Delta f = 1/T$ is the subcarrier separation in the frequency domain, and T is the OFDM symbol period. Assume that the channel state information $H^k_{i,j}(n)$ is known at the receiver, but not at the transmitter. In (3.97), $z^k_j(n)$ denotes the additive white complex Gaussian noise with zero mean and unit variance at the n-th subcarrier at receive antenna j in the k-th OFDM block. The factor $\sqrt{\rho/M_t}$ in (3.97) ensures that ρ is the average signal to noise ratio (SNR) at each receive antenna.

3.2.2   Performance criteria and maximum achievable diversity

In this subsection, we discuss the performance criteria for STF-coded MIMO-OFDM systems and determine the maximum achievable diversity order for such systems.

Using the notation

$$c_i((k-1)N+n) \triangleq c_i^k(n),$$

$$H_{i,j}((k-1)N+n) \triangleq H_{i,j}^k(n),$$

$$y_j((k-1)N+n) \triangleq y_j^k(n),$$

$$z_j((k-1)N+n) \triangleq z_j^k(n)$$

for $1 \leq k \leq K, 0 \leq n \leq N-1, 1 \leq i \leq M_t$ and $1 \leq j \leq M_r$, the received signal in (3.97) can be expressed as

$$y_j(m) = \sqrt{\frac{\rho}{M_t}} \sum_{i=1}^{M_t} c_i(m) H_{i,j}(m) + z_j(m) \tag{3.99}$$

for $m = 0, 1, \ldots, KN-1$. We further rewrite the received signal in vector form as

$$\mathbf{Y} = \sqrt{\frac{\rho}{M_t}} \mathbf{DH} + \mathbf{Z}, \tag{3.100}$$

where $\mathbf{D}$ is a $KNM_r \times KNM_t M_r$ matrix constructed from the STF codeword C in (3.95) as follows:

$$\mathbf{D} = I_{M_r} \otimes \begin{bmatrix} D_1 & D_2 & \cdots & D_{M_t} \end{bmatrix}, \tag{3.101}$$

where $\otimes$ denotes the tensor product, I_{M_r} is the identity matrix of size $M_r \times M_r$, and

$$D_i = \text{diag}\{c_i(0), \ c_i(1), \ \ldots, \ c_i(KN-1)\} \tag{3.102}$$

for any $i = 1, 2, \ldots, M_t$. The channel vector $\mathbf{H}$ of size $KNM_t M_r \times 1$ is formatted as

$$\mathbf{H} = [H_{1,1}^T \ \cdots \ H_{M_t,1}^T \ H_{1,2}^T \ \cdots \ H_{M_t,2}^T \ \cdots \ H_{1,M_r}^T \ \cdots \ H_{M_t,M_r}^T]^T, \tag{3.103}$$

where

$$H_{i,j} = [H_{i,j}(0) \ H_{i,j}(1) \ \cdots \ H_{i,j}(KN-1)]^T. \tag{3.104}$$

The received signal vector $\mathbf{Y}$ of size $KNM_r \times 1$ is given by

$$\mathbf{Y} = [y_1(0) \ \cdots \ y_1(KN-1) \ y_2(0) \ \cdots \ y_2(KN-1) \ \cdots \ y_{M_r}(0) \ \cdots \ y_{M_r}(KN-1)]^T, \tag{3.105}$$

and the noise vector $\mathbf{Z}$ has the same form as $\mathbf{Y}$, i.e.,

$$\mathbf{Z} = [z_1(0) \ \cdots \ z_1(KN-1) \ z_2(0) \ \cdots \ z_2(KN-1) \ \cdots \ z_{M_r}(0) \ \cdots \ z_{M_r}(KN-1)]^T. \tag{3.106}$$

Suppose that $\mathbf{D}$ and $\tilde{\mathbf{D}}$ are two matrices constructed from two different codewords C and $\tilde{C}$, respectively. Then, the pairwise error probability between $\mathbf{D}$ and $\tilde{\mathbf{D}}$ can be upper bounded as

$$\Pr(\mathbf{D} \to \tilde{\mathbf{D}}) \le \binom{2r-1}{r} \left(\prod_{i=1}^{r} \gamma_i \right)^{-1} \left(\frac{\rho}{M_t} \right)^{-r}, \qquad (3.107)$$

where r is the rank of $(\mathbf{D}-\tilde{\mathbf{D}})\mathbf{R}(\mathbf{D}-\tilde{\mathbf{D}})^{\mathrm{H}}$, $\gamma_1, \gamma_2, \ldots, \gamma_r$ are the nonzero eigenvalues of $(\mathbf{D}-\tilde{\mathbf{D}})\mathbf{R}(\mathbf{D}-\tilde{\mathbf{D}})^{\mathrm{H}}$, and $\mathbf{R} = \mathrm{E}\{\mathbf{H}\mathbf{H}^{\mathrm{H}}\}$ is the correlation matrix of $\mathbf{H}$. The superscript H stands for the complex conjugate and transpose of a matrix. Based on the upper bound on the pairwise error probability in (3.107), two general STF code performance criteria can be proposed as follows:

- *Diversity (rank) criterion:* The minimum rank of $(\mathbf{D} - \tilde{\mathbf{D}})\mathbf{R}(\mathbf{D} - \tilde{\mathbf{D}})^{\mathrm{H}}$ over all pairs of different codewords C and $\tilde{C}$ should be as large as possible.
- *Product criterion:* The minimum value of the product $\prod_{i=1}^{r} \gamma_i$ over all pairs of different codewords C and $\tilde{C}$ should be maximized.

In the case of spatially uncorrelated MIMO channels, i.e., the channel taps $\alpha_{i,j}^{k}(l)$ are independent for different transmit antenna index i and receive antenna index j, the correlation matrix $\mathbf{R}$ of size $KNM_tM_r \times KNM_tM_r$ becomes

$$\mathbf{R} = \mathrm{diag}\left(R_{1,1}, \ldots, R_{M_t,1}, R_{1,2}, \ldots, R_{M_t,2}, \ldots, R_{1,M_r}, \ldots, R_{M_t,M_r} \right), \quad (3.108)$$

where

$$R_{i,j} = \mathrm{E}\left\{ H_{i,j} H_{i,j}^{\mathrm{H}} \right\} \qquad (3.109)$$

is the correlation matrix of the channel frequency response from transmit antenna i to receive antenna j. Using the notation $w = \mathrm{e}^{-\mathrm{j}2\pi\Delta f}$, from (3.98), we have

$$H_{i,j} = (I_K \otimes W)A_{i,j}, \qquad (3.110)$$

where

$$W = \begin{bmatrix} 1 & 1 & \cdots & 1 \\ w^{\tau_0} & w^{\tau_1} & \cdots & w^{\tau_{L-1}} \\ \vdots & \vdots & \ddots & \vdots \\ w^{(N-1)\tau_0} & w^{(N-1)\tau_1} & \cdots & w^{(N-1)\tau_{L-1}} \end{bmatrix},$$

and

$$A_{i,j} = [\alpha_{i,j}^{1}(0)\ \alpha_{i,j}^{1}(1)\ \cdots\ \alpha_{i,j}^{1}(L-1)\ \cdots\ \alpha_{i,j}^{K}(0)\ \alpha_{i,j}^{K}(1)\ \cdots\ \alpha_{i,j}^{K}(L-1)]^{\mathrm{T}}.$$

Substituting (3.110) into (3.109), $R_{i,j}$ can be calculated as follows:

$$R_{i,j} = E\left\{ (I_K \otimes W)A_{i,j} A_{i,j}^{\mathrm{H}} (I_K \otimes W)^{\mathrm{H}} \right\}$$

$$= (I_K \otimes W)E\left\{ A_{i,j} A_{i,j}^{\mathrm{H}} \right\}(I_K \otimes W^{\mathrm{H}}).$$

With the assumptions that the path gains $\alpha_{i,j}^k(l)$ are independent for different paths and different pairs of transmit and receive antennas, and that the second order statistics of the time correlation is the same for all transmit and receive antenna pairs and all paths (i.e. the correlation values do not depend on i, j and l), we can define the time correlation at lag m as $r_T(m) = \mathrm{E}\{\alpha_{i,j}^k(l)\alpha_{i,j}^{k+m*}(l)\}$. Thus, the correlation matrix $\mathrm{E}\left\{A_{i,j}A_{i,j}^{\mathrm{H}}\right\}$ can be expressed as

$$\mathrm{E}\left\{A_{i,j}A_{i,j}^{\mathrm{H}}\right\} = R_T \otimes \Lambda, \tag{3.111}$$

where $\Lambda = \mathrm{diag}\{\delta_0^2, \delta_1^2, \cdots, \delta_{L-1}^2\}$, and R_T is the temporal correlation matrix of size $K \times K$, whose entry in the p-th row and the q-th column is given by $r_T(q-p)$ for $1 \leq p, q \leq K$. We can also define the frequency correlation matrix, R_F, as $R_F = \mathrm{E}\{H_{i,j}^k H_{i,j}^{k\,\mathrm{H}}\}$, where

$$H_{i,j}^k = [\, H_{i,j}^k(0), \ldots, H_{i,j}^k(N-1) \,]^{\mathrm{T}}.$$

Then, $R_F = W\Lambda W^{\mathrm{H}}$. As a result, we arrive at

$$\begin{aligned} R_{i,j} &= (I_K \otimes W)(R_T \otimes \Lambda)(I_K \otimes W^{\mathrm{H}}) \\ &= R_T \otimes (W\Lambda W^{\mathrm{H}}) = R_T \otimes R_F, \end{aligned} \tag{3.112}$$

yielding

$$\mathbf{R} = I_{M_tM_r} \otimes (R_T \otimes R_F). \tag{3.113}$$

Finally, combining (3.96), (3.101), (3.102), and (3.113), the expression for $(\mathbf{D} - \tilde{\mathbf{D}})\mathbf{R}(\mathbf{D} - \tilde{\mathbf{D}})^{\mathrm{H}}$ in (3.107) can be rewritten as

$$\begin{aligned} &(\mathbf{D} - \tilde{\mathbf{D}})\mathbf{R}(\mathbf{D} - \tilde{\mathbf{D}})^{\mathrm{H}} \\ &= I_{M_r} \otimes \left[\sum_{i=1}^{M_t} (D_i - \tilde{D}_i)(R_T \otimes R_F)(D_i - \tilde{D}_i)^{\mathrm{H}} \right] \\ &= I_{M_r} \otimes \left\{ \left[(C - \tilde{C})(C - \tilde{C})^{\mathrm{H}} \right] \circ (R_T \otimes R_F) \right\}, \end{aligned} \tag{3.114}$$

where $\circ$ denotes the Hadamard product. Let

$$\Delta \stackrel{\triangle}{=} (C - \tilde{C})(C - \tilde{C})^{\mathrm{H}}, \tag{3.115}$$

and $R \stackrel{\triangle}{=} R_T \otimes R_F$. Then, substituting (3.114) into (3.107), the pairwise error probability between C and $\tilde{C}$ can be upper bounded as

$$\Pr(C \to \tilde{C}) \leq \binom{2\nu M_r - 1}{\nu M_r} \left(\prod_{i=1}^{\nu} \lambda_i \right)^{-M_r} \left(\frac{\rho}{M_t} \right)^{-\nu M_r}, \tag{3.116}$$

where ν is the rank of $\Delta \circ R$, and $\lambda_1, \lambda_2, \ldots, \lambda_\nu$ are the nonzero eigenvalues of $\Delta \circ R$. The minimum value of the product $\prod_{i=1}^{\nu} \lambda_i$ over all pairs of distinct signals C and $\tilde{C}$ is termed as *coding advantage*, denoted by

$$\zeta_{\mathrm{STF}} = \min_{C \neq \tilde{C}} \prod_{i=1}^{\nu} \lambda_i. \tag{3.117}$$

As a consequence, we can formulate the performance criteria for STF codes as follows:

- *Diversity (rank) criterion:* The minimum rank of $\Delta \circ R$ over all pairs of distinct codewords C and $\tilde{C}$ should be as large as possible.
- *Product criterion:* The coding advantage or the minimum value of the product $\prod_{i=1}^{\nu} \lambda_i$ over all pairs of distinct signals C and $\tilde{C}$ should also be maximized.

If the minimum rank of $\Delta \circ R$ is ν for any pair of distinct STF codewords C and $\tilde{C}$, we say that the STF code achieves a *diversity order* of νM_r. For a fixed number of OFDM blocks K, number of transmit antennas M_t, and correlation matrices R_T and R_F, the *maximum achievable diversity* or *full diversity* is defined as the maximum diversity order that can be achieved by STF codes of size $KN \times M_t$.

According to the rank inequalities on Hadamard products and tensor products, we have

$$\text{rank}(\Delta \circ R) \leq \text{rank}(\Delta)\text{rank}(R_T)\text{rank}(R_F).$$

Since the rank of Δ is at most M_t and the rank of R_F is at most L, we obtain

$$\text{rank}(\Delta \circ R) \leq \min\{LM_t\text{rank}(R_T), KN\}. \tag{3.118}$$

Thus, the maximum achievable diversity is at most $\min\{LM_tM_r\text{rank}(R_T), KNM_r\}$. We can see from the following discussion that this upper bound can indeed be achieved. We also observe that if the channel stays constant over multiple OFDM blocks ($\text{rank}(R_T) = 1$), the maximum achievable diversity is $\min\{LM_tM_r, KNM_r\}$. In this case, STF coding basically reduces to SF coding and cannot provide additional diversity advantage compared to the SF coding approach.

Note that the above analytical framework includes ST and SF codes as special cases. If we consider only one subcarrier ($N = 1$), and one delay path ($L = 1$), the channel becomes a single-carrier, time-correlated, flat fading MIMO channel. The correlation matrix R simplifies to $R = R_T$, and the code design problem reduces to that of ST code design. In the case of coding over a single OFDM block ($K = 1$), the correlation matrix R becomes $R = R_F$, and the code design problem simplifies to that of SF codes.

3.2.3 Full-diversity STF code design methods

In this subsection, we review two STF code design methods to achieve the maximum achievable diversity order $\min\{LM_tM_r\text{rank}(R_T), KNM_r\}$. Without loss of generality, let us assume that the number of subcarriers, N, is not less than LM_t, so the maximum achievable diversity order is $LM_tM_r\text{rank}(R_T)$.

3.2.3.1 Repetition-based STF code design

In the previous section, a systematic approach was proposed to design full-diversity SF codes. Suppose that C_{SF} is a full-diversity SF code of size $N \times M_t$. A full-diversity STF code, C_{STF}, can be constructed by repeating C_{SF} K times (over K OFDM blocks) as follows:

$$C_{STF} = \mathbf{1}_{k \times 1} \otimes C_{SF}, \tag{3.119}$$

where $\mathbf{1}_{k \times 1}$ is an all one matrix of size $k \times 1$. Let

$$\Delta_{\text{STF}} = (C_{\text{STF}} - \tilde{C}_{\text{STF}})(C_{\text{STF}} - \tilde{C}_{\text{STF}})^{\text{H}}$$

and

$$\Delta_{\text{SF}} = (C_{\text{SF}} - \tilde{C}_{\text{SF}})(C_{\text{SF}} - \tilde{C}_{\text{SF}})^{\text{H}}.$$

Then we have

$$\Delta_{\text{STF}} = \left[\mathbf{1}_{k \times 1} \otimes (C_{\text{SF}} - \tilde{C}_{\text{SF}}) \right] \left[\mathbf{1}_{1 \times k} \otimes (C_{\text{SF}} - \tilde{C}_{\text{SF}})^{\text{H}} \right] = \mathbf{1}_{k \times k} \otimes \Delta_{\text{SF}}.$$

Thus,

$$\Delta_{\text{STF}} \circ R = (\mathbf{1}_{k \times k} \otimes \Delta_{\text{SF}}) \circ (R_{\text{T}} \otimes R_{\text{F}})$$
$$= R_{\text{T}} \otimes (\Delta_{\text{SF}} \circ R_{\text{F}}).$$

Since the SF code C_{SF} achieves full diversity in each OFDM block, the rank of $\Delta_{\text{SF}} \circ R_{\text{F}}$ is LM_{t}. Therefore, the rank of $\Delta_{\text{STF}} \circ R$ is $LM_{\text{t}}\text{rank}(R_{\text{T}})$, so C_{STF} in (3.119) is guaranteed to achieve a diversity order of $LM_{\text{t}}M_{\text{r}}\text{rank}(R_{\text{T}})$.

We can see that the maximum achievable diversity depends on the rank of the temporal correlation matrix R_{T}. If the fading channels are constant during K OFDM blocks, i.e., $\text{rank}(R_{\text{T}}) = 1$, the maximum achievable diversity order for STF codes (coding across several OFDM blocks) is the same as that for SF codes (coding within one OFDM block). Moreover, if the channel changes independently in time, i.e., $R_{\text{T}} = I_K$, the repetition structure of STF code C_{STF} in (3.119) is sufficient, but not necessary to achieve the full diversity. In this case,

$$\Delta \circ R = \text{diag}(\Delta_1 \circ R_{\text{F}}, \ \Delta_2 \circ R_{\text{F}}, \ \ldots, \ \Delta_K \circ R_{\text{F}}),$$

where $\Delta_k = (C_k - \tilde{C}_k)(C_k - \tilde{C}_k)^{\text{H}}$ for $1 \le k \le K$. Thus, in this case, the necessary and sufficient condition to achieve full diversity $KLM_{\text{t}}M_{\text{r}}$ is that each matrix $\Delta_k \circ R_{\text{F}}$ be of rank LM_{t} over all pairs of distinct codewords simultaneously for all $1 \le k \le K$.

The above repetition-based STF code design ensures full diversity at the price of the symbol rate decreasing by a factor of $1/K$ (over K OFDM blocks) compared to the symbol rate of the underlying SF code. The advantage of this approach is that *any* full-diversity SF code (block or trellis) can be used to design full-diversity STF codes.

3.2.3.2   Full-rate STF code design

Let us consider a STF code structure consisting of STF codewords C of size KN by M_{t}:

$$C = [C_1^{\text{T}} \ C_2^{\text{T}} \ \cdots \ C_K^{\text{T}}]^{\text{T}}, \tag{3.120}$$

where

$$C_k = \left[G_{k,1}^{\text{T}} \ G_{k,2}^{\text{T}} \ \cdots \ G_{k,P}^{\text{T}} \ \mathbf{0}_{N-P\Gamma M_{\text{t}}}^{\text{T}} \right]^{\text{T}} \tag{3.121}$$

for $k = 1, 2, \ldots, K$. In (3.121), $P = \lfloor N/(\Gamma M_{\text{t}}) \rfloor$, and each matrix $G_{k,p}$ ($1 \le k \le K, 1 \le p \le P$) is of size ΓM_{t} by M_{t}. The zero padding in (3.121) is used if the number of subcarriers N is not an integer multiple of ΓM_{t}. For each p ($1 \le p \le P$), we design the code matrices $G_{1,p}, G_{2,p}, \ldots, G_{K,p}$ jointly, but the design of G_{k_1,p_1} and G_{k_2,p_2}, $p_1 \neq p_2$, is independent of each other. For a fixed p $(1 \leq p \leq P)$, let

$$G_{k,p} = \sqrt{M_t}\, \mathrm{diag}\left(X_{k,1},\, X_{k,2}, \ldots, X_{k,M_t}\right), \quad k = 1, 2, \ldots, K \tag{3.122}$$

where $\mathrm{diag}(X_{k,1}, X_{k,2}, \ldots, X_{k,M_t})$ is a block diagonal matrix, in which $X_{k,i} = [x_{k,(i-1)\Gamma+1}\ \ x_{k,(i-1)\Gamma+2}\ \cdots\ x_{k,i\Gamma}]^T$, $i = 1, 2, \ldots, M_t$, and $x_{k,j}$, $j = 1, 2, \ldots, \Gamma M_t$, are complex symbols and will be specified later. The energy normalization condition is

$$E\left(\sum_{k=1}^{K}\sum_{j=1}^{\Gamma M_t} |x_{k,j}|^2\right) = K\Gamma M_t.$$

The symbol rate of the proposed scheme is $P\Gamma M_t/N$, ignoring the cyclic prefix. If N is a multiple of ΓM_t, the symbol rate is 1. If not, the rate is less than 1, but since usually N is much greater than ΓM_t, the symbol rate is very close to 1. We term full rate as one channel symbol per subcarrier per OFDM block period, so the proposed method can either achieve the full symbol rate, or it can perform very close to it. Note that this scheme includes the code design method proposed in [205] as a special case when $K = 1$.

The following theorem provides a sufficient condition for the STF codes described above to achieve a diversity order of $\Gamma M_t M_r \mathrm{rank}(R_T)$. For simplicity, let us use the notation $\mathbf{X} = [x_{1,1} \cdots x_{1,\Gamma M_t} \cdots x_{K,1} \cdots x_{K,\Gamma M_t}]$ and $\tilde{\mathbf{X}} = [\tilde{x}_{1,1} \cdots \tilde{x}_{1,\Gamma M_t} \cdots \tilde{x}_{K,1} \cdots \tilde{x}_{K,\Gamma M_t}]$. Moreover, for any $n \times n$ nonnegative definite matrix A, let us denote its eigenvalues in a non-increasing order as: $\mathrm{eig}_1(A) \geq \mathrm{eig}_2(A) \geq \cdots \geq \mathrm{eig}_n(A)$.

THEOREM 3.2.1 *For any STF code specified in (3.120)–(3.122), if $\prod_{k=1}^{K} \prod_{j=1}^{\Gamma M_t} |x_{k,j} - \tilde{x}_{k,j}| \neq 0$ for any pair of distinct symbols $\mathbf{X}$ and $\tilde{\mathbf{X}}$, the STF code achieves a diversity order of $\Gamma M_t M_r \mathrm{rank}(R_T)$, and the coding advantage is bounded by*

$$(M_t \delta_{\min})^{\Gamma M_t \mathrm{rank}(R_T)}\, \Phi \leq \zeta_{\mathrm{STF}} \leq (M_t \delta_{\max})^{\Gamma M_t \mathrm{rank}(R_T)}\, \Phi, \tag{3.123}$$

where

$$\delta_{\min} = \min_{\mathbf{X} \neq \tilde{\mathbf{X}}} \min_{1 \leq k \leq K, 1 \leq j \leq \Gamma M_t} |x_{k,j} - \tilde{x}_{k,j}|^2, \tag{3.124}$$

$$\delta_{\max} = \max_{\mathbf{X} \neq \tilde{\mathbf{X}}} \max_{1 \leq k \leq K, 1 \leq j \leq \Gamma M_t} |x_{k,j} - \tilde{x}_{k,j}|^2, \tag{3.125}$$

$$\Phi = |\det(Q_0)|^{M_t \mathrm{rank}(R_T)} \prod_{i=1}^{\mathrm{rank}(R_T)} \left(\mathrm{eig}_i(R_T)\right)^{\Gamma M_t}, \tag{3.126}$$

and

$$Q_0 = W_0 \mathrm{diag}(\delta_0^2, \delta_1^2, \ldots, \delta_{L-1}^2) W_0^H, \tag{3.127}$$

$$W_0 = \begin{bmatrix} 1 & 1 & \cdots & 1 \\ w^{\tau_0} & w^{\tau_1} & \cdots & w^{\tau_{L-1}} \\ \vdots & \vdots & \ddots & \vdots \\ w^{(\Gamma-1)\tau_0} & w^{(\Gamma-1)\tau_1} & \cdots & w^{(\Gamma-1)\tau_{L-1}} \end{bmatrix}_{\Gamma \times L}. \tag{3.128}$$

Furthermore, if the temporal correlation matrix R_T is of full rank, i.e., $\mathrm{rank}(R_T) = K$, the coding advantage is

$$\zeta_{\mathrm{STF}} = \delta\, M_t^{K\Gamma M_t}\, |\det(R_T)|^{\Gamma M_t}\, |\det(Q_0)|^{K M_t}, \tag{3.129}$$

where

$$\delta = \min_{\mathbf{X} \neq \tilde{\mathbf{X}}} \prod_{k=1}^{K} \prod_{j=1}^{\Gamma M_t} |x_{k,j} - \tilde{x}_{k,j}|^2. \tag{3.130}$$

Proof   Suppose that C and $\tilde{C}$ are two distinct STF codewords which are constructed from $G_{k,p}$ and $\tilde{G}_{k,p}$ ($1 \le k \le K$, $1 \le p \le P$), respectively. We would like to determine the rank of $\Delta \circ R$, where $\Delta = (C - \tilde{C})(C - \tilde{C})^{\mathrm{H}}$ and $R = R_T \otimes R_F$. For convenience, let

$$\mathbf{G}_p = \left[G_{1,p}^{\mathrm{T}}\ G_{2,p}^{\mathrm{T}}\ \cdots\ G_{K,p}^{\mathrm{T}} \right]^{\mathrm{T}}$$

for each $p = 1, 2, \ldots, P$. For two distinct codewords C and $\tilde{C}$, there exists at least one index p_0 ($1 \le p_0 \le P$) such that $\mathbf{G}_{p_0} \neq \tilde{\mathbf{G}}_{p_0}$. We may further assume that $\mathbf{G}_p = \tilde{\mathbf{G}}_p$ for any $p \neq p_0$ since the rank of $\Delta \circ R$ does not decrease if $\mathbf{G}_p \neq \tilde{\mathbf{G}}_p$ for some $p \neq p_0$ ([74], Corollary 3.1.3, p.149).

Note that the frequency correlation matrix R_F is a Toeplitz matrix. With the assumption that $\mathbf{G}_p = \tilde{\mathbf{G}}_p$ for any $p \neq p_0$, we observe that the nonzero eigenvalues of $\Delta \circ R$ are the same as those of $[(\mathbf{G}_{p_0} - \tilde{\mathbf{G}}_{p_0})(\mathbf{G}_{p_0} - \tilde{\mathbf{G}}_{p_0})^{\mathrm{H}}] \circ (R_T \otimes Q)$, where $Q = \{q_{i,j}\}_{1 \le i,j \le \Gamma M_t}$ is also a Toeplitz matrix whose entries are

$$q_{i,j} = \sum_{l=0}^{L-1} \delta_l^2 w^{(i-j)\tau_l}, \quad 1 \le i, j \le \Gamma M_t. \tag{3.131}$$

Note that Q is independent of the index p_0, i.e., it is independent of the position of $\mathbf{G}_{p_0} - \tilde{\mathbf{G}}_{p_0}$ in $C - \tilde{C}$. For any $1 \le k \le K$, we have

$$G_{k,p_0} - \tilde{G}_{k,p_0}$$
$$= \sqrt{M_t}\, \mathrm{diag}(X_{k,1} - \tilde{X}_{k,1},\ X_{k,2} - \tilde{X}_{k,2}, \ldots, X_{k,M_t} - \tilde{X}_{k,M_t})$$
$$= \sqrt{M_t}\, \mathrm{diag}(x_{k,1} - \tilde{x}_{k,1}, \ldots, x_{k,\Gamma M_t} - \tilde{x}_{k,\Gamma M_t})\left(I_{M_t} \otimes \mathbf{1}_{\Gamma \times 1}\right),$$

so the difference matrix between $\mathbf{G}_{p_0}$ and $\tilde{\mathbf{G}}_{p_0}$ is

$$\mathbf{G}_{p_0} - \tilde{\mathbf{G}}_{p_0} = \sqrt{M_t} \begin{bmatrix} \mathrm{diag}(x_{1,1} - \tilde{x}_{1,1}, \ldots, x_{1,\Gamma M_t} - \tilde{x}_{1,\Gamma M_t})\left(I_{M_t} \otimes \mathbf{1}_{\Gamma \times 1}\right) \\ \mathrm{diag}(x_{2,1} - \tilde{x}_{2,1}, \ldots, x_{2,\Gamma M_t} - \tilde{x}_{2,\Gamma M_t})\left(I_{M_t} \otimes \mathbf{1}_{\Gamma \times 1}\right) \\ \cdots \\ \mathrm{diag}(x_{K,1} - \tilde{x}_{K,1}, \ldots, x_{K,\Gamma M_t} - \tilde{x}_{K,\Gamma M_t})\left(I_{M_t} \otimes \mathbf{1}_{\Gamma \times 1}\right) \end{bmatrix}$$
$$= \sqrt{M_t}\, \mathrm{diag}(\mathbf{X} - \tilde{\mathbf{X}})\left[\mathbf{1}_{K \times 1} \otimes \left(I_{M_t} \otimes \mathbf{1}_{\Gamma \times 1}\right)\right],$$

where

$$\text{diag}(\mathbf{X} - \tilde{\mathbf{X}}) \overset{\triangle}{=} \text{diag}(x_{1,1} - \tilde{x}_{1,1}, \ldots, x_{1,\Gamma M_t} - \tilde{x}_{1,\Gamma M_t}, \ldots,$$
$$x_{K,1} - \tilde{x}_{K,1}, \ldots, x_{K,\Gamma M_t} - \tilde{x}_{K,\Gamma M_t}).$$

Thus, we have

$$\left[(\mathbf{G}_{p_0} - \tilde{\mathbf{G}}_{p_0})(\mathbf{G}_{p_0} - \tilde{\mathbf{G}}_{p_0})^H \right] \circ (R_T \otimes Q)$$
$$= M_t \left\{ \text{diag}(\mathbf{X} - \tilde{\mathbf{X}}) \left[\mathbf{1}_{K \times 1} \otimes \left(I_{M_t} \otimes \mathbf{1}_{\Gamma \times 1} \right) \right] \left[\mathbf{1}_{K \times 1} \otimes \left(I_{M_t} \otimes \mathbf{1}_{\Gamma \times 1} \right) \right]^H \right.$$
$$\left. \times \text{diag}(\mathbf{X} - \tilde{\mathbf{X}})^H \right\} \circ (R_T \otimes Q)$$
$$= M_t \left[\text{diag}(\mathbf{X} - \tilde{\mathbf{X}}) \left(\mathbf{1}_{K \times K} \otimes I_{M_t} \otimes \mathbf{1}_{\Gamma \times \Gamma} \right) \text{diag}(\mathbf{X} - \tilde{\mathbf{X}})^H \right] \circ (R_T \otimes Q)$$
$$= M_t \text{diag}(\mathbf{X} - \tilde{\mathbf{X}}) \left\{ R_T \otimes \left[\left(I_{M_t} \otimes \mathbf{1}_{\Gamma \times \Gamma} \right) \circ Q \right] \right\} \text{diag}(\mathbf{X} - \tilde{\mathbf{X}})^H$$
$$= M_t \text{diag}(\mathbf{X} - \tilde{\mathbf{X}}) \left(R_T \otimes I_{M_t} \otimes Q_0 \right) \text{diag}(\mathbf{X} - \tilde{\mathbf{X}})^H, \tag{3.132}$$

where $Q_0 = \{q_{i,j}\}_{1 \le i,j \le \Gamma}$ and $q_{i,j}$ is given by (3.131). In the above derivation, the second equality follows from the identities $[\mathbf{1}_{K \times 1} \otimes (I_{M_t} \otimes \mathbf{1}_{\Gamma \times 1})]^H = \mathbf{1}_{1 \times K} \otimes I_{M_t} \otimes \mathbf{1}_{1 \times \Gamma}$ and $(A_1 \otimes B_1)(A_2 \otimes B_2)(A_3 \otimes B_3) = (A_1 A_2 A_3) \otimes (B_1 B_2 B_3)$ ([74], p.251), and the third equality follows from a property of the Hadamard product ([74], p.304). From (3.132), we observe that if $\text{diag}(\mathbf{X} - \tilde{\mathbf{X}})$ is of full rank, i.e., $x_{k,j} - \tilde{x}_{k,j} \ne 0$ for any $1 \le k \le K$ and $1 \le j \le \Gamma M_t$, then the rank of $\left[(\mathbf{G}_{p_0} - \tilde{\mathbf{G}}_{p_0})(\mathbf{G}_{p_0} - \tilde{\mathbf{G}}_{p_0})^H \right] \circ (R_T \otimes Q)$ can be determined as $\text{rank}(R_T \otimes I_{M_t} \otimes Q_0)$, which is equal to $M_t \text{rank}(R_T) \text{rank}(Q_0)$. Similar to the correlation matrix R_F in (3.112), Q_0 can be expressed as

$$Q_0 = W_0 \text{diag}(\delta_0^2, \delta_1^2, \ldots, \delta_{L-1}^2) W_0^H,$$

where W_0 is defined in (3.128). Note that W_0 is a Γ by L matrix consisting of Γ rows of a Vandermonde matrix [74], so with $\tau_0 < \tau_1 < \cdots < \tau_{L-1}$, W_0 is nonsingular. Thus, Q_0 is of full rank (rank Γ). Therefore, if $\prod_{k=1}^{K} \prod_{j=1}^{\Gamma M_t} |x_{k,j} - \tilde{x}_{k,j}| \ne 0$, the rank of $\Delta \circ R$ is $\Gamma M_t \text{rank}(R_T)$.

The assumption that $\mathbf{G}_p = \tilde{\mathbf{G}}_p$ for any $p \ne p_0$ is also sufficient to calculate the coding advantage since the nonzero eigenvalues of $\Delta \circ R$ do not decrease if $\mathbf{G}_p \ne \tilde{\mathbf{G}}_p$ for some $p \ne p_0$ ([74], Corollary 3.1.3, p.149). Using the notation $v_0 = \Gamma M_t \text{rank}(R_T)$, the coding advantage can be calculated as

$$\zeta_{STF} = \min_{\mathbf{X} \ne \tilde{\mathbf{X}}} \prod_{i=1}^{v_0} \text{eig}_i \left(\left[(\mathbf{G}_{p_0} - \tilde{\mathbf{G}}_{p_0})(\mathbf{G}_{p_0} - \tilde{\mathbf{G}}_{p_0})^H \right] \circ (R_T \otimes Q) \right)$$

$$= \min_{\mathbf{X} \ne \tilde{\mathbf{X}}} \prod_{i=1}^{v_0} \text{eig}_i \left(M_t \text{diag}(\mathbf{X} - \tilde{\mathbf{X}}) \left(R_T \otimes I_{M_t} \otimes Q_0 \right) \text{diag}(\mathbf{X} - \tilde{\mathbf{X}})^H \right) \tag{3.133}$$

$$= \min_{\mathbf{X} \ne \tilde{\mathbf{X}}} \prod_{i=1}^{v_0} \theta_i M_t \text{eig}_i \left(R_T \otimes I_{M_t} \otimes Q_0 \right), \tag{3.134}$$

where the constants θ_i satisfy $\mathrm{eig}_{K\Gamma M_t}(\mathrm{diag}(\mathbf{X} - \tilde{\mathbf{X}})\mathrm{diag}(\mathbf{X} - \tilde{\mathbf{X}})^{\mathrm{H}}) \leq \theta_i \leq \mathrm{eig}_1(\mathrm{diag}(\mathbf{X}-\tilde{\mathbf{X}})\mathrm{diag}(\mathbf{X}-\tilde{\mathbf{X}})^{\mathrm{H}})$ for $i = 1, 2, \ldots, \nu_0$. In the above derivation, the second equality follows by (3.132), and the last equality follows by Ostrowski's theorem ([73], p.224). Since

$$\prod_{i=1}^{\nu_0} \mathrm{eig}_i \left(R_{\mathrm{T}} \otimes I_{M_t} \otimes Q_0 \right) = \prod_{i=1}^{\Gamma \mathrm{rank}(R_{\mathrm{T}})} \left(\mathrm{eig}_i (R_{\mathrm{T}} \otimes Q_0) \right)^{M_t}$$

$$= |\det(Q_0)|^{M_t \mathrm{rank}(R_{\mathrm{T}})} \prod_{i=1}^{\mathrm{rank}(R_{\mathrm{T}})} \left(\mathrm{eig}_i (R_{\mathrm{T}}) \right)^{\Gamma M_t},$$

and

$$\mathrm{eig}_1 \left(\mathrm{diag}(\mathbf{X} - \tilde{\mathbf{X}})\mathrm{diag}(\mathbf{X} - \tilde{\mathbf{X}})^{\mathrm{H}} \right) = \max_{\mathbf{X} \neq \tilde{\mathbf{X}}} \max_{1 \leq k \leq K, 1 \leq j \leq \Gamma M_t} |x_{k,j} - \tilde{x}_{k,j}|^2,$$

$$\mathrm{eig}_{K\Gamma M_t} \left(\mathrm{diag}(\mathbf{X} - \tilde{\mathbf{X}})\mathrm{diag}(\mathbf{X} - \tilde{\mathbf{X}})^{\mathrm{H}} \right) = \min_{\mathbf{X} \neq \tilde{\mathbf{X}}} \min_{1 \leq k \leq K, 1 \leq j \leq \Gamma M_t} |x_{k,j} - \tilde{x}_{k,j}|^2,$$

we have the lower and upper bounds in (3.123).

Finally, if R_{T} is of full rank, $\nu_0 = K\Gamma M_t$. From (3.133), the coding advantage is

$$\zeta_{\mathrm{STF}} = \min_{\mathbf{X} \neq \tilde{\mathbf{X}}} \det \left(M_t \, \mathrm{diag}(\mathbf{X} - \tilde{\mathbf{X}}) \left(R_{\mathrm{T}} \otimes I_{M_t} \otimes Q_0 \right) \mathrm{diag}(\mathbf{X} - \tilde{\mathbf{X}})^{\mathrm{H}} \right)$$

$$= M_t^{K\Gamma M_t} \det \left(R_{\mathrm{T}} \otimes I_{M_t} \otimes Q_0 \right) \min_{\mathbf{X} \neq \tilde{\mathbf{X}}} \prod_{k=1}^{K} \prod_{j=1}^{\Gamma M_t} |x_{k,j} - \tilde{x}_{k,j}|^2$$

$$= \delta \, M_t^{K\Gamma M_t} |\det(R_{\mathrm{T}})|^{\Gamma M_t} |\det(Q_0)|^{K M_t},$$

where δ is given by (3.130). Thus, we have proved Theorem 3.2.1 completely. ∎

From Theorem 3.2.1, we observe that with the code structure specified in (3.120)–(3.122), it is not difficult to achieve the maximum diversity order of $\Gamma M_t M_r \mathrm{rank}(R_{\mathrm{T}})$. The remaining problem is to design a set of complex symbol vectors, $\mathbf{X} = [x_{1,1} \cdots x_{1,\Gamma M_t} \cdots x_{K,1} \cdots x_{K,\Gamma M_t}]$, such that the coding advantage ζ_{STF} is as large as possible. One approach is to maximize $\delta_{\min}$ and $\delta_{\max}$ in (3.123) according to the lower and upper bounds of the coding advantage. Another approach is to maximize δ in (3.130). We follow the latter for two reasons. First, the coding advantage ζ_{STF} in (3.129) is determined by δ in closed form although this closed form only holds with the assumption that the temporal correlation matrix R_{T} is of full rank. Second, the problem of designing $\mathbf{X}$ to maximize δ is related to the problem of constructing signal constellations for Rayleigh fading channels. In the literature, δ is called the *minimum product distance* of the set of symbols $\mathbf{X}$.

We summarize some existing results on designing $\mathbf{X}$ in order to maximize the minimum product distance δ as follows. For simplicity, denote $\mathrm{L} = K\Gamma M_t$, and assume that Ω is a constellation such as QAM, PAM, and so on. The set of complex symbol vectors is obtained by applying a transform over a L-dimensional signal set Ω^L. Specifically,

$$\mathbf{X} = S \cdot \frac{1}{\sqrt{\mathrm{L}}} V(\theta_1, \theta_2, \ldots, \theta_L), \tag{3.135}$$

where $S = [s_1\ s_2\ \cdots\ s_L] \in \Omega^K$ is a vector of arbitrary channel symbols to be transmitted, and $V(\theta_1, \theta_2, \ldots, \theta_L)$ is a Vandermonde matrix with variables $\theta_1, \theta_2, \ldots, \theta_L$

$$V(\theta_1, \theta_2, \ldots, \theta_L) = \begin{bmatrix} 1 & 1 & \cdots & 1 \\ \theta_1 & \theta_2 & \cdots & \theta_L \\ \vdots & \vdots & \ddots & \vdots \\ \theta_1^{L-1} & \theta_2^{L-1} & \cdots & \theta_L^{L-1} \end{bmatrix}. \tag{3.136}$$

The optimum θ_l, $1 \leq l \leq \mathrm{L}$, have been specified for different L and Ω.

Example 3.4 If Ω is a QAM constellation, and $\mathrm{L} = 2^s$ ($s \geq 1$), the optimum θ_l's were given by

$$\theta_l = \mathrm{e}^{\mathrm{j}\frac{4l-3}{2L}\pi}, \quad l = 1, 2, \ldots, \mathrm{L}. \tag{3.137}$$

In case of $\mathrm{L} = 2^s \cdot 3^t$ ($s \geq 1, t \geq 1$), a class of θ_l's were given as

$$\theta_l = \mathrm{e}^{\mathrm{j}\frac{6l-5}{3L}\pi}, \quad l = 1, 2, \ldots, \mathrm{L}. \tag{3.138}$$

▲

For more details and other cases of Ω and L, readers can refer to [44, 14].

The STF code design discussed in this subsection achieves full symbol rate, which is much larger than that of the repetition coding approach. However, the maximum-likelihood decoding complexity of this approach is high. Its complexity increases exponentially with the number of OFDM blocks, K, while the decoding complexity of the repetition-coded STF codes increases only linearly with K. Fortunately, some sphere decoding methods [166] can be used to reduce the complexity.

3.2.3.3 STF code design examples and performance comparisons

In the following, we provide some full-diversity STF code design examples and show their performances. The OFDM modulation has $N = 128$ subcarriers, and the total bandwidth is 1 MHz. Thus, the OFDM block duration is 128 μs. Let us set the length of the cyclic prefix to 20 μs for all cases.

Performances of the repetition-coded STF codes

Let us consider a block code and a trellis code example. Assume that the communication system has $M_r = 1$ receive antenna. We consider a two-ray, equal power delay profile ($L = 2$), with a delay of 20 μs between the two rays. Each ray was modeled as a zero-mean, complex Gaussian random variable with variance 0.5.

The full-diversity STF block codes were obtained by repeating a full-diversity SF block code via (3.119) across $K = 1, 2, 3, 4$ OFDM blocks. The used full-diversity SF

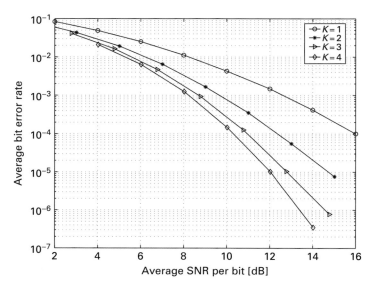

Fig. 3.15   The performance of the repetition block codes, with two transmit and one receive antennas.

block code for $M_t = 2$ transmit antennas was constructed from the Alamouti scheme with QPSK modulation via mapping. The spectral efficiency of the resulting STF code is $1, 0.5, 0.33, 0.25$ bit/s/Hz (omitting the cyclic prefix) for $K = 1, 2, 3, 4$, respectively. We consider the performance of the full-diversity STF block code without temporal correlation (R_T was an identity matrix). From Figure 3.15, we can see that, by repeating the SF code over multiple OFDM blocks, the achieved diversity order can be increased.

The simulated full-diversity STF trellis code was obtained from a full-diversity SF trellis code via (3.119) with $K = 1, 2, 3, 4$, respectively. The used full-diversity SF trellis code for $M_t = 3$ transmit antennas was constructed by applying the repetition mapping to the 16-state, QPSK ST trellis code given by [232]. Since the modulation was the same in all four cases, the spectral efficiency of the resulting STF codes were $1, 0.5, 0.33, 0.25$ bit/s/Hz (omitting the cyclic prefix) for $K = 1, 2, 3, 4$, respectively. Similarly to the previous case, we assumed that the channel changes independently from OFDM block to OFDM block. The obtained BER curves can be observed in Figure 3.16. As apparent from the figure, the STF codes ($K > 1$) achieved higher diversity order than the SF code ($K = 1$).

Performance of the full-rate STF codes

Let us consider a more realistic six-ray typical urban (TU) power delay profile, and simulate the fading channel with different temporal correlations. Assume that the fading channel is constant within each OFDM block period but varies from one OFDM block period to another according to a first-order Makovian model

$$\alpha_{i,j}^k(l) = \varepsilon\, \alpha_{i,j}^{k-1}(l) + \eta_{i,j}^k(l), \quad 0 \le l \le L - 1, \tag{3.139}$$

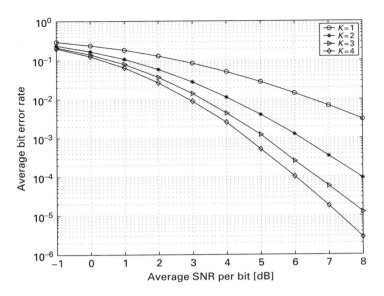

Fig. 3.16 The performance of the repetition trellis codes, with three transmit and one receive antennas.

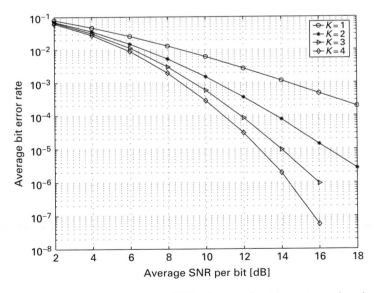

Fig. 3.17 The performance of the full-rate STF codes, $\varepsilon = 0$, with two transmit and one receive antennas.

where the constant ε $(0 \leq \varepsilon \leq 1)$ determines the amount the temporal correlation, and $\eta_{i,j}^{k}(l)$ is a zero-mean, complex Gaussian random variable with variance $\delta_l \sqrt{1 - \varepsilon^2}$. If $\varepsilon = 0$, there is no temporal correlation (independent fading), while if $\varepsilon = 1$, the channel stays constant over multiple OFDM blocks. Let us considered three temporal correlation scenarios: $\varepsilon = 0$, $\varepsilon = 0.8$, and $\varepsilon = 0.95$.

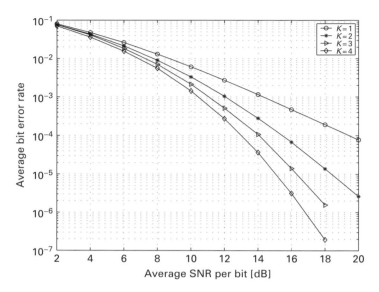

Fig. 3.18   The performance of the full-rate STF codes, $\varepsilon = 0.8$, with two transmit and one receive antennas.

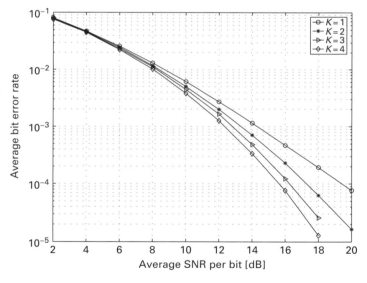

Fig. 3.19   The performance of the full-rate STF codes, $\varepsilon = 0.95$, with two transmit and one receive antennas.

The full-rate STF codes were constructed by (3.120)–(3.122) for $M_t = 2$ transmit antennas with $\Gamma = 2$. The set of complex symbol vectors $\mathbf{X}$ was obtained via (3.135) by applying Vandermonde transforms over a signal set Ω^{4K} for $K = 1, 2, 3, 4$. The Vandermonde transforms were determined for different K values according to (3.137) and (3.138). The constellation Ω was chosen to be BPSK. Thus, the spectral efficiency the resulting STF codes were 1 bit/s/Hz (omitting the cyclic prefix), which is independent of the number of jointly encoded OFDM blocks, K.

The performance of the full-rate STF codes are depicted in Figures 3.17–3.19 for the three different temporal correlation scenarios. From the figures, we observe that the diversity order of the STF codes increases with the number of jointly encoded OFDM blocks, K. However, the improvement of the diversity order depends on the temporal correlation. The performance gain obtained by coding across multiple OFDM blocks decreases as the correlation factor ε increases. For example, without temporal correlation ($\varepsilon = 0$), the STF code with $K = 4$ achieves an average BER of about 6.0×10^{-8} at a SNR of 16 dB. In case of the correlated channel model and $\varepsilon = 0.8$, the STF code with $K = 4$ has an average BER of only 3.0×10^{-6} at a SNR of 16 dB. Finally, in case of the correlated channel model and $\varepsilon = 0.95$, the STF code with $K = 4$ has an average BER of around 10^{-4} at a SNR of 16 dB.

3.3   Chapter summary and bibliographical notes

In this chapter, we considered MIMO broadband wireless communications. Specifically, we reviewed code design criteria for MIMO-OFDM systems and summarized some SF and STF code designs. We explored different coding approaches for MIMO-OFDM systems by taking into account all opportunities for performance improvement in the spatial, temporal, and the frequency domains in terms of the achievable diversity order. For SF coding, where the coding is applied within each OFDM block, we discussed two systematic SF code design methods that can guarantee to achieve the full diversity order of LM_tM_r, where the factor L comes from the frequency diversity due to the delay spread of the channel. For STF coding, where coding is applied over multiple OFDM blocks, we considered two STF code design methods by taking advantage of the SF code design methodology. The STF code design methods can guarantee the maximum achievable diversity order of LM_tM_rT, where T is the rank of the temporal correlation matrix of the fading channel. We observed that the performance of the SF and STF codes depends heavily on the channel power delay profile, and there is tradeoff between the diversity order and the spectral efficiency of the code.

Most early ST coding and modulation focused on MIMO systems with frequency non-selective (flat) fading channels to exploit spatial and temporal diversity in narrowband wireless communications. In case of broadband wireless communications with frequency selective fading channels, there is an additional frequency diversity. The first SF coding scheme was proposed in [3], in which previously existing ST codes were used by replacing the time domain with the frequency domain. The resulting SF codes could achieve only spatial diversity and were not guaranteed to achieve full (spatial and frequency) diversity. Later, similar schemes were described in [111, 13, 47]. The performance criteria for SF-coded MIMO-OFDM systems were derived in [125]. The maximum achievable diversity order was found to be the product of the number of transmit antennas, the number of receive antennas, and the number of delay paths. In [199], a systematic approach was proposed to design full-diversity SF codes from ST codes for arbitrary power delay profiles. It turns out that ST codes and SF codes are related, in the sense that ST codes achieving full (spatial) diversity in quasi-static flat fading environment can be used to construct SF codes that can achieve the maximum diversity available in frequency selective MIMO fading channels. More recently in [205], a SF code design approach was proposed that offers full symbol rate and guarantees full diversity for an arbitrary number of transmit antennas, any memoryless modulation method and arbitrary power delay profiles.

If longer decoding delay and higher decoding complexity are allowable, one may consider coding over several OFDM block periods, resulting in STF codes to exploit all of the spatial, temporal and frequency diversity. The STF coding strategy was first proposed in [46] for two transmit antennas and further developed in [126, 135, 121] for multiple transmit antennas. Both [46] and [121] assumed that the MIMO channel stays constant over multiple OFDM blocks; however, STF coding under this assumption cannot provide any additional diversity compared to the SF coding approach. In [135], an intuitive explanation on the equivalence between antennas and OFDM tones was presented from the viewpoint of channel capacity. In [126], the performance criteria for STF codes were derived, and an upper bound on the maximum achievable diversity order was established. However, there was no discussion in [126] whether the upper bound can be achieved or not, and the proposed STF codes were not guaranteed to achieve the full spatial, temporal, and frequency diversities. Later in [206], a systematic method was proposed to design full-diversity STF codes for MIMO-OFDM systems, in which an alternative performance analysis was obtained for STF-coded MIMO-OFDM systems that provides better insight for the STF code design.

Exercises

3.1 Show the PEP upper bound in (3.13), i.e., show that for any two distinct matrices $\mathbf{D}$ and $\tilde{\mathbf{D}}$, the pairwise error probability between $\mathbf{D}$ and $\tilde{\mathbf{D}}$ can be upper bounded as

$$P(\mathbf{D} \to \tilde{\mathbf{D}}) \leq \binom{2r-1}{r} \left(\prod_{i=1}^{r} \gamma_i \right)^{-1} \left(\frac{\rho}{M_t} \right)^{-r},$$

where r is the rank of $(\mathbf{D} - \tilde{\mathbf{D}})\mathbf{R}(\mathbf{D} - \tilde{\mathbf{D}})^H$, $\gamma_1, \gamma_2, \ldots, \gamma_r$ are the nonzero eigenvalues of $(\mathbf{D} - \tilde{\mathbf{D}})\mathbf{R}(\mathbf{D} - \tilde{\mathbf{D}})^H$, and $\mathbf{R} = E\{\mathbf{H}\mathbf{H}^H\}$ is the correlation matrix of $\mathbf{H}$.

3.2 Prove the result in Theorem 3.1.4: For any subcarrier permutation, the diversity product of the resulting SF code is

$$\zeta = \zeta_{\text{in}} \cdot \zeta_{\text{ex}},$$

and ζ_{in} and ζ_{ex} are the "intrinsic" and "extrinsic" diversity products defined in (3.60) and (3.73), respectively. Moreover, the "extrinsic" diversity product ζ_{ex} is upper bounded as:

(a) $\zeta_{\text{ex}} \leq 1$; and, more precisely,

(b) if we sort the power profile $\delta_0, \delta_1, \ldots, \delta_{L-1}$ in a non-increasing order as:
$\delta_{l_1} \geq \delta_{l_2} \geq \cdots \geq \delta_{l_L}$, then

$$\zeta_{ex} \leq \left(\prod_{i=1}^{\Gamma} \delta_{l_i}\right)^{\frac{1}{\Gamma}} \left|\prod_{m=1}^{M_t} \det(V_m V_m^{H})\right|^{\frac{1}{2\Gamma M_t}},$$

where equality holds when $\Gamma = L$. In this case, $\zeta_{ex} \leq \sqrt{L} \left(\prod_{i=1}^{\Gamma} \delta_{l_i}\right)^{\frac{1}{\Gamma}}$.

3.3 Prove the equation in (3.67), i.e., show that the product of the nonzero eigenvalues of $[\sigma(C - \tilde{C})\sigma(C - \tilde{C})^{H}] \circ R$, $\lambda_1, \lambda_2, \ldots, \lambda_{\Gamma M_t}$, can be calculated as

$$\prod_{k=1}^{\Gamma M_t} \lambda_k = \prod_{m=1}^{M_t} |\det(A_m)|.$$

3.4 Prove the result in Theorem 3.1.5: For the permutation specified in (3.81) with a separation factor μ, the "extrinsic" diversity product of the permuted SF code is

$$\zeta_{ex} = |\det(V_0 \Lambda V_0^{H})|^{\frac{1}{2\Gamma}},$$

where V_0 is specified in (3.83). Moreover, if $\Gamma = L$, the "extrinsic" diversity product ζ_{ex} can be calculated as

$$\zeta_{ex} = \left(\prod_{l=0}^{L-1} \delta_l\right)^{\frac{1}{L}} \left(\prod_{0 \leq l_1 < l_2 \leq L-1} \left|2 \sin\left(\frac{\mu(\tau_{l_2} - \tau_{l_1})\pi}{T}\right)\right|\right)^{\frac{1}{L}}.$$

3.5 Prove the result in (3.86): If $\Gamma = L$, the optimum separation factor μ_{op} defined in (3.85) can be determined as follows:

$$\mu_{op} = \arg \max_{1 \leq \mu \leq \lfloor N/\Gamma \rfloor} \prod_{0 \leq l_1 < l_2 \leq L-1} \left|\sin\left(\frac{\mu(\tau_{l_2} - \tau_{l_1})\pi}{T}\right)\right|,$$

which is independent to the power profile of the multipath channel.

3.6 (Simulation project) Consider a MIMO-OFDM system with $M_t = 2$ transmit and $M_r = 1$ receive antennas. It uses an SF code that is obtained by repeating each row of the Alamouti code twice as follows:

$$C = \begin{bmatrix} x_1 & x_2 \\ x_1 & x_2 \\ -x_2^* & x_1^* \\ -x_2^* & x_1^* \end{bmatrix}, \tag{E3.1}$$

in which x_1 and x_2 are QPSK symbols. Assume that the system bandwidth is $BW = 1\,MHz$ and the OFDM has $N = 128$ subcarriers. The SF code is applied over subcarriers within each OFDM block.

(a) Determine the duration of each OFDM block and spectral efficiency of the code without considering the cyclic prefix.

(b) Consider a two-ray, equal-power delay profile, with a delay of τ µs between the two rays. If the delay of the two rays is $\tau = 5\,\mu s$, simulate the system by Matlab and plot the bit error rate curve versus the signal-to-noise ratio. Repeat the simulation if $\tau = 20\,\mu s$. Compare the simulation results and explain your observations.

(c) Simulate the system by considering the TU six-ray channel model shown in Figure 3.5. Repeat the simulation if the system bandwidth is $BW = 4\,MHz$. Compare the simulation results and explain your observations.

3.7 (Simulation project)   Consider a MIMO-OFDM system with $M_t = 2$ transmit and $M_r = 1$ receive antennas. In this simulation, we compare performances of the SF code as shown in (E3.1) and the full-rate full-diversity SF code shown in (3.51).

(a) Design a proper full-rate full-diversity SF code based on (3.51) with the same spectral efficiency as the one shown in (E3.1).

(b) Assume that the system bandwidth is $BW = 1\,MHz$ and the OFDM has $N = 128$ subcarriers. Consider a two-ray, equal-power delay profile, with a delay of $\tau = 20\,\mu s$ between the two rays. Simulate both the SF codes and compare their bit error rate performances

(c) With the same channel assumption as in (b), compare performances of the two SF codes if rows of each code are randomly permuted among the $N = 128$ subcarriers.

(d) With the same channel assumption as in (b), determine the optimum permutation factor, i.e., the separation factor defined in (3.81), for the full-rate full-diversity SF code. Compare the performances of the full-rate code under three scenarios: without permutation, random permutation, and the optimum permutation.

(e) Repeat (b), (c), and (d) with the TU six-ray channel model shown in Figure 3.5 and a system bandwidth of $BW = 4\,MHz$.

Part II

Cooperative communications

4 Relay channels and protocols

In this chapter, we will discuss shortcomings of conventional point-to-point communications that led to the introduction of the new paradigm shift for wireless communications, i.e., cooperative communications. We will define what the relay channel is, and in what aspects it is different from the direct point-to-point channel. We will also describe several protocols that can be implemented at the relay channel, and discuss the performance of these protocols which will be assessed based on their outage probability and diversity gains.

4.1 Cooperative communications

In cooperative communications, independent paths between the user and the base station are generated via the introduction of a relay channel as illustrated in Figure 4.1. The relay channel can be thought of as an auxiliary channel to the direct channel between the source and destination. A key aspect of the cooperative communication process is the processing of the signal received from the source node done by the relay. These different processing schemes result in different cooperative communications protocol. Cooperative communications protocols can be generally categorized into fixed relaying schemes and adaptive relaying schemes. In fixed relaying, the channel resources are divided between the source and the relay in a fixed (deterministic) manner. The processing at the relay differs according to the employed protocol. In a fixed amplify-and-forward (AF) relaying protocol, the relay simply scales the received version and transmits an amplified version of it to the destination. Another possibility of processing at the relay node is for the relay to decode the received signal, re-encode it and then retransmit it to the receiver. This kind of relaying is termed a fixed decode-and-forward (DF) relaying protocol.

Fixed relaying has the advantage of easy implementation, but the disadvantage of low bandwidth efficiency. This is because half of the channel resources are allocated to the relay for transmission, which reduces the overall rate. This is true especially when the source–destination channel is not very bad, because in such a scenario a high percentage of the packets transmitted by the source to the destination could be received correctly by the destination and the relay's transmissions would be wasted. Adaptive relaying techniques, comprising selective and incremental relaying, try to overcome this problem.

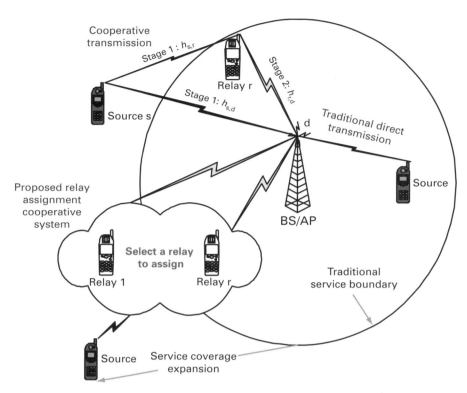

Fig. 4.1   Illustrating the difference between the direct and cooperative transmission schemes, and the coverage extension prospected by cooperative transmission.

In selective relaying, if the signal-to-noise ratio of the signal received at the relay exceeds a certain threshold, the relay performs decode-and-forward operation on the message. On the other hand, if the channel between the source and the relay has severe fading such that the signal-to-noise ratio is below the threshold, the relay idles. Moreover, if the source knows that the destination does not decode correctly, then the source may repeat to transmit the information to the destination or the relay may help forward information, which is termed as incremental relaying. In this case, a feedback channel from the destination to the source and the relay is necessary.

In this chapter, we discuss performance comparisons of some basic cooperation protocols. The performance metric used is the outage capacity defined in Section 1.2. Recall that the outage capacity given that a channel is required to support a transmission rate R is defined as

$$\Pr\left[I(x, y) \leq R\right],$$

where $I(x, y)$ is the the mutual information of a channel with input x and y is the channel output. Note that the mutual information is a random variable because the channel varies in a random way due to fading.

In practice, a device typically cannot listen and transmit simultaneously, otherwise transmitting signals will cause severe interference to incoming relatively weak received signals. Thus, a half-duplex constraint is assumed throughout the chapter, i.e, the relay cannot transmit and receive at the same time,

4.2   Cooperation protocols

In this chapter, we only consider a single relay helping a user (source) in the network forwarding information. The multiple relay scenarios will be considered in later chapters. A typical cooperation strategy can be modeled with two orthogonal phases, either in TDMA or FDMA, to avoid interference between the two phases:

- In phase 1, a source sends information to its destination, and the information is also received by the relay at the same time.
- In phase 2, the relay can help the source by forwarding or retransmitting the information to the destination.

Figure 4.2 depicts a general relay channel, where the source transmits with power P_1 and the relay transmits with power P_2. In this chapter, we will consider the special case where the source and the relay transmit with equal power P. Optimal power allocation is studied in the following chapters. In phase 1, the source broadcasts its information to both the destination and the relay. The received signals $y_{s,d}$ and $y_{s,r}$ at the destination and the relay, respectively, can be written as

$$y_{s,d} = \sqrt{P}\, h_{s,d}\, x + n_{s,d}, \tag{4.1}$$

$$y_{s,r} = \sqrt{P}\, h_{s,r}\, x + n_{s,r}, \tag{4.2}$$

in which P is the transmitted power at the source, x is the transmitted information symbol, and $n_{s,d}$ and $n_{s,r}$ are additive noise. In (4.1) and (4.2), $h_{s,d}$ and $h_{s,r}$ are the channel coefficients from the source to the destination and the relay, respectively. They are modeled as zero-mean, complex Gaussian random variables with variances $\delta_{s,d}^2$ and $\delta_{s,r}^2$, respectively. The noise terms $n_{s,d}$ and $n_{s,r}$ are modeled as zero-mean complex Gaussian random variables with variance N_0.

In phase 2, the relay forwards a processed version of the source's signal to the destination, and this can be modeled as

$$y_{r,d} = h_{r,d}\, q(y_{s,r}) + n_{r,d}, \tag{4.3}$$

where the function $q(\cdot)$ depends on which processing is implemented at the relay node.

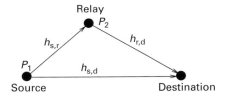

Fig. 4.2   A simplified cooperation model.

4.2.1   Fixed cooperation strategies

In fixed relaying, the channel resources are divided between the source and the relay in a fixed (deterministic) manner. The processing at the relay differs according to the employed protocols. The most common techniques are the fixed AF relaying protocol and the fixed relaying DF protocol.

4.2.1.1   Fixed amplify-and-forward relaying protocol

In a fixed AF relaying protocol, which is often simply called an AF protocol, the relay scales the received version and transmits an amplified version of it to the destination. The amplify-and-forward relay channel can be modeled as follows. The signal transmitted from the source x is received at both the relay and destination as

$$y_{s,r} = \sqrt{P} h_{s,r} x + n_{s,r}, \text{ and } y_{s,d} = \sqrt{P} h_{s,d} x + n_{s,d}, \tag{4.4}$$

where $h_{s,r}$ and $h_{s,d}$ are the channel fades between the source and the relay and destination, respectively, and are modeled as Rayleigh flat fading channels. The terms $n_{s,r}$ and $n_{s,d}$ denote the additive white Gaussian noise with zero-mean and variance N_0. In this protocol, the relay amplifies the signal from the source and forwards it to the destination ideally to equalize the effect of the channel fade between the source and the relay. The relay does that by simply scaling the received signal by a factor that is inversely proportional to the received power, which is denoted by

$$\beta_r = \frac{\sqrt{P}}{\sqrt{P|h_{s,r}|^2 + N_0}}. \tag{4.5}$$

The signal transmitted from the relay is thus given by $\beta_r y_{s,r}$ and has power P equal to the power of the signal transmitted from the source. To calculate the mutual information between the source and the destination, we need to calculate the total instantaneous signal-to-noise ratio at the destination. The SNR received at the destination is the sum of the SNRs from the source and relay links. The SNR from the source link is given by

$$\text{SNR}_{s,d} = \Gamma |h_{s,d}|^2, \tag{4.6}$$

where $\Gamma = P/N_0$.

In the following we calculate the received SNR from the relay link. In phase 2 the relay amplifies the received signal and forwards it to the destination with transmitted power P. The received signal at the destination in phase 2 according to (4.5) is given by

$$y_{r,d} = \frac{\sqrt{P}}{\sqrt{P|h_{s,r}|^2 + N_0}} h_{r,d} y_{s,r} + n_{r,d}, \tag{4.7}$$

where $h_{r,d}$ is the channel coefficient from the relay to the destination and $n_{r,d}$ is an additive noise. More specifically, the received signal $y_{r,d}$ in this case is

$$y_{r,d} = \frac{\sqrt{P}}{\sqrt{P|h_{s,r}|^2 + N_0}} \sqrt{P} h_{r,d} h_{s,r} x + n'_{r,d}, \tag{4.8}$$

where

$$n'_{r,d} = \frac{\sqrt{P}}{\sqrt{P|h_{s,r}|^2 + N_0}} h_{r,d} n_{s,r} + n_{r,d} \qquad (4.9)$$

Assume that the noise terms $n_{s,r}$ and $n_{r,d}$ are independent, then the equivalent noise $n'_{r,d}$ is a zero-mean, complex Gaussian random variable with variance

$$N'_0 = \left(\frac{P|h_{r,d}|^2}{P|h_{s,r}|^2 + N_0} + 1 \right) N_0. \qquad (4.10)$$

The destination receives two copies from the signal x through the source link and relay link. As discussed in Section 1.4, there are different techniques to combine the two signals. The optimal technique that maximize the overall signal-to-noise ratio is the maximal ratio combiner (MRC). Note that MRC combining requires a coherent detector that has knowledge of all channel coefficients. Also, as shown in (1.43), the signal-to-noise ratio at the output of the MRC is equal to the sum of the received signal-to-noise ratios from both branches.

With knowledge of the channel coefficients $h_{s,d}$, $h_{s,r}$, and $h_{r,d}$, the output of the MRC detector at the destination can be written as

$$y = a_1 y_{s,d} + a_2 y_{r,d}, \qquad (4.11)$$

The combining factors a_1 and a_2 should be designed to maximize the combined SNR. As discussed in Section 1.4, this can be solved by formulating an optimization problem and selecting these factors correspondingly. An easier way to design them is be resorting to signal space and detection theory principles. Since, the AWGN noise terms span the whole space, to minimize the noise effects the detector should project the received signals $y_{s,d}$ and $y_{s,r}$ to the desired signal spaces. Hence, $y_{s,d}$ and $y_{r,d}$ should be projected along the directions of $h_{s,d}$ and $h_{r,d}h_{s,r}$, respectively, after normalizing the noise variance terms in both received signals. Therefore, a_1 and a_2 are given by

$$a_1 = \frac{\sqrt{P} h^*_{s,d}}{N_0} \quad \text{and} \quad a_2 = \frac{\sqrt{\frac{P}{P|h_{s,r}|^2+N_0}} \sqrt{P}\, h^*_{s,r} h^*_{r,d}}{\left(\frac{P|h_{r,d}|^2}{P|h_{s,r}|^2+N_0} + 1 \right) N_0}. \qquad (4.12)$$

By assuming that the transmitted symbol x in (4.1) has average energy 1, the instantaneous SNR of the MRC output is

$$\gamma = \gamma_1 + \gamma_2, \qquad (4.13)$$

where

$$\gamma_1 = \frac{|a_1 \sqrt{(P)} h_{s,d}|^2}{|a_1|^2 N_0}$$
$$= P|h_{s,d}|^2 / N_0, \qquad (4.14)$$

and

$$\gamma_2 = \frac{\left|a_2 \dfrac{\sqrt{P}}{\sqrt{P|h_{s,r}|^2 + N_0}} \sqrt{P}\, h_{r,d}\, h_{s,r}\right|^2}{N_0'|a_2|^2}$$

$$= \frac{\dfrac{P^2}{P|h_{s,r}|^2 + N_0}|h_{s,r}|^2|h_{r,d}|^2}{\left(\dfrac{P|h_{r,d}|^2}{P_1|h_{s,r}|^2 + N_0} + 1\right)N_0}$$

$$= \frac{1}{N_0}\frac{P^2|h_{s,r}|^2|h_{r,d}|^2}{P|h_{s,r}|^2 + P|h_{r,d}|^2 + N_0}. \tag{4.15}$$

From the above, the instantaneous mutual information as a function of the fading coefficients for amplify-and-forward is given by

$$I_{AF} = \frac{1}{2}\log(1 + \gamma_1 + \gamma_2). \tag{4.16}$$

Substituting for the values of the SNR of both links, we can write the mutual information as

$$I_{AF} = \frac{1}{2}\log\left(1 + \Gamma|h_{s,d}|^2 + f(\Gamma|h_{s,r}|^2, \Gamma|h_{r,d}|^2)\right) \tag{4.17}$$

where

$$f(x, y) \triangleq \frac{xy}{x + y + 1}. \tag{4.18}$$

The outage probability can be obtained by averaging over the exponential channel gain distribution, as follows:

$$\Pr[I_{AF} < R] = E_{h_{s,d}, h_{s,r}, h_{r,d}}\left[\frac{1}{2}\log\left(1 + \Gamma|h_{s,d}|^2 + f(\Gamma|h_{s,r}|^2, \Gamma|h_{r,d}|^2)\right) < R\right] \tag{4.19}$$

Calculating the above integration, the outage probability at high SNR is given by

$$\Pr[I_{AF} < R] \simeq \left(\frac{\sigma_{s,r}^2 + \sigma_{r,d}^2}{2\sigma_{s,d}^2\left(\sigma_{s,r}^2\sigma_{r,d}^2\right)}\right)\left(\frac{2^{2R} - 1}{\Gamma}\right)^2, \tag{4.20}$$

where the multiplicative factor of 2 in $2R$ is because half of the bandwidth is lost in cooperation by allocating them to the relay. The outage expression decays as Γ^{-2}, which means that the AF protocol achieves diversity 2.

Example 4.1 In this example, we study the outage probability of amplify-and-forward protocol and compare it to the performance of direct transmission. The channel variance between the source and destination $\sigma_{s,d}$ is taken to be 1, while the channel variances for the source–relay or relay–destination channels are equal to 0.5. The noise variance is one.

In Figure 4.3, the outage probability is depicted versus SNR in dB for a fixed rate of 2 bps/Hz. As shown from the figure, diversity order two is achieved by AF as clear from the curve slope.

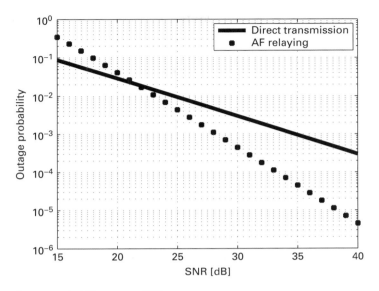

Fig. 4.3   Outage probability versus SNR.

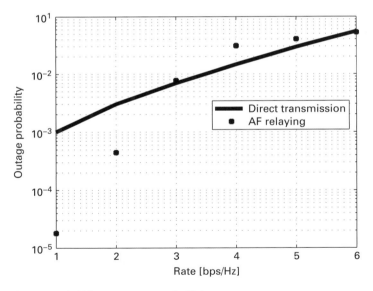

Fig. 4.4   Outage probability versus spectral efficiency.

In Figure 4.4, the outage probability is depicted versus spectral efficiency in bps/Hz for a fixed SNR of 40 dB. We have selected such high SNR to study the effect of increasing the transmission rate and isolate it from the SNR effects. As clear from the figure, the performance generally degrades with increasing R, but it degrades faster for AF because of the inherent loss in the spectral efficiency. At high enough R, direct transmission becomes more efficient than cooperation. ▲

4.2.1.2 Fixed decode-and-forward relaying protocol

Another processing possibility at the relay node is for the relay to decode the received signal, re-encode it, and then retransmit it to the receiver. This kind of relaying is termed as a *fixed* decode-and-forward (DF) scheme, which is often simply called a DF scheme without the confusion from the *selective* DF relaying scheme. If the decoded signal at the relay is denoted by $\hat{x}$, the transmitted signal from the relay can be denoted by $\sqrt{P}\hat{x}$, given that $\hat{x}$ has unit variance. Note that the decoded signal at the relay may be incorrect. If an incorrect signal is forwarded to the destination, the decoding at the destination is meaningless. It is clear that for such a scheme the diversity achieved is only one, because the performance of the system is limited by the worst link from the source–relay and source–destination. This will be illustrated through the following analysis.

Although fixed DF relaying has the advantage over AF relaying in reducing the effects of additive noise at the relay, it entails the possibility of forwarding erroneously detected signals to the destination, causing error propagation that can diminish the performance of the system. The mutual information between the source and the destination is limited by the mutual information of the weakest link between the source–relay and the combined channel from the source–destination and relay–destination. More specifically, the mutual information for decode-and-forward transmission in terms of the channel fades can be given by

$$I_{\text{DF}} = \frac{1}{2} \min \left\{ \log(1 + \Gamma |h_{\text{s,r}}|^2), \log(1 + \Gamma |h_{\text{s,d}}|^2 + \Gamma |h_{\text{r,d}}|^2) \right\}, \tag{4.21}$$

where the min operator in the above equation takes into account the fact that the relay only transmits if decoded correctly, and hence the performance is limited by the weakest link between the source–destination and source–relay.

The outage probability for the fixed DF relaying scheme is given by $\Pr[I_{\text{DF}} < R]$. Since log is a monotonic function, the outage event is equivalent to

$$\min \left\{ |h_{\text{s,r}}|^2, |h_{\text{s,d}}|^2 + |h_{\text{r,d}}|^2 \right\} < \frac{2^{2R} - 1}{\Gamma}. \tag{4.22}$$

The outage probability can be written as

$$\Pr[I_{\text{DF}} < R] = \Pr \left\{ |h_{\text{s,r}}|^2 < \frac{2^{2R} - 1}{\Gamma} \right\} \tag{4.23}$$

$$+ \Pr \left\{ |h_{\text{s,r}}|^2 > \frac{2^{2R} - 1}{\Gamma} \right\} \Pr \left\{ |h_{\text{s,d}}|^2 + |h_{\text{r,d}}|^2 < \frac{2^{2R} - 1}{\Gamma} \right\}.$$

Since the channel is Rayleigh fading, the above random variables are all exponential random variables with parameter one. Averaging over the channel conditions, the outage probability for decode-and-forward at high SNR is given by

$$\Pr[I_{\mathrm{DF}} < R] \simeq \frac{1}{\sigma_{\mathrm{s,r}}^2} \frac{2^{2R} - 1}{\Gamma}. \qquad (4.24)$$

From the above, fixed relaying has the advantage of easy implementation, but the disadvantage of low bandwidth efficiency. This is because half of the channel resources are allocated to the relay for transmission, which reduces the overall rate. This is true especially when the source–destination channel is not very bad, because under this scenario a high percentage of the packets transmitted by the source to the destination can be received correctly by the destination and the relay's transmissions are wasted.

Example 4.2 In this example, we study the outage probability of fixed DF relaying and compare it to the performance of direct transmission. The channel variance between the source and destination $\sigma_{\mathrm{s,d}}$ is taken to be 1, while the channel variances for the source–relay or relay–destination channels are equal to 0.5. The noise variance is one.

In Figure 4.5, the outage probability is depicted versus SNR in dB for a fixed rate of 2 bps/Hz. As shown from the figure, diversity order one is achieved by fixed DF relaying as is clear from the curve slope. This means that fixed DF relaying has no diversity gain.

In Figure 4.6, the outage probability is depicted versus spectral efficiency in bps/Hz for a fixed SNR of 40 dB. As is clear from the figure, the performance generally

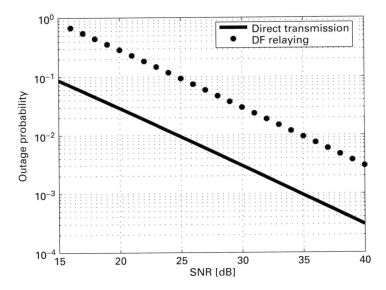

Fig. 4.5   Outage probability versus SNR.

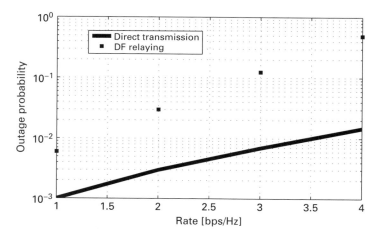

Fig. 4.6   Outage probability versus spectral efficiency.

degrades with increasing R. Again, fixed DF relaying does not have any advantage over direct transmission in terms of diversity.   ▲

4.2.1.3   Other cooperation strategies

Besides the two most common techniques for fixed relaying, there are other techniques, such as *compress-and-forward cooperation* and *coded cooperation*, which deserve some attention.

Compress-and-forward cooperation

The main difference between compress-and-forward and decode/amplify-and-forward is that while in the later the relay transmits a copy of the received message, in compress-and-forward the relay transmits a quantized and compressed version of the received message. Therefore, the destination node will perform the reception functions by combining the received message from the source node and its quantized and compressed version from the relay node.

The quantization and compression process at the relay node is a process of source encoding, i.e., the representation of each possible received message as a sequence of symbols. For clarity and simplicity, let us assume that these symbols are binary digits (bits). At the destination node, an estimate of the quantized and compressed message is obtained by decoding the received sequence of bits. This decoding operation simply involves the mapping of the received bits into a set of values that estimate the transmitted message. This mapping process normally involves the introduction of distortion (associated to the quantization and compression process), which can be considered as a form of noise.

In Section 1.2, we reviewed the concept of entropy and mentioned that the entropy of a random variable can be considered as the mean value of the information provided by all the outcomes of the random variable. In addition, the entropy provides a benchmark against which it is possible to evaluate the performance of source encoders. Next, and for the purpose of simplifying the presentation, let us consider that the source data is generated from a discrete memoryless source. For this setting, the entropy of the random variable being encoded at the source provides a lower bound on the average number of bits per source symbol (the source encoding rate) needed to encode the source. In this sense, the entropy provides a lower bound on the source encoding rate used at the relay node if in a peer-to-peer communication setup. The use of cooperation and the possibility of combining at the destination the messages from the source and the relay node, changes this point. Effectively, the information received at the destination from the source can be used, as side information, while decoding the message from the relay. This will allow for encoding at a lower source encoding rate. The following example further illustrates the operation of compress-and-forward.

Example 4.3 In this example we discuss a simplified model of a compress-and-forward scheme so as to highlight its main characteristics. Consider a source node that transmits a signal with four possible values: -1.5, 0.5, 0.5, and 1.5. With no cooperation, the signal is sent over an AWGN channel with standard deviation $\sigma = 0.5$. When using cooperation the source–relay and relay–destination channels are also assumed also AWGN with standard deviation $\sigma = 0.25$ and $\sigma = 0.6$, respectively.

Assume now that the signal equal to 1.5 is transmitted without cooperation. The setup is illustrated in Figure 4.7, which shows the figures associated with the channel. At the destination node, the value of 1.5 is decoded whenever the received signal exceeds a value of 1. Then, the probability of a successful reception equals

$$p_{\mathrm{nc}} = 1 - Q\left(\frac{1 - 1.5}{0.5}\right) = 0.84,$$

where

$$Q(x) = \int_x^\infty \frac{\exp\left(-\frac{x^2}{2}\right)}{\sqrt{2\pi}}\, \mathrm{d}x.$$

When using compress-and-forward, the relay transmits a quantized and compressed version of the received message. In this case, the message is the signal equal to 1.5. Since the destination node receives information from both the source and the relay, compression is not done in isolation from the information sent from the source to the destination node. As shown in Figure 4.7, the relay will encode the received signal as a binary 0 whenever the received signal is larger than 1 or in the interval $[-1, 0)$ and will encode the received signal as a binary 1 whenever the received signal is smaller than -1 or in the interval $[0, 1]$. Note that only one bit is necessary to encode the received signal, as opposed to two bits when the four possible inputs are equiprobable and there is no extra information at the receiver. The binary 0 or 1 resulting from the encoding process are transmitted as a signal with value $+1$ and -1, respectively. The threshold used to receive the message from the relay is equal to 0 (meaning that when the received signal is positive, a binary 0 is estimated as the encoded digit sent by the relay). At the destination node, the encoded message from the relay is decoded using the signal received from the source as side information. The decoding uses the following rules:

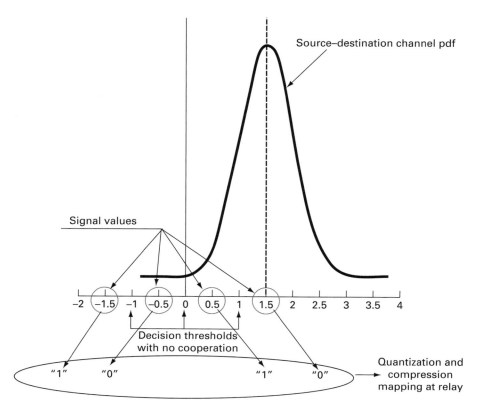

Fig. 4.7   Compress-and-forward example.

- If the signal from the relay is positive, a binary 0 is decoded and:
    - if the signal from the source is positive, the received signal becomes equal to 1.5;
    - if the signal from the source is negative, the received signal becomes equal to −0.5.
- If the signal from the relay is negative, a binary 1 is decoded and:
    - if the signal from the source is positive, the received signal becomes equal to 0.5;
    - if the signal from the source is negative, the received signal becomes equal to −1.5.

If we assume again that the transmitted signal equals 1.5, the probability of a successful reception is now equal to

$$p_{cf} = Q\left(\frac{0 - 1.5}{0.5}\right)\left(\left[Q\left(\frac{1 - 1.5}{0.25}\right) + Q\left(\frac{-1 - 1.5}{0.25}\right) - Q\left(\frac{0 - 1.5}{0.25}\right)\right](1 - q) \right.$$
$$\left. + \left(1 - \left[Q\left(\frac{1 - 1.5}{0.25}\right) + Q\left(\frac{-1 - 1.5}{0.25}\right) - Q\left(\frac{0 - 1.5}{0.25}\right)\right]\right)q\right)$$
$$= 0.93, \tag{4.25}$$

where

$$q = Q\left(\frac{0-1}{0.6}\right),$$

is the probability of successfully receiving the encoded message, a binary 0, from the relay. In 4.25, the first factor is the probability that the signal from the source is positive. This factor then multiplies two terms: the first being that the relay encodes a 0, which is received as such, and the second being the lucky coincidence that the relay incorrectly encodes a 1, but is received as a 0.

Although this example shows a simplified implementation of compress-and-forward, the numerical comparison of p_{nc} and p_{cf} shows in simple terms how cooperation improves the link quality. ▲

As a final note to compress-and-forward cooperation, we note that much of the source encoding operation done at the relay falls into the realm of the set of coding techniques known as distributed source coding, Sleppian–Wolf coding, or Wyner–Ziv coding.

Coded cooperation
Coded cooperation differs from the previous schemes in that the cooperation is implemented at the level of the channel coding subsystem. Note that both the amplify-and-forward and the decode-and-forward schemes presented earlier in this chapter were based on schemes where the relay repeats the bits sent by the source. In coded cooperation the relay sends incremental redundancy, which when combined at the receiver with the codeword sent by the source, results in a codeword with larger redundancy.

To understand coded cooperation, consider first the operation of a typical error correcting code. The encoder for an error correcting code takes a sequence of information-bearing symbols and applies mathematical operations in such a way that it generates a sequence of symbols containing not only the information present at the input sequence, but also redundant information. The redundancy in the codeword is used at the receiver to increase the chances of recovering the original information if errors have been introduced during the transmission process. While in some codewords the information and redundancy are encoded in such a way that they can only be separated through complete decoding, in many other codes the encoding and decoding operation can be done in such a way that it is possible to add redundancy to, or remove redundancy from, the codeword in a simple manner (such as through concatenation of new redundancy symbols or deletion of selected symbols). It is this second type of code that is used in coded cooperation.

In coded cooperation a codeword is transmitted in two parts, each using a different path or channel. Figure 4.8 summarizes one cycle of information transmission in coded cooperation. The main steps in the process are sequentially labeled with the circled numbers. One cycle of information transmission in coded cooperation is as follows. The cycle starts with a block of N_I information symbols being entered to a cyclic redundancy check (CRC) encoder at the source node (step 1). The result of the CRC encoder is then entered to a forward error correcting (FEC) code encoder (step 2), resulting

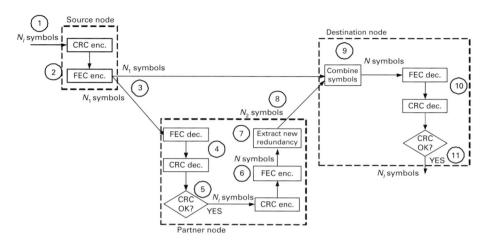

Fig. 4.8   A transmission cycle for coded cooperation.

in a codeword of N_1 symbols, i.e, with a rate $R_1 = N_1/N_I$. This codeword is then transmitted to a destination node and also is overheard by the source node cooperation partner node (step 3). After receiving the source transmission, the partner node decodes both the FEC and CRC codes (step 4). If the CRC does not reveal any error in the decoded message (step 5), the resulting N_I information symbols at the partner node are fed again into a CRC encoder. The output of the CRC encoder is then processed through an FEC encoder, resulting in a codeword of $N > N_1$ symbols (step 6), i.e, the channel encoding rate at the partner node is $R = N/N_I < R_1$.

The overall result of the processing at the partner node is a codeword generated with the same means as the codeword transmitted by the source node but containing $N_2 = N - N_1$ extra parity symbols, which are separated from the rest of the symbols (step 7). During a second phase of the communication process, the extra N_2 symbols are sent by the partner node to the destination (step 8). At the destination, the N_2 symbols from the partner node are combined with the N_1 symbols from the partner node to reconstitute the codeword with N symbols and rate R (step 9). This codeword is then decoded (step 10) and the original message is recovered if the FEC code was able to correct all the errors introduced during communication (step 11).

Note that the whole process can also be thought of as if the source node also performs channel encoding at a rate R and the transmitted codeword with N_1 symbols is generated by puncturing (deleting) N_2 symbols from the codeword with N symbols. Regardless of how the involved channel codes are generated, it is important to realize that the codeword transmitted by the source node belongs to a code that is weaker than the code used at the receiver. The code at the receiver is strengthen from that at the source node by combining the N_2 parity symbols received from the partner node. Also, it is important to note that the partner and the source operate over orthogonal channels so that at the same time that the source is transmitting its codeword, the partner is doing the same operation with its own data. Even more, while the partner is helping the source by sending N_2, parity symbols, the source is reciprocally helping the partner in the same way.

The operation of coded cooperation as described so far assumes that the partner is able to successfully decode the source message (which it knows after checking the CRC). If this is not the case, the partner is unable to generate N_2 extra parity symbols for the source and, instead, it generates and transmits N_2 parity symbols from its own data. Because of this, the weaker codeword transmitted by the source has to be a valid codeword that can be decoded when the partner is not sending the extra N_2 parity symbols.

Since in coded cooperation the operation of the source node and its partner is completely symmetric, let us denote in what follows the two cooperating nodes as node one and node two. The performance of coded cooperation is driven by the likelihood of four possible events:

(i) Both node one and node two succeed in decoding each other (weaker) channel code.

(ii) Both node one and node two fail in decoding each other (weaker) channel code.

(iii) Node one succeeds in decoding the channel code from node two but node two fails in decoding the channel code from node one. In this case, both nodes send the extra parity for node two, which can be combined together at the receiver using, for example, a maximum ratio combiner.

(iv) Node one succeeds in decoding the channel code from node two but node two fails in decoding the channel code from node one. In this case, both nodes send the extra parity for node two, which can be combined together at the receiver using, for example, a maximum ratio combiner.

As a consequence of this four-event dynamics, the outage probability in coded corporation is given by the mean outage probability among the four events.

4.2.2 Adaptive cooperation strategies

Fixed relaying suffers from deterministic loss in the transmission rate, for example, there is 50% loss in the spectral efficiency with transmissions in two phases. Moreover, fixed DF relaying suffers from the fact that the performance is limited by the weakest source–relay and relay–destination channels which reduces the diversity gains to one. To overcome this problem, adaptive relaying protocols can be developed to improve the inefficiency. We consider two strategies: selective DF relaying and incremental relaying.

4.2.2.1 Selective DF relaying

In a selective DF relaying scheme, if the signal-to-noise ratio of a signal received at the relay exceeds a certain threshold, the relay decodes the received signal and forwards the decoded information to the destination. On the other hand, if the channel between the source and the relay suffers a severe fading such that the signal-to-noise ratio falls below the threshold, the relay idles. Selective relaying improves upon the performance of fixed DF relaying, as the threshold at the relay can be determined to overcome the inherent problem in fixed DF relaying in which the relay forwards all decoded signals to the destination although some decoded signals are incorrect. For simplicity, selective

DF relaying is sometime simply called DF relaying without the confusion from fixed DF relaying, which is the case when we discuss selective DF relaying in later chapters in this book.

If the SNR in the source–relay link exceeds the threshold, the relay is likely able to decode the source's signal correctly. In this case, the SNR of the combined MRC signal at the destination is the sum of the received SNR from the source and the relay. Thus, the mutual information for selective DF relaying is given by

$$I_{\text{SDF}} = \begin{cases} \frac{1}{2}\log(1 + 2\Gamma|h_{\text{s,d}}|^2), & |h_{\text{s,r}}|^2 < g(\Gamma) \\ \frac{1}{2}\log(1 + \Gamma|h_{\text{s,d}}|^2 + \Gamma|h_{\text{r,d}}|^2), & |h_{\text{s,r}}|^2 \geq g(\Gamma) \end{cases} \qquad (4.26)$$

where $g(\Gamma) = (2^{2R} - 1)/\Gamma$.

The outage probability for selective relaying can be derived as follows. Using the law of total probability, conditioning on whether the relay forwards the source signal or not, we have

$$P[I_{\text{SDF}} < R] = P[I_{\text{SDF}} < R||h_{\text{s,r}}|^2 < g(\Gamma)]\Pr[|h_{\text{s,r}}|^2 < g(\Gamma)]$$
$$+ P[I_{\text{SDF}} < R||h_{\text{s,r}}|^2 > g(\Gamma)]\Pr[|h_{\text{s,r}}|^2 > g(\Gamma)].$$

From (4.26), the outage probability for selective DF relaying is given by

$$P[I_{\text{SDF}} < R] = P[\frac{1}{2}\log(1 + 2\Gamma|h_{\text{s,d}}|^2) < R||h_{\text{s,r}}|^2 < g(\Gamma)]\Pr[|h_{\text{s,r}}|^2 < g(\Gamma)]$$
$$+ P[\frac{1}{2}\log(1 + \Gamma|h_{\text{s,d}}|^2 + \Gamma|h_{\text{r,d}}|^2)||h_{\text{s,r}}|^2 > g(\Gamma)]\Pr[|h_{\text{s,r}}|^2 > g(\Gamma)].$$

From the above, the source can achieve diversity order two because the first term in the summation above is the product of two probabilities each account to diversity order one, and the second term in the summation is also a product of two terms the first of which has diversity order two. In other words, the selective relaying scheme can achieve diversity order two because in order for an outage event to happen, either both the source–destination and source–relay channels should be in outage, or the combined source–destination and relay–destination channel should be in outage. All the random variables in the above expression are independent exponential random variables, which makes the calculation of the outage probability straightforward, and the outage expression at high SNR is given by

$$\Pr[I_{\text{SDF}} < R] \simeq \left(\frac{\sigma_{\text{s,r}}^2 + \sigma_{\text{r,d}}^2}{2\sigma_{\text{s,d}}^2\left(\sigma_{\text{s,r}}^2\sigma_{\text{r,d}}^2\right)}\right)\left(\frac{2^{2R} - 1}{\Gamma}\right)^2, \qquad (4.27)$$

which has the same diversity gain as the AF case. This means that at high SNR, both selective DF relaying and AF relaying have the same diversity gain.

4.2.2.2   Incremental relaying

For incremental relaying, it is assumed that there is a feedback channel from the destination to the relay. The destination sends an acknowledgement to the relay if it was able to receive the source's message correctly in the first transmission phase, so the relay does not need to transmit. This protocol has the best spectral efficiency among the previously described protocols because the relay does not always need to transmit, and hence the second transmission phase becomes opportunistic depending on the channel state condition of the direct channel between the source and the destination. Nevertheless, incremental relaying achieves a diversity order of two as shown below.

In incremental relaying, if the source transmission in the first phase was successful, then there is no second phase and the source transmits new information in the next time slot. On the other hand, if the source transmission was not successful in the first phase, the relay can use any of the fixed relaying protocols to transmit the source signal from the first phase. We will focus here on the relay that utilizes the AF protocol, and in the following chapters the selective DF relaying protocol will be explored.

Note that the transmission rate is random in incremental relaying. If the first phase was successful, the transmission rate is R, while if the first transmission was in outage the transmission rate becomes $R/2$ as in fixed relaying. The outage probability can be calculated as follows:

$$\Pr[I_{\text{IR}} < R] = \Pr[I_{\text{D}} < R]\Pr[I_{\text{AF}} < \frac{R}{2}|I_{\text{D}} < R]. \tag{4.28}$$

The outage expressions for both direct transmission and AF relaying were computed before. As a function of the SNR and the rate R, the outage probability is

$$\Pr[I_{\text{IR}} < R] = \Pr\left[|h_{\text{s,d}}|^2 + \frac{1}{\Gamma}f(\Gamma|h_{\text{s,r}}|^2, \Gamma|h_{\text{r,d}}|^2) \leq g(\Gamma)\right], \tag{4.29}$$

where $g(\Gamma) = (2^R-1)/\Gamma$. The spectral efficiency is R if the source–destination channel is not in outage, and $R/2$ if the channel is not in outage. The average spectral efficiency is given by

$$\overline{R} = R\Pr[|h_{\text{s,r}}|^2 \geq g(\Gamma)] + \frac{R}{2}\Pr[|h_{\text{s,r}}|^2 < g(\Gamma)]$$

$$= \frac{R}{2}\left[1 + \exp\left(-\frac{2^R - 1}{\Gamma}\right)\right]. \tag{4.30}$$

For large SNR, we have

$$\Pr[I_{\text{IAF}} < R] \simeq \left(\frac{1}{\sigma_{\text{s,d}}^2}\frac{\sigma_{\text{s,r}}^2 + \sigma_{\text{r,d}}^2}{\sigma_{\text{s,r}}^2\sigma_{\text{r,d}}^2}\right)\left(\frac{2^{\overline{R}} - 1}{\Gamma}\right)^2. \tag{4.31}$$

Example 4.4 In this example, we study the outage probability of incremental relaying and compare it to the performance of direct transmission. The channel variance between the source and destination $\sigma_{\text{s,d}}$ is taken to be 1, while the channel variances for the source–relay or relay–destination channels are equal to 0.5. The noise variance is one.

In Figure 4.9, the outage probability is depicted versus SNR in dB for a fixed rate of 2 bps/Hz. As shown in the figure, a diversity order of two is achieved by incremental relaying, as is clear from the curve slope.

In Figure 4.10, the outage probability is depicted versus spectral efficiency in bps/Hz for a fixed SNR of 40 dB. Is is clear from the figure that the performance degrades with

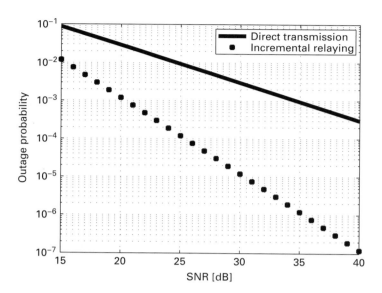

Fig. 4.9 Outage probability versus SNR.

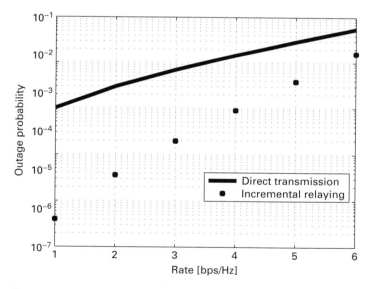

Fig. 4.10 Outage probability versus spectral efficiency.

increasing R, but it degrades faster for incremental relaying because of the inherent loss in the spectral efficiency. At high enough R, direct transmission becomes more efficient than cooperation.

Figures 4.11 and 4.12 compare the performance of the different relaying protocols discussed versus SNR and rate, respectively. The selective DF relaying has similar It is clear from both figures that incremental relaying has the best performance. This is

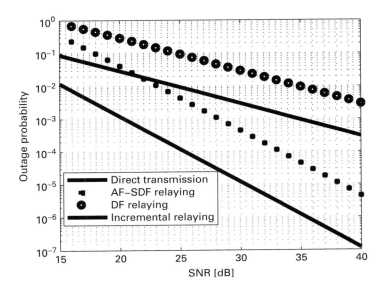

Fig. 4.11   Outage probability versus SNR for different relaying protocols.

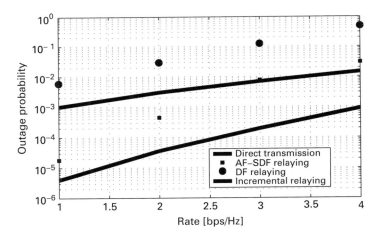

Fig. 4.12   Outage probability versus spectral efficiency for different relaying protocols.

because incremental relaying operates at a much higher spectral efficiency than the other relaying protocols and achieves a full diversity gain of two.

The simulation results also show that the fixed DF relaying offers only a diversity gain of one. In the following chapters, we consider only the selective DF relaying at the relay node and for convenience we refer to this scheme as DF relaying.   ▲

4.3   Hierarchical cooperation

Due to the spectrum scarcity, the design of future wireless systems strives to achieve high bandwidth efficiency, i.e., transmit more bits per unit channel resource (time and bandwidth). A good metric to measure the efficiency of a communication scheme is how the capacity of the network scales as the density of the users in the network increases. In this section, the impact of cooperative communications on the network capacity is studied.

4.3.1   Network model

Let us consider a network with n nodes that are uniformly distributed in a square of unit area. Each node in the network has traffic to transmit and can receive traffic from other nodes, i.e., each node is both a source and a destination. The sources and the destinations are paired arbitrarily and this pairing is fixed during the communication scenario. Each node has traffic $R(n)$ which is assumed to be the same for all nodes. The average transmit power budget is P.

The channel is assumed flat with bandwidth W Hz and the carrier frequency $f_c \gg W$. The channel between nodes i and k in the network is modeled as follows:

$$h_{ik}[m] = \sqrt{G} r_{ik}^{-\alpha/2} \exp(j\theta_{ik}[m]), \qquad (4.32)$$

where $\theta_{ik}[m]$ is a uniformly distributed random process that takes a value between 0 and 2π ($\theta_{ik} \propto U[0, 2\pi]$). The phase shift θ_{ik} is also assumed to be stationary and ergodic. The phase shifts at different nodes are mutually independent. Here r_{ik} denotes the distance between nodes i and k and it is assumed to be independent from the phase θ_{ik} and to be fixed during communications. Also, α denotes the path-loss exponent and G models the antenna gains at both transmitter and receiver.

The above channel model fits a far field assumption. Since a small position displacement of any node does not affect much the separation distance r_{ik}, the assumption that the distance r_{ik} is fixed is reasonable. On the other hand, the phase shift is measured relative to the wavelength, so any small displacement in the node location can change the phase shift of the received signal. The signal received by node i at time m is given by

$$y_i[m] = \sum_{k=1}^{n} h_{ik}[m] x_k[m] + z_i[m], \qquad (4.33)$$

where $x_k[m]$ is the signal transmitted by node k at time instant m, and $z_i[m]$ is the additive white Gaussian noise with variance N_0.

Since each node in the network has a traffic of $R(n)$ to be transmitted, the aggregate throughput in the network is given by

$$T(n) = nR(n). \qquad (4.34)$$

An interesting couple of questions that could be asked are what is the maximum aggregate throughput that can be achieved and how does it scale with the number of nodes? Furthermore, what communication schemes can achieve this aggregate throughput. It has been shown that the nearest neighbor multihop transmission has a network capacity that is upper bounded as follows [54]:

$$T(n) \leq O\left(\sqrt{n}\right). \tag{4.35}$$

This means that the traffic rate per link scales as

$$R(n) \leq O\left(\frac{1}{\sqrt{n}}\right), \tag{4.36}$$

hence the throughput per node drops as the density of the nodes increases. One can ask the question whether this network scaling is the best that we could get or it is a limitation of the nearest neighbor multihop transmission. It turns out that the network scaling in (4.35) is indeed a result of assuming nearest neighbor multihop transmission because this results in an interference limited scenario.

In the following section, a hierarchical cooperation protocol proposed in [142] that achieves linear capacity scaling is described.

4.3.2 Hierarchical cooperation protocol description

First, we consider an upper bound for the best network scaling that can be achieved for a unit area network with n nodes uniformly distributed.

THEOREM 4.3.1 *For a network with n nodes, the aggregate network throughput is upper bounded by*

$$T(n) \leq Kn \log n, \tag{4.37}$$

with probability approaching 1 as the number of nodes n goes to infinity, where K is a constant independent of n.

Proof To prove this upper bound, let us consider a single input multiple output (SIMO) system in which a source node s is transmitting to a destination node d and the rest of the nodes in the network are collocated with node d. The capacity of this channel is upper bounded by

$$R(n) \leq \log\left(1 + \frac{P}{N_0} \sum_{i=1, i=s}^{n} |h_{is}|^2\right)$$

$$= \log\left(1 + \frac{P}{N_0} \sum_{i=1, i=s}^{n} \frac{G}{r_{is}^{\alpha}}\right). \tag{4.38}$$

The right-hand side in the above inequality can be further upper bounded if the destination node is assumed to be the nearest neighbor for the source. The minimum distance between two nodes in a unit area where n nodes are uniformly distributed is larger than $1/n^{1+\delta}$ for $\delta > 0$. In the following this claim is verified.

Consider a node in the network, where the event that the nearest neighbor is at distance $1/n^{1+\delta}$ is equivalent to the event that an area equal to $\pi/n^{2+2\delta}$ is free of nodes. Denote this event by Φ, where the probability of Φ is given by

$$\Pr(\Phi) = \left(1 - \frac{\pi}{n^{2+2\delta}}\right)^{n-1}. \tag{4.39}$$

The above event is true for all nodes in the network with probability

$$\Pr\left(\text{minimum distance in the network is smaller than } \frac{1}{n^{1+\delta}}\right)$$
$$\leq n\left(1 - \frac{\pi}{n^{2+2\delta}}\right)^{n-1}, \tag{4.40}$$

where the above is true by the union bound. Therefore, the minimum distance between any two nodes in the network is given by

$$r_{\min} = \frac{1}{n^{1+\delta}}, \tag{4.41}$$

for $\delta > 0$ with high probability. Substituting the minimum distance in (4.38), we get

$$R(n) \leq \log\left(1 + \frac{PG}{N_0} n^{(1+\delta)\alpha+1}\right), \tag{4.42}$$

which concludes the theorem. ■

This means that the upper bound on the capacity scaling of a network with n nodes linearly scales with the number of nodes up to a logarithmic term. In other words, if there is a scheme that can achieve linear capacity scaling, this scheme is optimal.

Let us now describe the hierarchical cooperation protocol that is the focus of this section. The hierarchical scheme is based on clustering and long-range MIMO transmission across clusters. This can achieve two main goals:

- Using clustering one can achieve spatial reuse. This is because clusters that are well separated can work in parallel. The nodes operating in each cluster can lower their power accordingly to limit the interference to other clusters working in parallel.
- Using long-range MIMO we can achieve high spatial multiplexing gain.

The network is divided into clusters each with M nodes. Consider a specific node s within some cluster. Node s will first disseminate its bits to the other $M - 1$ nodes inside its cluster. The M nodes within the cluster can then form a distributed MIMO transmission and transmit the bits from source s to its destination, which might be in a different cluster. The nodes in destination's cluster form a MIMO receiver. Each node in the receiving cluster will then quantize the received observation and relay the information to the destination. In the following we describe these three phases in more detail.

4.3.2.1   Phase I: Setting up transmit cooperation

In this phase, each node distributes its data to the rest of the node inside its cluster. This is shown in Figure 4.13 in which source node s_1 distributes its bits to its neighbors within the cluster. There are nodes in all of the square areas in Figure 4.13, but only few of them are shown for clarity. Clusters work in parallel according to the 9-TDMA scheme depicted in Figure 4.14, which works as follows: between any two active clusters there

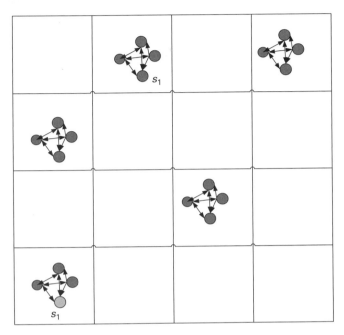

Fig. 4.13   Illustration of phase I. Each source distributes its bits to the nodes within its cluster. Adapted from [142].

Fig. 4.14   The 9-TDMA scheme. Squares with the same pattern can operate in parallel. Adapted from [142].

are at least two inactive clusters. Hence, the network needs nine time slots for all the nodes to perform phase I. This is clear from Figure 4.14, where the nodes with the same pattern are active simultaneously. Note that there are exactly nine patterns. This ensures that the interference received by any active cluster is bounded, as will be discussed later.

Let us focus on a specific source node. This source node will divide a block of length LM bits of its data into M blocks each of length L bits. The way that the source node disseminates these M blocks among the nodes inside its clusters depends on the relative location of the destination with respect to the source node. More specifically:

- if the source node and its destination are either in the same cluster or are not neighboring clusters then the source node keeps one block and the remaining $M - 1$ blocks are transmitted to the other $M - 1$ nodes located in the source's cluster;
- if the source node and its destination are neighboring clusters, then the cluster of the source is divided into two halves, one half located close to the cluster of the destination, and the other half away from it. The M blocks of the source node are distributed among the $M/2$ nodes in the half away from the destination.

Therefore, in the extreme case we need $2M$ sessions to deliver LM blocks.

To enable parallel operation of different clusters, we need to ensure that the interference from other clusters is bounded. Therefore, the transmit power is upper bounded by $PA_c^{\alpha/2}/M$, where A_c is the cluster's size and is given by

$$A_c = \frac{M}{n}. \tag{4.43}$$

Note that phase I resembles the original problem of communicating among n nodes in the unit area but scaled down to a smaller area with fewer nodes.

4.3.2.2   Phase II: MIMO transmission

In this phase, long-range MIMO transmission is implemented as depicted in Figure 4.15. Each node will independently encode the L bits received from the source node in its cluster, as in phase I. This is done by using a code with length C symbols in which the symbols have Gaussian distribution. The average transmit power is bounded above by $Pr_{s,d}^{\alpha}/M$, where $r_{s,d}$ is the distance separation between the centers of the clusters of the source and the destination.

The nodes in the destination's cluster quantize the signals they observe and store these quantized signals without decoding them. This MIMO transmission step requires C or $2C$ transmissions depending on the relative locations of the destination and the source as described in phase I. Completing this step for all of the nodes in the network thus requires $2nC$ transmissions at maximum.

4.3.2.3   Phase III: Cooperate to decode

In the previous phase, each observation is quantized into Q bits. Since each node received CM symbols from the previous phase, we end up by a traffic size of $CMQ \times M$ bits because we have M nodes within the clusters. Each node then needs to distribute this traffic to the nodes within its cluster and this can be done exactly using the scheme described in phase 1.

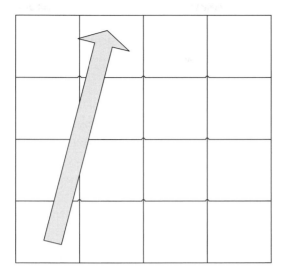

Fig. 4.15 Illustration of phase II showing Long-range MIMO transmission from the source's cluster to the destination's cluster.

Note that, depending on whether the source and the destination are in neighboring clusters or not, we will have $M/2$ nodes with $2C$ observations each, or M nodes with C observations each. An exception is when the destination node belongs to the same cluster as the source node. In this case, each node in the cluster has L bits of the original data and no MIMO observations because there is no phase II in this case. By assuming $L \leq CQ$ these exceptions can be ignored and all nodes can be assumed to have MIMO observations for simplicity.

4.3.3 Protocol analysis

In the following, the proof that a hierarchical protocol that consists of the previous three phases can achieve arbitrarily close to linear scaling is described. We first present a lemma that plays a fundamental rule in the proof.

Consider a path-loss exponent $\alpha > 2$ and a network with n nodes subject to interference from external sources. The received signal by node i is modeled as

$$y_i = \sum_{k=1}^{n} h_{ik}x_k + z_i + I_i. \tag{4.44}$$

Assume that $\{I_i, \ 1 \leq i \leq n\}$ is a collection of uncorrelated zero-mean stationary and ergodic random process with power P_{I_i} upper bounded by

$$P_{I_i} \leq K_I, \quad 1 \leq i \leq n, \tag{4.45}$$

for some constant $K_I > 0$ independent of n. We have the following result.

Lemma 4.3.2 *If there exists a scheme that can achieve an aggregate throughput of*

$$T(n) \geq K_1 n^b, \quad 0 \leq b < 1, \tag{4.46}$$

with probability at least $1 - \exp(-n^{c_1})$ *for some constant* c_1 *which is independent of n, and the per node average power budget required to achieve this scheme is upper-bounded by* P/n, *then one can construct another scheme for this network that achieve a higher aggregate throughput of*

$$T(n) \geq K_2 n^{\frac{1}{2-b}}, \tag{4.47}$$

and the failure rate of this second scheme is upper bounded by $\exp\left(-n^{C_2}\right)$. *Moreover, the per node average power required to achieve this scheme is also upper bounded by* P/n.

Proof   Let us divide the unit square into areas of A_c, each equal to

$$A_c = \frac{M}{n}. \tag{4.48}$$

Hence, each area has on average M nodes inside it. The proof of the lemma depends on a claim that the number of nodes inside each square equals M with high probability. To show this, let us calculate the probability that the number of nodes inside each square of area A_c is between $((1-\delta)A_c n, (1+\delta)A_c n)$ for some $\delta > 0$. The proof is a standard application of Chebyshev's inequality as follows.

The number of nodes in any area of size A_c is a sum of n i.i.d. Bernoulli random variables B_i such that $P(B_i = 1) = A_c$. We have

$$P\left(\sum_{i=1}^{n} B_i \geq (1+\delta)A_c n\right) = P\left(e^{s\sum_{i=1}^{n} B_i} \geq e^{s(1+\delta)A_c n}\right)$$

$$\leq \left(\mathbf{E}\left[e^{sB_1}\right]\right)^n e^{-s(1+\delta)A_c n}$$

$$= \left(e^s A_c + (1 - A_c)\right)^n e^{-s(1+\delta)A_c n}$$

$$\leq e^{-A_c n(s(1+\delta)-e^s+1)},$$

where the last inequality used the following lower bound

$$(1 - x) \leq \exp(-x). \tag{4.49}$$

Finally we have

$$P\left(\sum_{i=1}^{n} B_i \geq (1+\delta)A_c n\right) \leq e^{-A_c n \Lambda_+(\delta)} \tag{4.50}$$

where $\Lambda_+(\delta) = (1+\delta)\log(1+\delta)-\delta$, and this is achieved through choosing s to tighten the bound as follows:

$$s = \log(1 + \delta). \tag{4.51}$$

One can get a lower bound by following similar steps as above for negative B_i. This proves that the probability that the number of nodes inside the square A_c deviates from M decays exponentially with n. Hence, with high probability the number of nodes within a square A_c is equal to its mean M. In the following, it is always assumed that there exists M bits in each cluster.

Analysis of phase I

As described before, each node in a cluster divides its data into LM bits, and hence we have in total $LM \times M$ bits traffic for each cluster. For the special case where the source and the destination lie in neighbor clusters, the source will need to organize $2M$ sessions to transfer the information. Each node will scale its power proportional to the size of the cluster as $P(A_c)^{\alpha/2/M}$. The main reason for this is to allow parallel operation of the clusters. This leads to a bounded interference at each node (this is proven later in the exercises), then according to Lemma 4.3.2 all the conditions is satisfied to achieve an aggregate throughput of $K_1 M^b$ in each session for some $K_1 > 0$. This throughput also can be achieved with probability larger than $1 - \exp(-M^{c_1})$. By the union bound, and since we have at maximum $2M$ sessions in each cluster and n/M clusters, the throughput can be achieved with probability $1 - 2n \exp(-M^{c_1})$.

With this aggregate throughput, each session requires at most

$$T_{\text{session}} = \frac{LM}{K_1 M^b} = \frac{L}{K_1} M^{1-b} \tag{4.52}$$

channel uses. Since, we have at most $2M$ sessions, and taking into account the 9-TDMA scheme, the total number of channels uses required to complete phase I is given by

$$T_{\text{I}} = 9 \times 2M \times \frac{L}{K_1} M^{1-b} = \frac{18L}{K_1} M^{2-b} \tag{4.53}$$

channel uses to finish all the transmission required by the n nodes in the network within their clusters.

Analysis of phase II

In this phase, long-range MIMO transmission takes place. It is this step essentially that allows the protocol to achieve high capacity scaling. This is because with MIMO transmission, as discussed in earlier chapters, the throughput scales linearly with the minimum of the number of transmit and receive antennas.

In phase II, each source's information is transmitted to its destination while the other sources are silent. Hence, this resembles a TDMA scheme where all the sources in the network transmit in a serial way, but the major difference between this and plain TDMA is that while the source transmits all the nodes within its cluster cooperate by transmitting the bits they received from the source in the previous phase. This forms an M virtual antenna array at the source's cluster.

At the same time, all of the M nodes at the destination cluster cooperate to receive the transmitted information. This is done through each node storing a quantized version of the received information to forward later to the destination node, hence forming a distributed receive antenna array.

As described before, each node in the source's cluster encode the L bits received from the source in phase I into C symbols. This code is generated according to an average power constraint of $P r_{\text{s,d}}^{\alpha}/M$, where $r_{\text{s,d}}$ is the distance between the center of the source's cluster to the center of the destination's cluster.

In the special case in which the source and the destination belong to neighboring clusters, only half of the nodes in the source's cluster have the information. In this case, each node encodes the information into $2C$ symbols. Therefore, each node in the network requires either C or $2C$ channel uses to transmit its information. Since there are n nodes in the network, the maximum number of channel uses required to achieve phase II is given by

$$T_{\mathrm{II}} = 2nC. \tag{4.54}$$

Analysis of phase III

In this phase, each node delivers the observation it received from phase II to the intended destination. Either the M nodes in the clusters have C symbols or $M/2$ nodes have $2C$ symbols depending on whether the source and the destination belong to neighboring clusters or not. A node quantizes each symbol into Q bits. Therefore, at most we end up having a traffic of $2M \times QCM$ bits.

Similar to phase I, clusters work in parallel using the 9-TDMA scheme. This means that there are at least two idle clusters between any two active clusters. Since nodes within a cluster scale down their power proportionally to the cluster size, the interference from other active clusters can be shown to be bounded above by a constant independent of n.

Therefore, similar to phase I, the conditions to apply Lemma 4.3.2 are all satisfied. Hence, the number of channel uses required to complete this phase is given by $(2CQ/K_1) M^{2-b}$. Taking into account the 9-TDMA scheme, the total number of channel uses required for the n nodes in the network to complete phase III is given by

$$T_{\mathrm{III}} = \frac{18CQ}{K_1} M^{2-b}. \tag{4.55}$$

This is achieved with probability larger than $1 - 2ne^{-M^{c_1}}$.

The above depends on the quantization done at the nodes inside the clusters not introducing any distortion, although the nodes use a fixed quantizer of size Q bits. This is indeed possible and the observations can be encoded using a fixed number of bits, the rationale being that the observations have a bounded power that does not scale with M.

Combining all the results

Summarizing the results above, we have three phases and the total number of channel uses required to complete these phases has been computed. The total number of channel uses is given by

$$\begin{aligned} T_{\mathrm{t}} &= T_{\mathrm{I}} + T_{\mathrm{II}} + T_{\mathrm{III}} \\ &= \frac{18L}{K_1} M^{2-b} + 2Cn + \frac{18CQ}{K_1} M^{2-b}. \end{aligned} \tag{4.56}$$

Therefore, with high probability and bounded average power per node, the aggregate throughput is given by

$$T(n) = \frac{nML}{T_t} \geq K_2 n^{\frac{1}{2-b}},$$ (4.57)

for some $K_2 > 0$ independent of n. The above throughput is achieved by choosing the maximizing size for the cluster as

$$M = n^{\frac{1}{2-b}}.$$ (4.58)

This proves Lemma 4.3.2. ∎

4.3.4 Linear capacity scaling

In the previous subsection, the performance of the three-phase protocol was analyzed. To achieve the linear capacity scaling, the basic idea is to recursively apply the three-phase protocol to achieve higher throughput with each iteration. Figure 4.16 illustrates the idea of hierarchical cooperation in which each of the three phases is recursively applied. The recursion starts with smaller areas of the network and grows until it ultimately includes the whole network. The main result of the linear capacity scaling [142] is stated as follows for any path-loss exponent $\alpha \geq 2$.

THEOREM 4.3.3 *For any $\epsilon > 0$, there exists a constant $K_\epsilon > 0$ independent of n such that with high probability an aggregate throughput of*

$$T(n) \geq K_\epsilon n^{1-\epsilon}$$ (4.59)

is achievable in the network for all possible pairings between source and destination.

Proof Start with a TDMA scheme that satisfies Lemma 4.3.2. The aggregate throughput is $O(1)$, so $b = 0$. This is because each node takes exactly $1/n$ of the time. The failure probability of this scheme is 0, since there is no interference.

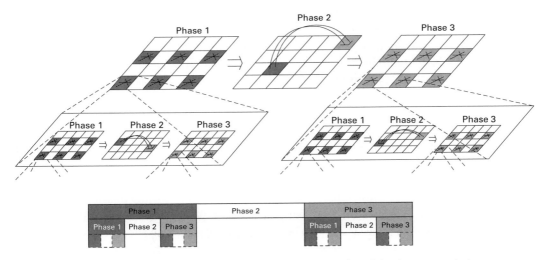

Fig. 4.16 Hierarchical cooperation. The figure illustrates the recursion of the three transmission phases © IEEE, 2007 [142].

Applying Lemma 4.3.2 recursively yields schemes that achieve higher throughput with each recursion. More precisely, starting with a TDMA scheme with $b = 0$ and applying Lemma 4.3.2 recursively h times, one gets a scheme achieving $\Theta\left(n^{h/(1+h)}\right)$ aggregate throughput.

Given any $\epsilon > 0$, we can now choose h such that

$$\frac{h}{1+h} \geq 1 - \epsilon, \tag{4.60}$$

and we get a scheme that achieves $\Theta\left(n^{1-\epsilon}\right)$ with high probability.   ∎

4.4   Chapter summary and bibliographical notes

In this chapter, we introduced the new notion of cooperative communications. Cooperative communications is a new communication paradigm that generalizes MIMO communications to much broader applications. In this new paradigm, the terminals dispersed in a wireless network cell can be thought of as distributed antennas. Through cooperation among these nodes, MIMO-like gains can be achieved by increasing the diversity gains of the system. Different protocols have been described to implement cooperation. We described the performance of these algorithms through calculating outage capacity and characterizing diversity gains. The performance of adaptive relaying techniques in general outperforms that of fixed relaying techniques because of the extra information utilized in implementing the protocols, for example, knowledge of the received SNR in selective relaying and the feedback from the destination in incremental relaying. On the other hand, fixed relaying techniques are simpler to implement compared to fixed relaying techniques due to the extra overhead needed to implement the former protocols. Therefore, it is ultimately a tradeoff between performance and complexity that the system designer must decide on.

In addition, in this chapter we have studied the capacity scaling in wireless ad hoc networks when cooperation is applied. It was shown that the hierarchical cooperation scheme proposed in [142] can achieve linear scaling in ad hoc networks. This means that as the number of nodes per unit area increases the throughput per node does not drop. This is an interesting result which shows that cooperation can overcome the interference limitation in wireless networks. The scheme that achieves the linear scaling is a recursive scheme that consists of three steps. The network is divided into clusters. In the first step, nodes within a cluster distribute their bits to the other nodes within the cluster. In the second phase, long-range MIMO transmission between clusters is implemented, and finally in the third phase local signal quantization and distribution within each cluster similar to phase 1 is implemented. Hence, the scheme makes use of two very important concepts: spatial reuse by allowing different clusters to operate in parallel, and long-range MIMO transmission that achieves linear throughput.

We provide some bibliographical notes as follows. The classical relay channel introduced by Van der Meulen [224] models a three-terminal communication channel. The relay channel, in general, is a channel with one transmitter, one receiver, and a number of intermediate nodes that act as relays to improve the system performance. The simplest relay channel contains one terminal, the relay, that listens to the signal transmitted by the source, processes it, and then transmits it to the destination. Cover and El Gamal [25] developed lower and upper bounds on the channel capacity for specific non-faded relay channel models. The lower and upper bound do not coincide in general except for some special cases, such as in the degraded relay channel. Later, several works have studied the capacity of the relay channels and developed coding strategies that can achieve the ergodic capacity of the channel under certain scenarios, see [99] and the references there in.

In [109] and [108], Laneman *et al.* proposed different cooperative diversity protocols and analyzed their performance in terms of outage behavior. The terms decode-and-forward and amplify-and-forward have been introduced in these two works. In decode-and-forward, each relay receives and decodes the signal transmitted by the source, and then forwards the decoded signal to the destination, which combines all the copies in a proper way. Amplify-and-forward is a simpler technique, in which the relay amplifies the received signal and then forwards it to the destination. Although the noise is amplified along with the signal in this technique, we still gain spatial diversity by transmitting the signal over two spatially independent channels. Compress-and-forward was studied in [25] and [99]. More information about the related distributed source coding technique can be found in [230]. Terminologies other than cooperative diversity are also used in the research community to refer to the same concept of achieving spatial diversity via forming virtual antennas. User cooperation diversity was introduced by Sendonaris *et al.* in [179] in which the authors implemented a two-user CDMA cooperative system, where both users are active and use orthogonal codes to avoid multiple access interference. Another technique to achieve diversity that incorporates error-control-coding into cooperation is coded cooperation, introduced by Hunter *et al.* in [78].

Capacity scaling in wireless networks has been studied by several researchers recently. The seminal work of Gupta and Kumar [54] has stimulated a lot of thinking in this problem. Gupta and Kumar showed that an aggregate throughput scaling of $\Theta(\sqrt{n})$ is an upper bound for what can be achievable by multihop transmission. The limitation in their model is that no cooperation is allowed between networks, and hence if a signal is not intended to be transmitted to a node it is treated as interference. Cooperation, on the other hand, benefits from the broadcast nature of the channel, and tries to take advantage of this aspect instead of treating the overheard signals as interference. In another recent work by Aeron and Saligrama [2], distributed collaborative schemes are proposed that lead to an increase in throughput over the traditional multihop schemes. The scheme proposed in this work achieves an aggregate throughput of $\Theta\left(n^{2/3}\right)$. The scheme that achieves the linear capacity scaling in wireless ad hoc networks when cooperation is applied, and the associated analysis studied in this chapter, was presented in [142].

Exercises

4.1   In (4.12), the MRC parameters were derived from a detection theory and signal space perspective. Another approach to solve for the MRC parameters is via solving an optimization problem with the cost function being the SNR at the output of the detector and the optimization variables being the MRC parameters.
        Assume that signal x is received at the destination via two independent paths as follows:

$$y_1 = h_1 x + n_1$$
$$y_2 = h_2 x + n_2$$

where h_1, h_2, n_1, and n_2 are all mutually independent complex normal Gaussian random variables with zero mean and variances $\sigma_1^2, \sigma_2^2, N_0 1$, and $N_0 2$. Find the optimal combining ratios a_1 and a_2 that maximize the output SNR of the detector.

4.2   In describing the different relaying protocols in this chapter, it was assumed that the relay already exists in some position and is helping the source node. In a real scenario, there might exist multiple relays that can help the source and the source needs to decide which relay to select. Therefore, studying the effect of relay position on the performance of various cooperation protocols is important.

(a) In AF relaying, assume the relay is located on the straight line joining the source and destination. Assume that the channel variance between two nodes depends on the distance d separating the two nodes according to an exponential path loss model as follows:

$$\sigma^2 = d^{-\alpha}. \tag{E4.1}$$

Find the outage probability of AF relaying as a function of the relay distance from the source $d_{s,r}$ assuming a unit distance between the source and destination, and that $d_{s,r}$ takes values between 0.1 and 0.9. Use a Matlab code to draw this outage probability as a function of the relay position.

(b) Repeat the previous calculations to DF, selective, and incremental relaying.

4.3   In proving that the schemes in phases I and III work, it was assumed that the 9-TDMA scheme and scaling the transmission power relative to the cluster size result in uncorrelated and bounded interference. Consider clusters of size M and area A_c. The clusters operate according to the 9-TDMA scheme shown in Figure 4.14 in a network of size n. The average power of each node is assumed to be constrained to $PA_c^{\alpha/2}/M$. For $\alpha > 2$, show that the interference power received by a node from the clusters operating in parallel is upper bounded by a constant K_I independent of n.

4.4   In phase III, each node in the destination cluster quantizes the MIMO received information into Q bits. The number of bits Q is assumed to be a fixed number independent of M. The argument is that this is sufficient to achieve linear throughput from MIMO transmission without introducing distortion because the received power in phase II by each node in the destination cluster is bounded below and above by constants P_1 and P_2, respectively, which are independent of M. Show that this claim is correct.

4.5   In this chapter, equal transmit power at both the source and the relay was assumed. Consider the case where the transmit power at the source and the relay are given by P_s and P_r.

(a) Find the outage probability for amplify-and-forward relaying as a function of P_s and P_r.

(b) Assuming an equal average power constraint as follows:

$$P_s + P_r = 2P, \tag{E4.2}$$

where P is the transmit power used for direct transmission. Find the optimal power allocation P_s and P_r to minimize the outage probability expression.

(c) Repeat the previous problem for selective decode-and-forward and incremental relaying.

(d) Find the outage probability for selective decode-and-forward relaying when two relays are available to help the source. What is the diversity gain achieved?

5 Cooperative communications with single relay

In this chapter, we consider single relay cooperative communications in wireless networks. We focus on the discussion of symbol error rate (SER) performance analysis and optimum power allocation for uncoded cooperative communications with either an amplify-and-forward (AF) or a selective decode-and-forward (DF)[1] cooperation protocol. In this chapter and the rest of this book, we simply call the *selective DF cooperation protocol* a *DF protocol* without confusing it with the *fixed DF relaying protocol*.

The chapter is organized as follows. First, we briefly describe a system model for cooperative communications with either DF or AF cooperation protocols. Second, we analyze the SER performance for DF cooperation systems, in which a closed-form SER formulation is obtained explicitly for systems with M-PSK and M-QAM modulations. An SER upper bound as well as an approximation are provided to reveal the asymptotic performance of the cooperative system. Based on the tight SER approximation, we can determine an asymptotic optimum power allocation for DF cooperation systems. Third, we consider the SER performance for AF cooperation systems, in which we first derive a simple closed-form moment-generating function (MGF) expression for the harmonic mean of two independent exponential random variables. Then, based on the simple MGF expression, closed-form SER formulations are given for AF cooperation systems with M-PSK and M-QAM modulations. We also provide a tight SER approximation to show the asymptotic performance of AF cooperation systems and to determine an optimum power allocation. Fourth, we discuss performance comparisons between the cooperation systems with DF and AF protocols and show some simulation examples. Finally, we consider a method of re-mapping signal constellation points at the relay to improve the SER performance of the DF protocol.

5.1   System model

A cooperative communication strategy with two phases is considered for wireless networks such as mobile ad hoc networks, sensor networks, or cellular networks. In both phases, users transmit signals through orthogonal channels by using TDMA, FDMA, or

---

[1] In this chapter, we consider that an ideal selective DF cooperation protocol is one in which a relay is able to detect whether the decoding at the relay is correct or not. If the decoding is correct, the relay forwards the decoded symbol to a destination, otherwise the relay does not send or remains idle.

CDMA. In this chapter, we focus on a two-user cooperation scheme. Specifically, user 1 sends information to its destination in phase 1, and user 2 also receives the information. User 2 helps user 1 to forward the information in phase 2. Similarly, when user 2 sends its information to its destination in phase 1, user 1 receives the information and forwards it to user 2's destination in phase 2. Due to the symmetry of the two users, we may consider only user 1's performance. Without loss of generality, we consider a concise model as shown in Figure 5.1, in which "source" denotes user 1 and "relay" represents user 2.

In Phase 1, the source broadcasts its information to both the destination and the relay. The received signals $y_{s,d}$ and $y_{s,r}$ at the destination and the relay, respectively, can be written as

$$y_{s,d} = \sqrt{P_1}\, h_{s,d}\, x + \eta_{s,d}, \tag{5.1}$$

$$y_{s,r} = \sqrt{P_1}\, h_{s,r}\, x + \eta_{s,r}, \tag{5.2}$$

in which P_1 is the transmitted power at the source, x is the transmitted information symbol, and $\eta_{s,d}$ and $\eta_{s,r}$ are additive noise. In (5.1) and (5.2), $h_{s,d}$ and $h_{s,r}$ are the channel coefficients from the source to the destination and the relay respectively. They are modeled as zero-mean, complex Gaussian random variables with variances $\delta_{s,d}^2$ and $\delta_{s,r}^2$, respectively. The noise terms $\eta_{s,d}$ and $\eta_{s,r}$ are modeled as zero-mean, complex Gaussian random variables with variance $\mathcal{N}_0$.

DF protocol

In Phase 2, for a DF cooperation protocol, if the relay is able to decode the transmitted symbol correctly, then the relay forwards the decoded symbol with power P_2 to the destination, otherwise the relay does not send or remains idle. The received signal at the destination in Phase 2 in this case can be modeled as

$$y_{r,d} = \sqrt{\tilde{P}_2}\, h_{r,d}\, x + \eta_{r,d}, \tag{5.3}$$

where $\tilde{P}_2 = P_2$ if the relay decodes the transmitted symbol correctly, otherwise $\tilde{P}_2 = 0$. In (5.3), $h_{r,d}$ is the channel coefficient from the relay to the destination, and it is modeled as a zero-mean, complex Gaussian random variable with variance $\delta_{r,d}^2$. The noise term $\eta_{r,d}$ is also modeled as a zero-mean complex Gaussian random variable with variance $\mathcal{N}_0$. Note that for analytical tractability of SER calculations, we assume here an ideal DF cooperation protocol that the relay is able to detect whether the transmitted symbol is decoded correctly or not. In practice, we may apply an SNR threshold at the relay. If

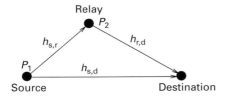

Fig. 5.1   A simplified cooperation model.

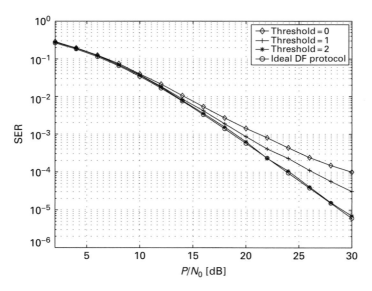

Fig. 5.2 Performance of the DF cooperation protocol with different SNR thresholds at the relay. It is assumed that $\delta_{s,d}^2 = 1$, $\delta_{s,r}^2 = 10$, $\delta_{r,d}^2 = 1$, $\mathcal{N}_0 = 1$, and $P_1 = P_2 = P/2$.

the received SNR at the relay is higher than the threshold, then the symbol has a high probability to be decoded correctly. For example, Figure 5.2 shows performances of a DF cooperation protocol with different SNR thresholds at the relay which are compared with that of the ideal DF protocol. We can see that if the SNR threshold at the relay is properly chosen (threshold = 2 in this case), the performance of the DF protocol is close to that of the ideal DF protocol and the detection error propagation at the relay is negligible. The ideal DF cooperation protocol considered here provides a performance benchmark for the DF protocol with SNR thresholds. Further discussion on threshold optimization at the relay can be found in later chapters.

AF protocol

For an AF cooperation protocol, in phase 2 the relay amplifies the received signal and forwards it to the destination with transmitted power P_2. The received signal at the destination in phase 2 is specified as

$$y_{r,d} = \frac{\sqrt{P_2}}{\sqrt{P_1 |h_{s,r}|^2 + \mathcal{N}_0}} h_{r,d} y_{s,r} + \eta_{r,d}, \tag{5.4}$$

where $h_{r,d}$ is the channel coefficient from the relay to the destination and $\eta_{r,d}$ is an additive noise, with the same statistics models as in (5.3), respectively. More specifically, by (5.2), the received signal $y_{r,d}$ in this case is

$$y_{r,d} = \frac{\sqrt{P_1 P_2}}{\sqrt{P_1 |h_{s,r}|^2 + \mathcal{N}_0}} h_{r,d} h_{s,r} x + \eta'_{r,d}, \tag{5.5}$$

where

$$\eta'_{r,d} = \frac{\sqrt{P_2}}{\sqrt{P_1|h_{s,r}|^2 + \mathcal{N}_0}} h_{r,d}\, \eta_{s,r} + \eta_{r,d} \qquad (5.6)$$

Assume that the noise terms $\eta_{s,r}$ and $\eta_{r,d}$ are independent, then the equivalent noise $\eta'_{r,d}$ is a zero-mean, complex Gaussian random variable with variance

$$\left(\frac{P_2|h_{r,d}|^2}{P_1|h_{s,r}|^2 + \mathcal{N}_0} + 1 \right) \mathcal{N}_0.$$

In both the DF and AF cooperation protocols, the channel coefficients $h_{s,d}$, $h_{s,r}$, and $h_{r,d}$ are assumed to be known at the receiver, but not at the transmitter. The destination jointly combines the received signal from the source in Phase 1 and that from the relay in phase 2, and detects the transmitted symbols by using maximum-ratio combining (MRC). In both protocols, we consider a total transmitted power P such as

$$P_1 + P_2 = P. \qquad (5.7)$$

Note that in the DF cooperation protocol, the power saving in the case of $\tilde{P}_2 = 0$ is negligible, since at high SNR, the chance that the relay incorrectly decodes the symbol is rare as we will see later in the performance analysis.

5.2 SER analysis for DF protocol

In this section, we consider the SER performance analysis for the DF cooperative communication systems. First, we introduce closed-form SER formulations for the systems with M-PSK and M-QAM[2] modulations, respectively. Then, we provide an SER upper bound as well as an approximation to reveal the asymptotic performance of the systems, in which the approximation is asymptotically tight at high SNR. Finally, based on the tight SER approximation, we are able to determine an asymptotic optimum power allocation for DF cooperation systems.

5.2.1 Closed-form SER analysis

With knowledge of the channel coefficients $h_{s,d}$ (between the source and the destination) and $h_{r,d}$ (between the relay and the destination), the destination detects the transmitted symbols by jointly combining the received signal $y_{s,d}$ (5.1) from the source and $y_{r,d}$ (5.3) from the relay. The combined signal at the MRC detector can be written as

$$y = a_1 y_{s,d} + a_2 y_{r,d}, \qquad (5.8)$$

in which the factors a_1 and a_2 are determined such that the SNR of the MRC output is maximized, and, from Section 1.4, they can be specified as

$$a_1 = \sqrt{P_1} h^*_{s,d}/\mathcal{N}_0, \quad a_2 = \sqrt{\tilde{P}_2} h^*_{r,d}/\mathcal{N}_0.$$

---

[2] In this chapter, QAM represents a square QAM constellation whose size is given by $M = 2^k$ with k even.

Assume that the transmitted symbol x in (5.1) and (5.3) has average energy 1, then the SNR of the MRC output is

$$\gamma = \frac{P_1|h_{s,d}|^2 + \tilde{P}_2|h_{r,d}|^2}{\mathcal{N}_0}. \tag{5.9}$$

First, let us review SER formulations for an uncoded system with M-PSK or M-QAM ($M = 2^k$ with k even) modulation [187], which are given by

$$\Psi_{PSK}(\rho) \triangleq \frac{1}{\pi} \int_0^{(M-1)\pi/M} \exp\left(-\frac{b_{PSK}\rho}{\sin^2\theta}\right) d\theta, \tag{5.10}$$

$$\Psi_{QAM}(\rho) \triangleq 4K Q(\sqrt{b_{QAM}\rho}) - 4K^2 Q^2(\sqrt{b_{QAM}\rho}), \tag{5.11}$$

in which ρ is SNR, $b_{PSK} = \sin^2(\pi/M)$, $K = 1 - (1/\sqrt{M})$, $b_{QAM} = 3/(M-1)$, and $Q(u) = (1/\sqrt{2\pi}) \int_u^\infty \exp(-t^2/2)dt$ is the Gaussian Q-function. Therefore, if M-PSK modulation is used in the DF cooperation system, with the instantaneous SNR γ in (5.9), the conditional SER of the system with the channel coefficients $h_{s,d}$, $h_{s,r}$ and $h_{r,d}$ can be written as

$$P_{PSK}^{h_{s,d},h_{s,r},h_{r,d}} = \Psi_{PSK}(\gamma). \tag{5.12}$$

If M-QAM ($M = 2^k$ with k even) signals are used in the system, the conditional SER of the system can also be expressed as

$$P_{QAM}^{h_{s,d},h_{s,r},h_{r,d}} = \Psi_{QAM}(\gamma). \tag{5.13}$$

It is easy to see that, in the case of QPSK or 4-QAM modulation, the conditional SERs in (5.12) and (5.13) are the same.

Note that in phase 2 we assume that if the relay decodes the transmitted symbol x from the source correctly, then the relay forwards the decoded symbol with power P_2 to the destination, i.e., $\tilde{P}_2 = P_2$; otherwise the relay does not send, i.e., $\tilde{P}_2 = 0$. If an M-PSK symbol is sent from the source, then at the relay, the chance of incorrect decoding is $\Psi_{PSK}(P_1|h_{s,r}|^2/\mathcal{N}_0)$, and the chance of correct decoding is $1 - \Psi_{PSK}(P_1|h_{s,r}|^2/\mathcal{N}_0)$. Similarly, if an M-QAM symbol is sent out at the source, then the chance of incorrect decoding at the relay is $\Psi_{QAM}(P_1|h_{s,r}|^2/\mathcal{N}_0)$, and the chance of correct decoding is $1 - \Psi_{QAM}(P_1|h_{s,r}|^2/\mathcal{N}_0)$.

Let us first focus on the SER performance analysis in the case of M-PSK modulation. Taking into account the two scenarios $\tilde{P}_2 = P_2$ and $\tilde{P}_2 = 0$, we further calculate the conditional SER in (5.12) as follows:

$$P_{PSK}^{h_{s,d},h_{s,r},h_{r,d}}$$

$$= \Psi_{PSK}(\gamma)|_{\tilde{P}_2=0}\Psi_{PSK}\left(\frac{P_1|h_{s,r}|^2}{\mathcal{N}_0}\right) + \Psi_{PSK}(\gamma)|_{\tilde{P}_2=P_2}\left[1 - \Psi_{PSK}\left(\frac{P_1|h_{s,r}|^2}{\mathcal{N}_0}\right)\right]$$

$$= \frac{1}{\pi^2} \int_0^{(M-1)\pi/M} \exp\left(-\frac{b_{PSK}P_1|h_{s,d}|^2}{\mathcal{N}_0\sin^2\theta}\right) d\theta \int_0^{(M-1)\pi/M} \exp\left(-\frac{b_{PSK}P_1|h_{s,r}|^2}{\mathcal{N}_0\sin^2\theta}\right) d\theta$$

$$+ \frac{1}{\pi} \int_0^{(M-1)\pi/M} \exp\left(-\frac{b_{PSK}(P_1|h_{s,d}|^2 + P_2|h_{r,d}|^2)}{\mathcal{N}_0 \sin^2 \theta} \right) d\theta$$

$$\times \left[1 - \frac{1}{\pi} \int_0^{(M-1)\pi/M} \exp\left(-\frac{b_{PSK} P_1 |h_{s,r}|^2}{\mathcal{N}_0 \sin^2 \theta} \right) d\theta \right]. \tag{5.14}$$

In the following, we average the conditional SER (5.14) over the Rayleigh fading channels $h_{s,d}$, $h_{s,r}$, and $h_{r,d}$ with variances $\delta_{s,d}^2$, $\delta_{s,r}^2$, and $\delta_{r,d}^2$, respectively. Since the fading channels $h_{s,d}$, $h_{s,r}$, and $h_{r,d}$ are independent of each other, and

$$\int_0^\infty \exp\left(-\frac{b_{PSK} P_1 z}{\mathcal{N}_0 \sin^2 \theta} \right) p_{|h|^2}(z) dz = \frac{1}{1 + \frac{b_{PSK} P_1 \delta_h^2}{\mathcal{N}_0 \sin^2 \theta}},$$

we are able to obtain the SER of the DF cooperation system with M-PSK modulation as follows:

$$P_{PSK} = F_1 \left(1 + \frac{b_{PSK} P_1 \delta_{s,d}^2}{\mathcal{N}_0 \sin^2 \theta} \right) F_1 \left(1 + \frac{b_{PSK} P_1 \delta_{s,r}^2}{\mathcal{N}_0 \sin^2 \theta} \right)$$

$$+ F_1 \left(\left(1 + \frac{b_{PSK} P_1 \delta_{s,d}^2}{\mathcal{N}_0 \sin^2 \theta} \right) \left(1 + \frac{b_{PSK} P_2 \delta_{r,d}^2}{\mathcal{N}_0 \sin^2 \theta} \right) \right)$$

$$\times \left[1 - F_1 \left(1 + \frac{b_{PSK} P_1 \delta_{s,r}^2}{\mathcal{N}_0 \sin^2 \theta} \right) \right], \tag{5.15}$$

where

$$F_1(x(\theta)) = \frac{1}{\pi} \int_0^{(M-1)\pi/M} \frac{1}{x(\theta)} d\theta. \tag{5.16}$$

For DF cooperation systems with M-QAM modulation, the conditional SER in (5.13) with the channel coefficients $h_{s,d}$, $h_{s,r}$, and $h_{r,d}$ can be similarly determined as

$$P_{QAM}^{h_{s,d},h_{s,r},h_{r,d}} = \Psi_{QAM}(\gamma)|_{\tilde{P}_2=0} \Psi_{QAM} \left(\frac{P_1|h_{s,r}|^2}{\mathcal{N}_0} \right)$$

$$+ \Psi_{QAM}(\gamma)|_{\tilde{P}_2=P_2} \left[1 - \Psi_{QAM} \left(\frac{P_1|h_{s,r}|^2}{\mathcal{N}_0} \right) \right]. \tag{5.17}$$

By substituting (5.11) into the above formulation and averaging it over the fading channels $h_{s,d}$, $h_{s,r}$, and $h_{r,d}$, as in the case of M-PSK modulation, the SER of the DF cooperation system with M-QAM modulation can be given by

$$P_{\text{QAM}} = F_2 \left(1 + \frac{b_{\text{QAM}} P_1 \delta_{\text{s,d}}^2}{2\mathcal{N}_0 \sin^2 \theta} \right) F_2 \left(1 + \frac{b_{\text{QAM}} P_1 \delta_{\text{s,r}}^2}{2\mathcal{N}_0 \sin^2 \theta} \right)$$

$$+ F_2 \left(\left(1 + \frac{b_{\text{QAM}} P_1 \delta_{\text{s,d}}^2}{2\mathcal{N}_0 \sin^2 \theta} \right) \left(1 + \frac{b_{\text{QAM}} P_2 \delta_{\text{r,d}}^2}{2\mathcal{N}_0 \sin^2 \theta} \right) \right)$$

$$\times \left[1 - F_2 \left(1 + \frac{b_{\text{QAM}} P_1 \delta_{\text{s,r}}^2}{2\mathcal{N}_0 \sin^2 \theta} \right) \right], \tag{5.18}$$

where

$$F_2(x(\theta)) = \frac{4K}{\pi} \int_0^{\pi/2} \frac{1}{x(\theta)} d\theta - \frac{4K^2}{\pi} \int_0^{\pi/4} \frac{1}{x(\theta)} d\theta. \tag{5.19}$$

In order to get the SER formulation in (5.18), two special properties of the Gaussian Q-function are needed as follows:

$$Q(u) = \frac{1}{\pi} \int_0^{\pi/2} \exp \left(-\frac{u^2}{2 \sin^2 \theta} \right) d\theta,$$

$$Q^2(u) = \frac{1}{\pi} \int_0^{\pi/4} \exp \left(-\frac{u^2}{2 \sin^2 \theta} \right) d\theta$$

for any $u \geq 0$. Note that for 4-QAM modulation,

$$F_2(x(\sin^2(\theta))) = \frac{2}{\pi} \int_0^{\pi/2} \frac{1}{x(\sin^2(\theta))} d\theta - \frac{1}{\pi} \int_0^{\pi/4} \frac{1}{x(\sin^2(\theta))} d\theta$$

$$= \frac{1}{\pi} \int_0^{\pi/2} \frac{1}{x(\sin^2(\theta))} d\theta + \frac{1}{\pi} \int_{\pi/4}^{\pi/2} \frac{1}{x(\sin^2(\theta))} d\theta$$

$$= \frac{1}{\pi} \int_0^{3\pi/4} \frac{1}{x(\sin^2(\theta))} d\theta,$$

which shows that the SER formulation in (5.18) for 4-QAM modulation is consistent with that in (5.15) for QPSK modulation.

5.2.2 SER upper bound and asymptotic approximation

Even though the closed-form SER formulations in (5.15) and (5.18) can be efficiently calculated numerically, they are very complex and it is hard to gain an insight into the system performance from these. In the following, we introduce a SER upper bound and SER approximation which are useful in demonstrating the asymptotic performance of the DF cooperation scheme.

THEOREM 5.2.1 *The SER of DF cooperation systems with M-PSK or M-QAM modulation can be upper bounded as*

$$P_s \leq \frac{(M-1)\mathcal{N}_0^2}{M^2} \cdot \frac{MbP_1\delta_{\text{s,r}}^2 + (M-1)bP_2\delta_{\text{r,d}}^2 + (2M-1)\mathcal{N}_0}{(\mathcal{N}_0 + bP_1\delta_{\text{s,d}}^2)(\mathcal{N}_0 + bP_1\delta_{\text{s,r}}^2)(\mathcal{N}_0 + bP_2\delta_{\text{r,d}}^2)}, \tag{5.20}$$

where $b = b_{\text{PSK}}$ for M-PSK signals and $b = b_{\text{QAM}}/2$ for M-QAM signals.

Proof In the case of M-PSK modulation, the closed-form SER expression is given in (5.15). By removing the negative term in (5.15), we have

$$P_{\text{PSK}} \leq F_1 \left(1 + \frac{b_{\text{PSK}} P_1 \delta_{\text{s,d}}^2}{\mathcal{N}_0 \sin^2 \theta} \right) F_1 \left(1 + \frac{b_{\text{PSK}} P_1 \delta_{\text{s,r}}^2}{\mathcal{N}_0 \sin^2 \theta} \right)$$
$$+ F_1 \left(\left(1 + \frac{b_{\text{PSK}} P_1 \delta_{\text{s,d}}^2}{\mathcal{N}_0 \sin^2 \theta} \right) \left(1 + \frac{b_{\text{PSK}} P_2 \delta_{\text{r,d}}^2}{\mathcal{N}_0 \sin^2 \theta} \right) \right), \qquad (5.21)$$

where

$$F_1(x(\theta)) = \frac{1}{\pi} \int_0^{(M-1)\pi/M} \frac{1}{x(\theta)} d\theta.$$

We observe that, in the right-hand side of the above inequality, all integrands have their maximum value when $\sin^2 \theta = 1$. Therefore, by substituting $\sin^2 \theta = 1$ into (5.21), we have

$$P_{\text{PSK}} \leq \frac{(M-1)^2}{M^2} \cdot \frac{\mathcal{N}_0^2}{(\mathcal{N}_0 + b_{\text{PSK}} P_1 \delta_{\text{s,d}}^2)(\mathcal{N}_0 + b_{\text{PSK}} P_1 \delta_{\text{s,r}}^2)}$$
$$+ \frac{M-1}{M} \cdot \frac{\mathcal{N}_0^2}{(\mathcal{N}_0 + b_{\text{PSK}} P_1 \delta_{\text{s,d}}^2)(\mathcal{N}_0 + b_{\text{PSK}} P_2 \delta_{\text{r,d}}^2)}$$
$$= \frac{(M-1)\mathcal{N}_0^2}{M^2} \cdot \frac{M b_{\text{PSK}} P_1 \delta_{\text{s,r}}^2 + (M-1) b_{\text{PSK}} P_2 \delta_{\text{r,d}}^2 + (2M-1)\mathcal{N}_0}{(\mathcal{N}_0 + b_{\text{PSK}} P_1 \delta_{\text{s,d}}^2)(\mathcal{N}_0 + b_{\text{PSK}} P_1 \delta_{\text{s,r}}^2)(\mathcal{N}_0 + b_{\text{PSK}} P_2 \delta_{\text{r,d}}^2)},$$

which validates the upper bound in (5.20) for M-PSK modulation. Similarly, in the case of M-QAM modulation, the SER in (5.18) can be upper bounded as

$$P_{\text{QAM}} \leq F_2 \left(1 + \frac{b_{\text{QAM}} P_1 \delta_{\text{s,d}}^2}{2\mathcal{N}_0 \sin^2 \theta} \right) F_2 \left(1 + \frac{b_{\text{QAM}} P_1 \delta_{\text{s,r}}^2}{2\mathcal{N}_0 \sin^2 \theta} \right)$$
$$+ F_2 \left(\left(1 + \frac{b_{\text{QAM}} P_1 \delta_{\text{s,d}}^2}{2\mathcal{N}_0 \sin^2 \theta} \right) \left(1 + \frac{b_{\text{QAM}} P_2 \delta_{\text{r,d}}^2}{2\mathcal{N}_0 \sin^2 \theta} \right) \right), \qquad (5.22)$$

where

$$F_2(x(\theta)) = \frac{4K}{\pi} \int_0^{\pi/2} \frac{1}{x(\theta)} d\theta - \frac{4K^2}{\pi} \int_0^{\pi/4} \frac{1}{x(\theta)} d\theta.$$

Note that the function $F_2(x(\theta))$ can be rewritten as

$$F_2(x(\theta)) = \frac{4K}{\pi \sqrt{M}} \int_0^{\pi/2} \frac{1}{x(\theta)} d\theta + \frac{4K^2}{\pi} \int_{\pi/4}^{\pi/2} \frac{1}{x(\theta)} d\theta, \qquad (5.23)$$

which does not contain negative term. It turns out that the integrands in (5.22) have their maximum value when $\sin^2 \theta = 1$. Thus, by substituting (5.23) and $\sin^2 \theta = 1$ into (5.22), we have

$$P_{\text{QAM}} \leq \left(\frac{2K}{\sqrt{M}} + K^2 \right)^2 \frac{\mathcal{N}_0^2}{(\mathcal{N}_0 + \frac{b_{\text{QAM}}}{2} P_1 \delta_{\text{s,d}}^2)(\mathcal{N}_0 + \frac{b_{\text{QAM}}}{2} P_1 \delta_{\text{s,r}}^2)}$$

$$+ \left(\frac{2K}{\sqrt{M}} + K^2 \right) \frac{\mathcal{N}_0^2}{(\mathcal{N}_0 + \frac{b_{\text{QAM}}}{2} P_1 \delta_{\text{s,d}}^2)(\mathcal{N}_0 + \frac{b_{\text{QAM}}}{2} P_2 \delta_{\text{r,d}}^2)}$$

$$= \frac{(M-1)\mathcal{N}_0^2}{M^2} \cdot \frac{M \frac{b_{\text{QAM}}}{2} P_1 \delta_{\text{s,r}}^2 + (M-1) \frac{b_{\text{QAM}}}{2} P_2 \delta_{\text{r,d}}^2 + (2M-1)\mathcal{N}_0}{(\mathcal{N}_0 + \frac{b_{\text{QAM}}}{2} P_1 \delta_{\text{s,d}}^2)(\mathcal{N}_0 + \frac{b_{\text{QAM}}}{2} P_1 \delta_{\text{s,r}}^2)(\mathcal{N}_0 + \frac{b_{\text{QAM}}}{2} P_2 \delta_{\text{r,d}}^2)},$$

in which $K = 1 - 1/\sqrt{M}$. Therefore, the upper bound in (5.20) holds for both M-PSK and M-QAM modulation. ∎

In the sequel, we provide an asymptotically tight SER approximation if all of the channel links $h_{\text{s,d}}$, $h_{\text{s,r}}$, and $h_{\text{r,d}}$ are available, i.e., $\delta_{\text{s,d}}^2 \neq 0$, $\delta_{\text{s,r}}^2 \neq 0$, and $\delta_{\text{r,d}}^2 \neq 0$. According to (5.15) and (5.18), let us denote the SER of DF cooperation systems with M-PSK or M-QAM modulation as

$$P_s = I_1(P_1/\mathcal{N}_0) + I_2(P_1/\mathcal{N}_0, P_2/\mathcal{N}_0), \qquad (5.24)$$

where

$$I_1(x) = F_i \left(1 + \frac{xb\delta_{\text{s,d}}^2}{\sin^2 \theta} \right) F_i \left(1 + \frac{xb\delta_{\text{s,r}}^2}{\sin^2 \theta} \right), \qquad (5.25)$$

$$I_2(x, y) = F_i \left(\left(1 + \frac{xb\delta_{\text{s,d}}^2}{\sin^2 \theta} \right) \left(1 + \frac{yb\delta_{\text{r,d}}^2}{\sin^2 \theta} \right) \right) \left[1 - F_i \left(1 + \frac{xb\delta_{\text{s,r}}^2}{\sin^2 \theta} \right) \right], \qquad (5.26)$$

in which $i = 1$ and $b = b_{\text{PSK}}$ for M-PSK modulation, and $i = 2$ and $b = b_{\text{QAM}}/2$ for M-QAM modulation. Then, we have the following results (leave proof as an exercise):

$$\lim_{x \to \infty} x^2 I_1(x) = \frac{A^2}{b^2 \delta_{\text{s,d}}^2 \delta_{\text{s,r}}^2}, \qquad (5.27)$$

$$\lim_{x, y \to \infty} xy I_2(x, y) = \frac{B}{b^2 \delta_{\text{s,d}}^2 \delta_{\text{r,d}}^2}, \qquad (5.28)$$

where, for M-PSK signals, $b = b_{\text{PSK}}$ and

$$A = \frac{M-1}{2M} + \frac{\sin \frac{2\pi}{M}}{4\pi}, \qquad (5.29)$$

$$B = \frac{3(M-1)}{8M} + \frac{\sin \frac{2\pi}{M}}{4\pi} - \frac{\sin \frac{4\pi}{M}}{32\pi}; \qquad (5.30)$$

while for M-QAM signals, $b = b_{\text{QAM}}/2$ and

$$A = \frac{M-1}{2M} + \frac{K^2}{\pi}, \qquad (5.31)$$

$$B = \frac{3(M-1)}{8M} + \frac{K^2}{\pi}. \qquad (5.32)$$

Therefore, for large enough x and y, we have the following asymptotically tight approximations:

$$I_1(x) \approx \frac{1}{x^2} \cdot \frac{A^2}{b^2 \delta_{s,d}^2 \delta_{s,r}^2}, \tag{5.33}$$

$$I_2(x, y) \approx \frac{1}{xy} \cdot \frac{B}{b^2 \delta_{s,d}^2 \delta_{r,d}^2}, \tag{5.34}$$

in which approximation errors become insignificant compared to the orders $1/x^2$ and $1/(xy)$ when x and y go to infinity. Replacing x and y in (5.33) and (5.34) with $P_1/\mathcal{N}_0$ and $P_2/\mathcal{N}_0$ respectively and then substituting the results into (5.24), we arrive at the following result.

THEOREM 5.2.2 *If all of the channel links $h_{s,d}$, $h_{s,r}$, and $h_{r,d}$ are available, i.e., $\delta_{s,d}^2 \neq 0$, $\delta_{s,r}^2 \neq 0$, and $\delta_{r,d}^2 \neq 0$, then when $P_1/\mathcal{N}_0$ and $P_2/\mathcal{N}_0$ go to infinity, the SER of the systems with M-PSK or M-QAM modulation can be tightly approximated as*

$$P_s \approx \frac{\mathcal{N}_0^2}{b^2} \cdot \frac{1}{P_1 \delta_{s,d}^2} \left(\frac{A^2}{P_1 \delta_{s,r}^2} + \frac{B}{P_2 \delta_{r,d}^2} \right), \tag{5.35}$$

where b, A, and B are specified in (5.29)–(5.32) for M-PSK and M-QAM signals, respectively.

In Figure 5.3, we compare the asymptotically tight approximation (5.35) and the SER upper bound (5.20) with the exact SER formulations (5.15) and (5.18) in the case of

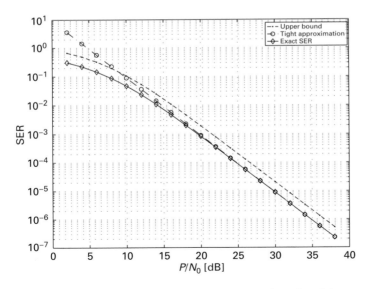

Fig. 5.3 Comparison of the exact SER formulation, the upper bound, and the asymptotically tight approximation for DF cooperation systems with QPSK or 4-QAM signals. It is assumed that $\delta_{s,d}^2 = \delta_{s,r}^2 = \delta_{r,d}^2 = 1$, $\mathcal{N}_0 = 1$, and $P_1 = P_2 = P/2$.

QPSK (or 4-QAM) modulation. In this case, the parameters b, A, and B in the upper bound (5.20) and the approximation (5.35) are specified as $b = 1$, $A = 3/8 + 1/4\pi$ and $B = 9/32 + 1/4\pi$. We can see that the upper bound (5.20) (dashed line with "·") is asymptotically parallel with the exact SER curve (solid line with "◇"), which means that they have the same diversity order. The approximation (5.35) (dashed line with "○") is loose at low SNR, but it is tight at reasonable high SNR. It merges with the exact SER curve at an SER of 10^{-3}. Both the SER upper bound and the approximation show the asymptotic performance of DF cooperation systems. Specifically, from the asymptotically tight approximation (5.35), we observe that the link between source and destination contributes diversity order one in the system performance. The term $A^2/P_1\delta_{s,r}^2 + B/P_2\delta_{r,d}^2$ also contributes diversity order one in the performance, but it depends on the balance of the two channel links from the source to the relay and from the relay to the destination. Therefore, the DF cooperation systems show an overall performance of diversity order two.

5.2.3   Optimum power allocation

Note that the SER approximation (5.35) is asymptotically tight at high SNR. In this subsection, we determine an asymptotic optimum power allocation for the DF cooperation protocol based on the asymptotically tight SER approximation. Specifically, we determine an optimum transmitted power P_1 that should be used at the source and P_2 at the relay for a fixed total transmission power $P_1 + P_2 = P$. We have the following result.

THEOREM 5.2.3   *In DF cooperation systems with M-PSK or M-QAM modulation, if all of the channel links $h_{s,d}$, $h_{s,r}$, and $h_{r,d}$ are available, i.e., $\delta_{s,d}^2 \neq 0$, $\delta_{s,r}^2 \neq 0$, and $\delta_{r,d}^2 \neq 0$, then, for sufficiently high SNR, the optimum power allocation is*

$$P_1 = \frac{\delta_{s,r} + \sqrt{\delta_{s,r}^2 + 8(A^2/B)\delta_{r,d}^2}}{3\delta_{s,r} + \sqrt{\delta_{s,r}^2 + 8(A^2/B)\delta_{r,d}^2}} P, \tag{5.36}$$

$$P_2 = \frac{2\delta_{s,r}}{3\delta_{s,r} + \sqrt{\delta_{s,r}^2 + (8A^2/B)\delta_{r,d}^2}} P, \tag{5.37}$$

where A and B are specified in (5.29)–(5.32) for M-PSK and M-QAM signals, respectively.

Proof   According to the asymptotically tight SER approximation (5.35), it is sufficient to minimize the following term $G(P_1, P_2)$ with the power constraint $P_1 + P_2 = P$ in order to optimize the asymptotic SER performance,

$$G(P_1, P_2) = \frac{1}{P_1\delta_{s,d}^2} \left(\frac{A^2}{P_1\delta_{s,r}^2} + \frac{B}{P_2\delta_{r,d}^2} \right).$$

By taking derivative in terms of P_1, we have

$$\frac{\partial G(P_1, P_2)}{\partial P_1} = \frac{1}{P_1 \delta_{s,d}^2} \left(-\frac{A^2}{P_1^2 \delta_{s,r}^2} + \frac{B}{P_2^2 \delta_{r,d}^2} \right) - \frac{1}{P_1^2 \delta_{s,d}^2} \left(\frac{A^2}{P_1 \delta_{s,r}^2} + \frac{B}{P_2 \delta_{r,d}^2} \right).$$

By setting the above derivation as 0, we come up with an equation as follows:

$$B \delta_{s,r}^2 (P_1^2 - P_1 P_2) - 2 A^2 \delta_{r,d}^2 P_2^2 = 0.$$

With the power constraint $P_1 + P_2 = P$, we can solve the above equation, which results in the optimum power allocation in (5.36) and (5.37). ■

The result in Theorem 5.2.3 is somewhat surprising since the asymptotic optimum power allocation does not depend on the channel link between source and destination, it depends only on the channel link between source and relay and that between relay and destination. Moreover, we can see that the optimum ratio of the transmitted power P_1 at the source over the total power P is less than 1 and larger than $1/2$, while the optimum ratio of the power P_2 used at the relay over the total power P is larger than 0 and less than $1/2$, i.e.,

$$\frac{1}{2} < \frac{P_1}{P} < 1 \quad \text{and} \quad 0 < \frac{P_2}{P} < \frac{1}{2}. \tag{5.38}$$

It means that we should always put more power at the source and less power at the relay. If the link quality between source and relay is much less than that between relay and destination, i.e., $\delta_{s,r}^2 << \delta_{r,d}^2$, then from (5.36) and (5.37), P_1 goes to P and P_2 goes to 0. It implies that we should use almost all of the power P at the source, and use few power at the relay. On the other hand, if the link quality between source and relay is much larger than that between relay and destination, i.e., $\delta_{s,r}^2 >> \delta_{r,d}^2$, then both P_1 and P_2 go to $P/2$. It means that we should put almost equal power at the source and the relay in this case.

The result in Theorem 5.2.3 may be interpreted as follows. With the assumption that all of the channel links $h_{s,d}$, $h_{s,r}$, and $h_{r,d}$ are available in the system, the cooperation strategy is expected to achieve a performance diversity of order two. The system is guaranteed to have a performance diversity of order one due to the channel link between source and destination. However, in order to achieve a diversity of order two, the channel link between source and relay and the channel link between relay and destination should be appropriately balanced. If the link quality between source and relay is bad, then it is difficult for the relay to correctly decode the transmitted symbol. Thus, the forwarding role of the relay is less important and it makes sense to put more power at the source.

On the other hand, if the link quality between source and relay is very good, the relay can always correctly decode the transmitted symbol, so the decoded symbol at the relay is almost the same as that at the source. We may consider the relay to be a copy of the source and put almost equal power on both. We want to emphasize that this interpretation is good only for the sufficiently high SNR scenario and under the assumption that all of the channel links $h_{s,d}$, $h_{s,r}$, and $h_{r,d}$ are available.

Example 5.1   In the case where the link quality between source and relay is the same as that between relay and destination, i.e., $\delta_{s,r}^2 = \delta_{r,d}^2$, the asymptotic optimum power allocation is given by

$$P_1 = \frac{1 + \sqrt{1 + 8A^2/B}}{3 + \sqrt{1 + 8A^2/B}} \, P, \qquad (5.39)$$

$$P_2 = \frac{2}{3 + \sqrt{1 + 8A^2/B}} \, P, \qquad (5.40)$$

where A and B depend on specific modulation signals. For example, if BPSK modulation is used, then $P_1 = 0.5931P$ and $P_2 = 0.4069P$; while if QPSK modulation is used, then $P_1 = 0.6270P$ and $P_2 = 0.3730P$. For 16-QAM, $P_1 = 0.6495P$ and $P_2 = 0.3505P$. We can see that the larger the constellation size, the more power should be put at the source.   ▲

Example 5.2   It is worth pointing out that, even though the asymptotic optimum power allocations in (5.36) and (5.37) are determined for high SNR, they also provide a good solution to a realistic moderate SNR scenario as in Figures 5.4–5.6, we plotted exact SER as a function of the ratio P_1/P for a DF cooperation system with QPSK modulation

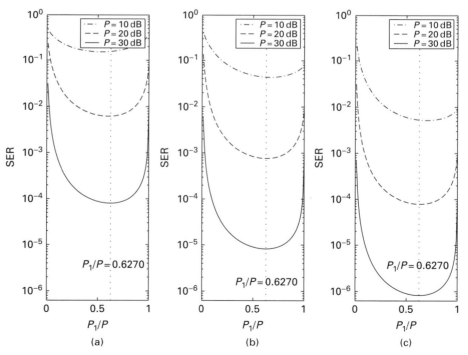

Fig. 5.4   SER of DF cooperation systems with $\delta_{s,r}^2 = 1$ and $\delta_{r,d}^2 = 1$: (a) $\delta_{s,d}^2 = 0.1$; (b) $\delta_{s,d}^2 = 1$; and (c) $\delta_{s,d}^2 = 10$. The asymptotic optimum power allocation is $P_1/P = 0.6270$ and $P_2/P = 0.3730$.

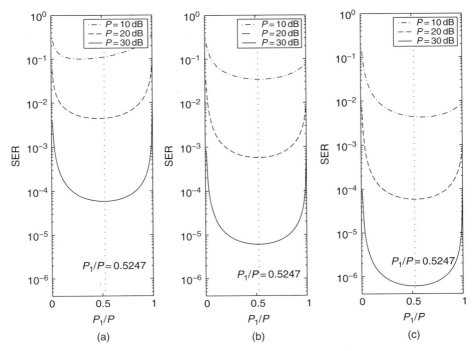

Fig. 5.5 SER of DF cooperation systems with $\delta_{s,r}^2 = 10$ and $\delta_{r,d}^2 = 1$: (a) $\delta_{s,d}^2 = 0.1$; (b) $\delta_{s,d}^2 = 1$; and (c) $\delta_{s,d}^2 = 10$. The asymptotic optimum power allocation is $P_1/P = 0.5247$ and $P_2/P = 0.4753$.

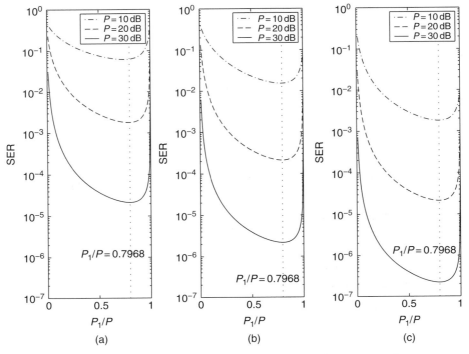

Fig. 5.6 SER of DF cooperation systems with $\delta_{s,r}^2 = 1$ and $\delta_{r,d}^2 = 10$: (a) $\delta_{s,d}^2 = 0.1$; (b) $\delta_{s,d}^2 = 1$; and (c) $\delta_{s,d}^2 = 10$. The asymptotic optimum power allocation is $P_1/P = 0.7968$ and $P_2/P = 0.2032$.

and different fading scenarios. In Figure 5.4, we considered the DF cooperation system with $\delta_{s,r}^2 = \delta_{r,d}^2 = 1$ and three different qualities of the channel link between source and destination: (a) $\delta_{s,d}^2 = 0.1$; (b) $\delta_{s,d}^2 = 1$; and (c) $\delta_{s,d}^2 = 10$. The asymptotic optimum power allocation in this case is $P_1/P = 0.6270$ and $P_2/P = 0.3730$. From the figures, we can see that the ratio $P_1/P = 0.6270$ almost provides the best performance for different total transmit power $P = 10, 20, 30$ dB and different channel variance $\delta_{s,d}^2$. We have the same observation in Figure 5.5 for a system with $\delta_{s,r}^2 = 10$ and $\delta_{r,d}^2 = 1$, and in Figure 5.6 for a system with $\delta_{s,r}^2 = 1$ and $\delta_{r,d}^2 = 10$. Note that for a system with fixed variances $\delta_{s,r}^2$ and $\delta_{r,d}^2$, the system performance improves by increasing the channel quality between source and destination, but the asymptotic optimum power allocation keeps the same. Moreover, the larger the channel quality ratio $\delta_{r,d}^2/\delta_{s,r}^2$, the larger the optimum power ratio P_1/P, i.e., the more power should be used at the source.   ▲

5.2.4   Some special scenarios

In the previous subsection, optimum power allocation was determined for DF cooperation systems where all the channel links $h_{s,d}$, $h_{s,r}$, and $h_{r,d}$ are available. In the following, we consider some special cases that some of the channel links are not available.

- If the channel link between relay and destination is not available, i.e., $\delta_{r,d}^2 = 0$, according to (5.15), the SER of the DF system with M-PSK modulation can be given by

$$P_{\text{PSK}} = F_1 \left(1 + \frac{b_{\text{PSK}} P_1 \delta_{s,d}^2}{N_0 \sin^2 \theta} \right) \leq \frac{A N_0}{b_{\text{PSK}} P_1 \delta_{s,d}^2}, \qquad (5.41)$$

where $F_1(\cdot)$ is defined in (5.16), and A is specified in (5.29). Similarly, from (5.18), the SER of the system with M-QAM modulation is

$$P_{\text{QAM}} = F_2 \left(1 + \frac{b_{\text{QAM}} P_1 \delta_{s,d}^2}{2 N_0 \sin^2 \theta} \right) \leq \frac{2 A N_0}{b_{\text{QAM}} P_1 \delta_{s,d}^2}, \qquad (5.42)$$

where $F_2(\cdot)$ is defined in (5.19), and A is specified in (5.31). From (5.41) and (5.42), we can see that for both M-PSK and M-QAM signals, the optimum power allocation is $P_1 = P$ and $P_2 = 0$. It means that we should use the direct transmission from source to destination in this case.
- If the channel link between source and relay is not available, i.e., $\delta_{s,r}^2 = 0$, from (5.15) and (5.18), the SER of the DF system with M-PSK or M-QAM modulation can be upper bounded as

$$P_s \leq \frac{2 A N_0}{b P_1 \delta_{s,d}^2},$$

where, for M-PSK modulation, $b = b_{\text{PSK}}$ and A is specified in (5.29), while for M-QAM modulation, $b = b_{\text{QAM}}/2$ and A is specified in (5.31). Therefore, the optimum power allocation in this case is $P_1 = P$ and $P_2 = 0$.

- If the channel link between source and destination is not available, i.e., $\delta_{\text{s,d}}^2 = 0$, according to (5.15) and (5.18), the SER of the DF system with M-PSK or M-QAM modulation can be given by

$$P_s = F_i \left(1 + \frac{b P_1 \delta_{\text{s,r}}^2}{\mathcal{N}_0 \sin^2 \theta} \right)$$
$$+ F_i \left(1 + \frac{b P_2 \delta_{\text{r,d}}^2}{\mathcal{N}_0 \sin^2 \theta} \right) \left[1 - F_i \left(1 + \frac{b P_1 \delta_{\text{s,r}}^2}{\mathcal{N}_0 \sin^2 \theta} \right) \right], \qquad (5.43)$$

in which $i = 1$ and $b = b_{\text{PSK}}$ for M-PSK modulation, and $i = 2$ and $b = b_{\text{QAM}}/2$ for M-QAM modulation. If $\delta_{\text{s,r}}^2 \neq 0$ and $\delta_{\text{r,d}}^2 \neq 0$, then by the same procedure as we obtained the SER approximation in (5.35), the SER in (5.43) can be asymptotically approximated as

$$P_s \approx \frac{A \mathcal{N}_0^2}{b^2} \left(\frac{1}{P_1 \delta_{\text{s,r}}^2} + \frac{1}{P_2 \delta_{\text{r,d}}^2} \right), \qquad (5.44)$$

where, for M-PSK modulation, $b = b_{\text{PSK}}$ and A is specified in (5.29), while for M-PSK modulation, $b = b_{\text{QAM}}/2$ and A is specified in (5.31). From (5.44), we can see that with the total power $P_1 + P_2 = P$, the optimum power allocation in this case is

$$P_1 = \frac{\delta_{\text{r,d}}}{\delta_{\text{s,r}} + \delta_{\text{r,d}}} P \qquad (5.45)$$

$$P_2 = \frac{\delta_{\text{s,r}}}{\delta_{\text{s,r}} + \delta_{\text{r,d}}} P \qquad (5.46)$$

for both M-PSK and M-QAM modulations.

Note that when the channel link between source and destination is not available (i.e., $\delta_{\text{s,d}}^2 = 0$), the system reduces to a two-hop communication scenario.

5.2.5 Simulation examples

In this subsection, we consider some examples to illustrate the theoretical analysis for DF cooperation systems with different modulation signals and different power allocation schemes. Assume that the variance of the noise is 1 (i.e., $\mathcal{N}_0 = 1$), and the variance of the channel link between source and destination is normalized as 1 (i.e., $\delta_{\text{s,d}}^2 = 1$). We consider two kinds of channel conditions: (a) $\delta_{\text{s,r}}^2 = 1$ and $\delta_{\text{r,d}}^2 = 1$; and (b) $\delta_{\text{s,r}}^2 = 1$ and $\delta_{\text{r,d}}^2 = 10$. We compare the SER simulation curves with the asymptotically tight SER approximation in (5.35). We also compare the performance of DF cooperation systems using the optimum power allocation scheme in Theorem 5.2.3 with that of systems using the equal power scheme, in which the total transmitted power is equally allocated at the source and at the relay ($P_1/P = P_2/P = 1/2$).

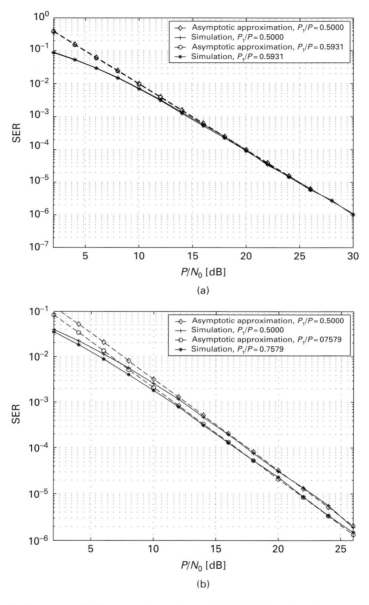

Fig. 5.7   Performance of DF cooperation systems with BPSK signals: optimum power allocation versus equal power scheme. (a) $\delta_{s,r}^2 = 1$ and $\delta_{r,d}^2 = 1$, (b) $\delta_{s,r}^2 = 1$ and $\delta_{r,d}^2 = 10$.

Figure 5.7 depicts the simulation results for DF cooperation systems with BPSK modulation. We can see that the SER approximations from (5.35) are tight at high SNR in all scenarios. From the figure we observe that, for $\delta_{s,r}^2 = 1$ and $\delta_{r,d}^2 = 1$, the performance of the optimum power allocation is almost the same as that of the equal power scheme, as shown in Figure 5.7(a). For $\delta_{s,r}^2 = 1$ and $\delta_{r,d}^2 = 10$ in Figure 5.7(b), the optimum power allocation scheme outperforms the equal power scheme with a performance improvement of about 1 dB. According to Theorem 5.2.3, the optimum power ratios are $P_1/P = 0.7579$ and $P_2/P = 0.2421$ in this case.

Figure 5.8 shows the simulation results for DF cooperation systems with QPSK modulation. For $\delta_{s,r}^2 = 1$ and $\delta_{r,d}^2 = 1$ in Figure 5.8(a), the optimum power ratios are

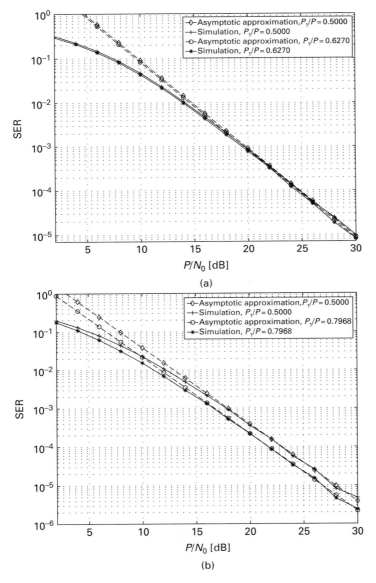

Fig. 5.8 Performance of DF cooperation systems with QPSK signals: optimum power allocation versus equal power scheme. (a) $\delta_{s,r}^2 = 1$ and $\delta_{r,d}^2 = 1$, (b) $\delta_{s,r}^2 = 1$ and $\delta_{r,d}^2 = 10$.

$P_1/P = 0.6270$ and $P_2/P = 0.3730$ by Theorem 5.2.3. From the figure we observe that the performance of the optimum power allocation is a little bit better than that of the equal power case, and the two SER approximations are consistent with the simulation curves at high SNR. For $\delta_{s,r}^2 = 1$ and $\delta_{r,d}^2 = 10$, the optimum power ratios are $P_1/P = 0.7968$ and $P_2/P = 0.2032$ according to Theorem 5.2.3. From Figure 5.8(b), we can see that the optimum power allocation scheme outperforms the equal power scheme with a performance improvement of about 1 dB. Note that if the ratio of the link quality $\delta_{r,d}^2/\delta_{s,r}^2$ becomes larger, we will observe a greater performance improvement of the optimum power allocation over the equal power case. In all of the above simulations, we can see that the SER approximation in (5.35) is asymptotically tight at high SNR.

5.3   SER analysis for AF protocol

In this section, we consider the SER performance of AF cooperative communication systems. First, we derive a simple closed-form MGF expression for the harmonic mean of two independent exponential random variables. Second, based on the simple MGF expression, closed-form SER formulations are given for AF cooperation systems with M-PSK and M-QAM modulations. Third, we provide an SER approximation, which is is tight at high SNR, to show the asymptotic performance of the systems. Finally, based on the tight approximation, we are able to determine an optimum power allocation for AF cooperation systems.

5.3.1   SER analysis by MGF approach

In the AF cooperation systems, the relay amplifies not only the received signal, but also the noise as shown in (5.4)–(5.6). The equivalent noise $\eta_{r,d}'$ at the destination in phase 2 is a zero-mean complex Gaussian random variable with variance

$$\left(\frac{P_2|h_{r,d}|^2}{P_1|h_{s,r}|^2 + \mathcal{N}_0} + 1 \right) \mathcal{N}_0.$$

Therefore, with knowledge of the channel coefficients $h_{s,d}$, $h_{s,r}$ and $h_{r,d}$, the output of the MRC detector at the destination can be written as

$$y = a_1 y_{s,d} + a_2 y_{r,d}, \tag{5.47}$$

where a_1 and a_2 are specified as

$$a_1 = \frac{\sqrt{P_1}h_{s,d}^*}{\mathcal{N}_0} \quad \text{and} \quad a_2 = \frac{\sqrt{\frac{P_1 P_2}{P_1|h_{s,r}|^2+\mathcal{N}_0}} \, h_{s,r}^* h_{r,d}^*}{\left(\frac{P_2|h_{r,d}|^2}{P_1|h_{s,r}|^2+\mathcal{N}_0} + 1 \right) \mathcal{N}_0}. \tag{5.48}$$

Note that to determine the factor a_2 in (5.48), we considered the equivalent received signal model in (5.5). By assuming that the transmitted symbol x in (5.1) has average energy 1, we know that the instantaneous SNR of the MRC output is

$$\gamma = \gamma_1 + \gamma_2, \tag{5.49}$$

where $\gamma_1 = P_1 |h_{s,d}|^2 / \mathcal{N}_0$, and

$$\gamma_2 = \frac{\frac{P_1 P_2}{P_1 |h_{s,r}|^2 + \mathcal{N}_0} |h_{s,r}|^2 |h_{r,d}|^2}{\left(\frac{P_2 |h_{r,d}|^2}{P_1 |h_{s,r}|^2 + \mathcal{N}_0} + 1 \right) \mathcal{N}_0}$$

$$= \frac{1}{\mathcal{N}_0} \frac{P_1 P_2 |h_{s,r}|^2 |h_{r,d}|^2}{P_1 |h_{s,r}|^2 + P_2 |h_{r,d}|^2 + \mathcal{N}_0}. \tag{5.50}$$

The instantaneous SNR γ_2 in (5.50) can be tightly upper bounded as

$$\gamma_2 \leq \tilde{\gamma}_2 \stackrel{\triangle}{=} \frac{1}{\mathcal{N}_0} \frac{P_1 P_2 |h_{s,r}|^2 |h_{r,d}|^2}{P_1 |h_{s,r}|^2 + P_2 |h_{r,d}|^2}, \tag{5.51}$$

which is the harmonic mean of two exponential random variables $P_1 |h_{s,r}|^2 / \mathcal{N}_0$ and $P_2 |h_{r,d}|^2 / \mathcal{N}_0$. If we approximate the SNR as $\gamma \approx \gamma_1 + \tilde{\gamma}_2$, then, according to (5.12) and (5.13), the conditional SER of AF cooperation systems with M-PSK and M-QAM modulations can be given as follows:

$$P_{PSK}^{h_{s,d}, h_{s,r}, h_{r,d}} \approx \frac{1}{\pi} \int_0^{(M-1)\pi/M} \exp\left(-\frac{b_{PSK}(\gamma_1 + \tilde{\gamma}_2)}{\sin^2 \theta} \right) d\theta, \tag{5.52}$$

$$P_{QAM}^{h_{s,d}, h_{s,r}, h_{r,d}} \approx 4K Q\left(\sqrt{b_{QAM}(\gamma_1 + \tilde{\gamma}_2)} \right) - 4K^2 Q^2 \left(\sqrt{b_{QAM}(\gamma_1 + \tilde{\gamma}_2)} \right), \tag{5.53}$$

where $b_{PSK} = \sin^2(\pi/M)$, $b_{QAM} = 3/(M-1)$, and $K = 1 - 1/\sqrt{M}$.

Let us denote the moment-generating function (MGF) of a random variable Z as

$$\mathcal{M}_Z(s) = \int_{-\infty}^{\infty} \exp(-sz) p_Z(z) dz, \tag{5.54}$$

for any real number s. By averaging over the Rayleigh fading channels $h_{s,d}$, $h_{s,r}$, and $h_{r,d}$ in (5.52) and (5.53), we obtain the SER of AF cooperation systems in terms of MGF $\mathcal{M}_{\gamma_1}(s)$ and $\mathcal{M}_{\tilde{\gamma}_2}(s)$ as follows:

$$P_{PSK} \approx \frac{1}{\pi} \int_0^{(M-1)\pi/M} \mathcal{M}_{\gamma_1}\left(\frac{b_{PSK}}{\sin^2 \theta} \right) \mathcal{M}_{\tilde{\gamma}_2}\left(\frac{b_{PSK}}{\sin^2 \theta} \right) d\theta, \tag{5.55}$$

$$P_{QAM} \approx \left[\frac{4K}{\pi} \int_0^{\pi/2} - \frac{4K^2}{\pi} \int_0^{\pi/4} \right] \mathcal{M}_{\gamma_1}\left(\frac{b_{QAM}}{2 \sin^2 \theta} \right) \mathcal{M}_{\tilde{\gamma}_2}\left(\frac{b_{QAM}}{2 \sin^2 \theta} \right) d\theta, \tag{5.56}$$

in which, for simplicity, we use the following notation

$$\left[\frac{4K}{\pi} \int_0^{\pi/2} - \frac{4K^2}{\pi} \int_0^{\pi/4} \right] x(\theta) d\theta \stackrel{\triangle}{=} \frac{4K}{\pi} \int_0^{\pi/2} x(\theta) d\theta - \frac{4K^2}{\pi} \int_0^{\pi/4} x(\theta) d\theta.$$

From (5.55) and (5.56), we can see that the remaining problem is to obtain the MGF $\mathcal{M}_{\gamma_1}(s)$ and $\mathcal{M}_{\tilde{\gamma}_2}(s)$. Since $\gamma_1 = P_1|h_{s,d}|^2/\mathcal{N}_0$ has an exponential distribution with parameter $\mathcal{N}_0/(P_1\delta_{s,d}^2)$, the MGF of γ_1 can be simply given by

$$\mathcal{M}_{\gamma_1}(s) = \frac{1}{1 + \frac{sP_1\delta_{s,d}^2}{\mathcal{N}_0}}. \tag{5.57}$$

However, it is not easy to get the MGF of $\tilde{\gamma}_2$, which is the harmonic mean of two exponential random variables $P_1|h_{s,r}|^2/\mathcal{N}_0$ and $P_2|h_{r,d}|^2/\mathcal{N}_0$. The MGF can be obtained by applying a Laplace transform and a solution was presented in [58] in terms of the hypergeometric function as follows:

$$\mathcal{M}_{\tilde{\gamma}_2}(s) = \frac{16\beta_1\beta_2}{3(\beta_1 + \beta_2 + 2\sqrt{\beta_1\beta_2} + s)^2} \left[\frac{4(\beta_1 + \beta_2)}{\beta_1 + \beta_2 + 2\sqrt{\beta_1\beta_2} + s} \right.$$
$$\times \; _2F_1\left(3, \frac{3}{2}; \frac{5}{2}; \frac{\beta_1 + \beta_2 - 2\sqrt{\beta_1\beta_2} + s}{\beta_1 + \beta_2 + 2\sqrt{\beta_1\beta_2} + s} \right)$$
$$\left. + \; _2F_1\left(2, \frac{1}{2}; \frac{5}{2}; \frac{\beta_1 + \beta_2 - 2\sqrt{\beta_1\beta_2} + s}{\beta_1 + \beta_2 + 2\sqrt{\beta_1\beta_2} + s} \right) \right], \tag{5.58}$$

in which $\beta_1 = \mathcal{N}_0/(P_1\delta_{s,r}^2)$, $\beta_2 = \mathcal{N}_0/(P_2\delta_{r,d}^2)$, and $_2F_1(\cdot, \cdot; \cdot; \cdot)$ is the hypergeometric function.[3] Because the hypergeometric function $_2F_1(\cdot, \cdot; \cdot; \cdot)$ is defined as an integral it is hard to use in an SER analysis aimed at revealing the asymptotic performance and optimizing the power allocation. Using an alternative approach, we introduce a simple closed-form solution for the MGF of $\tilde{\gamma}_2$ as shown in the next subsection.

5.3.2   Simple MGF expression for the harmonic mean

In this subsection, first we obtain a general result on the probability density function (pdf) for the harmonic mean of two independent random variables. Then, we are able to determine a simple closed-form MGF expression for the harmonic mean of two independent exponential random variables.

THEOREM 5.3.1   *Suppose that X_1 and X_2 are two independent random variables with pdf $p_{X_1}(x)$ and $p_{X_2}(x)$ defined for all $x \geq 0$, and $p_{X_1}(x) = 0$ and $p_{X_2}(x) = 0$ for $x < 0$. Then the pdf of $Z = \frac{X_1 X_2}{X_1 + X_2}$, the harmonic mean of X_1 and X_2, is*

$$p_Z(z) = z \int_0^1 \frac{1}{t^2(1-t)^2} \, p_{X_1}(\frac{z}{1-t}) \, p_{X_2}(\frac{z}{t}) dt \cdot U(z), \tag{5.59}$$

in which $U(z) = 1$ for $z \geq 0$ and $U(z) = 0$ for $z < 0$.

---

[3] A hypergeometric function with variables α, β, γ, and z is defined as [50]

$$_2F_1(\alpha, \beta; \gamma; z) = \frac{\Gamma(\gamma)}{\Gamma(\beta)\Gamma(\gamma - \beta)} \int_0^1 t^{\beta-1}(1-t)^{\gamma-\beta-1}(1-tz)^{-\alpha} dt,$$

where $\Gamma(\cdot)$ is the Gamma function.

The proof of Theorem 5.3.1 is omitted (as an exercise). Note that there is no specification for the distributions of the two independent random variables in Theorem 5.3.1. Suppose that X_1 and X_2 are two independent exponential random variables with parameters β_1 and β_2, respectively, i.e.,

$$p_{X_1}(x) = \beta_1\,e^{-\beta_1 x}\cdot U(x),$$
$$p_{X_2}(x) = \beta_2\,e^{-\beta_2 x}\cdot U(x).$$

Then, according to Theorem 5.3.1, the pdf of the harmonic mean $Z = \frac{X_1 X_2}{X_1+X_2}$ can be simply given as

$$p_Z(z) = z\int_0^1 \frac{\beta_1\beta_2}{t^2(1-t)^2}\,e^{-\left(\frac{\beta_1}{1-t}+\frac{\beta_2}{t}\right)z}\,dt\cdot U(z). \tag{5.60}$$

The pdf expression in (5.60) is critical in the following to obtain a simple closed-form MGF result for the harmonic mean Z.

Let us start calculating the MGF of the harmonic mean of two independent exponential random variables by substituting the pdf of Z (5.60) into the definition (5.54) as follows:

$$\mathcal{M}_Z(s) = \int_0^\infty e^{-sz} z \int_0^1 \frac{\beta_1\beta_2}{t^2(1-t)^2}\,e^{-\left(\frac{\beta_1}{1-t}+\frac{\beta_2}{t}\right)z}\,dt\,dz$$
$$= \int_0^1 \frac{\beta_1\beta_2}{t^2(1-t)^2}\left(\int_0^\infty z\,e^{-\left(\frac{\beta_1}{1-t}+\frac{\beta_2}{t}+s\right)z}\,dz\right)dt, \tag{5.61}$$

in which we switch the integration order. Since

$$\int_0^\infty z\,e^{-\left(\frac{\beta_1}{1-t}+\frac{\beta_2}{t}+s\right)z}\,dz = \left(\frac{\beta_1}{1-t}+\frac{\beta_2}{t}+s\right)^{-2},$$

the MGF in (5.61) can be determined as

$$\mathcal{M}_Z(s) = \int_0^1 \frac{\beta_1\beta_2}{\left[\beta_2+(\beta_1-\beta_2+s)t-st^2\right]^2}\,dt, \tag{5.62}$$

which is an integration of a quadratic trinomial and has a closed-form solution [50]. For notation simplicity, denote $\alpha = (\beta_1-\beta_2+s)/2$, then, for any $s>0$, we have

$$\int_0^1 \frac{1}{(\beta_2+2\alpha t - st^2)^2}\,dt$$
$$= \left.\frac{st-\alpha}{2(\beta_2 s+\alpha^2)(\beta_2+2\alpha t-st^2)}\right|_0^1$$
$$+ \left.\frac{s}{4(\beta_2 s+\alpha^2)^{\frac{3}{2}}}\ln\left|\frac{-st+\alpha-\sqrt{\beta_2 s+\alpha^2}}{-st+\alpha+\sqrt{\beta_2 s+\alpha^2}}\right|\right|_0^1$$
$$= \frac{\beta_2 s+\alpha(\beta_1-\beta_2)}{2\beta_1\beta_2(\beta_2 s+\alpha^2)} + \frac{s}{4(\beta_2 s+\alpha^2)^{\frac{3}{2}}}\ln\frac{\left(\beta_2+\alpha+\sqrt{\beta_2 s+\alpha^2}\right)^2}{\beta_1\beta_2}. \tag{5.63}$$

By substituting $\alpha = (\beta_1 - \beta_2 + s)/2$ into (5.63) and denoting $\Delta = 2\sqrt{\beta_2 s + \alpha^2}$, we obtain a simple closed-form MGF for the harmonic mean Z as follows:

$$\mathcal{M}_Z(s) = \frac{(\beta_1 - \beta_2)^2 + (\beta_1 + \beta_2)s}{\Delta^2} + \frac{2\beta_1\beta_2 s}{\Delta^3} \ln \frac{(\beta_1 + \beta_2 + s + \Delta)^2}{4\beta_1\beta_2}, \quad s > 0,$$

(5.64)

where $\Delta = \sqrt{(\beta_1 - \beta_2)^2 + 2(\beta_1 + \beta_2)s + s^2}$. We can see that if β_1 and β_2 go to zero, then Δ can be approximated as s. In this case, the MGF in (5.64) can be simplified as

$$\mathcal{M}_Z(s) \approx \frac{\beta_1 + \beta_2}{s} + \frac{2\beta_1\beta_2}{s^2} \ln \frac{s^2}{\beta_1\beta_2}.$$

(5.65)

Note that in (5.65), the second term goes to zero faster than the first term. As a result, the MGF in (5.65) can be further simplified as

$$\mathcal{M}_Z(s) \approx \frac{\beta_1 + \beta_2}{s}.$$

(5.66)

We summarize the above discussion in the following theorem.

THEOREM 5.3.2   *Let X_1 and X_2 be two independent exponential random variables with parameters β_1 and β_2 respectively. Then, the MGF of $Z = \frac{X_1 X_2}{X_1 + X_2}$ is*

$$\mathcal{M}_Z(s) = \frac{(\beta_1 - \beta_2)^2 + (\beta_1 + \beta_2)s}{\Delta^2} + \frac{2\beta_1\beta_2 s}{\Delta^3} \ln \frac{(\beta_1 + \beta_2 + s + \Delta)^2}{4\beta_1\beta_2}$$

(5.67)

for any $s > 0$, in which

$$\Delta = \sqrt{(\beta_1 - \beta_2)^2 + 2(\beta_1 + \beta_2)s + s^2}.$$

(5.68)

Furthermore, if β_1 and β_2 go to zero, then the MGF of Z can be approximated as

$$\mathcal{M}_Z(s) \approx \frac{\beta_1 + \beta_2}{s}.$$

(5.69)

We can see that the closed-form solution in (5.67) does not involve any integration. If X_1 and X_2 are i.i.d. exponential random variables with parameter β, then according to the result in Theorem 5.3.2, the MGF of $Z = \frac{X_1 X_2}{X_1 + X_2}$ can be simply given as

$$\mathcal{M}_Z(s) = \frac{2\beta}{4\beta + s} + \frac{4\beta^2 s}{\Delta_0^3} \ln \frac{2\beta + s + \Delta_0}{2\beta},$$

(5.70)

where $s > 0$ and $\Delta_0 = \sqrt{4\beta s + s^2}$. The approximation in (5.69) will provide a very simple solution for the SER calculations in (5.55) and (5.56), as shown in the next subsection.

5.3.3   Asymptotically tight approximation

Now let us apply the result of Theorem 5.3.2 to the harmonic mean of two random variables $X_1 = P_1 |h_{s,r}|^2 / N_0$ and $X_2 = P_2 |h_{r,d}|^2 / N_0$ as we considered in Section 4.1. They are two independent exponential random variables with parameters $\beta_1 = N_0/(P_1 \delta_{s,r}^2)$ and $\beta_2 = N_0/(P_2 \delta_{r,d}^2)$, respectively. With the closed-form MGF expression in Theorem

5.3.2, the SER formulations in (5.55) and (5.56) for AF systems with M-PSK and M-QAM modulations can be determined respectively as

$$P_{\text{PSK}} \approx \frac{1}{\pi} \int_0^{(M-1)\pi/M} \frac{1}{1 + \frac{b_{\text{PSK}}}{\beta_0 \sin^2 \theta}} \left\{ \frac{(\beta_1 - \beta_2)^2 + (\beta_1 + \beta_2)\frac{b_{\text{PSK}}}{\sin^2 \theta}}{\Delta^2} \right.$$
$$\left. + \frac{2\beta_1\beta_2 b_{\text{PSK}}}{\Delta^3 \sin^2 \theta} \ln \frac{(\beta_1 + \beta_2 + \frac{b_{\text{PSK}}}{\sin^2 \theta} + \Delta)^2}{4\beta_1\beta_2} \right\} d\theta, \tag{5.71}$$

$$P_{\text{QAM}} \approx \left[\frac{4K}{\pi} \int_0^{\pi/2} - \frac{4K^2}{\pi} \int_0^{\pi/4} \right] \frac{1}{1 + \frac{b_{\text{QAM}}}{2\beta_0 \sin^2 \theta}} \left\{ \frac{(\beta_1 - \beta_2)^2 + (\beta_1 + \beta_2)\frac{b_{\text{QAM}}}{2\sin^2 \theta}}{\Delta^2} \right.$$
$$\left. + \frac{\beta_1\beta_2 b_{\text{QAM}}}{\Delta^3 \sin^2 \theta} \ln \frac{(\beta_1 + \beta_2 + \frac{b_{\text{QAM}}}{2\sin^2 \theta} + \Delta)^2}{4\beta_1\beta_2} \right\} d\theta, \tag{5.72}$$

in which $\beta_0 = N_0/(P_1 \delta_{\text{s,d}}^2)$, $\beta_1 = N_0/(P_1 \delta_{\text{s,r}}^2)$, $\beta_2 = N_0/(P_2 \delta_{\text{r,d}}^2)$, and $\Delta^2 = (\beta_1 - \beta_2)^2 + 2(\beta_1 + \beta_2)s + s^2$ with $s = b_{\text{PSK}}/\sin^2 \theta$ for M-PSK modulation and $s = b_{\text{QAM}}/(2\sin^2 \theta)$ for M-QAM modulation.

We observe that it is hard to understand AF system performance based on the SER formulations in (5.71) and (5.72), even though they can be numerically calculated. In the following, we provide an asymptotically tight approximation for the SER formulations when all of the channel links $h_{\text{s,d}}$, $h_{\text{s,r}}$, and $h_{\text{r,d}}$ are available, i.e., $\delta_{\text{s,d}}^2 \neq 0$, $\delta_{\text{s,r}}^2 \neq 0$, and $\delta_{\text{r,d}}^2 \neq 0$. For simplicity of discussion, we denote

$$\int_{\Omega_1} x(\theta) \, d\theta \triangleq \frac{1}{\pi} \int_0^{(M-1)\pi/M} x(\theta) \, d\theta, \tag{5.73}$$

$$\int_{\Omega_2} x(\theta) \, d\theta \triangleq \left[\frac{4K}{\pi} \int_0^{\pi/2} - \frac{4K^2}{\pi} \int_0^{\pi/4} \right] x(\theta) \, d\theta, \tag{5.74}$$

for any function $x(\theta)$. Then, the SER formulations in (5.71) and (5.72) can be expressed as

$$P_s \approx \int_{\Omega_i} \frac{1}{1 + \frac{b P_1 \delta_{\text{s,d}}^2}{N_0 \sin^2 \theta}} \left\{ \frac{\left(\frac{N_0}{P_1 \delta_{\text{s,r}}^2} - \frac{N_0}{P_2 \delta_{\text{r,d}}^2} \right)^2 + \left(\frac{N_0}{P_1 \delta_{\text{s,r}}^2} + \frac{N_0}{P_2 \delta_{\text{r,d}}^2} \right) \frac{b}{\sin^2 \theta}}{\Delta^2} \right.$$
$$\left. + \frac{2b N_0^2 \ln \left[P_1 P_2 \delta_{\text{s,r}}^2 \delta_{\text{r,d}}^2 \left(\frac{N_0}{P_1 \delta_{\text{s,r}}^2} + \frac{N_0}{P_2 \delta_{\text{r,d}}^2} + \frac{b}{\sin^2 \theta} + \Delta \right)^2 / \left(4 N_0^2 \right) \right]}{P_1 P_2 \delta_{\text{s,r}}^2 \delta_{\text{r,d}}^2 \Delta^3 \sin^2 \theta} \right\} d\theta, \tag{5.75}$$

where $i = 1$ and $b = b_{\text{PSK}}$, for M-PSK modulation, and $i = 2$ and $b = b_{\text{QAM}}/2$ for M-QAM modulation, and

$$\Delta^2 = \left(\frac{\mathcal{N}_0}{P_1 \delta_{s,r}^2} - \frac{\mathcal{N}_0}{P_2 \delta_{r,d}^2} \right)^2 + 2 \left(\frac{\mathcal{N}_0}{P_1 \delta_{s,r}^2} + \frac{\mathcal{N}_0}{P_2 \delta_{r,d}^2} \right) \frac{b}{\sin^2 \theta} + \frac{b^2}{\sin^4 \theta}.$$

Let us separate the right-hand side of the above equation (5.75) as two parts as follows:

$$P_s \approx J_1(P_1/\mathcal{N}_0, P_2/\mathcal{N}_0) + J_2(P_1/\mathcal{N}_0, P_2/\mathcal{N}_0), \tag{5.76}$$

in which

$$J_1(x, y) = \int_{\Omega_i} \frac{1}{1 + \frac{x b \delta_{s,d}^2}{\sin^2 \theta}} \left\{ \frac{\frac{1}{x \delta_{s,r}^2} \left(\frac{1}{x \delta_{s,r}^2} - \frac{1}{y \delta_{r,d}^2} \right) + \frac{b}{x \delta_{s,r}^2 \sin^2 \theta}}{\Delta^2} \right.$$
$$\left. + \frac{2b \ln y}{x y \delta_{s,r}^2 \delta_{r,d}^2 \Delta^3 \sin^2 \theta} \right\} d\theta, \tag{5.77}$$

$$J_2(x, y) = \int_{\Omega_i} \frac{1}{1 + \frac{x b \delta_{s,d}^2}{\sin^2 \theta}} \left\{ \frac{-\frac{1}{y \delta_{r,d}^2} \left(\frac{1}{x \delta_{s,r}^2} - \frac{1}{y \delta_{r,d}^2} \right) + \frac{b}{y \delta_{r,d}^2 \sin^2 \theta}}{\Delta^2} \right.$$
$$\left. + \frac{2b \ln \left[x \delta_{s,r}^2 \delta_{r,d}^2 \left(\frac{1}{x \delta_{s,r}^2} + \frac{1}{y \delta_{r,d}^2} + \frac{b}{\sin^2 \theta} + \Delta \right)^2 \Big/ 4 \right]}{x y \delta_{s,r}^2 \delta_{r,d}^2 \Delta^3 \sin^2 \theta} \right\} d\theta, \tag{5.78}$$

in which $i = 1$ and $b = b_{PSK}$ for M-PSK modulation, and $i = 2$ and $b = b_{QAM}/2$ for M-QAM modulation. Then, we have the following results (leave the proof as an exercise)

$$\lim_{x, y \to \infty} x^2 J_1(x, y) = \frac{B}{b^2 \delta_{s,d}^2 \delta_{s,r}^2}, \tag{5.79}$$

$$\lim_{x, y \to \infty} x y J_2(x, y) = \frac{B}{b^2 \delta_{s,d}^2 \delta_{r,d}^2}, \tag{5.80}$$

where for M-PSK signals, $b = b_{PSK}$ and

$$B = \frac{3(M-1)}{8M} + \frac{\sin \frac{2\pi}{M}}{4\pi} - \frac{\sin \frac{4\pi}{M}}{32\pi}; \tag{5.81}$$

while for M-QAM signals, $b = b_{QAM}/2$ and

$$B = \frac{3(M-1)}{8M} + \frac{K^2}{\pi}. \tag{5.82}$$

Therefore, for large enough x and y, we have the following asymptotically tight approximations:

$$J_1(x, y) \approx \frac{1}{x^2} \cdot \frac{B}{b^2 \delta_{s,d}^2 \delta_{s,r}^2}, \tag{5.83}$$

$$J_2(x, y) \approx \frac{1}{xy} \cdot \frac{B}{b^2 \delta_{s,d}^2 \delta_{r,d}^2}, \tag{5.84}$$

in which approximation errors become insignificant compared to the orders $1/x^2$ and $1/xy$ when x and y go to infinity. Replacing x and y in (5.83) and (5.84) with $P_1/\mathcal{N}_0$ and $P_2/\mathcal{N}_0$, respectively, and substituting the results into (5.76), we arrive at the following result.

THEOREM 5.3.3 *If all of the channel links $h_{s,d}$, $h_{s,r}$, and $h_{r,d}$ are available, i.e., $\delta_{s,d}^2 \neq 0$, $\delta_{s,r}^2 \neq 0$, and $\delta_{r,d}^2 \neq 0$, then when $P_1/\mathcal{N}_0$ and $P_2/\mathcal{N}_0$ go to infinity, the SER of AF cooperation systems with M-PSK or M-QAM modulation can be tightly approximated as*

$$P_s \approx \frac{B\mathcal{N}_0^2}{b^2} \cdot \frac{1}{P_1 \delta_{s,d}^2} \left(\frac{1}{P_1 \delta_{s,r}^2} + \frac{1}{P_2 \delta_{r,d}^2} \right), \tag{5.85}$$

where b and B are specified in (5.81) and (5.82) for M-PSK signals and M-QAM signals, respectively.

Let us compare the SER approximations in (5.71), (5.72), and (5.85) with the SER simulation result in Figure 5.9 in the case of an AF cooperation systems with QPSK (or 4-QAM) modulation. It is easy to check that for both QPSK and 4-QAM modulations the parameters B in (5.81) and (5.82) are the same, in which $B = 9/32 + 1/4\pi$. We can see that the theoretical calculation in (5.71) or (5.72) matches the simulation curve, except for a small difference at low SNR, which is due to the approximation of the

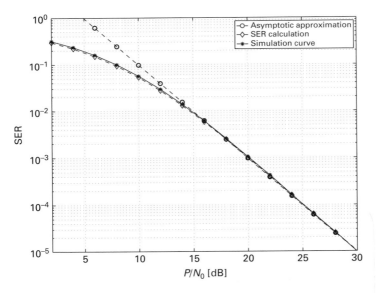

Fig. 5.9   Comparison of the SER approximations and the simulation result for an AF cooperation systems with QPSK or 4-QAM signals. It is assumed that $\delta_{s,d}^2 = \delta_{s,r}^2 = \delta_{r,d}^2 = 1$, $\mathcal{N}_0 = 1$, and $P_1/P = 2/3$ and $P_2/P = 1/3$.

SNR $\tilde{\gamma}_2$ in (5.51). Furthermore, the simple SER approximation in (5.85) is tight at high SNR, which is good enough to show the asymptotic performance of the AF cooperation system. From Theorem 5.3.3, we can see that the AF cooperation system also provides an overall performance of diversity order two, which is similar to that of DF cooperation systems.

5.3.4 Optimum power allocation

In this subsection we determine an asymptotic optimum power allocation for AF cooperation systems based on the tight SER approximation in (5.85) for sufficiently high SNR.

For a fixed total transmitted power $P_1 + P_2 = P$, we need to optimize P_1 and P_2 such that the asymptotically tight SER approximation in (5.85) is minimized. Equivalently, we try to minimize

$$G(P_1, P_2) = \frac{1}{P_1 \delta_{s,d}^2} \left(\frac{1}{P_1 \delta_{s,r}^2} + \frac{1}{P_2 \delta_{r,d}^2} \right).$$

By taking derivatives in terms of P_1, we have

$$\frac{\partial G(P_1, P_2)}{\partial P_1} = \frac{1}{P_1 \delta_{s,d}^2} \left(-\frac{1}{P_1^2 \delta_{s,r}^2} + \frac{1}{P_2^2 \delta_{r,d}^2} \right) - \frac{1}{P_1^2 \delta_{s,d}^2} \left(\frac{1}{P_1 \delta_{s,r}^2} + \frac{1}{P_2 \delta_{r,d}^2} \right).$$

By setting the above derivation as 0, we have

$$\delta_{s,r}^2 (P_1^2 - P_1 P_2) - 2\delta_{r,d}^2 P_2^2 = 0.$$

Together with the power constraint $P_1 + P_2 = P$, we can solve the above equation and arrive at the following result.

THEOREM 5.3.4 *For sufficiently high SNR, the optimum power allocation for AF cooperation systems with either M-PSK or M-QAM modulation is*

$$P_1 = \frac{\delta_{s,r} + \sqrt{\delta_{s,r}^2 + 8\delta_{r,d}^2}}{3\delta_{s,r} + \sqrt{\delta_{s,r}^2 + 8\delta_{r,d}^2}} P, \tag{5.86}$$

$$P_2 = \frac{2\delta_{s,r}}{3\delta_{s,r} + \sqrt{\delta_{s,r}^2 + 8\delta_{r,d}^2}} P. \tag{5.87}$$

From Theorem 5.3.4, we observe that the optimum power allocation for AF cooperation systems is not modulation-dependent, unlike that for DF cooperation systems in which the optimum power allocation depends on the specific M-PSK or M-QAM modulation as stated in Theorem 5.2.3. This is due to the fact that, in AF cooperation systems, the relay amplifies the received signal and forwards it to the destination regardless of what kind of received signal it is. In DF cooperation systems, the relay forwards information to the destination only if the relay correctly decodes the received signal, and the decoding at the relay requires specific modulation information, which results in the modulation-dependent optimum power allocation scheme.

On the other hand, the asymptotic optimum power allocation scheme in Theorem 5.3.4 for AF cooperation systems is similar to that in Theorem 5.2.3 for DF cooperation systems, in the sense that both of them do not depend on the channel link between source and destination, but instead depend only on the channel links between source and relay and between relay and destination. Similarly, we can see from Theorem 5.3.4 that the optimum ratio of the transmitted power P_1 at the source over the total power P is less than 1 and larger than $1/2$, while the optimum ratio of the power P_2 used at the relay over the total power P is larger than 0 and less than $1/2$. In general, the equal power strategy is not optimum. For example, if $\delta_{s,r}^2 = \delta_{r,d}^2$, then the optimum power allocation is $P_1 = P/3$ and $P_2 = P/3$.

5.3.5 Simulation examples

In this subsection, we consider some examples to illustrate the theoretical analysis for AF cooperation systems with different modulation signals and different power allocation schemes. We compare the performance of AF cooperation systems using the optimum power allocation scheme in Theorem 5.3.4 with that of the systems using the equal power scheme. We also compare the asymptotic tight SER approximation in (5.85) with the SER simulation curves. Assume that the variance of the noise is 1 (i.e., $\mathcal{N}_0 = 1$), and the variance of the channel link between source and destination is normalized as 1 (i.e., $\delta_{s,d}^2 = 1$).

Figure 5.10 provides the simulation results for AF cooperation systems with BPSK modulation. In case of $\delta_{s,r}^2 = 1$ and $\delta_{r,d}^2 = 1$ in Figure 5.10(a), we can see that the performance of the optimum power allocation is slightly better than that of the equal power case, in which the optimum power ratios are $P_1/P = 2/3$ and $P_2/P = 1/3$ according to Theorem 5.3.4. In the case of $\delta_{s,r}^2 = 1$ and $\delta_{r,d}^2 = 10$, the optimum power ratios are $P_1/P = 0.8333$ and $P_2/P = 0.1667$ according to Theorem 5.3.4. We observe from Figure 5.10(b) that the optimum power allocation scheme outperforms the equal power scheme with a performance improvement of more than 1.5 dB. Note that all SER approximations from (5.85) are respectively consistent with the simulation curves at reasonable high SNR.

The simulation results of AF cooperation systems with QPSK modulation are shown in Figure 5.11. In the case of $\delta_{s,r}^2 = 1$ and $\delta_{r,d}^2 = 1$ in Figure 5.11(a), the optimum power ratios in this case are $P_1/P = 2/3$ and $P_2/P = 1/3$ which are the same as those for the case of BPSK modulation. From the figure, we can see that the performance of the optimum power allocation is better than that of the equal power case, and the two SER approximations are consistent with the simulation curves at high SNR respectively. In case of $\delta_{s,r}^2 = 1$ and $\delta_{r,d}^2 = 10$, the optimum power ratios are $P_1/P = 0.8333$ and $P_2/P = 0.1667$ according to Theorem 5.3.4. From Figure 5.11(b), we observe that the optimum power allocation scheme outperforms the equal power scheme with a

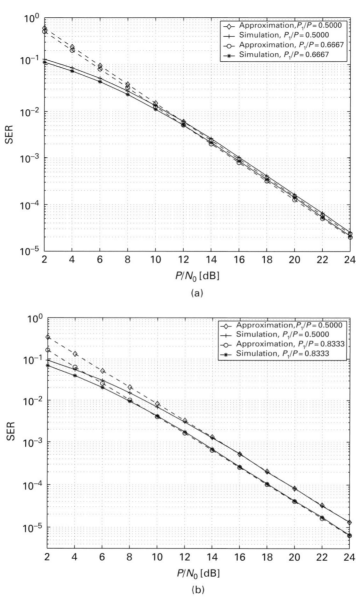

Fig. 5.10   Performance of AF cooperation systems with BPSK signals: optimum power allocation versus equal power scheme. (a) $\delta_{s,r}^2 = 1$ and $\delta_{r,d}^2 = 1$, (b) $\delta_{s,r}^2 = 1$ and $\delta_{r,d}^2 = 10$.

performance improvement of about 2 dB. If the ratio of the channel link quality $\delta_{r,d}^2/\delta_{s,r}^2$ increases, we expect to see a greater improvement in performance of the optimum power allocation over the equal power case. Moreover, from the figures we can see that in all of the above simulations, the SER approximations from (5.85) are tight enough at high SNR.

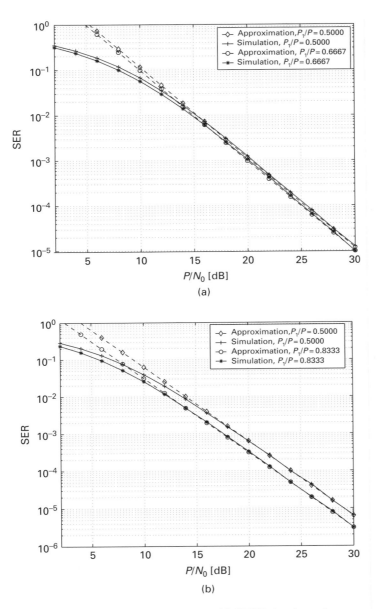

Fig. 5.11 Performance of AF cooperation systems with QPSK signals: optimum power allocation versus equal power scheme. (a) $\delta_{s,r}^2 = 1$ and $\delta_{r,d}^2 = 1$, (b) $\delta_{s,r}^2 = 1$ and $\delta_{r,d}^2 = 10$.

5.4 Comparison of DF and AF cooperation gains

Based on the asymptotically tight SER approximations and the optimum power allocation solutions established in the previous two sections, we consider in this section the overall cooperation gain and diversity order for DF and AF cooperation systems, respectively. Then, we compare the cooperation gain between the DF and AF cooperation protocols.

Let us first focus on the DF cooperation protocol. According to the asymptotically tight SER approximation (5.35) in Theorem 5.2.2, we know that for sufficiently high SNR, the SER performance of DF cooperation systems can be approximated as

$$P_s \approx \frac{\mathcal{N}_0^2}{b^2} \cdot \frac{1}{P_1 \delta_{s,d}^2} \left(\frac{A^2}{P_1 \delta_{s,r}^2} + \frac{B}{P_2 \delta_{r,d}^2} \right), \tag{5.88}$$

where A and B are specified in (5.29)–(5.32) for M-PSK and M-QAM signals, respectively. By substituting the asymptotic optimum power allocation (5.36) and (5.37) into (5.88), we have

$$P_s \approx \Delta_{\text{DF}}^{-2} \left(\frac{P}{\mathcal{N}_0} \right)^{-2}, \tag{5.89}$$

where

$$\Delta_{\text{DF}} = \frac{2\sqrt{2}\, b\delta_{s,d}\delta_{s,r}\delta_{r,d}}{\sqrt{B}} \frac{\left(\delta_{s,r} + \sqrt{\delta_{s,r}^2 + 8(A^2/B)\delta_{r,d}^2} \right)^{1/2}}{\left(3\delta_{s,r} + \sqrt{\delta_{s,r}^2 + 8(A^2/B)\delta_{r,d}^2} \right)^{3/2}}, \tag{5.90}$$

in which $b = b_{\text{PSK}}$ for M-PSK signals and $b = b_{\text{QAM}}/2$ for M-QAM signals. From (5.89), we can see that the DF cooperation systems can guarantee a performance diversity of order two. Note that the term Δ_{DF} in (5.90) depends only on the statistics of the channel links. We call it the *cooperation gain* of the DF cooperation systems, and it indicates the best performance gain that we are able to achieve through the DF cooperation protocol with any kind of power allocation. If the link quality between source and relay is much less than that between relay and destination, i.e., $\delta_{s,r}^2 << \delta_{r,d}^2$, then the cooperation gain is approximated as

$$\Delta_{\text{DF}} = \frac{b\delta_{s,d}\delta_{s,r}}{A}, \tag{5.91}$$

in which $A = (M-1)/2M + (\sin\frac{2\pi}{M})/4\pi \to 1/2$ (for large M) for M-PSK modulation, or $A = (M-1)/2M + K^2/\pi \to 1/2 + 1/\pi$ when M is large for M-QAM modulation. For example, in case of QPSK modulation, $A = 3/8 + 1/4\pi = 0.4546$.

On the other hand, if the link quality between source and relay is much larger than that between relay and destination, i.e., $\delta_{s,r}^2 >> \delta_{r,d}^2$, then the cooperation gain can be approximated as

$$\Delta_{\text{DF}} = \frac{b\delta_{s,d}\delta_{r,d}}{2\sqrt{B}}, \tag{5.92}$$

in which $B = 3(M-1)/8M + \left(\sin\frac{2\pi}{M}\right)/4\pi - \left(\sin\frac{4\pi}{M}\right)/32\pi \to 3/8$ (for large M) for M-PSK modulation, or $B = 3(M-1)/8M + K^2/\pi \to 3/8 + 1/\pi$ when M is large for M-QAM modulation. For example, in case of QPSK modulation, $B = 9/32 + 1/4\pi = 0.3608$.

Similarly, for the AF cooperation protocol, from the the asymptotically tight SER approximation (5.85) in Theorem 5.3.3, we can see that for sufficiently high SNR, the SER performance of AF cooperation systems can be approximated as

$$ P_s \approx \frac{B\mathcal{N}_0^2}{b^2} \cdot \frac{1}{P_1 \delta_{s,d}^2} \left(\frac{1}{P_1 \delta_{s,r}^2} + \frac{1}{P_2 \delta_{r,d}^2} \right), \tag{5.93} $$

where $b = b_{\text{PSK}}$ for M-PSK signals and $b = b_{\text{QAM}}/2$ for M-QAM signals, and B is specified in (5.81) and (5.82) for M-PSK and M-QAM signals respectively. By substituting the asymptotic optimum power allocation (5.86) and (5.87) into (5.93), we have

$$ P_s \approx \Delta_{\text{AF}}^{-2} \left(\frac{P}{\mathcal{N}_0} \right)^{-2}, \tag{5.94} $$

$$ \Delta_{\text{AF}} = \frac{2\sqrt{2}\, b \delta_{s,d} \delta_{s,r} \delta_{r,d}}{\sqrt{B}} \frac{\left(\delta_{s,r} + \sqrt{\delta_{s,r}^2 + 8\delta_{r,d}^2} \right)^{1/2}}{\left(3\delta_{s,r} + \sqrt{\delta_{s,r}^2 + 8\delta_{r,d}^2} \right)^{3/2}}, \tag{5.95} $$

which is called the *cooperation gain* of the AF cooperation systems, and indicates the best asymptotic performance gain of the AF cooperation protocol with the optimum power allocation scheme. From (5.94), we can see that AF cooperation systems can also guarantee a performance diversity of order two, which is similar to that of DF cooperation systems.

Note that both AF and DF cooperation systems are able to achieve a performance diversity of order two. In the following, we compare the cooperation gains of the two cooperation systems. Let us define a ratio $\lambda = \Delta_{\text{DF}}/\Delta_{\text{AF}}$ to indicate the performance gain of the DF cooperation protocol compared with the AF protocol. According to (5.90) and (5.95), we have

$$ \lambda = \left(\frac{\delta_{s,r} + \sqrt{\delta_{s,r}^2 + 8(A^2/B)\delta_{r,d}^2}}{\delta_{s,r} + \sqrt{\delta_{s,r}^2 + 8\delta_{r,d}^2}} \right)^{1/2} \left(\frac{3\delta_{s,r} + \sqrt{\delta_{s,r}^2 + 8\delta_{r,d}^2}}{\delta_{s,r} + \sqrt{3\delta_{s,r}^2 + 8(A^2/B)\delta_{r,d}^2}} \right)^{3/2}, \tag{5.96} $$

where A and B are specified in (5.29)–(5.32) for M-PSK and M-QAM signals, respectively.

- If the channel link quality between source and relay is much less than that between relay and destination, i.e., $\delta_{s,r}^2 << \delta_{r,d}^2$, then

$$ \lambda = \frac{\Delta_{\text{DF}}}{\Delta_{\text{AF}}} \to \frac{\sqrt{B}}{A}. \tag{5.97} $$

In the case of BPSK modulation, $A = 1/4$ and $B = 3/16$, so $\lambda = \sqrt{3} > 1$. In the case of QPSK modulation, $A = 3/8 + 1/4\pi$ and $B = 9/32 + 1/4\pi$, so $\lambda = 1.3214 > 1$. In the case of 8-PSK modulation, $A = 7/16 + \sqrt{2}/8\pi$ and $B = 21/64 + \sqrt{2}/8\pi - 1/32\pi$, so $\lambda = 1.2393 > 1$. In the case of 16-QAM modulation, $A = 15/32 + 9/16\pi$ and

$B = 45/128 + 9/16\pi$, so $\lambda = 1.1245 > 1$. In general, for M-PSK modulation (M large), $A = (M-1)/2M + \left(\sin\frac{2\pi}{M}\right)/4\pi \to 1/2$ and $B = 3(M-1)/8M + \left(\sin\frac{2\pi}{M}\right)/4\pi - \left(\sin\frac{4\pi}{M}\right)/32\pi \to 3/8$, so

$$\lambda \to \frac{\sqrt{6}}{2} \approx 1.2247 > 1.$$

For M-QAM modulation (M large), $A = (M-1)/2M + K^2/\pi \to 1/2 + 1/\pi$ and $B = 3(M-1)/8M + K^2/\pi \to 3/8 + 1/\pi$,

$$\lambda \to \frac{\sqrt{\frac{3}{8} + \frac{1}{\pi}}}{\frac{1}{2} + \frac{1}{\pi}} \approx 1.0175 > 1.$$

We can see that if $\delta_{s,r}^2 << \delta_{r,d}^2$, the cooperation gain of DF systems is always larger than that of AF systems for both M-PSK and M-QAM modulations. The advantage of the DF cooperation systems is more significant if M-PSK modulation is used.

- If the channel link quality between source and relay is much better than that between relay and destination, i.e., $\delta_{s,r}^2 >> \delta_{r,d}^2$, from (5.96) we have

$$\lambda = \frac{\Delta_{\text{DF}}}{\Delta_{\text{AF}}} \to 1.$$

This implies that if $\delta_{s,r}^2 >> \delta_{r,d}^2$, the performance of DF cooperation systems is almost the same as that of AF cooperation systems for both M-PSK and M-QAM modulations. Since the DF cooperation protocol requires decoding process at the relay, we may suggest the use of the AF cooperation protocol in this case to reduce the system complexity.

- If the channel link quality between source and relay is the same as that between relay and destination, i.e., $\delta_{s,r}^2 = \delta_{r,d}^2$, we have

$$\lambda = \left(\frac{1 + \sqrt{1 + 8(A^2/B)}}{4}\right)^{1/2} \left(\frac{6}{3 + \sqrt{1 + 8(A^2/B)}}\right)^{3/2}.$$

In the case of BPSK modulation, $A = 1/4$ and $B = 3/16$, so $\lambda \approx 1.1514 > 1$. In the case of QPSK modulation, $A = 3/8 + 1/4\pi$ and $B = 9/32 + 1/4\pi$, so $\lambda \approx 1.0851 > 1$. In the case of 8-PSK modulation, $A = 0.4938$ and $B = 0.3744$, so $\lambda = 1.0670 > 1$. In the case of 16-QAM modulation, $A = 0.6478$ and $B = 0.5306$, so $\lambda = 1.0378$. In general, for M-PSK modulation (M large), $A = (M-1)/2M + \left(\sin\frac{2\pi}{M}\right)/4\pi \to 1/2$ and $B = 3(M-1)/8M + \left(\sin\frac{2\pi}{M}\right)/4\pi - \left(\sin\frac{4\pi}{M}\right)/32\pi \to 3/8$, so

$$\lambda \to \left(\frac{1 + \sqrt{1 + 16/3}}{4}\right)^{1/2} \left(\frac{6}{3 + \sqrt{1 + 16/3}}\right)^{3/2} \approx 1.0635 > 1.$$

For M-QAM modulation (M large), $A = (M-1)/2M + K^2/\pi \rightarrow 1/2 + 1/\pi$ and $B = 3(M-1)/8M + K^2/\pi \rightarrow 3/8 + 1/\pi$, λ goes to

$$\left(\frac{1 + \sqrt{1 + 8(\frac{1}{2} + \frac{1}{\pi})^2/(\frac{3}{8} + \frac{1}{\pi})}}{4}\right)^{1/2} \left(\frac{6}{3 + \sqrt{1 + 8(\frac{1}{2} + \frac{1}{\pi})^2/(\frac{3}{8} + \frac{1}{\pi})}}\right)^{3/2}$$

which is approximately 1.0058. We can see that if the modulation size is large, the performance advantage of the DF cooperation protocol is negligible compared with the AF cooperation protocol. Actually, with QPSK modulation, the ratio of the cooperation gain is $\lambda \approx 1.0851$ which is already small.

From the above discussion, we can see that the performance of the DF cooperation protocol is always not less than that of the AF cooperation protocol. However, the performance advantage of the DF cooperation protocol is not significant unless (i) the channel link quality between the relay and the destination is much stronger than that between the source and the relay, and (ii) the constellation size of the signaling is small. These represent a tradeoff between the two cooperation protocols.

The complexity of the AF cooperation protocol is less than that of the DF cooperation protocol in which decoding process at the relay is required. For high-data-rate cooperative communications (with large modulation size), we may use the AF cooperation protocol to reduce the system complexity while the performance is comparable.

Example 5.3 In this example, we compare the performances of DF and AF cooperation systems with BPSK modulation, as shown in Figure 5.12. In the case of $\delta_{s,r}^2 = 1$ and $\delta_{r,d}^2 = 1$, the performance of the DF cooperation protocol is better than that of the AF protocol by about 1dB, as shown in Figure 5.12(a). In this case, the optimum power ratios for the DF cooperation protocol are $P_1/P = 0.5931$ and $P_2/P = 0.4069$ according to Theorem 5.2.3, while the optimum ratios for the AF protocol are $P_1/P = 2/3$ and $P_2/P = 1/3$ according to Theorem 5.3.4.

In the case of $\delta_{s,r}^2 = 1$ and $\delta_{r,d}^2 = 10$, from Figure 5.12(b) we can see that the DF cooperation protocol outperforms the AF protocol with an SER performance of about 2 dB. In this case, the optimum power ratios for the DF cooperation protocol are $P_1/P = 0.7579$ and $P_2/P = 0.2421$, while the optimum ratios for the AF protocol are $P_1/P = 0.8333$ and $P_2/P = 0.1667$. It seems that the larger the ratio of the channel link quality $\delta_{r,d}^2/\delta_{s,r}^2$, the greater the performance gain of the DF cooperation protocol compared with the AF protocol. However, the performance gain cannot be larger than $\lambda = \sqrt{3} \approx 2.4$ dB as shown in (5.97) in the case of BPSK modulation. ▲

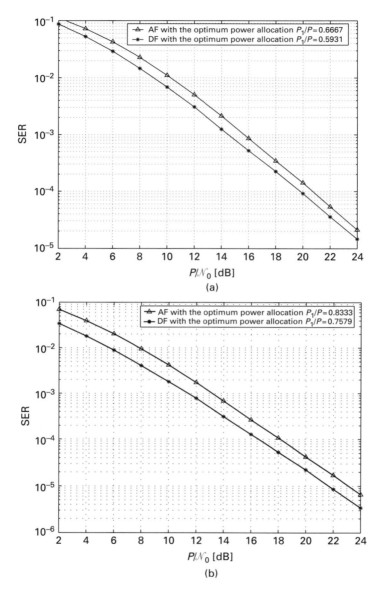

Fig. 5.12   Performance comparison of cooperation systems using either the AF or DF cooperation protocol with BPSK signals. (a) $\delta_{s,r}^2 = 1$ and $\delta_{r,d}^2 = 1$, (b) $\delta_{s,r}^2 = 1$ and $\delta_{r,d}^2 = 10$.

5.5   Trans-modulation in relay communications

In this section, we consider a method of re-mapping signal constellation points at the relay nodes to improve the destination SER performance of the DF cooperation protocol. The constellation reassignment method can be considered as some form of complex field coding by trying to increase the Euclidean distance between different transmitted symbols.

The transceiver model of the DF cooperation protocol was specified at the beginning of this chapter. We recall that if the relay node is able to decode correctly, it will help the source in phase 2; otherwise, it will remain idle. Specifically, the received signal at the destination in phase 2, due to the relay node transmission, is given by

$$y_{r,d} = \sqrt{P_r} h_{r,d} x_r + \eta_{r,d}, \tag{5.98}$$

where P_r is the relay transmitted power, $\eta_{r,d}$ denotes the additive white Gaussian noise at the destination, and $h_{r,d}$ is the channel coefficient from the relay node to the destination. Here x_r denotes the new constellation point transmitted from the relay node and is normalized such that $E\{|x_r|^2\} = 1$. For the repetition-based approach, the symbol x_r is the same as the transmitted symbol x_s from the source.

We derive in the following an expression of the pairwise symbol error rate (PSER) between two possible transmitted source symbols. This analysis gives a guideline on how to design the constellation points at the relay node to improve the system SER performance. The PSER at the destination node is defined as

$$\begin{aligned}
\Pr\{\mathbf{x}_1 \rightarrow \mathbf{x}_2\} = {} & \Pr\{\mathbf{x}_1 \rightarrow \mathbf{x}_2|\mathbf{x}_1, \text{ relay decodes erroneously}\} \\
& \times \Pr\{\text{relay decodes erroneously}\} \\
& + \Pr\{\mathbf{x}_1 \rightarrow \mathbf{x}_2|\mathbf{x}_1, \text{ relay decodes correctly}\} \\
& \times \Pr\{\text{relay decodes correctly}\}, \tag{5.99}
\end{aligned}$$

where $\mathbf{x}_1$ and $\mathbf{x}_2$ are two possible transmitted source symbols. The vector $\mathbf{x}_1 = [\sqrt{P_s} x_{s_1} \ \sqrt{P_r} x_{r_1}]^T$, where x_{s_1} is the source transmitted constellation point and x_{r_1} is the relay transmitted constellation point, similarly, $\mathbf{x}_2 = [\sqrt{P_s} x_{s_2} \ \sqrt{P_r} x_{r_2}]^T$. The PSER expression in (5.99) has two terms depending on the state of the relay node (whether or not it has decoded correctly). The first term corresponds to the case when the relay decodes erroneously. This term only depends on the constellation used at the source node while the second term clearly depends on the constellations used at the source and relay nodes. We consider the design of the constellation points at the relay node in order to minimize the PSER.

The use of a maximum likelihood (ML) detector at the receiver is assumed. We consider minimizing the term

$$\mathrm{PSER}_r = \Pr\{\mathbf{x}_1 \rightarrow \mathbf{x}_2|\mathbf{x}_1, \text{ relay decodes correctly}\},$$

which corresponds to the case when the relay correctly decodes the source symbol. Under our system model assumptions, the PSER_r of the ML detector can be expressed as

$$\mathrm{PSER}_r = E\{\Pr\{q < 0|\mathbf{x}_1, \text{ relay decodes correctly}\}\}, \tag{5.100}$$

where

$$q = \begin{bmatrix} \mathbf{z}_1^H & \mathbf{z}_2^H \end{bmatrix} \begin{bmatrix} \mathbf{I}_2 & \mathbf{0} \\ \mathbf{0} & -\mathbf{I}_2 \end{bmatrix} \begin{bmatrix} \mathbf{z}_1 \\ \mathbf{z}_2 \end{bmatrix}$$

in which $\mathbf{z}_1 = (\mathbf{diag}(\mathbf{x}_1) - \mathbf{diag}(\mathbf{x}_2))\mathbf{h} + \mathbf{n}$, $\mathbf{z}_2 = \mathbf{n}$, $\mathbf{h} = [h_{\mathrm{s,d}} \ h_{\mathrm{r,d}}]^T$, $\mathbf{n} = [\eta_{\mathrm{s,d}} \ \eta_{\mathrm{r,d}}]^T$, and $\mathbf{I}_2$ is the 2×2 identity matrix. It can be shown that the conditional pdf of q in (5.100), given the channel coefficients and given that $\mathbf{x}_1$ was transmitted and the relay decoded correctly, is Gaussian. The PSER$_r$ can be proved to be given by

$$\mathrm{PSER}_r = \mathrm{E}\left\{ Q\left(\sqrt{\frac{1}{2N_0} \left(P_s |h_{\mathrm{s,d}}|^2 |x_{\mathrm{s}_1} - x_{\mathrm{s}_2}|^2 + P_r |h_{\mathrm{r,d}}|^2 |x_{\mathrm{r}_1} - x_{\mathrm{r}_2}|^2 \right)} \right) \right\}, \quad (5.101)$$

where $Q(u) = (1/\sqrt{2\pi}) \int_u^\infty \exp(-t^2/2) \mathrm{d}t$ is the Gaussian Q-function. The expectation in (5.101) is with respect to the channel state information (CSI). Using the special property of the Gaussian Q-function as $Q(u) = 1/\pi \int_0^{\pi/2} \exp(-u^2/2\sin^2\theta)\mathrm{d}\theta$ and averaging over the exponential distribution of the squared magnitude of the channel gains, it can proved that the PSER$_r$ is given by

$$\mathrm{PSER}_r = \frac{1}{\pi} \int_0^{\pi/2} \frac{1}{\left(1 + \frac{P_s \delta_{\mathrm{s,d}}^2 |x_{\mathrm{s}_1} - x_{\mathrm{s}_2}|^2}{4N_0 \sin^2\theta}\right)} \cdot \frac{1}{\left(1 + \frac{P_r \delta_{\mathrm{r,d}}^2 |x_{\mathrm{r}_1} - x_{\mathrm{r}_2}|^2}{4N_0 \sin^2\theta}\right)} \mathrm{d}\theta. \quad (5.102)$$

An upper bound on PSER$_r$ can be obtained by neglecting the one term in the denominator of the terms inside the integration of (5.102). The PSER$_r$ can now be upper bounded as

$$\mathrm{PSER}_r \le \frac{3N_0^2}{\delta_{\mathrm{s,d}}^2 \delta_{\mathrm{r,d}}^2 P_s P_r |x_{\mathrm{s}_1} - x_{\mathrm{s}_2}|^2 |x_{\mathrm{r}_1} - x_{\mathrm{r}_2}|^2}. \quad (5.103)$$

In order to minimize the PSER, symbols that have adjacent constellation points in the source constellation are assigned nonadjacent constellation points in the relay constellation assignment and vice versa. So instead of using repetition at the relay node (i.e., using the same constellation as the source node) we can do constellation reassignment at the relay node to better separate the symbols to maximize the product $|x_{\mathrm{s}_1} - x_{\mathrm{s}_2}|^2 |x_{\mathrm{r}_1} - x_{\mathrm{r}_2}|^2$. The constellation reassignment scheme can improve the system PSER performance, and hence the SER performance, without increasing the complexity of the system.

The most common constellations used in communication systems are BPSK, QPSK, 16-QAM, and 64-QAM. For BPSK, constellation reassignment at the relay nodes is meaningless since the BPSK constellation has only two points. For QPSK, the performance gains of using constellation reassignment are not significant and we do not show the results for that case. For 16-QAM and 64-QAM constellations, one possible way to do the constellation reassignment at the relay nodes is to perform an exhaustive search over all possible relay constellation assignments and select the one that maximizes the minimum value of the product $|x_{\mathrm{s}_1} - x_{\mathrm{s}_2}|^2 |x_{\mathrm{r}_1} - x_{\mathrm{r}_2}|^2$ over all possible pairs of transmitted symbols. However, the exhaustive search is extremely complex–for example, for the single-relay case, the number of possible constellation assignments at the relay node is $16! = 2.0923 \times 10^{13}$ for the 16-QAM constellation and $64! = 1.2689 \times 10^{89}$ for the 64-QAM constellation. This renders the exhaustive search impractical for constellation reassignment at the relay node. Therefore, we resort to the use of heuristic approaches

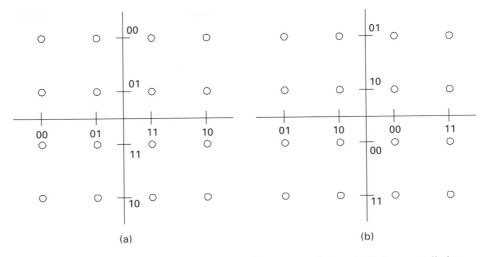

Fig. 5.13   Trans-modulation for 16-QAM constellation. (a) Source constellation, (b) Relay constellation.

for constellation reassignment at the relay nodes. For example, for the 16-QAM constellations in Figure 5.13, if we look at the source constellation as a 4×4 matrix, we first rearrange the rows and then the columns of that matrix to ensure that any two adjacent rows (columns) in the source constellation matrix are non-adjacent in the resulting matrix, which will be used as the relay constellation. This approach ensures that adjacent source constellation points are non-adjacent in the relay constellation assignment and, hence, improves the system SER performance.

Example 5.4   We present some simulation examples to show the performance gains using constellation reassignment. Figures 5.13 and 5.14 show the constellations used at the source and relay nodes for 16-QAM and 64-QAM constellations, respectively. The constellation assignment for the 64-QAM constellation is shown only along one (real) axis and the same reassignment is done along the other (complex) axis. Figure 5.15 shows the SER versus SNR, defined as $SNR = (P_s + P_r)/N_0$, for the 16-QAM and 64-QAM single-relay DAF system, where we assume equal power allocation between the source node and the relay node. The channel variance between the source and the destination is taken to be 1 in all cases. We consider two cases: relay close to source ($\delta_{s,r}^2 = 10$, $\delta_{r,d}^2 = 1$) and relay close to destination ($\delta_{s,r}^2 = 1$, $\delta_{r,d}^2 = 10$). From Figure 5.15, it is clear that for the 16-QAM constellation a gain of about 2 dB is achieved for the case where the relay is close to the source. For the 64-QAM constellation, we observe a gain of about 3 dB for the case where the relay is close to the source. For the case where the relay is close to the source, the relay will decode correctly with a high probability. Hence, the use of constellation reassignment at the relay node can greatly improve the system performance in this case compared to the repetition-based approach.

For the case where the relay is close to the destination, we observe that, for both 16-QAM and 64-QAM constellations, there is no significant performance gain using

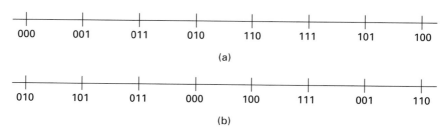

Fig. 5.14 Trans-modulation for 64-QAM constellation. (a) Source constellation (real axis), (b) Relay constellation (real axis).

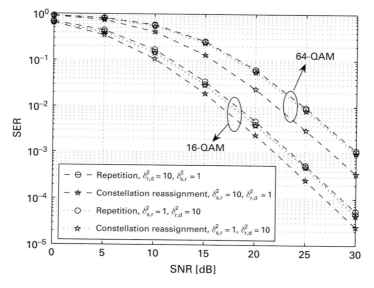

Fig. 5.15 SER for a single-relay DAF system using 16-QAM and 64-QAM constellations.

the constellation reassignment scheme as compared to the repetition-based scheme. In this case, and under relay transmission, the distance between two possible transmitted symbols will be dominated by the distance resulting from the relay constellation. Transmitted symbols are most likely to be mistaken with their adjacent symbols in the relay node constellation.

▲

5.6   Chapter summary and bibliographical notes

In this chapter, the SER performance analysis for the uncoded cooperation systems with either the DF or AF cooperation protocol was presented. For DF cooperation systems, a closed-form SER formulation explicitly for systems with PSK and QAM signals was developed. An SER upper bound as well as an approximation were also established to reveal the asymptotic performance of DF cooperation systems, in which the SER approximation is asymptotically tight at high SNR. Based on the asymptotically tight SER approximation, an optimum power allocation for DF cooperation systems was determined. For AF cooperation systems, in order to obtain a closed-form SER formulation, a simple closed-form MGF expression for the harmonic mean of two exponential random variables was obtained at first. By taking advantage of the simple MGF expression, an asymptotic tight SER approximation for AF cooperation systems with PSK and QAM signals was established. Also, based on the tight SER approximation, an optimum power allocation for AF cooperation systems was determined. Moreover, the performances of cooperation systems using the DF and AF protocols were compared. Finally, a method of re-mapping signal constellation points at the relay to improve the SER performance of the DF protocol was discussed.

Based on the theoretical and simulation results, several observations can be made. First, the equal power strategy [109] is good, but in general not optimum, in the cooperation systems using either the DF or AF protocol, and the optimum power allocation depends on the channel link quality. Second, in the case where all channel links are available in DF or AF cooperation systems, the optimum power allocation does not depend on the direct link between source and destination, but depends only on the channel links between source and relay and between relay and destination. Specifically, if the link quality between source and relay is much less than that between relay and destination, i.e., $\delta_{s,r}^2 << \delta_{r,d}^2$, then we should put the total power at the source and do not use the relay. On the other hand, if the link quality between source and relay is much larger than that between relay and destination, i.e., $\delta_{s,r}^2 >> \delta_{r,d}^2$, then the equal power strategy at the source and the relay tends to be optimum. Third, the performance of the cooperation systems using the DF protocol is better than those using the AF protocol. However, the performance gain varies with different modulation types. The larger the signal constellation size, the lower the performance gain. In the case of BPSK modulation, the performance gain cannot be larger than 2.4 dB; and for QPSK modulation, it cannot be larger than 1.2 dB. Therefore, for high data-rate cooperative communications (with large signal constellation size), we may use the AF cooperation protocol to reduce system complexity while maintaining a comparable performance.

The basic idea of cooperative communications can be traced back to the 1970s [224, 225], in which a basic three-terminal communication model was first introduced and studied by van der Meulen in the context of mutual information. A more thorough capacity analysis of the relay channel was provided later in [25], and there have been more recent works that further address the information-theoretic aspect of the relay channel, for example [99] on achievable capacity and coding strategies for wireless relay channels, [151] on the capacity region of a degraded Gaussian relay channel with multiple relay stages, [33] on the capacity of relay channels with orthogonal channels, and so on.

Recently, many efforts have been focused on the design of cooperative diversity protocols in order to combat the effects of severe fading in wireless channels. Specifically, in [108, 109], various cooperation protocols were proposed for wireless networks and extensive outage probability performance analysis has been presented for such coopera-tion systems. The concept of user cooperation diversity was also proposed in [179, 180], where a specific two-user cooperation scheme was investigated for CDMA systems and a substantial performance gain was demonstrated over the non-cooperative approach. In [92], a coded two-user cooperation scheme was proposed by taking advantage of the existing channel codes, in which the coded information of each user is divided into two parts: one part is transmitted by the user itself and the other sent by its cooperator.

Note that in order to analyze the SER performance of AF cooperation systems, we have to investigate the statistics of the harmonic mean of two random variables, which are related to the instantaneous SNR at the destination [58]. A moment-generating function of the harmonic mean of two independent exponential random variables was derived in [58] by applying the Laplace transform and the hypergeometric functions [50]. However, the result involves an integration of the hypergeometric functions and it is hard to use for analyzing AF cooperation systems. The simple closed-form MGF expression for the harmonic mean derived in this chapter is useful beyond the SER analysis. The analysis of AF and DF cooperation systems discussed in this chapter was published in [207] and [204]. Interested readers may refer to [168] for more details on trans-modulation.

Exercises

5.1 Prove the claims in (5.27) and (5.28), i.e., show the following results

$$\lim_{x \to \infty} x^2 I_1(x) = \frac{A^2}{b^2 \delta_{s,d}^2 \delta_{s,r}^2},$$

$$\lim_{x, y \to \infty} xy I_2(x, y) = \frac{B}{b^2 \delta_{s,d}^2 \delta_{r,d}^2},$$

where $I_1(x)$ and $I_2(x, y)$ are defined in (5.25) and (5.26), respectively, and the constants b, A, and B are specified in (5.29)–(5.32) for M-PSK and M-QAM modulation, respectively. Moreover, based on the above results, prove the claim in Theorem 5.2.2.

5.2 Let X be a random variable with pdf $p_X(x)$ for all $x \geq 0$ and $p_X(x) = 0$ for $x < 0$. Then, the pdf of $Y = 1/X$ is

$$p_Y(y) = \frac{1}{y^2} p_X \left(\frac{1}{y} \right) \cdot U(y).$$

5.3 Let X_1 and X_2 be two independent random variables with pdf $p_{X_1}(x)$ and $p_{X_2}(x)$ defined for all x. Then, the pdf of the sum $Y = X_1 + X_2$ is

$$p_Y(y) = \int_{-\infty}^{\infty} p_{X_1}(y - x) \, p_{X_2}(x) \mathrm{d}x,$$

which is the convolution of $p_{X_1}(x)$ and $p_{X_2}(x)$.

5.4 Prove the result in Theorem 5.3.1. Suppose that X_1 and X_2 are two independent random variables with pdf $p_{X_1}(x)$ and $p_{X_2}(x)$ defined for all $x \geq 0$, and $p_{X_1}(x) = 0$ and $p_{X_2}(x) = 0$ for $x < 0$. Then the pdf of $Z = \frac{X_1 X_2}{X_1 + X_2}$, the harmonic mean of X_1 and X_2, is

$$p_Z(z) = z \int_0^1 \frac{1}{t^2 (1-t)^2} \, p_{X_1}(\frac{z}{1-t}) \, p_{X_2}(\frac{z}{t}) dt \cdot U(z),$$

in which $U(z) = 1$ for $z \geq 0$ and $U(z) = 0$ for $z < 0$.

5.5 By applying the closed-form MGF expression in Theorem 5.3.2, show the SER formulations in (5.71) and (5.72) for AF systems with M-PSK and M-QAM modulations, respectively.

5.6 Prove the claims in (5.79) and (5.80), i.e., show the following results

$$\lim_{x, \, y \to \infty} x^2 \, J_1(x, y) = \frac{B}{b^2 \delta_{s,d}^2 \delta_{s,r}^2},$$

$$\lim_{x, \, y \to \infty} xy \, J_2(x, y) = \frac{B}{b^2 \delta_{s,d}^2 \delta_{r,d}^2},$$

where $J_1(x, y)$ and $J_2(x, y)$ are defined in (5.77) and (5.78), respectively, and the constants b and B are specified in (5.81) and (5.82) for M-PSK and M-QAM modulation, respectively.

5.7 Prove the results in Theorem 5.3.4, i.e., show that for sufficiently high SNR, the optimum power allocation for AF cooperation systems with either M-PSK or M-QAM modulation is

$$P_1 = \frac{\delta_{s,r} + \sqrt{\delta_{s,r}^2 + 8 \delta_{r,d}^2}}{3 \delta_{s,r} + \sqrt{\delta_{s,r}^2 + 8 \delta_{r,d}^2}} P,$$

$$P_2 = \frac{2 \delta_{s,r}}{3 \delta_{s,r} + \sqrt{\delta_{s,r}^2 + 8 \delta_{r,d}^2}} P.$$

5.8 (Simulation project) Consider a simplified network with a source node, a relay node and a destination node. The source–destination channel $h_{s,d}$, the source–relay channel $h_{s,r}$, and the relay–destination channel $h_{r,d}$ are modeled as independent zero-mean, complex Gaussian random variables with variances $\delta_{s,d}^2$, $\delta_{s,r}^2$ and $\delta_{r,d}^2$, respectively. Assume that $\delta_{s,r}^2 = 1$ and $\delta_{r,d}^2 = 1$. We compare performances of the AF and DF cooperation protocols with QPSK modulation.

(a) Determine the optimum power allocation at the source and the relay for the DF cooperation protocol.

(b) Determine the optimum power allocation at the source and the relay for the AF cooperation protocol.

(c) Simulate both the DF and AF protocols with corresponding optimum power allocations, and compare their performances.

(d) Repeat (a), (b), and (c) for the case of $\delta_{s,r}^2 = 1$ and $\delta_{r,d}^2 = 10$.

6   Multi-node cooperative communications

In the previous chapter, the symbol error rate performance of single-relay cooperative communications was analyzed for both the decode-and-forward and amplify-and-forward relaying strategies. This chapter builds upon the results in the previous chapter and generalizes the symbol error rate performance analysis to the multi-relay scenario.

Decode-and-forward relaying will be considered first, followed by the amplify-and-forward case. In both scenarios, exact and approximate expressions for the symbol error rate will be derived. The symbol error rate expressions are then used to characterize an optimal power allocation strategy among the relays and the source node.

6.1   Multi-node decode-and-forward protocol

We begin by presenting a class of cooperative decode-and-forward protocols for arbitrary N-relay wireless networks, in which each relay can combine the signal received from the source along with one or more of the signals transmitted by previous relays. Then, we focus on the performance of a general cooperation scenario and present an exact symbol error rate (SER) expressions for both M-ary phase shift keying (PSK) and quadrature amplitude modulation (QAM) signalling. We also consider an approximate expression for the SER of a general cooperation scenario that is shown to be tight at high enough SNR. Finally, we study optimal power allocation for the class of cooperative diversity schemes, where the optimality is determined in terms of minimizing the SER of the system.

6.1.1   System model and protocol description

We consider an arbitrary N-relay wireless network, where information is to be transmitted from a source to a destination. Due to the broadcast nature of the wireless channel, some relays can overhear the transmitted information and thus can cooperate with the source to send its data. The wireless link between any two nodes in the network is modeled as a Rayleigh fading narrowband channel with additive white Gaussian noise (AWGN). The channel fades for different links are assumed to be statistically independent. This is a reasonable assumption as the relays are usually spatially well separated. The additive noise at all receiving terminals is modeled as zero-mean, complex Gaussian random variables with variance N_0. For medium access, the relays are assumed to transmit over orthogonal channels, thus no inter-relay interference is considered in the signal model.

The cooperation strategy we are considering employs a selective decode-and-forward protocol at the relaying nodes. Each relay can measure the received SNR and forwards the received signal if the SNR is higher than some threshold. For mathematical tractability of symbol error rate calculations we assume the relays can judge whether the received symbols are decoded correctly or not and only forwards the signal if decoded correctly otherwise remains idle. This assumption will be shown via simulations to be very close to the performance of the practical scenario of comparing the received SNR to a threshold, specially when the relays operate in a high SNR regime, as for example when the relays are selected close to the source node. The rationale behind this is that when the relays are closer to the source node, or more generally operate in a high SNR regime, the channel fading becomes the dominant source of error, and hence measuring the received SNR gives a very good judgement on whether the received symbol can be decoded correctly or not with high probability. Various scenarios for the cooperation among the relays can be implemented. A general cooperation scenario, denoted as $\mathcal{C}(m)$ $(1 \leq m \leq N - 1)$, can be implemented in which each relay combines the signals received from the m previous relays along with that received from the source. The simplest scenario $\mathcal{C}(1)$ among the class of cooperative protocols is depicted in Figure 6.1, in which each relay combines the signal received from the previous relay and the source. The most complicated scenario $\mathcal{C}(N - 1)$ is depicted in Figure 6.2, in which each relay combines the signals received from all of the previous relays along with that from the source. This is the most sophisticated scenario and should provide the best performance in the class of cooperative protocols $\{\mathcal{C}(m)\}_{m=1}^{N-1}$, as in this case each relay utilizes the information from all previous phases of the protocol. In all of the considered cooperation scenarios, the destination coherently combines the signals received from the source and all of the relays. In the sequel, we focus on presenting the system model for a general cooperative scheme $\mathcal{C}(m)$ for any $1 \leq m \leq N - 1$.

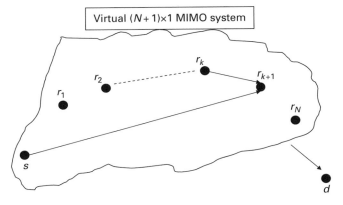

Fig. 6.1   Illustrating cooperation under $\mathcal{C}(1)$: the $(k + 1)$-th relay combines the signals received from the source and the k-th relay.

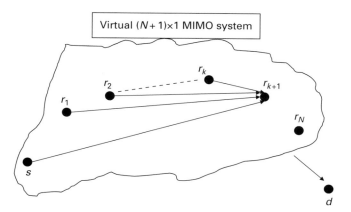

Fig. 6.2   Illustrating cooperation under $\mathcal{C}(N-1)$: the $(k+1)$-th relay combines the signals received from the source and all of the previous relays.

For a general scheme $\mathcal{C}(m)$, $1 \leq m \leq N-1$, each relay decodes the information after combining the signals received from the source and the previous m relays. The cooperation protocol has $(N+1)$ phases. In phase 1, the source transmits the information, and the received signal at the destination and the i-th relay can be modeled, respectively, as

$$y_{s,d} = \sqrt{P_0}h_{s,d}x + n_{s,d},$$
$$y_{s,r_i} = \sqrt{P_0}h_{s,r_i}x + n_{s,r_i}, \quad 1 \leq i \leq N, \tag{6.1}$$

where P_0 is the power transmitted at the source, x is the transmitted symbol with unit power, $h_{s,d} \sim CN(0, \sigma_{s,d}^2)$ and $h_{s,r_i} \sim CN(0, \sigma_{s,r_i}^2)$ are the channel fading coefficients between the source and the destination, and i-th relay, respectively, and $CN(\alpha, \sigma^2)$ denotes a circularly symmetric complex Gaussian random variable with mean α and variance σ^2. The terms $n_{s,d}$ and n_{s,r_i} denote the AWGN. In phase 2, if the first relay correctly decodes, it forwards the decoded symbol with power P_1 to the destination, otherwise it remains idle.

In general, during phase l, $2 \leq l \leq N$, the l-th relay combines the received signals from the source and the previous $\min\{m, l-1\}$ relays using a maximal-ratio-combiner (MRC) as follows[1]

$$y_{r_l} = \sqrt{P_0}h_{s,r_l}^* y_{s,r_l} + \sum_{i=\max(1,l-m)}^{l-1} \sqrt{\hat{P}_i}h_{r_i,r_l}^* y_{r_i,r_l}, \tag{6.2}$$

where $h_{r_i,r_l} \sim CN(0, \sigma_{r_i,r_l}^2)$ is the channel fading coefficient between the i-th and the l-th relays. In (6.2), y_{r_i,r_l} denotes the signal received at the l-th relay from the i-th relay, and can be modeled as

$$y_{r_i,r_l} = \sqrt{\hat{P}_i}h_{r_i,r_l}x + n_{r_i,r_l}, \tag{6.3}$$

---

[1] The $\max(1, l-m)$ function is used to make sure that if $l < m$ then the combining starts at the first relay.

where $\hat{P}_i$ is the power transmitted at relay i in phase $(i + 1)$, and $\hat{P}_i = P_i$ if relay i correctly decodes the transmitted symbol, otherwise $\hat{P}_i = 0$. The l-th relay uses y_{r_l} in (6.2) as the detection statistics. If relay l decodes correctly it transmits with power $\hat{P}_l = P_l$ in Phase $(l + 1)$, otherwise it remains idle. Finally, in phase $(N + 1)$, the destination coherently combines all of the received signals using an MRC as follows

$$y_d = \sqrt{P_0} h_{s,d}^* y_{s,d} + \sum_{i=1}^{N} \sqrt{\hat{P}_i} h_{r_i,d}^* y_{r_i,d}. \qquad (6.4)$$

For fair comparison purpose, the total transmitted power is fixed as $P_0 + \sum_{i=1}^{N} P_i = P$.

6.1.2 Exact SER performance analysis

In this subsection, we present the SER performance analysis for a general cooperative scheme $\mathcal{C}(m)$ for any $1 \leq m \leq N - 1$. Exact SER expressions of this general scheme are presented for systems with both M-PSK and M-QAM modulation. Some simulation examples are also discussed at the end of this subsection.

First, we introduce some terminologies that will be used throughout the analysis. For a given transmission, each relay can be in one of two states: either it decoded correctly or not. Let us define a $1 \times n$, $1 \leq n \leq N$, vector $\mathbf{S}_n$ to represent the states of the first n relays for a given transmission. The k-th entry of the vector $\mathbf{S}_n$ denotes the state of the k-th relay as follows:

$$S_n[k] = \begin{cases} 1 & \text{if relay } k \text{ correctly decodes,} \\ 0 & \text{otherwise,} \end{cases} \qquad 1 \leq k \leq n. \qquad (6.5)$$

Since the decimal value of the binary vector $\mathbf{S}_n$ can take on values from 0 to $2^n - 1$, for convenience we denote the state of the network by an integer decimal number. Let

$$\mathbf{B}_{x,n} = \left(B_{x,n}[1], B_{x,n}[2], \ldots, B_{x,n}[n] \right) \qquad (6.6)$$

be the $1 \times n$ binary representation of a decimal number x, with $B_{x,n}[1]$ being the most significant bit. So, $\mathbf{S}_N = \mathbf{B}_{x,N}$ indicates that the k-th relay, $1 \leq k \leq N$, is in state $S_N[k] = B_{x,N}[k]$.

We consider a general cooperation scheme $\mathcal{C}(m)$, $1 \leq m \leq N - 1$, in which the k-th $(1 \leq k \leq N)$ relay coherently combines the signals received from the source along with the signals received from the previous $\min\{m, k - 1\}$ relays. The state of each relay in this scheme depends on the states of the previous m relays, i.e., whether these relays decoded correctly or not. This is due to the fact that the number of signals received at each relay depends on the number of relays that decoded correctly from the previous m relays. Hence, the joint probability of the states is given by

$$P(\mathbf{S}_N) = P(S_N[1]) P(S_N[2] \mid S_N[1]) \cdots P(S_N[N] \mid S_N[N - 1], \ldots, S_N[N - m]). \qquad (6.7)$$

Conditioning on the network state, which can take 2^N values, the probability of error at the destination given the channel state information (CSI) can be calculated using the law of total probability as follows:

$$P_{e|CSI} = \sum_{i=0}^{2^N-1} \Pr(e \mid \mathbf{S}_N = \mathbf{B}_{i,N}) \Pr(\mathbf{S}_N = \mathbf{B}_{i,N}), \tag{6.8}$$

where e denotes the event that the destination decoded in error. The summation in the above equation is over all possible states of the network.

Now, let us compute the terms in (6.8). The destination collects the copies of the signal transmitted in the previous phases using an MRC (6.4). The resulting SNR at thedestination can be computed as

$$SNR_d = \frac{P_0 \mid h_{s,d} \mid^2 + \sum_{j=1}^N P_j B_{i,N}[j] \mid h_{r_j,d} \mid^2}{\mathcal{N}_o}, \tag{6.9}$$

where $B_{i,N}[j]$ takes value 1 or 0 and determines whether the j-th relay has decoded correctly or not. The k-th relay coherently combines the signals received from the source and the previous m relays. The resulting SNR can be calculated as

$$SNR_{r_k}^m = \frac{P_0 \mid h_{s,r_k} \mid^2 + \sum_{j=\max(1,k-m)}^{k-1} P_j B_{i,N}[j] \mid h_{r_j,r_k} \mid^2}{\mathcal{N}_o}. \tag{6.10}$$

If M-PSK modulation is used in the system, with instantaneous SNR γ, the SER with the channel state information is given by

$$P_{CSI}^{PSK} = \Psi_{PSK}(\gamma) \triangleq \frac{1}{\pi} \int_0^{(M-1)\pi/M} \exp\left(-\frac{b_{PSK}\gamma}{\sin^2(\theta)}\right) d\theta, \tag{6.11}$$

where $b_{PSK} = \sin^2(\pi/M)$. If M-QAM ($M = 2^k$ with k even) modulation is used in the system, the corresponding conditional SER can be expressed as

$$P_{CSI}^{QAM} = \Psi_{QAM}(\gamma) \triangleq 4CQ(\sqrt{b_{QAM}\gamma}) - 4C^2Q^2(\sqrt{b_{QAM}\gamma}), \tag{6.12}$$

in which $C = 1 - 1/\sqrt{M}$, $b_{QAM} = 3/(M-1)$, and $Q(x)$ is the complementary distribution function (CDF) of the Gaussian distribution.

Let us focus on computing the SER in the case of M-PSK modulation, and the same procedure is applicable for the case of M-QAM modulation. From (6.9), and for a given network state $\mathbf{S}_N = \mathbf{B}_{i,N}$, the conditional SER at the destination can be computed as

$$\Pr(e|\mathbf{S}_N = \mathbf{B}_{i,N}) = \Psi_{PSK}(SNR_d). \tag{6.13}$$

Denote the conditional probability that the k-th relay is in state $B_{i,N}[k]$ given the states of the previous m relays by $P_{k,i}^m$. From (6.10), this probability can be computed as follows:

$$\begin{aligned} P_{k,i}^m &\triangleq Pr(S_N[k] = B_{i,N}[k] \mid S_N[k-1] = B_{i,N}[k-1], \dots, S_N[k-m] \\ &= B_{i,N}[k-m]) \\ &= \begin{cases} \Psi_{PSK}(SNR_{r_k}^m), & \text{if } B_{i,N}[k] = 0, \\ 1 - \Psi_{PSK}(SNR_{r_k}^m), & \text{if } B_{i,N}[k] = 1. \end{cases} \end{aligned} \tag{6.14}$$

To compute the average SER, we need to average the probability in (6.8) over all channel realizations, i.e., $P_{SER}(m) = E_{CSI}[P_{e|CSI}]$. Using (6.7), (6.13), and (6.14), $P_{SER}(m)$ can be expanded as follows

$$P_{SER}(m) = \sum_{i=0}^{2^N-1} E_{CSI}\left[\Psi_{PSK}(SNR_d)\prod_{k=1}^{N}P_{k,i}^m\right]. \tag{6.15}$$

Since the channel fades between different pairs of nodes in the network are statistically independent by the virtue that different nodes are not co-located, the quantities inside the expectation operator in the above equation are functions of independent random variables, and thus can be further decomposed as

$$P_{SER}(m) = \sum_{i=0}^{2^N-1}\left\{E_{CSI}[\Psi_{PSK}(SNR_d)]\prod_{k=1}^{N}E_{CSI}[P_{k,i}^m]\right\}. \tag{6.16}$$

The above analysis is applicable to the M-QAM case by replacing the function $\Psi_{PSK}(\cdot)$ by $\Psi_{QAM}(\cdot)$.

Since the channels between the nodes are modeled as Rayleigh fading channels, the absolute norm square of any channel realization $h_{i,j}$ between any two nodes i and j in the network has an exponential distribution with mean $\sigma_{i,j}^2$. Hence, $E_{CSI}[\Psi_q(\gamma)]$ can be expressed as

$$E_{CSI}[\Psi_q(\gamma)] = \int_{\gamma} \Psi_q(\gamma)f(\gamma)d\gamma \tag{6.17}$$

where $f(\gamma)$ is the probability density function of the random variable γ, and $q = 1$ ($q = 2$) correspond to M-PSK (M-QAM), respectively. If γ is an exponentially distributed random variable with mean $\bar{\gamma}$, then it can be shown [187] that $E_{CSI}[\Psi_q(\gamma)]$ is given by

$$E_{CSI}[\Psi_q(\gamma)] = F_q\left(1 + \frac{b_q\bar{\gamma}}{\sin^2(\theta)}\right), \tag{6.18}$$

where $F_q(\cdot)$ and the constant b_q are defined as

$$F_1(x(\theta)) = \frac{1}{\pi}\int_0^{(M-1)\pi/M}\frac{1}{x(\theta)}d\theta, \qquad b_1 = b_{psk}$$

$$F_2(x(\theta)) = \frac{4C}{\pi}\int_0^{\pi/2}\frac{1}{x(\theta)}d\theta - \frac{4C^2}{\pi}\int_0^{\pi/4}\frac{1}{x(\theta)}d\theta, \qquad b_2 = \frac{b_{QAM}}{2}. \tag{6.19}$$

In order to get the above expressions, we use two special properties of the $Q(\cdot)$ function, specifically,

$$Q(x) = \frac{1}{\pi}\int_0^{\pi/2}\exp\left(-\frac{x^2}{2\sin^2(\theta)}\right)d\theta \tag{6.20}$$

$$Q^2(x) = \frac{1}{\pi}\int_0^{\pi/4}\exp\left(-\frac{x^2}{2\sin^2(\theta)}\right)d\theta \tag{6.21}$$

for $x \geq 0$.

Averaging over all the Rayleigh fading channel realizations, the SER at the destination for a given network state $\mathbf{B}_{i,N}$ is given by

$$E_{\text{CSI}}(\Psi_q(\text{SNR}_d)) = F_q\left[\left(1 + \frac{b_q P_0 \sigma_{\text{s,d}}^2}{\mathcal{N}_o \sin^2(\theta)}\right) \prod_{j=1}^{N}\left(1 + \frac{b_q B_{i,N}[j] P_j \sigma_{\text{r}_j,\text{d}}^2}{\mathcal{N}_o \sin^2(\theta)}\right)\right]. \quad (6.22)$$

Similarly, the probability that the k-th relay is in state $B_{i,N}[k]$ given the states of the
previous m relays is given by

$$E_{\text{CSI}}\left[P_{k,i}^m\right] = G_k^m(B_{i,N}[k]), \quad (6.23)$$

where $G_k^m(\cdot)$ is defined as

$$G_k^m(x) = \begin{cases} F_q\left[\left(1 + \frac{b_q P_0 \sigma_{\text{s,r}_k}^2}{\mathcal{N}_o \sin^2(\theta)}\right) \prod_{j=\max(1,k-m)}^{k-1}\left(1 + \frac{b_q B_{i,N}[j] P_j \sigma_{\text{r}_j,\text{r}_k}^2}{\mathcal{N}_o \sin^2(\theta)}\right)\right], \\ \text{if } x = 0, \\ 1 - F_q\left[\left(1 + \frac{b_q P_0 \sigma_{\text{s,r}_k}^2}{\mathcal{N}_o \sin^2(\theta)}\right) \prod_{j=\max(1,k-m)}^{k-1}\left(1 + \frac{b_q B_{i,N}[j] P_j \sigma_{\text{r}_j,\text{r}_k}^2}{\mathcal{N}_o \sin^2(\theta)}\right)\right], \\ \text{if } x = 1. \end{cases}$$
$$(6.24)$$

in which $F_q(\cdot)$ and the constant b_q are specified in (6.19). As a summary, the SER in
(6.16) of the cooperative multi-node system employing scenario $\mathcal{C}(m)$ with M-PSK or
M-QAM modulation can be determined from (6.22), (6.23), and (6.24) in the following
theorem.

THEOREM 6.1.1 *The SER of an N-relay decode-and-forward cooperative diversity
network utilizing protocol $\mathcal{C}(m)$, $1 \leq m \leq N-1$, and M-PSK or M-QAM modulation
is given by*

$$P_{\text{SER}}(m) = \sum_{i=0}^{2^N-1} F_q\left[\left(1 + \frac{b_q P_0 \sigma_{\text{s,d}}^2}{\mathcal{N}_o \sin^2(\theta)}\right) \prod_{j=1}^{N}\left(1 + \frac{b_q B_{i,N}[j] P_j \sigma_{\text{r}_j,\text{d}}^2}{N_0 \sin^2(\theta)}\right)\right]$$
$$\prod_{k=1}^{N} G_k^m(B_{i,N}[k]), \quad (6.25)$$

where the functions $F_q(\cdot)$ and $G_k^m(\cdot)$ are defined in (6.19) and (6.24), respectively.

Example 6.1   We illustrate the validity of the theoretical results by some simulation
examples. Let us focus on the cooperative protocol $\mathcal{C}(1)$. The number of relays is taken
to be $N = 1, 2, 3$, in addition to the source and the destination nodes. We considered
two simulation setups:

- In the first setup we simulate the SER performance under the assumption that the relay
  correctly judges whether the received signal is decoded correctly or not, i.e., no error
  propagation, which is the model analyzed in the chapter.

- In the second setup, we consider a more practical scenario in which each relay compares the instantaneous received SNR to a threshold and hence decides whether to forward the received signal or not, and thus error propagation is allowed (the threshold is taken equal to 3 dB here and is selected by experiment).

The relays are considered closer to the source than the destination. The channel variance depends on the distance l and propagation path-loss α as follows $\sigma^2 \propto l^{-\alpha}$, and $\alpha = 3$ in our simulations. The channel gains are as follows: $\sigma^2_{s,r_i} = 8\sigma^2_{s,d}$, and $\sigma^2_{r_i,d} = \sigma^2_{s,d}$. The noise variance is taken to be $N_0 = 1$. The total transmitted power in each case is considered fixed to P.

Figure 6.3 depicts the SER versus P/N_0 performance of cooperation scenario $\mathcal{C}(1)$ with QPSK. As shown in the figure, the performance curves of the two previously described simulation setups are very close for different numbers of relays. This validates that the model for selective relaying assumed for mathematical tractability has a performance close to that of practical selective relaying when comparing the SNR to a threshold. The intuition behind this is, as illustrated before, that when the relays in general operate in a high SNR regime, in this case the relays are closer to the source node, the error propagation from the relays becomes negligible, which is due to the fact that the channel outage event (the SNR is less than the threshold) becomes the dominating error event.

The performance of direct transmission without any relaying is also shown in Figure 6.3 as a benchmark for a no-diversity scheme. Moreover, the exact SER expression from the theorem is depicted as a benchmark. It is clear from the depicted figure that the analytical SER expression in (6.25) for scenario $\mathcal{C}(1)$ exactly matches the simulation results for each case. This confirms the theoretical analysis. The results also reveal

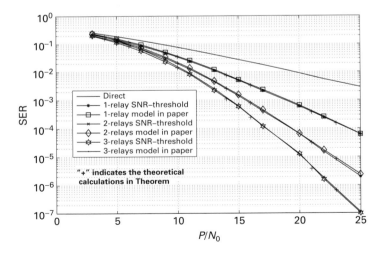

Fig. 6.3   SER versus SNR for two different scenarios. The first is the simulated SER for the model in which the relays know whether each symbol is decoded correctly or not. The second is the simulated SER for a practical scenario in which the relay forwards the decoded symbol based on comparing the received SNR with a threshold. Also the exact SER expression in (6.25) is plotted as "+." The cooperation protocol utilized is $\mathcal{C}(1)$ and the modulation scheme is QPSK.

that the cooperative diversity protocols can achieve full diversity gain in the number of cooperating terminals, which can be seen from the slopes of the performance curves, which become steeper with an increasing number of relays.   ▲

6.1.3   SER approximation

In the previous subsection, we provided exact expressions for the SER of a general cooperative scheme $\mathcal{C}(m)$, $1 \leq m \leq N - 1$, for arbitrary N-relay networks with either M-PSK or M-QAM modulation. The derived SER expressions, however, involve 2^N terms and integral functions. In this subsection, we provide approximate expressions for the SER performance of the class of cooperative diversity schemes. The approximation is derived at high SNR and yields simple expressions that can provide insights to help understand the factors that affect system performance, which helps in designing different network functions as power allocation, scheduling, routing, and node selection.

6.1.3.1   SER approximation for general cooperative protocol

One can see that any term in the exact SER formulation (6.25) in Theorem 6.1.1 consists of the product of two quantities:

- The first is

$$
F_q \left[\left(1 + \frac{b_q P_0 \sigma_{\mathrm{s,d}}^2}{\mathcal{N}_o \sin^2(\theta)} \right) \prod_{j=1}^{N} \left(1 + \frac{b_q B_{i,N}[j] P_j \sigma_{\mathrm{r}_j,\mathrm{d}}^2}{\mathcal{N}_o \sin^2(\theta)} \right) \right],   \tag{6.26}
$$

   which corresponds to the conditional SER at the destination for a given network state $\mathbf{B}_{i,N}$.

- The second is the probability of the network being in that state, and is given by $\prod_{k=1}^{N} G_k^m(B_{i,N}[k])$.

At high enough SNR, the probability of error $F_q(\cdot)$ is sufficiently small compared to 1, thus we can assume that $1 - F_q(\cdot) \simeq 1$. Hence, the only terms in the second quantity $\prod_{k=1}^{N} G_k^m(B_{i,N}[k])$ that will count are those corresponding to relays that have decoded in error. For convenience, we make the following definition: let $\Omega_i(n, m)$ denote the subset of nodes that decode correctly from node $\max(1, n - m)$ to node $n - 1$, when the network was in state $\mathbf{B}_{i,N}$. More specifically,

$$
\Omega_i(n, m) \triangleq \{ \mathrm{relay}\, j : s.t. B_{i,N}[j] = 1, \max(1, n - m) \leq j \leq n - 1. \}   \tag{6.27}
$$

Then, the SER formulation (6.25) in Theorem 6.1.1 can be approximated as

$$
P_{\mathrm{SER}}(m) \simeq \sum_{i=0}^{2^N - 1} F_q \left[\left(1 + \frac{b_q P_0 \sigma_{\mathrm{s,d}}^2}{\mathcal{N}_o \sin^2(\theta)} \right) \prod_{j=1}^{N} \left(1 + \frac{b_q B_{i,N}[j] P_j \sigma_{\mathrm{r}_j,\mathrm{d}}^2}{\mathcal{N}_o \sin^2(\theta)} \right) \right]
$$
$$
\cdot \prod_{k \in \Omega^c(N+1,N)} G_k^m(B_{i,N}[k]),   \tag{6.28}
$$

where Ω^c is the complementary set of Ω, i.e., the set of nodes that decoded erroneously.

First, we simplify the first term corresponding to the SER at the destination. Using the definition of F_q in (6.19), and ignoring all the 1's in $F_q(\cdot)$ in (6.28), the conditional SER at the destination for a given network state $\mathbf{B}_{i,N}$ can be approximated as

$$
F_q \left[\left(1 + \frac{b_q P_0 \sigma_{s,d}^2}{\mathcal{N}_o \sin^2(\theta)} \right) \prod_{j=1}^{N} \left(1 + \frac{b_q B_{i,N}[j] P_j \sigma_{r_j,d}^2}{\mathcal{N}_o \sin^2(\theta)} \right) \right]
$$
$$
\simeq \frac{\mathcal{N}_o^{1+|\Omega_i(N+1,N)|} g_q (1 + |\Omega_i(N+1,N)|)}{b_q^{1+|\Omega_i(N+1,N)|} P_0 \sigma_{s,d}^2 \prod_{j \in \Omega_i(N+1,N)} P_j \sigma_{r_j,d}^2}, \tag{6.29}
$$

where $|\Omega_i(N+1,N)|$ denotes the cardinality of the set $\Omega_i(N+1,N)$, i.e., the number of nodes that decode correctly, which also denotes the number of signal copies transmitted from the N relays to the destination at network state $\mathbf{B}_{i,N}$. The function $g_q(\cdot)$ in (6.29) is specified as

$$
g_q(x) = \begin{cases} \frac{1}{\pi} \int_0^{(M-1)\pi/M} \sin^{2x}(\theta) d\theta, & \text{for } M\text{-PSK}, q = 1 \\ \frac{4C}{\pi} \left[\int_0^{\pi/2} \sin^{2x}(\theta) d\theta - C \int_0^{\pi/4} \sin^{2x}(\theta) d\theta \right], & \text{for } M\text{-QAM}, q = 2. \end{cases} \tag{6.30}
$$

Let us write the transmitter powers allocated at the source and different relays as a ratio of the total available power P as follows: $P_0 = a_0 P$, and $P_i = a_i P$, $1 \le i \le N$, in which the power ratios are normalized as $a_0 + \sum_{i=1}^{N} a_i = 1$. One can then rewrite (6.29) in terms of the power allocation ratios as follows:

$$
F_q \left[\left(1 + \frac{b_q P_0 \sigma_{s,d}^2}{\mathcal{N}_o \sin^2(\theta)} \right) \prod_{j=1}^{N} \left(1 + \frac{b_q B_{i,N}[j] P_j \sigma_{r_j,d}^2}{\mathcal{N}_o \sin^2(\theta)} \right) \right]
$$
$$
\simeq \frac{(\mathcal{N}_o/P)^{1+|\Omega_i(N+1,N)|} g_q (1 + |\Omega_i(N+1,N)|)}{b_q^{1+|\Omega_i(N+1,N)|} a_0 \sigma_{s,d}^2 \prod_{j \in \Omega_i(N+1,N)} a_j \sigma_{r_j,d}^2}. \tag{6.31}
$$

Note that the SNR term $(\mathcal{N}_o/P)$ in (6.31) is of order $(1 + |\Omega_i(N+1,N)|)$. This is intuitively meaningful since the destination receives $(1 + |\Omega_i(N+1,N)|)$ copies of the signal, in which the term 1 is due to the copy from the source. Thus (6.31) decays as $\text{SNR}^{-(1+|\Omega_i(N+1,N)|)}$ at high SNR.

At the k-th relay, $1 \le k \le N$, the conditional SER for a given network state $\mathbf{B}_{i,N}$ can be similarly approximated as

$$
F_q \left[\left(1 + \frac{b_q P_0 \sigma_{s,r_k}^2}{\mathcal{N}_o \sin^2(\theta)} \right) \prod_{j=\max(1,k-m)}^{k-1} \left(1 + \frac{b_q B_{i,N}[j] P_j \sigma_{r_j,r_k}^2}{\mathcal{N}_o \sin^2(\theta)} \right) \right]
$$
$$
\simeq \frac{(\mathcal{N}_o/P)^{1+|\Omega_i(k,m)|} g_q (1 + |\Omega_i(k,m)|)}{b_q^{1+|\Omega_i(k,m)|} a_0 \sigma_{s,r_k}^2 \prod_{j \in \Omega_i(k,m)} a_j \sigma_{r_j,r_k}^2}, \tag{6.32}
$$

where $|\Omega_i(k,m)|$ is the number of relays that decodes correctly from the previous $\min(k-1,m)$ relays.

The SNR in the above expression is of order $1+ \mid \Omega_i(k,m) \mid$. From (6.32), the product $\prod_{k \in \Omega_i^c(N+1,N)} G_k^m(B_{i,N}[k])$ in (6.28) is given by

$$\prod_{k \in \Omega_i^c(N+1,N)} G_k^m(B_{i,N}[k]) = \prod_{k \in \Omega_i^c(N+1,N)} \frac{(\mathcal{N}_o/P)^{1+|\Omega_i(k,m)|} g_q(1+ \mid \Omega_i(k,m) \mid)}{b_q^{1+|\Omega_i(k,m)|} a_0 \sigma_{s,r_k}^2 \prod_{j \in \Omega_i(k,m)} a_j \sigma_{r_j,r_k}^2}, \tag{6.33}$$

in which the SNR is of order $\sum_{k \in \Omega_i^c(N+1,N)}(1+ \mid \Omega_i(k,m) \mid) = \mid \Omega_i^c(N+1,N) \mid + \sum_{k \in \Omega_i^c(N+1,N)} \mid \Omega_i(k,m) \mid$. Substituting (6.31) and (6.33) into (6.28), we get

$$P_{\text{SER}}(m) \simeq \sum_{i=0}^{2^N-1} \frac{(N_0/P)^{d_i} g_q(1+|\Omega_i(N+1,N)|)}{b_q^{d_i} a_0^{1+|\Omega_i^c(N+1,N)|} \sigma_{s,d}^2 \prod_{j \in \Omega_i(N+1,N)} a_j \sigma_{r_j,d}^2}$$
$$\cdot \frac{\prod_{k \in \Omega_i^c(N+1,N)} g_q(1+|\Omega_i(k,m)|)}{\prod_{k \in \Omega_i^c(N+1,N)} \sigma_{s,r_k}^2 \prod_{l \in \Omega_i(k,m)} a_l \sigma_{r_l,r_k}^2}, \tag{6.34}$$

where

$$d_i = 1+ \mid \Omega_i(N+1,N) \mid + \mid \Omega_i^c(N+1,N) \mid + \sum_{k \in \Omega_i^c(N+1,N)} \mid \Omega_i(k,m) \mid . \tag{6.35}$$

From (6.34), we can see that the SNR is of order d_i. Since $\mid \Omega_i(N+1,N) \mid + \mid \Omega_i^c(N+1,N) \mid = N$, the order d_i can be lower bounded as follows:

$$d_i = 1 + N + \sum_{k \in \Omega_i^c(N+1,N)} \mid \Omega_i(k,m) \mid \geq N+1, \tag{6.36}$$

in which the equality holds if and only if

$$\sum_{k \in \Omega_i^c(N+1,N)} \mid \Omega_i(k,m) \mid = 0. \tag{6.37}$$

Thus the smallest order of the SNR is $N + 1$.

The equality in (6.36) holds if and only if $\mid \Omega_i(k,m) \mid = 0$, for any $k \in \Omega_i^c(N+1,N)$, and $0 \leq i \leq 2^N - 1$. Essentially, this means that the equality in (6.36) is satisfied if and only if for each relay k that decodes erroneously, the m preceding relays must also have decoded erroneously. One can think of this condition as a chain rule, and this leads to the conclusion that the equality holds if and only if for each relay k that decodes in error all the previous relays must have decoded in error. As a result, the only network states that will contribute in the SER expression with terms of order $N + 1$ in the SNR are those of the form

$$\mathbf{S}_N = \mathbf{B}_{2^n-1,N}, \ 0 \leq n \leq N. \tag{6.38}$$

For example, a network state of the form $\mathbf{S}_N = [0, \dots, 0, 1, \dots, 1]$ will contribute to a term in the SER with SNR raised to the order $N + 1$, and a network state $\mathbf{S}_N = [0, \dots, 0, 1, 1, 0, 1, \dots, 1]$ will contribute to a term in the SER with SNR raised to an exponent larger than $(N + 1)$ depending on m. Therefore, only $N + 1$ states of the network have SER terms that decay as $1/\text{SNR}^{N+1}$ and the rest of the network states decay with faster rates, hence these $N + 1$ terms will dominant the SER expression at high enough SNR.

In order to write the approximate expression for the SER corresponding to these $N + 1$ terms, we need to note the following points that can be deduced from the above analysis. As described above, in order for the equality in (6.36) to hold, the following set of conditions must be satisfied:

- First, since for any relay that decodes erroneously all the previous m relays must have decoded in error, we have

$$\Omega_i(k, m) = \Phi, \tag{6.39}$$

  for all $k \in \Omega_i^c(N + 1, N)$, where Φ is the empty set.
- Second, for these $N + 1$ states that satisfy the equality in (6.36) the set $\Omega_i^c(N + 1, N)$ takes one of the following forms:

$$\Omega_i^c(N + 1, N) \in \{\Phi, \{1\}, \{1, 2\}, \dots, \{1, 2, \dots, N\}\}. \tag{6.40}$$

For example, $\Omega_i^c(N + 1, N) = \{1, 2, \dots, k\}$ denotes the state in which only the first k relays decoded erroneously. Accordingly, its cardinality, denoted by $| \Omega_i^c(N + 1, N) |$, takes one of the following values:

$$| \Omega_i^c(N + 1, N) | \in \{0, 1, 2, \dots, N\}. \tag{6.41}$$

Thus, only the $N + 1$ states determined from the above conditions will contribute to the SER expression at high SNR because they decay as $1/\text{SNR}^{N+1}$, which is the slowest decaying rate as seen from (6.36). From (6.34), (6.39), (6.40), and (6.41), the conditional SER for any of these states, e.g., $\Omega_i^c(N + 1, N) = \{1, 2, \dots, k\}$, can be determined as follows:

$$\text{SER}_k(m) = \frac{(\mathcal{N}_o/P)^{N+1} g_q(N - k + 1) g_q^k(1)}{b_q^{N+1} \sigma_{s,d}^2 a_o^{1+k} \prod_{j \in \Omega_i(N+1,N)} a_j \sigma_{r_j,d}^2 \prod_{l=1}^k \sigma_{s,r_l}^2}. \tag{6.42}$$

Summing the above expression over the $N + 1$ states in (6.40), we can further determine the approximate expression for the SER in the following theorem.

THEOREM 6.1.2 *At high enough SNR, the SER of an N relay decode-and-forward cooperative diversity network employing cooperation scheme $\mathcal{C}(m)$ and utilizing M-PSK or M-QAM modulation can be approximated by*

$$P_{\text{SER}}(m) \simeq \frac{(\mathcal{N}_o/P)^{N+1}}{b_q^{N+1} \sigma_{s,d}^2} \sum_{j=1}^{N+1} \frac{g_q(N - j + 2) g_q^{j-1}(1)}{a_0^j \prod_{i=j}^N a_i \sigma_{r_i,d}^2 \prod_{l=1}^{j-1} \sigma_{s,r_l}^2}. \tag{6.43}$$

A very important point to be noticed from the above theorem is that the approximated SER expression in (6.43) does not depend on m, the class parameter. Hence, the whole class of cooperative diversity protocols $\{\mathcal{C}(m)\}_{m=1}^{N-1}$ shares the same asymptotic performance at high enough SNR. The results obtained illustrate that utilizing the simplest scheme, namely, scenario $\mathcal{C}(1)$, results in the same asymptotic SER performance as the most sophisticated scheme, namely, $\mathcal{C}(N - 1)$. This motivates us to utilize scenario

$\mathcal{C}(1)$ as a cooperative protocol for multinode wireless networks employing decode-and-forward relaying. The simplicity behind scenario $\mathcal{C}(1)$ is due to the fact that it does not require each relay to estimate the CSI for all the previous relays as in scenario $\mathcal{C}(N-1)$. It only requires each relay to know the CSI to the previous relay and the destination, thus simplifying the channel estimation computations.

In the following, we determine roughly the savings in the computations needed for channel estimation when using scenario $\mathcal{C}(1)$ as opposed to scenario $\mathcal{C}(N-1)$ by computing the number of channels needed to be estimated in each case. The number of channels needed to be estimated in scenario $\mathcal{C}(1)$ is given by

$$n_{h,1} = 3N, \tag{6.44}$$

where N is the number of relays forwarding for the source. This value accounts for the $N+1$ channels estimated at the destination and $2N-1$ channels estimated by the N relays; the first relay estimates only one channel. In scenario $\mathcal{C}(N-1)$, the k-th relay estimates k channels, and thus the amount of computations for this case is given by

$$n_{h,N-1} = \frac{1}{2}\left[N^2 + 3N + 1\right]. \tag{6.45}$$

From (6.44) and (6.45), the savings in the computations needed for channel estimation when using scenario $\mathcal{C}(1)$ as opposed to scenario $\mathcal{C}(N-1)$ is given by

$$\frac{n_{h,1}}{n_{h,N-1}} = \frac{6N}{N^2 + 3N + 1}. \tag{6.46}$$

The above ratio approaches 0 in the limit as N tends to ∞. Hence, utilizing scenario $\mathcal{C}(1)$ will reduce the protocol complexity while having the same asymptotic performance as the best possible scenario.

6.1.3.2   Diversity order and cooperation gain

The philosophy behind employing cooperative diversity techniques in wireless networks is to form virtual MIMO systems from separated single-antenna terminals. The aim behind this is to emulate the performance gains that can be achieved in point-to-point communications when employing MIMO systems. Two well known factors that describe the performance of the system are the diversity order and coding gain of the transmit diversity scheme. To define these terms, the SER can be written in the following form:

$$P_{\text{SER}} \sim (\Delta \cdot \text{SNR})^{-d}. \tag{6.47}$$

The constant Δ which multiplies the SNR denotes the coding gain of the scheme, and the exponent d denotes the diversity order of the system. In the cooperative diversity schemes considered in this chapter, the relays simply repeat the decoded information, and thus we do not really have the notion of coding; although it can still be seen as a repetition coding scheme. Hence, we will denote the constant Δ that multiplies the SNR by the cooperation gain. From (6.43), the following observations can be deduced from the previous relation

- It is clear that the diversity order of the system is given by $d = N + 1$, which indicates that the cooperative diversity schemes described in Section 6.1.1 achieves full diversity order in the number of cooperating terminals; the source and the N relays.
- The cooperation gain of the system is given by

$$\Delta = \left[\frac{1}{b_q^{N+1} \sigma_{s,d}^2} \sum_{j=1}^{N+1} \frac{g_q (N - j + 2) g_q^{j-1}(1)}{a_0^j \prod_{i=j}^N a_i \sigma_{r_i,d}^2 \prod_{l=1}^{j-1} \sigma_{s,r_l}^2} \right]^{-1/(N+1)}. \tag{6.48}$$

Example 6.2 In this example, the performance of the presented protocols is compared against the approximate SER expressions. Without loss of generality, the channel gains are assumed to be unity and the noise variance is taken to be $N_0 = 1$. Figure 6.4 considers scenario $\mathcal{C}(1)$ and depicts the SER performance versus P/N_0 for QPSK signalling. The transmitting power P is fixed for different number of cooperating relays in the network. The results reveal that the derived approximations for the SER are tight at high enough SNR. Regarding scenario $\mathcal{C}(N - 1)$, we considered the $N = 3$ relays case. Figure 6.5 depicts the SER performance for QPSK and 16-QAM modulation. The results for scenario $\mathcal{C}(1)$ under the same simulation setup are included for comparison. It can be seen from the results that there is a very small gap between the SER performance of scenarios $\mathcal{C}(1)$ and $\mathcal{C}(N - 1)$, and that they almost merge together at high enough SNR. This confirms our observations that utilizing scenario $\mathcal{C}(1)$ can deliver the required SER performance for a fairly wide range of SNR. Hence, saving a lot in terms of channel estimation, thus computational complexity, requirements to implement the protocol. ▲

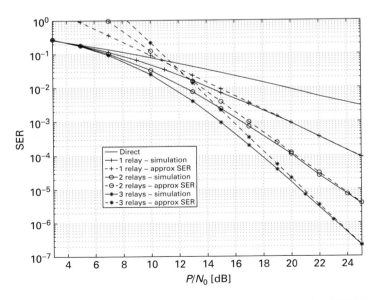

Fig. 6.4   Comparison between the approx. SER in (6.43), and the simulated SER for different numbers of relays. The cooperation protocol utilized is $\mathcal{C}(1)$ and the modulation scheme is QPSK.

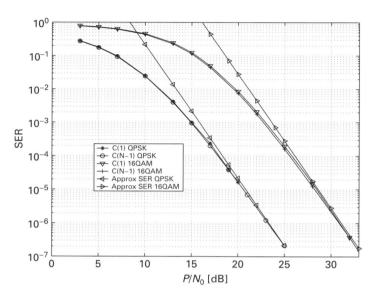

Fig. 6.5   Comparison between the performance of schemes $\mathcal{C}(1)$ and $\mathcal{C}(N-1)$ for both QPSK and 16-QAM modulation, $N = 3$.

6.1.3.3   Bandwidth efficiency versus diversity gain

Up to this point, we did not take into account the bandwidth (BW) efficiency as another important factor to determine the performance besides the SER. Increasing the number of relays reduces the BW efficiency of the system, as the source uses only a fraction of the total available degrees of freedom to transmit the information. There is a tradeoff between the diversity gain and the BW efficiency of the system, as higher diversity gain is usually translated into utilizing the available degrees of freedom to transmit more copies of the same message, which reduces the BW efficiency of the system. In order to have a fair comparison, we will fix the BW efficiency throughout the simulations. In order to achieve this, larger signal constellations are utilized with larger number of cooperating relays. For the direct transmission case, BPSK is used as a benchmark to achieve a bandwidth efficiency of 1 bit/channel use. QPSK is used with the $N = 1$ relay case, 8PSK with $N = 2$ relays and 16-QAM with $N = 3$ relays. In all of the aforementioned cases, the achieved BW efficiency is 1 bit/channel use. Figure 6.6 depicts the BER versus SNR per bit in dB for $N = 1, 2, 3$ relays along with the direct transmission case. The results reveal that, at low SNR, a lower number of nodes delivers a better performance due to the BW efficiency loss incurred through utilizing a larger number of cooperating nodes.

Another important point of concern is how the performance of cooperative diversity compares to that of time diversity without relaying under the same bandwidth efficiency. For example, if the target diversity gain is $N + 1$ then cooperation requires the employment of N relays, while in time diversity the source simply repeats the information

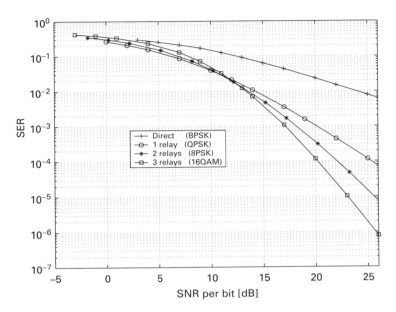

Fig. 6.6 BER performance comparison between different numbers of cooperating relays taking into account the BW efficiency, $\mathcal{C}(1)$.

for $N + 1$ successive time slots. Two factors can lead to cooperation yielding better performance than time diversity:

- The first is that the cooperation gain of cooperative diversity (6.48) can be considerably higher than that of time diversity if the propagation path loss is taken into account. This is because the relay nodes are usually closer to the destination node than the source itself, which results in a lower propagation path loss in the relay–destination links compared to the source–destination link. This is a natural gain offered by cooperation because of the distributed natural of the formed virtual array, and this is the same reason multihop communications offer more energy efficient transmission in general.
- The second factor that can lead to cooperation being a more attractive scheme than time diversity is that the spatial links between different nodes in the network fade independently, again because of the distributed nature of the formed virtual array, which leads to full diversity gain. In time diversity, however, full diversity gain is not guaranteed as there might be time correlation between successive time slots. This correlation is well modeled by a first-order Markov chain [226].

To illustrate these further we compare the SER performance of time and cooperative diversity in Figure 6.7. The desired diversity gain is 3. The time correlation factor for the first-order Markov model is taken to be equal to $\rho = 0.9, 0.7, 0.3, 0.1$. The two relays are taken in different positions as illustrated in the figure to illustrate different coding gains. It is clear from Figure 6.7 that cooperative diversity can offer better performance

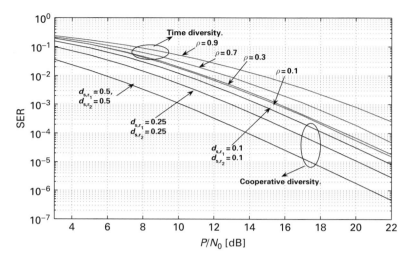

Fig. 6.7   Comparison between the SER performance of time diversity without any relaying and cooperative diversity. Two relays are utilized for cooperation and three time slots for time diversity. The first-order Markov model is utilized to account for time correlation, and the positions of different relays are depicted with the source–destination distance normalized.

than time diversity because of the higher possible coding gain that depends on the relay positions, and the degradation in the achieved performance of time diversity due to the correlation factor ρ.

6.1.4   Optimal power allocation

In this subsection, we discuss an optimal power allocation strategy for the multi-node cooperative scenarios considered in the previous subsections. The approximated SER formula derived in (6.43) is a function of the power allocated at the source and the N relays. For a fixed transmission power budget P, the power should be allocated optimally at the different nodes in order to minimize the SER.

Since the approximation in (6.43) is tight at high enough SNR, we use it to determine the asymptotic optimum power allocation. We also drop the parameter m since the asymptotic SER performance is independent of it. The SER can be written in terms of the power ratios allocated at the transmitting nodes as follows:

$$P_{\text{SER}} \simeq \left(\frac{P}{N_0}\right)^{-(N+1)} \frac{1}{b_q^{N+1}\sigma_{s,d}^2} \sum_{j=1}^{N+1} \frac{g_q(N-j+2)g_q^{j-1}(1)}{a_0^j \prod_{i=j}^N a_i \sigma_{r_i,d}^2 \prod_{k=1}^{j-1} \sigma_{s,r_k}^2}. \qquad (6.49)$$

The nonlinear optimization problem can be formulated as follows:

$$\mathbf{a}_{\text{opt}} = \arg \min_{\mathbf{a}} P_{\text{SER}} \qquad (6.50)$$

$$\text{s.t.} \quad a_i \geq 0 \ (0 \leq i \leq N), \quad \textstyle\sum_{i=0}^N a_i = 1,$$

where $\mathbf{a} = [a_0, a_1, \ldots, a_N]$ is the power allocating vector. The Lagrangian of this problem can be written as

$$L = P_{\text{SER}} + v \left(\sum_{i=0}^{N} a_i - 1 \right) - \sum_{j=0}^{N} \beta_j a_j \tag{6.51}$$

where the β's act as slack variables.

Although this nonlinear optimization problem should, in general, be solved numerically, there are some insights which can be drawn out of it. Applying first-order optimality conditions, we can show that the optimum power allocation vector $\mathbf{a}_{\text{opt}}$ must satisfy the following necessary conditions:

$$\frac{\partial P_{\text{SER}}}{\partial a_i} = \frac{\partial P_{\text{SER}}}{\partial a_j}, \quad i, j \in \{0, 1, 2, \ldots, N\}. \tag{6.52}$$

Next, we solve these equations simultaneously to get the relations between the optimal power allocations at different nodes. To simplify the notations, let μ_j denote the constant quantity inside the summation in (6.49), i.e.,

$$\mu_j = \left(\frac{P}{N_0} \right)^{-(N+1)} \frac{g_q (N - j + 2) g_q^{j-1}(1)}{b_q^{N+1} \sigma_{\text{s,d}}^2 \prod_{i=j}^{N} \sigma_{r_i,d}^2 \prod_{k=1}^{j-1} \sigma_{\text{s},r_k}^2}. \tag{6.53}$$

The derivative of the SER with respect to a_0 is given by

$$\frac{\partial P_{\text{SER}}}{\partial a_0} = \sum_{j=1}^{N+1} \frac{-j \mu_j}{a_0^{j+1} \prod_{i=j}^{N} a_i}, \tag{6.54}$$

while the derivative with respect to a_k, $1 \le k \le N$, is given by

$$\frac{\partial P_{\text{SER}}}{\partial a_k} = \sum_{j=1}^{k} \frac{-\mu_j}{a_0^j a_k \prod_{i=j}^{N} a_i}, \tag{6.55}$$

where the summation is to the k-th term only as a_k does not appear in the terms from $k + 1$ to N. Using (6.52), we equate the derivatives of the SER with respect to any two consecutive variables a_k and a_{k+1}, $1 \le k \le N - 1$, as follows:

$$\sum_{j=1}^{k} \frac{-\mu_j}{a_0^j a_k \prod_{i=j}^{N} a_i} = \sum_{j=1}^{k+1} \frac{-\mu_j}{a_0^j a_{k+1} \prod_{i=j}^{N} a_i}. \tag{6.56}$$

Rearranging the terms in the above equation we get

$$\frac{a_{k+1} - a_k}{a_k a_{k+1}} \sum_{j=1}^{k} \frac{\mu_j}{a_0^j \prod_{i=j}^{N} a_i} = \frac{\mu_{k+1}}{a_0^j a_{k+1} \prod_{i=k+1}^{N} a_i}. \tag{6.57}$$

Since both sides of the above equation are positive, we conclude that $a_{k+1} \ge a_k$ for any $k = 1, 2, \ldots, N$. Similarly, we can show that $a_o \ge a_k$ for all $1 \le k \le N$. Hence, solving the optimality conditions simultaneously we get the following relationships between the powers allocated at different nodes:

$$P_0 \ge P_N \ge P_{N-1} \ge \cdots \ge P_1. \tag{6.58}$$

The above set of inequalities demonstrates an important concept: power is allocated at different nodes according to the received signal quality at the node. We refer to the quality of the signal copy at a node as the reliability of the node, thus the more reliable the node the greater the power that is allocated to it. To further illustrate this concept, the $N + 1$ cooperative nodes form a virtual $(N + 1) \times 1$ MIMO system. The difference between this virtual array and a conventional point-to-point MIMO system is that in conventional point-to-point communications all the antenna elements at the transmitter are allocated at the same place and hence all the antenna elements can acquire the original signal. In a virtual array, the antenna elements constituting the array (the cooperating nodes) are not allocated at the same place and the channels among them are noisy. The source is the most reliable node as it has the original copy of the signal and thus it should be allocated the highest share of the power.

According to the cooperation protocol presented in this chapter, each relay combines the signal received from the source and the previous relays. As a result, each relay is more reliable than the previous relay, and hence the N-th relay is the most reliable node and is allocated the largest ratio of the power after the source, while the first relay is the least reliable and is allocated the smallest ratio of the transmitted power. Another important point to notice is that the channel quality of the direct link between the source and the destination $\sigma_{s,d}^2$ is a common factor in the μ_j's that appear in (6.57), hence the optimal power allocation does not depend on it.

To illustrate the effect of relay position on the values of the optimal power allocation ratios at the source and relay nodes, we consider a two-relay scenario in Figure 6.8. The two relays are taken in three different positions, close to the source, close to the destination, and in the middle between the source and the destination. In the first scenario, almost equal power allocation between the three nodes is optimal. When the relays are closer to the destination, more power is allocated to the source node, but still the second relay has a higher portion of the power relative to the first one. Similarly, in the last scenario the last relay has more power than the first one. These results show that the further

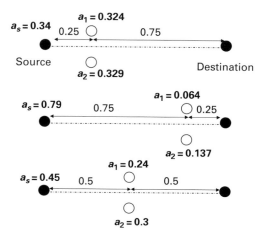

Fig. 6.8   Optimal power allocation for $N = 2$ relays under different relays positions.

the relays are from the source node the less the power that is allocated to the relays as they become less reliable, while as the relays become closer to the source, equal power allocation becomes near optimal.

There are a few special cases of practical interest that permit a closed-form solution for the optimization problem in (6.50), and they are discussed in the following examples

---

Example 6.3 For the $N = 1$ relay scenario, the optimization problem in (6.50) admits closed-form expression. The SER for this case is simply given as

$$P_{\text{SER}} = \left(\frac{P}{N_0}\right)^{-2} \frac{1}{b_q^2 \sigma_{s,d}^2} \left(\frac{g_q(2)}{a_0 a_1 \sigma_{r_1,d}^2} + \frac{g_q^2(1)}{a_0^2 \sigma_{s,r_1}^2}\right). \tag{6.59}$$

Solving the optimization problem for this case leads to the following solution for the optimal power allocation:

$$a_0 = \frac{\sigma_{s,r_1} + \sqrt{\sigma_{s,r_1}^2 + 8\frac{g_q^2(1)}{g_q(2)}\sigma_{r_1,d}^2}}{3\sigma_{s,r_1} + \sqrt{\sigma_{s,r_1}^2 + 8\frac{g_q^2(1)}{g_q(2)}\sigma_{r_1,d}^2}},$$

$$a_1 = \frac{2\sigma_{s,r_1}}{3\sigma_{s,r_1} + \sqrt{\sigma_{s,r_1}^2 + 8\frac{g_q^2(1)}{g_q(2)}\sigma_{r_1,d}^2}}. \tag{6.60}$$

▲

To study the effect of relay position on the optimal power allocation, we depict in Figure 6.9 the SER performance of a single relay scenario versus the power allocation

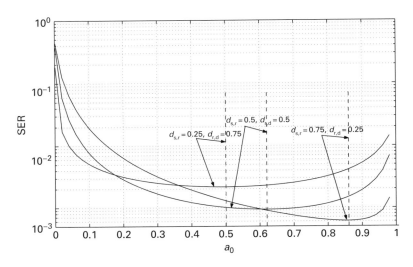

Fig. 6.9   SER versus power allocation ratio at the source node for different relay positions.

at the source node a_0 for different relay positions. The first observation that the figure reveals is that the SER performance is relatively flat around equal power allocation when the relay is not very close to the destination. Another observation to notice here is that as the relay becomes closer to the destination, the value of the optimal power allocation at the source node a_0 approaches 1, which means that as the relay node becomes less reliable more power should be allocated to the source node.

Example 6.4   The propagation path-loss and networks with linear topologies will be taken into account here. The channel attenuation between any two nodes $\sigma_{i,j}^2$ depends on the distance between these two nodes $d_{i,j}$ as follows: $\sigma_{i,j}^2 \propto d^{-\alpha}$, where α is the propagation constant. For a linear network topology, the most significant channel gains are for the channels between the source and the first relay σ_{s,r_1}^2, and between the last relay and the destination $\sigma_{r_N,d}^2$, the other channel gains are considerably smaller than these two channels. In the SER expression in (6.43), these two terms appear as a product in all the terms except the first and the last. Hence these two terms dominate the SER expression, and we can further approximate the SER in this case as follows:

$$P_{SER} \simeq \frac{(\mathcal{N}_o/P)^{N+1}}{b_q^{N+1}\sigma_{s,d}^2} \left[\frac{g_q(N+1)}{a_0 \prod_{i=1}^{N} a_i \sigma_{r_i,d}^2} + \frac{g_q^{N+1}(1)}{a_0^{N+1} \prod_{i=1}^{N} \sigma_{s,r_i}^2} \right]. \qquad (6.61)$$

Taking the power constraint into consideration, the Lagrangian of the above problem can be written as

$$L(\mathbf{a}) = \frac{\mu_1}{a_o \prod_{i=1}^{N} a_i} + \frac{\mu_{N+1}}{a_o^{N+1}} + \lambda \left(\sum_{i=0}^{N} a_i - 1 \right). \qquad (6.62)$$

Taking the partial derivatives of the Lagrangian with respect to a_j, $1 \leq j \leq N$, and equating with 0, we get

$$a_j = \frac{\mu_1}{\lambda a_o \prod_{i=1}^{N} a_i}. \qquad (6.63)$$

Thus, we deduce that the power allocated to all of the relays is the same. Let the constant κ be defined as follows:

$$\kappa = \frac{P_0 - P_j}{P_j}. \qquad (6.64)$$

From the above definition, along with the power constraint, we get

$$P_j = \frac{1}{1+\kappa+N} P,$$

$$P_0 = \frac{1+\kappa}{1+\kappa+N} P. \qquad (6.65)$$

To find the optimum value for κ, substitute (6.65) into the expression for the SER in (6.61) to get

$$P_{SER} \simeq \frac{\mu_1(1+\kappa)^N(1+\kappa+N)^{N+1} + \mu_{N+1}(1+\kappa+N)^{N+1}}{(1+\kappa)^{N+1}}. \qquad (6.66)$$

Differentiating (6.66) and equating to 0, we can find that the optimum κ satisfies the equation $\kappa(1 + \kappa)^N = A$, in which A is a constant given by

$$A = (N + 1)\frac{g_q^{N+1}(1)\prod_{i=1}^{N}\sigma_{r_i,d}^2}{g_q(N+1)\prod_{i=1}^{N}\sigma_{s,r_i}^2}. \tag{6.67}$$

▲

From the above analysis, the optimal power allocation for a linear network can be found in the following theorem.

THEOREM 6.1.3 *The optimal power allocation for a linear network that minimizes the SER expression in (6.61) is as follows:*

$$P_0 = \frac{1 + \kappa}{1 + \kappa + N}P, \qquad P_i = \frac{1}{1 + \kappa + N}P, \quad 1 \le i \le N, \tag{6.68}$$

where κ is found through solving the equation $\kappa(1 + \kappa)^N = A$, in which A is a constant given by $(N + 1)\frac{g^{N+1}(1)\prod_{i=1}^{N}\sigma_{r_i,d}^2}{g(N+1)\prod_{i=1}^{N}\sigma_{s,r_i}^2}$.

The above theorem agrees with the optimality conditions we found for the general problem in (6.58). Also, it shows an interesting property that in linear network topologies equal power allocation at the relays is asymptotically optimal.

Example 6.5 The cooperating relays can be chosen to be closer to the source than to the destination, in order for the $N + 1$ cooperating nodes to mimic a multi-input-single-output (MISO) transmit antenna diversity system. This case is of special interest because decode-and-forward relaying can be a capacity achieving scheme when the relays are taken to be closer to the source and it has the best performance compared to amplify-and-forward and compress-and-forward relaying in this case. In order to model this scenario in our SER formulation, we will consider the channel gains from the source to the relays to have higher gains than those from the relays to the destination, i.e., $\sigma_{s,r_i}^2 \gg \sigma_{r_i,d}^2$ for $1 \le i \le N$. Taking this into account, the approximate SER expression in (6.43) can be further approximated as

$$P_{\text{SER}} \simeq \frac{N_0^{N+1}g_q(N+1)}{b_q^{N+1}\sigma_{s,d}^2 P^{N+1}a_0\prod_{i=1}^{N}a_i\sigma_{r_i,d}^2}. \tag{6.69}$$

It is clear from the above equation that the SER depends equally on the power allocated to all nodes including the source, and thus the optimal power allocation strategy for this case is simply given by

$$P_0 = P_i = \frac{P}{N + 1}, \qquad 1 \le i \le N. \tag{6.70}$$

This result is intuitively meaningful as all the relays are located near to the source and thus they all have high reliability and are allocated equal power as if they form a conventional antenna array.   ▲

Example 6.6   Now we consider the opposite scenario in which all the relays are located near the destination. In this case the channels between the relays and the destination are of a higher quality, higher gain, than those between the source and the relays, i.e., $\sigma_{r_i,d}^2 \gg \sigma_{s,r_i}^2$ for $1 \leq i \leq N$. In this case the SER can be approximated as

$$P_{\text{SER}} \simeq \frac{N_0^{N+1} g(1)^{N+1}}{b_q^{N+1} \sigma_{s,d}^2 P_0^{N+1} \prod_{k=1}^{N} \sigma_{s,r_k}^2}. \tag{6.71}$$

The SER in the above equation is not a function of the power allocated at the cooperating relays, and thus the optimal power allocation in this case is simply $P_0 = P$, i.e., allocating all the available power at the source. This result is very interesting as it reveals a very important concepts: If the relays are located closer to the destination than to the transmitter then direct transmission can lead better performance than decode-and-forward relaying. This is also consistent with the results in [99] in which it was shown that the performance of the decode-and-forward strategy degrades significantly when the relays get closer to the destination. This result can be intuitively interpreted as follows: the further the relays are from the source the more noisy the channels between them and the less reliable the signals received by those relays, to the extent that we can not rely on them on forwarding copies of the signal to the destination.   ▲

6.1.4.1   Numerical examples

In this subsection, we present some numerical results to verify the analytical results for the optimal power allocation problem for the considered network topologies. The effect of the geometry on the channel links qualities is taken into consideration. We assume that the channel variance between any two nodes is proportional to the distance between them, more specifically $\sigma_{i,j}^2 \propto d_{i,j}^{-\alpha}$, where α is determined by the propagation environment is taken equal to 4 throughout our simulations. We provide comparisons between the optimal power allocation via exhaustive search to minimize the SER expression in (6.43), and optimal power allocation provided by the closed-form expressions provided in this section.

First, for the linear network topology, we consider a uniform linear network, i.e., $d_{s,r_1} = d_{r_1,r_2} = \cdots = d_{r_N,d}$. The variance of the direct link between the source and the destination is taken to be $\sigma_{s,d}^2 = 1$. Table 6.1 demonstrates the results for $N = 3$ relays. Second, for the case when all the relays are near the source, the channel links are taken to be: $\sigma_{s,r_i}^2 = \sigma_{r_i,r_j}^2 = 10$, while $\sigma_{s,d}^2 = \sigma_{r_i,d}^2 = 0.1$. Finally, for the case when all of the relays are near the destination, the channel link qualities are taken to be: $\sigma_{s,d}^2 = \sigma_{s,r_i}^2 = 0.1$, while $\sigma_{r_i,r_j}^2 = \sigma_{r_i,d}^2 = 10$. Table 6.2 illustrates the results for $N = 3$ relays for the two previous cases. In all of the provided numerical examples it is

Table 6.1 Comparison between optimal power allocation via exhaustive search and analytical results. $N=3$ relays, uniform network topology.

Exhaustive search	Analytical results
$P_0 = 0.31P$	$P_0 = 0.31P$
$P_1 = 0.23P$	$P_1 = 0.23P$
$P_2 = 0.23P$	$P_2 = 0.23P$
$P_3 = 0.23P$	$P_3 = 0.23P$

Table 6.2 Comparison between optimal power allocation via exhaustive search and analytical results. $N=3$ relays: (a) all relays near the source; (b) all relays near the destination.

(a) Exhaustive search	Analytical results	(b) Exhaustive search	Analytical results
$P_0 = 0.25P$	$P_0 = 0.25P$	$P_0 = 0.875P$	$P_0 = P$
$P_1 = 0.25P$	$P_1 = 0.25P$	$P_1 = 0.015P$	$P_1 = 0$
$P_2 = 0.25P$	$P_2 = 0.25P$	$P_2 = 0.035P$	$P_2 = 0$
$P_3 = 0.25P$	$P_3 = 0.25P$	$P_3 = 0.075P$	$P_3 = 0$

clear that the optimal power allocations obtained via exhaustive search agree with that via analytical results for all the considered scenarios. In addition, the numerical results show that the optimal power allocation obtained via exhaustive search has the same ordering as the one we got in (6.58).

6.2 Multi-node amplify-and-forward protocol

In the second part of this chapter, we focus on a multi-node amplify-and-forward strategy. Different from the decode-and-forward protocol, an amplify-and-forward protocol does not suffer from the error propagation problem because the relays do not perform any hard-decision operation on the received signal. We first describe the multi-node amplify-and-forward protocol in detail and then analyze its SER performance. Finally, we consider the outage probability of the multi-node amplify-and-forward protocol, as well as the mutual information of the system.

The multi-node amplify-and-forward system model is shown in Figure 6.10. In the following, we will consider two relaying strategies. In the first scenario, each relay forwards only the source's signal to the the destination, while in the second scenario each relay forwards a combined signal from the source and previous relays.

6.2.1 Source-only amplify-and-forward relaying

6.2.1.1 System model and protocol description

We first consider a source-only multi-node amplify-and-forward relaying strategy, i.e., each relay forwards only the source's signal to the the destination, in which cooperation

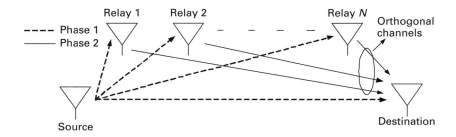

Fig. 6.10   Multi-node amplify-and-forward system model.

is done in two phases. In phase 1, the source broadcasts its information to the destination and N relay nodes. The received signals $y_{s,d}$ and y_{s,r_i} at the destination and the i-th relay can be written, respectively, as

$$y_{s,d} = \sqrt{P_0} h_{s,d} x + n_{s,d}, \tag{6.72}$$

$$y_{s,r_i} = \sqrt{P_0} h_{s,r_i} x + n_{s,r_i}, \tag{6.73}$$

for $i = 1, 2, \ldots, N$, in which P_0 is the transmitted source power, $n_{s,d}$ and n_{s,r_i} denote the additive white Gaussian noise at the destination and the i-th relay, respectively, and $h_{s,d}$ and h_{s,r_i} are the channel coefficients from the source to the destination and the i-th relay node, respectively. Each relay amplifies the received signal from the source and re-transmits it to the destination. The received signal at the destination node in phase 2 due to the i-th relay transmission is given as

$$y_{r_i,d} = \frac{\sqrt{P_i}}{\sqrt{P_0 |h_{s,r_i}|^2 + N_0}} h_{r_i,d} y_{s,r_i} + n_{r_i,d}, \tag{6.74}$$

where P_i is the i-th relay node power. The channel coefficients $h_{s,d}$, h_{s,r_i}, and $h_{r_i,d}$ are modeled as zero-mean, complex Gaussian random variables with variances $\sigma_{s,d}^2$, σ_{s,r_i}^2, and $\sigma_{r_i,d}^2$, respectively, at the receiving nodes but not at the the transmitting nodes. The noise terms are modeled as zero-mean, complex Gaussian random variables with variance N_0. Jointly combining the signal received from the source in phase 1 and those from the relays in phase 2, the destination detects the transmitted symbols by use of maximum-ratio combining.

6.2.1.2   SER performance analysis

In the following, we derive a closed form SER expression for the source-only amplify-and-forward cooperation protocol with M-PSK and square M-QAM signals. With the knowledge of the channel state information, the output of the MRC detector at the destination can be written as

$$y_d = \alpha_s y_{s,d} + \sum_{i=1}^{N} \alpha_i y_{r_i,d}, \tag{6.75}$$

where $\alpha_s = \sqrt{P_0} h_{s,d}^* / N_0$ and

$$\alpha_i = \frac{\sqrt{\frac{P_0 P_i}{P_0 |h_{s,r_i}|^2 + N_0}} h_{s,r_i}^* h_{r_i,d}^*}{\left(\frac{P_i |h_{r_i,d}|^2}{P_0 |h_{s,r_i}|^2 + N_0} + 1 \right) N_0}.$$

If we assume that the transmitted symbol x has an average energy of 1, then the SNR at the MRC detector output is

$$\text{SNR}_d = \gamma_s + \sum_{i=1}^{N} \gamma_i \tag{6.76}$$

where $\gamma_s = P_0 |h_{s,d}|^2 / N_0$, and

$$\gamma_i = \frac{1}{N_0} \frac{P_0 P_i |h_{s,r_i}|^2 |h_{r_i,d}|^2}{P_0 |h_{s,r_i}|^2 + P_i |h_{r_i,d}|^2 + N_0}. \tag{6.77}$$

The instantaneous SNR γ_i can be tightly upper bounded as

$$\tilde{\gamma}_i = \frac{1}{N_0} \frac{P_0 P_i |h_{s,r_i}|^2 |h_{r_i,d}|^2}{P_0 |h_{s,r_i}|^2 + P_i |h_{r_i,d}|^2}, \tag{6.78}$$

which is the harmonic mean of $P_0 |h_{s,r_i}|^2 / N_0$ and $P_i |h_{r_i,d}|^2 / N_0$. The SER of M-PSK and M-QAM conditional on the channel state information are defined before in (6.11) and (6.12), respectively.

Let us denote the MGF of a random variable Z as

$$M_Z(s) = \int_{-\infty}^{\infty} \exp(-sz) P_Z(z) dZ. \tag{6.79}$$

Averaging the conditional SER over the Rayleigh fading channels, the SER of the M-PSK signals and M-QAM signals can be given, respectively, as

$$P_{\text{SER}} \approx \frac{1}{\pi} \int_0^{(M-1)\pi/M} M_{\gamma_s}\left(\frac{b_{\text{PSK}}}{\sin^2 \theta}\right) \prod_{i=1}^{N} M_{\tilde{\gamma}_i}\left(\frac{b_{\text{PSK}}}{\sin^2 \theta}\right) d\theta, \tag{6.80}$$

$$P_{\text{SER}} \approx \frac{4C}{\pi} \int_0^{\pi/2} M_{\gamma_s}\left(\frac{b_{\text{QAM}}}{2\sin^2 \theta}\right) \prod_{i=1}^{N} M_{\tilde{\gamma}_i}\left(\frac{b_{\text{QAM}}}{2\sin^2 \theta}\right) d\theta$$

$$- \frac{4C^2}{\pi} \int_0^{\pi/4} M_{\gamma_s}\left(\frac{b_{\text{QAM}}}{2\sin^2 \theta}\right) \prod_{i=1}^{N} M_{\tilde{\gamma}_i}\left(\frac{b_{\text{QAM}}}{2\sin^2 \theta}\right) d\theta, \tag{6.81}$$

in which we use the SNR approximation $\tilde{\gamma}_i$ (6.78) instead of γ_i.

The MGF of γ_s, which is an exponential random variable, can be simply given by

$$M_{\gamma_s} = \frac{1}{1 + \frac{s P_0 \sigma_{s,d}^2}{N_0}}. \tag{6.82}$$

The challenge is how to get the MGF of $\tilde{\gamma}_i$. We follow the same approach of analyzing the SER performance of the single-relay amplify-and-forward protocol as presented in the previous chapter. The approach relies on a key result as follows. If X_1 and X_2 are two independent exponential random variables with parameters β_1 and β_2 respectively, and $Z = \frac{X_1 X_2}{X_1 + X_2}$ is the harmonic mean of X_1 and X_2, then the MGF of Z is

$$
\begin{aligned}
M_Z(s) &= \frac{(\beta_1 - \beta_2)^2 + (\beta_1 + \beta_2)s}{\Delta^2} \\
&\quad + \frac{2\beta_1\beta_2 s}{\Delta^3} \ln \frac{(\beta_1 + \beta_2 + s + \Delta)^2}{4\beta_1\beta_2},
\end{aligned}
\tag{6.83}
$$

where

$$
\Delta = \sqrt{(\beta_1 - \beta_2)^2 + 2(\beta_1 + \beta_2)s + s^2}.
$$

With $\beta_1 = N_0/P_0\sigma_{s,r_i}^2$ and $\beta_2 = N_0/P_i\sigma_{r_i,d}^2$, at high SNR, for any relay both β_1 and β_2 go to zero, and Δ goes to s. Thus, The MGF in (6.83) can be approximated as

$$
M_Z(s) \approx \frac{\beta_1 + \beta_2}{s} + \frac{2\beta_1\beta_2}{s^2} \ln \frac{s^2}{\beta_1\beta_2}.
\tag{6.84}
$$

At enough high SNR, the MGF can be further simplified as

$$
M_Z(s) \approx \frac{\beta_1 + \beta_2}{s}.
\tag{6.85}
$$

Substituting the above MGF approximation into (6.80), we arrive at the following result.

THEOREM 6.2.1   *At enough high SNR, the SER of the source-only amplify-and-forward cooperative protocol with N relay nodes employing M-PSK or M-QAM signals can be approximated as*

$$
P_{\text{SER}} \approx \frac{g(N)N_0^{N+1}}{b^{N+1}} \cdot \frac{1}{P_0\sigma_{s,d}^2} \prod_{i=1}^N \frac{P_0\sigma_{s,r_i}^2 + P_i\sigma_{r_i,d}^2}{P_0 P_i \sigma_{s,r_i}^2 \sigma_{r_i,d}^2},
\tag{6.86}
$$

where, in the case of M-PSK signals, $b = b_{\text{PSK}}$ and

$$
g(N) = \frac{1}{\pi} \int_0^{(M-1)\pi/M} \sin^{2(N+1)} \theta \, d\theta;
\tag{6.87}
$$

while, in the case of M-QAM signals, $b = b_{\text{QAM}}/2$ and

$$
g(N) = \frac{4C}{\pi} \int_0^{\pi/2} \sin^{2(N+1)} \theta \, d\theta - \frac{4C^2}{\pi} \int_0^{\pi/4} \sin^{2(N+1)} \theta \, d\theta.
\tag{6.88}
$$

It is clear that the source-only protocol achieves full diversity of order $N + 1$. We show the tightness of the SER approximated expression derived in the above theorem in the following example.

Example 6.7   Figures 6.11–6.13 show the performance of the relaying protocol with QPSK (4-QAM) modulation for one, two, and three relay nodes and various channel conditions, respectively. An equal power allocation between the source and the relay

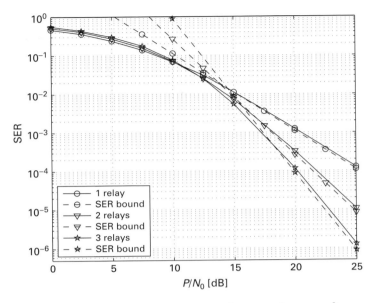

Fig. 6.11   SER performance for QPSK constellation ($\sigma_{s,d}^2 = 1$, $\sigma_{s,r_i}^2 = 1$, $\sigma_{r_i,d}^2 = 1$, equal power).

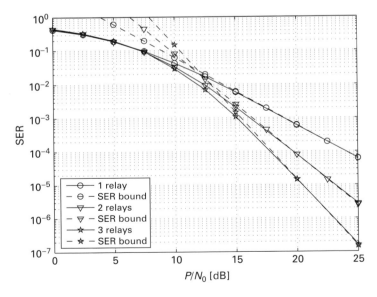

Fig. 6.12   SER performance for QPSK constellation with relays close to the source ($\sigma_{s,d}^2 = 1$, $\sigma_{s,r_i}^2 = 10$, $\sigma_{r_i,d}^2 = 1$, equal power).

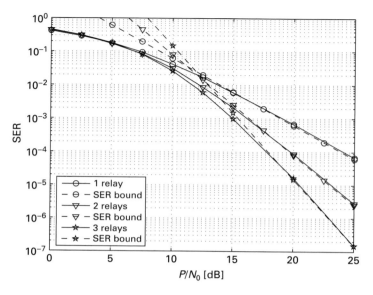

Fig. 6.13   SER performance for QPSK constellation with relays close to the destination
($\sigma_{s,d}^2 = 1$, $\sigma_{s,r_i}^2 = 1$, $\sigma_{r_i,d}^2 = 10$, equal power).

nodes is assumed. From the figure, it is clear that the derived SER approximation is
tight at high SNR for QPSK constellation.   ▲

6.2.2   MRC-based amplify-and-forward relaying

6.2.2.1   System model and protocol description

In this subsection, we address the MRC-based multi-node amplify-and-forward relaying
strategy. For simplicity of presentation, we will focus on the scenario of two relay nodes,
but the extension to N relay nodes is straightforward. With two relaying nodes, the
protocol is implemented in three phases as follows:

- In phase 1, the source broadcasts its information to the destination and the two relay
  nodes.
- In phase 2, the first relay helps the source by amplifying the source signal and sending
  it to the destination and the second relay.
- In phase 3, the second relay applies an MRC detector on the two received signals from
  the previous two phases and forwards the amplified MRC signal to the destination.

In general, we have $N + 1$ phases for a system with N relay nodes, in which each relay
applies an MRC detector on the signals that it receives from the source and all previous
relays. The MRC detector has the advantage of maximizing the SNR at the output of
the detector under the condition that the noise components of the combined signals are
uncorrelated.

Let us focus on the two-relay scenario. The received signals at the destination and the
relays in phase 1 are the same as those in the source-only amplify-and-forward scenario.

In phase 2, the received signal at the destination due to the first relay transmission is also the same as in the source-only scenario. However, the received signal at the second relay due to the first relay transmission is given as

$$y_{r_1,r_2} = \frac{\sqrt{P_1}}{\sqrt{P_0|h_{s,r_1}|^2 + N_0}} h_{r_1,r_2} y_{s,r_1} + n_{r_1,r_2}, \tag{6.89}$$

where the inter-relay channel, h_{r_1,r_2}, is modeled as a zero-mean, complex Gaussian random variable with variance $\sigma^2_{r_1,r_2}$. In phase 3, the second relay applies an MRC detector on the signals received in phases 1 and 2, and the output of the MRC detector can be written as

$$\tilde{y} = \tilde{\alpha}_s y_{s,r_2} + \tilde{\alpha}_1 y_{r_1,r_2}, \tag{6.90}$$

where $\tilde{\alpha}_s = \sqrt{P_0} h^*_{s,r_2}/N_0$ and

$$\tilde{\alpha}_1 = \frac{\sqrt{\frac{P_0 P_1}{P_0|h_{s,r_1}|^2 + N_0}} h^*_{s,r_1} h^*_{r_1,r_2}}{\left(\frac{P_1|h_{r_1,r_2}|^2}{P_0|h_{s,r_1}|^2 + N_0} + 1\right) N_0}.$$

The received signal at the destination in phase 3 is given by

$$y_{r_2,d} = \sqrt{P_2} h_{r_2,d} \frac{\tilde{y}}{\sqrt{K^2 + K}} + n_{r_2,d}, \tag{6.91}$$

where

$$K = \frac{\frac{P_0 P_1}{P_0|h_{s,r_1}|^2 + N_0} |h_{s,r_1}|^2 |h_{r_1,r_2}|^2}{\left(\frac{P_1|h_{r_1,r_2}|^2}{P_0|h_{s,r_1}|^2 + N_0} + 1\right) N_0} + \frac{P_0|h_{s,r_2}|^2}{N_0}. \tag{6.92}$$

Finally, the destination applies an MRC detector on the signals that it receives from all phases and jointly detects the information from the source.

6.2.2.2 Performance analysis

In this subsection, we try to gain some insight into the performance of the MRC-based amplify-and-forward protocol. We again focus on the two-relay scenario. The analysis can be similarly extended to the case of more relay nodes. With the knowledge of the channel state information, the output of the MRC detector in the two-relay case can be written as

$$y_d = \alpha_s y_{s,d} + \alpha_1 y_{r_1,d} + \alpha_2 y_{r_2,d}, \tag{6.93}$$

where α_s, α_1 are the same as the source-only amplify-and-forward protocol and α_2 is given by

$$\alpha_2 = \frac{\sqrt{P_2} h^*_{r_2,d} \frac{K}{\sqrt{K^2 + K}}}{P_2|h_{r_2,d}|^2 \frac{K}{K^2 + K} + N_0}.$$

The SER analysis of the protocol is very complicated and a close-form analysis is not tractable.

We may think the performance of the MRC-based protocol would be better than that of the source-only protocol since the relays put more efforts in forwarding signals. However, simulations show that this is not the case. The reason behind this is that the system suffers from the noise propagation problem. For the simple example of two relays network, the noise terms at the destination in phases 2 and 3 contain a contribution from the noise generated at the first relay, $n_{\mathrm{s,r_1}}$ in phase 1. So the noise components in the received signals during the several phases are no more uncorrelated and the MRC detector is no more optimal. This noise propagation problem causes a degradation in the SER performance of the protocol. The problem is more severe for increased number of relays because we will have more noise components that will propagate to the destination. The optimum receiver in this case is to apply a pre-whitening on the received signals and then apply the MRC detector. Although the source-only amplify-and-forward protocol is less complex than the MRC-based amplify-and-forward protocol, we will show that it can give approximately the same, if not better in some cases, SER performance. This is because the benefit we get from applying an MRC detector at each relay node is diminished by the noise propagation problem. To see the effect of the noise propagation on the MRC-based protocol, we consider two extreme scenarios for the two-relay network and compare the performance of the two protocols under these two scenarios:

(i) $|h_{\mathrm{r_1,d}}| = 0$: In this case, we do not have the noise propagation problem because the noise term $n_{\mathrm{s,r_1}}$ will be received only once in phase 3. In this case the SNR at the destination of the MRC-based protocol can be written as

$$\mathrm{SNR_{MRC}} \simeq \frac{P_0|h_{\mathrm{s,d}}|^2}{N_0} + \frac{K\,\frac{P_2|h_{\mathrm{r_2,d}}|^2}{N_0}}{K + \frac{P_2|h_{\mathrm{r_2,d}}|^2}{N_0}}. \tag{6.94}$$

Similarly, the SNR at the destination of the source-only protocol can be written as

$$\mathrm{SNR_{source\text{-}only}} \simeq \frac{P_0|h_{\mathrm{s,d}}|^2}{N_0} + \frac{\frac{P_0|h_{\mathrm{s,r_2}}|^2}{N_0}\cdot\frac{P_2|h_{\mathrm{r_2,d}}|^2}{N_0}}{\frac{P_0|h_{\mathrm{s,r_2}}|^2}{N_0} + \frac{P_2|h_{\mathrm{r_2,d}}|^2}{N_0}}. \tag{6.95}$$

Clearly, $\mathrm{SNR_{MRC}} > \mathrm{SNR_{source\text{-}only}}$ because $K > \frac{P_0|h_{\mathrm{s,r_2}}|^2}{N_0}$. Intuitively, because we do not have noise propagation in this case, it is better for relay 2 to combine the signals it receives from both the source and relay 1 using the MRC detector to maximize the SNR at its output instead of using the source signal only. Under this scenario the MRC-based protocol is better than the source-only protocol because it results in a higher SNR at the destination.

(ii) $|h_{\mathrm{r_1,d}}| >>$, $|h_{\mathrm{r_2,d}}| >>$, $|h_{\mathrm{r_1,r_2}}| >>$ and $|h_{\mathrm{s,r_2}}| = 0$: In this case, the relay–destination links can be approximated to be noise free. The output of the destination detector in the MRC-based protocol can be written as

$$y_{\mathrm{MRC}} \simeq \left(\frac{P_0|h_{\mathrm{s,d}}|^2}{N_0} + 2\frac{P_0|h_{\mathrm{s,r_1}}|^2}{N_0}\right)x + \frac{\sqrt{P_0}h_{\mathrm{s,d}}^*}{N_0}n_{\mathrm{s,d}} \tag{6.96}$$

$$+2\frac{\sqrt{P_0}h_{\mathrm{s,r_1}}^*}{N_0}n_{\mathrm{s,r_1}}, \tag{6.97}$$

because the signal at the first relay in phase 1 is transmitted twice in phases 2 and 3. The output of the destination detector in the source-only protocol can be written as

$$y_{\text{source-only}} \simeq \left(\frac{P_0 |h_{s,d}|^2}{N_0} + \frac{P_0 |h_{s,r_1}|^2}{N_0} \right) x \tag{6.98}$$

$$+ \frac{\sqrt{P_0} h_{s,d}^*}{N_0} n_{s,d} + \frac{\sqrt{P_0} h_{s,r_1}^*}{N_0} n_{s,r_1}. \tag{6.99}$$

Clearly, we have

$$\text{SNR}_{\text{source-only}} \simeq \frac{P_0 |h_{s,d}|^2}{N_0} + \frac{P_0 |h_{s,r_1}|^2}{N_0} \tag{6.100}$$

$$> \text{SNR}_{\text{MRC}} \simeq \frac{\left(\frac{P_0 |h_{s,d}|^2}{N_0} + 2 \frac{P_0 |h_{s,r_1}|^2}{N_0} \right)^2}{\frac{P_0 |h_{s,d}|^2}{N_0} + 4 \frac{P_0 |h_{s,r_1}|^2}{N_0}}. \tag{6.101}$$

In this case, the source-only protocol achieves better performance than the MRC-based protocol. This is because the MRC-based protocol combines the same signal twice (the signals received in phases 2 and 3). Thus, in this case, the noise propagation problem is highly severe and causes a high degradation in the system SER performance.

Although the above two scenarios occur with zero probability, they can give some insights about how the system performance is affected by the different channels magnitudes. From the above two scenarios, intuition suggests that the source-only protocol will give better performance than the MRC-based protocol if the relays become closer to the destination. Moreover, the MRC-based protocol requires more channels to estimate at each relay node, which increases the overhead in the network required for channel estimation. Finally, the MRC-based protocol may have a higher delay because each relay node must wait for the transmission of all of the previous relay nodes, while in the source-only amplify-and-forward protocol, all the relays can forward the source data at the same time using FDM, for example.

6.2.3 Outage analysis and optimum power allocation

In this subsection, we investigate the outage performance of the source-only multi-node amplify-and-forward relaying protocol and determine its optimum power allocation.

We recall that in the source-only relaying protocol, the received signal at the destination in phase 2 due to the i-th relay transmission is given by

$$y_{r_i,d} = h_{r_i,d} \beta_i y_{s,r_i} + n_{r_i,d}, \tag{6.102}$$

and β_i is a scaling factor that satisfies the power constraint

$$\beta_i \leq \sqrt{\frac{P_i}{P_0 |h_{s,r_i}|^2 + N_0}}. \tag{6.103}$$

The output of the MRC detector is given by

$$y_d = \alpha_s y_{s,d} + \sum_{i=1}^{N} \alpha_i y_{r_i,d}, \qquad (6.104)$$

where $\alpha_s = \sqrt{P_0} h_{s,d}^* / N_0$ and $\alpha_i = \frac{\sqrt{P_0} \beta_i h_{r_i,d}^* h_{s,r_i}^*}{(\beta_i^2 |h_{r_i,d}|^2 + 1) N_0}$. We can write y_d in terms of the source signal x as

$$y_d = \left(\frac{P_0 |h_{s,d}|^2}{N_0} + \sum_{i=1}^{N} \frac{P_0 \beta_i^2 |h_{r_i,d}|^2 |h_{s,r_i}|^2}{(\beta_i^2 |h_{r_i,d}|^2 + 1) N_0} \right) x + \frac{\sqrt{P_0} h_{s,d}^*}{N_0} n_{s,d}$$

$$+ \sum_{i=1}^{N} \frac{\sqrt{P_0} \beta_i h_{r_i,d}^* h_{s,r_i}^*}{(\beta_i^2 |h_{r_i,d}|^2 + 1) N_0} (n_{r_i,d} + h_{r_i,d} \beta_i n_{s,r_i}). \qquad (6.105)$$

The SNR at the MRC detector output is

$$\text{SNR}_d = \gamma_s + \sum_{i=1}^{N} \gamma_i \qquad (6.106)$$

where $\gamma_s = P_0 |h_{s,d}|^2 / N_0$, and

$$\gamma_i = \frac{P_0 \beta_i^2 |h_{r_i,d}|^2 |h_{s,r_i}|^2}{(\beta_i^2 |h_{r_i,d}|^2 + 1) N_0}. \qquad (6.107)$$

For simplicity, let us denote an $(N+1) \times 1$ vector

$$\mathbf{y} = [y_{s,d}, y_{r_1,d}, \ldots, y_{r_N,d}]^{\mathrm{T}}. \qquad (6.108)$$

Then, for given x and channel state information, the probability density function (pdf) of $\mathbf{y}$ follows an exponential distribution. Moreover, y_d is a sufficient statistic for x, i.e.,

$$p_{\mathbf{y}/x, y_d}(\mathbf{y}/x, y_d) = p_{\mathbf{y}/y_d}(\mathbf{y}/y_d), \qquad (6.109)$$

where $p_{\mathbf{y}/x, y_d}(\mathbf{y}/x, y_d)$ is pdf of $\mathbf{y}$ given x and y_d, and $p_{\mathbf{y}/y_d}(\mathbf{y}/y_d)$ is the pdf of $\mathbf{y}$ given y_d. Since y_d is a sufficient statistics for x, then the mutual information between x and $\mathbf{y}$ equals the mutual information between x and y_d, that is

$$I(x; y_d) = I(x; \mathbf{y}). \qquad (6.110)$$

Then the average mutual information for amplify-and-forward I_{AF} satisfies

$$I_{\text{AF}} \leq I(x; y_d) \leq \log \left(1 + \frac{P_0 |h_{s,d}|^2}{N_0} + \sum_{i=1}^{N} \frac{P_0 \beta_i^2 |h_{r_i,d}|^2 |h_{s,r_i}|^2}{(\beta_i^2 |h_{r_i,d}|^2 + 1) N_0} \right), \qquad (6.111)$$

with equality for x being a zero-mean, circularly symmetric complex Gaussian random variable. It is clear that (6.111) is increasing in β_i's, so to maximize the mutual information the constraint in (6.103) should be satisfied with equality, yielding

$$I_{\text{AF}} = \log \left(1 + |h_{s,d}|^2 \text{SNR}_{s,d} + \sum_{i=1}^{N} f \left(|h_{s,r_i}|^2 \text{SNR}_{s,r_i}, |h_{r_i,d}|^2 \text{SNR}_{r_i,d} \right) \right),$$

where $\text{SNR}_{\text{s,d}} = \text{SNR}_{\text{s},r_i} = P_0/N_0$ and $\text{SNR}_{r_i,\text{d}} = P_i/N_0$ for all $i = 1, 2, \ldots, N$, and

$$f(v, u) = \frac{uv}{u + v + 1}.$$

6.2.3.1 Outage analysis

The outage probability for spectral efficiency R is defined as

$$P_{\text{AF}}^{\text{out}}(R) = \Pr\left\{\frac{1}{N+1}I_{\text{AF}} < R\right\}, \tag{6.112}$$

and the $1/(N+1)$ factor comes from the fact that the relays help the source through N uses of orthogonal channels. Using the notation

$$\mathbf{p} = [P_0, P_1, P_2, \ldots, P_N]^{\text{T}}, \tag{6.113}$$

equation (6.112) can be rewritten as

$$P_{\text{AF}}^{\text{out}}(\mathbf{p}, R) = \Pr\left\{\left(\frac{P_0}{N_0}|h_{\text{s,d}}|^2 + \sum_{i=1}^{N} f\left(\frac{P_0}{N_0}|h_{\text{s},r_i}|^2, \frac{P_i}{N_0}|h_{r_i,\text{d}}|^2\right)\right) < \left(2^{(N+1)R} - 1\right)\right\}.$$

At high SNR, we can neglect the 1 in the denominator of the function $f(u, v)$, and it follows that the outage probability is given by

$$P_{\text{AF}}^{\text{out}}(\mathbf{p}, R) \simeq \Pr\left\{\left(\frac{P_0}{N_0}|h_{\text{s,d}}|^2 + \sum_{i=1}^{N} \frac{\frac{P_0}{N_0}|h_{\text{s},r_i}|^2 \frac{P_i}{N_0}|h_{r_i,\text{d}}|^2}{\frac{P_0}{N_0}|h_{\text{s},r_i}|^2 + \frac{P_i}{N_0}|h_{r_i,\text{d}}|^2}\right) < \left(2^{(N+1)R} - 1\right)\right\}.$$

Denote some random variables

$$w_1 = \frac{P_0}{N_0}|h_{\text{s,d}}|^2$$

$$w_{i+1} = \frac{\frac{P_0}{N_0}|h_{\text{s},r_i}|^2 \frac{P_i}{N_0}|h_{r_i,\text{d}}|^2}{\frac{P_0}{N_0}|h_{\text{s},r_i}|^2 + \frac{P_i}{N_0}|h_{r_i,\text{d}}|^2}, \quad \forall i \in [1, N]. \tag{6.114}$$

The outage probability is now given as

$$P_{\text{AF}}^{\text{out}}(\mathbf{p}, R) \simeq \Pr\left\{\sum_{j=1}^{N+1} w_j < (2^{(N+1)R} - 1)\right\}. \tag{6.115}$$

The random variable w_1 has an exponential distribution with rate

$$\lambda_1 = \frac{N_0}{P_0 \sigma_{\text{s,d}}^2}. \tag{6.116}$$

To calculate the outage probability in (6.115), we consider an approach based on approximating the harmonic mean of two exponential random variables by an exponential random variable.

Note that each w_j for $j \in [2, N + 1]$ is a harmonic mean of two exponential random variables. The CDF of w_j, $j = 2, \ldots, N + 1$ can be given as

$$P_{w_j}(w) = \Pr\left\{w_j < w\right\} = 1 - 2w\sqrt{\zeta_{j1}\zeta_{j2}}e^{-w(\zeta_{j1}+\zeta_{j2})}K_1(2w\sqrt{\zeta_{j1}\zeta_{j2}}), \tag{6.117}$$

where

$$\zeta_{j1} = \frac{N_0}{P_0 \sigma_{s,r_{j-1}}^2}$$

$$\zeta_{j2} = \frac{N_0}{P_{j-1} \sigma_{s,r_{j-1}}^2} \tag{6.118}$$

and $K_1(\cdot)$ is the first-order modified Bessel function of the second kind. The function $K_1(\cdot)$ can be approximated as $K_1(x) \simeq 1/x$ for small x, from which we can approximate the CDF of w_j at high SNR as

$$P_{w_j}(w) = \Pr\{w_j < w\} \simeq 1 - e^{-w(\zeta_{j1}+\zeta_{j2})}, \tag{6.119}$$

which is the CDF of an exponential random variable of rate

$$\lambda_j = \frac{N_0}{P_0 \sigma_{s,r_{j-1}}^2} + \frac{N_0}{P_{j-1} \sigma_{r_{j-1},d}^2}. \tag{6.120}$$

Define a random variable W as

$$W = \sum_{j=1}^{N+1} w_j, \tag{6.121}$$

the CDF of W, assuming the λ_i's to be distinct, can be obtained as

$$\Pr\{W \leq w\} \simeq \sum_{k=1}^{N+1} \left(\prod_{m=1,m\neq k}^{N+1} \frac{\lambda_m}{\lambda_m - \lambda_k} \right) (1 - e^{-\lambda_k w}). \tag{6.122}$$

The outage probability can be expressed in terms of the CDF of W as

$$P_{AF}^{out}(\mathbf{p}, R) \simeq \Pr\left\{ W \leq (2^{(N+1)R} - 1) \right\}. \tag{6.123}$$

The CDF of W can now be written as

$$\Pr\{W \leq w\} = \sum_{k=1}^{N+1} \left(\prod_{m=1,m\neq k}^{N+1} \frac{\lambda_m}{\lambda_m - \lambda_k} \right) \left(\sum_{n=1}^{N+1} (-1)^{n+1} \lambda_k^n \frac{w^n}{n!} \right)$$
$$+ \text{H.O.T.}, \tag{6.124}$$

where H.O.T. stands for the higher order terms. Rearranging the terms in (6.124) we get

$$\Pr\{W \leq w\} = \sum_{n=1}^{N+1} \left(\sum_{k=1}^{N+1} \left(\prod_{m=1,m\neq k}^{N+1} \frac{\lambda_m}{\lambda_m - \lambda_k} \right) \lambda_k^n \right) (-1)^{n+1} \frac{w^n}{n!}$$
$$+ \text{H.O.T.} \tag{6.125}$$

In order for the system to achieve a diversity of order $(N + 1)$, the coefficients of w^n have to be zero for all $n \in [1, N]$. This requirement can be reformulated in a matrix form as

$$
\begin{bmatrix}
\lambda_1 & \cdots & \lambda_{N+1} \\
\lambda_1^2 & \cdots & \lambda_{N+1}^2 \\
\vdots & \vdots & \vdots \\
\lambda_1^{N+1} & \cdots & \lambda_{N+1}^{N+1}
\end{bmatrix}
\underbrace{
\begin{bmatrix}
\prod_{m=2}^{N+1} \frac{\lambda_m}{\lambda_m - \lambda_1} \\
\prod_{m=1, m\neq2}^{N+1} \frac{\lambda_m}{\lambda_m - \lambda_2} \\
\vdots \\
\prod_{m=1}^{N} \frac{\lambda_m}{\lambda_m - \lambda_{N+1}}
\end{bmatrix}}_{\mathbf{q}}
=
\begin{bmatrix}
0 \\
0 \\
\vdots \\
c_1
\end{bmatrix}.
\tag{6.126}
$$

where $\underbrace{}_{\mathbf{V}}$

To prove (6.126), let us consider the following equations:

$$
\mathbf{V}\mathbf{a} = \underbrace{[0, 0, \ldots, 1]^{\mathrm{T}}}_{\mathbf{c}},
\tag{6.127}
$$

where c_1 is some non-zero constant, and $\mathbf{q} = c_1\mathbf{a}$. Note that as the columns of the $\mathbf{V}$ matrix are scaled versions of the columns of a Vandermonde matrix, i.e., it is a nonsingular matrix, the solution for the system of equations in (6.127) can be found as

$$
\mathbf{a} = \mathbf{V}^{-1}\mathbf{c} = \frac{1}{\det(\mathbf{V})} \mathrm{adj}(\mathbf{V})\mathbf{c}.
\tag{6.128}
$$

where $\mathrm{adj}(\mathbf{V})$ is the adjoint matrix of $\mathbf{V}$. The determinant of a Vandermonde matrix is given by

$$
\det
\begin{bmatrix}
1 & 1 & \cdots & 1 \\
\lambda_1 & \lambda_2 & \cdots & \lambda_{N+1} \\
\vdots & \vdots & \vdots & \vdots \\
\lambda_1^N & \lambda_2^N & \cdots & \lambda_{N+1}^N
\end{bmatrix}
= \prod_{k=1}^{N+1} \prod_{m>k}^{N+1} (\lambda_m - \lambda_k),
\tag{6.129}
$$

from which we can express the determinant of the $\mathbf{V}$ matrix as

$$
\det(\mathbf{V}) = \left(\prod_{j=1}^{N+1} \lambda_j \right) \prod_{k=1}^{N+1} \prod_{m>k}^{N+1} (\lambda_m - \lambda_k).
\tag{6.130}
$$

Due to the structure of the $\mathbf{c}$ vector, we are only interested in the last column of the $\mathrm{adj}(\mathbf{V})$ matrix. The i-th element of the $\mathbf{a}$ vector can be obtained as

$$
a_i = \frac{(-1)^{N+i-1} \left(\prod_{j=1, j\neq i}^{N+1} \lambda_j \right) \prod_{k=1, k\neq i}^{N+1} \prod_{m>k, m\neq i}^{N+1} (\lambda_m - \lambda_k)}{\left(\prod_{j=1}^{N+1} \lambda_j \right) \prod_{k=1}^{N+1} \prod_{m>k}^{N+1} (\lambda_m - \lambda_k)}
$$

$$
= \frac{(-1)^N}{\lambda_i} \prod_{j=1, j\neq i}^{N+1} \frac{1}{(\lambda_j - \lambda_i)}.
\tag{6.131}
$$

From (6.131), it is clear that $\mathbf{q} = c_1\mathbf{a}$ where

$$
c_1 = (-1)^N \prod_{i=1}^{N+1} \lambda_i.
\tag{6.132}
$$

Using the CDF of W defined in (6.125), The outage probability can now be expressed as

$$P_{AF}^{out}(\mathbf{p}, R) \simeq \Pr\left\{W < (2^{(N+1)R} - 1)\right\}$$

$$= \frac{1}{(N+1)!}\left(\prod_{i=1}^{N+1}\lambda_i\right)\left(2^{(N+1)R} - 1\right)^{N+1} + \text{H.O.T.}$$

(6.133)

Based on the λ_i's from (6.116), we get

$$P_{AF}^{out}(\mathbf{p}, R) \sim \frac{1}{(N+1)!}\cdot\frac{1}{P_0\sigma_{s,d}^2}\prod_{i=1}^{N}\frac{P_0\sigma_{s,r_i}^2 + P_i\sigma_{r_i,d}^2}{P_0 P_i \sigma_{s,r_i}^2 \sigma_{r_i,d}^2}\left(2^{(N+1)R} - 1\right)^{N+1} N_0^{N+1}.$$

(6.134)

From the expression in (6.134), let P be the total power and let $P_0 = a_0 P$ and $P_i = a_i P$ where $a_0 + \sum_{i=1}^{N} a_i = 1$, $a_0 > 0$, $a_i > 0$, $i = 1, \ldots, N$. Denote the SNR as $\text{SNR} = P/N_0$, the diversity order of the system, based on the outage probability, is defined as

$$d_{AF}^{out} = \lim_{\text{SNR}\to\infty} -\frac{\log P_{AF}^{out}(\text{SNR}, R)}{\log \text{SNR}} = N + 1.$$

(6.135)

So the system achieves a diversity of order $N + 1$, in terms of outage probability, for N relay nodes helping the source. For the special case of single relay node ($N = 1$) and let $\text{SNR} = P_0/N_0 = P_1/N_0$, we get

$$P_{AF}^{out}(\text{SNR}, R) \sim \frac{1}{2}\cdot\frac{1}{\sigma_{s,d}^2}\cdot\frac{\sigma_{s,r_1}^2 + \sigma_{r_1,d}^2}{\sigma_{s,r_1}^2 \sigma_{r_1,d}^2}\left(\frac{2^{2R} - 1}{\text{SNR}}\right)^2.$$

(6.136)

6.2.3.2   Optimal power allocation

The optimal power allocation is based on minimizing the outage probability expression in (6.134) under a total power constraint. Ignoring the constant factor from the outage probability, the optimization problem can be formulated as

$$\mathbf{p}_{opt} = \arg\min_{\mathbf{p}} \frac{1}{P_0^{N+1}}\prod_{i=1}^{N}\frac{P_0\sigma_{s,r_i}^2 + P_i\sigma_{r_i,d}^2}{P_i},$$

$$\text{s.t. } P_0 + \sum_{i=1}^{N} P_i \leq P, \quad P_i \geq 0 \quad \forall i,$$

(6.137)

where P is the maximum allowable total power for one symbol transmission.

It can be easily shown that the cost function in (6.137) is convex in $\mathbf{p}$ over the convex feasible set defined by the linear power constraints. The Lagrangian of this optimization problem can be written as

$$L = \frac{1}{P_0^{N+1}} \prod_{i=1}^{N} \frac{P_0 \sigma_{s,r_i}^2 + P_i \sigma_{r_i,d}^2}{P_i} \tag{6.138}$$

$$+ \tilde{\lambda} \left(P_0 + \sum_{i=1}^{N} P_i - P \right) + \sum_{i=1}^{N} \mu_i (0 - P_i), \tag{6.139}$$

where the μ's serve as the slack variables. To minimize the outage bound, it is clear that we must have $P_i > 0$ for all i. Then, the complementary slackness imply that $\mu_i = 0$ for all i. Denote an $N \times 1$ vector $\mathbf{a} = [a_0, a_1, \dots, a_N]$, where $a_0 = P_0/P$ and $a_i = P_i/P$ for any $i \in [1, N]$. Since the log function is a monotone function, the Lagrangian of the optimization problem in (6.137) can be given as

$$f = -\log a_0 + \sum_{i=1}^{N} \log \left(\frac{1}{a_i} \sigma_{s,r_i}^2 + \frac{1}{a_0} \sigma_{r_i,d}^2 \right) + \tilde{\lambda}(\mathbf{a}^{\mathsf{T}} \mathbf{1}_{N+1} - 1), \tag{6.140}$$

where $\mathbf{1}_{N+1}$ is an all 1 $(N+1) \times 1$ vector. Applying first-order optimality conditions, $\mathbf{a}_{\text{opt}}$ must satisfy

$$\frac{\partial f}{\partial a_0} = \frac{\partial f}{\partial a_i} = 0, \quad \forall i \in [1, N], \tag{6.141}$$

from which we get

$$\frac{1}{a_0} \left[1 + \sum_{j=1}^{N} \frac{\sigma_{r_j,d}^2}{\sigma_{r_j,d}^2 + \frac{a_0}{a_j} \sigma_{s,r_j}^2} \right] = \frac{1}{a_i} \left[\frac{\sigma_{s,r_i}^2}{\sigma_{s,r_i}^2 + \frac{a_i}{a_0} \sigma_{r_i,d}^2} \right]. \tag{6.142}$$

Since $a_0 > 0$ and $a_j > 0$, then we can easily show that $a_0 > a_i$, i.e., $P_0 > P_i$ $\forall i \in [1, N]$. This is due to the fact that the source power appears in all the SNR terms in (6.134) either through the source–destination direct link or through the harmonic mean of the source–relay and relay–destination links.

Using (6.141) we have

$$\frac{1}{a_j} \left[\frac{\sigma_{s,r_j}^2}{\sigma_{s,r_j}^2 + \frac{a_j}{a_0} \sigma_{r_j,d}^2} \right] = \frac{1}{a_i} \left[\frac{\sigma_{s,r_i}^2}{\sigma_{s,r_i}^2 + \frac{a_i}{a_0} \sigma_{r_i,d}^2} \right], \quad \forall i, j. \tag{6.143}$$

Define $c_i = a_i/a_0 = P_i/P_0$, $\forall i = 1, \dots, N$ and using (6.142), we get

$$\frac{\sigma_{r_i,d}^2}{\sigma_{s,r_i}^2} c_i^2 + c_i - c = 0, \quad \forall i \in [1, \dots, N], \tag{6.144}$$

for some constant c. From (6.142), c should satisfy the following equation

$$f(c) = c - \frac{1}{1 + \sum_{j=1}^{N} \frac{\sigma_{r_j,d}^2}{\sigma_{r_j,d}^2 + \frac{1}{c_j(c)} \sigma_{s,r_j}^2}} = 0. \tag{6.145}$$

Table 6.3 Optimal power allocation for one and two relays ($\sigma_{s,d}^2 = 1$ in all cases).

	$\sigma_{s,r_i}^2 = 10, \sigma_{r_i,d}^2 = 1$ Relays close to the source	$\sigma_{s,r_i}^2 = 1, \sigma_{r_i,d}^2 = 10$ Relays close to the destination
One relay	$P_0/P = 0.5393$ $P_1/P = 0.4607$	$P_0/P = 0.8333$ $P_1/P = 0.1667$
Two relays	$P_0/P = 0.3830$ $P_1/P = P_2/P = 0.3085$	$P_0/P = 0.75$ $P_1/P = P_2/P = 0.125$

Since $P_i < P_0$, $\forall i$, then $c_i < 1$, $\forall i$. Hence, using (6.144), we have $c \in (0, 1 + \min_i (\sigma_{r_i,d}^2/\sigma_{s,r_i}^2)$. Therefore, so we have reduced the $(N+1)$-dimensional problem to a single dimension search over the parameter c, which can be done using a simple numerical search or any other standard method such as the Newton's method. Convexity of both the cost function and the feasible set in (6.137) imply global optimality of the solution of (6.145) over the desired feasible set.

Example 6.8   Table 6.3 gives numerical results for the optimal power allocation for one and two relays helping the source. From the table, it is clear that equal power allocation is not optimal. As the relays get closer to the source, the equal power allocation scheme tends to be optimal. If the relays are close to the destination, the optimal power allocation can result in a significant performance improvement.   ▲

Example 6.9   In this example, the tightness of the derived outage probability approximation is studied for the system with one, two, and three relay nodes, respectively. Figure 6.14 shows the outage probability versus SNR$_{norm}$, defined as

$$\text{SNR}_{norm} = \frac{\text{SNR}}{2^R - 1}, \tag{6.146}$$

which is the SNR normalized by the minimum SNR required to achieve spectral efficiency R. In Figure 6.14, $R = 1$ is used. For the single-relay case, all the channel variances are taken to be 1. For the case of two relay nodes, all the channel variances are taken to be 1 except that between the source and the second relay which is taken to be $\sigma_{s,r_2}^2 = 10$; meaning that the second relay is close to the source. For the case of three relay nodes, all the channel variances are taken to be 1 except $\sigma_{s,r_2}^2 = 10$, and that between the source and the third relay is $\sigma_{s,r_3}^2 = 5$. From Figure 6.14 it is clear the the bound in (6.134) is tight at high SNR and that the multi-node amplify-and-forward protocol achieves full diversity of order $N + 1$ in terms of outage probability.   ▲

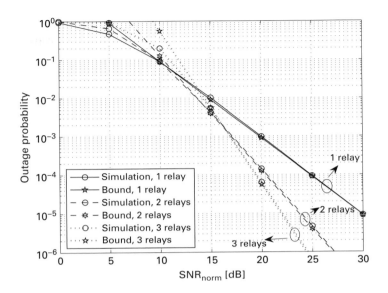

Fig. 6.14   Outage probability for a one-, two-, and three-nodes amplify-and-forward relay network.

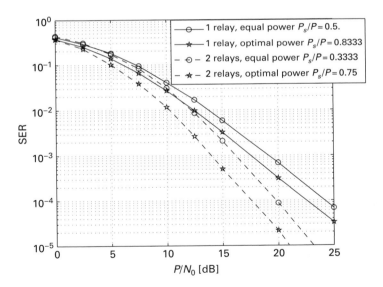

Fig. 6.15   Comparison of the SER for QPSK modulation using equal power allocation and the optimal power allocation for relays close to the destination.

Example 6.10   In this example, the gain of the optimal power allocation and the equal power allocation is compared. Figure 6.15 shows a comparison for a case that relays are close to the destination ($\sigma_{s,r_1}^2 = \sigma_{s,r_2}^2 = 1$, $\sigma_{r_1,d}^2 = \sigma_{r_2,d}^2 = 10$, and $\sigma_{s,d}^2 = 1$). From the figure, we can see that, using the optimal power allocation scheme, we can get about 1 dB improvement for the single relay case and about 2 dB improvement for the two relays case over the equal power assignment scheme.   ▲

6.3   Chapter summary and bibliographical notes

In this chapter, we discussed cooperative diversity protocols for multi-node wireless networks employing either decode-and forward or amplify-and-forward relaying strategy. In the first part of the chapter, we focused on a class of multi-node decode-and-forward relaying protocols. This class of protocols consists of schemes in which each relay can combine the signals arriving from an arbitrary but fixed number of previous relays along with that received from the source. We derived exact SER expressions for a general cooperation scheme for both M-PSK and M-QAM modulation. We also discussed SER approximations, which are shown to be tight at high enough SNR. The theoretical analysis reveals a very interesting result: this class of cooperative protocols shares the same asymptotic performance at high enough SNR. Thus the performance of a simple cooperation scenario in which each relay combines the signals arriving from the previous relay and the source is asymptotically exactly the same as that for the most complicated scenario in which each relay combines the signals arriving from all the previous relays and the source. The analysis also reveals that the described protocols achieve full diversity gain in the number of cooperating terminals. Moreover, we addressed the optimal power allocation problem and showed that the optimum power allocated at the nodes for an arbitrary network follow a certain ordering and do not depend on the quality of the direct link between the source and the destination.

In the second part of the chapter, we considered two kinds of multi-node amplify-and-forward relaying protocols. We compared the performances of the source-only amplify-and-forward protocol and the MRC-based amplify-and-forward protocol. Although the MRC-based protocol is more complex, it gives approximately the same performance as the simple source-only amplify-and-forward protocol due to the noise propagation problem. relays inter-channels which is difficult to achieve in wireless environment and represents a large overhead in the network. We provided an SER analysis for the source-only amplify-and-forward protocol that leads to a closed-form SER expression that can be written in the form of integral functions by directly substituting for the MGF functions. We also provided a simple SER approximation that is shown to be tight at high SNR. Furthermore, by forming an SER performance upper bound on any amplify-and-forward protocol, we showed that the source-only amplify-and-forward protocol achieves this bound if the relays are close to the source. Based on the SER upper bound, we determined an optimal power allocation between the source and the relays. We also provided an outage probability analysis for the source-only amplify-and-forward protocol and showed that the source-only amplify-and-forward protocol with N relay nodes.

The problem of SER performance analysis for multi-node amplify-and-forward relaying was studied in [152], in which the authors presented an approximate analysis and they considered the first term of the McLaurin series expansion of the SER around SNR $= 0$. The approach presented in this chapter has some similarities with the approach in [6] but differs in the way that the signal is normalized at each relay. In [6], the authors normalized the signal at the relays by a factor that can cause excessive increase in the signal peak in the case of deep fades. The approximations presented in this chapter are also tighter than the bound in [6]. Some protocols for multi node cooperative communications were presented in [156].

In [109], the outage probability of the single-relay amplify-and-forward network was obtained by considering the high SNR behavior of the outage probability based on the limiting behavior of the cumulative distribution function (CDF) of certain combinations of exponential random variables. We also considered the high SNR behavior of the outage probability of the multi-node source-only amplify-and-forward protocol. The case considered in [109] can be considered as a special case of the analysis presented in this chapter with $N = 1$. The study of multi-node cooperative communication can be found in [163, 176].

Exercises

6.1   The symbol error rate for a general decode-and-forward scenario $C(m)$ in which each relay combines the signals from the previous m relays and the source was analyzed for $1 \leq m \leq N$.

(a) Find the symbol error rate for the scenario $C(0)$ in whicheach relay only decodes the signal from the source.

(b) Find the asymptotic symbol error rate performance of $C(0)$ at high SNR. Does this scheme also achieve full diversity gains.

(c) Find the optimal power allocation in case of $C(0)$.

6.2   Find the symbol error rate performance for a time diversity scheme, in which the source transmits the same message in three consecutive time slots. Assume a first-order Markov model for the time correlation as follows

$$h_{s,d}(t + 1) = \rho h_{s,d}(t) + \sqrt{(1 - \rho^2)}n$$

where $h_{s,d}(t)$ is the channel gain at time t, and ρ is the correlation coefficient. The channel is modeled as a Rayleigh flat-fading channel with unit variance.

Plot the symbol error rate expression versus the SNR for different values of ρ.

6.3   Consider a symmetric network scenario in which all channels have identical statistics. Find the optimal power allocation among the relays and the source. Find the optimal power allocation for the following two scenarios:

(a) A three-relay decode-and-forward network in which a $C(1)$ scheme is used;

(b) a three-relay source-only amplify-and-forward network.

6.4   Write a Matlab code to compare the BPSK bit error rate performance of $N = 2$-relay $C(1)$ decode-and-forward protocol and $N = 2$-relay source-only amplify-and-forward protocols. Assume the source–destination channel variance to be equal to 1. Take $N = 2, 3, 4$. Consider the following three cases:
(a) All relays close to the source.
(b) All relays close to the destination.
(c) All relays are at mid-distance between the source and the destination.
For all of the previous cases assume the channel variance between any two relays to be equal to 1. What is the effects of increasing N to 3 on the performance gap between the two schemes?

6.5   Consider a multi-relay decode-and-forward protocol. For an average SNR of 20 dB per symbol and a spectral efficiency of 1 bps/Hz. Find the minimum number of relays to achieve a bit error rate of 10^{-4}. Assume a unit variance between the source and the destination and that all relays are close to the source with all inter-relay channel variance being equal to 1.
(a) What is the corresponding number of relays for amplify-and-forward relaying?
(b) Repeat for a bit error rate requirement of 10^{-6}.
(c) Find the minimum number of relays in all of the previous cases when the relays are closer to the destination.

6.6   Compare the bit error rate performance of a 2×1 Alamouti scheme and a one-relay decode-and-forward system under the following scenarios assuming an exponential propagation path-loss with exponent 3 and BPSK modulation. Assume the channel variance for a unit distance is normalized to 1.
(a) The relay is at the mid-point of the source–destination link with the source–destination separation equal to 1 and 2.
(b) The source–destination channel variance is 2 and the relay is located 0.25 from the source node.
(c) The source–destination separation distance is 2 and the relay is located 0.25 from the destination.

6.7   As shown in this chapter, cooperation provides both spatial diversity gains and power gains due to its multi-hop nature. Time diversity schemes on the other hand can only provide diversity gains. Consider a source–destination pair separated with distance d. Calculate the bit error rate (BER) of binary phase shift keying (BPSK) for the following two transmission schemes.
(a) A direct transmission scheme in which the source uses a two-repetition code under an average power constraint P per bit. In this repetition coding, a bit is transmitted twice in two consecutive time slots. Assume flat fading i.i.d. channel.
(b) A cooperative diversity scheme in which the source transmits in the first slot and the relay uses amplify-and-forward to forward the bit in the second time slot.

6.8   The relay state vector $\mathbf{S}_N$, which represents whether each relay has decoded correctly or not, was defined as

$$S_n[k] = \begin{cases} 1 & \text{if relay } k \text{ correctly decodes,} \\ 0 & \text{otherwise,} \end{cases} \qquad 1 \leq k \leq n.$$

Find the probability mass function of the vector $\mathbf{S}_N$.

6.9   The outage probability for the amplify-and-forward scenario was specified as

$$P_{\text{AF}}^{\text{out}}(\mathbf{p}, R) \simeq \Pr\left\{ W < (2^{(N+1)R} - 1) \right\}$$

$$= \frac{1}{(N+1)!} \left(\prod_{i=1}^{N+1} \lambda_i \right) \left(2^{(N+1)R} - 1 \right)^{N+1} + \text{H.O.T.}$$

in (6.133). Prove this expression.

7 Distributed space–time and space–frequency coding

The main drawback of the multi-node decode-and-forward (DF) protocol and the multi-node amplify-and-forward (AF) protocol, presented in Chapter 5, is the loss in the data rate as the number of relay nodes increases. The use of orthogonal subchannels for the relay node transmissions, either through TDMA or FDMA, results in a high loss of the system spectral efficiency. This leads to the use of what is known as distributed space–time coding (DSTC) and distributed space–frequency coding (DSFC), where relay nodes are allowed to simultaneously transmit over the same channel by emulating a space–time or a space–frequency code. The term *distributed* comes from the fact that the virtual multi-antenna transmitter is distributed between randomly located relay nodes. Employing DSTCs or DSFCs reduces the data rate loss due to relay nodes transmissions without sacrificing the system diversity order, as will be seen in this chapter.

7.1 Distributed space–time coding (DSTC)

In this section, the design of distributed space–time codes for wireless relay networks is considered. The two-hop relay network model depicted in Figure 7.1, where the system lacks a direct link from the source to destination node, is considered. Distributed space–time (space–frequency) coding can be achieved through node cooperation to emulate multiple antennas transmitter. First, the decode-and-forward protocol, in which each relay node decodes the symbols received from the source node before retransmission, is considered. A space–time code designed to achieve full diversity and maximum coding gain over MIMO channels is shown to achieve full diversity but does not necessarily maximize the coding gain if used with the decode-and-forward protocol. Next, the amplify-and-forward protocol is considered; each relay node can only perform simple operations such as linear transformation of the received signal and then amplify the signal before retransmission. A space–time code designed to achieve full diversity and maximum coding gain over MIMO channels is shown to achieve full diversity and maximum coding gain if used with the amplify-and-forward protocol.

Next, the design of DSTC that can mitigate the relay nodes synchronization errors is considered. Most of the work on cooperative transmission assume perfect synchronization between the relay nodes, which means that the relays' timings, carrier frequencies,

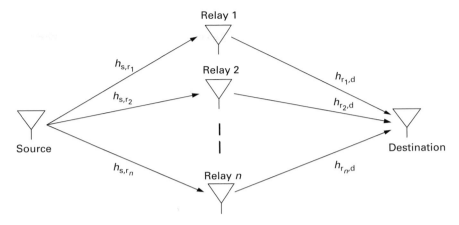

Fig. 7.1   Simplified system model for the two-hop distributed space–time codes.

and propagation delays are identical. Perfect synchronization is difficult to achieve among randomly located relay nodes. To simplify the synchronization in the network, a diagonal structure is imposed on the space–time code used. The diagonal structure of the code bypasses the perfect synchronization problem by allowing only one relay to transmit at any time slot (assuming TDMA). Hence, it is not necessary to synchronize simultaneous in-phase transmissions of randomly located relay nodes, which greatly simplifies the synchronization among the relay nodes.

7.1.1   DSTC with the DF protocol

In this section, the system model for DSTC with decode-and-forward cooperation protocol is presented, and a system performance analysis is provided.

7.1.1.1   System model
The source node is assumed to have n relay nodes assigned for cooperation as shown in Figure 7.1. The system has two phases given as follows. In phase 1, the source transmits data to the relay nodes with power P_1. The received signal at the k-th relay is modeled as

$$\mathbf{y}_{s,r_k} = \sqrt{P_1} h_{s,r_k} \mathbf{s} + \mathbf{v}_{s,r_k}, \quad k = 1, 2, \ldots, n, \tag{7.1}$$

where $\mathbf{s}$ is an $L \times 1$ transmitted data vector with a power constraint $||\mathbf{s}||_F^2 \leq L$, where $|| \cdot ||_F^2$ denotes the Frobenius norm and $h_{s,r_k} \sim \mathcal{CN}(0, \delta_{s,r_k}^2)$ denotes the channel gain between the source node and the k-th relay node. The channel gains from the source node to the relay nodes are assumed to be independent. All channel gains are fixed during the transmission of one data packet and can vary from one packet to another, i.e., a block flat-fading channel model is assumed. In (7.1), $\mathbf{v}_{s,r_i} \sim \mathcal{CN}(\mathbf{0}, N_0 \mathbf{I}_n)$ denotes additive white Gaussian noise (AWGN), where $\mathbf{I}_n$ denotes the $n \times n$ identity matrix.

The n relay nodes try to decode the received signals from the source node. Each relay node is assumed to be capable of deciding whether or not it has decoded correctly. If a relay node decodes correctly, it will forward the source data in the second phase of the cooperation protocol; otherwise, it remains idle. This can be achieved through the use of cyclic redundancy check (CRC) codes. This performance can also be approached by setting a SNR threshold at the relay nodes, and the relay will only forward the source data if the received SNR is larger than that threshold. For the analysis in this section, the relay nodes are assumed to be synchronized either by a centralized or a distributed algorithm.

In phase 2, the relay nodes that have decoded correctly re-encode the data vector $\mathbf{s}$ with a pre-assigned code structure. In the subsequent development, no specific code design will be assumed, instead a generic space–time (ST) code structure is considered. The ST code is distributed among the relays such that each relay will emulate a single antenna in a multiple-antenna transmitter. Hence, each relay will generate a column in the corresponding ST code matrix. Let $\mathbf{X}_r$ denote the $K \times n$ space–time code matrix with $K \geq n$. Column k of $\mathbf{X}_r$ represents the code transmitted from the k-th relay node. The signal received at the destination is given by

$$\mathbf{y_d} = \sqrt{P_2}\mathbf{X}_r\mathbf{D}_I\mathbf{h}_d + \mathbf{v}_d, \tag{7.2}$$

where

$$\mathbf{h}_d = \left[h_{r_1,d}, h_{r_2,d}, \ldots, h_{r_n,d}\right]^\mathsf{T} \tag{7.3}$$

is an $n \times 1$ channel gains vector from the n relays to the destination, $h_{r_k,d} \sim \mathcal{CN}(0, \delta_{r_k,d}^2)$, and P_2 is the relay node power. Without loss of generality, equal power allocation among the relay nodes is assumed. The channel gains from the relay nodes to the destination node are assumed to be statistically independent as the relays are spatially separated. The $K \times 1$ vector $\mathbf{v}_d \sim \mathcal{CN}(\mathbf{0}, N_0\mathbf{I}_K)$ denotes AWGN at the destination node.

The state of the k-th relay, i.e., whether it has decoded correctly or not, is denoted by the random variable I_k ($1 \leq k \leq n$), which takes values 1 or 0 if the relay decodes correctly or erroneously, respectively. Let

$$\mathbf{I} = [I_1, I_2, \ldots, I_n]^\mathsf{T} \tag{7.4}$$

denote the state vector of the relay nodes and n_I denote the number of relay nodes that have decoded correctly corresponding to a certain realization $\mathbf{I}$. The random variables I_k's are statistically independent as the state of each relay depends only on its channel conditions to the source node, which are independent from other relays. The matrix

$$\mathbf{D}_I = \mathbf{diag}(I_1, I_2, \ldots, I_n) \tag{7.5}$$

in (7.2) is defined as the state matrix of the relay nodes. An energy constraint is imposed on the generated ST code such that $||\mathbf{X}_r||_F^2 \leq L$, and this guarantees that the transmitted power per source symbol is less than or equal to $P_1 + P_2$.

7.1.1.2　Performance analysis

In this section, the pairwise error probability (PEP) performance analysis for the cooperation scheme described in Section 7.1.1.1 is provided. The diversity and coding gain achieved by the protocol are then analyzed.

One can see that the random variable I_k is a Bernoulli random variable. Therefore, the probability distribution of I_k is given by

$$I_k = \begin{cases} 0 & \text{with probability} = 1 - (1 - \text{SER}_k)^L \\ 1 & \text{with probability} = (1 - \text{SER}_k)^L, \end{cases}$$

where SER_k is the un-coded SER at the k-th relay node and is modulation dependent. For M-ary quadrature amplitude modulation (M-QAM, $M = 2^p$ with p even), the exact expression for SER_k can be shown to be upper-bounded by

$$\text{SER}_k \leq \frac{2 N_0 g}{b P_1 \delta_{s,r_k}^2}, \tag{7.6}$$

where $b = 3/(M - 1)$ and

$$g = \frac{4R}{\pi} \int_0^{\pi/2} \sin^2 \theta d\theta - \frac{4R^2}{\pi} \int_0^{\pi/4} \sin^2 \theta d\theta, \tag{7.7}$$

in which

$$R = 1 - \frac{1}{\sqrt{M}}. \tag{7.8}$$

The destination is assumed to have perfect channel state information as well as the relay nodes state vector. The destination applies a maximum likelihood (ML) receiver, which is based on a minimum distance rule. The conditional pairwise error probability (PEP) is given by

$$\Pr(\mathbf{X}_1 \rightarrow \mathbf{X}_2 | \mathbf{I}, \mathbf{h}_d)$$
$$= \Pr\left(\|\mathbf{y_d} - \sqrt{P_2}\mathbf{X}_1\mathbf{D_I}\mathbf{h}_d\|_F^2 > \|\mathbf{y_d} - \sqrt{P_2}\mathbf{X}_2\mathbf{D_I}\mathbf{h}_d\|_F^2 | \mathbf{I}, \mathbf{h}_d, \mathbf{X}_1 \text{ was transmitted}\right), \tag{7.9}$$

where $\mathbf{X}_1$ and $\mathbf{X}_2$ are two possible transmitted codewords. The conditional PEP can be expressed as quadratic form of a complex Gaussian random vector as

$$\Pr(\mathbf{X}_1 \rightarrow \mathbf{X}_2 | \mathbf{I}, \mathbf{h}_d) = \Pr(q < 0 | \mathbf{I}, \mathbf{h}_d), \tag{7.10}$$

where

$$q = \begin{bmatrix} \mathbf{z}_1^H & \mathbf{z}_2^H \end{bmatrix} \begin{bmatrix} \mathbf{I}_n & \mathbf{0} \\ \mathbf{0} & -\mathbf{I}_n \end{bmatrix} \begin{bmatrix} \mathbf{z}_1 \\ \mathbf{z}_2 \end{bmatrix},$$

$$\mathbf{z}_1 = \sqrt{P_2}(\mathbf{X}_1 - \mathbf{X}_2)\mathbf{D_I}\mathbf{h}_d + \mathbf{v}_d, \tag{7.11}$$

and

$$\mathbf{z}_2 = \mathbf{v}_d. \tag{7.12}$$

The random vectors $\mathbf{h}_d$ and $\mathbf{I}$ are mutually independent as they arise from independent processes. First, the conditional PEP was averaged over the channel realizations $\mathbf{h}_d$. By defining the signal matrix

$$\mathbf{C_I} = (\mathbf{X}_1 - \mathbf{X}_2)\mathbf{D_I}\text{diag}(\delta_{r_1,d}^2, \delta_{r_2,d}^2, \ldots, \delta_{r_n,d}^2)\mathbf{D_I}(\mathbf{X}_1 - \mathbf{X}_2)^H, \tag{7.13}$$

the conditional PEP in (7.10) can be tightly upper-bounded, using equation (2.13), by

$$\Pr\left(\mathbf{X}_1 \to \mathbf{X}_2|\mathbf{I}\right) \leq \frac{\left(\dfrac{2\Delta(\mathbf{I}) - 1}{\Delta(\mathbf{I}) - 1}\right) N_0^{\Delta(\mathbf{I})}}{P_2^{\Delta(\mathbf{I})} \prod_{i=1}^{\Delta(\mathbf{I})} \lambda_i^{\mathbf{I}}}, \qquad (7.14)$$

where $\Delta(\mathbf{I})$ is the number of nonzero eigenvalues of the signal matrix and $\lambda_i^{\mathbf{I}}$'s are the nonzero eigenvalues of the signal matrix corresponding to the state vector $\mathbf{I}$. The nonzero eigenvalues of the signal matrix are the same as the nonzero eigenvalues of the matrix

$$\Gamma\left(\mathbf{X}_1, \mathbf{X}_2\right) = \mathbf{diag}(\delta_{r_1,d}, \delta_{r_2,d}, \dots, \delta_{r_n,d})\mathbf{D}_{\mathbf{I}}\Phi\left(\mathbf{X}_1, \mathbf{X}_2\right)\mathbf{D}_{\mathbf{I}}\mathbf{diag}(\delta_{r_1,d}, \delta_{r_2,d}, \dots, \delta_{r_n,d}),$$

where

$$\Phi\left(\mathbf{X}_1, \mathbf{X}_2\right) = \left(\mathbf{X}_1 - \mathbf{X}_2\right)^{\mathrm{H}} \left(\mathbf{X}_1 - \mathbf{X}_2\right). \qquad (7.15)$$

The employed space–time scheme is assumed to achieve full diversity and maximum coding gain over MIMO channels, which means that the matrix $\Phi\left(\mathbf{X}_1, \mathbf{X}_2\right)$ is full rank of order n for any pair of distinct codewords $\mathbf{X}_1$ and $\mathbf{X}_2$. Achieving maximum coding gain means that the minimum of the products $\prod_{i=1}^{n} \lambda_i$, where the λ_i's are the eigenvalues of the matrix $\Phi\left(\mathbf{X}_1, \mathbf{X}_2\right)$, is maximized over all the pairs of distinct codewords.

Clearly, if the matrix $\Phi\left(\mathbf{X}_1, \mathbf{X}_2\right)$ has a rank of order n then the matrix $\Gamma\left(\mathbf{X}_1, \mathbf{X}_2\right)$ will have a rank of order $n_{\mathbf{I}}$, which is the number of relays that have decoded correctly. Equation (7.14) can now be rewritten as

$$\Pr\left(\mathbf{X}_1 \to \mathbf{X}_2|\mathbf{I}\right) \leq \frac{\left(\dfrac{2n_{\mathbf{I}} - 1}{n_{\mathbf{I}} - 1}\right) N_0^{n_{\mathbf{I}}}}{P_2^{n_{\mathbf{I}}} \prod_{i=1}^{n_{\mathbf{I}}} \lambda_i^{\mathbf{I}}}. \qquad (7.16)$$

Next, the conditional PEP is averaged over the relays' state vector $\mathbf{I}$. The dependence of the expression in (7.16) on $\mathbf{I}$ appears through the set of nonzero eigenvalues $\{\lambda_i^{\mathbf{I}}\}_{i=1}^{n_{\mathbf{I}}}$, which depends on the number of relays that have decoded correctly and their realizations. The state vector $\mathbf{I}$ of the relay nodes determines which columns from the ST code matrix are replaced with zeros and thus affect the resulting eigenvalues.

The probability of having a certain realization of $\mathbf{I}$ is given by

$$\Pr(\mathbf{I}) = \left(\prod_{k \in CR(\mathbf{I})} (1 - \mathrm{SER}_k)^L\right)\left(\prod_{k \in ER(\mathbf{I})} \left(1 - (1 - \mathrm{SER}_k)^L\right)\right), \qquad (7.17)$$

where $CR(\mathbf{I})$ is the set of relays that have decoded correctly and $ER(\mathbf{I})$ is the set of relays that have decoded erroneously corresponding to the $\mathbf{I}$ realization. For simplicity of presentation symmetry is assumed between all relays, i.e., is $\delta_{s,r_k}^2 = \delta_{s,r}^2$ and $\delta_{r_k,d}^2 = \delta_{r,d}^2$ for all k. Averaging over all realizations of the states of the relays, gives the PEP at high SNR as

$$\text{PEP} = \Pr(\mathbf{X}_1 \rightarrow \mathbf{X}_2)$$

$$\leq \sum_{k=0}^{n} \left((1 - \text{SER})^L\right)^k \left(1 - (1 - \text{SER})^L\right)^{n-k} \sum_{\mathbf{I}:\, n_{\mathbf{I}}=k} \frac{\binom{2k-1}{k-1} N_0^k}{P_2^k \prod_{i=1}^{k} \lambda_i^{\mathbf{I}}}, \qquad (7.18)$$

where SER is now the symbol error rate at any relay node due to the symmetry assumption.

The diversity order of a system determines the average rate with which the error probability decays at high enough SNR. In order to compute the diversity order of the system, the PEP in (7.18) is rewritten in terms of the SNR defined as $\text{SNR} = P/N_0$, where $P = P_1 + P_2$ is the transmitted power per source symbol. Let $P_1 = \alpha P$ and $P_2 = (1 - \alpha)P$, where $\alpha \in (0, 1)$. Substituting these definitions along with the SER expressions at the relay nodes from (7.6) into (7.18) and considering high SNR, the PEP can be upper-bounded as

$$\Pr(\mathbf{X}_1 \rightarrow \mathbf{X}_2) \leq \text{SNR}^{-n} \sum_{k=0}^{n} \left(\frac{2Lg}{ba\delta_{\text{s,r}}^2}\right)^{n-k} \sum_{\mathbf{I}:\, n_{\mathbf{I}}=k} \frac{\binom{2k-1}{k-1}}{(1-\alpha)^k \prod_{i=1}^{k} \lambda_i^{\mathbf{I}}}, \qquad (7.19)$$

where at high SNR

$$1 - (1 - \text{SER})^L \approx L\text{SER} \qquad (7.20)$$

and upper-bounding $1 - L\text{SER}$ by 1. The diversity gain is defined as

$$d = \lim_{\text{SNR} \rightarrow \infty} -\frac{\log(\text{PEP}(\text{SNR}))}{log(\text{SNR})}. \qquad (7.21)$$

Applying this definition to the PEP in (7.19), when the number of cooperating nodes is n, gives

$$d_{\text{DF}} = \lim_{\text{SNR} \rightarrow \infty} -\frac{\log(\text{PEP}(\text{SNR}))}{\log(\text{SNR})} = n. \qquad (7.22)$$

Hence, any code that is designed to achieve full diversity over MIMO channels will achieve full diversity in the distributed relay network if it is used in conjunction with the decode-and-forward protocol.

If full diversity is achieved, the coding gain is

$$C_{\text{DF}} = \left(\sum_{k=0}^{n} \left(\frac{2ng}{ba\delta_{\text{s,r}}^2}\right)^{n-k} \sum_{\mathbf{I}:\, n_{\mathbf{I}}=k} \frac{\binom{2k-1}{k-1}}{(1-\alpha)^k \prod_{i=1}^{k} \lambda_i^{\mathbf{I}}}\right)^{-\frac{1}{n}}, \qquad (7.23)$$

which is a term that does not depend on the SNR. To minimize the PEP bound the coding gain of the distributed space–time code needs to be maximized. This is different from the determinant criterion in the case of MIMO channels as discussed in Chapter 2. Hence, a space–time code designed to achieve full diversity and maximum coding gain over MIMO channels will achieve full diversity but not necessarily maximizing the coding gain if used in a distributed fashion with the decode-and-forward protocol. Intuitively, the difference is due to the fact that in the case of distributed space–time codes with decode-and-forward protocol, not all of the relays will always transmit their corresponding code matrix columns. The design criterion used in the case of distributed space–time codes makes sure that the coding gain is significant over all sets of possible relays that have decoded correctly. Although it is difficult to design codes to maximize the coding gain as given by (7.23), this expression gives insight on how to design good codes. The code design should take into consideration the fact that not all of the relays will always transmit in the second phase.

Example 7.1   In this example, a simple design of DSTC with the decode-and-forward protocol is considered. Consider the case of two relay nodes emulating the Alamouti scheme. In the first phase, the source broadcasts the data vector $\mathbf{s} = [s_1, s_2]^{\mathrm{T}}$, where s_1 and s_2 are carved from any QAM constellation. In phase 2, relay nodes emulate a two-antenna transmitter and transmit the 2×2 code given by

$$\begin{pmatrix} s_1 & s_2^* \\ s_2 & -s_1^* \end{pmatrix},$$

where the i-th row is transmitted by the i-th relay node. If a relay nodes decodes in error in phase 1, it will remain idle (i.e., transmits zeros) in phase 2.

We will have four possible states for the relay nodes after phase 1, each of which will depend on which relay(s) decoded correctly or erroneously after phase 1. The first state will correspond to the case when both relays erroneously decoded the source symbol. Under the assumption of having independent relay nodes channels, then the probability of error corresponding to that event will behave as $\propto 1/\mathrm{SNR}^2$ at high SNRs. The second state will correspond to the case when the first relay decodes correctly and the other relay decodes erroneously. The probability of that event will behave as $1/\mathrm{SNR}$ (which corresponds to the probability of error at the second relay node). In this state, relay 1 will transmit its part of the code while relay 2 remains idle. The probability of error at the destination node corresponding to that state will behave as $\propto 1/\mathrm{SNR}$ since we only a transmission from one relay node at the destination. Hence, the contribution of this state to the destination SER will behave as $\propto 1/\mathrm{SNR}^2$ at high SNRs (since it will be the multiplication of the probability of the state and the destination error corresponding to this state). The third state, which corresponds to the case when the first relay decodes erroneously and the second relay decodes correctly, will be similar to the second state. In the fourth state, which corresponds to both relays decoding correctly, the destination error will behave as $\propto 1/\mathrm{SNR}^2$ at high SNRs due to the use of Alamouti scheme as the coding scheme at the relay nodes. Therefore, the overall destination error expression will behave as $\propto 1/\mathrm{SNR}^2$ at high SNRs. In this case, closed from expressions for the SER can be derived (see Exercise 7.2).   ▲

From the above example, we can see that the Alamouti scheme, which achieves full diversity over MIMO channels, can be used in a distributed fashion to achieve full diversity over relay channels.

7.1.2 DSTC with the AF protocol

In this section, the distributed space–time coding based on the amplify-and-forward protocol is introduced. The relays do not perform any hard decision operation on the received data vectors. The system model is presented and a performance analysis is provided.

7.1.2.1 System model

The system has two phases as follows. In phase 1, if n relays are assigned for cooperation, the source transmits data to the relays with power P_1 and the signal received at the k-th relay is as modeled in (7.1) with $L = n$. Without loss of generality, symmetry of the relay nodes is assumed, i.e., $h_{s,r_k} \sim \mathcal{CN}(0, \delta_{s,r}^2)$, $\forall k$ and $h_{r_k,d} \sim \mathcal{CN}(0, \delta_{r,d}^2)$, $\forall k$. In the amplify-and-forward protocol, relay nodes do not decode the received signals. Instead, the relays can only amplify the received signal and perform simple operations such as permutations of the received symbols or other forms of *unitary* linear transformations. Let $\mathbf{A}_k$ denote the $n \times n$ unitary transformation matrix at the k-th relay node. Each relay will normalize the received signal by the factor $\sqrt{(P_2/n)/(P_1\delta_{s,r}^2 + N_0)}$ to satisfy a long-term power constraint. It can be easily shown that this normalization will give a transmitted power per symbol of $P = P_1 + P_2$.

The $n \times 1$ received data vector from the relays at the destination node can be modeled as

$$\mathbf{y_d} = \sqrt{\frac{P_2/n}{P_1\delta_{s,r}^2 + N_0}}\tilde{\mathbf{X}}_r\mathbf{h}_d + \mathbf{v}_d, \qquad (7.24)$$

where $\mathbf{h}_d = [h_{r_1,d}, h_{r_2,d}, \dots, h_{r_n,d}]^T$ is an $n \times 1$ vector channel gains from the n relays to the destination and $h_{r_i,d} \sim \mathcal{CN}(0, \delta_{r,d}^2)$, $\tilde{\mathbf{X}}_r$ is the $n \times n$ code matrix given by

$$\tilde{\mathbf{X}}_r = [h_{s,r_1}\mathbf{A}_1\mathbf{s}, h_{s,r_2}\mathbf{A}_2\mathbf{s}, \dots, h_{s,r_n}\mathbf{A}_n\mathbf{s}],$$

and $\mathbf{v}_d$ denotes additive white Gaussian noise. Each element of $\mathbf{v}_d$ has the distribution of

$$\mathcal{CN}\left(0, N_0\left(1 + \frac{P_2/n}{P_1\delta_{s,r}^2 + N_0}\sum_{i=1}^n |h_{r_i,d}|^2\right)\right),$$

and $\mathbf{v}_d$ accounts for both the noise propagated from the relay nodes as well as the noise generated at the destination. It can be easily shown that restricting the linear transformations at the relay nodes to be unitary causes the elements of the vector $\mathbf{v}_d$ to be independent.

Now, the received vector in (7.24) can be rewritten as

$$\mathbf{y_d} = \sqrt{\frac{P_2 P_1/n}{P_1\delta_{s,r}^2 + N_0}}\mathbf{X}_r\mathbf{h} + \mathbf{v}_d, \qquad (7.25)$$

where

$$\mathbf{h} = [h_{s,r_1}h_{r_1,d}, h_{s,r_2}h_{r_2,d}, \dots, h_{s,r_n}h_{r_n,d}]^T \qquad (7.26)$$

and

$$\mathbf{X}_r = [\mathbf{A}_1 \mathbf{s}, \mathbf{A}_2 \mathbf{s}, \ldots, \mathbf{A}_n \mathbf{s}] \tag{7.27}$$

plays the role of the space–time codeword.

7.1.2.2   Performance analysis

In this section, a pairwise error probability analytical framework is used to derive the code design criteria. The PEP of mistaking $\mathbf{X}_1$ by $\mathbf{X}_2$ can be upper-bounded by the following Chernoff bound:

$$\Pr(\mathbf{X}_1 \to \mathbf{X}_2) \leq$$

$$\mathrm{E}\left\{ \exp\left(-\frac{P_1 P_2/n}{4N_0 \left(P_1 \delta_{\mathrm{s,r}}^2 + N_0 + \frac{P_2}{n}\sum_{i=1}^{n} |h_{\mathrm{r}_i,\mathrm{d}}|^2 \right)} \mathbf{h}^{\mathrm{H}}(\mathbf{X}_1 - \mathbf{X}_2)^{\mathrm{H}}(\mathbf{X}_1 - \mathbf{X}_2)\mathbf{h} \right) \right\}. \tag{7.28}$$

Averaging over the source-to-relay channel coefficients, which are complex Gaussian random variables, gives

$$\Pr(\mathbf{X}_1 \to \mathbf{X}_2) \leq E_{h_{\mathrm{r}_1,\mathrm{d}},\ldots,h_{\mathrm{r}_n,\mathrm{d}}} \det^{-1}\left[\mathbf{I}_n + \frac{\delta_{\mathrm{s,r}}^2 P_1 P_2/n}{4N_0 \left(P_1 \delta_{\mathrm{s,r}}^2 + N_0 + \frac{P_2}{n}\sum_{i=1}^{n} |h_{\mathrm{r}_i,\mathrm{d}}|^2 \right)} \right.$$

$$\left. (\mathbf{X}_1 - \mathbf{X}_2)^{\mathrm{H}}(\mathbf{X}_1 - \mathbf{X}_2)\mathbf{diag}(|h_{\mathrm{r}_1,\mathrm{d}}|^2, |h_{\mathrm{r}_2,\mathrm{d}}|^2, \ldots, |h_{\mathrm{r}_n,\mathrm{d}}|^2) \right], \tag{7.29}$$

where $\mathbf{I}_n$ is the $n \times n$ identity matrix. Define the matrix

$$\mathbf{M} = \frac{\delta_{\mathrm{s,r}}^2 P_1 P_2/n}{4N_0 \left(P_1 \delta_{\mathrm{s,r}}^2 + N_0 + \frac{P_2}{n}\sum_{i=1}^{n} |h_{\mathrm{r}_i,\mathrm{d}}|^2 \right)} \Phi(\mathbf{X}_1, \mathbf{X}_2) \tag{7.30}$$

$$\mathbf{diag}(|h_{\mathrm{r}_1,\mathrm{d}}|^2, |h_{\mathrm{r}_2,\mathrm{d}}|^2, \ldots, |h_{\mathrm{r}_n,\mathrm{d}}|^2).$$

The bound in (7.29) can be written in terms of the eigenvalues of $\mathbf{M}$ as

$$\Pr(\mathbf{X}_1 \to \mathbf{X}_2) \leq E_{h_{\mathrm{r}_1,\mathrm{d}},\ldots,h_{\mathrm{r}_n,\mathrm{d}}} \frac{1}{\prod_{i=1}^{n}(1 + \lambda_{M_i})}, \tag{7.31}$$

where λ_{M_i} is the i-th eigenvalue of the matrix $\mathbf{M}$. If $P_1 = \alpha P$ and $P_2 = (1 - \alpha)P$, where P is the power per symbol for some $\alpha \in (0, 1)$ and define SNR $= P/N_0$, the eigenvalues of $\mathbf{M}$ increase with the increase of the SNR. Now assuming that the matrix $\mathbf{M}$ has full rank of order n, the following approximations hold at high SNR:

$$\prod_{i=1}^{n}(1+\lambda_{M_i})$$

$$\simeq 1 + \prod_{i=1}^{n}\lambda_{M_i}$$

$$= 1 + \left(\frac{\delta_{s,r}^2 P_1 P_2/n}{4N_0\left(P_1\delta_{s,r}^2 + N_0 + \frac{P_2}{n}\sum_{i=1}^{n}|h_{r_i,d}|^2\right)}\right)^n \prod_{i=1}^{n}\lambda_i \prod_{i=1}^{n}|h_{r_i,d}|^2$$

$$\simeq \prod_{i=1}^{n}\left(1 + \frac{\delta_{s,r}^2 P_1 P_2/n}{4N_0\left(P_1\delta_{s,r}^2 + N_0 + \frac{P_2}{n}\sum_{i=1}^{n}|h_{r_i,d}|^2\right)}\lambda_i|h_{r_i,d}|^2\right), \qquad (7.32)$$

where the λ_i's are the eigenvalues of the matrix $\Phi(\mathbf{X}_1, \mathbf{X}_2)$. The determinant of a matrix equals the product of the matrix eigenvalues and the determinant of the multiplication of two matrices equals the product of the individual matrices' determinants.

The PEP in (7.31) can now be approximated at high SNR as

$$\Pr(\mathbf{X}_1 \to \mathbf{X}_2)$$

$$\leq E_{h_{r_1,d},\dots,h_{r_n,d}} \frac{1}{\prod_{i=1}^{n}\left(1 + \frac{\delta_{s,r}^2 P_1 P_2/n}{4N_0\left(P_1\delta_{s,r}^2+N_0+\frac{P_2}{n}\sum_{i=1}^{n}|h_{r_i,d}|^2\right)}\lambda_i|h_{r_i,d}|^2\right)}. \qquad (7.33)$$

Consider now the term $h = \sum_{i=1}^{n}|h_{r_i,d}|^2$ in (7.33), which can be reasonably approximated as $\sum_{i=1}^{n}|h_{r_i,d}|^2 \approx n\delta_{r,d}^2$, especially for large n (by the strong law of large numbers). Averaging the expression in (7.33) over the exponential distribution of $|h_{r_i,d}|^2$ gives

$$\Pr(\mathbf{X}_1 \to \mathbf{X}_2) \leq \prod_{i=1}^{n}\left(\frac{(\delta_{s,r}^2\delta_{r,d}^2 P_1 P_2/n)\lambda_i}{4N_0\left(P_1\delta_{s,r}^2 + N_0 + P_2\delta_{r,d}^2\right)}\right)^{-1}$$

$$\times \prod_{i=1}^{n}\left[-\exp\left(-\frac{4N_0\left(P_1\delta_{s,r}^2 + N_0 + P_2\delta_{r,d}^2\right)}{(\delta_{s,r}^2\delta_{r,d}^2 P_1 P_2/n)\lambda_i}\right)\right.$$

$$\left.\times \mathbf{Ei}\left(-\frac{4N_0\left(P_1\delta_{s,r}^2 + N_0 + P_2\delta_{r,d}^2\right)}{(\delta_{s,r}^2\delta_{r,d}^2 P_1 P_2/n)\lambda_i}\right)\right], \qquad (7.34)$$

where $\mathbf{Ei}(.)$ is the exponential integral function defined as

$$\mathbf{Ei}(\mu) = \int_{-\infty}^{\mu}\frac{\exp(t)}{t}dt, \quad \mu < 0. \qquad (7.35)$$

The exponential integral function can be approximated as μ tends to 0 as

$$-\mathbf{Ei}(\mu) \approx \ln\left(-\frac{1}{\mu}\right), \quad \mu < 0. \qquad (7.36)$$

At high SNR (high P), we have

$$\exp\left(-\frac{4N_0\left(P_1\delta_{\mathrm{s,r}}^2 + N_0 + P_2\delta_{\mathrm{r,d}}^2\right)}{(\delta_{\mathrm{s,r}}^2\delta_{\mathrm{r,d}}^2 P_1 P_2/n)\lambda_i}\right) \approx 1, \qquad (7.37)$$

and using the approximation for the **Ei**(.) function provides the bound in (7.1.2.2) as

$$\Pr(\mathbf{X}_1 \to \mathbf{X}_2)$$

$$\leq \prod_{i=1}^{n}\left(\frac{(\delta_{\mathrm{s,r}}^2\delta_{\mathrm{r,d}}^2 P_1 P_2/n)\lambda_i}{4N_0\left(P_1\delta_{\mathrm{s,r}}^2 + P_2\delta_{\mathrm{r,d}}^2\right)}\right)^{-1} \prod_{i=1}^{n}\ln\left(\frac{(\delta_{\mathrm{s,r}}^2\delta_{\mathrm{r,d}}^2 P_1 P_2/n)\lambda_i}{4N_0\left(P_1\delta_{\mathrm{s,r}}^2 + P_2\delta_{\mathrm{r,d}}^2\right)}\right). \qquad (7.38)$$

Let $P_1 = \alpha P$ and $P_2 = (1-\alpha)P$, where P is the power per symbol, for some $\alpha \in (0, 1)$. With the definition of the SNR as $\mathrm{SNR} = P/N_0$, the bound in (7.38) can be given as

$$\Pr(\mathbf{X}_1 \to \mathbf{X}_2) \leq a_{\mathrm{AF}}\frac{1}{\prod_{i=1}^{n}\lambda_i}\mathrm{SNR}^{-n}\prod_{i=1}^{n}(\ln(\mathrm{SNR}) + \ln(C_i))$$

$$\simeq a_{\mathrm{AF}}\frac{1}{\prod_{i=1}^{n}\lambda_i}\mathrm{SNR}^{-n}(\ln(\mathrm{SNR}))^n, \qquad (7.39)$$

where

$$C_i = \frac{(\delta_{\mathrm{s,r}}^2\delta_{\mathrm{r,d}}^2\alpha(1-\alpha)/n)\lambda_i}{4\left(\alpha\delta_{\mathrm{s,r}}^2 + (1-\alpha)\delta_{\mathrm{r,d}}^2\right)}, \quad i = 1, \ldots, n,$$

are constant terms that do not depend on the SNR and a_{AF} is a constant that depends on the power allocation parameter α and the variances of the channels. The $\ln(C_i)$ terms are neglected at high SNRs resulting in the last bound in (7.39). The diversity order of the system can be calculated as

$$d_{\mathrm{AF}} = \lim_{\mathrm{SNR}\to\infty} -\frac{\log(\mathrm{PEP})}{\log(\mathrm{SNR})} = n. \qquad (7.40)$$

The system will achieve a full diversity of order n if the matrix $\mathbf{M}$ is full rank, that is the code matrix $\Phi(\mathbf{X}_1, \mathbf{X}_2)$ must be full rank of order n over all distinct pairs of codewords $\mathbf{X}_1$ and $\mathbf{X}_2$. It can be easily shown, following the same approach, that if the code matrix $\Phi(\mathbf{X}_1, \mathbf{X}_2)$ is rank deficient, then the system will not achieve full diversity. So any code that is designed to achieve full diversity over MIMO channels will achieve full diversity in the case of amplify-and-forward distributed space–time coding scheme.

If full diversity is achieved, the coding gain is given as

$$C_{\mathrm{AF}} = \left(a_{\mathrm{AF}}\frac{1}{\prod_{i=1}^{n}\lambda_i}\right)^{-\frac{1}{n}}.$$

To maximize the coding gain of the amplify-and-forward distributed space–time codes the product $\prod_{i=1}^{n}\lambda_i$ needs to be maximized, which is the same as the determinant criterion used over MIMO channels as presented in Chapter 2. So if a space–time code is designed to maximize the coding gain over MIMO channels, it will also maximize the coding gain if it can be used in a distributed fashion with the amplify-and-forward protocol.

7.1.3   Synchronization-aware distributed DSTC

In this section, the design of distributed space–time codes that relax the stringent synchronization requirement is considered. Most of the work on cooperative transmission assumed perfect synchronization between the relay nodes, which means that the relays' timings, carrier frequencies, and propagation delays are identical. To simplify the synchronization in the network a diagonal structure is imposed on the space–time code used (refer to the diagonal space–time codes presented in Section 2.1.2). Figure 7.2 shows the time frame structure for the conventional decode-and-forward (amplify-and-forward) distributed space–time codes and the diagonal distributed space–time codes (DDSTCs). The diagonal structure of the code bypasses the perfect synchronization problem by allowing only one relay to transmit at any time slot. Hence, synchronizing simultaneous in-phase transmissions of randomly distributed relay nodes is not necessary. This greatly simplifies the synchronization since nodes can maintain slot synchronization, which means that coarse slot synchronization is available. For example, any synchronization scheme that is used for TDMA systems can be employed to achieve synchronization in the network. However, fine synchronization is more difficult to be achieved. Guard intervals are introduced to ensure that the transmissions from different relays are not overlapped. One relay is allowed to consecutively transmit its part of the space–time code from different data packets. This allows the overhead introduced by the guard intervals to be neglected. Figure 7.3 shows the effect of propagation delay

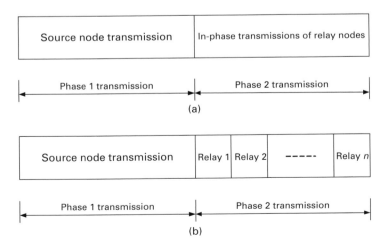

Fig. 7.2   Time frame structure for (a) decode-and-forward (amplify-and-forward)-based system, (b) DDSTC-based system.

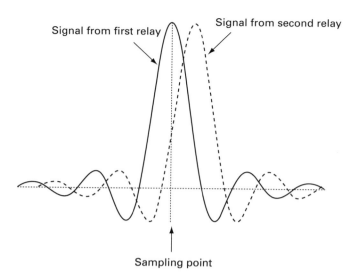

Fig. 7.3　Baseband signals (each is raised cosine pulse-shaped) from two relays at the receiver.

on the received signal from two relays. The sampling time in Figure 7.3 is the optimum sampling time for the first relay signal, but clearly it is not optimal for the second relay signal.

7.1.3.1 System model

In this subsection, the system model with n relay nodes, which helps the source by emulating a diagonal STC, is introduced. The system has two phases with the time frame structure shown in Figure 7.2(b). In phase 1, the received signals at the relay nodes are modeled as in (7.1) with $L = n$.

In phase 2, the k-th relay applies a linear transformation $\mathbf{t}_k$ to the received data vector, where $\mathbf{t}_k$ is an $1 \times n$ row vector, as

$$
\begin{aligned}
y_{r_k} &= \mathbf{t}_k \mathbf{y}_{\mathrm{s},r_k} \\
&= \sqrt{P_1} h_{\mathrm{s},r_k} \mathbf{t}_k \mathbf{s} + \mathbf{t}_k \mathbf{v}_{\mathrm{s},r_k} \\
&= \sqrt{P_1} h_{\mathrm{s},r_k} x_k + v_{r_k},
\end{aligned}
\tag{7.41}
$$

where $x_k = \mathbf{t}_k \mathbf{s}$ and $v_{r_k} = \mathbf{t}_k \mathbf{v}_{\mathrm{s},r_k}$. If the linear transformations are restricted to have unit norm, i.e., $||\mathbf{t}_k||^2 = 1$ for all k, then v_{r_k} is $\mathcal{CN}(0, N_0)$. The relay then multiplies y_{r_k} by the factor

$$
\beta_k \le \sqrt{\frac{P_2}{P_1 |h_{\mathrm{s},r_k}|^2}}
\tag{7.42}
$$

to satisfy a power constraint of $P = P_1 + P_2$ transmitted power per source symbol. The received signal at the destination due to the k-th relay transmission is given by

$$
\begin{aligned}
y_k &= h_{r_k,\mathrm{d}} \beta_k \sqrt{P_1} h_{\mathrm{s},r_k} x_k + h_{r_k,\mathrm{d}} \beta_k v_{r_k} + \tilde{v}_k \\
&= h_{r_k,\mathrm{d}} \beta_k \sqrt{P_1} h_{\mathrm{s},r_k} x_k + z_k, \qquad k = 1, \dots, n,
\end{aligned}
\tag{7.43}
$$

where $\tilde{v}_k$ is modeled as $\mathcal{CN}(0, N_0)$ and hence, z_k, given the channel coefficients is $\mathcal{CN}(0, (\beta_k^2 |h_{\mathrm{r}_k,\mathrm{d}}|^2 + 1)N_0)$, $k = 1, \ldots, n$.

7.1.3.2 Performance Analysis

In this subsection, the code design criterion of the DDSTC based on the PEP analysis is derived. In the following, the power constraint in (7.42) is set to be satisfied with equality.

Now, we start deriving a PEP upper bound to derive the code design criterion. Let σ_k^2 denote the variance of z_k in (7.43) and is given as

$$\sigma_k^2 = \left(\frac{P_2 |h_{\mathrm{r}_k,\mathrm{d}}|^2}{P_1 |h_{\mathrm{s},\mathrm{r}_k}|^2} + 1 \right) N_0, \quad k = 1, \ldots, n. \tag{7.44}$$

Then, define the codeword vector $\mathbf{x}$ from (7.41) as

$$\mathbf{x} = \underbrace{\left[\mathbf{t}_1^\mathrm{T}, \mathbf{t}_2^\mathrm{T}, \ldots, \mathbf{t}_n^\mathrm{T} \right]^\mathrm{T}}_{\mathbf{T}} \mathbf{s} = \mathbf{Ts}, \tag{7.45}$$

where $\mathbf{T}$ is an $n \times n$ linear transformation matrix. From $\mathbf{x}$ define the $n \times n$ code matrix $\mathbf{X} = \mathrm{diag}(\mathbf{x})$, which is a diagonal matrix with the elements of $\mathbf{x}$ on its diagonal. Let $\mathbf{y} = [y_1, y_2, \ldots, y_n]^\mathrm{T}$ denote the received data vector at the destination node as given from (7.43).

Using our system model assumptions, the pdf of $\mathbf{y}$ given the source data vector $\mathbf{s}$ and the channel state information (CSI) is given by

$$p(\mathbf{y}|\mathbf{s}, \mathrm{CSI}) = \left(\prod_{i=1}^n \frac{1}{\pi \sigma_i^2} \right) \exp \left(-\sum_{i=1}^n \frac{1}{\sigma_i^2} \left| y_i - \sqrt{\frac{P_1 P_2}{P_1 |h_{\mathrm{s},\mathrm{r}_i}|^2}} h_{\mathrm{s},\mathrm{r}_i} h_{\mathrm{r}_i,\mathrm{d}} x_i \right|^2 \right). \tag{7.46}$$

From which, the maximum likelihood (ML) decoder can be expressed as

$$\arg \max_{\mathbf{s} \in \mathcal{S}} p(\mathbf{y}|\mathbf{s}, \mathrm{CSI}) = \arg \min_{\mathbf{s} \in \mathcal{S}} \sum_{i=1}^n \frac{1}{\sigma_i^2} \left| y_i - \sqrt{\frac{P_1 P_2}{P_1 |h_{\mathrm{s},\mathrm{r}_i}|^2}} h_{\mathrm{s},\mathrm{r}_i} h_{\mathrm{r}_i,\mathrm{d}} x_i \right|^2, \tag{7.47}$$

where $\mathcal{S}$ is the set of all possible transmitted source data vectors.

The PEP of mistaking $\mathbf{X}_1$ by $\mathbf{X}_2$ can be upper-bounded as

$$\Pr(\mathbf{X}_1 \to \mathbf{X}_2) \leq \mathrm{E} \left\{ \exp \left(\lambda [\ln p(\mathbf{y}|\mathbf{s}_2) - \ln p(\mathbf{y}|\mathbf{s}_1)] \right) \right\}, \tag{7.48}$$

where $\mathbf{X}_1$ and $\mathbf{X}_2$ are the code matrices corresponding to the source data vectors $\mathbf{s}_1$ and $\mathbf{s}_2$, respectively. Equation (7.48) applies for any λ, which is a parameter that can be adjusted to get the tightest bound. Now, the PEP can be written as

$$\Pr(\mathbf{X}_1 \to \mathbf{X}_2) \leq \mathrm{E} \left\{ \exp \left(-\lambda \left[\sum_{i=1}^n \frac{1}{\sigma_i^2} \left(\sqrt{\frac{P_1 P_2}{P_1 |h_{\mathrm{s},\mathrm{r}_i}|^2}} h_{\mathrm{s},\mathrm{r}_i} h_{\mathrm{r}_i,\mathrm{d}} (x_{1i} - x_{2i}) z_i^* \right. \right. \right. \right.$$
$$\left. \left. \left. + \sqrt{\frac{P_1 P_2}{P_1 |h_{\mathrm{s},\mathrm{r}_i}|^2}} h_{\mathrm{s},\mathrm{r}_i}^* h_{\mathrm{r}_i,\mathrm{d}}^* (x_{1i} - x_{2i})^* z_i + \frac{P_1 P_2}{P_1 |h_{\mathrm{s},\mathrm{r}_i}|^2} |h_{\mathrm{s},\mathrm{r}_i}|^2 |h_{\mathrm{r}_i,\mathrm{d}}|^2 |x_{1i} - x_{2i}|^2 \right) \right] \right) \right\}, \tag{7.49}$$

where the expectation is over the noise and channel coefficients statistics and x_{ij} is the j-th element of the i-th code vector.

To average the expression in (7.49) over the noise statistics, define the receiver noise vector $\mathbf{z} = [z_1, z_2, \ldots, z_n]^{\mathrm{T}}$, where z_i's are as defined in (7.43). The pdf of $\mathbf{z}$ given the channel state information is given by

$$p(\mathbf{z}|CSI) = \left(\prod_{i=1}^{n} \frac{1}{\pi \sigma_i^2} \right) \exp\left(-\sum_{i=1}^{n} \frac{1}{\sigma_i^2} z_i z_i^* \right). \tag{7.50}$$

Taking the expectation in (7.49) over $\mathbf{z}$ given the channel coefficients yields

$$\Pr(\mathbf{X}_1 \rightarrow \mathbf{X}_2)$$

$$\leq \mathrm{E}\left\{ \exp\left(-\lambda(1-\lambda) \sum_{i=1}^{n} \frac{1}{\sigma_i^2} \frac{P_1 P_2}{P_1 |h_{\mathrm{s},\mathrm{r}_i}|^2} \left(|h_{\mathrm{s},\mathrm{r}_i}|^2 |h_{\mathrm{r}_i,\mathrm{d}}|^2 |x_{1i} - x_{2i}|^2 \right) \right) \right.$$

$$\left. \times \int_{\mathbf{z}} \left(\prod_{i=1}^{n} \frac{1}{\pi \sigma_i^2} \right) \exp\left(-\sum_{i=1}^{n} \frac{1}{\sigma_i^2} |z_i + \lambda \sqrt{\frac{P_1 P_2}{P_1 |h_{\mathrm{s},\mathrm{r}_i}|^2}} h_{\mathrm{s},\mathrm{r}_i} h_{\mathrm{r}_i,\mathrm{d}} (x_{1i} - x_{2i})|^2 \right) d\mathbf{z} \right\}$$

$$= \mathrm{E}\left\{ \exp\left(-\lambda(1-\lambda) \sum_{i=1}^{n} \frac{1}{\sigma_i^2} \frac{P_1 P_2}{P_1 |h_{\mathrm{s},\mathrm{r}_i}|^2} \left(|h_{\mathrm{s},\mathrm{r}_i}|^2 |h_{\mathrm{r}_i,\mathrm{d}}|^2 |x_{1i} - x_{2i}|^2 \right) \right) \right\}. \tag{7.51}$$

Choose $\lambda = 1/2$ that maximizes the term $\lambda(1-\lambda)$, i.e., minimizes the PEP upper bound. Substituting for σ_i^2's from (7.44), the PEP can be upper-bounded as

$$\Pr(\mathbf{X}_1 \rightarrow \mathbf{X}_2) \leq \mathrm{E}\left\{ \exp\left(-\frac{1}{4} \sum_{i=1}^{n} \frac{P_1 |h_{\mathrm{s},\mathrm{r}_i}|^2 P_2 |h_{\mathrm{r}_i,\mathrm{d}}|^2}{(P_1 |h_{\mathrm{s},\mathrm{r}_i}|^2 + P_2 |h_{\mathrm{r}_i,\mathrm{d}}|^2) N_0} |x_{1i} - x_{2i}|^2 \right) \right\}. \tag{7.52}$$

To get the expression in (7.52), let us define the variable

$$\gamma_i = \frac{P_1 |h_{\mathrm{s},\mathrm{r}_i}|^2 P_2 |h_{\mathrm{r}_i,\mathrm{d}}|^2}{(P_1 |h_{\mathrm{s},\mathrm{r}_i}|^2 + P_2 |h_{\mathrm{r}_i,\mathrm{d}}|^2) N_0}, \quad i = 1, \ldots, n,$$

which is the scaled harmonic mean[1] of the two exponential random variables $P_1 |h_{\mathrm{s},\mathrm{r}_i}|^2 / N_0$ and $P_2 |h_{\mathrm{r}_i,\mathrm{d}}|^2 / N_0$. Averaging the expression in (7.52) over the channel coefficients, the upper bound on the PEP can be expressed as

$$\Pr(\mathbf{X}_1 \rightarrow \mathbf{X}_2) \leq \prod_{i=1, x_{1i} \neq x_{2i}}^{n} M_{\gamma_i}\left(\frac{1}{4} |x_{1i} - x_{2i}|^2 \right), \tag{7.53}$$

where $M_{\gamma_i}(.)$ is the moment-generating function (MGF) of the random variable γ_i. Using the result in Chapter 4 (Equation (4.66)), the MGF for γ_i can be approximated at high enough SNR to be

$$M_{\gamma_i}(s) \simeq \frac{\zeta_i}{s}, \tag{7.54}$$

---

[1] The scaling factor is 1/2 since the harmonic mean of two numbers, g_1 and g_2, is $\frac{2g_1 g_2}{g_1 + g_2}$.

where

$$\zeta_i = \frac{N_0}{P_1 \delta_{s,r}^2} + \frac{N_0}{P_2 \delta_{r,d}^2}.$$

The PEP can now be upper-bounded as

$$\Pr(\mathbf{X}_1 \rightarrow \mathbf{X}_2)$$

$$\leq N_0^n \left(\prod_{i=1, x_{1i} \neq x_{2i}}^{n} \left(\frac{1}{P_1 \delta_{s,r}^2} + \frac{1}{P_2 \delta_{r,d}^2} \right) \right) \left(\prod_{i=1, x_{1i} \neq x_{2i}}^{n} \frac{1}{4} |x_{1i} - x_{2i}|^2 \right)^{-1}. \quad (7.55)$$

Let $P_1 = \alpha P$ and $P_2 = (1-\alpha)P$, where P is the power per symbol, for some $\alpha \in (0, 1)$ and define $SNR = P/N_0$. The diversity order d_{DDSTC} of the system is

$$d_{\text{DDSTC}} = \lim_{SNR \rightarrow \infty} -\frac{\log(\text{PEP})}{\log(\text{SNR})} = \min_{m \neq j} \text{rank}(\mathbf{X}_m - \mathbf{X}_j), \quad (7.56)$$

where $\mathbf{X}_m$ and $\mathbf{X}_j$ are two possible code matrices. To achieve a diversity order of n, the matrix $\mathbf{X}_m - \mathbf{X}_j$ should be of full rank for any $m \neq j$ (that is $x_{mi} \neq x_{ji}$ $\forall m \neq j$, $\forall i = 1, \dots, n$). Intuitively, if two code matrices exist for which the rank of the matrix $\mathbf{X}_m - \mathbf{X}_j$ is not n this means that they have at least one diagonal element that is the same in both matrices. Clearly, this element can not be used to decide between these two possible transmitted code matrices and hence, the diversity order of the system is reduced. This criterion implies that each element in the code matrix is unique to that matrix and any other matrix will have a different element at that same location and this is really the source of diversity. Furthermore, to minimize the PEP bound in (7.55) we need to maximize

$$\min_{m \neq j} \left(\prod_{i=1}^{n} |x_{mi} - x_{ji}|^2 \right)^{1/n}, \quad (7.57)$$

which is called the minimum product distance of the set of symbols $\mathbf{s} = [s_1, s_2, \dots, s_n]^T$. This is the same criteria used to design full-rate, full-diversity space–frequency codes in Chapter 3. We can use the design presented in Chapter 3 to design the DDSTC.

Example 7.2 In this example, the performance of the different schemes with two relays helping the source are compared. In the simulations, the variance of any source–relay or relay–destination channel is taken to be 1. Figure 7.4 shows the simulations for two decode-and-forward systems using the Alamouti scheme (DF Alamouti) and the diagonal STC (DF DAST), distributed space–time codes based on the linear dispersion (LD) space–time codes (LD-DSTC) [93] which are based on the AF scheme, and the DDSTC. In the LD-DSTC, the relay nodes apply linear transformations to the received data vectors. The linear dispersion matrices at the relay nodes are designed to guarantee that full diversity is achieved at the destination node. In Figure 7.4, the orthogonal distributed space–time codes (O-DSTC) in [94, 97] are simulated. The O-DSTC are similar to the LD-DSTC in the way the code is constructed. However, the design of the dispersion

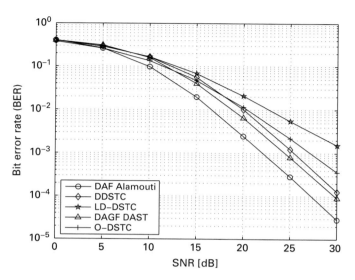

Fig. 7.4   BER for two relays with data rate 1 bit/sym.

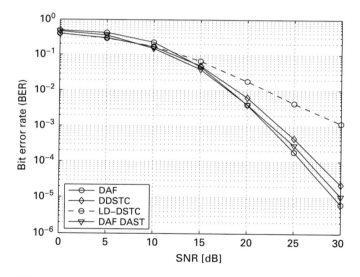

Fig. 7.5   BER for three relays with data rate 1 bit/sym.

matrices at the relay nodes in the case of O-DSTC is done in a way to reduce the complexity of the detector at the receiver. For the case of two relays, O-DSTC reduces to the Alamouti-based DSTC.

All of the above systems have a data rate of (1/2). QPSK modulation is used, which means that a rate of one transmitted bit per symbol (1 bit/sym) is achieved. Figure 7.5 shows the simulation results for two decode-and-forward systems using the $\mathcal{G}_3$ ST block code presented in Chapter 2 and the diagonal STC (DF DAST), LD-DSTC, and the DDSTC. For fair comparison the number of transmitted bits per symbol is fixed to be

1 bit/sym. The $\mathcal{G}_3$ ST block code has a data rate of (1/2) [211], which results in an overall system data rate of (1/3). Therefore, 8-PSK modulation is employed for the system that uses the $\mathcal{G}_3$ ST block code. For the other three systems QPSK modulation is used as these systems have a data rate of (1/2). Clearly, the decode-and-forward-based schemes outperform the amplify-and-forward-based schemes but this is under the assumption that each relay node can decide whether it has decoded correctly or not. Intuitively, the decode-and-forward will deliver signals that are less noisy to the destination. The noise is suppressed at the relay nodes by transmitting a noise-free version of the signal. The amplify-and-forward delivers more noise to the destination due to noise propagation from the relay nodes. However, the amplify-and-forward based schemes have the advantage of simple processing at the relay nodes, which makes the design of relay nodes simpler in this case compared to the decode-and-forward-based schemes.

▲

Example 7.3   In this example, the effect of the synchronization errors on the system BER performance is investigated. Figure 7.6 shows the case of having two relays helping the source and propagation delay mismatches of $T_2 = 0.2T$, $0.4T$, and $0.6T$, where T is the time slot duration. Raised cosine pulse-shaped waveforms were used with roll-off factor of 0.2 and QPSK modulation. Figure 7.7 shows the case of having three relays helping the source for different propagation delay mismatches. A decode-and-forward (DF) system using the $\mathcal{G}_3$ ST block code of [211] and the DDSTC were compared. For fair comparison the number of transmitted bits per symbol is fixed to be 1 bit/sym. Again, the $\mathcal{G}_3$ ST block code has a data rate of (1/2) [211], which results in an overall system data rate to be (1/3). Therefore, 8-PSK modulation is employed for the

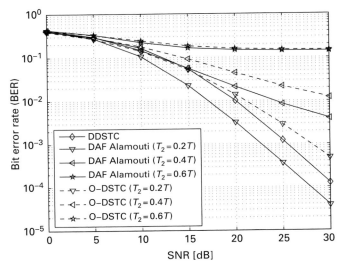

Fig. 7.6   BER performance with propagation delay mismatch: two-relay case.

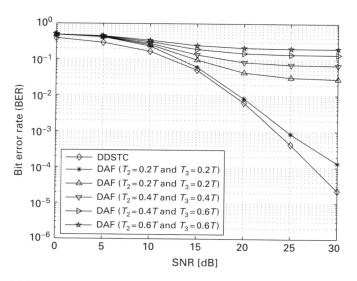

Fig. 7.7   BER performance with propagation delay mismatch: three-relays case.

system that uses the $\mathcal{G}_3$ ST block code. For the DDSTC, QPSK modulation is used as the system has a data rate of (1/2). Raised cosine pulse-shaped waveforms with a roll-off factor of 0.2 are used. It is clear that the synchronization errors can severely reduce the system BER performance. The DDSTC bypasses this problem by allowing only one relay transmission at any time slot. The diagonal structure of the code mitigates the effect of time synchronization mismatches.   ▲

7.2   Distributed space–frequency coding (DSFC)

In this section, we will consider the design of distributed space–frequency coding (DSFC) for broadband multipath fading channels to exploit the frequency diversity of the channel. The presence of multipaths in broadband channels provides another means for achieving diversity across the frequency axis. Exploiting the frequency axis diversity can highly improve the system performance by achieving higher diversity orders. The main problem for the wireless relay network is how to design space–frequency codes distributed among *spatially separated* relay nodes while guaranteeing to achieve full diversity at the destination node. The spatial separation of the relay nodes presents other challenges for the design of DSFCs such as time synchronization and carrier offset synchronization.

In this section, we will present some structures for distributed space–frequency codes (DSFCs) over wireless broadband relay networks. The presented DSFCs are designed to achieve the frequency and cooperative diversities of the wireless relay channels. The use of DSFCs with the decode-and-forward (DF) and amplify-and-forward (AF) protocols is considered. The code design criteria to achieve full diversity, based on the pairwise error probability (PEP) analysis, are derived. For DSFC with the DF protocol, a two-stage coding scheme, with source node coding and relay nodes coding, is presented. We derive sufficient conditions for the code structures at the source and relay nodes to achieve full diversity of order NL, where N is the number of relay nodes and L is the number of paths per channel. For the case of DSFC with the AF protocol, a structure for distributed space–frequency coding will be presented and sufficient conditions for that structure to achieve full diversity will then be derived.

7.2.1   DSFC with the DF protocol

In this section, the design and performance analysis for DSFCs with the DF protocol are presented. A two-stage structure is presented for the DSFCs with the DF protocol

7.2.1.1   System model

In this section, the system model is presented. The following notations are used: $\mathbf{diag}(\mathbf{y})$ is the $T \times T$ diagonal matrix with the elements of $\mathbf{y}$ on its diagonal, where $\mathbf{y}$ is a $T \times 1$ vector.

Without loss of generality, we assume a two-hop relay channel model, where there is no direct link from the source node to the destination node. The case when a direct link exists between the source node and the destination node will be discussed in Section 7.2.3. A schematic system model is depicted in Figure 7.8. The system is based on OFDM modulation with K subcarriers. The channel between the source node and the n-th relay node is modeled as a multipath fading channel with L paths as

$$h_{s,r_n}(\tau) = \sum_{l=1}^{L} \alpha_{s,r_n}(l)\delta(\tau - \tau_l), \tag{7.58}$$

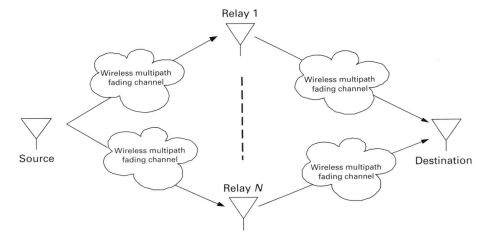

Fig. 7.8   Simplified system model for the distributed space–frequency codes.

where τ_l is the delay of the l-th path, $\delta(\cdot)$ is the Dirac delta function, and $\alpha_{s,r_n}(l)$ is the complex amplitude of the l-th path. The $\alpha_{s,r_n}(l)$'s are modeled as zero-mean, complex Gaussian random variables with variance $E\left[|\alpha_{s,r_n}(l)|^2\right] = \sigma^2(l)$. We assume symmetry between the relay nodes for simplicity of presentation; the analysis can be easily extended to the asymmetric case. The channels are normalized such that the channel variance $\sum_{l=1}^{L}\sigma^2(l) = 1$. A cyclic prefix is introduced to convert the multipath frequency-selective fading channels to flat fading subchannels.

The system has two phases as follows. In phase 1, the source node broadcasts the information to the N relays. The received signal in the frequency domain on the k-th subcarrier at the n-th relay node is given by

$$y_{s,r_n}(k) = \sqrt{P_1}H_{s,r_n}(k)s(k) + v_{s,r_n}(k), \quad k = 1, \ldots, K; \quad n = 1, \ldots, N, \quad (7.59)$$

where P_1 is the transmitted source node power, $H_{s,r_n}(k)$ is the channel attenuation of the source node to the n-th relay node channel on the k-th subcarrier, $s(k)$ is the transmitted source node symbol on the k-th subcarrier with $E\left\{|s(k)|^2\right\} = 1$, and $v_{s,r_n}(k)$ is the n-th relay node additive white Gaussian noise on the k-th subcarrier that is modeled as zero-mean circularly symmetric complex Gaussian random variable with variance $N_0/2$ per dimension. The subcarrier noise terms are statistically independent assuming that the time-domain noise samples are statistically independent and identically distributed.[2] In (7.59), $H_{s,r_n}(k)$ is given by

$$H_{s,r_n}(k) = \sum_{l=1}^{L}\alpha_{s,r_n}(l)e^{-j2\pi(k-1)\Delta f \tau_l}, \quad k = 1, \ldots, K, \quad (7.60)$$

where $\Delta f = 1/T$ is the subcarrier frequency separation and T is the OFDM symbol duration. We assume perfect channel state information at any receiving node but no channel information at transmitting nodes.

In phase 2, relays that have decoded correctly in phase 1 will forward the source node information. Each relay is assumed to be able to decide whether it has decoded correctly or not. This can be achieved through the use of error detecting codes such as the *cyclic redundancy* codes (CRC).

The transmitted $K \times N$ space–frequency (SF) codeword from the relay nodes is given by

$$\mathbf{C}_r = \begin{pmatrix} C_r(1,1) & C_r(1,2) & \cdots & C_r(1,N) \\ C_r(2,1) & C_r(2,2) & \cdots & C_r(2,N) \\ \vdots & \vdots & \ddots & \vdots \\ C_r(K,1) & C_r(K,2) & \cdots & C_r(K,N) \end{pmatrix}, \quad (7.61)$$

---

[2] FFT, which is used to transform the received data from the time-domain to the frequency-domain, can be represented by a unitary matrix multiplication. Unitary transformation of a Gaussian random vector, whose components are statistically independent and identically distributed, results in a Gaussian random vector with statistically independent and identically distributed components.

where $C_r(k, n)$ is the symbol transmitted by the n-th relay node on the k-th subcarrier. Note that $\mathbf{C}_r$ will be SF code transmitted by the relay nodes if all of them have decoded correctly in phase 1. The SF is assumed to satisfy the power constraint $||\mathbf{C}_r||_F^2 \leq K$.

The received signal at the destination node on the k-th subcarrier is given by

$$y_d(k) = \sqrt{P_2} \sum_{n=1}^{N} H_{r_n,d}(k) C_r(k, n) I_n + v_{r_n,d}(k), \qquad (7.62)$$

where P_2 is the relay node power, $H_{r_n,d}(k)$ is the attenuation of the channel between the n-th relay node and the destination node on the k-th subcarrier, $v_{r_n,d}(k)$ is the destination additive white Gaussian noise on the k-th subcarrier, and I_n is the state of the n-th relay. I_n will equal 1 if the n-th relay has decoded correctly in phase 1, otherwise, I_n will equal 0.

7.2.1.2 Performance analysis

It is now necessary to develop sufficient code design criteria for the DSFC to achieve full diversity of order $N \times L$. Unlike the case of MIMO space–frequency coding, we will need a two-stage coding to achieve full diversity at the destination node. Therefore, the presented DSFCs will have two stages of coding: the first stage is coding at the source node and the second stage is coding at the relay nodes. The transmitted source node code will be designed to guarantee a diversity of order L at any relay node, and this will in turn cause the presented DSFC to achieve full diversity of order $N \times L$, as will be shown later.

7.2.1.3 Source node coding

Due to the symmetry assumption, the pairwise error probability (PEP) is the same at any relay node. For two distinct transmitted source node symbols, $\mathbf{s}$ and $\tilde{\mathbf{s}}$, the PEP can be tightly upper-bounded as (refer to Chapter 3)

$$PEP(\mathbf{s} \rightarrow \tilde{\mathbf{s}}) \leq \binom{2v-1}{v} \left(\prod_{i=1}^{v} \lambda_i \right)^{-1} \left(\frac{P_1}{N_0} \right)^{-v} \qquad (7.63)$$

and v is the rank of the matrix $\mathbf{C} \circ \mathbf{R}$ where

$$\mathbf{C} = (\mathbf{s} - \tilde{\mathbf{s}})(\mathbf{s} - \tilde{\mathbf{s}})^H,$$

$$\mathbf{R} = E\left\{ \mathbf{H}_{s,r_n} \mathbf{H}_{s,r_n}^H \right\},$$

and

$$\mathbf{H}_{s,r_n} = [H_{s,r_n}(1), \ldots, H_{s,r_n}(K)]^T.$$

Here the λ_i's are the nonzero eigenvalues of the matrix $\mathbf{C} \circ \mathbf{R}$, where $\circ$ denotes the Hadamard product.

The correlation matrix, $\mathbf{R}$, of the channel impulse response can be found as

$$
\begin{aligned}
\mathbf{R} &= \mathrm{E}\left\{\mathbf{H}_{\mathrm{s},\mathrm{r}_n}\mathbf{H}_{\mathrm{s},\mathrm{r}_n}^{\mathrm{H}}\right\} \\
&= \mathbf{W}\mathrm{E}\left\{\alpha_{\mathrm{s},\mathrm{r}_n}\alpha_{\mathrm{s},\mathrm{r}_n}^{\mathrm{H}}\right\}\mathbf{W}^{\mathrm{H}} \\
&= \mathbf{W}\mathrm{diag}\{\sigma^2(1), \sigma^2(2), \ldots, \sigma^2(L)\}\mathbf{W}^{\mathrm{H}},
\end{aligned}
\tag{7.64}
$$

where

$$
\alpha_{\mathrm{s},\mathrm{r}_n} = [\alpha_{\mathrm{s},\mathrm{r}_n}(1), \alpha_{\mathrm{s},\mathrm{r}_n}(2), \ldots, \alpha_{\mathrm{s},\mathrm{r}_n}(L)]^{\mathrm{T}},
$$

$$
\mathbf{W} = \begin{pmatrix}
1 & 1 & \cdots & 1 \\
w^{\tau_1} & w^{\tau_2} & \cdots & w^{\tau_L} \\
\vdots & \vdots & \ddots & \vdots \\
w^{(K-1)\tau_1} & w^{(K-1)\tau_2} & \cdots & w^{(K-1)\tau_L}
\end{pmatrix},
$$

and $w = \mathrm{e}^{-\mathrm{j}2\pi\Delta f}$.

The coding at the source node is implemented to guarantee a diversity of order L, which is the maximum achievable diversity order at any relay node. The transmitted $K \times 1$ source node code is partitioned into subblocks of length L and each subblock will be designed to guarantee a diversity of order L at any relay node as will be seen later. Let $M = \lfloor K/L \rfloor$ denote the number of subblocks in the source node transmitted OFDM block. The transmitted $K \times 1$ source node code is given as

$$
\begin{aligned}
\mathbf{s} &= [s(1), s(2), \ldots, s(K)]^{\mathrm{T}} \\
&= [\mathbf{F}_1^{\mathrm{T}}, \mathbf{F}_2^{\mathrm{T}}, \ldots, \mathbf{F}_M^{\mathrm{T}}, \mathbf{0}_{K-ML}^{\mathrm{T}}]^{\mathrm{T}},
\end{aligned}
\tag{7.65}
$$

where

$$
\mathbf{F}_i = [F_i(1), \ldots, F_i(L)]^{\mathrm{T}}
$$

is the i-th subblock of dimension $L \times 1$. Zeros are padded if K is not an integer multiple of L. For any two distinct source codewords, $\mathbf{s}$ and $\tilde{\mathbf{s}} = [\tilde{\mathbf{F}}_1^{\mathrm{T}}, \tilde{\mathbf{F}}_2^{\mathrm{T}}, \ldots, \tilde{\mathbf{F}}_M^{\mathrm{T}}, \mathbf{0}_{K-ML}^{\mathrm{T}}]^{\mathrm{T}}$, at least one index p_0 exists for which $\mathbf{F}_{p_0}$ is not equal to $\tilde{\mathbf{F}}_{p_0}$.

Based on the presented structure of the transmitted code from the source node, sufficient conditions for the code to achieve a diversity of order L at the relay nodes are derived. We assume for $\mathbf{s}$ and $\tilde{\mathbf{s}}$ that $\mathbf{F}_p = \tilde{\mathbf{F}}_p$ for all $p \neq p_0$, which corresponds to the worst-case PEP. This does not decrease the rank of the matrix $\mathbf{C} \circ \mathbf{R}$.

Define the $L \times L$ matrix $\mathbf{Q} = \{q_{i,j}\}$ as

$$
q_{i,j} = \sum_{l=1}^{L}\sigma^2(l)w^{(i-j)\tau(l)}, \quad 1 \le i, j \le L.
$$

Note that the nonzero eigenvalues of the matrix $\mathbf{C} \circ \mathbf{R}$ are the same as those of the matrix $\left(\mathbf{F}_{po} - \tilde{\mathbf{F}}_{po} \right) \left(\mathbf{F}_{po} - \tilde{\mathbf{F}}_{po} \right)^{\mathrm{H}} \circ \mathbf{Q}$. Hence, we have

$$
\left(\mathbf{F}_{po} - \tilde{\mathbf{F}}_{po} \right) \left(\mathbf{F}_{po} - \tilde{\mathbf{F}}_{po} \right)^{\mathrm{H}} \circ \mathbf{Q}
$$
$$
= \left[\mathbf{diag} \left(\mathbf{F}_{po} - \tilde{\mathbf{F}}_{po} \right) \mathbf{1}_{L \times L} \mathbf{diag} \left(\mathbf{F}_{po} - \tilde{\mathbf{F}}_{po} \right)^{\mathrm{H}} \right] \circ \mathbf{Q}
$$
$$
= \mathbf{diag} \left(\mathbf{F}_{po} - \tilde{\mathbf{F}}_{po} \right) \mathbf{Q} \, \mathbf{diag} \left(\mathbf{F}_{po} - \tilde{\mathbf{F}}_{po} \right)^{\mathrm{H}} \qquad (7.66)
$$

where $\mathbf{1}_{L \times L}$ is the $L \times L$ matrix whose all elements are ones. The last equality follows from a property of the Hadamard product.

If all of the eignenvalues of the matrix $\left(\mathbf{F}_{po} - \tilde{\mathbf{F}}_{po} \right) \left(\mathbf{F}_{po} - \tilde{\mathbf{F}}_{po} \right)^{\mathrm{H}} \circ \mathbf{Q}$ are nonzero, then their product can be calculated as

$$
\det \left(\left(\mathbf{F}_{po} - \tilde{\mathbf{F}}_{po} \right) \left(\mathbf{F}_{po} - \tilde{\mathbf{F}}_{po} \right)^{\mathrm{H}} \circ \mathbf{Q} \right)
$$
$$
= \det \left(\mathbf{diag} \left(\mathbf{F}_{po} - \tilde{\mathbf{F}}_{po} \right) \right) \det (\mathbf{Q}) \det \left(\mathbf{diag} \left(\mathbf{F}_{po} - \tilde{\mathbf{F}}_{po} \right)^{\mathrm{H}} \right)
$$
$$
= \prod_{l=1}^{L} \left| F_{po}(l) - \tilde{F}_{po}(l) \right|^2 (\det(Q)). \qquad (7.67)
$$

The matrix $\mathbf{Q}$ is nonsingular. Hence, if the product $\prod_{l=1}^{L} \left| F_{po}(l) - \tilde{F}_{po}(l) \right|^2$ is nonzero over all possible pairs of distinct transmitted source codewords, $\mathbf{s}$ and $\tilde{\mathbf{s}}$, then a diversity of order L will be achieved at each relay node.

In phase 2, relays that have decoded correctly in phase 1 will forward the source node information. The received signal at the destination node on the k-th subcarrier is as given in (7.62). The state of the n-th relay node I_n is a Bernoulli random variable with a probability mass function (pmf) given by

$$
I_n = \begin{cases} 0 & \text{with probability} = \text{SER} \\ 1 & \text{with probability} = 1 - \text{SER}, \end{cases}
$$

where SER is the symbol error rate at the n-th relay node. Note that SER is the same for any relay node due to the symmetry assumption. If the transmitted code from the source node is designed such that the product $\prod_{l=1}^{L} \left| F_{po}(l) - \tilde{F}_{po}(l) \right|^2$ is nonzero, for at least one index p_0, over all the possible pairs of distinct transmitted source codewords, $\mathbf{s}$ and $\tilde{\mathbf{s}}$, then the SER at the n-th relay node can be upper-bounded as

$$
\text{SER} = \sum_{\mathbf{s} \in \mathcal{S}} \Pr\{\mathbf{s}\} \Pr\{\text{error given that } \mathbf{s} \text{ was transmitted}\}
$$
$$
\leq \sum_{\mathbf{s} \in \mathcal{S}} \Pr\{\mathbf{s}\} \sum_{\tilde{\mathbf{s}} \in \mathcal{S}, \tilde{\mathbf{s}} \neq \mathbf{s}} \text{PEP}(\mathbf{s} \to \tilde{\mathbf{s}})
$$
$$
\leq c \times \text{SNR}^{-L}, \qquad (7.68)
$$

where S is the set of all possible transmitted source codewords and c is a constant that does not depend on the SNR. The first inequality follows from the union upper bound and the second inequality follows from (7.63), where SNR is defined as $\text{SNR} = P_1/N_0$.

7.2.1.4   Relay Nodes Coding

Next, the design of the SF code at the relay nodes to achieve a diversity of order NL is considered. We will present SF codes constructed from the concatenation of block diagonal matrices, which is similar to the structure used in Chapter 3 to design full-rate, full-diversity space–frequency codes over MIMO channels. We will derive sufficient conditions for that code structure to achieve full diversity at the destination node.

Let $P = \lfloor K/NL \rfloor$ denote the number of subblocks in the transmitted OFDM block from the relay nodes. The transmitted $K \times N$ SF codeword from the relay nodes, if all relays decoded correctly, is given by

$$\mathbf{C}_r = [\mathbf{G}_1^\mathrm{T}, \mathbf{G}_2^\mathrm{T}, \dots, \mathbf{G}_P^\mathrm{T}, \mathbf{0}_{K-PLN}^\mathrm{T}]^\mathrm{T}, \tag{7.69}$$

where $\mathbf{G}_i$ is the i-th subblock of dimension $NL \times N$. Zeros are padded if K is not an integer multiple of NL. Each $\mathbf{G}_i$ is a block diagonal matrix that has the structure

$$\mathbf{G}_i = \begin{pmatrix} \mathbf{X}_{1_{L \times 1}} & \mathbf{0}_{L \times 1} & \cdots & \mathbf{0}_{L \times 1} \\ \mathbf{0}_{L \times 1} & \mathbf{X}_{2_{L \times 1}} & \cdots & \mathbf{0}_{L \times 1} \\ \vdots & \vdots & \ddots & \vdots \\ \mathbf{0}_{L \times 1} & \mathbf{0}_{L \times 1} & \cdots & \mathbf{X}_{N_{L \times 1}} \end{pmatrix} \tag{7.70}$$

and let

$$\mathbf{X} = [\mathbf{X}_1^\mathrm{T}, \mathbf{X}_2^\mathrm{T}, \cdots, \mathbf{X}_N^\mathrm{T}] = [x(1), x(2), \cdots, x(NL)].$$

For two distinct transmitted source codewords, $\mathbf{s}$ and $\tilde{\mathbf{s}}$, and a given realization of the relays states $\mathbf{I} = [I_1, I_2, \dots, I_n]^\mathrm{T}$, the conditional PEP can be tightly upper-bounded as

$$\text{PEP}(\mathbf{s} \rightarrow \tilde{\mathbf{s}}/\mathbf{I}) \leq \left(\frac{2\kappa - 1}{\kappa} \right) \left(\prod_{i=1}^{\kappa} \lambda_i \right)^{-1} \left(\frac{P_2}{N_0} \right)^{-\kappa}, \tag{7.71}$$

and κ is the rank of the matrix $\mathbf{C}(\mathbf{I}) \circ \mathbf{R}$ where

$$\mathbf{C}(\mathbf{I}) = (\mathbf{C}_r - \tilde{\mathbf{C}}_r)\,\text{diag}(\mathbf{I})(\mathbf{C} - \tilde{\mathbf{C}}_r)^\mathrm{H}.$$

For two source codewords, $\mathbf{s}$ and $\tilde{\mathbf{s}}$, at least one index p_0 exists for which $\mathbf{G}_{p_0} \neq \tilde{\mathbf{G}}_{p_0}$. We assume for $\mathbf{s}$ and $\tilde{\mathbf{s}}$ that $\mathbf{G}_p = \tilde{\mathbf{G}}_p$ for all $p \neq p_0$. As for the source node coding case, this does not decrease the rank of the matrix $\mathbf{C}(\mathbf{I}) \circ \mathbf{R}$ that corresponds to any realization $\mathbf{I}$ of the relays states.

Define the $NL \times NL$ matrix $\mathbf{S} = \{s_{i,j}\}$ as

$$s_{i,j} = \sum_{l=1}^{L} \sigma^2(l) w^{(i-j)\tau(l)}, \quad 1 \leq i, j \leq NL.$$

Note that the nonzero eigenvalues of the matrix $\mathbf{C}(\mathbf{I}) \circ \mathbf{R}$ are the same as the nonzero eigenvalues of the matrix $\left(\mathbf{G}_{p_0}(\mathbf{I}) - \tilde{\mathbf{G}}_{p_0}(\mathbf{I}) \right) \left(\mathbf{G}_{p_0}(\mathbf{I}) - \tilde{\mathbf{G}}_{p_0}(\mathbf{I}) \right)^\mathrm{H} \circ \mathbf{S}$, where $\mathbf{G}_{p_0}(\mathbf{I})$ is formed from $\mathbf{G}_{p0}$ by setting the columns corresponding to the relays that have decoded erroneously to zeros. Hence,

$$\left(\mathbf{G}_{p0}(\mathbf{I}) - \tilde{\mathbf{G}}_{p0}(\mathbf{I})\right)\left(\mathbf{G}_{p0}(\mathbf{I}) - \tilde{\mathbf{G}}_{p0}(\mathbf{I})\right)^{\mathrm{H}} \circ \mathbf{S}$$

$$= \left(\mathbf{diag}(\mathbf{X} - \tilde{\mathbf{X}})\,(\mathbf{diag}(\mathbf{I}) \otimes \mathbf{1}_{L\times 1})\,(\mathbf{diag}(\mathbf{I}) \otimes \mathbf{1}_{L\times 1})^{\mathrm{H}}\,\mathbf{diag}(\mathbf{X} - \tilde{\mathbf{X}})^{\mathrm{H}}\right) \circ \mathbf{S}$$

$$= \left(\mathbf{diag}(\mathbf{X} - \tilde{\mathbf{X}})\,(\mathbf{diag}(\mathbf{I}) \otimes \mathbf{1}_{L\times L})\,\mathbf{diag}(\mathbf{X} - \tilde{\mathbf{X}})^{\mathrm{H}}\right) \circ \mathbf{S}$$

$$= \mathbf{diag}(\mathbf{X} - \tilde{\mathbf{X}})\left[(\mathbf{diag}(\mathbf{I}) \otimes \mathbf{1}_{L\times L}) \circ \mathbf{S}\right]\mathbf{diag}(\mathbf{X} - \tilde{\mathbf{X}})^{\mathrm{H}}, \tag{7.72}$$

where the second and the third equalities follow from the properties of the tensor and Hadamard products.

Let

$$n_{\mathbf{I}} = \sum_{n=1}^{N} I_n$$

denote the number of relays that have decoded correctly corresponding to a realization $\mathbf{I}$ of the relays states. Using (7.72), the product of the nonzero eigenvalues of the matrix $\mathbf{C}(\mathbf{I}) \circ \mathbf{R}$ can be found as

$$\prod_{i=1}^{\kappa} \lambda_i = \left(\prod_{i=1,\,i\in\mathcal{I}}^{NL} |x(i) - \tilde{x}(i)|^2\right) \cdot (\det(\mathbf{S}_0))^{n_{\mathbf{I}}} \tag{7.73}$$

where $\mathcal{I}$ is the index set of symbols that are transmitted from the relays that have decoded correctly corresponding to the realization $\mathbf{I}$ and

$$\mathbf{S}_0 = \{s_{i,j}\}, \ 1 \le i, j \le L.$$

The result in (7.73) is based on the assumption that the product $\prod_{i=1,\,i\in\mathcal{I}}^{NL} |x(i) - \tilde{x}(i)|^2$ is nonzero. The first product in (7.73) is over $n_{\mathbf{I}}L$ terms. The matrix $\mathbf{S}_0$ is always full rank of order L. Hence, designing the product $\prod_{i=1,\,i\in\mathcal{I}}^{NL} |x(i) - \tilde{x}(i)|^2$ to be nonzero will guarantee a rate of decay, at high SNR, of the conditional PEP as $\mathrm{SNR}^{-n_{\mathbf{I}}L}$, where SNR is now defined as $\mathrm{SNR} = P_2/N_0$. To guarantee that this rate of decay, $\mathrm{SNR}^{-n_{\mathbf{I}}L}$, is always achieved irrespective of the state realization $\mathbf{I}$ of the relay nodes then the product $\prod_{i=1}^{NL} |x(i) - \tilde{x}(i)|^2$ should be nonzero. Hence, designing the product $\prod_{i=1}^{NL} |x(i) - \tilde{x}(i)|^2$ to be nonzero for any pair of distinct source codewords is a sufficient condition for the conditional PEP to decay as $\mathrm{SNR}^{-n_{\mathbf{I}}L}$ for any realization $\mathbf{I}$, where $n_{\mathbf{I}}$ is the number of relays that have decoded correctly corresponding to $\mathbf{I}$.

Now, we calculate the PEP at the destination node for our DSFC structure. Let c_r denote the number of relays that have decoded correctly. Then c_r follows a Binomial distribution as

$$\Pr\{c_r = k\} = \binom{N}{k}(1 - \mathrm{SER})^k \mathrm{SER}^{N-k}, \tag{7.74}$$

where SER is the symbol error rate at the relay nodes. The destination PEP is given by

$$\text{PEP}(\mathbf{s} \rightarrow \tilde{\mathbf{s}}) = \sum_{\mathbf{I}} \Pr\{\mathbf{I}\}\text{PEP}(\mathbf{s} \rightarrow \tilde{\mathbf{s}}/\mathbf{I})$$

$$= \sum_{k=0}^{N} \Pr\{c_r = k\} \sum_{\{\mathbf{I}:n_{\mathbf{I}}=k\}} \text{PEP}(\mathbf{s} \rightarrow \tilde{\mathbf{s}}/\mathbf{I})$$

$$= \sum_{k=0}^{N} \binom{N}{k} (1 - \text{SER})^k \text{SER}^{N-k} \sum_{\{\mathbf{I}:n_{\mathbf{I}}=k\}} \text{PEP}(\mathbf{s} \rightarrow \tilde{\mathbf{s}}/\mathbf{I}), \qquad (7.75)$$

Using the upper bound on the SER at the relay nodes given in (7.68) and the expression for the conditional PEP at the destination node in (7.71), and upper-bounding $(1 - \text{SER})$ by 1, it can be shown that

$$\text{PEP}(\mathbf{s} \rightarrow \tilde{\mathbf{s}}) \leq \text{constant} \times \text{SNR}^{-NL}. \qquad (7.76)$$

Hence, our structure for DSFCs with two-stage coding at the source node and the relay nodes achieves a diversity of order NL, which is the rate of decay of the PEP at high SNR.

7.2.2   DSFC with the AF protocol

In this section, the design and performance analysis for DSFCs with the AF protocol are presented. A structure is presented and sufficient conditions for that structure to achieve full diversity are then derived for some special cases.

7.2.2.1   System model

In this section, we describe the system model for DSFC with the AF protocol. The received signal model at the relay nodes and the channel gains are modeled as in Section 7.2.1. The transmitted data from the source node is parsed into subblocks of size $NL \times 1$. Let $P = \lfloor K/NL \rfloor$ denote the number of subblocks in the transmitted OFDM block. The transmitted $K \times 1$ source codeword is given by

$$\mathbf{s} = [s(1), s(2), \dots, s(K)]^{\mathrm{T}} = [\mathbf{B}_1^{\mathrm{T}}, \mathbf{B}_2^{\mathrm{T}}, \dots, \mathbf{B}_P^{\mathrm{T}}, \mathbf{0}_{K-\text{PLN}}^{\mathrm{T}}]^{\mathrm{T}}, \qquad (7.77)$$

where $\mathbf{B}_i$ is the i-th subblock of dimension $NL \times 1$. Zeros are padded if K is not an integer multiple of NL. For each subblock, $\mathbf{B}_i$, the n-th relay only forwards the data on L subcarriers. For example, relay 1 will only forward $[\mathbf{B}_i(1), \dots, \mathbf{B}_i(L)]$ for all i's and send zeros on the remaining set of subcarriers. In general, the n-th relay will only forward $[\mathbf{B}_i((n-1)L + 1), \dots, \mathbf{B}_i((n-1)L + L)]$ for all i's.

At the relay nodes, each node will normalize the received signal on the subcarriers that it will forward before retransmission and send zeros on the remaining set of subcarriers. If the k-th subcarrier is to be forwarded by the n-th relay, the relay will normalize the received signal on that subcarrier by the factor

$$\beta(k) = \sqrt{\frac{1}{P_1 |H_{s,r_n}(k)|^2 + N_0}}.$$

The relay nodes will use OFDM modulation for transmission to the destination node. At the destination node, the received signal on the k-th subcarrier, assuming it was forwarded by the n-th relay, is given by

$$y(k) = H_{r_n,d}(k)\sqrt{P_2}\left(\sqrt{\frac{1}{P_1|H_{s,r_n}(k)|^2 + N_0}}\left(\sqrt{P_1}H_{s,r_n}(k)s(k) + v_{s,r_n}(k)\right)\right)$$
$$+ v_{r_n,d}(k),$$
(7.78)

where P_2 is the relay node power, $H_{r_n,d}(k)$ is the attenuation of the channel between the n-th relay node and the destination node on the k-th subcarrier, and $v_{s,r_n}(k)$ is the destination noise on the k-th subcarrier. The $v_{r_n,d}(k)$'s are modeled as zero mean, circularly symmetric complex Gaussian random variables with a variance of $N_0/2$ per dimension.

7.2.2.2 Performance analysis

In this section, the PEP of the DSFC with the AF protocol is presented. Based on the PEP analysis, code design criteria are derived.

The received signal at destination on the k-th subcarrier given by (7.78) can be rewritten as

$$y(k) = H_{r_n,d}(k)\sqrt{P_2}\left(\sqrt{\frac{1}{P_1|H_{s,r_n}(k)|^2 + N_0}}\sqrt{P_1}H_{s,r_n}(k)s(k)\right) + z_{r_n,d}(k),$$
(7.79)

where $z_{r_n,d}(k)$ accounts for the noise propagating from the relay node as well as the destination noise. $z_{r_n,d}(k)$ follows a circularly symmetric complex Gaussian random variable with a variance

$$\delta_z^2(k) = \left(\frac{P_2|H_{r_n,d}(k)|^2}{P_1|H_{s,r_n}(k)|^2 + N_0} + 1\right)N_0.$$

The probability density function of $z_{r_n,d}(k)$ given the channel state information (CSI) is given by

$$p(z_{r_n,d}(k)/\text{CSI}) = \frac{1}{\pi\delta_z^2(k)}\exp\left(-\frac{1}{\delta_z^2(k)}|z_{r_n,d}(k)|^2\right).$$
(7.80)

The receiver applies a *maximum likelihood* (ML) detector to the received signal, which is given as

$$\hat{s} = \arg\min_{s}\sum_{k=1}^{K}\frac{1}{\delta_z^2(k)}\left|y(k) - \frac{\sqrt{P_1 P_2}H_{s,r_n}(k)H_{r_n,d}(k)}{\sqrt{P_1|H_{s,r_n}(k)|^2 + N_0}}s(k)\right|^2,$$
(7.81)

where the n index (which is the index of the relay node) is adjusted according to the k index (which is the index of the subcarrier).

Now, sufficient conditions for the presented code structure to achieve full diversity are derived. The pdf of a received vector

$$\mathbf{y} = [y(1), y(2), \ldots, y(K)]^{\mathrm{T}}$$

given that the codeword **s** was transmitted is given by

$$p(\mathbf{y}/\mathbf{s}, \text{CSI})$$

$$= \left(\prod_{k=1}^{K} \frac{1}{\pi \delta_z^2(k)} \right) \exp \left(\sum_{k=1}^{K} -\frac{1}{\delta_z^2(k)} \left| y(k) - \frac{\sqrt{P_1 P_2} H_{\text{s},\text{r}_n}(k) H_{\text{r}_n,\text{d}}(k)}{\sqrt{P_1 |H_{\text{s},\text{r}_n}(k)|^2 + N_0}} \mathbf{s}(k) \right|^2 \right).$$

(7.82)

The PEP of mistaking **s** by **s̃** can be upper-bounded as

$$\text{PEP}(\mathbf{s} \to \tilde{\mathbf{s}}) \leq \text{E} \left\{ \exp \left(\lambda [\ln p(\mathbf{y}/\tilde{\mathbf{s}}) - \ln p(\mathbf{y}/\mathbf{s})] \right) \right\}, \tag{7.83}$$

and the relation applies for any λ, which can selected to get the tightest bound. Any two distinct codewords **s** and

$$\tilde{\mathbf{s}} = [\tilde{\mathbf{B}}_1, \tilde{\mathbf{B}}_2, \dots, \tilde{\mathbf{B}}_p]^{\text{T}}$$

will have at least one index p_0 such that $\tilde{\mathbf{B}}_{p_0} \neq \mathbf{B}_{p_0}$. We will assume that **s** and **s̃** will have only one index p_0 such that $\tilde{\mathbf{B}}_{p_0} \neq \mathbf{B}_{p_0}$, which corresponds to the worst case PEP.

Averaging the PEP expression in (7.83) over the noise distribution given in (7.80) we get

$$\text{PEP}(\mathbf{s} \to \tilde{\mathbf{s}}) \leq \text{E} \left\{ \exp \left(-\lambda(1 - \lambda) \sum_{n=1}^{N} \sum_{l=1}^{L} \right. \right.$$

$$\left(\frac{P_1 |H_{\text{s},\text{r}_n}(J + (n-1)L + l)|^2 P_2 |H_{\text{r}_n,\text{d}}(J + (n-1)L + l)|^2}{\left(P_1 |H_{\text{s},\text{r}_n}(J + (n-1)L + l)|^2 + P_2 |H_{\text{r}_n,\text{d}}(J + (n-1)L + l)|^2 + N_0 \right) N_0} \right)$$

$$\times \left| \mathbf{B}_{p_0}((n-1)L + l) - \tilde{\mathbf{B}}_{p_0}((n-1)L + l) \right|^2 \right) \right\}, \tag{7.84}$$

where $J = (p_0 - 1)NL$. Take $\lambda = 1/2$ to minimize the upper bound in (7.84), hence, we get

$$\text{PEP}(\mathbf{s} \to \tilde{\mathbf{s}}) \leq \text{E} \left\{ \exp \left(-\frac{1}{4} \sum_{n=1}^{N} \sum_{l=1}^{L} \right. \right.$$

$$\left(\frac{P_1 |H_{\text{s},\text{r}_n}(J + (n-1)L + l)|^2 P_2 |H_{\text{r}_n,\text{d}}(J + (n-1)L + l)|^2}{\left(P_1 |H_{\text{s},\text{r}_n}(J + (n-1)L + l)|^2 + P_2 |H_{\text{r}_n,\text{d}}(J + (n-1)L + l)|^2 + N_0 \right) N_0} \right)$$

$$\times \left| \mathbf{B}_{p_0}((n-1)L + l) - \tilde{\mathbf{B}}_{p_0}((n-1)L + l) \right|^2 \right) \right\}. \tag{7.85}$$

At high SNR, we have

$$\frac{P_1 |H_{\text{s},\text{r}_n}(k)|^2 P_2 |H_{\text{r}_n,\text{d}}(k)|^2}{\left(P_1 |H_{\text{s},\text{r}_n}(k)|^2 + P_2 |H_{\text{r}_n,\text{d}}(k)|^2 + N_0 \right) N_0} \approx \frac{P_1 |H_{\text{s},\text{r}_n}(k)|^2 P_2 |H_{\text{r}_n,\text{d}}(k)|^2}{\left(P_1 |H_{\text{s},\text{r}_n}(k)|^2 + P_2 |H_{\text{r}_n,\text{d}}(k)|^2 \right) N_0},$$

which is the scaled harmonic mean of the source–relay and relay–destination SNRs on the k-th subcarrier.

The scaled harmonic mean of two non-negative numbers, a_1 and a_2, can be upper- and lower-bounded as

$$\frac{1}{2} \min(a_1, a_2) \leq \frac{a_1 a_2}{a_1 + a_2} \leq \min(a_1, a_2). \tag{7.86}$$

Using the lower-bound in (7.86) the PEP in (7.85) can be further upper-bounded as

$$\text{PEP}(\mathbf{s} \rightarrow \tilde{\mathbf{s}}) \leq \text{E} \left\{ \exp \left(-\frac{1}{8} \sum_{n=1}^{N} \sum_{l=1}^{L} \min \left(\frac{P_1}{N_0} |H_{s,r_n}((p_0 - 1)NL \right. \right. \right.$$
$$+ (n-1)L + l)|^2, \frac{P_2}{N_0} |H_{r_n,d}((p_0 - 1)NL + (n-1)L + l)|^2 \Bigg)$$
$$\left. \left. \left| \mathbf{B}_{p_0}((n-1)L + l) - \tilde{\mathbf{B}}_{p_0}((n-1)L + l) \right|^2 \right) \right\}. \tag{7.87}$$

If $P_2 = P_1$ and SNR is defined as P_1/N_0, then the PEP is now upper-bounded as

$$\text{PEP}(\mathbf{s} \rightarrow \tilde{\mathbf{s}}) \leq \text{E} \left\{ \exp \left(-\frac{1}{8} \sum_{n=1}^{N} \sum_{l=1}^{L} \min \left(\text{SNR}|H_{s,r_n}((p_0 - 1)NL \right. \right. \right.$$
$$+ (n-1)L + l)|^2, \text{SNR}|H_{r_n,d}((p_0 - 1)NL + (n-1)L + l)|^2 \Bigg)$$
$$\left. \left. \left| \mathbf{B}_{p_0}((n-1)L + l) - \tilde{\mathbf{B}}_{p_0}((n-1)L + l) \right|^2 \right) \right\}. \tag{7.88}$$

7.2.2.3 PEP Analysis for $L = 1$

The case of L equal to 1 corresponds to a flat, frequency nonselective fading channel. The PEP in (7.88) is now given by

$$\text{PEP}(\mathbf{s} \rightarrow \tilde{\mathbf{s}}) \leq \text{E} \left\{ \exp \left(-\frac{1}{8} \sum_{n=1}^{N} \min \left(\text{SNR}|H_{s,r_n}((p_0 - 1)NL \right. \right. \right.$$
$$+ (n-1)L + 1)|^2, \text{SNR}|H_{r_n,d}((p_0 - 1)NL + (n-1)L + 1)|^2 \Bigg)$$
$$\left. \left. \left| \mathbf{B}_{p_0}((n-1)L + 1) - \tilde{\mathbf{B}}_{p_0}((n-1)L + 1) \right|^2 \right) \right\}. \tag{7.89}$$

The random variables $\text{SNR}|H_{s,r_n}(k)|^2$ and $\text{SNR}|H_{r_n,d}(k)|^2$ follow an exponential distribution with rate $1/\text{SNR}$ for all k. The minimum of two exponential random variables is an exponential random variable with rate that is the sum of the two random variables rates. Hence, $\min \left(\text{SNR}|H_{s,r_n}(k)|^2, \text{SNR}|H_{r_n,d}(k)|^2 \right)$ follows an exponential distribution with rate $2/\text{SNR}$.

The PEP upper bound is now given by

$$
\text{PEP}(\mathbf{s} \to \tilde{\mathbf{s}}) \le \prod_{n=1}^{N} \frac{1}{1 + \frac{1}{16}\text{SNR}\left|\mathbf{B}_{p_0}((n-1)L+1) - \tilde{\mathbf{B}}_{p_0}((n-1)L+1)\right|^2}. \quad (7.90)
$$

At high SNR, we can neglect 1 besides the SNR dependent term in the denominator of (7.90). Hence, the PEP can now be upper-bounded as

$$
\text{PEP}(\mathbf{s} \to \tilde{\mathbf{s}})
$$

$$
\lesssim \left(\frac{1}{16}\text{SNR}\right)^{-N} \left(\prod_{n=1}^{N} \left|\mathbf{B}_{p_0}((n-1)L+1) - \tilde{\mathbf{B}}_{p_0}((n-1)L+1)\right|^2\right)^{-1}. \quad (7.91)
$$

The result in (7.91) is under the assumption that the product

$$
\prod_{n=1}^{N} \left|\mathbf{B}_{p_0}((n-1)L+1) - \tilde{\mathbf{B}}_{p_0}((n-1)L+1)\right|^2
$$

is nonzero. Clearly, if that product is nonzero, then the system will achieve a diversity of order NL, where L is equal to 1 in this case. From the expression in (7.91) the coding gain of the space–frequency code is maximized when the product

$$
\min_{\mathbf{s} \ne \tilde{\mathbf{s}}} \prod_{n=1}^{N} \left|\mathbf{B}_{p_0}((n-1)L+1) - \tilde{\mathbf{B}}_{p_0}((n-1)L+1)\right|^2
$$

is maximized, which is the minimum product distance.

7.2.2.4 PEP analysis for $L=2$

The PEP in (7.88) can now be given as

$$
\text{PEP}(\mathbf{s} \to \tilde{\mathbf{s}}) \le \text{E}\left\{\exp\left(-\frac{1}{8}\sum_{n=1}^{N}\sum_{l=1}^{2}\min\left(\text{SNR}|H_{\text{s},\text{r}_n}((p_0-1)NL \right.\right.\right.
$$

$$
\left. + (n-1)L+l)|^2, \text{SNR}|H_{\text{r}_n,\text{d}}((p_0-1)NL+(n-1)L+l)|^2\right)
$$

$$
\left.\left.\left|\mathbf{B}_{p_0}((n-1)L+l) - \tilde{\mathbf{B}}_{p_0}((n-1)L+l)\right|^2\right)\right\}, \quad (7.92)
$$

where $L = 2$. The analysis in this case is more involved since the random variables appearing in (7.92) are correlated. Signals transmitted from the same relay node on different subcarriers will experience correlated channel attenuations.

As a first step in deriving the code design criterion, we prove that the channel attenuations, $|H_{\text{s},\text{r}_n}(k_1)|^2$ and $|H_{\text{s},\text{r}_n}(k_2)|^2$ for any $k_1 \ne k_2$, have a bivariate Gamma distribution as their joint pdf. The same applies for $|H_{\text{r}_n,\text{d}}(k_1)|^2$ and $|H_{\text{r}_n,\text{d}}(k_2)|^2$ for any $k_1 \ne k_2$. The proof of this result is given in the Appendix to this chapter.

To evaluate the expectation in (7.92) we need the expression for the joint pdf of the two random variables

$$
M_1 = \min\left(\text{SNR}|H_{\text{s},\text{r}_n}(k_1)|^2, \text{SNR}|H_{\text{r}_n,\text{d}}(k_1)|^2\right)
$$

and

$$M_2 = \min \left(\text{SNR} |H_{\text{s},\text{r}_n}(k_2)|^2, \text{SNR} |H_{\text{r}_n,\text{d}}(k_2)|^2 \right)$$

for some $k_1 \neq k_2$. Although M_1 and M_2 can be easily seen to be marginally exponential random variables, they are not jointly Gamma distributed. Define the random variables $X_1 = \text{SNR} |H_{\text{s},\text{r}_n}(k_1)|^2$, $X_2 = \text{SNR} |H_{\text{s},\text{r}_n}(k_2)|^2$, $Y_1 = \text{SNR} |H_{\text{r}_n,\text{d}}(k_1)|^2$, and $Y_2 = \text{SNR} |H_{\text{r}_n,\text{d}}(k_2)|^2$. All of these random variables are marginally exponential with rate $1/SNR$. Under the assumptions of our channel model, the pairs (X_1, X_2) and (Y_1, Y_2) are independent. Hence, the joint pdf of (X_1, X_2, Y_1, Y_2), using the result in the Appendix, is given by

$$
\begin{aligned}
&f_{X_1,X_2,Y_1,Y_2}(x_1, x_2, y_1, y_2) \\
&= f_{X_1,X_2}(x_1, x_2) f_{Y_1,Y_2}(y_1, y_2) \\
&= \frac{1}{\text{SNR}^2 (1 - \rho_{x_1 x_2})(1 - \rho_{y_1 y_2})} \exp\left(-\frac{x_1 + x_2}{\text{SNR}(1 - \rho_{x_1 x_2})} \right) \\
&\quad \cdot I_0 \left(\frac{2\sqrt{\rho_{x_1 x_2}}}{\text{SNR}(1 - \rho_{x_1 x_2})} \sqrt{x_1 x_2} \right) \exp\left(-\frac{y_1 + y_2}{\text{SNR}(1 - \rho_{y_1 y_2})} \right) \\
&\quad \cdot I_0 \left(\frac{2\sqrt{\rho_{y_1 y_2}}}{\text{SNR}(1 - \rho_{y_1 y_2})} \sqrt{y_1 y_2} \right) U(x_1) U(x_2) U(y_1) U(y_2),
\end{aligned}
\tag{7.93}
$$

where $I_0(\cdot)$ is the modified Bessel function of the first kind of order zero and $U(\cdot)$ is the Heaviside unit step function. $\rho_{x_1 x_2}$ is the correlation coefficient between X_1 and X_2 and similarly, $\rho_{y_1 y_2}$ is the correlation coefficient between Y_1 and Y_2.

The joint cumulative distribution function (cdf) of the pair (M_1, M_2) can be computed as

$$
\begin{aligned}
&F_{M_1,M_2}(m_1, m_2) \\
&\triangleq \Pr[M_1 \leq m_1, M_2 \leq m_2] \\
&= \Pr[\min(X_1, Y_1) \leq m_1, \min(X_2, Y_2) \leq m_2] \\
&= 2 \int_{y_1=0}^{m_1} \int_{x_1=y_1}^{\infty} \int_{y_2=0}^{m_2} \int_{x_2=y_2}^{\infty} f_{X_1,X_2}(x_1, x_2) f_{Y_1,Y_2}(y_1, y_2) dy_1 dx_1 dy_2 dx_2 \\
&\quad + 2 \int_{y_1=0}^{m_1} \int_{x_1=y_1}^{\infty} \int_{x_2=0}^{m_2} \int_{y_2=x_2}^{\infty} f_{X_1,X_2}(x_1, x_2) f_{Y_1,Y_2}(y_1, y_2) dy_1 dx_1 dx_2 dy_2,
\end{aligned}
\tag{7.94}
$$

where we have used the symmetry assumption of the source–relay and relay–destination channels.

The joint pdf of (M_1, M_2) can now be given as

$$
\begin{aligned}
f_{M_1,M_2}(m_1, m_2) &= \frac{\partial^2}{\partial m_1 \partial m_2} F_{M_1,M_2}(m_1, m_2) \\
&= 2 f_{Y_1,Y_2}(m_1, m_2) \int_{x_1=m_1}^{\infty} \int_{x_2=m_2}^{\infty} f_{X_1,X_2}(x_1, x_2) dx_1 dx_2 \\
&\quad + 2 \int_{x_1=m_1}^{\infty} \int_{y_2=m_2}^{\infty} f_{X_1,X_2}(x_1, m_2) f_{Y_1,Y_2}(m_1, y_2) dx_1 dx_2.
\end{aligned}
\tag{7.95}
$$

To get the PEP upper bound in (7.92) we need to calculate the expectation

$$
\mathrm{E}\left\{\exp\left(-\frac{1}{8}\left(M_1\left|\mathbf{B}(k_1)-\tilde{\mathbf{B}}(k_1)\right|^2+M_2\left|\mathbf{B}(k_2)-\tilde{\mathbf{B}}(k_2)\right|^2\right)\right)\right\}
$$
$$
=\int_{m_1=0}^{\infty}\int_{m_2=0}^{\infty}\exp\left(-\frac{1}{8}\left(m_1\left|\mathbf{B}(k_1)-\tilde{\mathbf{B}}(k_1)\right|^2+m_2\left|\mathbf{B}(k_2)-\tilde{\mathbf{B}}(k_2)\right|^2\right)\right)
$$
$$
f_{M_1,M_2}(m_1,m_2)\mathrm{d}m_1\mathrm{d}m_2.
$$

(7.96)

At high enough SNR, we have

$$
I_0\left(\frac{2\sqrt{\rho_{x_1x_2}}}{\mathrm{SNR}(1-\rho_{x_1x_2})}\sqrt{x_1x_2}\right)\approx 1.
$$

Using this approximation, the PEP upper bound can be approximated at high SNR as

$$
\mathrm{PEP}(\mathbf{s}\to\tilde{\mathbf{s}})\lesssim\left(\prod_{m=1}^{2N}\left|\mathbf{B}_{p_0}(m)-\tilde{\mathbf{B}}_{p_0}(m)\right|^2\right)^{-1}\left(\frac{1}{16}(1-\rho)\mathrm{SNR}\right)^{-2N},\qquad(7.97)
$$

where $\rho=\rho_{x_1x_2}=\rho_{y_1y_2}$. Again, full diversity is achieved when the product

$$
\prod_{m=1}^{2N}\left|\mathbf{B}_{p_0}(m)-\tilde{\mathbf{B}}_{p_0}(m)\right|^2
$$

is nonzero. The coding gain of the space–frequency code is maximized when the product

$$
\min_{\mathbf{s}\neq\tilde{\mathbf{s}}}\prod_{m=1}^{2N}\left|\mathbf{B}_{p_0}(m)-\tilde{\mathbf{B}}_{p_0}(m)\right|^2
$$

is maximized.

The analysis becomes highly involved for any $L\geq 3$. It is very difficult to get closed form expressions in this case due to the correlation among the summed terms in (7.88) for which there is no closed form pdf expressions, similar to (7.93).

Example 7.4   In this example, we investigate the performance of the presented SDFCs. Figure 7.9 shows the case of a simple two-ray, $L=2$, channel model with a delay of $\tau=5\,\mu\mathrm{s}$ between the two rays. The two rays have equal powers, i.e., $\sigma^2(1)=\sigma^2(2)$. The number of subcarriers is $K=128$ with a system bandwidth of 1 MHz. We use BPSK modulation and Vandermonde-based linear transformations. Figure 7.9 shows the SER of the presented DSFCs versus the SNR defined as SNR $=(P_1+P_2)/N_0$, and we use $P_1=P_2$, i.e., equal power allocation between the source and relay nodes. We simulated three cases: all channel variances are ones, relays close to source, and relays close to destination. For the case of relays close to source, the variance of any source–relay channel is taken to be 10 and the variance of any relay–destination channel is taken to be 1. For the case of relays close to destination, the variance of any source–relay channel is taken to be 1 and the variance of any relay–destination channel is taken to be 10. From Figure 7.9, it is clear that DSFCs with the DF protocol have a better

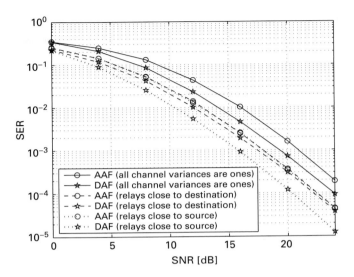

Fig. 7.9   SER for DSFCs for BPSK modulation, $L = 2$, and delay $= [0, 5\,\mu\text{s}]$ versus SNR.

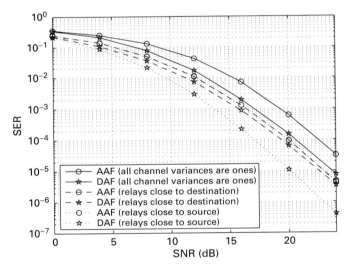

Fig. 7.10   SER for DSFCs for BPSK modulation, $L = 2$, and delay $= [0, 20\,\mu\text{s}]$ versus SNR.

performance than DSFCs with the AF protocol. The reason is that DSFCs with the DF protocol deliver a less noisy code to the destination node as compared to DSFCs with the AF protocol, where noise propagation results from the transmissions of the relay nodes. Decoding at the relay nodes, in the DF protocol, has the effect of removing the noise before retransmission to the destination node. As can be seen from Figure 7.9, a gain of about 3 dB is achieved, for the case of relays close to the source, by employing DSFCs with the DF protocol as compared to DSFCs with the AF protocol. Figure 7.10 shows the case of a simple two-ray, $L = 2$, with a delay of $\tau = 20\,\mu\text{s}$ between the two rays. The simulation setup is the same as that used in Figure 7.9. From Figure 7.10, it is clear that DSFCs with the DF protocol have a better performance than DSFCs with the AF protocol.

▲

7.2.3   Code design and remarks

For DSFCs with the DF and AF protocols, The code design criteria is to minimize the minimum product distance. Therefore, We can use the design presented in Chapter 3. Here we summarize some remarks related to our presented DSFCs.

- *Remark 1*: In our problem formulation, we have considered a two-hop system model that lacks a direct link from the source node to the destination node. If such a direct link between the source node and the destination node exists, then the destination node can use its received signal from the source node to help recovering the source symbols. Assuming that the channel from the source node to the destination node has L paths, it can be shown that our presented DSFCs, with the presented coding at the source node and the relay nodes for both the DF and AF protocols, achieve a diversity of order $(N + 1)L$.
- *Remark 2*: The presented DSFCs with the DF protocol can be easily modified to achieve full diversity for the asymmetric case where the number of paths per fading channel is not the same for all channels. Let L_{s,r_n} denote the number of paths of the channel between the source node and n-th relay and $L_{r_n,d}$ denote the number of paths of the channel between the n-th relay node and the destination node. The presented DSFC can be easily modified to achieve a diversity d of order

$$d = \sum_{n=1}^{N} \min(L_{s,r_n},\ L_{r_n,d}),$$

which can be easily shown to be the maximal achievable diversity order. This maximal diversity order can be achieved, for example, by designing the codes at the source node and relay nodes using

$$L = \max_n\ \min(L_{s,r_n},\ L_{r_n,d}).$$

- *Remark 3*: The presented construction for the design of DSFCs can be easily generalized to the case of multi-antenna nodes, where any node may have more than one antenna. Each antenna can be treated as a separate relay node and the analysis presented before directly applies.
- *Remark 4*: As mentioned before, the presence of the cyclic prefix in the OFDM transmission provides a mean for combating the relays synchronization mismatches. Hence, our presented DSFCs, which are based on OFDM transmission, are robust against synchronization mismatches within the duration of the cyclic prefix.

7.3 Chapter summary and bibliographical notes

The design of distributed space–time codes in wireless relay networks is considered for different schemes, which vary in the processing performed at the relay nodes. For the decode-and-forward distributed space–time codes, any space–time code that is designed to achieve full diversity over MIMO channels can achieve full diversity under the assumption that the relay nodes can decide whether they have decoded correctly or not. A code that maximizes the coding gain over MIMO channels is not guaranteed to maximize the coding gain in the decode-and-forward distributed space–time coding. This is due to the fact that not all of the relays will always transmit their code columns in the second phase. The code design criteria for the amplify-and-forward distributed space–time codes were also considered. In this case, a code designed to achieve full diversity over MIMO channels will also achieve full diversity. Furthermore, a code that maximizes the coding gain over MIMO channels will also maximize the coding gain in the amplify-and-forward distributed space–time scheme.

Then, the design of DDSTC for wireless relay networks was studied. In DDSTC, the diagonal structure of the code was imposed to simplify the synchronization between randomly located relay nodes. Synchronization mismatches between the relay nodes causes inter-symbol interference, which can highly degrade the system performance. DDSTC relaxes the stringent synchronization requirement by allowing only one relay to transmit at any time slot. The code design criterion for the DDSTC based on minimizing the PEP was derived and the design criterion is found to be maximizing the minimum product distance. This is the same criterion used for designing DAST codes and full-rate, full-diversity space frequency codes.

Next, the design of distributed space–frequency codes (DSFCs) was considered for the wireless multipath relay channels. The use of DSFCs can greatly improve system performance by achieving higher diversity orders by exploiting the multipath diversity of the channel as well as the cooperative diversity. We have considered the design of DSFCs with the DF and AF cooperation protocols. For the case of DSFCs with the DF protocol, we have presented a two-stage coding scheme: source node coding and relay nodes coding. We have derived sufficient conditions for the presented code structure to achieve full diversity of order NL where N is the number of relay nodes and L is the number of multipaths per channel. For the case of DSFCs with the AF protocol, we have derived sufficient conditions for the presented code structure to achieve full diversity of order NL for the special cases of $L = 1$ and $L = 2$.

The presented DSFCs are robust against the synchronization errors caused by the relays timing mismatches and propagation delays due to the presence of the cyclic prefix in the OFDM transmission.

Several works have considered the design of distributed space–time codes. It was proposed in [108] to use relay nodes to form a virtual multi-antenna transmitter to achieve diversity. In addition, an outage analysis was presented for the system.

Some works have considered the application of the existing space–time codes in a distributed fashion for the wireless relay network [193, 7, 8, 93]. In [193], space–time block codes were used in a completely distributed fashion. Each relay node transmits a randomly selected column from the space–time code matrix. This system achieves a diversity of order one, as the signal-to-noise (SNR) tends to infinity, limited by the probability of having all of the relay nodes selecting to transmit the same column of the space–time code matrix. In [7], distributed space–time coding based on the Alamouti scheme and amplify-and-forward cooperation protocol was analyzed. An expression for the average symbol error rate (SER) was derived. In [8], a performance analysis of the gain of using cooperation among nodes was considered assuming that the number of relays available for cooperation is a Poisson random variable. The authors compared the performance of different distributed space–time codes designed for the MIMO channels under this assumption. In [93], the performance of the linear dispersion (LD) space–time codes of [59] was analyzed when used for distributed space–time coding in wireless relay networks.

For the case of DSFCs, a design of DSFCs was considered in [116], where a system that employs the DF protocol was used. It was assumed that all of the relay nodes will always decode correctly, which is not always true, especially over wireless fading channels. Such proposed DSFCs mitigate synchronization errors due to the timing mismatchs of relay nodes and propagation delays by employing OFDM transmission. The presence of the cyclic prefix in OFDM modulation can mitigate synchronization errors of relay nodes. This follows directly due to the use of OFDM transmission and the same applies for our proposed DSFCs.

Further information and details can be found in [169, 170, 171, 172, 174, 175, 178].

Appendix

Consider the two random variables $H_{s,r_n}(k_1)$ and $H_{s,r_n}(k_2)$, we will assume without loss of generality that $\tau_1 = 0$, i.e., the delay of the first path is zero. $H_{s,r_n}(k_1)$ is given by

$$H_{s,r_n}(k_1) = \alpha_{s,r_n}(1) + \alpha_{s,r_n}(2)e^{-j2\pi(k_1-1)\Delta f \tau_2} = \Re\left(H_{s,r_n}(k_1)\right) + j\Im\left(H_{s,r_n}(k_1)\right),$$
(7.98)

where $\Re(x)$, and $\Im(x)$ are the real, and imaginary parts of x, respectively. From (7.98) we have

$$\Re\left(H_{s,r_n}(k_1)\right) = \Re(\alpha_{s,r_n}(1)) + \Re(\alpha_{s,r_n}(2))\cos(2\pi(k_1-1)\Delta f \tau_2)$$
$$+ \Im(\alpha_{s,r_n}(2))\sin(2\pi(k_1-1)\Delta f \tau_2)$$
$$\Im\left(H_{s,r_n}(k_1)\right) = \Im(\alpha_{s,r_n}(1)) + \Im(\alpha_{s,r_n}(2))\cos(2\pi(k_1-1)\Delta f \tau_2)$$
$$- \Re(\alpha_{s,r_n}(2))\sin(2\pi(k_1-1)\Delta f \tau_2).$$
(7.99)

Based on the channel model presented in Section 7.2.1.1 both $\Re\left(H_{s,r_n}(k_1)\right)$ and $\Im\left(H_{s,r_n}(k_1)\right)$ are zero-mean Gaussian random variables with variance $1/2$. The correlation coefficient, ρ_{ri}, between $\Re\left(H_{s,r_n}(k_1)\right)$ and $\Im\left(H_{s,r_n}(k_1)\right)$ can be calculated as

$$\rho_{ri} = E\left\{\Re\left(H_{s,r_n}(k_1)\right)\Im\left(H_{s,r_n}(k_1)\right)\right\} = 0.$$
(7.100)

Hence, $H_{s,r_n}(k_1)$ is a circularly symmetric complex Gaussian random variable with variance $1/2$ per dimension and the same applies for $H_{s,r_n}(k_2)$. To get the joint probability distribution of $|H_{s,r_n}(k_1)|^2$ and $|H_{s,r_n}(k_2)|^2$, we can use the standard techniques of transformation of random variables. Using transformation of random variables and the fact that both $H_{s,r_n}(k_1)$ and $H_{s,r_n}(k_2)$ are circularly symmetric complex Gaussian random variables, it can be shown that $X_1 = |H_{s,r_n}(k_1)|^2$ and $X_2 = |H_{s,r_n}(k_2)|^2$ are jointly distributed according to a bivariate Gamma distribution with pdf [131, 188]

$$f_{X_1,X_2}(x_1, x_2) = \frac{1}{1 - \rho_{x_1 x_2}} \exp\left(-\frac{x_1 + x_2}{1 - \rho_{x_1 x_2}}\right) I_0\left(\frac{2\sqrt{\rho_{x_1 x_2}}}{1 - \rho_{x_1 x_2}} \sqrt{x_1 x_2}\right) U(x_1) U(x_2),$$

(7.101)

where $I_0(\cdot)$ is the modified Bessel function of the first kind of order zero and $U(\cdot)$ is the Heaviside unit step function [50]. $\rho_{x_1 x_2}$ is the correlation between $|H_{s,r_n}(k_1)|^2$ and $|H_{s,r_n}(k_2)|^2$ and it can calculated as

$$\rho_{x_1,x_2} = \frac{\text{Cov}(X_1, X_2)}{\sqrt{\text{Var}(X_1)\,\text{Var}(X_2)}}.$$

(7.102)

Following tedious computations, it can be shown that

$$\rho_{x_1,x_2} = \frac{1}{2} + 2\sigma^2(1)\sigma^2(2)\cos(2\pi(k_2 - k_1)\Delta f \tau_2),$$

(7.103)

where the last equation applies under the assumption of having $\sigma^2(1) + \sigma^2(2) = 1$ and both, $\sigma^2(1)$ and $\sigma^2(2)$, are nonzeros. From (7.103) it is clear that $0 \leq \rho_{x_1,x_2} \leq 1$.

Exercises

7.1 Define the SNR as SNR $= P/N_0$, where $P = P_1 + P_2$ is the transmitted power per source symbol. Let $P_1 = \alpha P$ and $P_2 = (1 - \alpha)P$, where $\alpha \in (0, 1)$. Use the SER expression at the relay nodes from (7.6) and use (7.18) to prove the expression in (7.19).

7.2 In this problem, the case of two relays helping the source to forward its information using the Alamouti scheme is considered. The system model is as presented in Section 7.1.1 and the data transmission is as presented in Example 7.1. Each relay is assumed to able to decide whether or not it has decoded correctly in the first phase

(a) Prove the following results:

(i) For M-PSK modulation, the destination SER is given by

$$P_{\text{PSK}} = 4F_1\left(1 + \frac{b_{\text{PSK}} P_1 \delta_{s,d}^2}{N_0 \sin^2 \theta}\right)\left[F_1\left(1 + \frac{b_{\text{PSK}} P_1 \delta_{s,r}^2}{N_0 \sin^2 \theta}\right)\right]^2$$

$$+ 4F_1\left(\left(1 + \frac{b_{\text{PSK}} P_1 \delta_{s,d}^2}{N_0 \sin^2 \theta}\right)\left(1 + \frac{b_{\text{PSK}} P_2 \delta_{r,d}^2}{N_0 \sin^2 \theta}\right)\right)$$

$$\times F_1\left(1+\frac{b_{\text{PSK}}P_1\delta_{\text{s,r}}^2}{N_0\sin^2\theta}\right)\left[1-2F_1\left(1+\frac{b_{\text{PSK}}P_1\delta_{\text{s,r}}^2}{N_0\sin^2\theta}\right)\right]$$

$$+F_1\left(\left(1+\frac{b_{\text{PSK}}P_1\delta_{\text{s,d}}^2}{N_0\sin^2\theta}\right)\left[\left(1+\frac{b_{\text{PSK}}P_2\delta_{\text{r,d}}^2}{N_0\sin^2\theta}\right)\right]^2\right)$$

$$\times\left[1-2F_1\left(1+\frac{b_{\text{PSK}}P_1\delta_{\text{s,r}}^2}{N_0\sin^2\theta}\right)\right]^2, \tag{E7.1}$$

where

$$F_1(x(\theta))=\frac{1}{\pi}\int_0^{(M-1)\pi/M}\frac{1}{x(\theta)}\,d\theta. \tag{E7.2}$$

(ii) For the M-QAM modulation, the destination SER is given by

$$P_{\text{QAM}}=4F_2\left(1+\frac{b_{\text{QAM}}P_1\delta_{\text{s,d}}^2}{2N_0\sin^2\theta}\right)\left[F_2\left(1+\frac{b_{\text{QAM}}P_1\delta_{\text{s,r}}^2}{2N_0\sin^2\theta}\right)\right]^2$$

$$+4F_2\left(\left(1+\frac{b_{\text{QAM}}P_1\delta_{\text{s,d}}^2}{2N_0\sin^2\theta}\right)\left(1+\frac{b_{\text{QAM}}P_2\delta_{\text{r,d}}^2}{2N_0\sin^2\theta}\right)\right)$$

$$\times F_2\left(1+\frac{b_{\text{QAM}}P_1\delta_{\text{s,r}}^2}{2N_0\sin^2\theta}\right)\left[1-2F_2\left(1+\frac{b_{\text{QAM}}P_1\delta_{\text{s,r}}^2}{2N_0\sin^2\theta}\right)\right]$$

$$+F_2\left(\left(1+\frac{b_{\text{QAM}}P_1\delta_{\text{s,d}}^2}{2N_0\sin^2\theta}\right)\left[\left(1+\frac{b_{\text{QAM}}P_2\delta_{\text{r,d}}^2}{2N_0\sin^2\theta}\right)\right]^2\right)$$

$$\times\left[1-2F_2\left(1+\frac{b_{\text{QAM}}P_1\delta_{\text{s,r}}^2}{2N_0\sin^2\theta}\right)\right]^2, \tag{E7.3}$$

where

$$F_2(x(\theta))=\frac{4K}{\pi}\int_0^{\pi/2}\frac{1}{x(\theta)}\,d\theta-\frac{4K^2}{\pi}\int_0^{\pi/4}\frac{1}{x(\theta)}\,d\theta. \tag{E7.4}$$

P_1 denotes the source node power and each relay transmits using a power of $P_2/2$ in each time slot of the second phase.

(b) Prove that if all of the channel links are available, i.e., $\delta_{\text{s,d}}^2\neq 0$, $\delta_{\text{s,r}}^2\neq 0$, and $\delta_{\text{r,d}}^2\neq 0$, the SER of the Alamouti-based cooperative diversity can be upper-bounded as follows:

(i) For M-PSK

$$P_{\text{PSK}}\leq\frac{N_0^3}{b_{\text{PSK}}^3}\left[4(g(2))^3\frac{1}{P_1^3\delta_{\text{s,d}}^2\delta_{\text{s,r}}^4}+4g(4)g(2)\right.$$

$$\left.\times\frac{1}{P_1^2P_2\delta_{\text{s,d}}^2\delta_{\text{r,d}}^2\delta_{\text{s,r}}^2}+g(6)\frac{1}{P_1P_2^2\delta_{\text{s,d}}^2\delta_{\text{r,d}}^4}\right], \tag{E7.5}$$

where $g(n)=\frac{1}{\pi}\int_0^{(M-1)\pi/M}\sin^n\theta\,d\theta$.

(ii) For M-QAM

$$P_{\text{QAM}} \le \frac{N_0^3}{(b_{\text{QAM}}/2)^3} \left[4(f(2))^3 \frac{1}{P_1^3 \delta_{s,d}^2 \delta_{s,r}^4} + 4f(4)f(2) \right.$$

$$\left. \times \frac{1}{P_1^2 P_2 \delta_{s,d}^2 \delta_{r,d}^2 \delta_{s,r}^2} + f(6) \frac{1}{P_1 P_2^2 \delta_{s,d}^2 \delta_{r,d}^4} \right],$$

$$\text{(E7.6)}$$

where $f(n) = \frac{4K}{\pi} \int_0^{\pi/2} \sin^n \theta d\theta - \frac{4K^2}{\pi} \int_0^{\pi/4} \sin^n \theta d\theta.$

7.3　In this problem, the AF-based system presented in Section 7.1.2 is considered. The probability density function (pdf) of $\mathbf{y_d}$, given a certain code $\mathbf{X_m}$ and $\mathbf{h}$, is given by

$$P(\mathbf{y_d}/\mathbf{X_m}, \mathbf{h})$$

$$= \frac{1}{\left[\pi N_0 \left(1 + \frac{P_2/K_n}{P_1 \delta_{s,r}^2 + N_0} \sum_{i=1}^n |h_{r_i,d}|^2 \right) \right]^{K_n}}$$

$$\times \exp \left(- \frac{\left(\mathbf{y_d} - \sqrt{\frac{P_2 P_1/K_n}{P_1 \delta_{s,r}^2 + N_0}} \mathbf{X_m H} \right)^{\mathbf{H}} \left(\mathbf{y_d} - \sqrt{\frac{P_2 P_1/K_n}{P_1 \delta_{s,r}^2 + N_0}} \mathbf{X_m H} \right)}{N_0 \left(1 + \frac{P_2/K_n}{P_1 \delta_{s,r}^2 + N_0} \sum_{i=1}^n |h_{r_i,d}|^2 \right)} \right). \quad \text{(E7.7)}$$

Use (E7.7) and (7.48) to prove the PEP upper bound given by (7.28).

7.4　Prove (7.29).

7.5　(Computer simulation) Returning to Exercise 7.1, the validity of the derived expressions will be checked. In all of the simulations, set the variance of the channel between the source and destination to 1, i.e., $\delta_{s,d}^2 = 1$.

　　(a) Simulate the cooperation system for QPSK modulation scheme (4-QAM). Take the channel variance between the source and relays to be 1, i.e., $\delta_{s,r}^2 = \delta_{r,d}^2 = 1$. Use the $P_1 = 2P_2 = P/2$ power assignment, where P is a total power constraint. Plot the simulated SER, exact SER expression, and SER upper bound versus P/N_0. Comment on the validity of the derived expressions.

　　(b) Repeat part (a) for the case of a 16-QAM constellation. Comment on the validity of the derived expressions.

7.6　Write a Matlab code to compare the symbol error rate performance of a two-relay decode-and-forward protocol and a two-relay distributed space–time coding system that utilizes an Alamouti scheme. Assume a Rayleigh flat-fading channel with unit variance, and a propagation path-loss with an exponent of 3.5. Experiment with different source–destination separations with the relays located at the middle between the source and destination.

7.7　Write a Matlab code to simulate a three-relay DSFC for the case of a two-ray, $L = 2$, channel model with a delay of $\tau = 5 \mu s$ between the two rays. The two rays have equal powers, i.e., $\sigma^2(1) = \sigma^2(2)$. The number of subcarriers is $K = 128$ with a system bandwidth of 1 MHz. Use BPSK modulation and Vandermonde-based linear transformations.

8 Relay selection: when to cooperate and with whom

In this chapter, we present a cooperative protocol based on the *relay-selection* technique using the availability of the partial channel state information (CSI) at the source and the relays. The main objective of this scheme is to achieve higher bandwidth efficiency while guaranteeing the same diversity order as in a conventional cooperative scheme. Two cooperation scenarios are addressed: a single-relay scenario and a multi-relay scenario. In the single-relay scenario, the focus is to answer the question: *"When to cooperate?"* The rationale behind this protocol is that there is no need for the relay to forward the source's information if the direct link, between the source and destination, is of high quality. It turns out that an appropriate metric to represent the relay's ability to help is a modified version of the harmonic mean function of its source–relay and relay–destination instantaneous channel gains. The source decides when to cooperate by taking the ratio between the source–destination channel gain and the relay's metric and comparing it with a threshold, which is referred to as the *cooperation threshold* . If this ratio is greater than or equal to the cooperation threshold, then the source sends its information to the destination directly without the need for the relay. Otherwise, the source employs the relay in forwarding its information to the destination as in the conventional cooperative scheme.

An extension to the multi-node cooperative scenario with arbitrary N relays is considered as well. In the multi-node scenario, the focus is to answer two questions: *"When to cooperate?"* and *"Who to cooperate with?"* The user picks only one relay to cooperate with, in case it needs help. This optimal relay is the one which has the maximum instantaneous value of the relay's metric among the N relays. By determining that instantaneous optimal relay, the user can determine whether to utilize it or not, in a similar fashion to the single-relay scenario. For the symmetric scenario, it is shown that full diversity order is guaranteed and obtain the optimum power allocation which minimizes the SER performance. Finally, we discuss tradeoff curves between the bandwidth efficiency and the SER, which are utilized to determine the optimum cooperation threshold.

8.1 Motivation and relay-selection protocol

In this section, we review the system model of the conventional single-relay decode-and-forward cooperative scenario along with the SER results. This is to help illustrate the motivation behind choosing a modified harmonic mean function of the source–relay and relay–destination channels gain as an appropriate metric to represent the relay's ability to help the source. This motivation is further discussed in Section 8.1.2. Finally, we consider the multi-node relay-selection decode-and-forward cooperative scenario.

8.1.1 Conventional cooperation scenario

We start by considering the conventional single-relay decode-and-forward cooperative scheme where we denote the source as s, the destination as d and the relay as r. The received symbols at the destination and relay after the first transmission phase can be modeled as

$$y_{s,d} = \sqrt{P_1}\, h_{s,d}\, x + \eta_{s,d}, \tag{8.1}$$

$$y_{s,r} = \sqrt{P_1}\, h_{s,r}\, x + \eta_{s,r}, \tag{8.2}$$

where P_1 is the source transmitted power, x is the transmitted information symbol, and $\eta_{s,d}$ and $\eta_{s,r}$ are additive noises. Also, $h_{s,d}$ and $h_{s,r}$ are the source–destination and source–relay channel gains, respectively.

The relay decides whether to forward the received information or not according to the quality of the received signal. If the relay decodes the received symbol correctly, then it forwards the decoded symbol to the destination in the second phase, otherwise it remains idle. The received symbol at the destination from the relay is written as

$$y_{r,d} = \sqrt{\tilde{P}_2}\, h_{r,d}\, x + \eta_{r,d}\ ,$$

where $\tilde{P}_2 = P_2$ if the relay decodes the symbol correctly, otherwise $\tilde{P}_2 = 0$, $\eta_{r,d}$ is an additive noise, and $h_{r,d}$ is the relay–destination channel coefficient. The destination applies maximal-ratio combining (MRC) for the received signals from the source and the relay. The output of the MRC is

$$y = \frac{\sqrt{P_1}\, h_{s,d}^*}{N_0}\, y_{s,d} + \frac{\sqrt{\tilde{P}_2}\, h_{r,d}^*}{N_0}\, y_{r,d}\ . \tag{8.3}$$

The channel coefficients $h_{s,d}$, $h_{s,r}$, and $h_{r,d}$ are modeled as zero-mean, complex Gaussian random variables with variances $\delta_{s,d}^2$, $\delta_{s,r}^2$, and $\delta_{r,d}^2$, respectively. The noise terms $\eta_{s,d}$, $\eta_{s,r}$, and $\eta_{r,d}$ are modeled as zero-mean complex Gaussian random variables with variance N_0.

It has been shown previously in Chapter 5 that the SER for M-PSK signalling can be upper bounded as

$$\Pr(e) \le \frac{N_0^2}{b^2} \cdot \frac{A^2\, P_2\, \delta_{r,d}^2 + B\, P_1\, \delta_{s,r}^2}{P_1^2\, P_2\, \delta_{s,d}^2\, \delta_{s,r}^2\, \delta_{r,d}^2}, \tag{8.4}$$

where $b = \sin^2(\pi/M)$,

$$A = \frac{1}{\pi} \int_0^{\frac{(M-1)\pi}{M}} \sin^2\theta \, d\theta = \frac{M-1}{2M} + \frac{\sin(\frac{2\pi}{M})}{4\pi} ,$$

$$\text{and } B = \frac{1}{\pi} \int_0^{\frac{(M-1)\pi}{M}} \sin^4\theta \, d\theta = \frac{3(M-1)}{8M} + \frac{\sin(\frac{2\pi}{M})}{4\pi} - \frac{\sin(\frac{4\pi}{M})}{32\pi} . \quad (8.5)$$

Moreover, it was shown that the SER upper bound in (8.4) is tight at high enough SNR.

8.1.2   Relay-selection criterion

In this subsection, we deduce a relay-selection criterion from the SER expression in (8.4). Let $\gamma \triangleq P/N_0$ denote the signal-to-noise ratio (SNR), where $P = P_1 + P_2$ is the total power. Hence, (8.4) can be written as

$$\Pr(e) \le (CG \, \gamma)^{-2} , \quad (8.6)$$

where CG denotes the coding gain and it is equal to

$$CG = b^2 \, \delta_{s,d}^2 \left(\frac{\delta_{s,r}^2 \, \delta_{r,d}^2}{q_1 \, \delta_{r,d}^2 + q_2 \, \delta_{s,r}^2} \right) , \quad (8.7)$$

where

$$q_1 = \frac{A^2}{r^2} , \quad q_2 = \frac{B}{r \, (1-r)} , \quad (8.8)$$

and $r \triangleq P_1/P$ is referred to as *power ratio*. If we consider choosing one relay among a set of N relays, we will choose the relay that maximizes the coding gain in (8.7). Note that the coding gain depends on the source–relay and relay–destination channel variances, which are assumed to be fixed during the whole transmission time. Therefore, only one relay will be chosen to help the source during the whole transmission process and the resulting diversity order will be two.

On the other hand, if the instantaneous source–relay and relay–destination channel gains are available at the relay, then this extra information can be utilized in order to benefit from the multi-node diversity inherent in the system, which results in full diversity gain equal to $N + 1$, as we will show later. Therefore, we define the metric for each relay to be the modified harmonic mean function of its source–relay and relay–destination channels gain as

$$\beta_i = \mu_H(q_1 \, \beta_{r_i,d}, q_2 \, \beta_{s,r_i}) = \frac{2 \, q_1 \, q_2 \, \beta_{r_i,d} \, \beta_{s,r_i}}{q_1 \, \beta_{r_i,d} + q_2 \, \beta_{s,r_i}} , \quad for \; i = 1, 2, \ldots, N . \quad (8.9)$$

where $\beta_{s,r_i} = |h_{s,r_i}|^2$, $\beta_{r_i,d} = |h_{r_i,d}|^2$, and $\mu_H(.)$ denotes the standard harmonic mean function. This relay's metric is a scaled version of the coding gain (8.7) after replacing the channel variances with the instantaneous channel gains. The relay's metric β_i (8.9) gives an instantaneous indication of the relay's ability to cooperate with the source.

8.1.3 Single-relay decode-and-forward protocol

We consider that the channels between the nodes present a flat quasi-static fading, hence the channel coefficients are assumed to be constant during a complete frame, and can vary independently from one frame to another. It is assumed that the channels are reciprocal as in the time division duplex (TDD) mode, hence the relay knows its instantaneous source–relay and relay–destination channel gains. The relay calculates the instantaneous metric, denoted by β_m, given by

$$\beta_m = \mu_{\mathrm{H}}(q_1 \, \beta_{\mathrm{r,d}}, q_2 \, \beta_{\mathrm{s,r}}) \triangleq \frac{2 \, q_1 \, q_2 \, \beta_{\mathrm{s,r}} \, \beta_{\mathrm{r,d}}}{q_1 \, \beta_{\mathrm{r,d}} + q_2 \, \beta_{\mathrm{s,r}}}, \tag{8.10}$$

where $\beta_{\mathrm{s,r}} = |h_{\mathrm{s,r}}|^2$, $\beta_{\mathrm{r,d}} = |h_{\mathrm{r,d}}|^2$. Then it sends its metric value through a feedback channel to the source. The source is assumed to know its instantaneous source–destination channel gain, which is then compared with the relay's metric to determine whether to cooperate with this relay or not. Finally, the source sends a control signal to determine if it is going to cooperate or not. We assume here that the channels vary slowly, so that the overhead resulting from the transmission of the relay's metric and source's decision is negligible.

The transmission protocol can be described as follows. In the first phase, the source computes the ratio $\beta_{\mathrm{s,d}}/\beta_m$ and compares it to the cooperation threshold α. If $\beta_{\mathrm{s,d}}/\beta_m \geq \alpha$, then the source decides to use direct transmission only. This mode will be referred to as the *direct-transmission* mode. Let

$$\phi = \{ \, \beta_{\mathrm{s,d}} \geq \alpha \, \beta_m \, \}$$

be the event of direct transmission. The received symbol at the destination can then be modeled as

$$y_{\mathrm{s,d}}^{\phi} = \sqrt{P} \, h_{\mathrm{s,d}} \, x + \eta_{\mathrm{s,d}}, \tag{8.11}$$

where P is the total transmitted power, x is the transmitted symbol, $h_{\mathrm{s,d}}$ is the source–destination channel coefficient, and $\eta_{\mathrm{s,d}}$ is an additive noise.

On the other hand, if $\beta_{\mathrm{s,d}}/\beta_m < \alpha$, then the source employs the relay to transmit its information as in the conventional decode-and-forward cooperative protocol. This mode will be called the *relay-cooperation* mode and can be described as follows. In the first phase, the source broadcasts its symbol to both the relay and the destination. The received symbols at the destination and the relay can be modeled as

$$y_{\mathrm{s,d}}^{\phi^c} = \sqrt{P_1} \, h_{\mathrm{s,d}} \, x + \eta_{\mathrm{s,d}}, \tag{8.12}$$

$$y_{\mathrm{s,r}}^{\phi^c} = \sqrt{P_1} \, h_{\mathrm{s,r}} \, x + \eta_{\mathrm{s,r}}, \tag{8.13}$$

where P_1 is the source transmitted power, $h_{\mathrm{s,r}}$ is the source–relay channel coefficient, $\eta_{\mathrm{s,r}}$ is an additive noise, and ϕ^c denotes the complement of the event ϕ.

The relay decodes the received symbol and re-transmits the decoded symbol if correctly decoded in the second phase, otherwise it remains idle. The received symbol at the destination from the relay is written as

$$y_{\mathrm{r,d}}^{\phi^c} = \sqrt{\tilde{P}_2} \, h_{\mathrm{r,d}} \, x + \eta_{\mathrm{r,d}}, \tag{8.14}$$

where $\tilde{P}_2 = P_2$ if the relay decodes the symbol correctly, otherwise $\tilde{P}_2 = 0$, $h_{r,d}$ is the relay–destination channel coefficient, and $\eta_{r,d}$ is an additive noise. Power is distributed between the source and the relay subject to the power constraint $P_1 + P_2 = P$.

The channel coefficients $h_{s,d}$, $h_{s,r}$, and $h_{r,d}$ are modeled as zero-mean complex Gaussian random variables with variances $\delta_{s,d}^2$, $\delta_{s,r}^2$, and $\delta_{r,d}^2$, respectively. The noise terms, $\eta_{s,d}$, $\eta_{s,r}$, and $\eta_{r,d}$, are modeled as zero-mean, complex Gaussian random variables with equal variance N_0.

8.2   Performance analysis

In this section, first we calculate the probability of the direct-transmission and relay-cooperation modes for the single-relay relay-selection decode-and-forward cooperative scenario. Then, they are used to obtain an approximate expression of the bandwidth efficiency and an upper bound on the SER performance.

8.2.1   Average bandwidth efficiency analysis

We derive the average achievable bandwidth efficiency as follows. Let $\beta_{i,j} = |h_{i,j}|^2$ represent the source–destination, source–relay, and relay–destination channel gains. We know that the $\beta_{i,j}$ is exponentially distributed with parameter $1/\delta_{i,j}^2$, i.e.,

$$ p_{\beta_{i,j}} \left(\beta_{i,j} \right) = \frac{1}{\delta_{i,j}^2} \exp \left(-\frac{\beta_{i,j}}{\delta_{i,j}^2} \right) U \left(\beta_{i,j} \right) , \tag{8.15} $$

is the probability density function (PDF) of $\beta_{i,j}$, where $U(\cdot)$ is the unit-step function. The PDF and the cumulative distribution function (CDF) of β_m in (8.10), denoted by $p_{\beta_m}(\cdot)$ and $P_{\beta_m}(\cdot)$, respectively, can be written as (leave proof as an exercise)

$$ p_{\beta_m} \left(\beta_m \right) = \frac{\beta_m}{2 \, t_1^2} \exp \left(-\frac{t_2}{2} \beta_m \right) \left(t_1 \, t_2 \, K_1 \left(\frac{\beta_m}{t_1} \right) + 2 \, K_0 \left(\frac{\beta_m}{t_1} \right) \right) U \left(\beta_m \right) , $$

$$ P_{\beta_m} \left(\beta_m \right) = 1 - \frac{\beta_m}{t_1} \exp \left(-\frac{t_2}{2} \beta_m \right) K_1 \left(\frac{\beta_m}{t_1} \right) , \tag{8.16} $$

where

$$ t_1 = \sqrt{q_1 q_2 \delta_{s,r}^2 \delta_{r,d}^2} , \quad t_2 = \frac{1}{q_2 \delta_{s,r}^2} + \frac{1}{q_1 \delta_{r,d}^2} . \tag{8.17} $$

In (8.16), $K_0(x)$ and $K_1(x)$ are the zero-order and first-order modified Bessel functions of the second kind, respectively, defined in [[1], (9.6.21–22)] as

$$ K_0(x) = \int_0^\infty \cos \left(x \, \sinh (t) \right) \, \mathrm{d}t = \int_0^\infty \frac{\cos (x \, t)}{\sqrt{t^2 + 1}} \, \mathrm{d}t , $$

$$ K_1(x) = \int_0^\infty \sin \left(x \, \sinh (t) \right) \, \sinh (t) \, \mathrm{d}t , \tag{8.18} $$

for $x > 0$.

The probability of direct-transmission mode can be obtained as follows:

$$\Pr(\phi) = \Pr\left(\beta_{\text{s,d}} \geq \alpha \, \beta_m\right) = \int_0^\infty P_{\beta_m}\left(\frac{\beta_{\text{s,d}}}{\alpha}\right) p_{\beta_{\text{s,d}}}\left(\beta_{\text{s,d}}\right) \, d\beta_{\text{s,d}}$$

$$= 1 - \frac{1}{\alpha t_1 \delta_{\text{s,d}}^2} \int_0^\infty \beta_{\text{s,d}} \exp\left(-\left(\frac{1}{\delta_{\text{s,d}}^2} + \frac{t_2}{2\alpha}\right)\beta_{\text{s,d}}\right) K_1\left(\frac{\beta_{\text{s,d}}}{\alpha t_1}\right) \, d\beta_{\text{s,d}},$$

$$\tag{8.19}$$

where we used (8.15) and (8.16). $K_1(.)$ can be approximated as

$$K_1(x) \approx \frac{1}{x} \, . \tag{8.20}$$

Substituting (8.20) into (8.19), the probability of the direct transmission is approximated as

$$\Pr(\phi) \approx 1 - \frac{1}{\delta_{\text{s,d}}^2} \int_0^\infty \exp\left(-\left(\frac{1}{\delta_{\text{s,d}}^2} + \frac{t_2}{2\alpha}\right)\beta_{\text{s,d}}\right) \, d\beta_{\text{s,d}}$$

$$\approx \frac{t_2 \, \delta_{\text{s,d}}^2}{2\alpha + t_2 \, \delta_{\text{s,d}}^2} \, , \tag{8.21}$$

and the probability of the relay-cooperation mode is

$$\Pr(\phi^c) = 1 - \Pr(\phi) \approx \frac{2\alpha}{2\alpha + t_2 \, \delta_{\text{s,d}}^2} \, . \tag{8.22}$$

Since the bandwidth efficiency of the direct-transmission mode is 1 symbols per channel use (SPCU), and that of the relay-cooperation mode is 1/2 SPCU, the average bandwidth efficiency can be written as

$$R = \Pr(\phi) + \frac{1}{2}\Pr(\phi^c) \, . \tag{8.23}$$

Substituting (8.21) and (8.22) into (8.23) and substituting t_2 as in (8.17), we get the following theorem.

THEOREM 8.2.1 *The bandwidth efficiency of the relay-selection decode-and-forward cooperative scenario, utilizing a single relay, can be approximated as*

$$R \approx \frac{\alpha + \left(\frac{1-r}{B\,\delta_{\text{s,r}}^2} + \frac{r}{A^2\,\delta_{\text{r,d}}^2}\right) r\,\delta_{\text{s,d}}^2}{2\alpha + \left(\frac{1-r}{B\,\delta_{\text{s,r}}^2} + \frac{r}{A^2\,\delta_{\text{r,d}}^2}\right) r\,\delta_{\text{s,d}}^2} \quad \text{SPCU}, \tag{8.24}$$

where SPCU stands for symbols per channel use.

8.2.2   SER analysis and upper bound

We obtain the SER of the single-relay decode-and-forward scheme for M-PSK signalling. The probability of symbol error, or SER, is defined as

$$\Pr(e) = \Pr(e|\phi) \cdot \Pr(\phi) + \Pr(e|\phi^c) \cdot \Pr(\phi^c) \, , \tag{8.25}$$

where $\Pr(e|\phi) \cdot \Pr(\phi)$ represents the SER of the direct-transmission mode and $\Pr(e|\phi^c) \cdot \Pr(\phi^c)$ represents the relay-cooperation mode SER. The SER of the direct-transmission mode can be calculated as follows. First, the instantaneous direct-transmission SNR is

$$\gamma^\phi = \frac{P \, \beta_{s,d}}{N_0} \, . \tag{8.26}$$

The conditional direct-transmission SER can be written as

$$\Pr(e|\text{phi}, \beta_{s,d}) = \Psi(\gamma^\phi) \triangleq \frac{1}{\pi} \int_0^{\frac{(M-1)\pi}{M}} \exp\left(-\frac{b \, \gamma^\phi}{\sin^2 \theta}\right) \, d\theta \, . \tag{8.27}$$

where $b = \sin^2(\pi/M)$. Since

$$
\begin{aligned}
\Pr(e|\phi) \, \Pr(\phi) &= \int_0^\infty \Pr(e|\phi, \beta_{s,d}) \, \Pr(\phi|\beta_{s,d}) \, p_{\beta_{s,d}}(\beta_{s,d}) \, d\beta_{s,d} \\
&= \int_0^{\frac{(M-1)\pi}{M}} \frac{1}{\pi} \int_0^\infty \frac{1}{\delta_{s,d}^2} \exp\left(-\left(\frac{b \, P}{N_0 \sin^2 \theta} + \frac{1}{\delta_{s,d}^2}\right) \beta_{s,d}\right) \\
&\quad \times \left(1 - \frac{\beta_{s,d}}{\alpha t_1} \exp\left(-\frac{t_2}{2\alpha} \beta_{s,d}\right) K_1\left(\frac{\beta_{s,d}}{\alpha \, t_1}\right)\right) \, d\beta_{s,d} \, d\theta,
\end{aligned}
\tag{8.28}
$$

where we used (8.15), (8.16), and (8.27). Substituting (8.20) into (8.28), we get

$$
\begin{aligned}
\Pr(e|\phi) \, \Pr(\phi) &\approx F_1\left(1 + \frac{b \, P \, \delta_{s,d}^2}{N_0 \sin^2 \theta}\right) - F_1\left(1 + \frac{t_2 \delta_{s,d}^2}{2\alpha} + \frac{b \, P \, \delta_{s,d}^2}{N_0 \sin^2 \theta}\right) \\
&\approx \frac{t_2 \, \delta_{s,d}^2}{2\alpha} F_1\left(\left(1 + \frac{b \, P \, \delta_{s,d}^2}{N_0 \, \sin^2 \theta}\right)\left(1 + \frac{t_2 \, \delta_{s,d}^2}{2\alpha} + \frac{b \, P \, \delta_{s,d}^2}{N_0 \, \sin^2 \theta}\right)\right),
\end{aligned}
\tag{8.29}
$$

where

$$F_1(x(\theta)) = \frac{1}{\pi} \int_0^{\frac{(M-1)\pi}{M}} \frac{1}{x(\theta)} \, d\theta.$$

We obtain the SER of the relay-cooperation mode as follows. For the relay-cooperation mode, maximal-ratio combining (MRC) is applied at the destination. The output of the MRC can be written as

$$y^{\phi^c} = \frac{\sqrt{P_0} \, h_{s,d}^*}{N_0} \, y_{s,d}^{\phi^c} + \frac{\sqrt{\tilde{P}_1} \, h_{r,d}^*}{N_0} \, y_{r,d}^{\phi^c} \, . \tag{8.30}$$

The instantaneous SNR of the MRC output can be written as

$$\gamma^{\phi^c} = \frac{P_0 \beta_{s,d} + \tilde{P}_1 \beta_{r,d}}{N_0} \, . \tag{8.31}$$

We can easily see that the conditional SER of the relay-cooperation mode can be written as

$$\Pr(e|\phi^c, \beta_{s,d}, \beta_{s,r}, \beta_{r,d}) = \Psi(\gamma^{\phi^c})|_{\tilde{P}_1=0}\, \Psi\left(\frac{P_0\beta_{s,r}}{N_0}\right)$$
$$+ \Psi(\gamma^{\phi^c})|_{\tilde{P}_1=P_1}\left(1 - \Psi\left(\frac{P_0\beta_{s,r}}{N_0}\right)\right). \tag{8.32}$$

Let us denote $\Pr(A|\phi^c, \beta_{s,d}, \beta_{s,r}, \beta_{r,d}) \triangleq \Psi(\gamma^{\phi^c})\Psi(\frac{P_0\beta_{s,r}}{N_0})$ and $\Pr(B|\phi^c, \underline{\beta}) \triangleq \Psi(\gamma^{\phi^c})$, in which $\underline{\beta} \triangleq [\beta_{s,d}, \beta_{s,r}, \beta_{r,d}]$. Then

$$\Pr(A|\phi^c, \beta_{s,d}, \beta_{s,r}, \beta_{r,d})$$
$$= \frac{1}{\pi}\int_{\theta_1=0}^{\frac{(M-1)\pi}{M}} \exp\left(-\frac{b\,P_0}{N_0\,\sin^2\theta_1}\beta_{s,d}\right) \exp\left(-\frac{b\,\tilde{P}_1}{N_0\,\sin^2\theta_1}\beta_{r,d}\right) d\theta_1$$
$$\times \frac{1}{\pi}\int_{\theta_2=0}^{\frac{(M-1)\pi}{M}} \exp\left(-\frac{b\,P_0}{N_0\sin^2\theta_2}\beta_{s,r}\right) d\theta_2, \tag{8.33}$$

so we have

$$\Pr(A|\phi^c)\Pr(\phi^c) = \int_{\underline{\beta}} \Pr(A/\phi^c, \underline{\beta})\,\Pr(\phi^c/\underline{\beta})\,p_{\underline{\beta}}(\underline{\beta})\,d\underline{\beta} \tag{8.34}$$

Moreover, since β_m is a function of $\beta_{s,r}$ and $\beta_{r,d}$ as given by (8.10), so

$$\Pr(\phi^c|\underline{\beta}) = \Pr(\beta_{s,d} < \alpha\beta_m/\beta_{s,d}, \beta_{s,r}, \beta_{r,d}) = U(\alpha\beta_m - \beta_{s,d}). \tag{8.35}$$

Substituting (8.33) and (8.35) into (8.34), we get

$$\Pr(A|\phi^c)\,Pr(\phi^c)$$
$$= \int_{\underline{\beta}} \frac{1}{\pi^2}\int_{\theta_1=0}^{\frac{(M-1)\pi}{M}} \int_{\theta_2=0}^{\frac{(M-1)\pi}{M}} \exp\left(-P_0\,C(\theta_1)\,\beta_{s,d}\right)\exp\left(-\tilde{P}_1\,C(\theta_1)\,\beta_{r,d}\right)$$
$$\times \exp\left(-P_0\,C(\theta_2)\,\beta_{s,r}\right) U(\alpha\beta_m - \beta_{s,d})\,p_{\underline{\beta}}(\underline{\beta})\,d\theta_2\,d\theta_1\,d\underline{\beta}, \tag{8.36}$$

where $C(\theta) = b/(N_0\sin^2\theta)$. Since $\beta_{s,d}$, $\beta_{s,r}$, and $\beta_{r,d}$ are statistically independent, thus

$$p_{\underline{\beta}}(\underline{\beta}) = p_{\beta_{s,d}}(\beta_{s,d})\,p_{\beta_{s,r}}(\beta_{s,r})\,p_{\beta_{r,d}}(\beta_{r,d}) = p_{\beta_{s,d}}(\beta_{s,d})\,p_{\tilde{\underline{\beta}}}(\tilde{\underline{\beta}}), \tag{8.37}$$

$$\Pr(A|\phi^c)\,\Pr(\phi^c)$$
$$= \int_{\tilde{\underline{\beta}}} \frac{1}{\pi^2}\int_{\theta_1=0}^{\frac{(M-1)\pi}{M}} \int_{\theta_2=0}^{\frac{(M-1)\pi}{M}} \frac{1 - \exp\left(-\left(P_0\,C(\theta_1) + \frac{1}{\delta_{s,d}^2}\right)\alpha\beta_m\right)}{1 + P_0\,C(\theta_1)\delta_{s,d}^2}$$
$$\times \exp\left(-\tilde{P}_1\,C(\theta_1)\beta_{r,d}\right)\exp\left(-P_0\,C(\theta_2)\beta_{s,r}\right) p_{\tilde{\underline{\beta}}}(\tilde{\underline{\beta}})\,d\theta_2\,d\theta_1\,d\tilde{\underline{\beta}}. \tag{8.38}$$

It is difficult to get an exact expression of (8.38) for β_m defined in (8.10). Thus, we obtain an upper bound via the worst-case scenario. We replace $\beta_{s,r}$ and $\beta_{r,d}$ in (8.38) by their worst-case values in terms of β_m. Then, we average (8.38) over β_m only. It follows from (8.10) that

$$\frac{1}{\beta_m} = \frac{1}{2\ q_2\ \beta_{\mathrm{s,r}}} + \frac{1}{2\ q_1\ \beta_{\mathrm{r,d}}}, \tag{8.39}$$

thus, $\beta_m \le 2\ q_2\ \beta_{\mathrm{s,r}}$ and $\beta_m \le 2\ q_1\ \beta_{\mathrm{r,d}}$. If we replace $\beta_{\mathrm{s,r}}$ and $\beta_{\mathrm{r,d}}$ by their worst-case values in terms of β_m as

$$\beta_{\mathrm{s,r}} \longrightarrow \frac{\beta_m}{2\ q_2}, \quad \beta_{\mathrm{r,d}} \longrightarrow \frac{\beta_m}{2\ q_1}. \tag{8.40}$$

Then, the probability in (8.38) can be upper bounded as

$$\Pr(A|\phi^c)\ \Pr(\phi^c)$$

$$\le \frac{1}{\pi^2} \int_{\theta_1=0}^{\frac{(M-1)\pi}{M}} \frac{1}{1 + P_0\ C(\theta_1)\delta_{\mathrm{s,d}}^2} \int_{\theta_2=0}^{\frac{(M-1)\pi}{M}} M_{\beta_m}\left(\frac{\tilde{P}_1\ C(\theta_1)}{2\ q_1} + \frac{P_0\ C(\theta_2)}{2\ q_2} \right)$$

$$- M_{\beta_m}\left(\left(P_0\ C(\theta_1) + \frac{1}{\delta_{\mathrm{s,d}}^2} \right)\alpha + \frac{\tilde{P}_1\ C(\theta_1)}{2\ q_1} + \frac{P_0\ C(\theta_2)}{2\ q_2} \right) \mathrm{d}\theta_2\mathrm{d}\theta_1, \tag{8.41}$$

where $M_{\beta_m}(.)$ is the moment-generation function (MGF) of β_m. It can be shown that for two independent exponential random variables with parameters λ_1 and λ_2, the MGF of their harmonic mean function can be written as (leave proof as an exercise)

$$M_{\beta_m}(\gamma) = E_{\beta_m}\left(\exp(-\gamma\ \beta_m) \right)$$

$$= \frac{16\ \lambda_1\ \lambda_2}{3\ (\lambda_1 + \lambda_2 + 2\ \sqrt{\lambda_1\ \lambda_2} + \gamma)^2} \left(\frac{4\ (\lambda_1 + \lambda_2)\ {}_2F_1\left(3, \frac{3}{2}; \frac{5}{2}; \frac{\lambda_1 + \lambda_2 - 2\ \sqrt{\lambda_1\ \lambda_2} + \gamma}{\lambda_1 + \lambda_2 + 2\ \sqrt{\lambda_1\ \lambda_2} + \gamma} \right)}{(\lambda_1 + \lambda_2 + 2\ \sqrt{\lambda_1\ \lambda_2} + \gamma)} \right.$$

$$\left. + {}_2F_1\left(2, \frac{1}{2}; \frac{5}{2}; \frac{\lambda_1 + \lambda_2 - 2\ \sqrt{\lambda_1\ \lambda_2} + \gamma}{\lambda_1 + \lambda_2 + 2\ \sqrt{\lambda_1\ \lambda_2} + \gamma} \right) \right) \tag{8.42}$$

where $E_{\beta_m}(.)$ represents the statistical average with respect to β_m and ${}_2F_1(., .; .; .)$ is the Gauss' hypergeometric function as defined on page 172. Following similar steps as done in (8.33)–(8.41), we have (leave proof as an exercise)

$$\Pr(B|\phi^c)\ \Pr(\phi^c)$$

$$\le \frac{1}{\pi} \int_{\theta=0}^{\frac{(M-1)\pi}{M}} \frac{M_{\beta_m}\left(\frac{\tilde{P}_1\ C(\theta)}{2\ q_1} \right) - M_{\beta_m}\left(\left(P_0\ C(\theta) + \frac{1}{\delta_{\mathrm{s,d}}^2} \right)\alpha + \frac{\tilde{P}_1\ C(\theta)}{2\ q_1} \right)\mathrm{d}\theta}{1 + P_0\ C(\theta)\delta_{\mathrm{s,d}}^2}. \tag{8.43}$$

Note that the unconditional SER of the relay-cooperation mode in (8.32) can be rewritten as

$$\Pr(e|\phi^c)\ \Pr(\phi^c)$$

$$= \left(\Pr(A|\phi^c)|_{\tilde{P}_1=0} - \Pr(A|\phi^c)|_{\tilde{P}_1=P_1} + \Pr(B|\phi^c)|_{\tilde{P}_1=P_1} \right) \Pr(\phi^c),$$

$$\le \left(\Pr(A|\phi^c)|_{\tilde{P}_1=0} + \Pr(B|\phi^c)|_{\tilde{P}_1=P_1} \right) \Pr(\phi^c). \tag{8.44}$$

Based on (8.41) and (8.43), we get an upper bound as

$$\Pr(e|\phi^c)\,\Pr(\phi^c)$$

$$\leq \frac{1}{\pi}\int_{\theta=0}^{\frac{(M-1)\pi}{M}} \frac{M_{\beta_m}\!\left(\frac{P_1\,C(\theta)}{2\,q_1}\right)}{1+P_0\,C(\theta)\delta_{s,d}^2}\,d\theta$$

$$+\frac{1}{\pi^2}\int_{\theta_1=0}^{\frac{(M-1)\pi}{M}} \frac{1}{1+P_0\,C(\theta_1)\delta_{s,d}^2}\int_{\theta_2=0}^{\frac{(M-1)\pi}{M}} M_{\beta_m}\!\left(\frac{P_0\,C(\theta_2)}{2\,q_2}\right)d\theta_2\,d\theta_1.$$

$$(8.45)$$

Therefore, an upper bound on the total SER can be obtained as

$$\Pr(e)\leq \frac{t_2\,\delta_{s,d}^2}{2\alpha}\,F_1\!\left(\left(1+\frac{b\,P\,\delta_{s,d}^2}{N_0\,\sin^2\theta}\right)\left(1+\frac{t_2\,\delta_{s,d}^2}{2\alpha}+\frac{b\,P\,\delta_{s,d}^2}{N_0\,\sin^2\theta}\right)\right)$$

$$+\frac{1}{\pi}\int_{\theta_1=0}^{\frac{(M-1)\pi}{M}} \frac{\left(M_{\beta_m}\!\left(\frac{b\,P_1}{2\,q_1\,N_0\,\sin^2\theta_1}\right)+\frac{1}{\pi}\int_{\theta_2=0}^{\frac{(M-1)\pi}{M}} M_{\beta_m}\!\left(\frac{b\,P_0}{2\,q_2\,N_0\,\sin^2\theta_2}\right)d\theta_2\right)}{1+\frac{b\,P_0\,\delta_{s,d}^2}{N_0\,\sin^2\theta_1}}\,d\theta_1.$$

$$(8.46)$$

The above SER upper bound expression is in terms of the MGF M_{β_m}, which is mathematically intractable. In order to show that full diversity order is achieved, we derive an upper bound on the SER performance at high SNR as follows. We recall that an approximation to the MGF of two independent exponential random variables at high enough SNR can be obtained as (see Chapter 5)

$$M_{\beta_m}(\gamma)\approx \frac{q_1\,\delta_{r,d}^2+q_2\,\delta_{s,r}^2}{2\,\gamma}.$$

$$(8.47)$$

Then, substituting (8.47) into (8.46) and neglecting all terms added to $\left(b\,P\,\delta_{s,d}^2\right)/\left(N_0\,\sin^2\theta\right)$ and $\left(b\,P_0\,\delta_{s,d}^2\right)/\left(N_0\,\sin^2\theta\right)$ at high SNR, we get

$$\Pr(e)\leq \frac{t_2}{2\alpha}\,\frac{N_0^2\,B}{b^2\,P^2\,\delta_{s,d}^2}+\left(\frac{N_0}{b}\right)^2\frac{\left(q_1\,\delta_{r,d}^2+q_2\,\delta_{s,r}^2\right)\left(\frac{q_1\,B}{P_1}+\frac{q_2\,A^2}{P_0}\right)}{P_0\,\delta_{s,d}^2}$$

$$=\left(\frac{N_0}{b}\right)^2\left(\frac{B\,t_2}{2\,\alpha\,\delta_{s,d}^2\,P^2}+\frac{2\,q_2\,A^2\,P}{\delta_{s,d}^2}\left(\frac{A^2\,\delta_{r,d}^2\,P_1+B\,\delta_{s,r}^2\,P_0}{P_0^3\,P_1}\right)\right).$$

Note that $(q_1\,B)/P_1=(q_2\,A^2)/P_0$ after substituting q_1 and q_2 by their expressions in (8.8), and using the equality $P_0\,(1-r)=P_1\,r$. Thus,

$$\Pr(e)\leq \frac{B}{b^2\,\delta_{s,d}^2}\left(\frac{t_2}{2\,\alpha}+\frac{2\,A^2\,(q_1\,\delta_{r,d}^2+q_2\,\delta_{s,r}^2)}{r^2\,(1-r)}\right)\cdot\left(\frac{1}{\gamma}\right)^2,$$

$$(8.48)$$

where $\gamma=P/N_0$ is the SNR. Equation (8.48) shows that diversity order two is achieved as long as $\alpha>0$. Replacing q_1 and q_2 by their expressions defined in (8.8) and t_2 by its expression defined in (8.17), we obtain the following theorem.

THEOREM 8.2.2   *At high SNR, $\gamma = P/N_0$, the SER of the single-relay relay-selection decode-and-forward cooperative scheme is upper bounded as* $\Pr(e) \leq (\mathrm{CG} \cdot \gamma)^{-2}$, *where* CG *denotes the coding gain and is equal to*

$$
\mathrm{CG} = \sqrt{\frac{b^2\, \delta_{\mathrm{s,d}}^2}{B\left(\dfrac{\frac{r(1-r)}{B\, \delta_{\mathrm{s,r}}^2} + \frac{r^2}{A^2\, \delta_{\mathrm{r,d}}^2}}{2\,\alpha} + \dfrac{2\, A^2 \left(\frac{A^2\, \delta_{\mathrm{r,d}}^2}{r^2} + \frac{B\, \delta_{\mathrm{s,r}}^2}{r(1-r)}\right)}{r^2\, (1-r)}\right)}}\,.
\tag{8.49}
$$

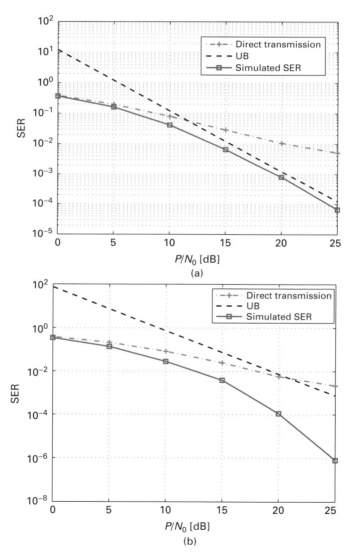

Fig. 8.1   SER simulated, SER upper bound, and direct transmission curves for single-relay relay-selection decode-and-forward cooperative scheme with QPSK modulation, $\alpha = 1$, and $r_o = 0.5$ for (a) unity channel variances and (b) $\delta_{\mathrm{s,d}}^2 = 1$, $\delta_{\mathrm{s,r}}^2 = 10$, and $\delta_{\mathrm{r,d}}^2 = 1$.

Example 8.1 This example is to demonstrate the single-relay relay-selection cooperative communications scheme by assuming $\alpha = 1$, $r_o = 0.5$, $N_0 = 1$, and QPSK signalling for (a) Unity channel variances and (b) $\delta_{s,d}^2 = 1$, $\delta_{s,r}^2 = 10$, and $\delta_{r,d}^2 = 1$. Figure 8.1(a) depicts the simulated SER curves for the single-relay relay-selection decode-and-forward cooperative scheme with unity channel variances. Also, we plot the direct transmission curve which achieves diversity order one, to show the advantage of using the cooperative scenario. Figure 8.1(b) shows the simulated SER curve for single-relay relay-selection decode-and-forward cooperative scheme when the source–relay channel is stronger, $\delta_{s,r}^2 = 10$. We can see that the SER upper bound achieves full diversity order, which guarantees that the actual SER performance has full diversity order as well.   ▲

8.3 Multi-node scenario

In this section, we discuss the multi-node relay-selection decode-and-forward cooperative scenario and determine when the source needs to cooperate and which relay to cooperate with. Let us consider a cooperative communication system with N relays, as shown in Figure 8.2, which consists of a source, s, its destination, d, and N relays. Each relay receives the symbols from its previous relays, applies MRC on the received output, and re-transmits if correctly decoded. This protocol achieves full diversity order as shown in Chapter 6. However, it requires $N + 1$ phases and the bandwidth efficiency is $R = 1/N + 1$ SPCU. Therefore, the objective of the relay-selection scenario is to increase the bandwidth efficiency, while achieving full diversity order.

   In the multi-node relay selection cooperative scenarios, there are two main questions to be answered:

- The first question is how to determine the optimal relay to cooperate with. The answer comes from the motivation described earlier. The modified harmonic mean function

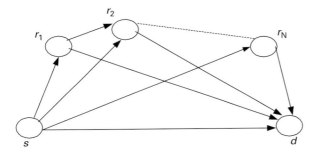

Fig. 8.2   Multi-node cooperative communication system.

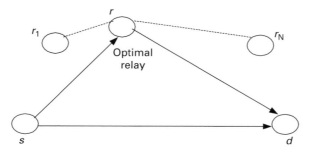

Fig. 8.3   Multi-node relay-selection cooperative communication system behaves like a single-relay selection system with the optimal relay.

of the source–relay and relay–destination channel gains is an appropriate measure on how much help a relay can offer. Thus, the optimal relay is the relay with the maximum modified harmonic mean function of its source–relay and relay–destination channel gains among all the N relays. With this optimal relay being decided, the system consists of the source, the destination, and the optimal relay, as shown in Figure 8.3.

• The second question is how to determine when to cooperate. Its answer is the same as that of the single-relay case described before. The source compares the source–destination channel gain with the optimal relay's metric to decide whether to cooperate or not. Therefore, the system model of the multi-node relay-selection cooperative scenario is the same as that of the single relay scenario utilizing the optimal relay, which is described by (8.11)–(8.14).

We assume that the channels are reciprocal, hence each relay knows its source–relay and relay–destination channel gains and calculates its metric. Then, each relay sends this metric to the source through a feedback channel. Furthermore, we assume that the source knows its source–destination channel gain. Thus, the source computes the ratio between its source–destination channel gain and the maximum metric of the relays, and compares this ratio to the cooperation threshold. If this ratio is greater than the cooperation threshold, then the source does not cooperate with any of the relays and sends its information directly to the destination in single phase. Otherwise, the source cooperates with one relay, which has the maximum metric, in two phases. Finally, the source sends a control signal to the destination and the relays to indicate which mode it is going to use. The relay, which has the maximum metric, will be referred to as the *optimal relay*. This procedure is repeated every time the channel gains vary. We assume that the channel gains vary slowly so that the overhead resulting from sending the relays's metrics is negligible. We should note here that the source and the relays are not required to know the phase information of their channels.

8.3.1   Bandwidth efficiency

We derive the achievable bandwidth efficiency of the multi-node relay-selection decode-and-forward cooperative scheme as follows. The metric for each relay is defined in

(8.9). We denote the PDF and CDF of β_i for $i = 1, 2, \ldots, N$ as $p_{\beta_i}(.)$ and $P_{\beta_i}(.)$, respectively. They are as defined in (8.16) with t_1 and t_2 being replaced by

$$t_{1,i} = \sqrt{q_1 \, q_2 \, \delta_{s,r_i}^2 \, \delta_{r_i,d}^2} \quad \text{and} \quad t_{2,i} = \frac{1}{q_2 \, \delta_{s,r_i}^2} + \frac{1}{q_1 \, \delta_{r_i,d}^2},$$

respectively. The optimum relay will have a metric which is equal to

$$\beta_{\max} = \max\{ \beta_1, \beta_2, \ldots, \beta_N \} . \tag{8.50}$$

The CDF of $\beta_{\max}$ can be written as

$$P_{\beta_{\max}}(\beta) = \Pr(\beta_1 \leq \beta, \beta_2 \leq \beta, \ldots, \beta_N \leq \beta) = \prod_{i=1}^{N} P_{\beta_i}(\beta)$$

$$= \prod_{i=1}^{N} \left(1 - \frac{\beta_i}{t_{1,i}} \exp(-\frac{t_{2,i}}{2} \beta) \, K_1(\frac{\beta}{t_{1,i}}) \right) , \tag{8.51}$$

and the PDF of $\beta_{\max}$ is written as

$$p_{\beta_{\max}}(\beta) = \frac{\partial P_{\beta_{\max}}(\beta)}{\partial \beta} = \sum_{j=1}^{N} p_{\beta_j}(\beta) \left(\prod_{i=1,i \neq j}^{N} P_{\beta_i}(\beta) \right)$$

$$\approx \sum_{j=1}^{N} p_{\beta_j}(\beta) \left(\prod_{i=1,i \neq j}^{N} \left(1 - \exp(-\frac{t_{2,i}}{2} \beta) \right) \right) , \tag{8.52}$$

where we have used the approximation in (8.20). The expression in (8.52) is complex and will lead to more complex and intractable expressions. Without loss of generality, we consider the symmetric scenario where all the relays have the same source–relay and relay–destination channel variances, i.e.,

$$\delta_{s,r_i}^2 = \delta_{s,r}^2 \quad , \quad \delta_{r_i,d}^2 = \delta_{r,d}^2 , \text{ for } i = 1, 2, \ldots, N . \tag{8.53}$$

According to (8.51) and (8.52), the CDF and PDF of $\beta_{\max}$ can be written as

$$P_{\beta_{\max}}(\beta) = \left(1 - \frac{\beta}{t_1} \exp\left(-\frac{t_2}{2} \beta\right) K_1\left(\frac{\beta}{t_1}\right) \right)^{N}$$

$$p_{\beta_{\max}}(\beta) = N \left(1 - \frac{\beta}{t_1} \exp\left(-\frac{t_2}{2} \beta\right) K_1\left(\frac{\beta}{t_1}\right) \right)^{N-1} p_{\beta_m}(\beta) , \tag{8.54}$$

where $p_{\beta_m}(.)$ is as defined in (8.16) and t_1 and t_2 are as defined in (8.17). The probability of the direct-transmission mode can be obtained as follows:

$$\Pr(\phi) = \Pr(\beta_{s,d} \geq \alpha \, \beta_{\max})$$

$$= \sum_{n=0}^{N} \binom{N}{n}(-1)^n \frac{1}{(\alpha t_1)^n \delta_{s,d}^2}$$

$$\times \int_0^\infty \beta_{s,d}^n \exp\left(-\left(\frac{1}{\delta_{s,d}^2} + \frac{t_2 \, n}{2\alpha}\right) \beta_{s,d}\right) \left(K_1\left(\frac{\beta_{s,d}}{\alpha t_1}\right)\right)^n d\beta_{s,d}$$

$$\approx \sum_{n=0}^{N} \binom{N}{n}(-1)^n \frac{2\alpha}{2\alpha + t_2 \, \delta_{s,d}^2 \, n} \, , \tag{8.55}$$

where we used (8.15), (8.20), and (8.54) to get (8.55). The probability of relay-cooperation mode is obtained using (8.22). Finally, we obtain the average bandwidth efficiency using (8.23), which leads to the following theorem.

THEOREM 8.3.1   *The bandwidth efficiency of the multi-node relay-selection decode-and-forward symmetric cooperative scenario, employing N relays, is approximated as*

$$R \approx \frac{1}{2}\left(1 + \sum_{n=0}^{N} \binom{N}{n}(-1)^n \frac{2\alpha}{2\alpha + \left(\frac{1-r}{B \, \delta_{s,r}^2} + \frac{r}{A^2 \, \delta_{r,d}^2}\right) r \, \delta_{s,d}^2 \, n}\right) \quad \text{SPCU.} \tag{8.56}$$

Example 8.2   This example is to plot the bandwidth efficiency given in Theorem 8.3.1 versus the number of relays for unity channel variances, $\alpha = 1$, and $r = 0.5$. Figure 8.4 depicts the bandwidth efficiency of the relay-selection with $\alpha = 1$ and $r = 0.5$ and

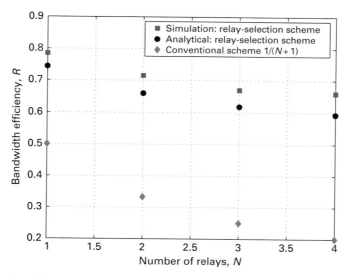

Fig. 8.4   Bandwidth efficiency dependence on the number of relays with QPSK modulation and unity channel variances, $\alpha = 1$, and $r = 0.5$.

the conventional cooperative schemes for different number of relays and unity channel variances. It is clear that the bandwidth efficiency decreases down to 0.5 as N increases, because the probability of the direct-transmission mode decreases down to 0 as N goes to ∞. Furthermore, we plot the bandwidth efficiency of the conventional cooperative scheme, $R_{\mathrm{conv}} = 1/N + 1$ SPCU, to show the significant increasing in the bandwidth efficiency of the relay-selection cooperative scenario over the conventional cooperative scheme.

$\blacktriangle$

8.3.2   SER analysis and upper bound

Before determining the SER of the multi-node relay-selection decode-and-forward symmetric cooperative scenario, we mention the following lemma, which is used to prove full diversity later.

Lemma 8.3.1   *For any x, y, and N*

$$\sum_{n=0}^{N} \binom{N}{n} (-1)^n \frac{1}{x+n\,y} = \frac{(N)!\,y^N}{\prod_{n=0}^{N}(x+n\,y)}, \tag{8.57}$$

Proof   First, we prove that

$$\sum_{n=0}^{N} \frac{\binom{N}{n}(-1)^n}{1+n\,z} = \frac{N!\,z^N}{\prod_{n=0}^{N}(1+n\,z)}. \tag{8.58}$$

Let $A(z) = \sum_{n=0}^{N} \frac{\binom{N}{n}(-1)^N}{1+n\,z}$, $B(z) = \prod_{n=0}^{N}(1+n\,z)$, and $G(z) = A(z)\cdot B(z)$. The order of $G(z)$ is N, thus it can be written as $G(z) = \sum_{i=0}^{N} g_i\,z^i$ where $g_i = \frac{1}{i!}\frac{\partial^i G(z)}{\partial z^i}|_{z=0}$. It can be easily shown that

$$\frac{\partial^i A(z)}{\partial z^i} = (-1)^i\,i!\sum_{n=0}^{N} \frac{\binom{N}{n}(-1)^n\,n^i}{(1+n\,z)^{i+1}}. \tag{8.59}$$

Using the identity obtained in [[50], (0.154, 3–4)] As

$$\sum_{n=0}^{N} \binom{N}{n}(-1)^n\,n^i = \begin{cases} 0, & 0 \le i < N \\ (-1)^N\,N!, & i = N \end{cases} \tag{8.60}$$

we get

$$\frac{\partial^i A(z)}{\partial z^i}\bigg|_{z=0} = (-1)^i\,i!\left(\sum_{n=0}^{N}\binom{N}{n}(-1)^n\,n^i\right) = \begin{cases} 0, & 0 \le i < N \\ (N!)^2, & i = N \end{cases} \tag{8.61}$$

Since, $A(z)|_{z=0} = 0$ and $B(z)|_{z=0} = 1$, thus

$$g_i = \frac{1}{i!} \frac{\partial^i \left(A(z) \cdot B(z) \right)}{\partial z^i} \bigg|_{z=0}$$

$$= \frac{1}{i!} \frac{\partial^i A(z)}{\partial z^i} \bigg|_{z=0} \cdot B(z) \bigg|_{z=0}$$

$$= \begin{cases} 0, & 0 \leq i < N. \\ \\ N!, & i = N. \end{cases} \tag{8.62}$$

Thus,

$$A(z) = \frac{G(z)}{B(z)} = \frac{N! \, z^N}{\prod_{n=0}^{N}(1 + n \, z)}, \tag{8.63}$$

which proves (8.58). Replacing $z = y/x$ in (8.58), we obtain Lemma 8.3.1.

The SER derivation in the multi-node system is similar to those in the single-relay system. The major difference is the usage of the PDF and CDF of $\beta_{\max}$ defined in (8.54) instead of the PDF and CDF of β_m defined in (8.16). The total SER can be obtained as a summation of the SER of both the direct-transmission and relay-cooperation modes as in (8.25). First, we obtain the SER of the direct-transmission mode similar to the single-relay case (8.28) as follows:

$$\Pr(e|\phi) \Pr(\phi) \approx \sum_{n=0}^{N} \binom{N}{n}(-1)^n F_1 \left(1 + \frac{t_2 \, \delta_{s,d}^2 \, n}{2\alpha} + \frac{b \, P}{N_0 \, \sin^2 \theta} \, \delta_{s,d}^2 \right)$$

$$= N! \left(\frac{t_2 \, \delta_{s,d}^2}{2\alpha} \right)^N F_1 \left(\prod_{n=0}^{N}(1 + \frac{t_2 \, \delta_{s,d}^2 \, n}{2\alpha} + \frac{b \, P}{N_0 \, \sin^2 \theta} \, \delta_{s,d}^2) \right), \tag{8.64}$$

where we have applied (8.15), (8.20), (8.54), and Lemma 8.3.1. As for the relay-cooperation mode, its SER is similar to that of the single-relay system, but using the MGF of $\beta_{\max}$. The MGF of $\beta_{\max}$ can be written as

$$M_{\beta_{\max}}(\gamma) = E_{\beta_{\max}} \left(\exp(-\gamma \, \beta_{\max}) \right)$$

$$= N \sum_{n=0}^{N-1} \binom{N-1}{n} (-1)^n \, M_{\beta_m} \left(\gamma + \frac{n \, t_2}{2} \right), \tag{8.65}$$

where we have applied (8.20).

Similar to the single-relay scenario, we can obtain an upper bound on the total SER as (leave proof as an exercise)

$$\Pr(e) \leq N! \left(\frac{N_0}{b \, P} \right)^{N+1} \left(\frac{t_2}{2\alpha} \right)^N \frac{I(2N+2)}{\delta_{s,d}^2} + N! \left(\frac{N_0}{b} \right)^{N+1} t_2^{N-1} \frac{\left(q_1 \, \delta_{r,d}^2 + q_2 \, \delta_{s,r}^2 \right)}{P_0 \, \delta_{s,d}^2}$$

$$\times \left(\left(\frac{q_1}{P_1} \right)^N I(2 \, N + 2) + \left(\frac{q_2}{P_0} \right)^N A \, I(2 \, N) \right), \tag{8.66}$$

where

$$
I(p) = \frac{1}{\pi} \int_{\theta=0}^{\frac{(M-1)\pi}{M}} \sin^p \theta \, d\theta. \tag{8.67}
$$

For the single-relay case $N = 1$, (8.66) and (8.48) are the same. Replacing q_1 and q_2 by their expressions defined in (8.8), t_2 by its expression defined in (8.17), and using $P_0 = r \, P$ and $P_1 = (1 - r) \, P$, we get the following theorem which shows that full diversity order of $N + 1$ is achieved. ∎

THEOREM 8.3.2 *At high SNR* $\gamma = P/N_0$, *the SER of the multi-node relay-selection decode-and-forward symmetric cooperative scenario, utilizing N relays, is upper bounded by*

$$
\Pr(e) \leq (CG \cdot \gamma)^{-(N+1)} \tag{8.68}
$$

where

$$
CG = \left(\frac{N! \left(\frac{r(1-r)}{B \, \delta_{s,r}^2} + \frac{r^2}{A^2 \, \delta_{r,d}^2} \right)^{N-1}}{b^{N+1} \, \delta_{s,d}^2} \right)^{-\frac{1}{(N+1)}} \left(\frac{\left(\frac{r(1-r)}{B \, \delta_{s,r}^2} + \frac{r^2}{A^2 \, \delta_{r,d}^2} \right) I(2\,N + 2)}{(2\,\alpha)^N} \right.
$$

$$
\left. + \frac{\left(\frac{A^2 \delta_{r,d}^2}{r^2} + \frac{B \delta_{s,r}^2}{r(1-r)} \right) \left(A^{2N} \, I(2\,N + 2) + B^N \, A \, I(2\,N) \right)}{r^{N+1} \, (1 - r)^N} \right)^{-\frac{1}{(N+1)}}.
$$

$$
\tag{8.69}
$$

8.4   Optimum power allocation

In this section, an analytical expression of the optimum power allocation is derived, and bandwidth efficiency-SER tradeoff curves are shown to obtain the optimum cooperation threshold. We clarify that as α increases, the probability of choosing the relay-cooperation mode increases. Therefore, the bandwidth efficiency and the SER, given by (8.56) and (8.66), respectively, decrease monotonically with α. In addition, the bandwidth efficiency is a monotonically increasing or decreasing function of the power ratio r, depending on the channel variances. On the contrary, there exists an optimum power ratio r^*, which minimizes the SER. We determine the optimum power allocation as follows.

In the direct-transmission mode, all the power is transmitted through the source–destination channel. In the relay-cooperation mode, we determine the optimum powers P_1 and P_2 which minimize the SER upper bound expression in (8.66) subject to constraint $P_1 + P_2 = P$. Substituting (8.8) into (8.66), we can approximate the optimization problem as

$$\min_{P_1} \left(\frac{A^2 \, \delta_{r,d}^2}{r^2 \, P_1} + \frac{B \, \delta_{s,r}^2}{r \, (1-r) \, P_1} \right) \cdot \left(\left(\frac{A^2}{r^2 \, P_2} \right)^N I(2\,N+2) \right.$$

$$\left. + \left(\frac{B}{r \, (1-r) \, P_1} \right)^N A \, I(2\,N) \right)$$

$$\text{s.t. } P_1 + P_2 = P \, . \tag{8.70}$$

By substituting $r = P_1/P$ into (8.70), we get

$$\min_{P_1} \frac{\left(A^{2N+2} \, I(2\,N+2) + A^3 \, B^N \, I(2\,N) \right) \delta_{r,d}^2}{P_1^{2\,N+3} \, P_2^N}$$

$$+ \frac{\left(A^{2N} \, B \, I(2\,N+2) + A \, B^{N+1} \, I(2\,N) \right) \delta_{s,r}^2}{P_1^{2\,N+2} \, P_2^{N+1}}$$

$$\text{s.t. } P_1 + P_2 = P \, . \tag{8.71}$$

Solving (8.71) using the standard Lagrangian method results in the following theorem (leave proof as an exercise).

THEOREM 8.4.1 *The optimum power allocation of the multi-node relay-selection decode-and-forward symmetric cooperative scenario, employing N relays, is obtained as*

$$P_1 = \frac{1 - \frac{N \, X_1}{2\,(N+1)\,X_2} + \sqrt{1 + \frac{(N+2)\,X_1}{(N+1)\,X_2} + \left(\frac{N \, X_1}{2\,(N+1)\,X_2} \right)^2}}{2 - \frac{N \, X_1}{2\,(N+1)\,X_2} + \sqrt{1 + \frac{(N+2)\,X_1}{(N+1)\,X_2} + \left(\frac{N \, X_1}{2\,(N+1)\,X_2} \right)^2}} \, P \, ,$$

$$P_2 = \frac{1}{2 - \frac{N \, X_1}{2\,(N+1)\,X_2} + \sqrt{1 + \frac{(N+2)\,X_1}{(N+1)\,X_2} + \left(\frac{N \, X_1}{2\,(N+1)\,X_2} \right)^2}} \, P \, , \tag{8.72}$$

where

$$X_1 = \left(A^{2N+2} \, I(2\,N+2) + A^3 \, B^N \, I(2\,N) \right) \delta_{r,d}^2 \, ,$$

$$X_2 = \left(A^{2N} \, B \, I(2\,N+2) + A \, B^{N+1} \, I(2\,N) \right) \delta_{s,r}^2 \, .$$

Theorem 8.4.1 shows that the optimum power allocation does not depend on the source–destination channel variance. It depends basically on the parameter M and the source–relay and relay–destination channel variances. If $\delta_{r,d}^2 \gg \delta_{s,r}^2$ then P_1 goes to P and P_2 goes to zero. Intuitively, this is because the source–relay link is of bad quality. Thus, it is reasonable to send the total power through the source–destination channel. In addition, if $\delta_{s,r}^2 \gg \delta_{r,d}^2$ then P_1 goes to $P/2$ and P_2 goes to $P/2$ as well, which is expected because if the source–relay channel is so good, then the symbols will be received correctly by the relay with high probability. Thus, the relay will be almost the same as the source, thus both source and relay share the power equally.

The obtained optimum power ratio will be used to get the optimum cooperation threshold as follows. Figure 8.5(a) and (b) depicts the bandwidth efficiency-SER tradeoff curves for different number of relays at SNR equal to 20 dB and 25 dB, respectively. This tradeoff is the achievable bandwidth efficiency and SER for different values of cooperation threshold. At a certain SER value, the maximum achievable bandwidth efficiency while guaranteeing full diversity order, can be obtained through Figure 8.5. In Figure 8.5, it is shown that the tradeoff achieved using four relays is the best among the

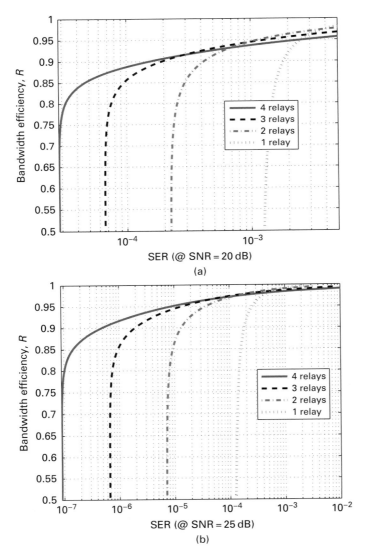

Fig. 8.5   Bandwidth efficiency versus SER at (a) SNR = 20 dB, (b) SNR = 25 dB.

Table 8.1 Single-relay optimum values using the (CG*R) optimization criterion.

$\delta_{s,d}^2$	$\delta_{s,r}^2$	$\delta_{r,d}^2$	$r = P_1/P$	α	Bandwidth efficiency (R)	Coding gain (CG)
1	1	1	0.6902	0.55	0.8624	0.247
1	1	10	0.7487	0.09	0.9075	0.1613
1	10	1	0.6697	0.14	0.9443	0.0939

plotted curves at low SER region. Moreover, it is clear in Figure 8.5(a) that the SER is almost constant at 2×10^{-5} while the bandwidth efficiency increases from 0.5 to 0.8 SPCU for four relays. Thus, an increase in the bandwidth efficiency of about 60% can be achieved with the same SER performance.

In the sequel, we consider three different channel-variances cases as follows. Case 1 which corresponds to the unity channel variances, where $\delta_{s,d}^2 = \delta_{s,r}^2 = \delta_{r,d}^2 = 1$ and it is represented at the first row of Table 8.1. Case 2 expresses a stronger relay–destination channel $\delta_{r,d}^2 = 10$, while case 3 expresses a stronger source–relay channel $\delta_{s,r}^2 = 10$. Cases 2 and 3 are represented at the second and third rows of Table 8.1, respectively. Table 8.1 shows the optimum values of the power allocation ratio (8.72) for the three cases. Since we aim at maximizing both the coding gain and the bandwidth efficiency, we choose as an example an optimization metric, which is the product of the coding gain and bandwidth efficiency, to find the optimum cooperation threshold. This optimization metric can be written as

$$\max_{\alpha} \ \mathrm{CG} \cdot R \ ,$$

where R and CG are obtained from (8.56) and (8.69), respectively.

Example 8.3 This example considers the CG*R optimization metric and indicate the optimum cooperation threshold for (a) single-relay scenario using various channel variances and (b) optimum cooperation thresholds for different number of relays with unity channel variances. The CG*R curves are plotted in Figure 8.6(a) for different channel variances. It is shown that the optimum cooperation threshold for the three different cases are $\alpha_o = 0.55$, $\alpha_o = 0.09$, and $\alpha_o = 0.14$, respectively. These values of cooperation thresholds result in bandwidth efficiencies equal to $R_o = 0.8624$, $R_o = 0.9075$, and $R_o = 0.9443$ SPCU, respectively. Notably for $\delta_{r,d}^2 = 10$, the optimum power ratio is $r_o = 0.7487$, which is greater than $r_o = 0.6902$ for $\delta_{r,d}^2 = 1$; $\delta_{s,r}^2 = 1$ in both cases. This is in agreement with the conclusion that more power should be put for P_1 if $\delta_{r,d}^2 \gg \delta_{s,r}^2$.

As shown in Figure 8.6(b), increasing the number of relays affects the optimum cooperation threshold values according to the CG*R optimization criterion. Table 8.2 describes the effect of changing the number of relays on the power ratio, cooperation threshold, bandwidth efficiency, and the coding gain using the unity channel-variances

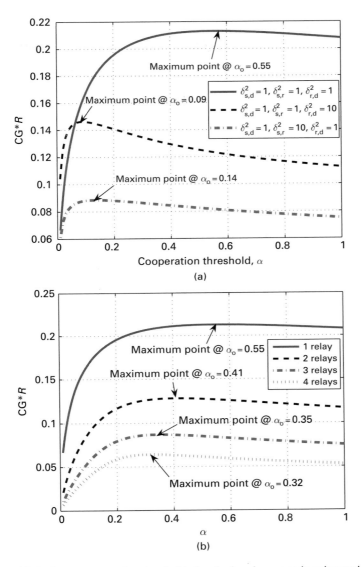

Fig. 8.6   (a) Optimum cooperation thresholds for single-relay scenario using various channel variances, (b) optimum cooperation thresholds for different number of relays with unity channel variances.

case. A few comments on Table 8.2 are as follows: (1) The optimum power ratio is slightly decreasing with the number of relays. Because, increasing the number of relays will increase the probability of finding a better relay, which can receive the symbols from the source more correctly. Thus, it can send with almost equal power with the source. (2) The bandwidth efficiency is slightly decreasing with increasing the number of relays, because the probability that the source–destination channel is better than all the relays' metrics goes to 0 as N goes to ∞.   ▲

Table 8.2 CG*R multi-node optimum values for unity channel variances.

N	Power ratio (r)	Cooperation threshold (α)	Bandwidth efficiency (R)	Coding gain (CG)
1	0.6902	0.55	0.8624	0.247
2	0.6826	0.41	0.8397	0.1512
3	0.6787	0.35	0.8297	0.1046
4	0.6764	0.32	0.82	0.0776

Example 8.4 In this example, the SER of multi-node relay-selection cooperative communications are considered for one, two, and three relays considering unity channel variances, $N_0 = 1$, and QPSK signalling. Figure 8.7 depicts the SER performance employing one, two, and three relays for unity channel variances. We plot the simulated SER curves using the optimum power ratios and the optimum cooperation thresholds obtained in Table 8.2. Moreover, we plot the SER upper bounds obtained in (8.66). It was shown in Theorem 8.3.2 that these upper bounds achieve full diversity order. It is obvious that the simulated SER curves are bounded by these upper bounds, hence they achieve full diversity order as well. The direct-transmission SER curve is plotted as well to show the effect of employing the relays in a cooperative way.

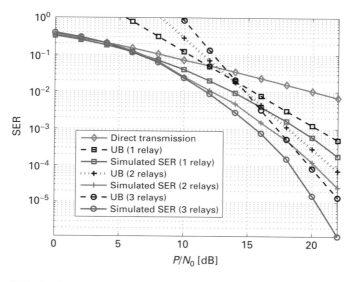

Fig. 8.7   SER simulated with optimum power ratio and SER upper bound curves for multi-node relay-selection decode-and-forward cooperative scheme with QPSK modulation and unity channel variances.

Moreover, the simulated bandwidth efficiencies are 0.8973, 0.8805, and 0.8738 employing one, two, and three relays, respectively. These results are slightly higher than the analytical results shown in Table 8.2.   ▲

8.5 Chapter summary and bibliographical notes

In this chapter, we have discussed single-relay and multi-node relay-selection decode-and-forward cooperative scenarios, which utilize the partial CSI available at the source and the relays to achieve higher bandwidth efficiency and guarantee full diversity order. In the single-relay system, it was shown that the modified harmonic mean function of the source–relay and relay–destination channels gain gives a proper measure on how much help a relay can offer. An approximate expression of the achievable bandwidth efficiency and an upper bound on the SER performance were derived. It turns out that the bandwidth efficiency is increased to 0.9 SPCU when the source–relay channel is relatively strong. Moreover, it was also shown that full diversity order is guaranteed as long as there is a positive probability of having the relay-cooperation mode.

In the multi-node system using arbitrary N relays, we considered the optimal relay as the one which has the maximum instantaneous modified harmonic mean function of its source–relay and relay–destination channels gain among the N relays. For the symmetric scenario, we have presented the approximate expression of the achievable bandwidth efficiency, which decreases with increasing the number of employed relays. Furthermore, we have also derived the SER upper bound, which proves that full diversity order is guaranteed as long as there is a positive probability of having the relay-cooperation mode. It was shown that the bandwidth efficiency is boosted up from 0.2 to 0.77 SPCU for $N = 4$ relays and unity channel variances case. Moreover, we determined the optimum power allocation for the multi-node relay-selection scheme, which minimizes the SER. As for the optimum cooperation threshold, the bandwidth efficiency-SER tradeoff curves was also shown to determine the optimal cooperation threshold.

There are various protocols proposed to choose the best relay among a collection of available relays in the literature. In [237], the authors proposed to choose the best relay depending on its geographic position, based on the geographic random forwarding (GeRaF) protocol proposed in [240]. In GeRaF, the source broadcasts its data to a collection of nodes and the node that is closest to the destination is chosen in a distributed manner to forward the source's data to the destination. In [128], the authors considered a best-select relay scheme in which only the relay, which has received the transmitted data from the source correctly and has the highest mean signal-to-noise ratio (SNR) to the destination node, is chosen to forward the source's data. In [83], a relay-selection scheme for single-relay decode-and-forward cooperative systems was proposed. In this scheme, the source decides whether to employ the relay in forwarding its information or not, depending on the instantaneous values of the source–destination and source–relay channel gains. While [83] considered a single relay scenario, the multi-node case was considered in [84]. More information on the topics studied in this chapter can be found in [82].

Exercises

8.1   Prove that the PDF and CDF of the variable β_m specified in (8.10) can be determined as

$$p_{\beta_m}(\beta_m) = \frac{\beta_m}{2\,t_1^2}\,\exp\left(-\frac{t_2}{2}\,\beta_m\right)\left(t_1\,t_2\,K_1\left(\frac{\beta_m}{t_1}\right) + 2\,K_0\left(\frac{\beta_m}{t_1}\right)\right)U\,(\beta_m),$$

$$P_{\beta_m}(\beta_m) = 1 - \frac{\beta_m}{t_1}\,\exp\left(-\frac{t_2}{2}\,\beta_m\right)K_1\left(\frac{\beta_m}{t_1}\right),$$

where $t_1 = \sqrt{q_1 q_2 \delta_{s,r}^2 \delta_{r,d}^2}$, $t_2 = 1/q_2\delta_{s,r}^2 + 1/q_1\delta_{r,d}^2$, and $K_0(x)$ and $K_1(x)$ are the zero-order and first-order modified Bessel functions of the second kind, respectively, defined as

$$K_0(x) = \int_0^\infty \cos\left(x\,\sinh\,(t)\right)\,dt = \int_0^\infty \frac{\cos\,(x\,t)}{\sqrt{t^2+1}}\,dt\ ,$$

$$K_1(x) = \int_0^\infty \sin\left(x\,\sinh\,(t)\right)\,\sinh\,(t)\,dt\ ,$$

for $x > 0$.

8.2   Show that the MGF of the variable β_m specified in (8.10) can be written as

$$M_{\beta_m}(\gamma) = E_{\beta_m}\left(\exp(-\gamma\,\beta_m)\right)$$

$$= \frac{16\,\lambda_1\,\lambda_2}{3\,(\lambda_1 + \lambda_2 + 2\,\sqrt{\lambda_1\,\lambda_2} + \gamma)^2}$$

$$\times\left(\frac{4\,(\lambda_1 + \lambda_2)\,{}_2F_1\left(3,\frac{3}{2};\frac{5}{2};\frac{\lambda_1+\lambda_2-2\,\sqrt{\lambda_1\,\lambda_2}+\gamma}{\lambda_1+\lambda_2+2\,\sqrt{\lambda_1\,\lambda_2}+\gamma}\right)}{(\lambda_1 + \lambda_2 + 2\,\sqrt{\lambda_1\,\lambda_2} + \gamma)}\right.$$

$$\left. + {}_2F_1\left(2,\frac{1}{2};\frac{5}{2};\frac{\lambda_1+\lambda_2-2\,\sqrt{\lambda_1\,\lambda_2}+\gamma}{\lambda_1+\lambda_2+2\,\sqrt{\lambda_1\,\lambda_2}+\gamma}\right)\right)$$

where $E_{\beta_m}(.)$ represents the statistical average with respect to β_m and ${}_2F_1(.,.;.;.)$ is the Gauss' hypergeometric function.

8.3   Show the probability upper bound in (8.43), i.e.,

$$\Pr(B|\phi^c)\,\Pr(\phi^c)$$

$$\leq \frac{1}{\pi}\int_{\theta=0}^{\frac{(M-1)\pi}{M}}\frac{M_{\beta_m}\left(\frac{\tilde{P}_1\,C(\theta)}{2\,q_1}\right) - M_{\beta_m}\left((P_0\,C(\theta) + \frac{1}{\delta_{s,d}^2})\alpha + \frac{\tilde{P}_1\,C(\theta)}{2\,q_1}\right)d\theta}{1 + P_0\,C(\theta)\delta_{s,d}^2}.$$

8.4   Prove the result in Theorem 8.4.1, i.e., the optimum power allocation of the multi-node relay-selection decode-and-forward symmetric cooperative scenario with N relays is given by

$$P_1 = \frac{1 - \frac{N X_1}{2(N+1) X_2} + \sqrt{1 + \frac{(N+2) X_1}{(N+1) X_2} + (\frac{N X_1}{2(N+1) X_2})^2}}{2 - \frac{N X_1}{2(N+1) X_2} + \sqrt{1 + \frac{(N+2) X_1}{(N+1) X_2} + (\frac{N X_1}{2(N+1) X_2})^2}} P,$$

$$P_2 = \frac{1}{2 - \frac{N X_1}{2(N+1) X_2} + \sqrt{1 + \frac{(N+2) X_1}{(N+1) X_2} + (\frac{N X_1}{2(N+1) X_2})^2}} P,$$

where

$$X_1 = \left(A^{2N+2} \, I(2N+2) + A^3 \, B^N \, I(2N) \right) \delta_{r,d}^2,$$

$$X_2 = \left(A^{2N} \, B \, I(2N+2) + A \, B^{N+1} \, I(2N) \right) \delta_{s,r}^2.$$

8.5   In this exercise, we will consider an alternative relay selection metric and shown its effect on the single-relay relay-selection decode-and-forward scheme. The relay's metric is represented by its source–relay channel gain. Based on this metric, the source determines whether to cooperate with the relay or not. Let $\phi = \{ \beta_{s,d} \geq \alpha \, \beta_{s,r} \}$ be the event of direct transmission. The transmission protocol can be modelled as in (8.11)–(8.14). The channel coefficients $h_{s,d}, h_{s,r}$, and $h_{r,d}$ are modeled as zero-mean complex Gaussian random variables with variances $\delta_{s,d}^2, \delta_{s,r}^2$, and $\delta_{r,d}^2$, respectively. The noise terms $\eta_{s,d}, \eta_{s,r}$, and $\eta_{r,d}$ are modeled as zero-mean, complex Gaussian random variables with equal variance N_0.

(a) Show that the probability of direct transmission is given by

$$\Pr(\phi) = \Pr(\beta_{s,d} \geq \alpha \beta_{s,r}) = \frac{\delta_{s,d}^2}{\delta_{s,d}^2 + \alpha \, \delta_{s,r}^2}. \tag{E8.1}$$

(b) Based on (8.23), show that the bandwidth efficiency is given by

$$R = \frac{2 \, \delta_{s,d}^2 + \alpha \, \delta_{s,r}^2}{2 \, \delta_{s,d}^2 + 2 \, \alpha \, \delta_{s,r}^2}. \tag{E8.2}$$

(c) For the direct-transmission mode and using M-PSK modulation, the conditional direct-transmission SER is given by (8.27). By averaging (8.27) over $\beta_{s,d}$, show that the direct-transmission SER is given by

$$\Pr(e|\phi) \, Pr(\phi)$$
$$= F_1 \left(1 + \frac{b P \delta_{s,d}^2}{N_0 \sin^2 \theta} \right) - F_1 \left(1 + \frac{\delta_{s,d}^2}{\alpha \delta_{s,r}^2} + \frac{b P \delta_{s,d}^2}{N_0 \sin^2 \theta} \right), \tag{E8.3}$$

where

$$F_1 \left(x(\theta) \right) = \frac{1}{\pi} \int_0^{\frac{(M-1)\pi}{M}} \frac{1}{x(\theta)} d\theta. \tag{E8.4}$$

(d) For the relay-cooperation mode, maximum ratio combining (MRC) [17] is applied at the destination. The output of the MRC can be written as in (8.30). The conditional SER of the relay-cooperation mode is given by (8.32). By averaging (8.32) over the exponentially distributed random variables $\beta_{s,d}$, $\beta_{s,r}$, and $\beta_{r,d}$, show that

$$
\begin{aligned}
\Pr(e/\phi^c)\Pr(\phi^c) = {} & F_2\left(\frac{1}{\pi^2}, P_1, 0, \frac{(M-1)\pi}{M}, \frac{(M-1)\pi}{M}\right) \\
& + F_1\left(\left(1 + \frac{\delta_{s,d}^2}{\alpha\delta_{s,r}^2} + \frac{bP_1\delta_{s,d}^2}{N_0\sin^2\theta}\right)\left(1 + \frac{bP_2\delta_{r,d}^2}{N_0\sin^2\theta}\right)\right) \\
& - F_2\left(\frac{1}{\pi^2}, P_1, P_2, \frac{(M-1)\pi}{M}, \frac{(M-1)\pi}{M}\right),
\end{aligned}
\tag{E8.5}
$$

where

$$
F_2\left(C, P_1, P_2, \theta_1, \theta_2\right) = C \int_0^{\theta_1}\int_0^{\theta_2} \frac{1}{\left(1 + \frac{b\,P_1\delta_{s,r}^2}{N_0\sin^2\theta_2}\right)}
$$
$$
\frac{d\theta_2\,d\theta_1}{\left(1 + \frac{\delta_{s,d}^2}{\alpha\delta_{s,r}^2} + \frac{b\,P_1\delta_{s,d}^2}{N_0}\left(\frac{1}{\sin^2\theta_1} + \frac{1}{\alpha\sin^2\theta_2}\right)\right)\left(1 + \frac{b\,P_2\delta_{r,d}^2}{N_0\sin^2\theta_1}\right)}.
\tag{E8.6}
$$

(e) Obtain a complete SER expression for the M-PSK signalling based on (c) and (d).

8.6 For the communication system described in Exercise 8.5, consider the ratio between the SER and the bandwidth efficiency (SER/R) as an optimization metric to determine the optimum cooperation threshold and power ratio.

(a) At SNR=10 dB and using a unity channel-variances case $\delta_{s,d}^2 = \delta_{s,r}^2 = \delta_{r,d}^2 = 1$ with QPSK modulation, plot a 3-dimensional curve for the SER/R, and verify that the optimum values, which minimize the SER/R optimization metric, are $\alpha_o = 0.35$ and $r_o = 0.52$. Moreover, plot the SER of the relay-selection scheme and compare the simulated SER with the analytical SER, given by the summation of (E8.3) and (E8.5). Verify that the bandwidth efficiency resulting from the simulations is 0.870 37 SPCU.

(b) Consider the multi-node case where arbitrary N relays are available. Each relay's metric is represented by its source–relay channel gain. Plot the SER of this system assuming two and three relays. From the simulation results, does this relay's metric guarantees full diversity order? Explain.

8.7 Consider a relay metric, which is given by the minimum of the relay's source–relay and relay–destination channel gains, i.e., the relay's metric can be given by

$$
\beta_i = \min(|h_{s,r_i}|^2, |h_{r_i,d}|^2), \quad for\ i = 1, 2, \ldots, N.
\tag{E8.7}
$$

For simplicity, consider that the cooperation threshold is 1 and the power ratio is 0.5. Hence, the direct transmission event is denoted by $\phi = \{\ \beta_{s,d} \geq \alpha\beta_i\ \}$. Plot the SER for system with one, two, and three relays assuming unity channel variances for all the channels, $r = 0.5$, $\alpha = 1$, and QPSK signalling. Based on the simulation results, does this relay's metric guarantee full diversity order? Explain.

9 Differential modulation for cooperative communications

In wireless communication systems with a single antenna, the channel capacity can be very low and the bit error rate high when fading occurs. Various techniques can be utilized to mitigate fading, e.g., robust modulation, coding and interleaving, error-correcting coding, equalization, and diversity. Different kinds of diversities such as space, time, frequency, or any combination of them are possible. Among these diversity techniques, space diversity is of special interest because of its ability to improve performance without sacrificing delay and bandwidth efficiency. Recently, space diversity has been intensively investigated in point-to-point wireless communication systems by the deployment of a MIMO concept together with efficient coding and modulation schemes. In recent years, cooperative communication [109] has been proposed as an alternative communication system that explores MIMO-like diversity to improve link performance without the requirement of additional antennas. However, most of existing works on MIMO systems and cooperative communications are designed based on an assumption that the receivers have full knowledge of the channel state information (CSI). In this case, the schemes must incorporate reliable multi-channel estimation, which inevitably increases the cost of frequent retraining and the number of estimated parameters to the receivers. Although the channel estimates may be available when the channel changes slowly comparing with the symbol rate, they may not be possibly acquired in a fast-fading environment. To develop practical schemes that omit such CSI requirements, we consider in this chapter differential modulations for cooperative communications.

We first briefly review differential modulation for non-cooperative or direct wireless communication systems. Then, we consider differential modulation for cooperative communication systems employing the decode-and-forward (DF) protocol. We discuss system design, performance analysis, and performance optimization. Finally, we consider differential modulation for the amplify-and-forward (AF) cooperative communication protocol.

9.1 Differential modulation

In conventional single-antenna systems, non-coherent modulation is useful when the knowledge of CSI is not available. The non-coherent modulation simplifies the receiver structure by omitting channel estimation and carrier or phase tracking. In this chapter, although we focus on the differential M-ary phase-shift-keying (DMPSK) scheme, the concept of differential modulation schemes covered in subsequent sections can also be applied to differential M-ary quadrature amplitude modulation (DMQAM).

In the DMPSK scheme, information is modulated through the phase difference among two consecutive symbols. Specifically, for a data rate of R bits per channel use, the DPMSK signal constellation comprises $M = 2^R$ symbols. Each symbol $m \in 0 \leq M - 1$ is generated by the m-th root of unity: $v_m = e^{j2\pi m/M}$. Define s^τ as the differentially encoded symbol to be transmitted at time τ. The differential modulator transmits $s^\tau = 1$ at the first symbol period, i.e., at $\tau = 0$. For subsequent symbol periods, the differential transmission follows a recursive multiplication rule:

$$s^\tau = v_m s^{\tau-1}. \tag{9.1}$$

If we let h^τ denote the fading coefficient and n^τ represent additive noise, then the received signal for DMPSK modulation can be written as

$$y^\tau = \sqrt{\rho} h^\tau s^\tau + w^\tau. \tag{9.2}$$

The detection rule for differential demodulation can be obtained by the following derivations. We first rewrite (9.2) as

$$y^\tau = \sqrt{\rho} h^\tau s^\tau + w^\tau = \sqrt{\rho} h^\tau v_m s^{\tau-1} + w^\tau.$$

We also knew from (9.2) that

$$\sqrt{\rho} h^{\tau-1} s^{\tau-1} = y^{\tau-1} - w^{\tau-1}. \tag{9.3}$$

By assuming that fading channel keeps constant over two symbol periods, i.e., $h^{\tau-1} \approx h^\tau$, and substitute (9.3) into (9.2), we have

$$y^\tau = y^{\tau-1} v_m - v_m w^{\tau-1} + w^\tau.$$
$$= y^{\tau-1} v_m + \sqrt{2} \tilde{w}^\tau, \tag{9.4}$$

where $\tilde{w}^\tau \triangleq (1/\sqrt{2})(w^\tau - v_m w^{\tau-1})$ is a complex Gaussian random variable with zero mean and unit variance. From (9.4), by treating $y^{\tau-1}$ as a known channel information and v_m as a transmitted symbol, the expression in (9.4) can be considered as a received signal for coherent detection. In this way, the optimum detector is simply the minimum distance detector. Therefore, the differential decoder estimates the transmit information using the following decoding rule:

$$\hat{m} = \arg \min_{m \in 0, 1, \dots M-1} |y^\tau - v_m y^{\tau-1}|^2. \tag{9.5}$$

We can see from (9.5) that the efficient differential demodulation for an information symbol relies on two consecutive received symbols, and the decoding error does not propagate to subsequent decoded information symbols. In terms of performance, since the noise power has twice the variance of its coherent counterpart, the differential scheme looses 3 dB performance in comparison to its coherent counterpart. For convenience in subsequence derivations of optimum differential detector, (9.5) is further simplified to

$$\hat{m} = \arg \min_{m \in 0,1,\ldots M-1} |y^\tau - v_m y^{\tau-1}|^2,$$

$$= \arg \min_{m \in 0,1,\ldots M-1} y^\tau (y^\tau)^* - (v_m)^* (y^{\tau-1})^* y^\tau - v_m y^{\tau-1} (y^\tau)^* + y^{\tau-1} (y^{\tau-1})^*,$$

$$= \arg \max_{m \in 0,1,\ldots M-1} (v_m)^* (y^{\tau-1})^* y^\tau + v_m y^{\tau-1} (y^\tau)^*,$$

$$= \arg \max_{m \in 0,1,\ldots M-1} \mathrm{Re}\{(v_m)^* (y^{\tau-1})^* y^\tau\}, \tag{9.6}$$

where $(\cdot)^*$ denotes the complex conjugate and $\mathrm{Re}(\cdot)$ represents the real part. The third equality in (9.6) follows the fact that $y^\tau (y^\tau)^*$ and $y^{\tau-1}(y^{\tau-1})^*$ are the same for any index m.

9.2   Differential modulations for DF cooperative communications

9.2.1   DiffDF with single-relay systems

In this section, we consider a differential cooperative scheme employing the DF pro-tocol, known as DiffDF, for systems with a single relay. In particular, the relay helps forward the source symbol only if the symbol is correctly decoded. At the destination, the received signal from the relay is combined with that from the source only if its amplitude is larger than a threshold; otherwise only the received signal from the source is used for the detection. A properly designed threshold allows the destination to make a judgement whether the signal from the relay contains the information such that the signals from the source and the relay can be efficiently combined and jointly decoded. The bit error rate (BER) performance of the threshold-based differential DF scheme is derived for differential DMPSK modulation. A tight approximate BER formulation as well as the BER upper and lower bounds are provided.

9.2.1.1   Signal model and protocol description for the DiffDF scheme

We present in this section the threshold-based differential scheme for DF cooperation systems. Consider a two-user cooperation system as shown in Figure 9.1 in which signal transmission involves two transmission phases. A user who sends information directly to the destination is considered to be a source node. The other user who helps forwarding the information from the source node is a relay node. In phase 1, the source differentially encodes its information and then broadcasts the encoded symbol to the destination. Due to the broadcasting nature of the wireless networks, the relay is also able to receive the transmitted symbol from the source. In phase 2, while the source is silent, the relay differentially decodes the received signal from the source. If the relay correctly decodes the transmitted symbol, the relay differentially re-encodes the information, and then forwards the encoded symbol to the destination. Otherwise, the relay does not send or remains idle. In both phases, we assume that all users transmit their signals through orthogonal channels by the use of existing schemes such as TDMA, frequency division multiple access (FDMA), or code division multiple access (CDMA).

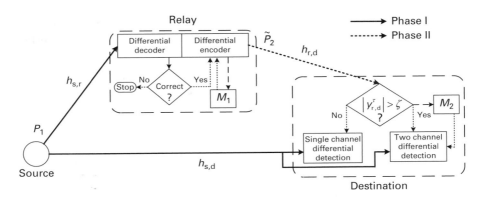

Fig. 9.1   The threshold-based differential scheme for decode-and-forward cooperative communications.

In differential modulation, the information is conveyed in the difference of the phases of two consecutive symbols. The set of information symbols to be transmitted by the source can be given by $v_m = e^{j\varphi_m}$, where $\{\phi_m\}_{m=0}^{M-1}$ is a set of M information phases. In case of differential M-ary phase shift keying (DMPSK), ϕ_m is specified as $\varphi_m = 2\pi m/M$ for $m = 0, 1, \ldots, M - 1$. Instead of directly transmitting the information as in coherent transmission [109], [204], the source node differentially encodes the information symbol v_m as

$$x^\tau = v_m x^{\tau-1}, \tag{9.7}$$

where τ is the time index, and x^τ is the differentially encoded symbol to be transmitted at time τ. After differential encoding, the source sends out the symbol x^τ with transmitted power P_1 to the destination and the relay. The corresponding received signals at the destination and the relay can be expressed as

$$y_{s,d}^\tau = \sqrt{P_1} h_{s,d}^\tau x^\tau + w_{s,d}^\tau,$$
$$y_{s,r}^\tau = \sqrt{P_1} h_{s,r}^\tau x^\tau + w_{s,r}^\tau,$$

where $h_{s,d}^\tau$ and $h_{s,r}^\tau$ are fading coefficients at the source–destination link and the source–relay link, respectively, and $w_{s,d}^\tau$ and $w_{s,r}^\tau$ are additive noise. Both channel coefficients $h_{s,d}^\tau$ and $h_{s,r}^\tau$ are modeled as zero-mean, complex Gaussian random variables with variances $\sigma_{s,d}^2$ and $\sigma_{s,r}^2$, i.e., $\mathcal{CN}(0, \sigma_{s,d}^2)$ and $\mathcal{CN}(0, \sigma_{s,r}^2)$, respectively. Each of the additive noise terms is modeled as $\mathcal{CN}(0, \mathcal{N}_0)$, where $\mathcal{N}_0$ is the noise power spectral density.

In phase 2, the relay differentially decodes the transmitted symbol from the source. Two consecutive received signals, $y_{s,r}^{\tau-1}$ and $y_{s,r}^\tau$, are required to recover the transmitted information at each symbol period. By assuming that the channel coefficient $h_{s,r}^\tau$ is almost constant over two symbol periods, the differential decoder at the relay decodes based on following the decision rule:

$$\hat{m} = \arg \max_{m=0,1,\ldots,M-1} \mathrm{Re}\left\{ \left(v_m y_{s,r}^{\tau-1} \right)^* y_{s,r}^\tau \right\},$$

in which the CSI is not required. In the relay-cooperation mode, the relay decides whether to forward the received information or not according to the quality of the received signal. For mathematical tractability, we assume that the relay can judge whether the decoded information is correct or not.[1] If the relay incorrectly decodes the received signal, such incorrectly decoded symbol is discarded, and the relay does not send any information. Otherwise, the relay differentially re-encodes the correctly decoded information symbol and forwards it to the destination.

In general, successful differential decoding requires that the encoder differentially encodes each information symbol with the previously transmitted symbol. For example, if the information symbols are sent every time slot, then the information symbol to be transmitted at time τ is differentially encoded with the transmitted symbol at time $\tau - 1$. In this differential DF scheme, since the information symbols at the relay are transmitted only if they are correctly decoded, the transmission time of the previously transmitted symbol can be any time before the current time τ. We denote such previous transmission time as $\tau - k$, $k \geq 1$, i.e., $\tau - k$ is the latest time that the relay correctly decodes the symbol before time τ. In order to perform successful differential en/decoding, we let a memory M_1 at the relay (see Figure 9.1) store the transmitted symbol at time $\tau - k$. Note that having a memory M_1 does not increase the system complexity compared to the conventional differential system. The difference is that the memory in this DiffDF scheme stores the transmitted symbol at time $\tau - k$ instead of time $\tau - 1$ as does the conventional differential scheme.

The differentially re-encoded signal at the relay in phase 2 can be expressed as

$$\tilde{x}^\tau = v_m \tilde{x}^{\tau-k}, \tag{9.8}$$

where $\tilde{x}^\tau$ is the differentially encoded symbol at the relay at time τ. We can see from (9.7) and (9.8) that the differentially encoded symbols $\tilde{x}^\tau$ at the relay and x^τ at the source convey the same information symbol v_m. However, the two encoded symbols can be different since the relay differentially encodes the information symbol v_m with the symbol in the memory, which may not be $x^{\tau-1}$ as used at the source. After differential re-encoding, the relay sends the symbol $\tilde{x}^\tau$ to the destination with transmitted power P_2, and then stores the transmitted symbol $\tilde{x}^\tau$ in the memory M_1 for subsequent differential encoding. The received signal at the destination from the relay in phase 2 can be expressed as

$$y_{r,d}^\tau = \begin{cases} \sqrt{P_2} h_{r,d}^\tau \tilde{x}^\tau + w_{r,d}^\tau & : \text{if relay correctly decodes } (\tilde{P}_2^\tau = P_2); \\ w_{r,d}^\tau & : \text{if relay incorrectly decodes } (\tilde{P}_2^\tau = 0), \end{cases} \tag{9.9}$$

where $h_{r,d}^\tau$ is the channel coefficient from the relay to the destination and $w_{r,d}^\tau$ is an additive noise. We assume that $h_{r,d}^\tau$ is $\mathcal{CN}(0, \sigma_{r,d}^2)$ distributed, and $w_{r,d}^\tau$ is $\mathcal{CN}(0, N_0)$ distributed.

The received signals at the destination comprise the received signal from the source in phase 1 and that from the relay in phase 2. As discussed previously, in phase 2, the relay may forward the information or remain idle. Without knowledge of the CSI, the

---

[1] Practically, this can be done at the relay by applying a simple SNR threshold test on the received data, although it can lead to some error propagation. Nevertheless, for practical ranges of operating SNR, the event of error propagation can be assumed negligible.

destination is unable to know whether the received signal from the relay contains the information or not. In order for the destination to judge whether to combine the signals from the source–destination and relay–destination links, we use a decision threshold ζ at the destination node (see Figure 9.1). We consider the received signal with amplitude $|y_{r,d}^\tau|$ greater than the threshold ζ as a high-potential information bearing signal to be used for further differential detection.

Particularly, if the amplitude of the received signal from the relay is not greater than the decision threshold, i.e., $|y_{r,d}^\tau| \leq \zeta$, the destination estimates the transmitted symbol based only on the received signal from the direct link. On the other hand, if $|y_{r,d}^\tau| > \zeta$, the received signal from the source and that from the relay are combined together, and then the combined output is jointly differentially decoded. Note that, in order to successfully decode, the differential detector requires the previously received signal, which serves as a CSI estimate. Since the received signal from the relay may contain the transmitted symbol or only noise, we use a memory M_2 at the destination as shown in Figure 9.1 to store the previously received signal at the relay–destination link that tends to contain the information. An ideal situation is to let the memory store the received signal $y_{r,d}^\tau$ only when the signal contains the transmitted symbol; however, such information is not available since the destination does not have the knowledge of the CSI. So, to efficiently decode the received signal from the relay, the memory M_2 is used to store the received signal $y_{r,d}^\tau$ whose amplitude is greater than the decision threshold. If the threshold is properly designed, then the signal in the memory M_2 corresponds to the received signal from the relay that carries the encoded symbol stored in the memory M_1.

With an assumption that the channel coefficients stay almost constant for several symbol periods, the signal in the memory M_2 serves as a channel estimate of the relay–destination link, which can then be used for efficient differential decoding at the destination. When we combine all received signals for joint detection, the combined signal prior to the differential decoding is

$$y = \begin{cases} a_1(y_{s,d}^{\tau-1})^* y_{s,d}^\tau + a_2(y_{r,d}^{\tau-l})^* y_{r,d}^\tau & \text{if } |y_{r,d}^\tau| > \zeta; \\ (y_{s,d}^{\tau-1})^* y_{s,d}^\tau & \text{if } |y_{r,d}^\tau| \leq \zeta, \end{cases} \tag{9.10}$$

where a_1 and a_2 are combining weight coefficients, and $\tau - l$ ($l \geq 1$) represents the time index of the latest signal in memory M_2, i.e., $y_{r,d}^{\tau-l}$ is the most recent received signal from the relay whose amplitude is larger than the threshold. Note that different combining weights, a_1 and a_2, result in different system performances. In Section 9.2.1.2, the BER performance is derived in case of $a_1 = 1/2N_0$ and $a_2 = 1/2N_0$. The use of these combining weights maximizes the signal-to-noise ratio (SNR) of the combiner output when the destination is able to differentially decode the signals from both source and relay. Based on the combined signal in (9.10), the decoder at the destination jointly differentially decodes the transmitted information symbol by using the following decision rule:

$$\hat{m} = \arg \max_{m=0,1,\ldots,M-1} \text{Re}\left\{v_m^* y\right\}.$$

9.2.1.2   Performance Analysis

This section considers BER performance of the threshold-based differential DF scheme employing DMPSK modulation. First, we classify different scenarios that lead to different instantaneous SNRs at the combiner output of the destination. Next, the probability that each scenario occurs is determined. Then, we evaluate an average BER performance by taking into account all the possible scenarios.

Classification of different scenarios

There are different scenarios that result in different SNRs at the combiner output. Recall that if the amplitude of received signal from the relay is larger than the decision threshold ($|y_{r,d}^\tau| > \zeta$), then the destination jointly decodes the received signals from the source $\left(y_{s,d}^\tau\right)$ and the relay $\left(y_{r,d}^\tau\right)$. Otherwise, only the received signal $y_{s,d}^\tau$ from the source is used for differential detection. Therefore, we can classify the scenarios into two major groups, namely the scenarios that $|y_{r,d}^\tau| > \zeta$ and $|y_{r,d}^\tau| \le \zeta$. In case that $|y_{r,d}^\tau| \le \zeta$, the SNR can be simply determined based on the received signal from the direct link. On the other hand, if $|y_{r,d}^\tau| > \zeta$, the SNR at the combiner output depends not only on the received signals from the direct link but also on that from the relay link.

According to the decision rule in (9.10), if $|y_{r,d}^\tau| > \zeta$, the performance of the differential decoder relies on the received signals from the source at the current time τ ($y_{s,d}^\tau$) and the previous time $\tau - 1$ $\left(y_{s,d}^{\tau-1}\right)$ as well as the received signal from the relay at the current time τ $\left(y_{r,d}^\tau\right)$ and that stored in the memory $\left(y_{r,d}^{\tau-l}\right)$. The received signals $y_{r,d}^\tau$ and $y_{r,d}^{\tau-l}$ from the relay may or may not contain the information, depending on the correctness of the decoded symbol at the relay at time τ and time $\tau - l$. If the relay decodes the symbol correctly, then the relay sends the encoded symbol with transmitted power P_2. Thus, based on the transmitted power at the relay at time τ and time $\tau - l$, we can further classify the scenarios into four different categories, namely:

(a) $\tilde{P}_2^\tau = P_2$ and $\tilde{P}_2^{\tau-l} = P_2$;
(b) $\tilde{P}_2^\tau = P_2$ and $\tilde{P}_2^{\tau-l} = 0$;
(c) $\tilde{P}_2^\tau = 0$ and $\tilde{P}_2^{\tau-l} = P_2$;
(d) $\tilde{P}_2^\tau = 0$ and $\tilde{P}_2^{\tau-l} = 0$.

In case (a), the received signals $y_{r,d}^\tau$ and $y_{r,d}^{\tau-l}$ convey the symbols $\tilde{x}^\tau$ and $\tilde{x}^{\tau-l}$, respectively. Since the relay differentially encodes the information symbol from the source with the symbol $\tilde{x}^{\tau-k}$ in the memory M_1, the SNR at the combiner output also depends on whether the received signal $y_{r,d}^{\tau-l}$ used for decoding corresponds to the symbol $\tilde{x}^{\tau-k}$ used for encoding. Therefore, the scenarios under case (a) can be further separated into two cases, namely $l = k$ and $l \ne k$. All of these six possible scenarios are summarized in Figure 9.2.

In order to facilitate the BER analysis in the subsequent subsection, we define six different scenarios by $\Phi_i, i = 1, 2, \ldots, 6$ as follows. We let the first scenario Φ_1 be the scenario that the amplitude of the received signal is not larger than the threshold ζ, i.e.,

$$\Phi_1 \triangleq \left\{|y_{r,d}^\tau| \le \zeta\right\}. \tag{9.11}$$

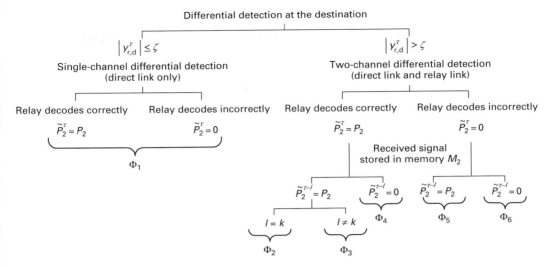

Fig. 9.2   Two possible differential detection techniques at the destination: single-channel differential detection or two-channel differential detection, with six possible scenarios based on the currently received signal and the signal stored in memory M_2.

When Φ_1 occurs, the destination does not combine the received signal from the relay with that from the source. The second scenario Φ_2 is defined as the case that the amplitude of the received signal is larger than the threshold, the relay transmitted powers at both time τ and $\tau - l$ are equal to P_2, and the received signal $y_{r,d}^{\tau-l}$ stored in memory M_2 conveys the information symbol $\tilde{x}^{\tau-k}$ stored in memory M_1. Specifically, Φ_2 can be written as

$$\Phi_2 \triangleq \left\{ |y_{r,d}^{\tau}| > \zeta, \tilde{P}_2^{\tau} = P_2, \tilde{P}_2^{\tau-l} = P_2, l = k \right\}. \tag{9.12}$$

The scenario Φ_3 is similar to the scenario Φ_2 excepts that the received signal $y_{r,d}^{\tau-l}$ does not contain the symbol $\tilde{x}^{\tau-k}$, i.e., $l \neq k$. We express this scenario as

$$\Phi_3 \triangleq \left\{ |y_{r,d}^{\tau}| > \zeta, \tilde{P}_2^{\tau} = P_2, \tilde{P}_2^{\tau-l} = P_2, l \neq k \right\}. \tag{9.13}$$

The scenarios Φ_4 to Φ_6 correspond to the scenarios that $|y_{r,d}^{\tau}| > \zeta$ and either $\tilde{P}_2^{\tau}$ or $\tilde{P}_2^{\tau-l}$ is zero. Under the scenario Φ_4, the transmitted power $\tilde{P}_2^{\tau}$ is P_2 whereas the transmitted power $\tilde{P}_2^{\tau-l}$ is 0. We express the scenario Φ_4 as

$$\Phi_4 \triangleq \left\{ |y_{r,d}^{\tau}| > \zeta, \tilde{P}_2^{\tau} = P_2, \tilde{P}_2^{\tau-l} = 0 \right\}. \tag{9.14}$$

Under the scenario Φ_5, the transmitted power $\tilde{P}_2^{\tau}$ is 0 whereas the transmitted power $\tilde{P}_2^{\tau-l}$ is P_2, i.e., $\Phi_5 \triangleq \left\{ |y_{r,d}^{\tau}| > \zeta, \tilde{P}_2^{\tau} = 0, \tilde{P}_2^{\tau-l} = P_2 \right\}$. The scenario Φ_6 corresponds to the scenario that both $\tilde{P}_2^{\tau}$ and $\tilde{P}_2^{\tau-l}$ are zeros. We define the scenario Φ_6 as

$$\Phi_6 \triangleq \left\{ |y_{r,d}^{\tau}| > \zeta, \tilde{P}_2^{\tau} = 0, \tilde{P}_2^{\tau-l} = 0 \right\}. \tag{9.15}$$

For subsequent performance derivation, we denote $P_{\text{BER}}^h |_{\Phi_i}$ as the conditional BER, given a scenario Φ_i and a set of channel realizations for the source–destination link, the source–relay link, and the relay–destination link. We also denote $P_r^h(\Phi_i)$ as the chance that the scenario Φ_i occurs given a set of channel realizations. Accordingly, the conditional BER of the differential DF scheme can be expressed as

$$P_{\text{BER}}^h = \sum_{i=1}^{6} P_{\text{BER}}^h |_{\Phi_i} P_r^h(\Phi_i). \tag{9.16}$$

By averaging (9.16) over all channel realizations, the average BER of the DiffDF scheme is given by

$$P_{\text{BER}} = \sum_{i=1}^{6} \text{E}\left[P_{\text{BER}}^h |_{\Phi_i} P_r^h(\Phi_i) \right] = \sum_{i=1}^{6} P_{\text{BER}}^{(i)}, \tag{9.17}$$

where $\text{E}[\cdot]$ represents the expectation operation.

In the following subsections, we determine the chance that each scenario occurs ($P_r^h(\Phi_i)$), the conditional BER ($P_{\text{BER}}^h |_{\Phi_i}$), and finally obtain the average BER (P_{BER}) for the differential DF scheme.

Probability of occurrence

We first note that for a specific channel realization, the amplitude of the received signal $y_{\text{r,d}}^\tau$ depends on the relay transmitted power $\tilde{P}_2^\tau$, which in turn relies on the correctness of the decoded symbol at the relay. With DMPSK signals, the chance of incorrect decoding at the relay, and hence the chance of $\tilde{P}_2^\tau = 0$, can be obtained from the conditional symbol error rate (SER) of the transmission from the source to the relay. The SER of DMPSK signals is [145]:

$$P_r^h\left(\tilde{P}_2^\tau = 0\right) = \Psi(\gamma_{\text{s,r}}^\tau) \triangleq \frac{1}{\pi} \int_0^{(M-1)\pi/M} \exp\left[-g(\phi)\gamma_{\text{s,r}}^\tau\right] d\phi, \tag{9.18}$$

where

$$\gamma_{\text{s,r}}^\tau = \frac{P_1 |h_{\text{s,r}}^\tau|^2}{N_0} \tag{9.19}$$

is the instantaneous SNR per symbol at the relay due to the transmitted symbol from the source, and

$$g(\phi) = \frac{\sin^2(\pi/M)}{1 + \cos(\pi/M)\cos(\phi)}. \tag{9.20}$$

The chance that the relay forwards the symbol with transmitted power $\tilde{P}_2^\tau = P_2$ is determined by the chance of correct decoding at the relay, hence

$$P_r^h\left(\tilde{P}_2^\tau = P_2\right) = 1 - \Psi(\gamma_{\text{s,r}}^\tau), \tag{9.21}$$

where $\Psi(\gamma_{\text{s,r}}^\tau)$ is specified in (9.18).

Consider the scenario Φ_1 in which the amplitude of the received signal $y_{\text{r,d}}^\tau$ is not greater than the decision threshold. The chance that Φ_1 occurs can be written as

$$P_r^h(\Phi_1) = P_r^h\left(|y_{r,d}^\tau| \leq \varsigma \mid \tilde{P}_2^\tau = 0\right)\Psi(\gamma_{s,r}^\tau)$$
$$+ P_r^h\left(|y_{r,d}^\tau| \leq \varsigma \mid \tilde{P}_2^\tau = P_2\right)\left[1 - \Psi(\gamma_{s,r}^\tau)\right]. \tag{9.22}$$

The conditional probabilities, $P_r^h(|y_{r,d}^\tau| \leq \varsigma \mid \tilde{P}_2^\tau = 0)$ and $P_r^h(|y_{r,d}^\tau| \leq \varsigma \mid \tilde{P}_2^\tau = P_2)$ can be obtained from the cumulative distribution function (CDF) of the random variable $|y_{r,d}^\tau|$. The received signal $y_{r,d}^\tau$ is a complex Gaussian random variable with mean $\sqrt{\tilde{P}_2}h_{r,d}^\tau\tilde{x}^\tau$ and variance $\mathcal{N}_0$, i.e., $y_{r,d}^\tau \sim \mathcal{CN}\left(\sqrt{\tilde{P}_2}h_{r,d}^\tau\tilde{x}^\tau, \mathcal{N}_0\right)$. If $\tilde{P}_2^\tau = 0$, which results from incorrect decoding at the relay, the received signal $y_{r,d}^\tau$ is simply a zero-mean, complex Gaussian random variable with variance $\mathcal{N}_0$, i.e., $y_{r,d}^\tau \sim \mathcal{CN}(0, \mathcal{N}_0)$, and its amplitude is Rayleigh distributed. Hence, the conditional probability that $|y_{r,d}^\tau|$ is not greater than the decision threshold given that the relay does not send information can be expressed as

$$P_r^h\left(|y_{r,d}^\tau| \leq \varsigma \mid \tilde{P}_2^\tau = 0\right) = 1 - \exp(-\varsigma^2/\mathcal{N}_0). \tag{9.23}$$

If $\tilde{P}_2^\tau = P_2$, which corresponds to the case of correct decoding at the relay, then the received signal $y_{r,d}^\tau$ is Gaussian distributed with mean $\sqrt{P_2}h_{r,d}^\tau\tilde{x}^\tau$ and variance $\mathcal{N}_0$. In this case, $|y_{r,d}^\tau|^2$ can be viewed as a summation of two squared Gaussian random variables with means

$$m_1 = \mathrm{Re}\left\{\sqrt{P_2}h_{r,d}^\tau\tilde{x}^\tau\right\}, \quad m_2 = \mathrm{Im}\left\{\sqrt{P_2}h_{r,d}^\tau\tilde{x}^\tau\right\},$$

and a common variance $\sigma^2 = \mathcal{N}_0/2$. We observe that $|y_{r,d}^\tau|^2$ is non-central chi-square distributed with parameter $s^2 = m_1^2 + m_2^2$ [146]. In case of DMPSK signals, the parameter s can be determined as

$$s = \sqrt{P_2|h_{r,d}^\tau\tilde{x}^\tau|^2} = \sqrt{P_2|h_{r,d}^\tau|^2}, \tag{9.24}$$

where the second equality results from the fact that each DMPSK symbol has unit energy. From (9.24), the conditional probability that $|y_{r,d}^\tau|$ is not greater than the decision threshold given that the relay sends the information with transmitted power P_2 can be written as

$$P_r^h\left(|y_{r,d}^\tau| \leq \varsigma \mid \tilde{P}_2^\tau = P_2\right) = 1 - \mathcal{M}\left(P_2|h_{r,d}^\tau|^2, \varsigma\right), \tag{9.25}$$

where

$$\mathcal{M}\left(P_2|h_{r,d}^\tau|^2, \varsigma\right) \triangleq Q_1\left(\sqrt{\frac{P_2|h_{r,d}^\tau|^2}{\mathcal{N}_0/2}}, \frac{\varsigma}{\sqrt{\mathcal{N}_0/2}}\right), \tag{9.26}$$

in which $Q_1(\alpha, \beta)$ is the Marcum Q-function [188]:

$$Q_1(\alpha, \beta) = \int_\beta^\infty \lambda\exp\left[-\left(\frac{\lambda^2 + \alpha^2}{2}\right)\right]I_0(\alpha\lambda)d\lambda, \tag{9.27}$$

and $I_0(\cdot)$ is the zeroth-order modified Bessel function of the first kind. By substituting (9.23) and (9.25) into (9.22), we can express the chance that Φ_1 occurs as

$$P_r^h(\Phi_1) = \left(1 - e^{(-\zeta^2/N_0)}\right) \Psi(\gamma_{s,r}^\tau) + \left(1 - \mathcal{M}\left(P_2|h_{r,d}^\tau|^2, \zeta\right)\right)\left[1 - \Psi(\gamma_{s,r}^\tau)\right]. \quad (9.28)$$

The rest of the scenarios, Φ_2 to Φ_6, are related to the situation when the amplitude of the received signal from the relay, $|y_{r,d}^\tau|$, is greater than the decision threshold ζ. In these scenarios, both the currently received signal $(y_{r,d}^\tau)$ and that stored in the memory M_2 $(y_{r,d}^{\tau-l})$ are used for differential detection at the destination. In the DiffDF scheme, the memory M_2 stores only the received signal from the relay whose amplitude is larger than the threshold. This implies that the amplitude $|y_{r,d}^{\tau-l}|$ is larger than the threshold. Therefore, the chance that each of the scenarios Φ_2 to Φ_6 happens is conditioned on the event that $|y_{r,d}^{\tau-l}| > \zeta$. From (9.12), the chance that the scenario Φ_2 occurs is given by

$$P_r^h(\Phi_2) = P_r^h\left(|y_{r,d}^\tau| > \zeta, \tilde{P}_2^\tau = P_2, \tilde{P}_2^{\tau-l} = P_2, l = k \Big| |y_{r,d}^{\tau-l}| > \zeta\right). \quad (9.29)$$

Since the events at time $\tau - l$ are independent of the events at time τ, $P_r^h(\Phi_2)$ can be written as a product of the probabilities:

$$P_r^h(\Phi_2) = P_r^h\left(|y_{r,d}^\tau| > \zeta, \tilde{P}_2^\tau = P_2\right) P_r^h\left(\tilde{P}_2^{\tau-l} = P_2, l = k \Big| |y_{r,d}^{\tau-l}| > \zeta\right). \quad (9.30)$$

The first term on the right-hand side of (9.30) represents the probability that the relay transmits the decoded symbol with power P_2 and that the received signal from the relay is larger than the threshold. This term can be expressed as

$$P_r^h\left(|y_{r,d}^\tau| > \zeta, \tilde{P}_2^\tau = P_2\right) = P_r^h\left(|y_{r,d}^\tau| > \zeta \Big| \tilde{P}_2^\tau = P_2\right) P_r^h\left(\tilde{P}_2^\tau = P_2\right). \quad (9.31)$$

In (9.31), the chance that the amplitude of the received signal from the relay is larger than the threshold given that the relay sends the information can be obtained from (9.25) as

$$P_r^h\left(|y_{r,d}^\tau| > \zeta \Big| \tilde{P}_2^\tau = P_2\right) = \mathcal{M}\left(P_2|h_{r,d}^\tau|^2, \zeta\right). \quad (9.32)$$

Therefore, using the results in (9.21) and (9.32), (9.31) can be rewritten as

$$P_r^h\left(|y_{r,d}^\tau| > \zeta, \tilde{P}_2^\tau = P_2\right) = \mathcal{M}\left(P_2|h_{r,d}^\tau|^2, \zeta\right)\left(1 - \Psi(\gamma_{s,r}^\tau)\right). \quad (9.33)$$

The second term on the right-hand side of (9.30) represents the chance that the relay transmits with power $\tilde{P}_2^{\tau-l} = P_2$, and the received signal $y_{r,d}^{\tau-l}$ stored in the memory M_2 conveys the information symbol $\tilde{x}^{\tau-k}$ stored in the memory M_1. We can calculate it as

$$P_r^h\left(\tilde{P}_2^{\tau-l} = P_2, l = k \Big| |y_{r,d}^{\tau-l}| > \zeta\right)$$
$$= \sum_{k \geq 1} P_r^h\left(\tilde{P}_2^{\tau-k} = P_2 \Big| |y_{r,d}^{\tau-k}| > \zeta\right) P_r^h\left(|y_{r,d}^{\tau-k}| > \zeta\right)$$
$$\times \prod_{i=1}^{k-1} P_r^h\left(|y_{r,d}^{\tau-i}| \leq \zeta\right) P_r^h\left(\tilde{P}_2^{\tau-i} = 0 \Big| |y_{r,d}^{\tau-i}| \leq \zeta\right). \quad (9.34)$$

Based on (9.33), the term $P_r^h(\tilde{P}_2^{\tau-k} = P_2||y_{r,d}^{\tau-k}| > \varsigma)P_r^h(|y_{r,d}^{\tau-k}| > \varsigma)$ in (9.34) can be evaluated as $\mathcal{M}(P_2|h_{r,d}^{\tau-k}|^2, \varsigma)(1 - \Psi(\gamma_{s,r}^{\tau-k}))$, which can be approximated by $\mathcal{M}(P_2|h_{r,d}^{\tau}|^2, \varsigma)(1 - \Psi(\gamma_{s,r}^{\tau}))$ if the channels stay almost constant for several time slots. By applying Bayes' rule and using the results in (9.18) and (9.23), we can express the product term in (9.34) as

$$\prod_{i=1}^{k-1} P_r^h\left(|y_{r,d}^{\tau-i}| \leq \varsigma\right) P_r^h\left(\tilde{P}_2^{\tau-i} = 0\big||y_{r,d}^{\tau-i}| \leq \varsigma\right)$$

$$= \prod_{i=1}^{k-1} P_r^h\left(|y_{r,d}^{\tau-i}| \leq \varsigma\big|\tilde{P}_2^{\tau-i} = 0\right) P_r^h\left(\tilde{P}_2^{\tau-i} = 0\right)$$

$$\approx \left[(1 - e^{-\varsigma^2/N_0})\Psi(\gamma_{s,r}^{\tau})\right]^{k-1}, \tag{9.35}$$

where the resulting approximation comes from approximating $P_r^h(\tilde{P}_2^{\tau-i} = 0) = \Psi(\gamma_{s,r}^{\tau-i})$ by $\Psi(\gamma_{s,r}^{\tau})$ for all i. Accordingly, we can approximate (9.34) as

$$P_r^h\left(\tilde{P}_2^{\tau-l} = P_2, l = k\big||y_{r,d}^{\tau-l}| > \varsigma\right)$$

$$\approx \mathcal{M}(P_2|h_{r,d}^{\tau}|^2, \varsigma)(1 - \Psi(\gamma_{s,r}^{\tau}))\sum_{k\geq 1}\left[(1 - e^{-\varsigma^2/N_0})\Psi(\gamma_{s,r}^{\tau})\right]^{k-1}$$

$$= \frac{\mathcal{M}\left(P_2|h_{r,d}^{\tau}|^2, \varsigma\right)(1 - \Psi(\gamma_{s,r}^{\tau}))}{(1 - e^{-\varsigma^2/N_0})\Psi(\gamma_{s,r}^{\tau})}. \tag{9.36}$$

Hence, by substituting (9.33) and (9.36) into (9.30), the chance that the scenario Φ_2 happens can be approximated by

$$P_r^h(\Phi_2) \approx \frac{\mathcal{M}^2\left(P_2|h_{r,d}^{\tau}|^2, \varsigma\right)(1 - \Psi(\gamma_{s,r}^{\tau}))^2}{(1 - e^{-\varsigma^2/N_0})\Psi(\gamma_{s,r}^{\tau})}. \tag{9.37}$$

Next, we consider the scenario Φ_3, which is similar to the scenario Φ_2 except that the received signal $y_{r,d}^{\tau-l}$ stored in the memory M_2 at the destination does not convey the symbol $\tilde{x}^{\tau-k}$ stored in the memory M_2 at the relay. The chance that the scenario Φ_3 happens can be given by

$$P_r^h(\Phi_3) = P_r^h\left(|y_{r,d}^{\tau}| > \varsigma, \tilde{P}_2^{\tau} = P_2, \tilde{P}_2^{\tau-l} = P_2, l \neq k\big||y_{r,d}^{\tau-l}| > \varsigma\right). \tag{9.38}$$

Observe that the scenarios Φ_2 and Φ_3 are disjoint. Thus, the chance that the scenario Φ_3 happens can be obtained from $P_r^h(\Phi_2)$ as

$$P_r^h(\Phi_3) = P_r^h(\Phi_2 \cup \Phi_3) - P_r^h(\Phi_2), \tag{9.39}$$

where

$$P_r^h(\Phi_2 \cup \Phi_3) = P_r^h\left(|y_{r,d}^{\tau}| > \varsigma, \tilde{P}_2^{\tau} = P_2, \tilde{P}_2^{\tau-l} = P_2, ||y_{r,d}^{\tau-l}| > \varsigma\right).$$

Since the signals at time τ and time $\tau - l$ are independent, we can express $P_r^h(\Phi_2 \cup \Phi_3)$ as

$$P_r^h(\Phi_2 \cup \Phi_3) = P_r^h\left(|y_{r,d}^\tau| > \varsigma, \tilde{P}_2^\tau = P_2\right) P_r^h\left(\tilde{P}_2^{\tau-l} = P_2 \big| |y_{r,d}^{\tau-l}| > \varsigma\right). \quad (9.40)$$

By applying Bayes' rule, the second term on the right-hand side of (9.40) is given by

$$P_r^h\left(\tilde{P}_2^{\tau-l} = P_2 \big| |y_{r,d}^{\tau-l}| > \varsigma\right) = \frac{P_r^h\left(|y_{r,d}^{\tau-l}| > \varsigma, \tilde{P}_2^{\tau-l} = P_2\right)}{P_r^h\left(|y_{r,d}^{\tau-l}| > \varsigma\right)}. \quad (9.41)$$

The numerator on the right-hand side of (9.41) is in the same form as (9.33) with τ replaced by $\tau - l$, whereas the denominator can be calculated by using the concept of total probability:

$$P\left(|y_{r,d}^{\tau-l}| > \varsigma\right) = P\left(|y_{r,d}^{\tau-l}| > \varsigma \big| \tilde{P}_2^{\tau-l} = P_2\right) P\left(\tilde{P}_2^{\tau-l} = P_2\right)$$
$$+ P\left(|y_{r,d}^{\tau-l}| > \varsigma \big| \tilde{P}_2^{\tau-l} = 0\right) P\left(\tilde{P}_2^{\tau-l} = 0\right). \quad (9.42)$$

In (9.42), the chance that the received signal $|y_{r,d}^{\tau-l}|$ is greater than the decision threshold given that the relay does not send information can be obtained from (9.23) as

$$P_r^h\left(|y_{r,d}^{\tau-l}| > \varsigma \big| \tilde{P}_2^\tau = 0\right) = \exp(-\varsigma^2/N_0). \quad (9.43)$$

Substitute (9.18), (9.33), and (9.43) into (9.42) resulting in

$$P\left(|y_{r,d}^{\tau-l}| > \varsigma\right) = \mathcal{M}\left(P_2|h_{r,d}^{\tau-l}|^2, \varsigma\right)\left(1 - \Psi(\gamma_{s,r}^{\tau-l})\right) + \exp(-\varsigma^2/N_0)\Psi(\gamma_{s,r}^{\tau-l})$$
$$\triangleq \Gamma(P_1|h_{s,r}^{\tau-l}|^2, P_2|h_{r,d}^{\tau-l}|^2). \quad (9.44)$$

By assuming that the channel coefficients $h_{i,j}^{\tau-l} \approx h_{i,j}^\tau$ for any (i, j) link, then from (9.33), (9.42), and (9.44), we can determine the chance $P_r^h(\Phi_2 \cup \Phi_3)$ in (9.40) as

$$P_r^h(\Phi_2 \cup \Phi_3) = \frac{\mathcal{M}^2\left(P_2|h_{r,d}^\tau|^2, \varsigma\right)\left(1 - \Psi(\gamma_{s,r}^\tau)\right)^2}{\Gamma(P_1|h_{s,r}^\tau|^2, P_2|h_{r,d}^\tau|^2)}. \quad (9.45)$$

Similarly, we can obtain the following results (proofs left as exercises):

$$P_r^h(\Phi_3) = \mathcal{M}^2\left(P_2|h_{r,d}^\tau|^2, \varsigma\right)\left(1 - \Psi(\gamma_{s,r}^\tau)\right)^2$$
$$\times \left(\frac{1}{\Gamma(P_1|h_{s,r}^\tau|^2, P_2|h_{r,d}^\tau|^2)} - \frac{1}{(1 - e^{-\varsigma^2/N_0})\Psi(\gamma_{s,r}^\tau)}\right), \quad (9.46)$$

$$P_r^h(\Phi_4) = P_r^h(\Phi_5) = \frac{\mathcal{M}\left(P_2|h_{r,d}^\tau|^2, \varsigma\right)\exp(-\varsigma^2/N_0)\Psi(\gamma_{s,r}^\tau)\left(1 - \Psi(\gamma_{s,r}^\tau)\right)}{\Gamma(P_1|h_{s,r}^\tau|^2, P_2|h_{r,d}^\tau|^2)}, \quad (9.47)$$

$$P_r^h(\Phi_6) = \frac{\exp(-2\varsigma^2/N_0)\Psi(\gamma_{s,r}^\tau)}{\Gamma(P_1|h_{s,r}^\tau|^2, P_2|h_{r,d}^\tau|^2)}. \quad (9.48)$$

Average BER analysis

In this subsection, we provide the conditional BER given that each scenario Φ_i for $i = 1, 2, \ldots, 6$ happens. From the resulting conditional BER together with the chance that each scenario occurs as derived in the previous subsection, we then provide the average BER formulation for the differential DF scheme.

Conditional BER of each scenario   When the scenario Φ_1 occurs, the destination estimates the transmitted symbol v_m from the source by using only the received signal $y_{s,d}^\tau$ from the source. The conditional BER for the scenario Φ_1 is

$$P_{BER}^h|_{\Phi_1} = \Omega_1(\gamma_1) \triangleq \frac{1}{4\pi} \int_{-\pi}^{\pi} f_1(\theta) \exp\left[-\alpha(\theta)\gamma_1\right] d\theta, \tag{9.49}$$

where γ_i represents the instantaneous SNR given that the scenario Φ_i occurs, and

$$f_1(\theta) = \frac{1 - \beta^2}{1 + 2\beta \sin\theta + \beta^2}, \tag{9.50}$$

$$\alpha(\theta) = \frac{b^2}{2 \log_2 M}(1 + 2\beta \sin\theta + \beta^2), \tag{9.51}$$

in which M is the constellation size [188]. For the scenario Φ_1, the instantaneous SNR γ_1 is specified as $\gamma_1 = P_1|h_{s,d}^\tau|^2/N_0$. In (9.50) and (9.51), the parameter $\beta = a/b$ is a constant whose value depends on constellation size. For example, $a = 10^{-3}$ and $b = \sqrt{2}$ for DBPSK modulation, and $a = \sqrt{2 - \sqrt{2}}$ and $b = \sqrt{2 + \sqrt{2}}$ for DQPSK modulation [188]. The values of a and b for larger constellation sizes can be obtained by using the result in [146].

The scenarios Φ_2 to Φ_6 correspond to the case that the destination combines the received signal $y_{s,d}^\tau$ from the source and $y_{r,d}^\tau$ from the relay. The conditional BER for these scenarios depends on the combining weight coefficients a_1 and a_2 (see (9.10)). Under the scenario Φ_2, the received signals from the source and the relay can be expressed as

$$y_{s,d}^\tau = v_m y_{s,d}^{\tau-1} + \tilde{w}_{s,d}^\tau, \tag{9.52}$$

$$y_{r,d}^\tau = v_m y_{r,d}^{\tau-l} + \tilde{w}_{r,d}^\tau, \tag{9.53}$$

where $\tilde{w}_{s,d}^\tau$ and $\tilde{w}_{r,d}^\tau$ are the additive noise terms, and each of them is zero-mean Gaussian distributed with variance $2N_0$. Since the noise terms of the received signals from both direct link and relay link have the same mean and variance, the SNR of the combiner output under the scenario Φ_2 can be maximized by choosing the weighting coefficients $a_1 = a_2 = 1/2N_0$. Such weighting coefficients lead to an optimum two-branch differential detection with equal gain combining. With weighting coefficients $a_1 = a_2 = 1/2N_0$, the conditional BER can be determined as follows [188]:

$$P_{\text{BER}}^h | \Phi_2 = \Omega_2(\gamma_2) \triangleq \frac{1}{16\pi} \int_{-\pi}^{\pi} f_2(\theta) \exp\left[-\alpha(\theta)\gamma_2\right] d\theta, \tag{9.54}$$

where

$$f_2(\theta) = \frac{b^2(1-\beta^2)[3+\cos(2\theta)-(\beta+\frac{1}{\beta})\sin(\theta)]}{2\alpha(\theta)}, \tag{9.55}$$

and $\alpha(\theta)$ is specified in (9.51). The instantaneous SNR γ_2 is given by

$$\gamma_2 = \frac{P_1 |h_{\text{s,d}}^\tau|^2}{N_0} + \frac{P_2 |h_{\text{r,d}}^\tau|^2}{N_0}. \tag{9.56}$$

For the remaining scenarios, namely Φ_3 to Φ_6, the destination also combines the received signal from the source and that from the relay. However, the use of two-channel differential detection for these four cases is not guaranteed to be optimum since either the received signals $y_{\text{r,d}}^\tau$ or $y_{\text{r,d}}^{\tau-l}$ from the relay contain only noise. Up to now, the conditional BER formulation for DMPSK with arbitrary-weighted combining has not been available in the literature. For analytical tractability of the analysis, we resort to an approximate BER, in which the signal from the relay is considered as noise when Φ_3 to Φ_6 occur. As we will show in the succeeding section, the analytical BER obtained from this approximation is very close to the simulation results. The approximate conditional BER for the scenarios Φ_i, $i = 3, \ldots, 6$, are $P_{\text{BER}}^h | \Phi_i \approx \Omega_2(\gamma_i)$, where

$$\gamma_i = \frac{P_1 |h_{\text{s,d}}^\tau|^2}{N_0 + N_i/(P_1 |h_{\text{s,d}}^\tau|^2/N_0)}, \tag{9.57}$$

in which N_i represents the noise power that comes from the relay link given that the scenario Φ_i occurs. The noise power N_i depends on the received signal from the relay at the current time τ $(y_{\text{r,d}}^\tau)$ and that stored in the memory $(y_{\text{r,d}}^{\tau-l})$. Under the scenario Φ_3, N_3 is given by $(P_2 |h_{\text{r,d}}|^2 + N_0)^2/N_0$. Under the scenarios Φ_4 and Φ_5, the relay does not send the information at either time τ or time $\tau - l$. These two scenarios result in the same noise power such that $N_4 = N_5 = P_2 |h_{\text{r,d}}|^2 + N_0$. The last scenario, i.e., Φ_6, corresponds to the case when both $y_{\text{r,d}}^\tau$ and $y_{\text{r,d}}^{\tau-l}$ does not contain any information symbol. The noise power N_6 is equal to the noise variance N_0.

Average BER In what follows, we determine the average BER

$$P_{\text{BER}}^{(i)} \triangleq E\left[P_{\text{BER}}^h | \Phi_i \, P_r^h(\Phi_i)\right]$$

for each scenario. The average BER for each scenario, Φ_i, $i = 1, 2, \ldots, 6$, can be determined as follows (proofs left as exercises):

- For Φ_1:

$$P_{\text{BER}}^{(1)} = F_1\left(1 + \frac{\alpha(\theta)P_1\sigma_{s,d}^2}{N_0}\right)\left[\left(1 - e^{-\zeta^2/N_0}\right)G\left(1 + \frac{g(\phi)P_1\sigma_{s,r}^2}{N_0}\right)\right.$$

$$+ \left(1 - \frac{1}{\sigma_{r,d}^2}\int_0^\infty \mathcal{M}(P_2q, \zeta)\,e^{-q/\sigma_{r,d}^2}dq\right)$$

$$\left.\times\left(1 - G\left(1 + \frac{g(\phi)P_1\sigma_{s,r}^2}{N_0}\right)\right)\right], \tag{9.58}$$

where $g(\phi)$ and $f_1(\theta)$ are specified in (9.20) and (9.50), respectively, and

$$G(c(\phi)) = \frac{1}{\pi}\int_0^{(M-1)\pi/M} \frac{1}{c(\phi)}d\phi, \tag{9.59}$$

$$F_1(c(\theta)) = \frac{1}{4\pi}\int_{-\pi}^{\pi} \frac{f_1(\theta)}{c(\theta)}d\theta, \tag{9.60}$$

in which $c(\theta)$ is an arbitrary function of θ.

- For Φ_2:

$$P_{\text{BER}}^{(2)} \approx \frac{1}{\sigma_{r,d}^2}\int_0^\infty s_2(q)\mathcal{M}^2(P_2q, \zeta)\,e^{-q/\sigma_{r,d}^2}dz$$

$$\times\frac{1}{\sigma_{s,r}^2}\int_0^\infty \frac{\left(1 - \Psi\left(\frac{P_1u}{N_0}\right)\right)^2}{1 - (1 - e^{-\zeta^2/N_0})\Psi\left(\frac{P_1u}{N_0}\right)}e^{-u/\sigma_{s,r}^2}du, \tag{9.61}$$

where

$$s_2(q) = \frac{1}{16\pi}\int_{-\pi}^{\pi} \frac{f_2(\theta)}{1 + \frac{\alpha(\theta)P_1\sigma_{s,d}^2}{N_0}}e^{-\alpha(\theta)P_2q/N_0}d\theta. \tag{9.62}$$

- For Φ_3:

$$P_{\text{BER}}^{(3)} \approx \frac{1}{\sigma_{s,r}^2}\int_0^\infty\left[\frac{1}{4\pi}\int_{-\pi}^{\pi} f_2(\theta)s_3(u, \theta)d\theta\right]e^{-u/\sigma_{s,r}^2}du, \tag{9.63}$$

where

$$s_3(u, \theta) = \frac{1}{\sigma_{s,d}^2\sigma_{r,d}^2}\int_0^\infty\int_0^\infty \exp\left(-\frac{\alpha(\theta)P_1z}{N_0 + \frac{(P_2q+N_0)^2}{P_1z}} - \frac{z}{\sigma_{s,d}^2} - \frac{q}{\sigma_{r,d}^2}\right)$$

$$\times\left(1 - \Psi\left(\frac{P_1u}{N_0}\right)\right)^2 \times \mathcal{M}^2(P_2q, \zeta)$$

$$\times\left(\frac{1}{\Gamma(P_1u, P_2q)} - \frac{1}{1 - (1 - e^{-\zeta^2/N_0})\Psi\left(\frac{P_1u}{N_0}\right)}\right)dq\,dz. \tag{9.64}$$

- For Φ_4 and Φ_5:

$$P_{\text{BER}}^{(4)} = P_{\text{BER}}^{(5)} \approx \frac{1}{\sigma_{s,r}^2}\int_0^\infty\left[\frac{1}{4\pi}\int_{-\pi}^{\pi} f_2(\theta)s_4(u, \theta)d\theta\right]e^{-u/\sigma_{s,r}^2}du, \tag{9.65}$$

where

$$s_4(u, \theta) = \frac{1}{\sigma_{s,d}^2 \sigma_{r,d}^2} \int_0^\infty \int_0^\infty \exp\left(-\frac{\alpha(\theta) P_1 z}{N_0 + \frac{P_2 q + N_0}{P_1 z / N_0}} - \frac{\zeta^2}{N_0} - \frac{z}{\sigma_{s,d}^2} - \frac{q}{\sigma_{r,d}^2}\right)$$

$$\times \mathcal{M}(P_2 q, \zeta) \frac{\Psi\left(\frac{P_1 u}{N_0}\right)\left(1 - \Psi\left(\frac{P_1 u}{N_0}\right)\right)}{\Gamma(P_1 u, P_2 q)} dq \, dz. \qquad (9.66)$$

- For Φ_6:

$$P_{\text{BER}}^{(6)} \approx \exp(-2\zeta^2/N_0) \left[\frac{1}{16\pi}\int_{-\pi}^{\pi} f_2(\theta) s_6(\theta) d\theta\right]$$

$$\times \frac{1}{\sigma_{s,r}^2}\frac{1}{\sigma_{r,d}^2}\int_0^\infty \int_0^\infty \frac{\Psi^2(P_1 u / N_0)}{\Gamma(P_1 u, P_2 q)}\exp\left(-\frac{u}{\sigma_{s,r}^2} - \frac{q}{\sigma_{r,d}^2}\right) dq \, du, \qquad (9.67)$$

where

$$s_6(\theta) = \frac{1}{\sigma_{s,d}^2}\int_0^\infty \exp\left(-\frac{\alpha(\theta) P_1 z}{N_0 + N_0/(P_1 z / N_0)} - \frac{z}{\sigma_{s,d}^2}\right) dz. \qquad (9.68)$$

Finally, the average BER, P_{BER}, can be determined by summing together the average BER $P_{\text{BER}}^{(i)}$ for $i = 1, 2, \ldots, 6$:

$$P_{\text{BER}} = P_{\text{BER}}^{(1)} + P_{\text{BER}}^{(2)} + P_{\text{BER}}^{(3)} + 2P_{\text{BER}}^{(4)} + P_{\text{BER}}^{(6)}, \qquad (9.69)$$

in which $P_{\text{BER}}^{(i)}$ are specified in (9.58), (9.61), (9.64), (9.65), and (9.67), respectively.

9.2.1.3   Upper bound and lower bound

Since the closed-form BER result of the threshold-based differential DF scheme in the previous section involves integrals, it is hard to understand the system performance based on the result. We provide in this section, therefore, the expressions for the BER upper and lower bounds. We show through simulation results that when the power allocation and the decision threshold are properly designed, the BER upper and lower bounds are close to the simulated BER.

To obtain a BER upper bound, we first note that the conditional BER for each case, $P_{\text{BER}}^h|\Phi_i$, is at most $1/2$. In addition, if the threshold is properly designed, then the chance that the scenarios Φ_3 to Φ_6 happen are small compared to the chance that the scenarios Φ_1 and Φ_2 happen. Therefore, the BER of the differential DF cooperative scheme can be obtained by bounding the conditional BER $P_{\text{BER}}^h|\Phi_3$, $P_{\text{BER}}^h|\Phi_4$, $P_{\text{BER}}^h|\Phi_5$, and $P_{\text{BER}}^h|\Phi_6$ by $1/2$. The resulting BER upper bound can be expressed as

$$P_{\text{BER}} \leq P_{\text{BER}}^{(1)} + P_{\text{BER}}^{(2)} + \frac{1}{2}\{P_r^h(\Phi_3) + P_r^h(\Phi_4) + P_r^h(\Phi_5) + P_r^h(\Phi_6)\}, \qquad (9.70)$$

where $P_{\text{BER}}^{(1)}$ and $P_{\text{BER}}^{(2)}$ are determined in (9.58) and (9.61), and $P_r^h(\Phi_i)$, $i = 3, 4, 5$, and 6, are given in (9.38), (9.47), and (9.48), respectively.

Next, we determine a BER lower bound as follows. Since the exact expressions of the chances that the scenarios Φ_2 and Φ_3 occur involve the approximation that the channel coefficients are constant for several symbol periods, and the BER formulation given that the scenario Φ_3 happens is currently unavailable, it is hard to obtain the exact BER of $P_{\text{BER}}^{(2)}$ and $P_{\text{BER}}^{(3)}$. As we will show later in this section, if the power ratio and the threshold are properly designed, the chance that the scenario Φ_3 occurs tends to be small compared to the chance that the scenario Φ_2 occurs. In addition, the conditional BER under the scenario Φ_3 ($P_{\text{BER}}^h|_{\Phi_3}$) is larger than that under the scenario Φ_2 ($P_{\text{BER}}^h|_{\Phi_2}$). Therefore, the BER lower bound of the DiffDF scheme can be obtained by bounding the conditional BER $P_{\text{BER}}^h|_{\Phi_3}$ with $P_{\text{BER}}^h|_{\Phi_2}$. In this way, the BER under the scenarios Φ_2 and Φ_3 can be lower bounded as

$$P_{\text{BER}}^{(2)} + P_{\text{BER}}^{(3)} \geq \mathrm{E}\big[P_{\text{BER}}^h|_{\Phi_2} P_r^h(\Phi_2 \cup \Phi_3)\big], \tag{9.71}$$

where $P_r^h(\Phi_2 \cup \Phi_3)$, which is evaluated in (9.45), is the chance that the scenarios Φ_2 and Φ_3 occur. According to (9.37) and (9.54), the average BER in (9.71) can be expressed as

$$P_{\text{BER}}^{(2)} + P_{\text{BER}}^{(3)} \geq \mathrm{E}\Bigg[\Omega_2\left(\frac{P_1|h_{\text{s,d}}^\tau|^2}{N_0} + \frac{P_2|h_{\text{r,d}}^\tau|^2}{N_0}\right) \\ \times \frac{\mathcal{M}^2\left(P_2|h_{\text{r,d}}^\tau|^2, \zeta\right)\left(1 - \Psi\left(\frac{P_1|h_{\text{s,r}}^\tau|^2}{N_0}\right)\right)^2}{\Gamma(P_1|h_{\text{s,r}}^\tau|^2, P_2|h_{\text{r,d}}^\tau|^2)}\Bigg]. \tag{9.72}$$

By averaging over the channel realization, we obtain

$$P_{\text{BER}}^{(2)} + P_{\text{BER}}^{(3)} \geq \frac{1}{\sigma_{\text{s,r}}^2} \int_0^\infty \left[\frac{1}{16\pi} \int_{-\pi}^{\pi} \frac{f_2(\theta)s(u,\theta)}{1 + \alpha(\theta)\frac{P_1\sigma_{\text{s,d}}^2}{N_0}}d\theta\right] \exp(-u/\sigma_{\text{s,r}}^2)du.$$

$$\triangleq \mathrm{LB}\{P_{\text{BER}}^{(2)} + P_{\text{BER}}^{(3)}\} \tag{9.73}$$

where

$$s(u,\theta) = \frac{1}{\sigma_{\text{r,d}}^2} \int_0^\infty \frac{\mathcal{M}^2\left(P_2 q, \zeta\right)\left(1 - \Psi\left(\frac{P_1 u}{N_0}\right)\right)^2}{\Gamma(P_1 u, P_2 q)} \exp\left(-\frac{\alpha(\theta)P_2 q}{N_0} - \frac{q}{\sigma_{\text{r,d}}^2}\right)dq. \tag{9.74}$$

Since the exact BER formulations under the scenarios 4, 5, and 6 are currently unavailable, and the chances that these three scenarios happen are small at high SNR, we further lower bound the BER $P_{\text{BER}}^{(4)}$, $P_{\text{BER}}^{(5)}$, and $P_{\text{BER}}^{(6)}$ by 0. As a result, the BER of the differential DF scheme can be lower bounded by

$$P_{\text{BER}} \geq P_{\text{BER}}^{(1)} + \mathrm{LB}\{P_{\text{BER}}^{(2)} + P_{\text{BER}}^{(3)}\}, \tag{9.75}$$

where $P_{\text{BER}}^{(1)}$ is determined in (9.58).

Example 9.1   This example compares the BER approximation (9.69), the BER upper bound (9.70), and the BER lower bound (9.75) with the simulated performance in case of DQPSK modulation. In Figure 9.3(a)–(c), we consider the differential DF cooperation system with $\sigma_{\text{s,d}}^2 = \sigma_{\text{s,r}}^2 = \sigma_{\text{r,d}}^2 = 1$, i.e., all the channel links have the same

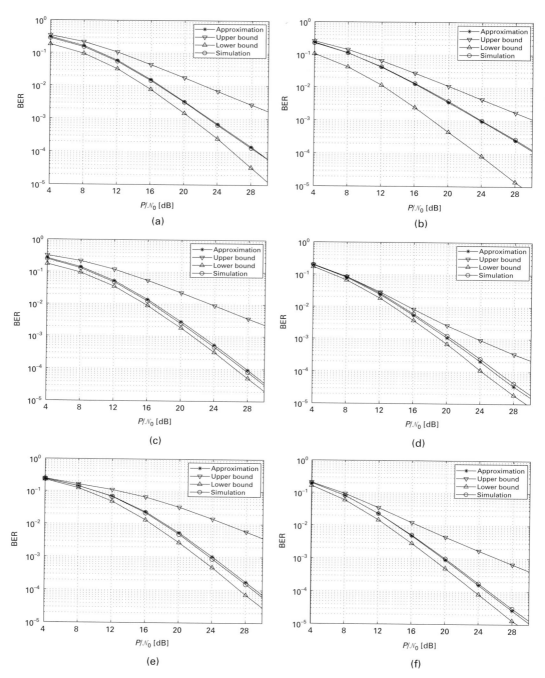

Fig. 9.3 DQPSK: BER performance versus $P/\mathcal{N}_0$ for $\sigma_{s,d}^2 = 1$, $\sigma_{s,r}^2 = 1$. (a) $P_1 = 0.5P$, $P_2 = 0.5P$, $\zeta = 1$, $\sigma_{r,d}^2 = 1$; (b) $P_1 = 0.8P$, $P_2 = 0.2P$, $\zeta = 1$, $\sigma_{r,d}^2 = 1$; (c) $P_1 = 0.8P$, $P_2 = 0.2P$, $\zeta = 2$, $\sigma_{r,d}^2 = 1$; (d) $P_1 = 0.5P$, $P_2 = 0.5P$, $\zeta = 1$, $\sigma_{r,d}^2 = 10$; (e) $P_1 = 0.5P$, $P_2 = 0.5P$, $\zeta = 2$, $\sigma_{r,d}^2 = 10$; (f) $P_1 = 0.8P$, $P_2 = 0.2P$, $\zeta = 2$, $\sigma_{r,d}^2 = 10$.

qualities. From the figures, we can see that the approximated BER closely matches the simulated BER, and both the approximated and simulated BER lie between the BER upper and lower bounds. Moreover, the system performance depends on the power allocation and the threshold. By choosing proper power allocation and threshold, not only the BER performance improves, but also the lower bound is closer to the simulated performance.

For example, considering a system with threshold of $\zeta = 1$ and total transmit power $P_1 + P_2 = P$ where P_1 and P_2 are transmitted power of the source and the relay, respectively. By changing the power allocation from $P_1/P = 0.5$ to $P_1/P = 0.8$ the BER performance is improved by 1 dB at a BER of 10^{-4}, while the performance gap between the simulated BER and the BER lower bound is reduced by 2 dB at the same BER.

This result follows the fact that when the threshold ζ is appropriately chosen, the scenarios Φ_3 to Φ_6 occur with much smaller probability than the scenarios Φ_1 and Φ_2. Even though the BER under each of the scenarios Φ_3 to Φ_6 is larger than that under the scenarios Φ_1 or Φ_2, the average BER $P_{\mathrm{BER}}^{(i)}$, $i = 3, \ldots, 6$ are smaller. The BER P_{BER} is dominated by the BER under scenarios Φ_1 and Φ_2. Therefore, the performance gaps between the bounds and the approximate BER is small if the threshold is properly designed.

For a system with $\sigma_{\mathrm{s,d}}^2 = 1$, $\sigma_{\mathrm{s,r}}^2 = 1$, and $\sigma_{\mathrm{r,d}}^2 = 10$, the same observation can be obtained in Figure 9.3(d)–(f). At a threshold of $\zeta = 1$, the performance can be improved by allocating more power at the source and less power at the relay. It illustrates that the channel link between source and relay and the channel link between relay and destination should be balanced in order to achieve a performance diversity of two.

Interestingly, by choosing a proper threshold, the performance can be significantly improved, regardless of the power allocation. For instance, increasing the threshold from 1 to 2, the performance of the DiffDF scheme with equal power allocation improves 4 dB at a BER of 10^{-4}. At the threshold of $\zeta = 2$, by changing the power allocation from $P_1/P = 0.5$ to $P_1/P = 0.8$, the system performance is further improved by only 0.5 dB at the same BER. This implies that the effect of the threshold dominates; the performance does not severely depend on the power allocation after the threshold is properly designed.

Another observation obtained from Figure 9.3 is that the threshold depends on the channel link qualities. To be specific, the threshold should be increased as the link quality between the relay and the destination increases. For example, under the scenario $\sigma_{\mathrm{s,d}}^2 = \sigma_{\mathrm{s,r}}^2 = 1$, the threshold $\zeta = 1$ results in superior performance in case of $\sigma_{\mathrm{r,d}}^2 = 1$, while the threshold $\zeta = 2$ leads to better performance in case of $\sigma_{\mathrm{r,d}}^2 = 10$. This observation can be explained as follows. When the link quality between the relay and the destination is good, i.e., the channel variance is high, the received signal from the relay tends to have large energy if it carries the information. As a result, by increasing the threshold from 1 to 2, we reduce the chance that the received signal whose amplitude is larger than the threshold contains no information. Thus, with a threshold of 2, the received signals from the relay and the destination are efficiently combined, and hence result in better performance. ▲

9.2.1.4   Optimum decision threshold and power allocation

As we observe in the above examples in the previous section, the choice of power allocation and threshold ζ affect the performance a lot; we determine in this section the optimum decision threshold and the optimum power allocation for the differential DF cooperation system based on the tight BER approximation in (9.69). To simplify the notation, let us denote $r = P_1/P$ as the power ratio of the transmitted power at the source (P_1) over the total power (P). For a fixed total transmitted power $P_1 + P_2 = P$, we will jointly optimize the threshold ζ and the power ratio r such that the tight BER approximation in (9.69) is minimized. The optimization problem can be formulated as

$$\min_{\zeta, r} \ P_{\text{BER}}(\zeta, r), \tag{9.76}$$

where $P_{\text{BER}}(\zeta, r)$ represents the BER approximation with $P_1 = rP$ and $P_2 = (1-r)P$.

Figure 9.4 shows the BER performance of the DiffDF scheme with DQPSK signals as a function of power allocation and threshold. In Figure 9.4(a)–(c), we consider the case when the channel variances of all communication links are equal, i.e., $\sigma_{s,d}^2 = \sigma_{s,r}^2 = \sigma_{r,d}^2 = 1$. The BER approximation is plotted in Figure 9.4(a) and its cross sections are shown in Figure 9.4(b) and (c) together with the simulated BER curves. Based on the approximated BER in Figure 9.4(a), the jointly optimum power allocation and decision threshold are $r = 0.7$ and $\zeta = 1$. Figure 9.4(b) compares the cross sectional curve of the approximated BER with $r = 0.7$ with the simulated BER. We can see that the approximate BER is close to the simulated BER.

Furthermore, the DiffDF scheme with any threshold less than 1.5 yields almost the same BER performance, and the performance significantly degrades as the threshold increases above 1.5. This is because as the threshold increases, the chances that the scenarios Φ_4 to Φ_6 occur increases, and hence the average BER is dominated by the BER under these scenarios. Figure 9.4(c) depicts the approximated and simulated BER curves as functions of power allocation in case of the decision threshold $\zeta = 1$. We can obviously see that the power ratio of $r = 0.7$ results in the optimum performance.

In Figure 9.4(d)–(f), we consider the case of channel variances $\sigma_{s,d}^2 = 1$, $\sigma_{s,r}^2 = 1$, $\sigma_{r,d}^2 = 10$. Figure 9.4(d) depicts the BER of the DiffDF scheme as a function of the power allocation r and the decision threshold ζ. In this scenario, we can see that the jointly optimum power allocation and decision threshold are $r = 0.8$ and $\zeta = 1.7$. Figure 9.4(e) shows a cross-sectional curve of the approximate BER in Figure 9.4(d) at $r = 0.8$ together with the simulated BER performance of the cooperation system with the same power allocation. We can see that the approximated BER closely matches to the simulated BER for every threshold value. According to both the simulated and approximated BER, the optimum threshold for this case is about 1.7. We show in Figure 9.4(f) a comparison of the approximated and simulated BER with a decision threshold $\zeta = 1.7$ under different power allocations. Clearly, the approximated BER follows the same trend as the simulated BER, and the optimum power allocation is $r = 0.8$ for the differential DF system with a decision threshold $\zeta = 1.7$.

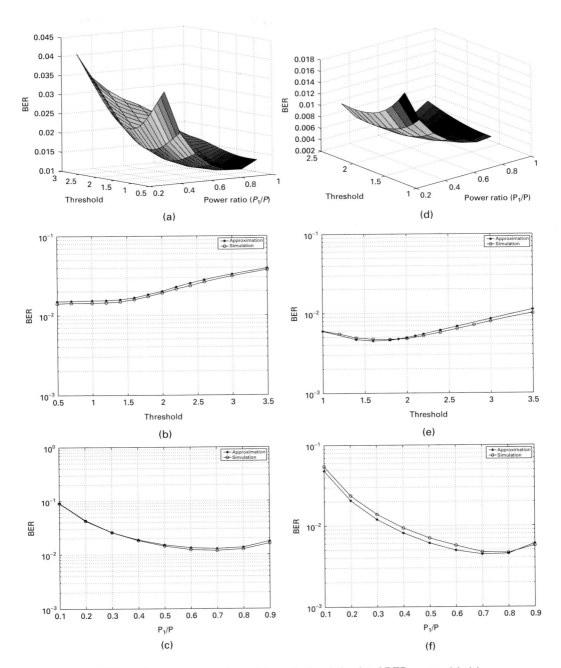

Fig. 9.4   DQPSK: performance comparison of theoretical and simulated BER curves. (a)–(c)
$\sigma_{s,d}^2 = \sigma_{s,r}^2 = \sigma_{r,d}^2 = 1$; (d)–(f) $\sigma_{s,d}^2 = 1$, $\sigma_{s,r}^2 = 1$, $\sigma_{r,d}^2 = 10$: (a) jointly optimum ζ and r; (b)
vary ζ, $P_1 = 0.7P$ and $P_2 = 0.3P$; (c) vary r, $\zeta = 1$; (d) jointly optimum ζ and r; (e) vary ζ,
$P_1 = 0.8P$ and $P_2 = 0.2P$; (f) vary r, $\zeta = 1.7$.

9.2.1.5   Examples for single-relay DiffDF scheme

In what follows, a two-user cooperation system employing the DF protocol is considered. The channel fading coefficients are modeled according to the Jakes' model (see Section 1.1.5) with the Doppler frequency $f_D = 75\,\text{Hz}$ and normalized fading parameter $f_D T_s = 0.0025$ where T_s is the sampling period. The noise variance is $\mathcal{N}_0 = 1$. The DQPSK modulation is used in all examples. We plot the BER performance curves as functions of $P/\mathcal{N}_0$, where P is the total transmitted power. We assume that the power allocation at the source node and relay node are fixed at $P_1 + P_2 = P$.

Example 9.2   This example compares the performance of the threshold-based differential DF scheme to that of the differential DF scheme without threshold at the destination and that of the differential DF scheme where the relay always forwards the decoded symbols to the destination. We consider the system with power allocation $P_1 = 0.5P$, $P_2 = 0.5P$, and channel variances $\sigma_{s,d}^2 = 1$, $\sigma_{s,r}^2 = 1$, and $\sigma_{r,d}^2 = 10$. We can see that the threshold-based differential DF scheme outperforms the other two schemes. The reason is that a decoding error at the relay tends to result in an error at the destination. Hence, the performance of the differential DF scheme in which all the decoded symbols at the relay are forwarded is worse than that of the DiffDF scheme. In particular, the performance degradation of 11 dB can be observed at a BER of 10^{-3}. Adding a threshold at the destination can reduce the chance that the incorrectly decoded signal from the relay is combined to the signal from the source. Therefore, the DiffDF scheme yields superior performance to the differential DF scheme without threshold.

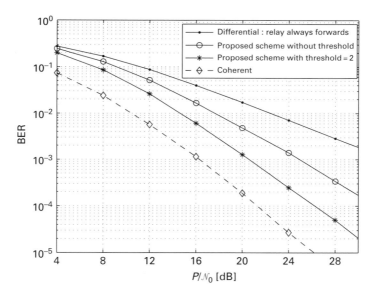

Fig. 9.5   DQPSK with or without CRC at relay node, and with or without threshold at destination node, $P_1 = P_2 = 0.5P$, $\sigma_{s,d}^2 = 1$, $\sigma_{s,r}^2 = 1$, $\sigma_{r,d}^2 = 10$.

As seen in Figure 9.5, the DiffDF scheme yields a gain of about 4 dB at a BER of 10^{-3} compared to the scheme without threshold. We also compare the performance of the coherent DF scheme in which the relay forwards only the correctly decoded symbols. Such scheme is the coherent counterpart of both the DiffDF scheme and the differential DF scheme without threshold. The DiffDF scheme shows a 5 dB performance gap in comparison to its coherent counterpart, but the differential scheme without a decision threshold looses about 9 dB in comparison to its coherent counterpart at a BER of 10^{-3}. ▲

Example 9.3 We now look at the BER performances of the DiffDF scheme with different thresholds. In Figure 9.6(a), the power allocation is $P_1 = 0.7P$ and $P_2 = 0.3P$, and the channel variances are $\sigma_{s,d}^2 = \sigma_{s,r}^2 = \sigma_{r,d}^2 = 1$. We can see that the DiffDF scheme achieves a performance diversity of two at high SNR for any threshold. However, the performance degrades as the threshold increases.

Figure 9.6(b) shows the BER performance in for a power allocation $P_1 = 0.7P$ and $P_2 = 0.3P$, and channel variances $\sigma_{s,d}^2 = 1$, $\sigma_{s,r}^2 = 1$, and $\sigma_{r,d}^2 = 10$. Obviously, different thresholds result in different BER performances. The threshold of $\zeta = 1.7$ provides the best performance in this scenario. Furthermore, if the threshold is too small, e.g. $\zeta = 1$, not only the BER performance degrades but also the diversity order is less than two. This is because when the threshold is small, the destination tends to combine the signals from both the relay and the destination. As a result, the incorrect decoding at the relay leads to a significant performance degradation at the destination. ▲

Example 9.4 Next we study effect of power allocation on BER performance of the differential DF scheme with a fixed threshold. In Figure 9.7(a), we consider the cooperation system with channel variances $\sigma_{s,d}^2 = \sigma_{s,r}^2 = \sigma_{r,d}^2 = 1$ and a threshold $\zeta = 2$. We can see that the power ratios $r = P_1/P = 0.5, 0.6$, and 0.7 yield almost the same performances. When the power ratio increases to $r = 0.9$, the performance degradation is about 2 dB at a BER of 10^{-4} compared to the equal power allocation scheme. This is due to the fact that at $r = 0.9$, only low power is allocated at the relay. Consequently, even though the received signal from the relay carries information, there is a high chance that its amplitude is smaller than the threshold, and hence the detection is based only on the received signal from the direct link.

Figure 9.7(b) depicts the performance in the case of channel variances $\sigma_{s,d}^2 = 1$, $\sigma_{s,r}^2 = 1$, and $\sigma_{r,d}^2 = 10$, and a threshold $\zeta = 1$. We can see that the performance improves as the power ratio increases from $r = 0.5$ to $r = 0.9$. The reason is that the relay–destination link is of high quality while the threshold is small. With only low power at the relay, the amplitude of the received signal from the relay can be larger than the threshold. Therefore, by allocating more power at the source, we not only

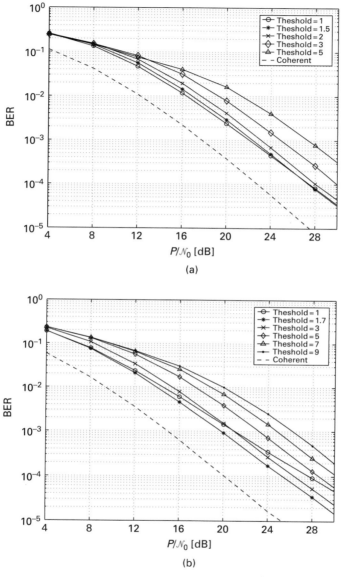

Fig. 9.6   DQPSK: different thresholds with fixed power allocation. (a) $r = 0.7$, $\sigma_{s,d}^2 = \sigma_{s,r}^2 = \sigma_{r,d}^2 = 1$; (b) $r = 0.8$, $\sigma_{s,d}^2 = \sigma_{s,r}^2 = 1$, $\sigma_{r,d}^2 = 10$.

increase the chance of correct decoding at the relay, but also increase the SNR of the MRC output. Based on the numerical results in Figure 9.4(d), the optimum power ratio for this scenario is $r = 0.9$ at an SNR of $P/N_0 = 16\,\text{dB}$. Clearly, the simulation results in Figure 9.7(b) agree with the numerical results at the SNR of 16 dB. Moreover, Figure 9.7(b) illustrates that the power ratio of $r = 0.9$ results in optimum performance

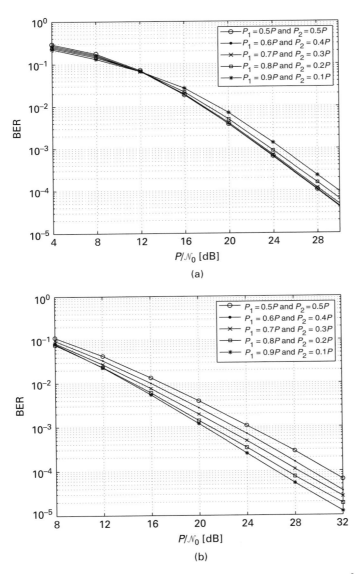

Fig. 9.7 DQPSK: different power allocations with fixed threshold. (a) $\zeta = 2$, $\sigma_{s,d}^2 = \sigma_{s,r}^2 = \sigma_{r,d}^2 = 1$; (b) $\zeta = 1$, $\sigma_{s,d}^2 = \sigma_{s,r}^2 = 1$, $\sigma_{r,d}^2 = 10$.

for the entire SNR range. At the threshold $\zeta = 1$, the differential DF scheme with optimum power allocation achieves an improvement of about 5 dB over that with equal power allocation at a BER of 10^{-4}. ▲

Example 9.5 Here we compare the performances of the differential DF scheme with different power allocation and decision threshold. We consider the case of $\sigma_{s,d}^2 = \sigma_{s,r}^2 =$

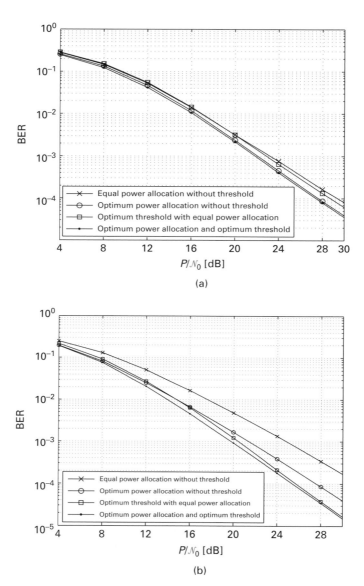

Fig. 9.8   DQPSK: different power allocations and different thresholds. (a) $\sigma_{s,d}^2 = \sigma_{s,r}^2 = \sigma_{r,d}^2 = 1$;
(b) $\sigma_{s,d}^2 = \sigma_{s,r}^2 = 1$, $\sigma_{r,d}^2 = 10$.

$\sigma_{r,d}^2 = 1$ in Figure 9.8(a), and the case of $\sigma_{s,d}^2 = \sigma_{s,r}^2 = 1$, and $\sigma_{r,d}^2 = 10$ in Figure 9.8(b).
From both figures, it is clear that the DiffDF scheme with jointly optimum power allocation and optimum threshold yields the best performance over the entire SNR range. In the case of equal link qualities, i.e., $\sigma_{s,d}^2 = \sigma_{s,r}^2 = \sigma_{r,d}^2 = 1$, optimum power allocation and optimum threshold yields a performance improvement of 2 dB at a BER of 10^{-4} compared to the scheme with equal power allocation and without threshold.

In addition, if the power allocation is optimum, the scheme without threshold yields almost the same performance as that with optimum threshold. When the quality of the relay–destination link is very good, e.g. $\sigma_{r,d}^2 = 10$, the use of optimum threshold is more important than the use of optimum power allocation at high SNR. Specifically, by properly choosing the threshold, the differential DF scheme achieves almost the same performance for any power allocation at high SNR.

As we can see from Figure 9.8(b), in case of equal power allocation, using the optimum threshold leads to an improvement gain of more than 5 dB over the scheme without threshold at a BER of 10^{-4}. With optimum threshold, the performance difference between the differential DF scheme with optimum power allocation and that with equal power allocation is only about 0.5 dB at a BER of 10^{-4}. ▲

9.2.2 DiffDF with multi-relay systems

This section considers DiffDF schemes in multi-node cooperative networks. Due to their low complexity implementations, the multi-node DiffDF scheme can be deployed in sensor and ad-hoc networks in which multi-node signal transmissions are necessary for reliable communications among nodes. In the DiffDF scheme, each relay decodes the received signal and it forwards only correctly decoded symbols to the destination. A number of decision thresholds that correspond to the number of relays are used at the destination to efficiently combines received signals from each relay–destination link with that from the direct link. The BER performance of DiffDF schemes is analyzed and optimum power allocation and threshold is provided to further improve the system performance. In addition, BER approximation and a tractable BER lower bound are provided. Then, power allocation and thresholds are jointly optimized.

9.2.2.1 Signal models for multi-node DiffDF scheme

For the DiffDF scheme, as in Figure 9.9, each relay decodes each received signal and then forwards only correctly decoded symbols to the destination. In order to take advantage of the DiffDF protocol, by which only correctly decoded symbols at each relay are forwarded with a certain amount of power to the destination, a decision threshold (ζ_i) is used at the destination to allow only high potential information bearing signals from each of the i-th relay links to be combined with that from the direct link before being differentially decoded.

With N cooperative relays in the network, signal transmissions comprise $N + 1$ phases. The first phase belongs to direct transmission, and the remaining N phases are for signal transmission for each of the N relays. The signal models for each of the $N+1$ transmission phases are as follows.

In phase 1, suppose that differential M-ary phase shift keying (DMPSK) modulation is used, the modulated information at the source is $v_m = e^{j\phi_m}$, where $\phi_m = 2\pi m/M$ for $m = 0, 1, \ldots, M - 1$, and M is the constellation size. The source differentially encodes v_m by $x^\tau = v_m x^{\tau-1}$, where τ is the time index, and x^τ is the differentially encoded symbol to be transmitted at time τ. After that, the source transmits x^τ with transmitted

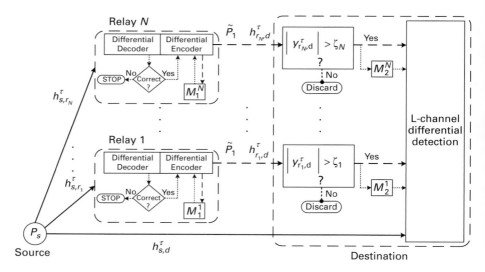

Fig. 9.9   System description of the multi-node differential DF scheme.

power P_s to the destination. Due to the broadcasting nature of the wireless network, the information can also be received by each of the N relays. The corresponding received signals at the destination and the i-th relay, for $i = 1, 2, \ldots, N$, can be expressed as

$$y_{s,d}^\tau = \sqrt{P_s} h_{s,d}^\tau x^\tau + \eta_{s,d}^\tau,$$

and

$$y_{s,r_i}^\tau = \sqrt{P_s} h_{s,r_i}^\tau x^\tau + \eta_{s,r_i}^\tau.$$

where $h_{s,d}^\tau$ and h_{s,r_i}^τ represent the channel coefficients from the source to the destination and from the source to the i-th relay, respectively. The channel coefficients $h_{s,d}^\tau$ and h_{s,r_i}^τ are modeled as complex Gaussian random variables with zero means and variances $\sigma_{s,d}^2$ and σ_{s,r_i}^2, respectively. The terms $\eta_{s,d}^\tau$ and η_{s,r_i}^τ are additive white Gaussian noise at the destination and the i-th relay, respectively. Both of these noise terms are modeled as zero-mean, complex Gaussian random variables with the same variance of $\mathcal{N}_0$.

In phases 2 to $N + 1$, depending on the cooperation protocol under consideration, each of the N relays forwards the decoded signal to the destination. In this cooperation system, each relay forwards only correctly decoded symbols to the destination, i.e., $\tilde{P}_i = P_i$ when the i-th relay decodes correctly, and $\tilde{P}_i = 0$ otherwise. As shown in Figure 9.9, N decision thresholds are used at the destination to allow only high potential information bearing signals from each of the i-th relays to be combined with that from the direct link before being differentially decoded.

Specifically, in phases 2 to $N + 1$, each of the i-th relays differentially decodes the received signal from the source by using the decision rule:

$$\hat{m} = \arg \max_{m=0,1,\ldots,M-1} \mathrm{Re}\{(v_m y_{s,r_i}^{\tau-1})^* y_{s,r_i}^\tau\}.$$

Here, we assume an ideal relay that can make a judgement on the decoded information as to whether it is correct or not. If any of the i-th relays incorrectly decodes, the incorrectly decoded symbol is discarded. Otherwise, the i-th relay differentially re-encodes the information symbol as $\tilde{x}^\tau = v_m \tilde{x}^{\tau - k_i}$, where k_i represents the time index that the i-th relay correctly decodes before time τ. Then, $\tilde{x}^\tau$ is forwarded to the destination with transmitted power $\tilde{P}_i = P_i$. After that, $\tilde{x}^\tau$ is stored in a memory, represented by M_1^i in Figure 9.9, for subsequent differential encoding. Note that the time index $\tau - k_i$ in $\tilde{x}^{\tau - k_i}$ can be any time before time τ depending on the decoding result in the previous time. The received signal at the destination in phases 2 to $N + 1$ can be expressed as

$$
y_{r_i,d}^\tau = \begin{cases} \sqrt{\tilde{P}_i} h_{r_i,d}^\tau \, \tilde{x}^\tau + \eta_{r_i,d}^\tau, & \text{if relay correctly decodes } (\tilde{P}_i = P_i); \\ \eta_{r_i,d}^\tau, & \text{Otherwise } (\tilde{P}_i = 0), \end{cases}
\tag{9.77}
$$

where $i = 1, 2, \ldots, N$, $h_{r_i,d}^\tau$ denotes the channel coefficient between the i-th relay and the destination, and $\eta_{r_i,d}^\tau$ represents an additive noise.

Since the perfect knowledge of CSI is not available at each time instant, the destination does not know when the received signal from the i-th relay contains the information. For each i-th relay–destination link, a decision threshold ζ_i is used at the destination to make a decision as to whether to combine $y_{r_i,d}^\tau$ with the received signal from the direct link. Specifically, if $|y_{r_i,d}^\tau| \le \zeta_i$ for all i, where $|x|$ denotes the absolute value of x, the destination estimates the transmitted symbol based only on the received signal from the direct link. However, if $|y_{r_i,d}^\tau| > \zeta_i$ for any i, the received signals from the source and from the i-th relay are combined for joint decoding. In this way, the combined signal at the destination can be written as

$$
y^{\text{DF}} = w_s^{\text{DF}} (y_{s,d}^{\tau-1})^* y_{s,d}^\tau + \sum_{i=1}^{N} w_i^{\text{DF}} I_{\zeta_i}[|y_{r_i,d}^\tau|] (y_{r_i,d}^{\tau-l_i})^* y_{r_i,d}^\tau.
\tag{9.78}
$$

where w_s^{DF} and w_i^{DF} are combining weights. In (9.78), $y_{r_i,d}^{\tau-l_i}$ ($l_i \ge 1$) is the most recent received signal from the i-th relay with $|y_{r_i,d}^{\tau-l_i}| > \zeta_i$. It is stored in a memory, represented by M_2^i in Figure 9.9, at the destination. The function $I_{\zeta_i}[|y_{r_i,d}^\tau|]$ in (9.78) represents an indicator function in which

$$
I_{\zeta_i}[|y_{r_i,d}^\tau|] = \begin{cases} 1, & \text{if } |y_{r_i,d}^\tau| > \zeta_i; \\ 0, & \text{otherwise.} \end{cases}
$$

After signal combining, the destination jointly differentially decodes the transmitted information by

$$
\hat{m} = \arg \max_{m=0,1,\ldots,M-1} \text{Re}\left\{ v_m^* y^{\text{DF}} \right\}.
$$

Note that using different combining weights (w_s^{DF} and w_i^{DF}) results in different system performances. In the following, we use $w_s^{\text{DF}} = w_i^{\text{DF}} = 1/(2N_0)$ which maximizes the signal-to-noise ratio (SNR) at the combiner output.

9.2.2.2   BER Analysis

BER analysis of the multi-node DiffDF scheme is considered in this subsection. First, different SNR scenarios are characterized according to the received signal $y^\tau_{r_i,d}$, threshold ζ_i, and memories M^i_1 and M^i_2. Then, the probability of occurrence is provided for each of these SNR scenarios. After that, the average BER is derived based on the probability of occurrence and the combined SNR for each scenario. Finally, a tractable BER lower bound is provided.

Characterization of different SNR scenarios

At the destination, different combined SNRs may occur based on the received signal $(y^\tau_{r_i,d})$, the threshold (ζ_i), and the signals stored in memory M^i_1 and M^i_2. In this way, the destination encounters six possible SNR scenarios at each relay–destination link, and we characterize each of them as follows. For a given network state j, we denote s^i_j as an integer number that represents an SNR scenario at the i-th relay–destination link, i.e., $s^i_j \in \{1, 2, 3, 4, 5, 6\}$. A set of joint events $\Phi^i_{s^i_j}$ for $i = 1, 2, \ldots, N$ will be related to each of the scenarios s^i_j. Specifically, when $s^i_j = 1$,

$$\Phi^i_1 \triangleq \left\{ |y^\tau_{r_i,d}| \le \zeta_i \right\}$$

represents a joint event that received signals from the i-th relay link are not greater than the thresholds. We characterize

$$\Phi^i_2 \triangleq \left\{ |y^\tau_{r_i,d}| > \zeta_i, \tilde{P}^\tau_i = P_i, \tilde{P}^{\tau-l_i}_i = P_i, l_i = k_i \right\}$$

as a joint event including $|y^\tau_{r_i,d}| > \zeta_i$, the relay correctly decodes at time τ and $\tau - l_i$, and the information symbols at time k_i and l_i in memories M^i_1 and M^i_2 are the same. The remaining scenarios are

$$\Phi^i_3 \triangleq \left\{ |y^\tau_{r_i,d}| > \zeta_i, \tilde{P}^\tau_i = P_i, \tilde{P}^{\tau-l_i}_i = P_i, l_i \ne k_i \right\},$$

$$\Phi^i_4 \triangleq \left\{ |y^\tau_{r_i,d}| > \zeta_i, \tilde{P}^\tau_i = P_i, \tilde{P}^{\tau-l_i}_i = 0 \right\},$$

$$\Phi^i_5 \triangleq \left\{ |y^\tau_{r_i,d}| > \zeta_i, \tilde{P}^\tau_i = 0, \tilde{P}^{\tau-l_i}_i = P_i \right\},$$

$$\Phi^i_6 \triangleq \left\{ |y^\tau_{r_i,d}| > \zeta_i, \tilde{P}^\tau_i = 0, \tilde{P}^{\tau-l_i}_i = 0 \right\}.$$

They are interpreted in a similar way to Φ^i_2.

Probability of occurrence for each SNR scenario

To determine the probability of occurrence for each scenario, we first find that the probability that the i-th relay forwards information with transmitted power $\tilde{P}_i = 0$ due to incorrect decoding is related to the symbol error rate of DMPSK modulation as

$$\Psi(\gamma^{DF}_{s,r_i}) = \frac{1}{\pi} \int_0^{(M-1)\pi/M} \exp\left[-g(\phi)\gamma^{DF}_{s,r_i}\right] d\phi, \tag{9.79}$$

where $\gamma^{DF}_{s,r_i} = P_s|h^\tau_{s,r_i}|^2/N_0$ represents an instantaneous SNR at the i-th relay, and $g(\phi)$ is given by (9.20). Accordingly, the probability of correct decoding at the i-th relay (or probability of forwarding with transmitted power $\tilde{P}_i = P_i$) is $1 - \Psi(\gamma^{DF}_{s,r_i})$. Therefore, the chance that Φ_1^i occurs is determined by the weighted sum of conditional probabilities given that $\tilde{P}_i = P_i$ or 0, so that

$$
\begin{aligned}
P_r^{h,\mathrm{DF}}(\Phi_1^i) &= P_r^{h,\mathrm{DF}}\left(|y_{r_i,d}^\tau| \leq \zeta_i \mid \tilde{P}_i^\tau = 0\right)\Psi(\gamma_{s,r_i}^\tau) \\
&\quad + P_r^{h,\mathrm{DF}}\left(|y_{r_i,d}^\tau| \leq \zeta_i \mid \tilde{P}_i^\tau = P_i\right)\left[1 - \Psi(\gamma_{s,r_i}^\tau)\right] \\
&= \left(1 - \exp(-\zeta_i^2/N_0)\right)\Psi(\gamma_{s,r_i}^\tau) + \left(1 - \mathcal{M}\left(P_i|h_{r_i,d}^\tau|^2, \zeta_i\right)\right)\left[1 - \Psi(\gamma_{s,r_i}^\tau)\right],
\end{aligned}
\tag{9.80}
$$

where

$$
\mathcal{M}\left(P_i|h_{r_i,d}^\tau|^2, \zeta_i\right) \triangleq Q_1\left(\sqrt{P_i|h_{r_i,d}^\tau|^2/(N_0/2)}, \zeta_i/\sqrt{N_0/2}\right),
\tag{9.81}
$$

in which $Q_1(\alpha, \beta)$ is the Marcum Q-function [188](proof left as an exercise).

According to the definition of each SNR scenario in Section 9.2.2.2, the chance that each of the scenarios Φ_2 to Φ_6 happens is conditioned on an event that $|y_{r_i,d}^{\tau-l_i}| > \zeta_i$. Since the events at time $\tau - l_i$ and time τ are independent, then the probability that Φ_2^i occurs is given by

$$
\begin{aligned}
P_r^{h,\mathrm{DF}}(\Phi_2^i) &= P_r^{h,\mathrm{DF}}\left(|y_{r_i,d}^\tau| > \zeta_i, \tilde{P}_i^\tau = P_i\right) P_r^{h,\mathrm{DF}}\left(\tilde{P}_i^{\tau-l_i} = P_i, l_i = k_i \,\Big|\, |y_{r_i,d}^{\tau-l_i}| > \zeta_i\right). \\
&\approx \frac{\mathcal{M}^2\left(P_i|h_{r_i,d}^\tau|^2, \zeta_i\right)\left(1 - \Psi(\gamma_{s,r_i}^\tau)\right)^2}{1 - (1 - e^{-\zeta_i^2/N_0})\Psi(\gamma_{s,r_i}^\tau)}.
\end{aligned}
\tag{9.82}
$$

Next, the chance that the scenario Φ_3^i happens can be written as

$$
P_r^{h,\mathrm{DF}}(\Phi_3^i) = P_r^{h,\mathrm{DF}}(\Phi_2^i \cup \Phi_3^i) - P_r^{h,\mathrm{DF}}(\Phi_2^i),
\tag{9.83}
$$

where

$$
\begin{aligned}
P_r^{h,\mathrm{DF}}(\Phi_2^i \cup \Phi_3^i) &\triangleq P_r^{h,\mathrm{DF}}(|y_{r_i,d}^\tau| > \zeta_i, \tilde{P}_i^\tau = P_i, \tilde{P}_i^{\tau-l_i} = P_i, ||y_{r_i,d}^{\tau-l_i}| > \zeta_i) \\
&= P_r^{h,\mathrm{DF}}\left(|y_{r_i,d}^\tau| > \zeta_i, \tilde{P}_i^\tau = P_i\right) \\
&\quad \times \frac{P_r^{h,\mathrm{DF}}\left(|y_{r_i,d}^{\tau-l_i}| > \zeta_i, \tilde{P}_i^{\tau-l_i} = P_i\right)}{P_r^{h,\mathrm{DF}}\left(|y_{r_i,d}^{\tau-l_i}| > \zeta_i\right)},
\end{aligned}
\tag{9.84}
$$

Substituting (9.82) and (9.84) into (9.83), after some manipulations, we have

$$
\begin{aligned}
P_r^{h,\mathrm{DF}}(\Phi_3^i) &= \mathcal{M}^2\left(P_i|h_{r_i,d}^\tau|^2, \zeta_i\right)\left(1 - \Psi(\gamma_{s,r_i}^\tau)\right)^2 \\
&\quad \times \left(\frac{1}{\Gamma(P_s|h_{s,r_i}^\tau|^2, P_i|h_{r_i,d}^\tau|^2)} - \frac{1}{\left(1 - (1 - e^{-\zeta_i^2/N_0})\Psi(\gamma_{s,r_i}^\tau)\right)}\right),
\end{aligned}
\tag{9.85}
$$

in which $\Gamma(P_s|h_{s,r_i}^{\tau-l_i}|^2, P_i|h_{r_i,d}^{\tau-l_i}|^2)$ is defined as an expression that results from applying the concept of total probability[195] to $P_r^{h,\mathrm{DF}}\left(|y_{r_i,d}^{\tau-l_i}| > \zeta_i\right)$:

$$
\begin{aligned}
P_r^{h,\mathrm{DF}}\left(|y_{r_i,d}^{\tau-l_i}| > \zeta_i\right) &= \mathcal{M}(P_i|h_{r_i,d}^{\tau-l_i}|^2, \zeta_i)\left(1 - \Psi(\gamma_{s,r_i}^{\tau-l_i})\right) + e^{-\zeta_i^2/N_0}\Psi(\gamma_{s,r_i}^{\tau-l_i}) \\
&\triangleq \Gamma(P_s|h_{s,r_i}^{\tau-l_i}|^2, P_i|h_{r_i,d}^{\tau-l_i}|^2).
\end{aligned}
\tag{9.86}
$$

With the assumption of almost constant channels at time τ and $\tau - l_i$, we have $P_r^{h,\mathrm{DF}}(\Phi_4^i) = P_r^{h,\mathrm{DF}}(\Phi_5^i)$, i.e., scenarios Φ_4^i and Φ_5^i occur with the same probability. Following the calculation steps used in (9.84), we have

$$P_r^{h,\mathrm{DF}}(\Phi_4^i) = P_r^{h,\mathrm{DF}}(\Phi_5^i) = \frac{\mathcal{M}\left(P_i|h_{r_i,d}^\tau|^2, \zeta_i\right)e^{-\zeta_i^2/N_0}\Psi(\gamma_{s,r_i}^\tau)\left(1 - \Psi(\gamma_{s,r_i}^\tau)\right)}{\Gamma(P_s|h_{s,r_i}^\tau|^2, P_i|h_{r_i,d}^\tau|^2)}. \quad (9.87)$$

Finally, the chance that scenario Φ_6^i occurs can be determined as

$$P_r^{h,\mathrm{DF}}(\Phi_6^i) = \frac{P_r^{h,\mathrm{DF}}\left(|y_{r_i,d}^\tau| > \zeta_i, \tilde{P}_i^\tau = 0\right)P_r^{h,\mathrm{DF}}\left(|y_{r_i,d}^{\tau-l_i}| > \zeta_i, \tilde{P}_i^{\tau-l_i} = 0\right)}{P_r^{h,\mathrm{DF}}\left(|y_{r_i,d}^{\tau-l_i}| > \zeta_i\right)}$$

$$= e^{-2\zeta_i^2/N_0}\Psi(\gamma_{s,r_i}^\tau)\left(\frac{1}{\Gamma(P_s|h_{s,r_i}^\tau|^2, P_i|h_{r_i,d}^\tau|^2)}\right). \quad (9.88)$$

Approximated BER expression for the DiffDF scheme

We know from Section 9.2.2.2 that each relay contributes six possible SNR scenarios at the destination. For a network with N relays, there are a total of 6^N numbers of network states. We denote $S_j \triangleq [s_j^1 \ s_j^2 \ \cdots \ s_j^N]$ as an $1 \times N$ matrix of a network state j, where $s_j^i \in \{1, 2, \ldots, 6\}$. Accordingly, the average BER can be expressed as

$$P_b^{\mathrm{DF}} = \mathrm{E}\left[P_b^{h,\mathrm{DF}}\right] = \sum_{j=1}^{6^N}\mathrm{E}\left[\left(P_b^{h,\mathrm{DF}}|_{S_j}\right)\prod_{i=1}^{N}P_r^{h,\mathrm{DF}}(\Phi_{s_j^i}^i)\right], \quad (9.89)$$

where $P_r^{h,\mathrm{DF}}(\Phi_{s_j^i}^i)$ for each s_j^i is specified in (9.80)–(9.88), $P_b^{h,\mathrm{DF}}|_{S_j}$ represents a conditional BER for a given S_j, and $\mathrm{E}[\cdot]$ denotes the expectation operator.

Since it is difficult to find a closed-form solution for the BER in (9.89), we further simplify (9.89) by separating a set of all possible network states, denoted by $\mathbb{S}$, into two disjoint subsets as $\mathbb{S} = \mathbb{S}^{1,2} \cup (\mathbb{S}^{1,2})^c$, where $\mathbb{S}^{1,2}$ denotes all possible network states that every element in the network state S_j is either one or two, and $(\mathbb{S}^{1,2})^c$ denotes the remaining possible network states. Note that the cardinality of $\mathbb{S}^{1,2}$ and $(\mathbb{S}^{1,2})^c$ are $|\mathbb{S}^{1,2}| = 2^N$ and $|(\mathbb{S}^{1,2})^c| = 6^N - 2^N$, respectively. In this way, we can express the average BER (9.89) as

$$P_b^{\mathrm{DF}} = \sum_{j=1}^{2^N}\mathrm{E}\left[\left(P_b^{h,\mathrm{DF}}|_{S_j \in \mathbb{S}^{1,2}}\right)\prod_{i=1}^{N}P_r^{h,\mathrm{DF}}(s_j^i)\right]$$

$$+ \sum_{j=1}^{6^N-2^N}\mathrm{E}\left[\left(P_b^{h,\mathrm{DF}}|_{S_j \in (\mathbb{S}^{1,2})^c}\right)\prod_{i=1}^{N}P_r^{h,\mathrm{DF}}(s_j^i)\right],$$

$$\triangleq P_{b,1}^{\mathrm{DF}} + P_{b,2}^{\mathrm{DF}}. \quad (9.90)$$

The first term in the right-hand side of (9.90), $P_{b,1}^{\mathrm{DF}}$, results from the cases where every element in the network state S_j is either one or two; the second term, $P_{b,2}^{\mathrm{DF}}$, results from the remaining cases. These two terms can be determined as follows.

First, for notational convenience, let us denote $L(S_j)$ as the number of combining branches. By definition, we can express $L(S_j)$ as

$$L(S_j) = \sum_{i=1}^{N} \hat{L}(s_j^i) + 1, \tag{9.91}$$

where $\hat{L}(s_j^i) = 0$ when $s_j^i = 1$, and $\hat{L}(s_j^i) = 1$ otherwise. Note that the addition of 1 in (9.91) corresponds to the contribution of the signal from the direct link.

Next, consider the case that every element in the network state S_j is either one or two. Therefore, the conditional BER $P_b^{h,DF}|_{S_j \in \mathbb{S}^{1,2}}$ can be obtained from the multi-branch differential detection of DMPSK signals as

$$P_b^{h,DF}\Big|_{S_j \in \mathbb{S}^{1,2}} = \frac{1}{2^{2L(S_j)}\pi} \int_{-\pi}^{\pi} f\left(\theta, \beta, L(S_j)\right) \exp\left[-\alpha(\theta)\gamma_{S_j \in \mathbb{S}^{1,2}}^{DF}\right] d\theta$$

$$\triangleq \Lambda(\gamma_{S_j \in \mathbb{S}^{1,2}}^{DF}), \tag{9.92}$$

in which $\alpha(\theta)$ and $f(\theta, \beta, L(S_j))$ are specified in (9.51) and (9.140), respectively. The term $\gamma_{S_j \in \mathbb{S}^{1,2}}^{DF}$ is the SNR at the combined output, which is given by

$$\gamma_{S_j \in \mathbb{S}^{1,2}}^{DF} = \frac{P_s |h_{s,d}^{\tau}|^2}{N_0} + \sum_{i=1}^{N} \frac{(s_j^i - 1)P_i |h_{r_i,d}^{\tau}|^2}{N_0}, \qquad s_j^i \in \{1, 2\}, \ \forall i. \tag{9.93}$$

Then, the conditional BER $P_b^{h,DF}|_{S_j \in (\mathbb{S}^{1,2})^c}$ for the remaining cases can be found as follows. Since up to now the conditional BER formulation for DMPSK with arbitrary-weighted combining has not been available in the literature, $P_b^{h,DF}|_{S_j \in (\mathbb{S}^{1,2})^c}$ cannot be exactly determined. For analytical tractability of the analysis, we resort to an approximated BER, in which the signal from the relay i is considered as noise when any scenario from Φ_3^i to Φ_6^i occurs. As we will show in the following section, the analytical BER obtained from this approximation is close to the simulation results. The conditional BER for these cases can be approximated as

$$P_b^{h,DF}\Big|_{S_j \in (\mathbb{S}^{1,2})^c} \approx \Lambda\left(\gamma_{S_j \in (\mathbb{S}^{1,2})^c}^{DF}\right),$$

where

$$\gamma_{S_j \in (\mathbb{S}^{1,2})^c}^{DF} = \frac{P_s |h_{s,d}^{\tau}|^2 + \sum_{i=1}^{N} I_2[s_j^i] P_i |h_{r_i,d}^{\tau}|^2}{N_0 + \hat{N}_0}, \qquad s_j^i \in \{1, 2, \ldots, 6\}, \ \forall i, \tag{9.94}$$

in which

$$\hat{N}_0 \triangleq \left(\sum_{i=1}^{N} (1 - I_2[s_j^i]) \mathcal{N}_{s_j^i}\right) \Bigg/ \left(P_s |h_{s,d}^{\tau}|^2/N_0 + \sum_{i=1}^{N} I[s_j^i] P_i |h_{r_i,d}^{\tau}|^2 N_0\right),$$

and $\mathcal{N}_{s_j^i}$ depends on s_j^i as follows:

$$\mathcal{N}_{s_j^i} = \begin{cases} (P_i |h_{r_i,d}^{\tau}|^2 + N_0)^2/N_0, & \text{when } s_j^i = 2; \\ P_i |h_{r_i,d}^{\tau}|^2 + N_0, & \text{when } s_j^i = 3, 4, 5; \\ N_0, & \text{when } s_j^i = 6. \end{cases}$$

In (9.94), $I_2[s_j^i]$ is defined as an indicator function based on the occurrence of s_j^i such that $I_2[s_j^i] = 1$ when $s_j^i = 2$, and $I_2[s_j^i] = 0$ when $s_j^i = 1, 3, 4, 5, 6$.

From the above results, $P_{b,2}^{DF}$ in (9.90) can be approximated as (proof left as an exercise)

$$P_{b,2}^{DF}$$
$$\approx \sum_{j=1}^{6^N - 2^N} \frac{1}{4L(S_j)\pi} \int_{-\pi}^{\pi} f\left(\theta, \beta, L(S_j)\right) \mathrm{E}\left[e^{\left(-\alpha(\theta)\gamma_{S_j \in (\mathbb{S}1,2)^c}^{DF}\right)} \prod_{i=1}^{N} P_r^{h,DF}(s_j^i)\right]d\theta, \quad (9.95)$$

where

$$\mathrm{E}\left[e^{\left(-\alpha(\theta)\gamma_{S_j \in (\mathbb{S}1,2)^c}^{DF}\right)} \prod_{i=1}^{N} P_r^{h,DF}(s_j^i)\right]$$
$$= \underbrace{\int \cdots \int}_{2N+1 \text{ folds}} \exp\left(-\alpha(\theta)\gamma_{S_j \in (\mathbb{S}1,2)^c}^{DF}\right)$$
$$\times \prod_{i=1}^{N} P_r^{h,DF}(s_j^i) f(\varepsilon_1) f(\varepsilon_2) \cdots f(\varepsilon_{2N+1}) d\varepsilon_1 d\varepsilon_2 \cdots d\varepsilon_{2N+1}, \quad (9.96)$$

in which $\gamma_{S_j \in (\mathbb{S}1,2)^c}^{DF}$ is given in (9.94), and $\prod_{i=1}^{N} P_r^{h,DF}(s_j^i)$ is calculated using (9.80)–(9.88). We can see from (9.95) that the evaluation of $P_{b,2}^{DF}$ involves at most $(2N+2)$-fold integration. Although $P_{b,2}^{DF}$ can be numerically determined, the calculation time is prohibitively long even for a cooperation system with a small number of relays.

Now we determine $P_{b,1}^{DF}$ in (9.90) as follows. The term $\prod_{i=1}^{N} P_r^{h,DF}(s_j^i)$ is a product of the probabilities of occurrence of scenarios 1 and 2, and can be expressed as

$$\prod_{i=1}^{N} P_r^{h,DF}(s_j^i) = \prod_{i=1}^{N}\left[(2 - s_j^i)P_r^{h,DF}(\Phi_1^i) + (s_j^i - 1)P_r^{h,DF}(\Phi_2^i)\right]. \quad (9.97)$$

Substitute (9.92), (9.93), and (9.97) into the expression of $P_{b,1}^{DF}$ in (9.90), and then average over all CSIs, resulting in

$$P_{b,1}^{DF} \approx \sum_{j=1}^{2^N}\left(\frac{1}{2^{2L(S_j)}\pi}\right)\int_{-\pi}^{\pi}\left(\frac{f(\theta, \beta, L(S_j))}{1 + \alpha(\theta)P_s\sigma_{s,d}^2/N_0}\right)\prod_{i=1}^{N}\left[(2 - s_j^i)X + (s_j^i - 1)Y\right]d\theta, \quad (9.98)$$

where we denote $X \triangleq \mathrm{E}\left[P_r^{h,DF}(\Phi_1^i)\right]$, which can be determined as

$$X = (1 - e^{-\xi_i^2/N_0})G\left(1 + \frac{g(\phi)P_s\sigma_{s,r_i}^2}{N_0}\right)$$
$$+ \left(1 - \int_0^{\infty}\frac{M(P_iq, \xi_i)}{\sigma_{r_i,d}^2}e^{-q/\sigma_{r_i,d}^2}dq\right)\left(1 - G(1 + \frac{g(\phi)P_s\sigma_{s,r_i}^2}{N_0})\right), \quad (9.99)$$

in which

$$G(c(\phi)) \triangleq \frac{1}{\pi} \int_0^{(M-1)\pi/M} [c(\phi)]^{-1} d\phi, \tag{9.100}$$

and

$$Y \triangleq E\left[\exp(-\alpha(\theta) P_i |h_{r_i,d}^\tau|^2 / \mathcal{N}_0) \cdot P_r^{h,\mathrm{DF}}(\Phi_2^i)\right] \frac{1}{\pi} \int_0^{(M-1)\pi/M} [c(\phi)]^{-1} d\phi, \tag{9.101}$$

which can be further approximated as

$$Y \approx \frac{1}{\sigma_{s,r_i}^2 \sigma_{r_i,d}^2} \int_0^\infty \int_0^\infty \frac{\mathcal{M}^2(P_i q, \zeta_i)(1 - \Psi(P_s u / \mathcal{N}_0))^2}{1 - (1 - e^{-\zeta_i^2 / \mathcal{N}_0})\Psi(P_s u / \mathcal{N}_0)}$$
$$\times e^{-(\alpha(\theta)\frac{P_i}{\mathcal{N}_0} + \frac{1}{\sigma_{r_i,d}^2})q} e^{-u/\sigma_{s,r_i}^2} dq \, du. \tag{9.102}$$

Substituting (9.95) and (9.98) into (9.90), we finally obtain the average BER of the multi-node DiffDF scheme.

To gain a more insightful understanding, we further determine a BER lower bound of the multi-node DiffDF scheme as follows. Since the exact BER formulations under the scenarios Φ_4^i, Φ_5^i, and Φ_6^i are currently unavailable, and the chances that these three scenarios happen are small at high SNR, we lower bound the BER from these scenarios by zero. Also, we lower bound the BER under the scenario Φ_3^i by that under Φ_2^i; this allows us to express the lower bound in terms of $P_r^h(\Phi_2^i \cup \Phi_3^i)$ (instead of $P_r^h(\Phi_3^i)$ or $P_r^h(\Phi_2^i)$) which can be obtained without any approximation. In this way, the BER of the multi-node DiffDF scheme can be lower bounded by

$$P_b^{lb,\mathrm{DF}}$$
$$\approx \sum_{j=1}^{2^N} \left(\frac{1}{2^{2L(S_j)}\pi}\right) \int_{-\pi}^{\pi} \left(\frac{f(\theta, \beta, L(S_j))}{1 + \alpha(\theta) P_s \sigma_{s,d}^2 / \mathcal{N}_0}\right) \prod_{i=1}^{N} \left[(2 - s_j^i)X + (s_j^i - 1)\hat{Y}\right] d\theta, \tag{9.103}$$

where X is given in (9.99) and

$$\hat{Y} \triangleq E\left[\exp(-\alpha(\theta) P_i |h_{r_i,d}^\tau|^2 / \mathcal{N}_0) \cdot P_r^{h,\mathrm{DF}}(\Phi_2^i \cup \Phi_3^i)\right]$$
$$= \frac{1}{\sigma_{s,r_i}^2} \int_0^\infty s(u, \theta) e^{-u/\sigma_{s,r_i}^2} du,$$

in which

$$s(u, \theta) \triangleq \int_0^\infty \frac{\mathcal{M}^2(P_i q, \zeta_i)(1 - \Psi(P_s u/\mathcal{N}_0))^2}{\sigma_{r_i,d}^2 \Gamma(P_s u, P_i q)} e^{-\left(\alpha(\theta)\frac{P_i}{\mathcal{N}_0} + \frac{1}{\sigma_{r_i,d}^2}\right)q} dq. \qquad (9.104)$$

We will show through numerical evaluation that the BER lower bound (9.103) is very close to the simulated performance.

9.2.2.3   Optimizing power allocation and thresholds

This subsection considers the performance improvement of the DiffDF scheme through the joint optimization of power allocation and thresholds based on the BER lower bound (9.103). Specifically, for a fixed total power $P = P_s + \sum_{i=1}^N P_i$, we jointly optimize the threshold ζ_i, the power allocation at the source $a_s = P_s/P$, and the power allocation at each of the i-th relays $a_i = P_i/P$ with an objective to minimize the BER lower bound (9.103):

$$(\{\hat{\zeta}_i\}_{i=1}^N, \hat{a}_s, \{\hat{a}_i\}_{i=1}^N) = \arg \min_{\{\zeta_i\}_{i=1}^N, a_s, \{a_i\}_{i=1}^N} P_b^{lb,\text{DF}}(\{\zeta_i\}_{i=1}^N, a_s, \{a_i\}_{i=1}^N), \qquad (9.105)$$

where $P_b^{lb,\text{DF}}(\{\zeta_i\}_{i=1}^N, a_s, \{a_i\}_{i=1}^N)$ results from substituting $P_s = a_s P$ and $P_i = a_i P$ into (9.103). However, joint optimization in (9.105) involves $2N + 1$ dimensional searching, which includes $N + 1$ power allocation ratios and N decision thresholds. To make the optimization problem tractable and to gain some insights into the optimum power allocation and the optimum thresholds, each relay is assumed to be allocated with the same transmitted power, and the decision thresholds are assumed to be the same at the destination for each relay–destination link. Accordingly, the source is allocated with power $a_s = P_s/P$ and every relay is allocated with power $a_i = (1 - a_s)/N$. Hence, the search space for this optimization problem reduces to two-dimensional searching over a_s and ζ:

$$(\zeta^*, a_s^*) = \arg \min_{\zeta, a_s} P_b^{lb,\text{DF}}(\zeta, a_s, \{a_i\}_{i=1}^N), \qquad (9.106)$$

where $P_b^{lb,\text{DF}}(\zeta, a_s, \{a_i\}_{i=1}^N)$ results from substituting $\zeta_i = \zeta$, $P_s = a_s P$, and $P_i = (1 - a_s)P/N$ into (9.103).

Table 9.1 summarizes theobtained power allocation and thresholds based on the optimization problem (9.106). The DBPSK and DQPSK cooperation systems with two relays are considered, and different channel variances are used to investigate power allocation and thresholds for different cooperation network setups. From the results in Table 9.1, even though the obtained power allocation is sub-optimum, it provides some insights into how much power should be allocated to improve system performance. In particular, as the channel quality of the relay–destination links increases, the threshold should be increased and more power should be allocated at the source to maintain link reliability.

For example, if all the channel links are of the same quality, about half of the transmitted power should be allocated at the source and the optimum threshold is 0.4. On the other hand, if the channel link between each relay and the destination is very good, then the optimum power allocation at the source increases to about 70% of the transmitted

Table 9.1 DiffDF: optimum power allocation and thresholds for a cooperation system with two relays.

$\left[\sigma_{s,d}^2, \sigma_{s,r_i}^2, \sigma_{r_i,d}^2\right]$	DBPSK	DQPSK
	$[a_s, a_1, a_2, \zeta]$	$[a_s, a_1, a_2, \zeta]$
$[1, 1, 1]$	$[0.50, 0.25, 0.25, 0.4]$	$[0.52, 0.24, 0.24, 0.4]$
$[1, 10, 1]$	$[0.44, 0.28, 0.28, 0.4]$	$[0.40, 0.30, 0.30, 0.4]$
$[1, 1, 10]$	$[0.68, 0.16, 0.16, 1.6]$	$[0.70, 0.15, 0.15, 1.8]$

power, and the optimum threshold increases to 1.6 and 1.8 for DBPSK and DQPSK modulations, respectively.

9.2.2.4 Examples for the multi-node DiffDF scheme

Let us see some simulation performances of the multi-node DiffDF scheme with DBPSK and DQPSK modulations. Let us consider the scenarios where two or three relays ($N = 2$ or 3) are in the networks. The channel coefficients follow the Jakes' model (see Section 1.1.5) with Doppler frequency $f_D = 75\,\text{Hz}$ and normalized fading parameter $f_D T_s = 0.0025$, where T_s is the sampling period. The noise variance is assumed to be one ($N_0 = 1$). The average BER curves are plotted as functions of P/N_0.

Example 9.6 We first compare the BER lower bound with the simulated performance. Let us consider a DQPSK cooperation system with two relays. All nodes are allocated with equal power. The decision threshold is set at $\zeta = 1$ and the channel variances are $\sigma_{s,d}^2 = \sigma_{s,r_i}^2 = \sigma_{r_i,d}^2 = 1$ for all i. We can see from Figure 9.10(a) that the BER lower bound yields the same diversity order as that from the simulated performance even though there is a performance gap of 2 dB between these two curves. Also in the figure, the performance of the multi-node DiffDF scheme is 5 dB away from the performance with coherent detection at a BER of 10^{-3}. An interesting observation is that when the transmitted powers are optimally allocated ($P_s = 0.6P$, $P_i = 0.2P$) at a fixed threshold of $\zeta = 1$, as shown in Figure 9.10(b), the performance gap between the simulated performance and the BER lower bound is reduced to about 1 dB at a BER of 10^{-3}.

In addition, the performance of the DiffDF scheme is closer to the performance of coherent detection scheme. From the two figures, we can see that the performance gap between the differential detection and coherent detection is larger than 3 dB. The reason is that the power allocation and the combining weights used in the proposedmulti-node DiffDF scheme is not optimum, hence the maximum SNR of the combined signal cannot be achieved. However, when we discard the scenarios that the signals from the relays do not contain any information, the resulting BER lower bound is about 3 dB away from the performance with coherent detection. ▲

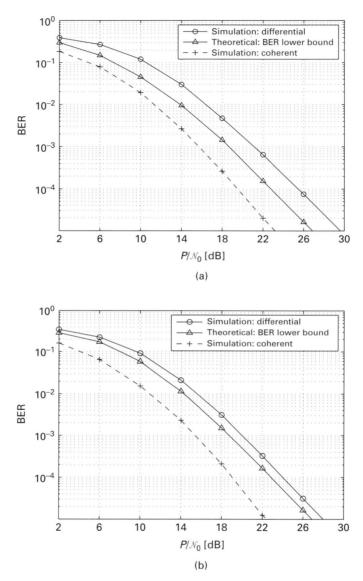

Fig. 9.10   DiffDF scheme with DQPSK: $\sigma_{s,d}^2 = \sigma_{s,r_i}^2 = \sigma_{r_i,d}^2 = 1$, and $\zeta = 1$: (a) equal power allocation and (b) optimum power allocation.

Example 9.7   Next we show the performance of the DiffDF scheme with DQPSK modulation for different number of relays. The channel variances are $\sigma_{s,d}^2 = \sigma_{s,r_i}^2 = \sigma_{r_i,d}^2 = 1$ for all i. All nodes are allocate with equal power, and the threshold at the destination is fixed at $\zeta = 1$. We can see from

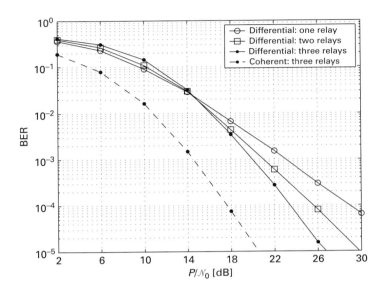

Fig. 9.11 DiffDF with DQPSK : different number of relays, equal power allocation, threshold $= 1$, and $\sigma^2_{s,d} = \sigma^2_{s,r_i} = \sigma^2_{r_i,d} = 1$.

Figure 9.11 that diversity order increases when higher numbers of relays are used. We observe a performance improvement of about 3.5 dB at a BER of 10^{-4} when the number of relays increases from one to two relays. An additional 2 dB gain at the same BER is obtained when the system increases from two to three relays. We also observe a performance gap of about 5.5 dB at a BER of 10^{-3} between the DiffDF scheme and its coherent counterpart for a cooperation system with three relays. ▲

Example 9.8 Figure 9.12(a) shows the effect of using different thresholds on the performance of the multi-node DiffDF scheme, by considering a DBPSK cooperation system with three relays, and all nodes are allocated with equal power. The channel variances are $\sigma^2_{s,d} = \sigma^2_{s,r_i} = 1$, and $\sigma^2_{r_i,d} = 10$ for all i. Clearly, different thresholds result in different performance. Specifically, the multi-node DiffDF scheme with $\zeta = 2$ provides the best performance under this simulation scenario. When $\zeta = 1$, not only BER deteriorates but also the diversity order reduces. Hence, an appropriate decision threshold should be employed such that the DiffDF scheme yields reasonable good performance. Comparing the simulated performance when $\zeta = 2$ with the coherent cooperative scheme without threshold, we observe a performance gap of about 6 dB between the two performance curves at a BER of 10^{-3}. This performance gap is large because in the DiffDF scheme, the CSIs are not available at the receivers, and the destination does not know whether the relay transmits or not. ▲

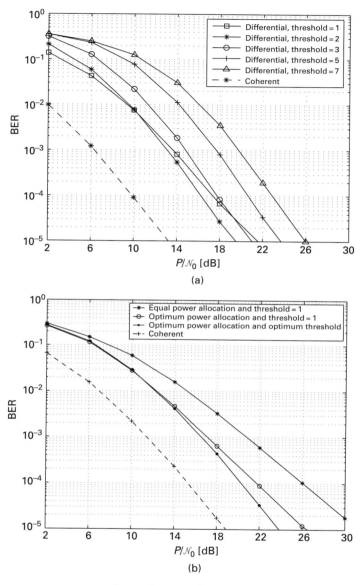

Fig. 9.12   DiffDF scheme with $\sigma_{s,d}^2 = \sigma_{s,r_i}^2 = 1, \sigma_{r_i,d}^2 = 10$: (a) DBPSK: three relays, fixed power allocation but different thresholds and (b) DQPSK: two relays, different power allocation and thresholds.

Example 9.9   Finally, Figure 9.12(b) shows the performance improvement when power allocation and decision thresholds are jointly optimized. It assumes a DQPSK cooperation system with two relays. The channel variances are $\sigma_{s,d}^2 = \sigma_{s,r_i}^2 = 1$, and $\sigma_{r_i,d}^2 = 10$ for all i. In this scenario, the optimum power allocation is $a_s = 0.70$, $a_1 = 0.15$, and

$a_2 = 0.15$, and the optimum threshold is $\zeta = 1.8$. We can see that the performance curve with optimum power allocation and threshold significantly improves from that with equal power allocation and an arbitrary decision threshold ($\zeta = 1$ in this case). A performance gain of 4–5 dB is observed at a BER of 10^{-3}–10^{-4}.

Also in the figure, we compare the performance of optimum power allocation and threshold with that of optimum power allocation ($a_s = 0.8$, $a_1 = 0.1$, and $a_2 = 0.1$) but an arbitrary threshold ($\zeta = 1$). We can see that jointly optimizing power allocation and threshold leads to a gain of about 1–2.5 dB over the scheme with optimum power allocation but arbitrary threshold at BER range between 10^{-3} and 10^{-5}. Note that the performance of the DiffDF scheme with optimum power allocation and threshold is 5 dB away from that of coherent detection. ▲

9.3 Differential modulation for AF cooperative communications

We now consider differential modulation schemes for amplify-and-forward cooperative communications, *DiffAF* schemes in short, and we start from a single relay scheme.

9.3.1 DiffAF with single-relay systems

This section considers a DiffAF protocol in a two-user cooperative communications system. The scheme efficiently combines signals from all branches, in which only long term average of the received signals is required. As a performance benchmark, we provide an exact BER formulation and its simple bounds for the optimum-combining cooperation system with DMPSK signals. Based on the theoretical BER benchmark, the optimum power allocation can be obtained, and it is used to further improve the performance of the DiffAF scheme.

9.3.1.1 Signal model and protocol description for the DiffAF scheme

We consider a two-user cooperative communications system employing the AF protocol. Each user can be a source node that sends its information to the destination, or it can be a relay node that helps transmit the other user's information. Basically, signal transmission can be separated into two phases. In phase 1, the source node transmits the information to its destination. Due to the broadcasting nature of wireless networks, this information is also received by the relay node. In phase 2, while the source node is silent, the relay node amplifies the received signal and forwards it to the destination. In both phases, the signals of all users are transmitted through orthogonal channels by using existing schemes such as TDMA, FDMA, or CDMA.

In differential transmission, the information is conveyed in the difference of the phases of two consecutive symbols. Specifically, information symbols to be broadcasted by the source are given by $v_m = e^{j\phi_m}$, where $\{\phi_m\}_{m=0}^{M-1}$ is a set of M information phases. In the case of DMPSK, ϕ_m can be specified as $\phi_m = 2\pi m/M$ for $m = 0, 1, \ldots, M-1$.

Instead of directly transmitting the information as in coherent transmission, the source node differentially encodes the information symbol v_m as

$$x^\tau = v_m x^{\tau-1}, \tag{9.107}$$

where τ is the time index, and x^τ is the differentially encoded symbol to be transmitted at time τ.

In phase 1, the source sends out the symbol x^τ with transmit power P_1. The corresponding received signals at the destination and the relay nodes can be expressed as

$$y_{s,d}^\tau = \sqrt{P_1} h_{s,d}^\tau x^\tau + w_{s,d}^\tau,$$

and

$$y_{s,r}^\tau = \sqrt{P_1} h_{s,r}^\tau x^\tau + w_{s,r}^\tau,$$

respectively. Here, $h_{s,d}^\tau$ and $h_{s,r}^\tau$ represent the channel coefficients from the source to the destination and from the source to the relay, respectively, whereas $w_{s,d}^\tau$ and $w_{s,r}^\tau$ are additive noise.

In phase 2, the relay amplifies the received signal and forwards it to the destination with transmit power P_2. The received signal at the destination can be modeled as

$$y_{r,d}^\tau = \sqrt{\tilde{P}_2} h_{r,d}^\tau y_{s,r}^\tau + w_{r,d}^\tau, \tag{9.108}$$

where $h_{r,d}^\tau$ is the channel coefficient from the relay to the destination, and $w_{r,d}^\tau$ is additive noise. In a Rayleigh fading environment, the channel coefficients $h_{s,d}^\tau$, $h_{s,r}^\tau$, and $h_{r,d}^\tau$ can be modeled as independent zero-mean complex Gaussian random variables with variances $\sigma_{s,d}^2$, $\sigma_{s,r}^2$, and $\sigma_{r,d}^2$, respectively. All of the fading coefficients are unknown to either the transmitter or the receiver, and they are assumed to be constant over two symbol periods. The noise $w_{s,d}^\tau$, $w_{s,r}^\tau$, and $w_{r,d}^\tau$ are modeled as independent complex Gaussian random variables with zero means and variances $\mathcal{N}_0$. To ensure that the average transmit power of the relay node is P_2, the normalized power $\tilde{P}_2$ is specified as

$$\tilde{P}_2 = \frac{P_2}{P_1 \sigma_{s,r}^2 + \mathcal{N}_0}, \tag{9.109}$$

where the variance $\sigma_{s,r}^2$ can be obtained from the long-term average of the received signals. The normalized power in (9.109) differs from its coherent counterpart in that the latter uses instantaneous fading amplitude, i.e., $|h_{s,r}|^2$, instead of $\sigma_{s,r}^2$.

At the destination, the received signal from the source through direct transmission in phase 1 ($y_{s,d}^\tau$) and that from the relay in phase 2 ($y_{r,d}^\tau$) are combined together, and then the combined output is differentially decoded. The combined signal prior to differential decoding is

$$y = a_1 \left(y_{s,d}^{\tau-1}\right)^* y_{s,d}^\tau + a_2 \left(y_{r,d}^{\tau-1}\right)^* y_{r,d}^\tau, \tag{9.110}$$

where the coefficients a_1 and a_2 are given by

$$a_1 = \frac{1}{\mathcal{N}_0}, \quad a_2 = \frac{P_1 \sigma_{s,r}^2 + \mathcal{N}_0}{\mathcal{N}_0(P_1 \sigma_{s,r}^2 + P_2 \sigma_{r,d}^2 + \mathcal{N}_0)}, \tag{9.111}$$

which are determined to maximized the SNR of the combined signal. Without acquiring CSI, the decoder use the sufficient statistics given in (9.110), and its decision rule for decoding information symbol follows:

$$\hat{m} = \arg \max_{m = 0,1,\dots,M-1} \mathrm{Re}\left\{v_m^* y\right\}.$$

9.3.1.2 Performance analysis and discussions

While the exact BER formulation that is applicable for multi-channel differential scheme with arbitrary-weight combining as in (9.111) is currently not available, in this section we discuss a BER formulation for the case of optimum combining weights, which are specified in the following:

$$\hat{a}_1 = \frac{1}{N_0}, \quad \hat{a}_2 = \frac{P_1 \sigma_{s,r}^2 + N_0}{N_0 (P_1 \sigma_{s,r}^2 + P_2 |h_{r,d}|^2 + N_0)}. \tag{9.112}$$

Even though these optimum weights are not practical for the DiffAD scheme because the instantaneous channel amplitude at the relay–destination link, i.e., $|h_{r,d}|^2$, is assumed unknown to the receiver, the performance evaluation based on optimum weights (9.112) can be used as a performance benchmark for the DiffAF scheme.

With the optimum weight $\hat{a}_1$ and $\hat{a}_2$, the instantaneous SNR per bit of the optimum combiner output can be calculated as

$$\gamma = \gamma_1 + \gamma_2, \tag{9.113}$$

where

$$\gamma_1 = \frac{P_1 |h_{s,d}|^2}{N_0}, \tag{9.114}$$

and

$$\gamma_2 = \frac{P_1 P_2 |h_{s,r}|^2 |h_{r,d}|^2}{N_0 \left(P_1 \sigma_{s,r}^2 + P_2 |h_{r,d}|^2 + N_0\right)}. \tag{9.115}$$

To simplify the notation, we omit the time index τ in this section for convenience in the derivation.

Similar to the DiffDF scheme, the conditional BER of the DiffAF is given by

$$P_{b|\gamma} = \frac{1}{16\pi} \int_{-\pi}^{\pi} f(\theta) \exp\left[-\alpha(\theta)\gamma\right] d\theta, \tag{9.116}$$

where

$$f(\theta) = \frac{b^2 (1 - \beta^2) \left[3 + \cos(2\theta) - (\beta + 1/\beta) \sin\theta\right]}{2\alpha(\theta)}, \tag{9.117}$$

and

$$\alpha(\theta) = \frac{b^2 (1 + 2\beta \sin\theta + \beta^2)}{2}, \tag{9.118}$$

in which $\beta = a/b$ denotes a constant parameter in which a and b depend on the modulation size [188]. Specifically, $a = 10^{-3}$ and $b = \sqrt{2}$ for DBPSK modulation, and

$a = \sqrt{2 - \sqrt{2}}$ and $b = \sqrt{2 + \sqrt{2}}$ for DQPSK modulation [188]. The values of a and b for larger modulation sizes can be obtained by using the result in [146].

By averaging the conditional BER over the Rayleigh fading channels, $h_{s,d}$, $h_{s,r}$, and $h_{r,d}$, we have an average BER formulation in terms of MGF functions as follows:

$$P_b = \frac{1}{16\pi} \int_{-\pi}^{\pi} f(\theta) \mathcal{M}_{\gamma_1}(\theta) \, \mathcal{M}_{\gamma_2}(\theta) \, d\theta, \tag{9.119}$$

where

$$\mathcal{M}_{\gamma_i}(\theta) = \int_{-\infty}^{+\infty} e^{-\alpha(\theta)\lambda} p_{\gamma_i}(\lambda) d\lambda, \tag{9.120}$$

denotes the MGF of the SNR γ_i for $i = 1, 2$, evaluating at $\alpha(\theta)$, and $f(\theta)$ and $\alpha(\theta)$ are specified in (9.117) and (9.118), respectively. Observe that the exact evaluation of the BER in (9.141) involves triple integration, which is hard to obtain. In what follows, we provide a simplified exact BER formulation that involve only double integration. The BER formulation in (9.119) can be further calculated as follows.

The simplified BER formulation can be obtained by calculating $\mathcal{M}_{\gamma_1}(\theta)$ and $\mathcal{M}_{\gamma_2}$ separately as follows. For Rayleigh fading channels, $|h_{s,d}|^2$, $|h_{s,r}|^2$, and $|h_{r,d}|^2$ are independent exponential random variables with parameters $1/\sigma_{s,d}^2$, $1/\sigma_{s,r}^2$, and $1/\sigma_{r,d}^2$, respectively. Based on (9.120), $\mathcal{M}_{\gamma_1}(\theta)$ can be expressed as (proof left as an exercise)

$$\mathcal{M}_{\gamma_1}(\theta) = \frac{1}{1 + k_{s,d}(\theta)}, \tag{9.121}$$

where

$$k_{s,d}(\theta) \triangleq \alpha(\theta) P_1 \sigma_{s,d}^2 / \mathcal{N}_0. \tag{9.122}$$

Observe that $\mathcal{M}_{\gamma_2}(\theta)$ depends on both $|h_{s,r}|^2$ and $|h_{r,d}|^2$. By averaging over $|h_{s,r}|^2$, and after some manipulations, we can express $\mathcal{M}_{\gamma_2}(\theta)$ as (proof left as an exercise)

$$\mathcal{M}_{\gamma_2}(\theta)$$

$$= \frac{1}{1 + k_{s,r}(\theta)} \left(1 + \frac{k_{s,r}(\theta)}{1 + k_{s,r}(\theta)} \frac{P_1 \sigma_{s,r}^2 + \mathcal{N}_o}{P_2} \frac{1}{\sigma_{r,d}^2} \int_0^\infty \frac{\exp\left(-u/\sigma_{r,d}^2\right)}{u + R(\theta)} du \right), \tag{9.123}$$

where

$$R(\theta) \triangleq \frac{P_1 \sigma_{s,r}^2 + \mathcal{N}_o}{P_2 \left(1 + k_{s,r}(\theta)\right)}, \tag{9.124}$$

in which $k_{s,r}(\theta)$ is given by

$$k_{s,r}(\theta) \triangleq \alpha(\theta) P_1 \sigma_{s,r}^2 / \mathcal{N}_0, \tag{9.125}$$

Finally, by substituting (9.121) and (9.123) into (9.119), we obtain a BER expression that involves only double integration. In what follows, we can further obtain single-integral BER lower bound, single-integral BER upper bound, and their corresponding simple BER approximations that involve no integration.

The BER expression in (9.119) can be upper bounded by using the upper bound of $\mathcal{M}_{\gamma_2}(\theta)$ in (9.123). We will bound the integration term in (9.123) with a constant $Z_{\max}$ that is not a function of θ. The value of $Z_{\max}$ can be found by replacing $R(\theta)$ in the integrand with its minimum value. We can see from (9.125) and (9.124) that $R(\theta)$ reaches its minimum value when $B(\theta)$ attains its maximum at $\theta = \pi/2$, i.e., $B(\theta) \leq -b^2(1+\beta)^2/2$. Accordingly, the minimum value of $R(\theta)$ is given by

$$R(\theta) \geq \frac{P_1 \sigma_{s,r}^2 + \mathcal{N}_o}{P_2} \left[1 - \frac{P_1 \sigma_{s,r}^2 b^2 (1+\beta)^2}{2\mathcal{N}_0} \right]^{-1} \triangleq R_{\min}. \tag{9.126}$$

Substituting $R(\theta)$ in (9.123) with $R_{\min}$, the BER upper bound is given by

$$P_b \leq \frac{1}{16\pi} \int_{-\pi}^{\pi} F(\theta) \frac{1}{[1 + k_{s,d}(\theta)][1 + k_{s,r}(\theta)]}$$
$$\times \left(1 + \frac{P_1 \sigma_{s,r}^2 + 1}{P_2 \sigma_{r,d}^2} \frac{k_{s,r}(\theta)}{1 + k_{s,r}(\theta)} Z_{\max} \right) d\theta, \tag{9.127}$$

in which

$$Z_{\max} = \int_0^{\infty} \exp(-u/\sigma_{r,d}^2) \left[u + R_{\min} \right]^{-1} du \tag{9.128}$$

is a constant that can be easily obtained for any given values of $\sigma_{r,d}^2$ and $R_{\min}$. The upper bound in (9.127) involves only single integration, and it is simpler than the exact BER provided in (9.119).

To gain further insight, we further simplify the BER upper bound in (9.127) as follows. For high enough SNR, we can ignore all 1's in the denominator of (9.127). This results in

$$P_b \leq \left(1 + \frac{P_1 \sigma_{s,r}^2 + 1}{P_2 \sigma_{r,d}^2} Z_{\max} \right) \frac{1}{16\pi} \int_{-\pi}^{\pi} \frac{f(\theta)}{k_{s,d}(\theta) k_{s,r}(\theta)} d\theta$$
$$= \frac{(P_1 \sigma_{s,r}^2 + 1) Z_{\max} + P_2 \sigma_{r,d}^2}{P_1^2 P_2 \sigma_{s,d}^2 \sigma_{s,r}^2 \sigma_{r,d}^2} \mathcal{N}_0^2 C(\beta, \theta), \tag{9.129}$$

where

$$C(\beta, \theta) = \frac{1}{8\pi b^4} \int_{-\pi}^{\pi} \frac{(1 - \beta^2)[3 + \cos(2\theta) - (\beta + 1/\beta)\sin\theta]}{(1 + 2\beta \sin\theta + \beta^2)^3} d\theta \tag{9.130}$$

is a constant that depends on the modulation size.

The BER lower bound can be obtained in a similar way to the BER upper bound. We will briefly provide details of the derivations. In this case, $R(\theta)$ reaches its maximum value when $B(\theta)$ attains its minimum at $\theta = -\pi/2$, i.e., $B(\theta) \geq -b^2(1-\beta)^2/2$. Accordingly, the maximum value of $R(\theta)$ is given by

$$R(\theta) \leq \frac{P_1 \sigma_{s,r}^2 + \mathcal{N}_o}{P_2} \left[1 + \frac{P_1 \sigma_{s,r}^2 b^2 (1-\beta)^2}{2\mathcal{N}_0} \right]^{-1} \triangleq R_{\max}. \tag{9.131}$$

Therefore, the lower bound is given by

$$P_b \geq \frac{1}{16\pi} \int_{-\pi}^{\pi} f(\theta) \frac{1}{[1 + k_{s,d}(\theta)][1 + k_{s,r}(\theta)]}$$
$$\times \left(1 + \frac{P_1 \sigma_{s,r}^2 + 1}{P_2 \sigma_{r,d}^2} \frac{k_{s,r}(\theta)}{1 + k_{s,r}(\theta)} Z_{min}\right) d\theta, \tag{9.132}$$

in which

$$Z_{min} = \int_0^{\infty} \exp(-u/\sigma_{r,d}^2)[u + R_{max}]^{-1} du$$

is a constant that can be easily obtained for any given values of $\sigma_{r,d}^2$ and R_{max}. We will show in the examples that the lower bound is tight in the high SNR region.

We further simplify the BER lower bound in (9.132) for high enough SNR by ignoring all 1's in the denominator of (9.132). The resulting BER approximation of (9.132) can be expressed as

$$P_b \approx \frac{\left(P_1 \sigma_{s,r}^2 + 1\right) Z_{min} + P_2 \sigma_{r,d}^2}{P_1^2 P_2 \sigma_{s,d}^2 \sigma_{s,r}^2 \sigma_{r,d}^2} \mathcal{N}_0^2 C(\beta, \theta), \tag{9.133}$$

where $C(\beta, \theta)$ is specified in (9.130). The approximated BER in (9.133) is tight at high SNRs.

Based on the analysis in this section, we will show in the next section that the BER expression, and its upper and lower bounds, together with their simple BER approximations, provide performance curves as a performance benchmark of the DiffAF scheme. Moreover, the optimum power allocation based on this BER expression can be used to further improve the performance of our propose scheme.

9.3.1.3   Examples for single-relay DiffAF scheme

Let us see some examples for the two-user cooperation systems employing the AF protocol with DQPSK signals. The channel fading coefficients are assumed to be independent between communication links, but are time correlated according to the Jakes' model as in Section 1.1.5, in which the Doppler frequency is $f_D = 75$ Hz and the normalized fading parameter is $f_D T_s = 0.0025$, where T_s is the sampling period. The noise variance is assumed to be unity ($\mathcal{N}_0 = 1$). We plot the performance curves in terms of average BER versus $P/\mathcal{N}_0$, where P is the total transmit power. We assume that the power allocation at the source and relay nodes is fixed at $P_1 + P_2 = P$.

Example 9.10   First let us compare simulated performances of the DiffAF scheme to various transmission techniques, including differential direct transmission, and their coherent counterparts. For fair comparison, we simulate the direct transmission schemes with DBPSK signals. Here, the channel variances at the relay link are chosen as $\sigma_{s,d}^2 = \sigma_{s,r}^2 = \sigma_{r,d}^2 = 1$, and the power ratios are given by $P_1 = 0.7P$ and $P_2 = 0.3P$. From Figure 9.13, it is apparent that the DiffAF scheme achieves higher diversity

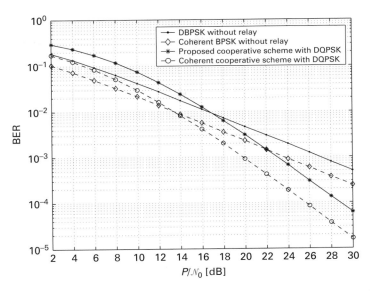

BER

P/N_0 [dB]

Fig. 9.13 Cooperative communication system with DQPSK signals, $\sigma_{s,d}^2 = \sigma_{s,r}^2 = \sigma_{r,d}^2 = 1$, $P_1/P = 0.7$, and $P_2/P = 0.3$.

orders than the DBPSK with direct transmission at high SNRs. We observe a performance gain of about 4 dB at a BER of 10^{-3}. In addition, at SNRs higher than 21 dB,
the DiffAF scheme provides significant performance improvement over that of the
directly transmitted BPSK with coherent detection. Therefore, the differential cooperative scheme can be a viable candidate for exploiting the inherent spatial diversity
of virtual antenna arrays in wireless networks with low complexity and simple implementation. Also in Figure 9.13, we can observe that the performance of the DiffAF
scheme is 3 dB away from its coherent counterpart at the high SNR region, which is as
expected. ▲

Example 9.11 In Figure 9.14, we compare the performance of the DiffAF scheme
and the optimum-combining scheme in case of equal power allocation, i.e.,
$P_1 = P_2 = 0.5P$. We can see that when $\sigma_{s,d}^2 = \sigma_{s,r}^2 = \sigma_{r,d}^2 = 1$, the performance of
both schemes are comparable, and both performance curves are close to the analytical
BER curve with optimum weights. When $\sigma_{s,d}^2 = \sigma_{s,r}^2 = 1$ and $\sigma_{r,d}^2 = 10$, we observe
that the performance of the DiffAF scheme and the optimum optimum-combining
scheme are about 0.7 dB different at a BER of 10^{-3}. However, the performance curve
of the optimum-combining scheme are the same as the theoretical curve with optimum
weights. Hence, the equal power allocation strategy does not always provide the best
performance. Moreover, the results illustrate that we can use the performance curve
with optimum weights as a benchmark for the performance of the DiffAF scheme. ▲

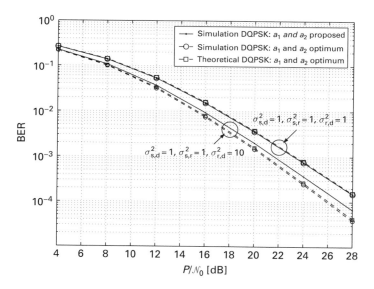

Fig. 9.14   Performance comparison of the DiffAF scheme and that with optimum weights, $P_1 = 0.5P$, $P_2 = 0.5P$.

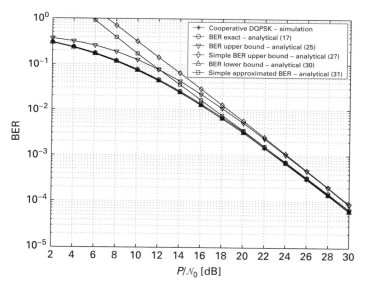

Fig. 9.15   Asymptotically tight curve of the BER formulations with optimum weights to the simulated performance of the DiffAF scheme with DQPSK signals, $\sigma^2_{s,d} = \sigma^2_{s,r} = \sigma^2_{r,d} = 1$; $P_1 = 0.7P$, $P_2 = 0.3P$.

Example 9.12   We illustrate in Figure 9.15 the numerically evaluated BER with optimum weights in comparison to simulated performance of the DiffAF scheme. For $\sigma^2_{s,d} = \sigma^2_{s,r} = \sigma^2_{r,d} = 1$, and $P_1 = 0.7P$ and $P_2 = 0.3P$, we can see that the BER expression with optimum weights in (9.119) is close to the simulated performance of the DiffAF scheme. The upper bound (9.127) and its simple approximation (9.129) are asymptotically parallel with the BER curve with optimum weights. The lower bound in (9.132) is asymptotically tight to the simulated performance and the BER curve using optimum weights. There is only a small difference at low SNRs. The simple approximated BER in (9.133) is loose at low SNRs, but appears to be asymptotically tight for a reasonably high range of SNRs. ▲

Example 9.13 Now let us use the BER formulation with optimum weights to numerically find the optimum power allocation at an SNR of 20 dB. We find that both the exact BER with optimum weights and its approximated BER formulations yield the same optimum power allocation. In Figure 9.16, for the case where all the channel variances are unity, we show the performance comparison of the simulated performance of the DiffAF scheme, the DiffAF scheme with optimum weights, and the analytical BER formulation with optimum weights; there is no performance gap observed in the figure. For the case where $\sigma_{s,d}^2 = \sigma_{s,r}^2 = 1$, and $\sigma_{r,d}^2 = 10$, the optimum powers at the source node and the relay node are $\hat{P}_1 = 0.8P$ and $\hat{P}_2 = 0.2P$, respectively. We can see in Figure 9.16 that the obtained optimum power allocation can significantly reduce the performance gap between the simulated performance of the DiffAF scheme and that with optimum weights. Furthermore, the two curves are close to the analytical BER curve using optimum weights.

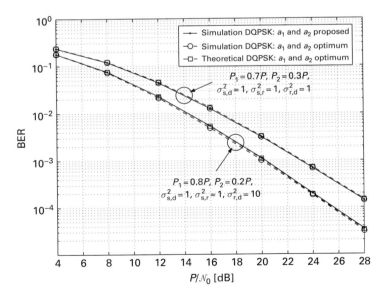

Fig. 9.16 Cooperative communication system with DQPSK signals for different power allocation schemes, and different channel variances.

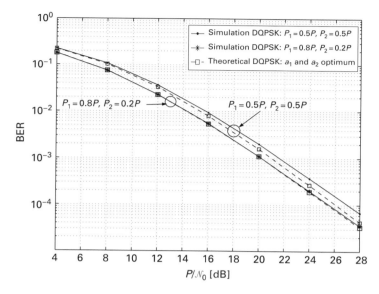

Fig. 9.17   Performance comparison of the optimum power allocation scheme and the equal power allocation scheme, $\sigma^2_{s,d} = \sigma^2_{s,r} = 1$ and $\sigma^2_{r,d} = 10$.

  This interesting result provides a key concept in allocating power among users. Specifically, in order to balance the link qualities, lower power should be allocated to the link with the largest channel variance, while higher power should be put in the link with the smallest channel variance. As can be seen in Figure 9.17, with a power loading of $\hat{P}_1$ and $\hat{P}_2$, there is a performance improvement of about 1.4 dB at a BER of 10^{-3} over the scheme with equal power allocation.   ▲

9.3.2   Multi-node DiffAF scheme

  This section considers DiffAF cooperative communications in multi-node cooperative networks. In this scheme, the destination in the multi-node DiffAF scheme requires only the long-term average of the received signals to efficiently combine the signals from all communications links. The BER performance of the multi-node DiffAF scheme is analyzed and the optimum power allocation is provided to further improve the system performance. An exact BER formulation is provided based on optimum combining weights for MDPSK modulation. The obtained BER formulation serves as a performance benchmark of the multi-node DiffAF scheme. In addition, BER upper bounds and simple BER approximations are provided. One of the tight BER approximations allows us to optimize the power allocation through a simple single-dimensional search.

9.3.2.1   Signal model and protocol description

  We consider a multi-node cooperative wireless network with a source and N relays as shown in Figure 9.18. For the multi-node DiffAF scheme, each relay amplifies each

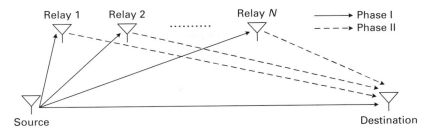

Fig. 9.18 Multi-node differential AF scheme.

received signal from the source and then forwards the amplified signal to the destination. With N cooperative relays in the network, signal transmissions for the multi-node DiffAF scheme comprise $N + 1$ phases. The first phase belongs to direct transmission, and the remaining N phases are for signal transmission for each of the N relays. The signal models for each of the $N + 1$ transmission phases are as follows.

In phase 1, suppose that differential M-ary phase shift keying (DMPSK) modulation is used, the modulated information at the source is $v_m = e^{j\phi_m}$, where $\phi_m = 2\pi m/M$ for $m = 0, 1, \ldots, M - 1$, and M is the constellation size. The source differentially encodes v_m by $x^\tau = v_m x^{\tau-1}$, where τ is the time index, and x^τ is the differentially encoded symbol to be transmitted at time τ. After that, the source transmits x^τ with transmitted power P_s to the destination. Due to the broadcasting nature of the wireless network, the information can also be received by each of the N relays. The corresponding received signals at the destination and the i-th relay, for $i = 1, 2, \ldots, N$, can be expressed as

$$y_{s,d}^\tau = \sqrt{P_s} h_{s,d}^\tau x^\tau + \eta_{s,d}^\tau,$$

and

$$y_{s,r_i}^\tau = \sqrt{P_s} h_{s,r_i}^\tau x^\tau + \eta_{s,r_i}^\tau.$$

where $h_{s,d}^\tau$ and h_{s,r_i}^τ represent channel coefficients from the source to the destination and from the source to the i-th relay, respectively. In this paper, $h_{s,d}^\tau$ and h_{s,r_i}^τ are modeled as complex Gaussian random variables with zero mean and variances $\sigma_{s,d}^2$ and σ_{s,r_i}^2, respectively. The terms $\eta_{s,d}^\tau$ and η_{s,r_i}^τ are additive white Gaussian noise at the destination and the i-th relay, respectively. Both of these noise terms are modeled as zero-mean, complex Gaussian random variables with the same variance of $\mathcal{N}_0$.

The received signal for the multi-node DiffAF scheme in phases 2 to $N+1$, is given by

$$y_{r_i,d}^\tau = \frac{\sqrt{P_i}}{\sqrt{P_s \sigma_{s,r_i}^2 + \mathcal{N}_0}} h_{r_i,d}^\tau y_{s,r_i}^\tau + \eta_{r_i,d}^\tau, \tag{9.134}$$

where P_i represents the transmitted power at the i-th relay, and $h_{r_i,d}^\tau$ denotes the channel coefficient at time τ at the i-th relay–destination link. We model $h_{r_i,d}^\tau$ as a zero-mean, complex Gaussian random variable with variance $\sigma_{r_i,d}^2$. In (9.134), P_i is normalized by $P_s \sigma_{s,r_i}^2 + \mathcal{N}_0$, and hence the i-th relay requires only the channel variance between the source and the i-th relay (σ^2_{s,r_i}) rather than its instantaneous value. In practice, σ^2_{s,r_i} can be obtained through long-term averaging of the received signals at the i-th relay.

Finally, the received signals from the source and all of the relays are combined at the destination, giving

$$y^{AF} = w_s^{AF}\left(y_{s,d}^{\tau-1}\right)^* y_{s,d}^{\tau} + \sum_{i=1}^{N} w_i^{AF}\left(y_{r_i,d}^{\tau-1}\right)^* y_{r_i,d}^{\tau}, \tag{9.135}$$

where

$$w_s^{AF} = \frac{1}{N_0}, \quad w_i^{AF} = \frac{P_s\sigma^2_{s,r_i} + N_0}{N_0(P_s\sigma^2_{s,r_i} + P_i\sigma^2_{r_i,d} + N_0)}$$

are combining weights that maximize the combiner output. Here, the channel $\sigma^2_{r_i,d}$ and σ^2_{s,r_i} are assumed to be available at the destination. Note that σ^2_{s,r_i} and $\sigma^2_{r_i,d}$ can be obtained through long-term averaging of the received signals at the i-th relay and the destination, respectively. In practice, σ^2_{s,r_i} can be forwarded from each of the i-th relays to the destination over a reliable channel link. Accordingly, without acquiring perfect CSI, the combined signal (9.135) is differentially decoded using the detection rule

$$\hat{m} = \arg \max_{m = 0,1,\dots,M-1} \mathrm{Re}\left\{v_m^* y^{AF}\right\}.$$

9.3.2.2   BER analysis

The BER formulation based on arbitrary combining weights, . i.e., w_s^{AF} and w_i^{AF} in (9.135), is currently not available in the literature. For mathematical tractability, we provide an alternative BER analysis based on the following optimum combining weights:

$$\hat{w}_s^{AF} = \frac{1}{N_0}, \quad \hat{w}_i^{AF} = \frac{P_s\sigma^2_{s,r_i} + N_0}{N_0(P_s\sigma^2_{s,r_i} + P_i|h_{r_i,d}^{\tau}|^2 + N_0)},$$

which are determined to maximized the SNR of the combined signal. Note that $\hat{w}_i^{AF}$ requires instantaneous channel information which is not available in the differential modulation; however, the BER analysis based on these optimum combining weights serves as the BER performance benchmark of the multi-node DiffAF scheme. As will be shown in Section 9.3.2.4, the multi-node DiffAF scheme with optimum power allocation yields very close performance to the BER benchmark. Using the optimum combining weights $\hat{w}_s^{AF}$ and $\hat{w}_i^{AF}$, an instantaneous SNR at the combiner output can be written as

$$\gamma^{AF} = \gamma_s^{AF} + \sum_{i=1}^{N} \gamma_i^{AF}, \tag{9.136}$$

where

$$\gamma_s^{AF} \triangleq \frac{P_s|h_{s,d}^{\tau}|^2}{N_0}, \tag{9.137}$$

and

$$\gamma_i^{AF} \triangleq \frac{P_s P_i |h_{s,r_i}^\tau|^2 |h_{r_i,d}^\tau|^2}{N_0(P_s \sigma_{s,r_i}^2 + P_i |h_{r_i,d}^\tau|^2 + N_0)}. \tag{9.138}$$

Accordingly, the conditional BER expression for the multi-node DiffAF scheme can be given by

$$P_{b|\gamma^{AF}}^{AF} = \frac{1}{2^{2L}\pi} \int_{-\pi}^{\pi} f(\theta, \beta, L) \exp[-\alpha(\theta)\gamma^{AF}] d\theta, \tag{9.139}$$

where $\alpha(\theta)$ is specified in (9.118) and

$$f(\theta, \beta, L) = \frac{b^2}{2\alpha(\theta)} \sum_{l=1}^{L} \binom{2L-1}{L-1} \Theta(\beta, \theta, l), \tag{9.140}$$

in which

$$\Theta(\beta, \theta, l) = \left[(\beta^{-l+1} - \beta^{l+1}) \cos((l-1)(\theta + \frac{\pi}{2})) - (\beta^{-l+2} - \beta^l) \cos(l(\theta + \frac{\pi}{2})) \right].$$

Here, $L = N+1$, and $\beta = a/b$ in which $a = 10^{-3}$ and $b = \sqrt{2}$ for DBPSK modulation, and $a = \sqrt{2 - \sqrt{2}}$ and $b = \sqrt{2 + \sqrt{2}}$ for DQPSK modulation [188]. The value of β for higher constellation sizes can be found in [146]. Averaging the conditional BER (9.139) over the Rayleigh distributed random variables by the use of the MGF functions, we have

$$P_b^{AF} = \frac{1}{2^{2(N+1)}\pi} \int_{-\pi}^{\pi} f(\theta, \beta, N+1) \mathcal{M}_{\gamma_s^{AF}}(\theta) \prod_{i=1}^{N} \mathcal{M}_{\gamma_i^{AF}}(\theta) d\theta, \tag{9.141}$$

where

$$\mathcal{M}_{\gamma_\mu^{AF}}(\theta) \triangleq \int_{-\infty}^{+\infty} e^{-\alpha(\theta)\lambda} p_{\gamma_\mu^{AF}}(\lambda) d\lambda,$$

denotes the MGF of γ_μ^{AF} in which $\mu \in \{s, 1, \ldots, N\}$. In (9.141), $\mathcal{M}_{\gamma_s^{AF}}(\theta)$ is obtained by an integration over an exponential random variable $|h_{s,d}|^2$ such that

$$\mathcal{M}_{\gamma_s^{AF}}(\theta) = \frac{1}{1 + k_{s,d}(\theta)}, \tag{9.142}$$

in which $k_{s,d}(\theta) \triangleq \alpha(\theta) P_s \sigma_{s,d}^2 / N_0$. The MGF $\mathcal{M}_{\gamma_i^{AF}}(\theta)$, however, is obtained through double integration of each γ_i^{AF} over two exponential random variables $|h_{s,r_i}|^2$ and $|h_{r_i,d}|^2$. After some manipulation, we have

$$\mathcal{M}_{\gamma_i^{AF}}(\theta) = \frac{1}{\sigma_{r_i,d}^2} \int_0^\infty \Omega_i(\theta) \exp\left(-\frac{u}{\sigma_{r_i,d}^2}\right) du, \tag{9.143}$$

where

$$\Omega_i(\theta) = \frac{N_0 (P_i u + P_s \sigma_{s,r_i}^2 + N_0)}{N_0 (P_i u + P_s \sigma_{s,r_i}^2 + N_0) + \alpha(\theta) P_s P_i \sigma_{s,r_i}^2 u}.$$

Denoting $k_{s,r_i}(\theta) \triangleq \alpha(\theta) P_s \sigma_{s,r_i}^2 / N_0$ and $\hat{k}_{s,r_i}(\theta) \triangleq P_i(1 + k_{s,r_i}(\theta))$, we can further simplify it as

$$\Omega_i(\theta) = \frac{1}{1 + k_{s,r_i}(\theta)} + \frac{1 - 1/\left[1 + k_{s,r_i}(\theta)\right]}{1 + \hat{k}_{s,r_i}(\theta) u / \left(P_s \sigma_{s,r_i}^2 + N_0\right)}, \tag{9.144}$$

such that (9.143) can be expressed as

$$\mathcal{M}_{\gamma_i^{AF}}(\theta) = \frac{1}{1 + k_{s,r_i}(\theta)}\left(1 + \frac{k_{s,r_i}(\theta)}{1 + k_{s,r_i}(\theta)} \frac{P_s \sigma_{s,r_i}^2 + N_0}{P_i} \frac{1}{\sigma_{r_i,d}^2} Z_i(\theta)\right), \tag{9.145}$$

where

$$\begin{aligned} Z_i(\theta) &= \int_0^\infty \frac{\exp\left(-u/\sigma_{r_i,d}^2\right)}{u + \hat{R}_i(\theta)} du \\ &= -e^{R_i(\theta)}\left[\mathcal{E} + \ln R_i(\theta) + \int_0^{R_i(\theta)} \frac{\exp(-t) - 1}{t} dt\right], \end{aligned} \tag{9.146}$$

in which

$$\hat{R}_i(\theta) \triangleq \left(P_s \sigma_{s,r_i}^2 + N_0\right) / \left(P_i\left[1 + k_{s,r_i}(\theta)\right]\right)$$

and $R_i(\theta) \triangleq \hat{R}_i(\theta)/\sigma_{r_i,d}^2$. Note that the last expression in (9.146) is obtained by applying results from [50] (p.358: Eq. (3.352.4) and p.934: Eq. (8.212.1)), and $\mathcal{E} \triangleq 0.577\,215\,664\,90...$ represents the Euler's constant (proof left as an exercise). Hence, by substituting (9.142) and (9.145) into (9.141), the average BER formulation is

$$\begin{aligned} P_b^{AF} &= \frac{1}{2^{2(N+1)}\pi} \int_{-\pi}^\pi \frac{f(\theta, \beta, N+1)}{1 + k_{s,d}(\theta)} \\ &\quad \times \prod_{i=1}^N \frac{1}{1 + k_{s,r_i}(\theta)}\left(1 + \frac{k_{s,r_i}(\theta)Z_i(\theta)}{1 + k_{s,r_i}(\theta)} \frac{P_s \sigma_{s,r_i}^2 + N_0}{P_i \sigma_{r_i,d}^2}\right) d\theta. \end{aligned} \tag{9.147}$$

Although P_b^{AF} can be calculated numerically, it is difficult to get some insights because it involves double integration. In the following we provide a single-integral BER upper bound, a simple BER upper bound without integration, and two tight BER approximations.

We first determine the BER upper bound and its simple expression as follows. From (9.145)–(9.147), we can see that the BER upper bound can be obtained by lower bound- ing $\hat{R}_i(\theta)$ in the denominator of the integrand of $Z_i(\theta)$ in (9.146). By substituting

$\theta = \pi/2$ in (9.118), $\alpha(\theta)$ is upper bounded by $\alpha(\theta) \leq (b^2(1 + \beta)^2)/2$. Therefore, $\hat{R}_i(\theta)$ is lower bounded by

$$\hat{R}_i(\theta) \geq \frac{P_s \sigma_{s,r_i}^2 + N_0}{P_i} \left[1 + \frac{P_s \sigma_{s,r_i}^2 b^2 (1 + \beta)^2}{2N_0} \right]^{-1} \triangleq \hat{R}_{i,\min}. \tag{9.148}$$

Substituting $R_i(\theta) = \hat{R}_{i,\min}$ into (9.146) results in an upper bound on $Z_i(\theta)$, i.e., $Z_i(\theta) \leq Z_{i,\max}$ where

$$Z_{i,\max} = -e^{R_{i,\min}} \left[\mathcal{E} + \ln R_{i,\min} + \int_0^{R_{i,\min}} \frac{\exp(-t) - 1}{t} dt \right], \tag{9.149}$$

in which $R_{i,\min} \triangleq \hat{R}_{i,\min}/\sigma_{r_i,d}^2$. Note that for any given channel variances σ_{s,r_i}^2 and $\sigma_{r_i,d}^2$, and power allocation, $Z_{i,\max}$ can be determined numerically. Therefore, the BER upper bound can be obtained by replacing $Z_i(\theta)$ in (9.147) by $Z_{i,\max}$, giving

$$P_b^{AF} \leq \frac{1}{2^{2(N+1)}\pi} \int_{-\pi}^{\pi} \frac{f(\theta, \beta, N+1)}{1 + k_{s,d}(\theta)}$$
$$\times \prod_{i=1}^{N} \frac{1}{1 + k_{s,r_i}(\theta)} \left(1 + \frac{k_{s,r_i}(\theta) Z_{i,\max}}{1 + k_{s,r_i}(\theta)} \frac{P_s \sigma_{s,r_i}^2 + N_0}{P_i \sigma_{r_i,d}^2} \right) d\theta. \tag{9.150}$$

We further simplify the right-hand side of (9.150) to obtain a simple BER upper bound and to get some insights into the achievable diversity order. We focus on the high SNR region so that all 1's in the denominator of (9.150) can be discarded. After some manipulation, the simple BER upper bound can be expressed as

$$P_b^{AF} \leq \frac{C(\beta, N+1) N_0^{N+1}}{P_s \sigma_{s,d}^2} \prod_{i=1}^{N} \frac{P_i \sigma_{r_i,d}^2 + (P_s \sigma_{s,r_i}^2 + N_0) Z_{i,\max}}{P_s P_i \sigma_{s,r_i}^2 \sigma_{r_i,d}^2}, \tag{9.151}$$

where

$$C(\beta, N+1) = \frac{1}{2^{2(N+1)}\pi} \int_{-\pi}^{\pi} \frac{f(\theta, \beta, N+1)}{\alpha^{N+1}(\theta)} d\theta \tag{9.152}$$

is a constant that depends on modulation size and number of relays, and $\alpha(\theta)$ and $f(\theta, \beta, N+1)$ are specified in (9.118) and (9.140), respectively. The BER upper bound (9.151) reveals that when N relays are available in the network, the multi-node DiffAF scheme achieves a diversity order of $N + 1$, as specified in the exponent of the noise variance.

In the following derivations, we determine two BER approximations in which one of them is an asymptotically tight simple BER approximation. We first note that $\alpha(\theta)$ in (9.118) can be lower bounded by $\alpha(\theta) \geq \alpha(-\pi/2) = (b^2(1 - \beta)^2)/2$ such that the upper bound on $R_i(\theta)$ can be written as

$$\hat{R}_i(\theta) \leq \frac{P_s \sigma_{s,r_i}^2 + N_0}{P_i} \left[1 + \frac{P_s \sigma_{s,r_i}^2 b^2 (1 - \beta)^2}{2N_0} \right]^{-1} \triangleq \hat{R}_{i,\max}. \tag{9.153}$$

By substituting $\hat{R}_i(\theta) = \hat{R}_{i,\max}$ into (9.146), we obtain a lower bound on $Z_i(\theta)$, denoted by $Z_{i,\min}$:

$$
Z_{i,\min} = -e^{R_{i,\max}} \left[\mathcal{E} + \ln R_{i,\max} + \int_0^{R_{i,\max}} \frac{\exp(-t) - 1}{t} dt \right],
\tag{9.154}
$$

in which $R_{i,\max} \triangleq \hat{R}_{i,\max}/\sigma_{r_i,d}^2$ and $\mathcal{E}$ is the Euler's constant. Then, replacing $Z_i(\theta)$ in (9.147) by $Z_{i,\min}$, we obtain a BER approximation

$$
P_b^{\mathrm{AF}} \gtrsim \frac{1}{2^{2(N+1)}\pi} \int_{-\pi}^{\pi} \frac{f(\theta, \beta, N+1)}{1 + k_{s,d}(\theta)}
$$

$$
\times \prod_{i=1}^{N} \frac{1}{1 + k_{s,r_i}(\theta)} \left(1 + \frac{k_{s,r_i}(\theta) Z_{i,\min}}{1 + k_{s,r_i}(\theta)} \frac{P_s \sigma_{s,r_i}^2 + \mathcal{N}_0}{P_i \sigma_{r_i,d}^2} \right) d\theta.
\tag{9.155}
$$

Furthermore, by ignoring all 1's in the denominator of (9.155), we obtain a simple BER approximation

$$
P_b^{\mathrm{AF}} \approx \frac{C(\beta, N+1)\mathcal{N}_0^{N+1}}{P_s \sigma_{s,d}^2} \prod_{i=1}^{N} \frac{P_i \sigma_{r_i,d}^2 + (P_s \sigma_{s,r_i}^2 + \mathcal{N}_0) Z_{i,\min}}{P_s P_i \sigma_{s,r_i}^2 \sigma_{r_i,d}^2},
\tag{9.156}
$$

where $C(\beta, N+1)$ and $Z_{i,\min}$ are specified in (9.152) and (9.154), respectively. We can see from the exponent of the noise variance in (9.156) that the achievable diversity order is $N+1$. As will be shown in the simulation results, these two BER approximations are tight in the high SNR region.

9.3.2.3 Optimum power allocation

Next, we formulate an optimization problem to minimize the BER under a fixed total transmitted power, $P = P_s + \sum_{i=1}^{N} P_i$. Based on the simple BER approximation (9.156), the optimization problem can be formulated as

$$
\arg \min_{P_s, \{P_i\}_{i=1}^N} \left\{ \frac{C(\beta, N+1)\mathcal{N}_0^{N+1}}{P_s \sigma_{s,d}^2} \prod_{i=1}^{N} \frac{P_i \sigma_{r_i,d}^2 + (P_s \sigma_{s,r_i}^2 + \mathcal{N}_0) Z_{i,\min}}{P_s P_i \sigma_{s,r_i}^2 \sigma_{r_i,d}^2} \right\},
$$

$$
\text{s.t. } P_s + \sum_{i=1}^{N} P_i \leq P, \; P_i \geq 0, \forall i,
\tag{9.157}
$$

Although (9.157) can be solved numerically, it is difficult to gain insights from numerical solutions. To further simplify the problem, we consider the high SNR region where $\hat{R}_{i,\max}$ in (9.153) can be approximated as

$$
\hat{R}_{i,\max} \approx (2\mathcal{N}_0)/(b^2(1-\beta)^2 P_i).
$$

Then we can rewrite $Z_{i,\min}$ in (9.154) as

$$
Z_{i,\min} \approx \int_0^{\infty} \frac{\exp(-u/\sigma_{r_i,d}^2)}{u + 2\mathcal{N}_0/(b^2(1-\beta)^2 P_s c_i)} du
$$

$$
= -e^{B_{c_i}} \left[\mathcal{E} + \ln B_{c_i} + \int_0^{B_{c_i}} \frac{\exp(-t) - 1}{t} dt \right],
\tag{9.158}
$$

where $c_i = P_i/P_s$ and $B_{c_i} \triangleq \hat{B}_{c_i}/\sigma^2_{r_i,d}$ in which $\hat{B}_{c_i} = 2N_0/(b^2(1-\beta)^2 P_s c_i)$. Since the integration term in the last expression of (9.158) is small compared to $\mathcal{E} + \ln B_{c_i}$, it can be neglected without significant effect on the power allocation. Hence $Z_{i,\min}$ can be further approximated by

$$Z_{i,\min} \approx -e^{B_{c_i}} \left(\mathcal{E} + \ln B_{c_i} \right). \tag{9.159}$$

As will be shown later, the obtained optimum power allocation based on the approximated $Z_{i,\min}$ in (9.159) yields almost the same performance as that with exact $Z_{i,\min}$ as specified in (9.154).

Substituting (9.159) into (9.157), the optimization problem can be simplified to

$$\arg\min_{P_s,\{P_i\}_{i=1}^N} \left\{ \frac{1}{P_s^{N+1}} \prod_{i=1}^N \frac{P_i \sigma^2_{r_i,d} - P_s \sigma^2_{s,r_i} e^{B_{c_i}} \left(\mathcal{E} + \ln B_{c_i} \right)}{P_i} \right\},$$

$$\text{s.t. } P_s + \sum_{i=1}^N P_i \leq P, \quad P_i \geq 0, \quad \forall i. \tag{9.160}$$

By taking the logarithm of the Lagrangian of (9.160) and letting $c_i = P_i/P_s$, we obtain

$$\mathcal{G} = -(N+1)\ln P_s + \lambda(\mathbf{c}^\mathsf{T}\mathbf{1} - P/P_s)$$

$$- \sum_{i=1}^N \ln c_i + \sum_{i=1}^N \ln \left(c_i \sigma^2_{r_i,d} - \sigma^2_{s,r_i} e^{B_{c_i}} \left(\mathcal{E} + \ln B_{c_i} \right) \right), \tag{9.161}$$

in which $\mathbf{c} = [1, c_1, \ldots, c_N]^\mathsf{T}$ is an $N \times 1$ vector, and $\mathbf{1}$ denotes an $N \times 1$ vector with all ones. By differentiating (9.161) with respect to c_i and P_s and equating the results to zero, we have

$$\frac{\partial \mathcal{G}}{\partial c_i} = \lambda - \frac{1}{c_i} + \frac{\sigma^2_{r_i,d} + \sigma^2_{s,r_i} \left[\frac{B_{c_i} e^{B_{c_i}}}{c_i} \left(\mathcal{E} + \ln B_{c_i} \right) + \frac{e^{B_{c_i}}}{c_i} \right]}{c_i \sigma^2_{r_i,d} - \sigma^2_{s,r_i} e^{B_{c_i}} \left(\mathcal{E} + \ln B_{c_i} \right)} = 0, \tag{9.162}$$

and

$$\frac{\partial \mathcal{G}}{\partial P_s} = -\frac{(N+1)}{P_s} + \lambda \frac{P}{P_s^2} + \sum_{i=1}^N \frac{\sigma^2_{s,r_i} \frac{e^{B_{c_i}}}{P_s} \left[\frac{B_{c_i}}{P_s} \left(\mathcal{E} + \ln B_{c_i} \right) + 1 \right]}{c_i \sigma^2_{r_i,d} - \sigma^2_{s,r_i} e^{B_{c_i}} \left(\mathcal{E} + \ln B_{c_i} \right)} = 0, \tag{9.163}$$

respectively. From (9.163), we can find that

$$\lambda = (N+1)\frac{P_s}{P} - \frac{1}{P} \sum_{i=1}^N \frac{\sigma^2_{s,r_i} \Upsilon e^{B_{c_i}} \left(\mathcal{E} + \ln B_{c_i} + \frac{1}{B_{c_i}} \right)}{c_i \left[c_i \sigma^2_{r_i,d} - \sigma^2_{s,r_i} e^{B_{c_i}} \left(\mathcal{E} + \ln B_{c_i} \right) \right]}, \tag{9.164}$$

in which we denote

$$\Upsilon \triangleq (2N_0)/(b^2(1-\beta^2)\sigma^2_{r_i,d}).$$

Observe from (9.158) that B_{c_i} can be re-expressed as

$$B_{c_i} = (2N_0)/(b^2(1-\beta)^2 \sigma^2_{r_i,d} P q c_i),$$

where $q \triangleq P_s/P$ for $q \in (0, 1)$. Then, substituting (9.164) into (9.162), we have

$$(N+1)q - \frac{1}{c_i} + \frac{\sigma_{r_i,d}^2 + \sigma_{s,r_i}^2 \left[\frac{\Upsilon}{Pqc_i^2} e^{B_{c_i}} \left(\mathcal{E} + \ln B_{c_i} \right) + \frac{e^{B_{c_i}}}{c_i} \right]}{c_i \sigma_{r_i,d}^2 - \sigma_{s,r_i}^2 e^{B_{c_i}} \left(\mathcal{E} + \ln B_{c_i} \right)} - Q(c_i, q) = 0, \quad (9.165)$$

in which

$$Q(c_i, q) \triangleq \frac{1}{P} \sum_{i=1}^{N} \frac{\sigma_{s,r_i}^2 \Upsilon e^{B_{c_i}} \left(\mathcal{E} + \ln B_{c_i} + \frac{1}{B_{c_i}} \right)}{c_i \left[c_i \sigma_{r_i,d}^2 - \sigma_{s,r_i}^2 e^{B_{c_i}} \left(\mathcal{E} + \ln B_{c_i} \right) \right]}. \quad (9.166)$$

The optimum solution of (9.160) can be obtained by finding the c_i that satisfies

$$(N+1)q - \frac{1}{c_i} + \frac{\sigma_{r_i,d}^2 + \sigma_{s,r_i}^2 \left[\frac{\Upsilon}{Pqc_i^2} e^{B_{c_i}} \left(\mathcal{E} + \ln B_{c_i} \right) + \frac{e^{B_{c_i}}}{c_i} \right]}{c_i \sigma_{r_i,d}^2 - \sigma_{s,r_i}^2 e^{B_{c_i}} \left(\mathcal{E} + \ln B_{c_i} \right)} - Q(c_i, q) = 0,$$

$$(9.167)$$

in which $Q(c_i, q)$ is specified in (9.166).

Given a specific $q \triangleq P_s/P$ for $q \in (0, 1)$, we can find the corresponding c_i, denoted by $c_i(q)$ that satisfies (9.167). The optimum power allocation can then be obtained by finding the $q = \hat{q}$ that satisfies

$$1 + \sum_{i=1}^{N} c_i(\hat{q}) - \frac{1}{\hat{q}} = 0. \quad (9.168)$$

The resulting optimum power allocation for the source is $a_s = \hat{q}$. Since $c_i(\hat{q}) = P_i/P_s = a_i/\hat{q}$, then the optimum power allocation for each of the i-th relay is $a_i = \hat{q} c_i(\hat{q})$ for $i = 1, 2, \ldots, N$.

Optimum power allocation for single-relay DiffAF systems

For single-relay systems, the optimization problem (9.165) and (9.168) is reduced to finding q such that

$$(N+1)q - \frac{1}{c_1} + \frac{\sigma_{r_1,d}^2 + \sigma_{s,r_1}^2 \Upsilon e^{B_{c_1}} \left[\frac{1}{Pq} \left(\mathcal{E} + \ln B_{c_1} \right) + \frac{1}{B_{c_1}} \right]}{c_1^2 [c_1 \sigma_{r_1,d}^2 - \sigma_{s,r_1}^2 e^{B_{c_1}} \left(\mathcal{E} + \ln B_{c_1} \right)]}$$

$$- \frac{1}{P} \frac{\sigma_{s,r_1}^2 \Upsilon e^{B_{c_1}} \left(\mathcal{E} + \ln B_{c_1} + \frac{1}{B_{c_1}} \right)}{c_1 \left[c_1 \sigma_{r_1,d}^2 - \sigma_{s,r_1}^2 e^{B_{c_1}} \left(\mathcal{E} + \ln B_{c_1} \right) \right]} = 0,$$

$$\text{and} \quad 1 + c_1(q) - \frac{1}{q} = 0, \quad (9.169)$$

which can be simply solved by any single-dimensional search techniques. In this way, the complexity of the optimization problem can be greatly reduced, while the resulting optimum power allocation is close to that from exhaustive search in [62]. For example, Table 9.2 compares the optimum power allocation from the exhaustive search in [62] and that from solving the low-complexity optimization problem in (9.169). The results in Table 9.2 are obtained at a reasonable high SNR region, e.g., 20 or 30 dB. The

Table 9.2 DiffAF: optimum power allocation for a cooperation system with a relay based on exhaustive search and approximate closed-form formulation (9.169).

$\left[\sigma_{s,d}^2, \sigma_{s,r_i}^2, \sigma_{r_i,d}^2\right]$	DBPSK $[a_s, a_1]$		DQPSK $[a_s, a_1]$	
	Exhaustive	Approximate	Exhaustive	Approximate
$[1, 1, 1]$	$[0.66, 0.34]$	$[0.66, 0.34]$	$[0.70, 0.30]$	$[0.69, 0.31]$
$[1, 10, 1]$	$[0.54, 0.46]$	$[0.54, 0.46]$	$[0.54, 0.46]$	$[0.54, 0.46]$
$[1, 1, 10]$	$[0.80, 0.20]$	$[0.79, 0.21]$	$[0.80, 0.20]$	$[0.78, 0.22]$

DBPSK or DQPSK modulations are used, and $\left[\sigma_{s,d}^2, \sigma_{s,r_i}^2, \sigma_{r_i,d}^2\right]$ represents a vector containing channel variances of the source–destination link, the source–relay link, and the relay–destination link, respectively. We can see from the table that the optimum power allocation based on (9.169) is very close to that from the numerical search, for any relay location. There is a difference of only about $1 - 2\%$ in the obtained results between these two methods.

Optimum power allocation for multi-relay DiffAF systems

For multi-relay systems, (9.165) and (9.168) can be used to find the optimum power allocation. Nevertheless, the optimization based on (9.165) and (9.168) involves an $(N + 1)$-dimensional search because $Q(c_i, q)$ in (9.165) contains the power allocation of each relay inside the summation. To reduce the complexity of the search space, we remove the summation inside $Q(c_i, q)$ such that the approximate $Q(c_i, q)$ depends only on the c_i of interest. Therefore, an optimum power allocation can be approximately obtained by finding q such that

$$(N+1)q - \frac{1}{c_i} + \frac{\sigma_{r_i,d}^2 + \sigma_{s,r_i}^2 \Upsilon e^{B_{c_i}} \left[\frac{1}{Pq}\left(\mathcal{E} + \ln B_{c_i}\right) + \frac{1}{B_{c_i}}\right]}{c_i^2 \left[c_i \sigma_{r_i,d}^2 - \sigma_{s,r_i}^2 e^{B_{c_i}}\left(\mathcal{E} + \ln B_{c_i}\right)\right]}$$
$$- \frac{1}{P} \frac{\sigma_{s,r_i}^2 \Upsilon e^{B_{c_i}}\left(\mathcal{E} + \ln B_{c_i} + \frac{1}{B_{c_i}}\right)}{c_i \left[c_i \sigma_{r_i,d}^2 - \sigma_{s,r_i}^2 e^{B_{c_i}}\left(\mathcal{E} + \ln B_{c_i}\right)\right]} = 0,$$

and $\quad 1 + \sum_{i=1}^{N} c_i(q) - \frac{1}{q} = 0, \forall i, i = 1, \ldots, N.$ $\qquad(9.170)$

From (9.170), the optimum power allocation that involves an $(N+1)$-dimensional search is reduced to a single-dimensional search over the parameter q, $q \in (0, 1)$. Table 9.3(a) for DBPSK signaling and Table 9.3(b) for DQPSK signaling summarize the numerical search results from the multi-dimensional search based on (9.157) in comparison to those from the approximated one-dimensional search using (9.170). Based on the optimization problem (9.170), the searching time for optimum power allocation can be greatly reduced, while the obtained power allocation is very close to that from solving (9.157) using the multi-dimensional search.

Table 9.3 Multi-node DiffAF: optimum power allocation for a cooperation system with two relays based on exhaustive search and approximate formulation (9.170) : (a) DBPSK signaling and (b) DQPSK signaling.

(a)

$\left[\sigma_{s,d}^2, \sigma_{s,r_i}^2, \sigma_{r_i,d}^2\right]$	DBPSK $[a_s, a_1, a_2]$	
	Exhaustive	Approximate
$[1, 1, 1]$	$[0.46, 0.33, 0.21]$	$[0.49, 0.26, 0.26]$
$[1, 10, 1]$	$[0.44, 0.28, 0.28]$	$[0.47, 0.27, 0.27]$
$[1, 1, 10]$	$[0.65, 0.21, 0.14]$	$[0.65, 0.17, 0.17]$

(b)

$\left[\sigma_{s,d}^2, \sigma_{s,r_i}^2, \sigma_{r_i,d}^2\right]$	DQPSK $[a_s, a_1, a_2]$	
	Exhaustive	Approximate
$[1, 1, 1]$	$[0.48, 0.33, 0.19]$	$[0.50, 0.25, 0.25]$
$[1, 10, 1]$	$[0.40, 0.30, 0.30]$	$[0.39, 0.31, 0.30]$
$[1, 1, 10]$	$[0.66, 0.21, 0.13]$	$[0.67, 0.16, 0.16]$

From the results in Tables 9.2 and 9.3, we can also observe that, for any channel link qualities, more power should be allocated to the source in order to maintain link reliability. This observation holds true for both DBPSK and DQPSK modulations. When the channel link qualities between the source and the relays are good (e.g., $[\sigma_{s,d}^2, \sigma_{s,r_i}^2, \sigma_{r_i,d}^2]$ = $[1, 10, 1,]$), the system replicates the multiple transmit antenna system. Therefore, almost equal power should be allocated to the source and all the relays. However, more power should be allocated to the source so that the transmit information can reach the relays, and the remaining power is allocated to the relays.

The results also show that if there are two relays in the networks, almost the same amount of power as the source should be allocated to the first relay, and the remaining power should be allocated to the second relay. In the case of $[\sigma_{s,d}^2, \sigma_{s,r_i}^2, \sigma_{r_i,d}^2]$ = $[1, 1, 10]$, which indicates that the channel qualities between the relays and the destination are good. When comparing to the case with $[\sigma_{s,d}^2, \sigma_{s,r_i}^2, \sigma_{r_i,d}^2] = [1, 1, 1]$, more power should be allocated at the source, while less power should be put at the relays. The reason is that the channel quality at the source–destination link is lower than that of the relay–destination links, so more power is required at the source to balance the qualities of all possible links so that the system can provide reliable communications.

9.3.2.4   Examples for the multi-relay DiffAF scheme

Let us consider some examples for the multi-node DiffAF scheme with DBPSK and DQPSK modulations. We consider the scenarios where two or three relays ($N = 2$ or 3) are in the networks. The channel coefficients follow the Jakes' model with Doppler frequency $f_D = 75\,\text{Hz}$ and normalized fading parameter $f_D T_s = 0.0025$, where T_s is the sampling period. The noise variance is assumed to be one ($\mathcal{N}_0 = 1$). The average BER curves are plotted as functions of $P/\mathcal{N}_0$.

Example 9.14 Let us see the performance of the multi-node DiffAF scheme with DBPSK modulation for a network with two relays. Figure 9.19(a) is observed under equal channel variances, i.e., $\sigma_{s,d}^2 = \sigma_{s,r_i}^2 = \sigma_{r_i,d}^2 = 1$, and equal power allocation strategy ($P_s = P_1 = P_2 = P/3$). We can see that the exact theoretical BER benchmark well matches the simulated BER curve. In addition, the BER upper bound, the simple BER upper bound, and the two simple BER approximations are tight to the simulated curve at high SNR. The BER curve for coherent detection is also shown in the figure; we observe a performance gap of about 4 dB between the multi-node DiffAF scheme and its coherent counterpart at a BER of 10^{-3}.

We also illustrate in Figure 9.19(b) the BER performance of the multi-node DiffAF scheme with DQPSK modulation when using a different number of relays (N). The simulation scenario is the same as that of Figure 9.19(a), and we consider two possible numbers of relays, namely $N = 2$ and $N = 3$. It is apparent that the multi-node DiffAF scheme achieves higher diversity orders as N increases. Specifically, as N increases from 2 to 3, we observe a gain of about 1.7–2 dB at a BER of 10^{-3}. This observation confirms our theoretical analysis in Section 9.3.2.2. Also in the figure, the exact theoretical BER curves for $N = 2$ and $N = 3$ are tight to the corresponding simulated curves. In addition, the performance curves of the multi-node DiffAF scheme are about 4 dB away from their coherent counterparts.   ▲

Example 9.15 This example shows the BER performance of the multi-node DiffAF scheme with optimum power allocation in comparison to that with equal power allocation. We consider the multi-node DiffAF scheme with DQPSK modulation for a network with two relays. The channel variances are $\sigma_{s,d}^2 = \sigma_{r_i,d}^2 = 1$ and $\sigma_{s,r_i}^2 = 10$, and the optimum power allocation is $[0.39, 0.31, 0.30]$ (from Table 9.3). Figure 9.20(a) shows that when all relays are close to the source, i.e., $\sigma_{s,r_i}^2 = 10$, the multi-node DiffAF scheme with optimum power allocation yields again of about 0.6 dB a gain of over the scheme with equal power allocation at a BER of 10^{-3}. We observe a small performance gain in this scenario because the signals at the relays are as good as the signal at the source. Therefore, the scheme with equal power allocation yields almost as good a performance under optimal power allocation, so that there is a small room for improvement. Also in the figure, the exact theoretical BER curves are provided for both power allocation schemes, and they closely match to their corresponding simulated curves.

Figure 9.20(b) shows the BER performance of the optimum power allocation scheme for a DQPSK cooperation system with two relays. The channel variances are $\sigma_{s,d}^2 = \sigma_{s,r_i}^2 = 1$ and $\sigma_{r_i,d}^2 = 10$, which corresponds to a scenario where all relays are close to the destination. The optimum power allocation for this scenario is $[0.67, 0.16, 0.16]$ (from Table 9.3). We observe that the performance with optimum power allocation is

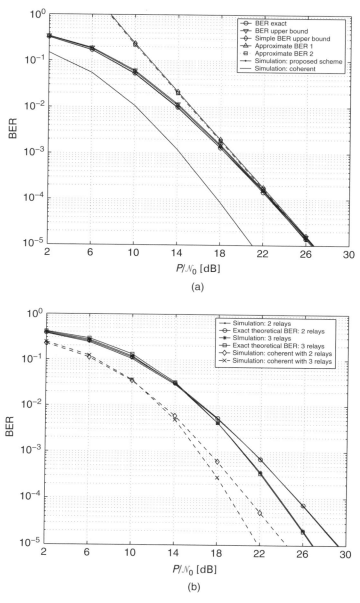

Fig. 9.19   Multi-node DiffAF scheme with equal power allocation strategy and $\sigma_{s,d}^2 = \sigma_{s,r_i}^2 = \sigma_{r_i,d}^2 = 1$. (a) DBPSK: two relays; (b) DQPSK: two and three relays.

about 2 dB superior to that with equal power allocation at a BER of 10^{-3}. In this scenario, we observe a larger performance gain than the case of $\sigma_{s,d}^2 = \sigma_{r_i,d}^2 = 1$ and $\sigma_{s,r_i}^2 = 10$ in Figure 9.20(a). The reason is that using equal power allocation in this case leads to low quality of the received signals at the relays, and thus causes higher chance

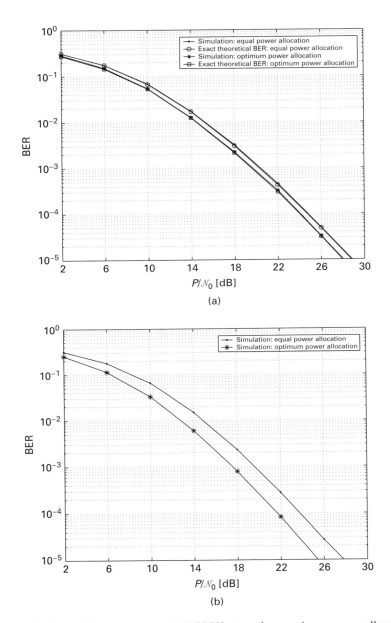

Fig. 9.20 Multi-Node DiffAF scheme with DQPSK: two relays, optimum power allocation strategy. (a) $\sigma_{s,d}^2 = 1, \sigma_{s,r_i}^2 = 10, \sigma_{r_i,d}^2 = 1$; (b) $\sigma_{s,d}^2 = \sigma_{s,r_i}^2 = 1$, and $\sigma_{r_i,d}^2 = 10$.

of decoding error at the destination based on the combined signal from cooperative links. With optimum power allocation, more power is allocated to the source, and consequently the quality of the received signals at the relays is improved. This results in more reliable combined signal at the destination, hence yielding better system performance. ▲

9.4   Chapter summary and bibliographical notes

In the first part of this chapter, differential decode-and-forward schemes (DiffDF) for cooperative communications are presented. Both single-relay and multi-relay cooperation systems are given. By allowing the relay to forward only the correctly decoded symbols and introducing a decision threshold at the destination node, the DiffDF scheme efficiently combines the signals from the direct and the relay links.

In case of the single-relay DiffDF scheme, BER analysis for DMPSK modulation is categorized into six different scenarios that lead to different instantaneous SNRs at the combiner output of the destination. A tight BER approximation is also provided. Based on the tight BER approximation, the optimum decision threshold and power allocation are jointly determined. Both theoretical and simulation examples reveal that the optimum threshold and optimum power allocation rely on the qualities of the channel links. When the quality of the relay–destination link is much larger than the other links, i.e., $\sigma_{s,d}^2 = \sigma_{s,r}^2 = 1$ and $\sigma_{r,d}^2 = 10$, then the decision threshold is more important than the power allocation at high SNR. For instance, in case of DQPSK signals with equal power allocation, using the optimum threshold resulted in an improvement gain of more than 5 dB over the scheme without threshold at a BER of 10^{-4}. By further using the optimum power allocation, the performance improvement is about 0.5 dB at the same BER. Simulation examples also showed that the DiffDF scheme with DQPSK signals provides a performance improvement of 11 dB at a BER of 10^{-3} over the differential DF scheme where the relay always forwards the decoded symbols.

In the case of the multi-node DiffDF scheme, a multi-node scenario follows the assumption that each of N cooperative relays forwards only correctly decoded symbols to the destination. Decision thresholds are used at the destination to efficiently combine the signals from each relay–destination link with that from the direct link. An approximated BER analysis for DMPSK is provided, and a low-complexity BER lower bound is derived. The BER lower bound is very close to the simulated performance under some scenarios. While jointly optimizing power allocation and thresholds based on the BER lower bound introduces $(2N + 1)$-dimensional searching, the search space is reduced by assuming that the same power is used at each relay and the same threshold is used at the destination. Numerical results reveal that more power should be allocated at the source, with the rest allocated to the relays. In addition, a larger threshold should be used when the relays are close to the destination. Simulation examples show that the diversity gain of the multi-node DiffDF scheme increases with the number of relays. For a DBPSK cooperation system, the multi-node DiffDF scheme with different thresholds leads to a performance improvement of up to 6 dB at a BER of 10^{-4}. In the case of the DQPSK cooperation system, the multi-node DiffDF scheme with joint optimum power allocation and optimum threshold achieves a gain of about 4–5 dB over the scheme with equal power allocation and a unit threshold at a BER of $10^{-3}-10^{-4}$.

The second part of the chapter presents differential modulation for the amplify-and-forward cooperation protocol (DiffAF) for single-relay and multi-relay systems. For the single-relay DiffAF scheme, the DiffAF scheme with DQPSK signals yielded a performance improvement of 4 dB at a BER of 10^{-3} over that of the DBPSK direct transmission scheme. In comparison to the coherent detection without relay, the DiffAF scheme provided a practical alternative with lower complexity and simpler implementation. In addition, simulation examples showed that the performance of the DiffAF scheme was superior to that of direct transmission with coherent detection at SNRs higher than 21 dB. This is due to the fact that the cooperative communications provide more diversity gain than the direct transmission schemes. While the BER analysis of the DiffAF scheme is not available currently, we provided the exact BER expression based on optimum combining weights, and it is considered to be a performance benchmark for the DiffAF scheme. By using the obtained optimum power allocation based on the provided BER expression, the DiffAF scheme is able to achieve comparable performance to the scheme with optimum weights in any channel variances of all links. Moreover, the performance of the scheme with optimum power strategy outperforms the equal power scheme by about 1.4 dB at a BER of 10^{-3}.

In the case of the multi-node DiffAF scheme, the BER expression for DMPSK modulation based on optimum combining weights is given; it is considered as a performance benchmark of the multi-node DiffAF scheme. BER upper bounds and BER approximations are provided; they are tight to the simulated performance, especially at high SNR. The theoretical BER reveals that the diversity order of the multi-node DiffAF scheme is $N + 1$, where N is the number of relays, and this is confirmed by the simulation results. There is a gain of about 1.7–2 dB at a BER of 10^{-3} when N increases from 2 to 3. The BER approximation is further simplified; based on the approximate BER, power allocation can be optimized by using a low-complexity single-dimensional search. Simulation examples show that when all relays are close to the source, the multi-node DiffAF scheme obtains a gain of about 0.6 dB over the scheme with equal power allocation at a BER of 10^{-3}. When all relays are close to the destination, the performance with optimum power allocation achieves an improvement of about 2 dB over that with equal power allocation.

The study of MIMO channel capacity was presented in [215] and many works on MIMO systems assuming full CSI available at the transmitter have been given, for example in [212, 215, 109, 5, 213, 213, 198, 3, 111, 199, 46, 135, 206]. In conventional single-antenna systems, noncoherent modulation is useful when the knowledge of CSI is not available. The noncoherent modulation simplifies the receiver structure by omitting channel estimation and carrier or phase trackings. Some examples of the noncoherent modulation techniques are noncoherent frequency shift keying (NFSK) and differential modulation [146]. Among these modulation techniques, the differential scheme is preferred to NFSK because it provides a better performance at the same operating SNR. Two classes of differential modulation schemes are available: differential M-ary quadrature amplitude modulation (DMQAM) and differential M-ary phase shift keying (DMPSK). In the DMQAM scheme, information is modulated through the amplitude difference between two consecutive symbols. The DMPSK, however, modulates information through the phase difference among two consecutive symbols. For a data rate of R bits per channel use, DPMSK signal constellation contains $M = 2^R$ symbols. Each of the symbol $m \in 0 \leq M - 1$ is generated by the m-th root of unity: $v_m = e^{j2\pi m/M}$.

To learn more about this topic, interested readers should also refer to [66, 67].

Exercises

9.1   Assume that the Rayleigh fading channels $h_{s,d}^\tau$, $h_{s,r}^\tau$, and $h_{r,d}^\tau$ are independent of each other. Prove the average BER $P_{BER}^{(i)}$, $i = 1, 2, \ldots, 6$, which are specified in (9.58), (9.61), (9.64), (9.65), and (9.67), respectively.

9.2   Prove the probabilities for different scenarios specified in (9.46)–(9.88), respectively, i.e.,

$$P_r^h(\Phi_3) = \mathcal{M}^2\left(P_2|h_{r,d}^\tau|^2, \zeta\right)\left(1 - \Psi(\gamma_{s,r}^\tau)\right)^2$$

$$\times \left(\frac{1}{\Gamma(P_1|h_{s,r}^\tau|^2, P_2|h_{r,d}^\tau|^2)} - \frac{1}{(1 - e^{-\zeta^2/N_0})\Psi(\gamma_{s,r}^\tau)}\right),$$

$$P_r^h(\Phi_4) = P_r^h(\Phi_5)$$

$$= \frac{\mathcal{M}\left(P_2|h_{r,d}^\tau|^2, \zeta\right)\exp(-\zeta^2/N_0)\Psi(\gamma_{s,r}^\tau)\left(1 - \Psi(\gamma_{s,r}^\tau)\right)}{\Gamma(P_1|h_{s,r}^\tau|^2, P_2|h_{r,d}^\tau|^2)},$$

$$P_r^h(\Phi_6) = \frac{\exp(-2\zeta^2/N_0)\Psi(\gamma_{s,r}^\tau)}{\Gamma(P_1|h_{s,r}^\tau|^2, P_2|h_{r,d}^\tau|^2)}.$$

9.3   Prove the probability approximations in (9.95) and (9.98) for the differential DF cooperative scheme.

9.4   For the differential AF cooperative scheme, prove the moment generation functions $\mathcal{M}_{\gamma_1}(\theta)$ and $\mathcal{M}_{\gamma_2}(\theta)$, specified in (9.121) and (9.123), respectively.

9.5   Prove the moment generation function $\mathcal{M}_{\gamma_i^{AF}}(\theta)$, specified in (9.145).

9.6   (Simulation project) Consider a simplified model with a source node, a relay node, and a destination node. The source–destination channel $h_{s,d}$, the source–relay channel $h_{s,r}$, and the relay–destination channel $h_{r,d}$ are modeled as independent zero-mean, complex Gaussian random variables with variances $\delta_{s,d}^2$, $\delta_{s,r}^2$, and $\delta_{r,d}^2$, respectively. Assume that $\delta_{s,d}^2 = \delta_{s,r}^2 = \delta_{r,d}^2 = 1$, and DQPSK modulation is used. In this project, we compare performances of the threshold-based differential DF cooperation scheme with choices of power allocations and decision threshold at the destination.

(a)   Simulate the threshold-based differential DF scheme with an equal power allocation scheme ($P_1 = P_2$) and a decision threshold of $\zeta = 1$. Compares the simulated BER performance with the BER approximation (9.69), the BER upper bound (9.70), and the BER lower bound (9.75).

(b)   With the equal power allocation ($P_1 = P_2$), simulate the differential DF scheme without threshold at the destination and the differential DF scheme that the relay always forwards the decoded symbols to the destination. Compare their BER performances with that of the threshold-based differential DF scheme simulated in (a).

(c)   Simulate the threshold-based differential DF scheme with different thresholds (for example, $\zeta = 0.5, 1$ or 2). Compare the BER performances and determine a best threshold.

(d) Repeat (a), (b), and (c) for an unequal power allocation ($P_1/P_2 = 9$).

9.7   (Simulation project) We consider a differential AF cooperative scheme with a single-relay node. The source–destination channel $h_{s,d}$, the source–relay channel $h_{s,r}$, and the relay–destination channel $h_{r,d}$ are modeled as independent zero-mean, complex Gaussian random variables with variances $\delta_{s,d}^2$, $\delta_{s,r}^2$, and $\delta_{r,d}^2$, respectively. Assume that $\delta_{s,d}^2 = \delta_{s,r}^2 = \delta_{r,d}^2 = 1$, and DQPSK modulation and an equal power allocation scheme ($P_1 = P_2$) are used.

(a) Simulate the differential AF cooperative scheme, the direct differential scheme, and their coherent counterparts, respectively. Compare their performances and explain.

(b) Simulate the differential AF scheme with an equal weight combining, the weight combining proposed in (9.111), and the ideal maximum ratio combining (9.112), respectively, and compare their performances.

(c) Repeat (a) and (b) for an unequal power allocation ($P_1/P_2 = 4$).

10 Energy efficiency in cooperative sensor networks

In the previous chapters, the gains of cooperative diversity were established under the ideal model of negligible listening and computing power. In sensor networks, and depending on the type of motes used, the power consumed in receiving and processing may constitute a significant portion of the total consumed power. Cooperative diversity can provide gains in terms of savings in the required transmit power in order to achieve a certain performance requirement because of the spatial diversity it adds to the system. However, if one takes into account the extra processing and receiving power consumption at the relay and destination nodes required for cooperation, then there is obviously a tradeoff between the gains in the transmit power and the losses due to the receive and processing powers when applying cooperation. Hence, there is a tradeoff between the gains promised by cooperation, and this extra overhead in terms of the energy efficiency of the system should be taken into consideration in the design of the network.

In this chapter the gains of cooperation under this extra overhead are studied. Moreover, some practical system parameters, such as the power amplifier loss, the quality of service (QoS) required, the relay location, and the optimal number of relays, are considered. Two communications architectures are considered, direct transmission and cooperative transmission. The performance metric for comparison between the two architectures is the energy efficiency of the communication scheme. More specifically, for both architectures the optimal total power consumption to achieve certain QoS requirements is computed, and the cooperation gain is defined as the ratio between the power required for direct transmission and cooperation. When this ratio is smaller than one, this indicates that direct transmission is more energy efficient, and that the extra overhead induced by cooperation overweighs its gains in the transmit power. Comparisons between optimal power allocation at the source and relay nodes and equal power allocation are demonstrated. The results reveal that, under some scenarios, equal power allocation is almost equivalent to optimal power allocation. The effect of relay location on the performance is investigated to provide guidelines for relay assignment algorithms.

10.1 System model

Consider a single source–destination pair separated by a distance $r_{s,d}$. The number of potential relays available to help the source is N. This is illustrated in Figure 10.1, where the distances between source and relay i, and relay i and destination are $r_{s,i}$ and $r_{i,d}$,

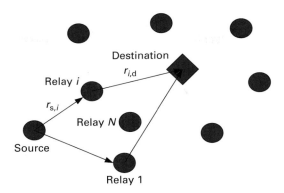

Relay i

Destination

$r_{i,d}$

$r_{s,i}$

Relay N

Source

Relay 1

Fig. 10.1 System model

respectively, and $i \in \{1, 2, \ldots, N\}$. First we analyze the performance of the single relay scenario, and later we extend the results for arbitrary finite N.

We compare the performance of two communication scenarios. In the first scenario only direct transmission between the source and destination nodes is allowed, and this accounts for conventional direct transmission. In direct transmission, if the channel link between the source and destination encounters a deep fade or strong shadowing for example, then the communication between these two nodes fails. Moreover, if the channel is slowly varying, which is the case in sensor networks due to the stationarity or limited mobility of the nodes, then the channel might remain in the deep fade state for long time (strong time correlation), hence conventional automatic repeat request (ARQ) might not help in this case.

In the second communication scenario, we consider a two phase cooperation protocol. In the first phase, the source transmits a signal to the destination, and due to the broadcast nature of the wireless medium the relay can overhear this signal. If the destination receives the packet from this phase correctly, then it sends back an acknowledgement (ACK) and the relay just idles. On the other hand, if the destination can not decode the received packet correctly, then it sends back a negative acknowledgement (NACK). In this case, if the relay was able to receive the packet correctly in the first phase, then it forwards it to the destination. So the idea behind this cooperation protocol is to introduce a new ARQ in another domain, which is the spatial domain, as the links between different pairs of nodes in the network fade independently. The assumptions of high temporal correlation and independence in the spatial domain will be verified through experiments as discussed in Section 10.4.

Next the wireless channel and system models are described. We consider a sensor network in which the link between any two nodes in the network is subject to narrowband Rayleigh fading, propagation path-loss, and additive white Gaussian noise (AWGN). The channel fades for different links are assumed to be statistically mutually independent. This is a reasonable assumption as the nodes are usually spatially well separated. For medium access, the nodes are assumed to transmit over orthogonal channels, thus no mutual interference is considered in the signal model. All nodes in the network are assumed to be equipped with single-element antennas, and transmission at all nodes is constrained to the half-duplex mode, i.e., any terminal cannot transmit and receive simultaneously.

The power consumed in a transmitting or receiving stage is described as follows. If a node transmits with power P, only $P(1 - \alpha)$ is actually utilized for RF transmission, where $(1 - \alpha)$ accounts for the efficiency of the RF power amplifier which generally has a non-linear gain function. The processing power consumed by a transmitting node is denoted by P_c. Any receiving node consumes P_r power units to receive the data. The values of the parameters α, P_r, P_c are assumed the same for all nodes in the network and are specified by the manufacturer. Following, we describe the received signal model for both direct and cooperative transmissions.

First, we describe the received signal model for the direct transmission mode. In the direct transmission scheme, which is employed in current wireless networks, each user transmits their signal directly to the next node in the route which we denote as the destination d here. The signal received at the destination d from source user s, can be modeled as

$$y_{s,d} = \sqrt{P_s^D(1 - \alpha)r_{s,d}^{-\gamma}} h_{s,d} x + n_{s,d}, \qquad (10.1)$$

where P_s^D is the transmission power from the source in the direct communication scenario, x is the transmitted data with unit power, $h_{s,d}$ is the channel fading gain between the two terminals s and d. The channel fade of any link is modeled as a zero mean circularly symmetric complex Gaussian random variable with unit variance. In (10.1), γ is the path loss exponent, and $r_{s,d}$ is the distance between the two terminals. The term $n_{s,d}$ in (10.1) denotes additive noise; the noise components throughout the paper are modeled as white Gaussian noise (AWGN) with variance N_0.

Second, we describe the signal model for cooperative transmission. The cooperative transmission scenario comprises two phases as illustrated before. The signals received from the source at the destination d and relay 1 in the first stage can be modeled respectively as,

$$y_{s,d} = \sqrt{P_s^C(1 - \alpha)r_{s,d}^{-\gamma}} h_{s,d} x + n_{s,d}, \qquad (10.2)$$

$$y_{s1} = \sqrt{P_s^C(1 - \alpha)r_{s,1}^{-\gamma}} h_{s,1} x + n_{s,1}, \qquad (10.3)$$

where P_s^C is the transmission power from the source in the cooperative scenario. The channel gains $h_{s,d}$ and $h_{s,1}$ between the source–destination and source–relay are modeled as zero-mean circular symmetric complex Gaussian random variables with zero mean. If the SNR of the signal received at the destination from the source falls below the threshold β, the destination broadcasts a NACK. In this case, if the relay was able to receive the packet from the source correctly in the first phase, it forwards the packet to the destination with power P_1

$$y_{1,d} = \sqrt{P_1(1 - \alpha)r_{1,d}^{-\gamma}} h_{1,d} x + n_{1,d}. \qquad (10.4)$$

Cooperation results in additional spatial diversity by introducing this artificial multipath through the relay link. This can enhance the transmission reliability against wireless channel impairmens as fading, but will also result in extra receiving and processing power. In the next section, we discuss this in more detail.

10.2 Performance analysis and optimum power allocation

We characterize the system performance in terms of outage probability. Outage is defined as the event that the received SNR falls below a certain threshold β, hence, the probability of outage $\mathcal{P}_O$ is defined as

$$\mathcal{P}_O = \mathcal{P}(\text{SNR} \leq \beta). \tag{10.5}$$

If the received SNR is higher than the threshold β, the receiver is assumed to be able to decode the received message with negligible probability of error. If an outage occurs, the packet is considered lost. The SNR threshold β is determined according to the application and the transmitter/receiver structure. For example, larger values of β is required for applications with higher QoS requirements. Also increasing the complexity of transmitter and/or receiver structure, for example applying strong error coding schemes, can reduce the value of β for the same QoS requirements.

Based on the derived outage probability expressions, we can formulate a constrained optimization problem to minimize the total consumed power subject to a given outage performance. We then compare the total consumed power for the direct and cooperative scenarios to quantify the energy savings, if any, gained by applying cooperative transmission.

10.2.1 Direct transmission

As discussed before, the outage is defined as the event that the received SNR falls below a predefined threshold which we denoted by β. From the received signal model in (10.1), the received SNR from a user at a distance $r_{s,d}$ from the destination is given by

$$\text{SNR}(r_{s,d}) = \frac{|h_{s,d}|^2 \, r_{s,d}^{-\gamma} \, P_s^D(1 - \alpha)}{N_0}, \tag{10.6}$$

where $|h_{s,d}|^2$ is the magnitude square of the channel fade and follows an exponential distribution with unit mean; this follows because of the Gaussian zero mean distribution of $h_{s,d}$. Hence, the outage probability for the direct transmission mode $\mathcal{P}_{OD}$ can be calculated as

$$\mathcal{P}_{OD} = \mathcal{P}\left(\text{SNR}(r_{s,d}) \leq \beta\right) = 1 - \exp\left(-\frac{N_0 \gamma r_{s,d}^\gamma}{(1-\alpha)P_s^D}\right). \tag{10.7}$$

The total transmitted power P_{tot}^D for the direct transmission mode is given by

$$P_{tot}^D = P_s^D + P_c + P_r, \tag{10.8}$$

where P_s^D is the power consumed at the RF stage of the source node, P_c is the processing power at the source node, and P_r is the receiving power at the destination. The requirement is to minimize this total transmitted power subject to the constraint that we meet a certain QoS requirement that the outage probability is less than a given outage requirement, which we denote by $\mathcal{P}_{\text{out}}^*$. Since both the processing and receiving powers are fixed, the only variable of interest is the transmitting power P_{s}^{D}.

The optimization problem can be formulated as follows:

$$\min_{P_{\text{s}}^{\text{D}}} P_{\text{tot}}^{\text{D}}, \tag{10.9}$$

$$\text{s.t.} \ \mathcal{P}_{\text{OD}} \leq \mathcal{P}_{\text{out}}^*. \tag{10.10}$$

The outage probability $\mathcal{P}_{\text{OD}}$ is a decreasing function in the power P_{s}^{D}. Substituting $\mathcal{P}_{\text{out}}^*$ in the outage expression in (10.7), we get after some simple arithmetics that the optimal transmitting power is given by

$$P_{\text{s}}^{\text{D}*} = -\frac{\beta N_0 r_{\text{s,d}}^{\gamma}}{(1-\alpha)\ln(1-\mathcal{P}_{\text{out}}^*)}. \tag{10.11}$$

The minimum total power required for direct transmission in order to achieve the required QoS requirement is therefore given by

$$P_{\text{tot}}^* = P_{\text{c}} + P_{\text{r}} - \frac{\beta N_0 r_{\text{s,d}}^{\gamma}}{(1-\alpha)\ln(1-\mathcal{P}_{\text{out}}^*)}. \tag{10.12}$$

In the next subsection we formulate the optimal power allocation problem for the cooperative communication scenario.

10.2.2   Cooperative transmission

For the optimal power allocation problem in cooperative transmission, we consider two possible scenarios.

- In the first scenario, the relay is allowed to transmit with different power than the source and hence the optimization space is two-dimensional: source and relay power allocations. The solution for this setting provides the minimum possible total consumed power. However, the drawback of this setting is that the solution for the optimization problem is complex and might not be feasible to implement in sensor nodes.
- The second setting that we consider is constraining the source and relay nodes to transmit with equal power. This is much easier to implement as the optimization space is one dimensional in this case, moreover, a relaxed version of the optimization problem can render a closed form solution.

Clearly the solution of the equal power allocation problem provides a suboptimal solution to the general case in which we allow different power allocations at the source and the relay. It is interesting then to investigate the conditions under which these two power allocation strategies have close performance.

First, we characterize the optimal power allocations at the source and relay nodes. Consider a source–destination pair that are $r_{\text{s,d}}$ units distance. Let us compute the conditional outage probability for given locations of the source and the helping relay. As discussed before, cooperative transmission encompasses two phases. Using (10.2), the SNR received at the destination d and relay 1 from the source s in the first phase are given by

$$\text{SNR}_{\text{s,d}} = \frac{|h_{\text{s,d}}|^2 \, r_{\text{s,d}}^{-\gamma} \, P_{\text{s}}^C (1 - \alpha)}{N_0}, \tag{10.13}$$

$$\text{SNR}_{\text{s,1}} = \frac{|h_{\text{s,1}}|^2 \, r_{\text{s,1}}^{-\gamma} \, P_{\text{s}}^C (1 - \alpha)}{N_0}. \tag{10.14}$$

While from (10.4), the SNR received at the destination from the relay in the second phase is given by

$$\text{SNR}_{\text{1,d}} = \frac{|h_{\text{1,d}}|^2 \, r_{\text{1,d}}^{-\gamma} \, P_1 (1 - \alpha)}{N_0}. \tag{10.15}$$

Note that the second phase of transmission is only initiated if the packet received at the destination from the first transmission phase is not correctly received. The terms $|h_{\text{s,d}}|^2$, $|h_{\text{s,1}}|^2$, and $|h_{\text{1,d}}|^2$ are mutually independent exponential random variables with unit mean.

The outage probability of the cooperative transmission $\mathcal{P}_{\text{OC}}$ can be calculated as follows

$$\begin{aligned}
\mathcal{P}_{\text{OC}} &= \mathcal{P}\left((\text{SNR}_{\text{s,d}} \leq \beta) \cap (\text{SNR}_{\text{s,1}} \leq \beta)\right) \\
&\quad + \mathcal{P}\left((\text{SNR}_{\text{s,d}} \leq \beta) \cap (\text{SNR}_{\text{l,d}} \leq \beta) \cap (\text{SNR}_{\text{s,1}} > \beta)\right) \\
&= \left(1 - f(r_{\text{s,d}}, P_{\text{s}}^C)\right)\left(1 - f(r_{\text{s,1}}, P_{\text{s}}^C)\right) \\
&\quad + \left(1 - f(r_{\text{s,d}}, P_{\text{s}}^C)\right)\left(1 - f(r_{\text{l,d}}, P_l)\right) f(r_{\text{s,1}}, P_{\text{s}}^C), \tag{10.16}
\end{aligned}$$

where

$$(x, y) = \exp\left(-\frac{N_0 \beta x^\gamma}{y(1 - \alpha)}\right). \tag{10.17}$$

The first term in the above expression corresponds to the event that both the source–destination and the source–relay channels are in outage, and the second term corresponds to the event that both the the source–destination and the relay–destination channels are in outage while the source–relay channel is not. The above expression can be simplified as follows

$$\mathcal{P}_{\text{OC}} = \left(1 - f(r_{\text{s,d}}, P_{\text{s}}^C)\right)\left(1 - f(r_{\text{l,d}}, P_l) f(r_{\text{s,1}}, P_{\text{s}}^C)\right). \tag{10.18}$$

The total average consumed power for cooperative transmission to transmit a packet is given by

$$\begin{aligned}
\text{E}[P_{\text{tot}}^C] &= (P_{\text{s}}^C + P_{\text{c}} + 2P_{\text{r}})\mathcal{P}(\text{SNR}_{\text{s,d}} \geq \beta) \\
&\quad + (P_{\text{s}}^C + P_{\text{c}} + 2P_{\text{r}})\mathcal{P}(\text{SNR}_{\text{s,d}} < \beta)\mathcal{P}(\text{SNR}_{\text{s,1}} < \beta) \\
&\quad + (P_{\text{s}}^C + P_1 + 2P_{\text{c}} + 3P_{\text{r}})\mathcal{P}(\text{SNR}_{\text{s,d}} < \beta)\mathcal{P}(\text{SNR}_{\text{s,1}} > \beta),
\end{aligned}$$

where the first term in the right-hand side corresponds to the event that the direct link in the first phase is not in outage, therefore, the total consumed power is only given by that of the source node, and the 2 in front of the received power term P_r is to account for the relay receiving power. The second term in the summation corresponds to the event that both the direct and the source–relay links are in outage, hence the total consumed power is still given as in the first term. The last term in the total summation accounts for the event that the source–destination link is in outage while the source–relay link is not, and hence we need to account for the relay transmitting and processing powers, and the extra receiving power at the destination. Using the Rayleigh fading channel model, the average total consumed power can be given as follows

$$
\begin{aligned}
P_{\text{tot}}^C = {} & (P_s^C + P_c + 2P_r) f(r_{s,d}, P_s^C) \\
& + (P_s^C + P_c + 2P_r)\left(1 - f(r_{s,d}, P_s^C)\right)\left(1 - f(r_{s,l}, P_s^C)\right) \\
& + (P_s^C + P_1 + 2P_c + 3P_r)\left(1 - f(r_{s,d}, P_s^C)\right) \times f(r_{s,l}, P_s^C).
\end{aligned}
$$

(10.19)

We can formulate the power minimization problem in a similar way to (10.10) with the difference that there are two optimization variables in the cooperative transmission mode, namely, the transmit powers P_s^C and P_1 at the source and relay nodes respectively. The optimization problem can be stated as follows

$$
\min_{P_s^C, P_1} P_{\text{tot}}^C(P_s^C, P_1),
$$

(10.20)

$$
\text{s.t. } \mathcal{P}_{\text{OC}}(P_s^C, P_1) \leq \mathcal{P}_{\text{out}}^*.
$$

(10.21)

This optimization problem is nonlinear and does not admit a closed form solution. Therefore we resort to numerical optimization techniques in order to solve for this power allocation problem at the relay and source nodes, and the results are shown in the simulations section.

In the above formulation we considered optimal power allocation at the source and relay node in order to meet the outage probability requirement. The performance attained by such an optimization problem provides a benchmark for the cooperative transmission scheme. However, in a practical setting, it might be difficult to implement such a complex optimization problem at the sensor nodes. A more practical scenario would be that all the nodes in the network utilize the same power for transmission. Denote the equal transmission power in this case by P_{CE}; the optimization problem can then be formulated as

$$
\min_{P_{\text{CE}}} P_{\text{tot}}^C(P_{\text{CE}}),
$$

$$
\text{s.t. } \mathcal{P}_{\text{OC}}(P_{\text{CE}}) \leq \mathcal{P}_{\text{out}}^*.
$$

Beside being a one-dimensional optimization problem that can be easily solved, the problem can be relaxed to render a closed form solution. Note that at enough high SNR the following approximation holds $\exp(-x) \simeq (1 - x)$; where x here is proportional to $1/SNR$.

Using the above approximation in (10.19), and after some mathematical manipulation, the total consumed power can be approximated as follows

$$P_{\text{tot}}^C \simeq P_{\text{CE}} + P_{\text{c}} + 2P_{\text{r}} + (P_{\text{CE}} + P_{\text{c}} + P_{\text{r}}) \frac{k_1}{P_{\text{CE}}} - (P_{\text{CE}} + P_{\text{c}} + P_{\text{r}}) \frac{k_1 k_2}{P_{\text{CE}}^2}.$$

Similarly, the outage probability can be written as follows

$$\mathcal{P}_{\text{OC}} \simeq \frac{k_1 k_2}{P_{\text{CE}}^2} + \frac{k_1 k_3}{P_{\text{CE}}^2} - \frac{k_1 k_2 k_3}{P_{\text{CE}}^3}, \tag{10.22}$$

where $k_1 = \beta N_0 r_{\text{s,d}}^{\gamma}/(1 - \alpha)$, $k_2 = \beta N_0 r_{\text{s,l}}^{\gamma}/(1 - \alpha)$, and $k_3 = \beta N_0 r_{\text{l,d}}^{\gamma}/(1 - \alpha)$. This is a constrained optimization problem in one variable and its Lagrangian is given by

$$\frac{\partial P_{\text{tot}}^C}{\partial P_{\text{CE}}} + \lambda \frac{\partial \mathcal{P}_{\text{OC}}}{\partial P_{\text{CE}}} = 0, \tag{10.23}$$

where the derivatives of the total power consumption P_{tot}^C and the outage probability $\mathcal{P}_{\text{OC}}$ with respect to the transmit power P_{CE} are given by

$$\frac{\partial P_{\text{tot}}^C}{\partial P_{\text{CE}}} = 1 + \frac{k_1 k_2 - (P_{\text{c}} + P_{\text{r}})k_1}{P_{\text{CE}}^2} + \frac{2k_1 k_2 (P_{\text{c}} + P_{\text{r}})}{P_{\text{CE}}^3}; \tag{10.24}$$

$$\frac{\partial \mathcal{P}_{\text{OC}}}{\partial P_{\text{CE}}} = \frac{-2(k_1 k_2 + k_1 k_3)}{P_{\text{CE}}^3} + \frac{3k_1 k_2 k_3}{P_{\text{CE}}^4}, \tag{10.25}$$

respectively. Substituting the derivatives in (10.25) into the Lagrangian in (10.23), and doing simple change of variables $1/P_{\text{CE}} = x$, the Lagrangian can be written in the following simple polynomial form

$$1 + (k_1 k_2 - (P_{\text{c}} + P_{\text{r}})k_1)x^2 + 2(k_1 k_2 (P_{\text{c}} + P_{\text{r}}))$$
$$- \lambda(k_1 k_2 + k_1 k_3))x^3 + 3\lambda k_1 k_2 k_3 x^4 = 0,$$

under the outage constraint

$$(k_1 k_2 + k_1 k_3) x^2 - k_1 k_2 k_3 x^3 = \mathcal{P}_{\text{out}}^*. \tag{10.26}$$

The constraint equation above is only a polynomial of order three, so it can be easily solved and we can find the root that minimizes the cost function.

10.3 Multi-relay scenario

In this section, we extend the study to the case when there is more than one potential relay. Let N be the number of relays assigned to help a given source. The cooperation protocol then works as an N-stage ARQ protocol as follows. The source node transmits its packets to the destination and the relays try to decode this packet. If the destination does not decode the packet correctly, it sends a NACK that can be heard by the relays. If the first relay is able to decode the packet correctly, it forwards the packet with power P_1 to the destination. If the destination does not receive correctly again, then it sends a NACK and the second candidate relay, if it received the packet correctly, forwards the source's packet to the destination with power P_2. This is repeated until the destination gets the packet correctly or the N trials corresponding to the N relays are exhausted.

We model the status of any relay by 1 or 0, corresponding to whether the relay received the source's packet correctly or not, respectively. Writing the status of all the relays in a column vector results in a $N \times 1$ vector whose entries are either 0 or 1. Hence, the decimal number representing this $N \times 1$ vector can take any integer value between 0 and $2^N - 1$. Denote this vector by S_k, where $k \in \{0, 1, 2, \dots, 2^N - 1\}$.

For a given status of the N relays, an outage occurs if and only if the links between the relays that decoded correctly and the destination are all in outage. Denote the set of the relays that received correctly by $\chi(S_k) = \{i : S_k(i) = 1, 1 \le i \le N\}$, and $\chi^C(S_k)$ as the set of relays that have not received correctly, i.e., $\chi^C(S_k) = \{i : S_k(i) = 0, 1 \le i \le N\}$. The conditional probability of outage given the relays status S_k is thus given by

$$\mathcal{P}_{\mathrm{OC}|S_k} = P\left(\mathrm{SNR}_{\mathrm{s,d}} \le \beta \bigcap_{j \in \chi(S_k)}^{N} (\mathrm{SNR}_{jd} \le \beta)\right), \tag{10.27}$$

The total outage probability is thus given by

$$\mathcal{P}_{\mathrm{OC}} = \sum_{k=0}^{2^N - 1} P(S_k)\mathcal{P}_{\mathrm{OC}|S_k}. \tag{10.28}$$

We then need to calculate the probability of the set S_k, which can then be written as

$$P(S_k) = P\left(\bigcap_{i \in \chi(S_k)} (\mathrm{SNR}_{si} \ge \beta) \bigcap_{j \in \chi^c(S_k)} (\mathrm{SNR}_{sj} \le \beta)\right). \tag{10.29}$$

The average outage probability expression can thus be given by

$$\mathcal{P}_{\mathrm{OC}} = \sum_{k=0}^{2^N - 1} \left(1 - f(r_{\mathrm{s,d}}, P_{\mathrm{s}}^C)\right) \prod_{j \in \chi(S_k)} \left(1 - f(r_{jd}, P_j)\right) f(r_{sj}, P_{\mathrm{s}}^C) \tag{10.30}$$

$$\cdot \prod_{j \in \chi^C(S_k)} \left(1 - f(r_{sj}, P_{\mathrm{s}}^C)\right). \tag{10.31}$$

where P_j, $j \in \{1, 2, \dots, N\}$, is the power allocated to the j-th relay.

Next we compute the average total consumed power for the N-relays scenario. First, we condition on some relays' status vector $\chi(S_k)$:

$$\mathrm{E}\left[P_{\mathrm{tot}}^C\right] = \mathrm{E}\left[\mathrm{E}\left[P_{\mathrm{tot}}^C|\chi(S_k)\right]\right] = \sum_{k=0}^{2^N - 1} P\left(\chi(S_k)\right) \mathrm{E}\left[P_{\mathrm{tot}}^C|\chi(S_k)\right]. \tag{10.32}$$

For a given $\chi(S_k)$, we can further condition on whether the source get the packet through from the first trial or not. This event happens with probability $f(r_{\mathrm{s,d}}, P_{\mathrm{s}}^C)$, and the consumed power in this case is given by

$$P_{\mathrm{tot}}^{C,1} = P_{\mathrm{s}}^C + (N + 1)P_{\mathrm{r}} + P_{\mathrm{c}}; \tag{10.33}$$

The complementary event that the source failed to transmit its packet from the direct transmission phase happens with probability $1 - f(r_{s,d}, P_s^C)$, and this event can be further divided into two mutually exclusive events. The first is when the first $|\chi(S_k)| - 1$ relays from the set $\chi(S_k)$ fails to forward the packet and this happens with probability $\prod_{i=1}^{|\chi(S_k)|-1} (1 - f(r_{i,d}, P_i))$ and the corresponding consumed power is given by

$$P_{tot}^{C,2} = P_s^C + (N + 1 + |\chi(S_k)|) P_r + (|\chi(S_k)| + 1) P_c + \sum_{n=1}^{|\chi(S_k)|} P_{\chi(S_k)(n)}; \quad (10.34)$$

The second is when one of the intermediate relays in the set $\chi(S_k)$ successfully forwards the packet and this happens with probability $\prod_{m=1}^{j-1} (1 - f(r_{m,d}, P_m)) f(r_{j,d}, P_j)$ if this intermediate relay was relay number j, and the corresponding power is given by

$$P_{tot}^{C,3,j} = P_s^C + (N + 1 + j) P_r + (1 + j) P_c + \sum_{i=1}^{j} P_{\chi(S_k)(i)}. \quad (10.35)$$

From (10.32), (10.33), (10.34), and (10.35), the average total consumed power can be given by

$$E\left[P_{tot}^C\right] = \sum_{k=0}^{2^N-1} P(\chi(S_k)) \left\{ f(r_{s,d}, P_s^C) P_{tot}^{C,1} \right.$$

$$+ \left(1 - f(r_{s,d}, P_s^C)\right) \left[\prod_{i=1}^{|\chi(S_k)|-1} (1 - f(r_{i,d}, P_i)) P_{tot}^{C,2} \right.$$

$$\left. + \sum_{j=1}^{|\chi(S_k)|-1} \prod_{m=1}^{j-1} (1 - f(r_{m,d}, P_m)) f(r_{j,d}, P_j) P_{tot}^{C,3,j} \right] \right\}.$$

The optimization problem can then be written as

$$\min_{\mathbf{P}} P_{tot}^C(\mathbf{P}),$$

$$\text{s.t. } \mathcal{P}_{OC}(\mathbf{P}) \leq \mathcal{P}_{out}^*.$$

where $\mathbf{P} = \left[P_s^C, P_1, P_2, \ldots, P_N \right]^T$.

10.4 Experimental results

In the system model, it is assumed that the channel independence between the following links: the source–relay link, the source–destination link, and the relay–destination link. Moreover, a strong motivation for applying cooperative transmission instead of ARQ in the time domain, is the assumption of high temporal correlation which results in delay and requires performing interleaving at the transmitter side. In this section, we show the results for a set of experiments to justify these two fundamental assumptions.

The experiments are set up as follows. We have three wireless nodes in the experiments, one of them acts as the sender and the other two act as receivers. Each wireless node is a computer equipped with a IEEE 802.11g wireless card, specifically, we utilized three LINKSYS wireless-G USB network adaptors. The sender's role is to broadcast data packets with a constant rate, while the two receivers' role is to decode the packets and record which packet is erroneous. The traffic rate is 100 packets per second, and the size of each packet is 554 bytes (including packet headers). The two receivers are placed together, with the distance between them being 20 cm. The distance between the transmitter and the receiver is around 5 m. The experiments have been mainly conducted in office environments. The experiments results, which are illustrated next, have revealed two important observations: the channels exhibit strong time correlation for each receiver, while there is negligible dependence between the two receivers. Figure 10.2 illustrates one instantiation of the experiments. The first figure illustrates the results obtained at the first receiver and the second figure is for the second receiver.

For each figure, the horizontal axis denotes the sequence number of the first 100 000 packets, and the vertical axis denotes whether a packet is erroneous or not. First, from these results we can see that packet errors exhibit strong correlation in time. For example, for the first receiver, most erroneous packets cluster at around 22nd second and around 83rd second. Similar observations also hold for the second receiver. If we take a further look at the results we can see that in this set of experiments the duration for the cluster is around 2 s. To help better understand the time correlation of erroneous packets, we have also used a two-state Markov chain to model the channel, as illustrated in

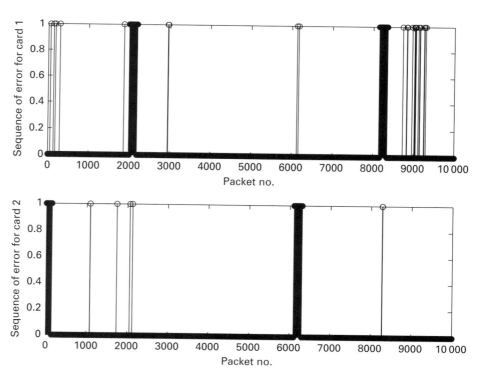

Fig. 10.2   Sequence of packet errors at the two utilized wireless cards.

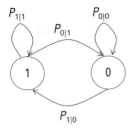

Fig. 10.3   Modelling the channel by a two (on–off) state Markov chain to study the time correlation.

Figure 10.3. In this model "1" denotes that the packet is correct, and "0" denotes that the packet is erroneous. $P_{i|j}$ denotes the transition probability from state i to state j, that is, the probability to reach state j given the previous state is i. The following transition probabilities have been obtained after using the experimental results to train the model: $P_{1|0} = 0.03$, $P_{1|1} = 0.999$, $P_{0|0} = 0.97$, $P_{0|1} = 0.001$. These results also indicate strong time correlation. For example, given the current received packet is erroneous, the probability that the next packet is also erroneous is around $P_{0|0} = 0.97$.

   Now we take a comparative look at the results obtained at the two receivers. From these results we can see that although there exists slight correlation in packet errors between the two receivers, it is almost negligible. To provide more concrete evidence of independence, we have estimated the correlation between the two receivers using the obtained experiment results. Specifically, we have measured the correlation coefficient between the received sequences at the two receivers and we found that the correlation coefficient is almost 0 which indicates a strong spatial independence between the two receivers.

10.4.1   Numerical examples

As discussed in the previous sections, there are different system parameters that can control whether we can gain from cooperation or not. Among which are the received power consumption, the processing power, the SNR threshold, the power amplifier loss, and the relative distances between the source, relay, and destination.

   In order to understand the effect of each of these parameters, we are going to study the performance of cooperative and direct transmission when varying one of these parameters and fixing the rest. This is described in more details in the following. In all of the numerical examples, the aforementioned parameters take the following values when considered fixed: $\alpha = 0.3$, $\beta = 10$, $N_0 = 10^{-3}$, $P_c = 10^{-4}$ W, $P_r = 5 \times 10^{-5}$, QoS $= P_{\text{out}}^* = 10^{-4}$. We define the cooperation gain as the ratio between the total power required for direct transmission to achieve a certain QoS, and the total power required by cooperation to achieve the same QoS.

Example 10.1   We study the effect of varying the receive power P_r as depicted in Figure 10.4. We plot the cooperation gain versus the distance between the source and the destination for different values of receive power $P_r = 10^{-4}, 5 \times 10^{-5}, 10^{-5}$ W.

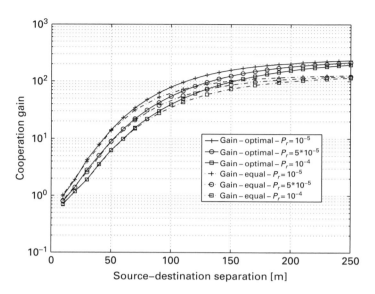

Fig. 10.4   Cooperation gain versus the source–destination distance for different values of received power consumption.

At source–destination distances below 20 m, the results reveal that direct transmission is more energy efficient than cooperation, i.e., the overhead in receive and processing power due to cooperation outweighs its gains in saving the transmit power. For $r_{s,d} > 20$ m, the cooperation gain starts increasing as the transmit power starts constituting a significant portion of the total consumed power. This ratio increases until the transmit power is the dominant part of the total consumed power and hence the cooperation gain starts to saturate.

In the plotted curves, the solid lines denote the cooperation gain when utilizing optimal power allocation at the source and the relay, while the dotted curves denote the gain for equal power allocation. For $r_{s,d} \leq 100$ m, both optimal power allocation and equal power allocation almost yield the same cooperation gain. For larger distances, however, a gap starts to appear between optimal and equal power allocation. The rationale behind these observations is that at small distances the transmit power is a small percentage of the total consumed power and hence optimal and equal power allocation almost have the same behavior, while at larger distances, transmit power plays a more important role and hence a gap starts to appear.   ▲

Example 10.2   In this example, we study the effect of changing the SNR threshold β as depicted in Figure 10.5. The distance between source and destination $r_{s,d}$ is fixed to 100 m. It is clear that the cooperation gain increases with increasing β, and that for the considered values of the system parameters, equal power allocation provides almost the same gains as optimal power allocation. In Figure 10.6 we study the effect of the power amplifier loss α.

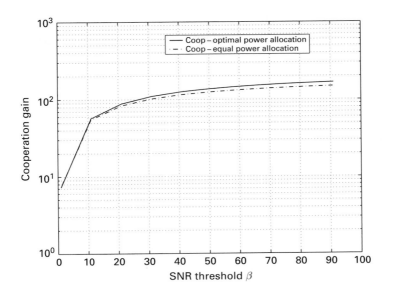

Fig. 10.5   Cooperation gain versus the SNR threshold β.

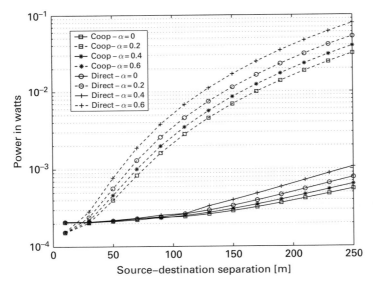

Fig. 10.6   Optimal power consumption for both cooperation and direct transmission scenarios for different values of power amplifier loss α.

In this case, we plot the total consumed power for cooperation and direct transmission versus distance for different values of α. Again below 20m separation between the source and the destination, direct transmission provides better performance over cooperation. It can also be seen from the plotted curves that the required power for direct

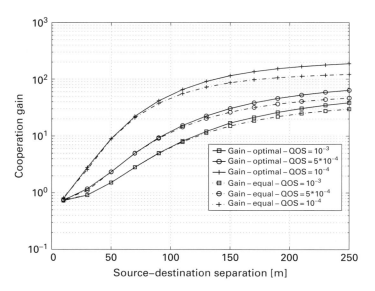

Fig. 10.7   Cooperation gain versus the source–destination distance for different values of QoS.

transmission is more sensitive to variations in α than the power required for cooperation. The reason is that the transmit power constitutes a larger portion in the total
consumed power in direct transmission than in cooperation, and hence the effect of α
is more significant. The QoS, measured by the required outage probability, has similar
behavior and the results are depicted in Figure (10.7).   ▲

Example 10.3   In this example, we study the effect of varying the relay location. We
consider three different positions for the relay, close to the source, in the middle between
the source and the destination, and close to the destination. In particular, the relay posi-
tion is taken equal to $(r_{s,1} = 0.2r_{s,d}, r_{1,d} = 0.8r_{s,d})$, $(r_{s,1} = 0.5r_{s,d}, r_{1,d} = 0.5r_{s,d})$, and
$(r_{s,1} = 0.8r_{s,d}, r_{1,d} = 0.2r_{s,d})$.

Figures 10.8 and 10.9 depict the power required for cooperation and direct transmission versus $r_{s,d}$ for equal power and optimal power allocation, respectively. In the equal
power allocation scenario, the relay in the middle gives the best results, and the other
two scenarios, relay close to source and relay close to destination provide the same
performance.

This can be expected because for the equal power allocation scenario the problem
becomes symmetric in the source–relay and relay–destination distances. For the optimal
power allocation scenario depicted in Figure 10.9, the problem is no more symmetric because different power allocation is allowed at the source and relay. In this case,
numerical results show that the closer the relay to the source the better the performance.
The intuition behind this is that when the relay is closer to the source, the source–relay
channel is very good and almost error-free.

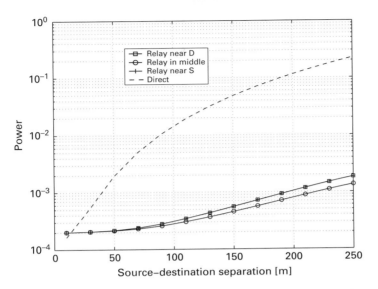

Fig. 10.8   Optimal consumed power versus distance for different relay locations for equal power allocation at source and relay.

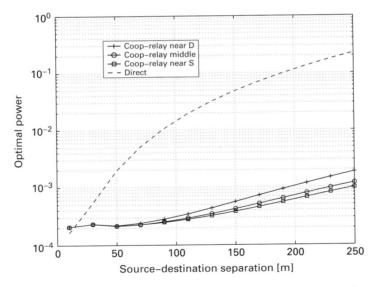

Fig. 10.9   Optimal consumed power versus distance for different relay locations for optimal power allocation at source and relay.

From both figures, it is also clear that for small source–destination separation $r_{s,d}$, equal and optimal power allocation almost provide the same cooperation gain while for larger $r_{s,d}$ optimal power allocation provides more gain. Another important observation is that at small distances below 100 m, the location of the relay does not affect

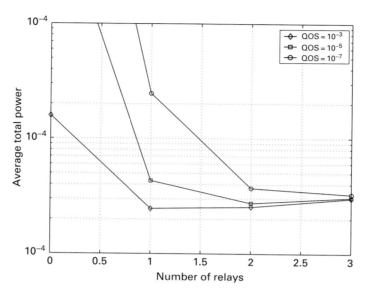

Fig. 10.10   Optimal Consumed Power versus number of relays for different values of required outage probability.

the performance much. This makes the algorithms required to select a relay in cooperative communications simpler to implement for source–destination separations in this range. Finally, the threshold behavior below 20 m still appears where direct transmission becomes more energy efficient.

                                                                    ▲

Example 10.4   Figure 10.10 depicts the multiple relays scenario for different values of outage probability $\mathcal{P}_{out}^*$. The results are depicted for a source–destination distance of 100 m, and for $N = 0, 1, 2, 3$ relays, where $N = 0$ refers to direct transmission. As shown in Figure 10.10, for small values of required outage probability, one relay is more energy efficient than two or three relays. As we increase the required QoS, reflected by $\mathcal{P}_{out}^*$, the optimal number of relays increases. Hence, our analytical framework can also provide guidelines to determining the optimal number of relays under any given scenario.

                                                                    ▲

10.5   Chapter summary and bibliographical notes

In this chapter, the energy efficiency of cooperation in wireless networks is studied under a practical setting where the extra overhead of cooperation is taken into account. The approach taken was to formulate a constrained optimization problem to minimize the total consumed power under a given QoS requirement. The numerical results reveal that for short distance separations between the source and the destination, e.g., below

a threshold of 20 m, the overhead of cooperation outweighs its gains and direct transmission is more efficient. Above that threshold, cooperation gains can be achieved. It
was also shown that simple equal power allocation at the source and the relay achieves
almost the same gains as optimal power allocation at these two nodes for distances
below 100 m, for the specific parameters used.

Furthermore, choosing the optimal relay location for cooperation plays an important
role when the source–destination separation exceeds 100 m, and the best relay location
depends on the power allocation scheme, whether optimal or equal. The results can
also be used to provide guidelines in determining the optimal number of relays for any
given communication setup; increasing the number of relays is not always beneficial. In
summary, caution must be taken before applying cooperative communications to sensor networks, in particular whether we should apply cooperation or not, whether equal
power allocation is good enough, how to choose a partner or a relay for cooperation,
and how many relays should be assigned to help the source. Further discussion on these
topics can be found in [161] and [154]

Exercises

10.1 Derive from scratch the average outage probability expression in (10.18).
  (a) First, find an expression for the unconditional outage probability given the
      channel state information;
  (b) second, find the average outage expression by averaging over the channel
      statistics.
  Assume that different channel pairs fade independently.

10.2 In this chapter, incremental-relaying with decode-and-forward was utilized at
  the relay. Solve the optimization problem for the equal power allocation scenario
  when amplify-and-forward is used instead with incremental-relaying.
  (a) What is the impact of using amplify-and-forward at the relay on the system
      performance?
  (b) Compare the cooperation gain when using amplify-and-forward versus
      that of decode-and-forward for a fixed source–destination separation and
      different relay location on the line joining the source and the destination.
  (c) Now fix the relay location to be always at the middle between the source and
      the destination, and vary the source–destination separation.

10.3 Formulate a relaxed version of the power allocation optimization problem in
  (10.21). Find the corresponding Lagrangian at high signal-to-noise ratio.

10.4 Write down closed form expressions for the outage probability and total power
  consumption for the two-relay scenario. Consider arbitrary relay locations and
  independent channel fading among all the links.

10.5 Given two binary sequences that represent the packet errors at two receivers.
  (a) Write a Matlab code that estimates the transition probabilities for the two-
      state Markov model that describes the binary sequences.
  (b) Write a Matlab code to estimate the cross-correlation between the two
      sequences.

10.6 As shown in the results in this chapter, direct transmission might be more energy efficient than cooperation depending on the relay location and source–destination separation. Let us define the cooperation region to be the set of relay locations that results in cooperation being more energy efficient than direct transmission.

    (a) For a source–destination separation of 100 m. Write Matlab code to specify the cooperation region and draw it. Consider optimal power allocation in your calculation.

    (b) Find the cooperation region for the above scenario for a source–destination separation of 500 m and 1000 m.

    (c) Repeat the above assuming equal power allocation at the source and the relay.

10.7 A cooperative communication system is a generalization of a MIMO system when the antennas are distributed in a geographic area. Compare the performance of a 2×1 MIMO system that utilizes an Alamouti scheme and a single-relay cooperative communication system considering the overhead of cooperation. Can cooperation be more energy efficient than a MIMO system? If yes, determine under what conditions. In your calculations consider a propagation path-loss exponent of 2.5, 3, 3.5, and 4. What is the effect of increasing the propagation path-loss on the performance of the two systems?

10.8 Repeat the previous problem for a 1×2 MIMO system that utilizes a maximal ratio combiner at the receiver. Is there a difference, and why?

Part III

Cooperative networking

11 Cognitive multiple access via cooperation

Despite the promised gains of cooperative communication demonstrated in many previous works, the impact of cooperation at higher network levels is not yet completely understood. In the previous chapters, it was assumed that the user always has a packet to transmit, which is not generally true in a wireless network. For example, in a network, most of the sources are bursty in nature, which leads to periods of silence in which the users may have no data to transmit. Such a phenomenon may affect important system parameters that are relevant to higher network layers, for example, buffer stability and packet delivery delay. We focus on the multiple access layer in this chapter. One can ask many important questions now. Can we design cooperation protocols that take these higher layer network features into account? Can the gains promised by cooperation at the physical layer be applied to the multiple access layer? More specifically, what is the impact of cooperation on important multiple access performance metrics such as stable throughput region and packet delivery delay?

In this chapter, we try to address all of these important questions to demonstrate the possible gains of cooperation at the multiple access layer. A slotted time division multiple access (TDMA) framework in which each time slot is assigned only to one terminal, i.e., orthogonal multiple access is considered. If a user does not have a packet to transmit in his time slot, then this time slot is not utilized. These unutilized time slots are wasted channel resources that could be used to enhance the system performance. In fact, the concept of cognitive radio has been introduced recently to allow the utilization of unused channel resources by enabling the operation of a secondary system overlapping with the original system.

This chapter presents a cognitive multiple access strategy with the concept of cooperation. The relay is cognitive in that it tries to "*smartly*" utilize the periods of source silence to cooperate with other terminals in the network, i.e., to increase the reliability of communications against random channel fades. In particular, when the relay senses the channel for empty time slots, the slots are then used to help other users in the network by forwarding any of their packets that were lost in previous transmissions. Thus the protocol has cooperative cognitive aspects in the sense that the unused channel resources are being utilized by the relay to cooperate with other users in the network. It should be pointed out that the cooperative protocol does not result in any bandwidth loss because there are no channel resources reserved for the relay to cooperate. We demonstrate later that this important feature of the protocol can lead to significant gains, particularly in high spectral efficiency regimes.

Two protocols to implement this new cooperative cognitive multiple access (CCMA) strategy are developed. The first protocol is CCMA within a single frame (CCMA-S), where the relay keeps a lost packet for no more than one time frame and then drops the packet if unable to deliver it successfully to the destination. The dropped packet then has to be retransmitted by the originating user. It turns out that in this protocol the relay's queue is always bounded, and that the terminals queues are interacting. To analyze the stability of the system's queues we resort to a stochastic dominance approach. The interaction between the queues arise due to the role of the relay in enabling cooperation. To analyze the stability of CCMA-S, we present a dominant system to resolve this interaction.

The second protocol, named CCMA-multiple-enhanced (CCMA-Me), differs in the way the relay handles the lost packets. In CCMA-Me, if a packet is captured by the relay and not by the destination, then this packet is removed from the corresponding user's queue and it becomes the relay's responsibility to deliver it to the destination. The term enhanced refers to the design of the protocol, as the relay only helps the users with inferior channel gains, which are reflected in the distances from the terminals to the destination. Unlike CCMA-S, the size of the relay's queue can possibly be unbounded and so its stability must be taken into consideration.

In addition to characterizing the stable throughput region, an analysis for the queuing delay performance is presented. Delay is an important performance measure and network parameter that may affect the tradeoff between the rate and reliability of communication. We consider a symmetric two-user scenario when analyzing the delay performance of the protocols.

11.1   System model

Let us consider the uplink of a TDMA system. The network consists of a finite number $M < \infty$ of source terminals numbered $1, 2, \ldots, M$, a relay node l,[1] and a destination node d, see Figure 11.1. Let $\mathcal{T} = \{\mathcal{M}, l\}$ denote the set of transmitting nodes, where $\mathcal{M} = \{1, 2, \ldots, M\}$ is the set of source terminals, and $\mathcal{D} = \{l, d\}$ denotes the set of receiving nodes or possible destinations. For simplicity of presentation, in the following we use terminal to refer to a source terminal.

First, we describe the queuing model for the multiple access channel. Each of the M terminals and the relay l have an infinite buffer for storing fixed length packets. The channel is slotted, and a slot duration is equal to a packet duration. The arrival process at any terminal's queue is independent identically distributed (i.i.d.) from one slot to another, and the arrival processes are independent from one terminal to another. The arrival process at the i-th queue ($i \in \{1, 2, \ldots, M\}$) is assumed stationary with mean λ_i. Terminals access the channel by dividing the channel resources, time in this case, among them, hence, each terminal is allocated a fraction of the time. Let $\Omega = [\omega_1, \omega_2, \ldots, \omega_M]$

---

[1] We use l here to denote the relay in order not to confuse with r, which denotes distance.

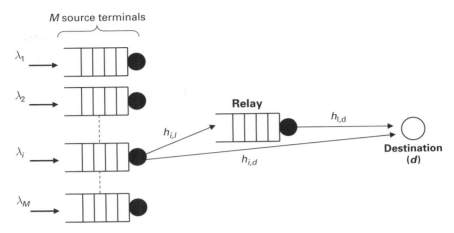

Fig. 11.1   Network and channel model.

denote a resource-sharing vector, where $\omega_i \geq 0$ is the fraction of the time allocated to terminal $i \in \mathcal{M}$, or it can represent the probability that terminal i is allocated the whole time slot. The later notation is used as it allows to consider fixed duration time slots with continuous values of the resource-sharing vector. A time frame is defined as M consecutive time slots. The set of all feasible resource-sharing vectors is specified as follows:

$$F \triangleq \left\{ \Omega = [\omega_1, \omega_2, \ldots, \omega_M] \in \Re_+^M : \sum_{i \in \mathcal{M}} \omega_i \leq 1 \right\}. \tag{11.1}$$

A fundamental performance measure of a communication network is the stability of its queues. Stability can be loosely defined as having a certain quantity of interest kept bounded. In our case, we are interested in the queue size and the packet delivery delay to be bounded. More rigorously, stability can be defined as follows. Denote the queue sizes of the transmitting nodes at any time t by the vector $\mathbf{Q}^t = [Q_i^t, i \in \mathcal{T}]$. Let us adopt the following definition of stability:

DEFINITION 11.1.1   *Queue $i \in \mathcal{T}$ of the system is stable, if*

$$\lim_{t \to \infty} \Pr\left[Q_i^t < x \right] = F(x) \quad and \quad \lim_{x \to \infty} F(x) = 1. \tag{11.2}$$

If

$$\lim_{x \to \infty} \lim_{t \to \infty} \inf \Pr\left[Q_i^t < x \right] = 1, \tag{11.3}$$

the queue is called substable [209].

From the definition, if a queue is stable then it is also substable. If a queue is not substable, then it is unstable. An arrival rate vector $[\lambda_1, \lambda_2, \ldots, \lambda_M]$ is said to be stable if there exists a resource-sharing vector $\Omega \in F$ such that all the queues in $\mathcal{T} = \{\mathcal{M}, l\}$ are stable. The multi-dimensional stochastic process $\mathbf{Q}^t$ can be easily shown to be an irreducible and aperiodic discrete-time Markov process with a countable number of states and state space $\in Z_+^{M+1}$. For such a Markov chain, the process is stable if and only if a positive probability exists for every queue being empty [127], i.e.,

$$\lim_{t \to \infty} \Pr[Q_i(t) = 0] > 0, \qquad i \in \mathcal{T}. \tag{11.4}$$

If the arrival and service processes of a queuing system are strictly stationary, then one can apply Loynes' theorem to check for stability conditions [124]. This theorem states that if the arrival process and the service process of a queuing system are strictly stationary, and the average arrival rate is less than the average service rate, then the queue is stable; if the average arrival rate is greater than the average service rate then the queue is unstable.

Next, we describe the physical channel model. The wireless channel between any two nodes in the network is modeled as a Rayleigh narrowband flat-fading channel with additive Gaussian noise. The transmitted signal also suffers from propagation path loss that causes the signal power to attenuate with distance. The signal received at a receiving node $j \in \mathcal{D}$ from a transmitting node $i \in \mathcal{T}$ at time t can be modeled as

$$y_{i,j}^t = \sqrt{Gr_{i,j}^{-\gamma}} h_{i,j}^t x_i^t + n_{i,j}^t, \quad i \in \mathcal{T}, \ j \in \mathcal{D}, \ i \neq j, \tag{11.5}$$

where G is the transmitting power, assumed to be the same for all transmitting terminals, $r_{i,j}$ denotes the distance between the two nodes i, j, γ is the path loss exponent, and $h_{i,j}^t$ captures the channel fading coefficient at time t and is modeled as an i.i.d. zero-mean, circularly symmetric complex Gaussian random process with unit variance. The term x_i^t denotes the transmitted packet with average unit power at time t, and $n_{i,j}^t$ denotes i.i.d. additive white Gaussian noise with zero mean and variance N_0. Since the arrival, channel gains, and additive noise processes are assumed to be stationary, we can drop the index t without loss of generality. We consider the scenario in which the fading coefficients are known to the appropriate receivers, but are not known at the transmitters.

In this chapter, we characterize the success and failure of packet reception by outage events and outage probability, which is defined as follows. For a target signal-to-noise (SNR) ratio β, if the received SNR as a function of the fading realization h is given by $\mathrm{SNR}(h)$, then the outage event O is the event that $\mathrm{SNR}(h) < \beta$, and $\Pr[\mathrm{SNR}(h) < \beta]$ denotes the outage probability, also defined in previous chapters. The SNR threshold β is a function of different parameters in the communication system; it is a function of the application, data rate, signal-processing applied at the encoder/decoder sides, error-correction codes, and other factors. For example, varying the data rate and fixing all other parameters, the required SNR threshold β to achieve certain system performance is a monotonically increasing function of the data rate. Also, increasing the signal-processing and encode/decoder complexity in the physical layer reduces the required SNR threshold β for a required system performance.

For the channel model in (11.5), the received SNR of a signal transmitted between two terminals i and j can be specified as follows:

$$\mathrm{SNR}_{i,j} = \frac{|h_{i,j}|^2 r_{i,j}^{-\gamma} G}{N_0}, \tag{11.6}$$

where $\mid h_{i,j} \mid^2$ is the magnitude channel gain square and has an exponential distribution with unit mean. The outage event for a SNR threshold β is equivalent to

$$O_{i,j} = \{h_{i,j} : \mathrm{SNR}_{i,j} < \beta\} = \left\{ h_{i,j} :\mid h_{i,j} \mid^2 < \frac{\beta N_0 r_{i,j}^\gamma}{G} \right\} \qquad (11.7)$$

Accordingly, the probability of outage is given by

$$\Pr\left[O_{i,j}\right] = \Pr\left[\mid h_{i,j} \mid^2 < \frac{\beta N_0 r_{i,j}^\gamma}{G}\right] = 1 - \exp\left(-\frac{\beta N_0 r_{i,j}^\gamma}{G}\right), \qquad (11.8)$$

where the above follows from the exponential distribution of the received SNR. Since we will use the above expression frequently in our subsequent analysis, and for compactness of representation, we will use the following notation to denote the success probability (no outage) at SNR threshold β:

$$f_{i,j} = \exp\left(-\frac{\beta N_0 r_{i,j}^\gamma}{G}\right). \qquad (11.9)$$

11.2   Cooperative cognitive multiple access (CCMA) protocols

In a TDMA system without relays, if a terminal does not have a packet to transmit, its time slot remains idle, i.e., the slot becomes a wasted channel resource. We consider the possibility of utilizing these wasted channel resources by employing a relay. In this section, we introduce the cognitive multiple access strategy based on employing relays in the wireless network. Furthermore, we develop two protocols to implement this new approach. We assume that the relay can sense the communication channel to detect empty time slots and we assume that the errors and delay in the packet acknowledgement feedback are negligible.

First, let us describe the multiple access strategy:

- Due to the broadcast nature of the wireless medium, the relay can listen to the packets transmitted by the terminals to the destination.
- If the packet is not received correctly by the destination, the relay stores this packet in its queue, given that it was able to decode this packet correctly. Thus, the relay's queue contains packets that have not been transmitted successfully by the terminals.
- At the beginning of each time slot, the relay listens to the channel to check whether the time slot is empty (not utilized for packet transmission) or not.
- If the time slot is empty the relay will retransmit the packet at the head of its queue, hence utilizing a channel resource that was previously wasted in a TDMA system without a relay.

Moreover, this introduces spatial diversity in the network as the channel fades between different nodes in the network are independent. In the following we present two different protocols to implement the proposed cognitive multiple access approach. The protocols are cognitive in the sense that the system introduces a relay in the network that tries to detect unused channel resources and uses them to help other terminals by forwarding packets lost in previous transmissions.

11.2.1   CCMA-single frame

The first protocol that we present is cooperative cognitive multiple access within a single frame duration or (CCMA-S). The characteristic feature of CCMA-S is that any terminal keeps its lost packet in its queue until it is captured successfully at the destination. CCMA-S operates according to the following rules:

- Each terminal transmits the packet at the head of its queue in its assigned slot, if the terminal's queue is empty the slot is free.
- If destination receives a packet successfully, it sends an ACK which can be heard by both the terminal and the relay. If the destination does not succeed in receiving the packet correctly but the relay does, then the relay stores this packet at the end of its queue. The corresponding terminal still keeps the lost packet at the head of its queue
- The relay senses the channel, and at each empty time slot the relay transmits the packet at the head of its queue, if its queue is nonempty. If the transmitted packet is received correctly by the destination it sends an ACK and the corresponding terminal removes this packet from its queue.
- If the relay does not succeed in delivering a packet to the destination during a time frame starting from the time it received this packet, then the relay drops the packet from its queue. In this case the corresponding terminal becomes responsible for delivering the packet to the destination.

Following are some important remarks on the above protocol. According to the above description of CCMA-S, the relay's queue has always a finite number of packets (at most it has M backlogged packets). This follows because, according to the protocol, the relay can have at most one packet from each terminal. Thus the stability of the system is only determined by the stability of the terminals' queues. Secondly, successful service of a packet in a frame depends on whether the other terminals have idle time slots or not. Therefore individual terminals' queues are *interacting*.

11.2.2   CCMA-multiple frames

In this section, we describe the implementation of protocol CCMA-M. The main difference between protocols CCMA-S and CCMA-M is in the role of the relay and the behavior of the terminals' regarding their backlogged packets. More specifically, a terminal removes a packet from its queue if it is received successfully by either the destination or the relay. CCMA-M operates according to the following rules:

- Each terminal transmits the packet at the head of its queue in its assigned time slot. If the queue is empty the time slot is free.

- If a packet is received successfully by either the destination or the relay, the packet is removed from the terminal's queue (the relay needs to send an ACK if one is not heard by the destination in this case).
- If a packet is not received successfully by both the relay and the destination, the corresponding terminal retransmits this packet in its next assigned time slot.
- At each sensed empty time slot, the relay retransmits the packet at the head of its queue.

One can now point out the differences between the queues in CCMA-S and CCMA-M:

- The size of the relay's queue can possibly grow in CCMA-M as it can have more than one packet from each terminal; however, it can not exceed size M in CCMA-S.
- The terminal's queues in CCMA-M are not interacting as in CCMA-S. This is because the terminal removes the packets which were received correctly by the relay or the destination. In other words, servicing the queue of any terminal depends only on the channel conditions from that terminal to the destination and relay, and does not depend on the status of the other terminals' queues.
- The stable throughput region of CCMA-M requires studying the stability of both the terminals' and the relay's queues.

11.3 Stability analysis

The aim of this section is to characterize the stable throughput region of the presented cooperation protocols. Furthermore, we show comparison for the stable throughput regions of TDMA without relaying, ALOHA, selection decode-and-forward, and incremental decode-and-forward.

11.3.1 Stability analysis of CCMA-S

11.3.1.1 The two-terminals case

In CCMA-S, we observed in the previous section that the relay's queue size is always finite, hence it is always stable. For the two-terminals case, the queues evolve as a two-dimensional Markov-chain in the first quadrant. From the protocol description in the previous section, one can observe that the queues in CCMA-S are interacting. In other words, the transition probabilities differ according to whether the size of the queues are empty or not. For example, if one of the two terminals' queues was empty for a long time, then the relay serves the lost packets from the other terminal more often. On the other hand, if one of the two terminals queues never empties, then the other terminal will never get served by the relay. Studying stability conditions for interacting queues is a difficult problem that has been addressed for ALOHA systems [149, 127]. The concept of dominant systems can be adopted to help finding bounds on the stability region.

To analyze the stability of CCMA-S, we develop a dominant system to decouple the interaction of the originating terminals from the role of the relay in cooperation.

Using our developed dominant system we are able to characterize the stability region of CCMA-S for a fixed resource-sharing vector, and hence the whole stability region. The following lemma states the stability region of CCMA-S for a fixed resource sharing vector $[\omega_1, \omega_2]$.

Lemma 11.3.1   *The stability region of CCMA-S for a fixed resource-sharing vector $[\omega_1, \omega_2]$ is given by $\mathcal{R}(S_1) \bigcup \mathcal{R}(S_2)$ where*

$$\mathcal{R}(S_1) = \left\{ [\lambda_1, \lambda_2] \in \mathbf{R}_+^2 : \lambda_2 < h(\lambda_1; w_1, w_2, f_{1,d}, f_{2,d}, f_{2,l}, f_{1,d}), \textit{for } \lambda_1 < \omega_1 f_{1,d}. \right\}$$

(11.10)

and

$$\mathcal{R}(S_2) = \left\{ [\lambda_1, \lambda_2] \in \mathbf{R}_+^2 : \lambda_1 < h(\lambda_2; w_2, w_1, f_{2,d}, f_{1,d}, f_{1,l}, f_{1,d}), \textit{for } \lambda_2 < \omega_2 f_{2,d}. \right\}$$

(11.11)

where

$$h(x; \alpha_1, \alpha_2, \alpha_3, \alpha_4, \alpha_5, \alpha_6) = \alpha_2\alpha_4 + \alpha_1\alpha_2 \left(1 - \frac{x}{\alpha_1\alpha_3} \right) (1 - \alpha_3)\alpha_5\alpha_6.$$ (11.12)

Proof   The proof of Lemma 11.3.1 depends on constructing a dominant system that decouples the interaction between the queues and thus renders the analysis tractable. By dominance, we mean that the queues in the dominant system stochastically dominate the queues in the original CCMA-S system, i.e., with the same initial conditions for queue sizes in both the original and dominant systems, the queue sizes in the dominant system are not smaller than those in the original system.

We define the dominant system for CCMA-S as follows. For $j \in \{1, 2\}$, define S_j as follows:

- arrivals at queue $i \in \mathcal{M}$ in S_j are the same as in CCMA-S;
- the channel realizations $h_{k,l}$, where $k \in \mathcal{T}$ and $l \in \mathcal{D}$, are identical for both S_j and CCMA-S;
- time slots assigned to user $i \in \mathcal{M}$ are identical for both S_j and CCMA-S;
- the noise generated at the receiving ends of both systems is identical;
- the packets successfully transmitted by the relay for user j are not removed from user j's queue in S_j.

The above definition of the dominant system implies that queue j evolves exactly as in a TDMA system without a relay. If both the dominant system and the original CCMA-S started with the same initial queue sizes, then the queues in system S_j are always not shorter than those in CCMA-S. This follows because a packet successfully transmitted for queue j in S_j is always successfully transmitted from the corresponding queue in CCMA-S. However, the relay can succeed in forwarding some packets from queue j in CCMA-S in the empty time slots of the other terminal. This implies that queue j empties more frequently in CCMA-S and therefore the other terminal is better served in CCMA-S compared to S_j. Consequently, stability conditions for the dominant system S_j ($j \in \{1, 2\}$) are sufficient for the stability of the original CCMA-S system. In the following, we first derive the sufficient conditions for stability of CCMA-S.

Consider system S_1 in which the relay only helps terminal 2, and terminal 1 acts exactly as in a TDMA system. In order to apply Loynes' theorem, we require the arrival and service processes for each queue to be stationary. The queue size for terminal $i \in \{1, 2\}$ in system S_1 at time t, denoted by $Q_i^t(S_1)$, evolves as follows:

$$Q_i^{t+1}(S_1) = \left(Q_i^t(S_1) - Y_i^t(S_1)\right)^+ + X_i^t(S_1), \tag{11.13}$$

where $X_i^t(S_1)$ represents the number of arrivals in slot t and is a stationary process by assumption with finite mean $E\left[X_i^t(S_1)\right] = \lambda_i$. The function $()^+$ is defined as $(x)^+ = \max(x, 0)$. $Y_i^t(S_1)$ denotes the possible (*virtual*) departures from queue i at time t; by virtual we mean that $Y_i^t(S_1)$ can be equal to 1 even if $Q_i^t(S_1 = 0)$. We assume that departures occur before arrivals, and the queue size is measured at the beginning of the slot. For terminal $i = 1$, the service process can be modeled as

$$Y_1^t(S_1) = \mathbf{1}\left[A_1^t \cap \overline{O_{1,d}^t}\right], \tag{11.14}$$

where $\mathbf{1}[\cdot]$ is the indicator function, A_1^t denotes the event that slot t is assigned to terminal 1, and $\overline{O_{1,d}^t}$ denotes the complement of the outage event between terminal 1 and the destination d at time t.[2] Due to the stationarity assumption of the channel gain process $\{h_{i,d}^t\}$, and using the outage expression in (11.8), the probability of this event is given by $\Pr\left[\overline{O_{1,d}^t}\right] = f_{1,d}$. From the above, it is clear that the service process $Y_1^t(S_1)$ is stationary and has a finite mean given by

$$E\left[Y_1^t(S_1)\right] = \omega_1 f_{1,d}, \tag{11.15}$$

where $E[\cdot]$ denotes statistical expectation. According to Loynes theorem, stability of queue 1 in the dominant system S_1 is achieved if the following condition holds:

$$\lambda_1 < \omega_1 f_{1,d}. \tag{11.16}$$

Consider now queue 2 in system S_1. The difference between the evolution of this queue and queue 1 is in the definition of the service process $Y_2^t(S_1)$. A packet from queue 2 can be served in a time slot in either one of the two following events:

- if the time slot belongs to queue 2 and the associated channel $h_{2,d}^t$ is not in outage; or
- the time slot belongs to queue 1, queue 1 is empty, in the previous time slot there was a successful "*maybe virtual*" reception of a packet at the relay from terminal 2, and the relay–destination channel is not in outage.

This can be modeled as

$$Y_2^t(S_1) = \mathbf{1}\left[A_2^t \cap \overline{O_{2,d}^t}\right] + \mathbf{1}\left[A_1^t \cap \{Q_1^t(S_1)\} = 0\} \right.$$
$$\left. \cap A_2^{t-1} \cap \overline{O_{2,1}^{t-1}} \cap \overline{O_{2,d}^{t-1}} \cap \overline{O_{1,d}^t}\right] \tag{11.17}$$

---

[2] $\overline{(\cdot)}$ denotes the complement of the event.

where $\{Q_1^t(S_1) = 0\}$ denotes the event that terminal's 1 queue is empty in time slot t. The two indicator functions in the right-hand side of equation (11.17) are mutually exclusive, hence, the average rate of the service process is given by

$$E\left[Y_2^t(S_1^1)\right]$$

$$= \omega_2 f_{2,d} + \omega_1 \Pr\left[\{Q_1^t(S_1^1) = 0\}\right] \omega_2 \left(1 - \Pr[O_{2l}]\right) \Pr[O_{2,d}] \left(1 - \Pr[O_{1,d}]\right),$$

(11.18)

where $\Pr[O_{i,j}]$ is the probability of outage between nodes i and j.

Using (11.16) and the fact that, at steady state, the fraction of time a queue is occupied equals the ratio of the arrival rate and service rate, the probability that queue 1 is empty is given by

$$\Pr\left[\{Q_1^t(S_1^1) = 0\}\right] = 1 - \frac{\lambda_1}{\omega_1 f_{1,d}}.$$

(11.19)

Using the expression of the outage probability in (11.8) and Loynes conditions for stability [124], the stability condition for queue 2 in the dominant system S_1 is given by

$$\lambda_2 < \omega_2 f_{2,d} + \omega_1 \omega_2 \left(1 - \frac{\lambda_1}{\omega_1 f_{1,d}}\right)\left(1 - f_{2,d}\right) f_{2,l} f_{1,d}.$$

(11.20)

Both conditions (11.16) and (11.20) represent the stability region for system S_1 for a specific resource-sharing vector (ω_1, ω_2) pair. Call this region $\mathcal{R}(S_1)$. Using parallel arguments for the dominant system S_2, we can characterize the stability region $\mathcal{R}(S_2)$ for this system by the following pair of inequalities:

$$\lambda_2 < \omega_2 f_{2,d},$$

$$\lambda_1 < \omega_1 f_{1,d} + \omega_2 \omega_1 \left(1 - \frac{\lambda_2}{\omega_2 f_{2,d}}\right)\left(1 - f_{1,d}\right) f_{1,l} f_{1,d}.$$

(11.21)

Since the stability conditions for a dominant system are sufficient for the stability of CCMA-S, any point inside the regions $\mathcal{R}(S_1)$ and $\mathcal{R}(S_2)$ can be achieved by the original system CCMA-S, hence $\mathcal{R}(S_1) \bigcup \mathcal{R}(S_2)$ is a subset from the stability region of CCMA-S for a fixed resource sharing pair (ω_1, ω_2). This region is depicted in Figure 11.2.

Up to this point have we only proved the sufficient conditions for the stability of CCMA-S in Lemma 11.3.1. To prove the necessary conditions, we have to prove the *indistinguishability* of the dominant and original systems at saturation. The argument is as follows. Consider the dominant system S_1 whose stability region is characterized by the pair of inequalities (11.16), (11.20). Note that if queue 2 does not empty, packets of user 1 are always dropped by the relay in both S_1 and CCMA-S, and both systems become identical. In S_1, for $\lambda_1 < \omega_1 f_{1,d}$, if $\lambda_2 > \omega_2 f_{2,d} + \omega_1 \omega_2 \left(1 - \lambda_1/\omega_1 f_{1,d}\right)\left(1 - f_{2,d}\right) f_{2,l} f_{1,d}$ then using same argument as before Q_1^t is stable and Q_2^t is unstable by Loynes, i.e., $\lim_{t \to \infty} Q_2^t \to \infty$ almost surely. If Q_2^t tends to infinity almost surely, i.e., does not empty, then system S_1 and CCMA-S are identical, and if both systems are started from the same initial conditions, then on a set of sample paths of positive probability Q_2^t in CCMA-S never returns to zero for $t \geq 0$. Hence Q_2^t

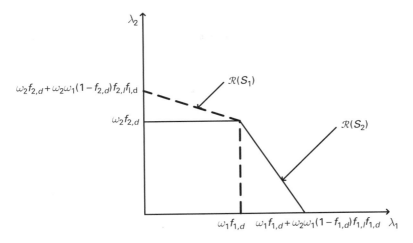

Fig. 11.2   Stable throughput region for system CCMA-S for a fixed resource-sharing vector (ω_1, ω_2) given by $\mathcal{R}(S_1) \bigcup \mathcal{R}(S_2)$.

in CCMA-S tends to infinity with positive probability, i.e., CCMA-S is also unstable. This means that the boundary for the stability region of the dominant system is also a boundary for the stability region of the original CCMA-S system. Thus, conditions for stability of the dominant system is sufficient and necessary for stability of the original system. This completes the proof of Lemma 11.3.1. ∎

The whole stability region for system CCMA-S can be determined by taking the union over all feasible resource-sharing vectors as follows:

$$\mathcal{R}(\text{CCMA-S}) = \bigcup_{\Omega \in F} \left\{ \mathcal{R}_1(S^1) \bigcup \mathcal{R}_2(S^1) \right\}. \tag{11.22}$$

We give a complete characterization of the stability region of CCMA-S in the following theorem.

THEOREM 11.3.1 *The stability region for a two-user CCMA-S system is given by*

$$\mathcal{R}(\text{CCMA-S}) = \left\{ [\lambda_1, \lambda_2] \in \mathbf{R}_+^2 : \lambda_2 < \max\left[g_1(\lambda_1), g_2(\lambda_1)\right] \right\} \tag{11.23}$$

where the functions $g_1(\cdot)$ and $g_2(\cdot)$ are defined as follows:

$$g_1(\lambda_1) = \begin{cases} K_2 \left(\frac{\lambda_1 + f_{1,d}}{2 f_{1,d}} - \frac{f_{2,d}}{2K_2} \right)^2 - \frac{K_2 \lambda_1}{f_{1,d}} + f_{2,d}, & 0 \le \lambda_1 \le f_{1,d} - \frac{f_{1,d} f_{2,d}}{K_2}, \\ f_{2,d} - \frac{f_{2,d}}{f_{1,d}} \lambda_1, & f_{1,d} - \frac{f_{1,d} f_{2,d}}{K_2} < \lambda_1 \le f_{1,d}. \end{cases} \tag{11.24}$$

And the function $g_2(\cdot)$ is specified as

$$g_2(\lambda_1) = \begin{cases} f_{2,d} - \frac{f_{2,d}}{f_{1,d}} \lambda_1, & 0 \le \lambda_1 < \frac{f_{1,d}^2}{(1 - f_{1,d}) f_{1,l} f_{1,d}}, \\ \frac{f_{1,d} f_{2,d}}{K_1} + f_{2,d} - 2 f_{2,d} \sqrt{\frac{\lambda_1}{K_1}}, & \frac{f_{1,d}^2}{(1 - f_{1,d}) f_{1,l} f_{1,d}} \le \lambda_1 \le \lambda_1^*. \end{cases} \tag{11.25}$$

where

$$\lambda_1^* = \begin{cases} f_{1,d}, & (1 - f_{1,d})f_{1,l}f_{1,d} < f_{1,d}, \\ \frac{1}{4K_1}(f_{1,d} + K_1)^2, & f_{1,d} \leq (1 - f_{1,d})f_{1,l}f_{1,d}. \end{cases} \tag{11.26}$$

and $K_i = (1 - f_{i,d})f_{i,l}f_{1,d}$, $i \in \{1, 2\}$.

Proof The stable throughput region of CCMA-S for a fixed resource-sharing vector (ω_1, ω_2) is specified in Lemma 11.3.1. In order to find the whole stability region of the protocol, we need to take the union over all possible values of (ω_1, ω_2) in F. One method to characterize this union is to solve a constrained optimization problem to find the maximum feasible λ_2 corresponding to each feasible λ_1. For a fixed λ_1, the maximum stable arrival rate for queue 2 is given by solving the following optimization problem:

$$\max_{w_1, w_2} \lambda_2 = w_2 f_{2,d} + w_1 w_2 K_2 - \frac{\lambda_1 w_2 K_2}{f_{1,d}}, \tag{11.27}$$

$$\text{s.t. } w_1 + w_2 \leq 1, \quad \lambda_1 \leq w_1 f_{1,d}, \tag{11.28}$$

where $K_i = (1 - f_{2,d})f_{2,l}f_{1,d}$, and $i \in \{1, 2\}$. To put this problem in a standard form, we can equivalently write it as the minimization of $-\lambda_2$. The Lagrangian of this optimization problem is given by

$$\begin{bmatrix} -w_2 K_2 \\ -w_1 K_2 + \frac{\lambda_1 K_2}{f_{1,d}} - f_{2,d} \end{bmatrix} + u_1 \begin{bmatrix} 1 \\ 1 \end{bmatrix} + u_2 \begin{bmatrix} -f_{1,d} \\ 0 \end{bmatrix} = 0, \tag{11.29}$$

where u_1, u_2 are the complementary slackness variables that are nonnegative. Solving for the complementary slackness variables we get the following relation between u_1 and u_2:

$$u_1 = w_2 K_2 + u_2 f_{1,d}. \tag{11.30}$$

This shows that $u_1 > 0$, i.e., the first constraint ($w_1 + w_2 = 1$) in (11.28) is met with equality.

Substituting $w_2 = 1 - w_1$ in the cost function in (11.28), we get

$$\lambda_2 = (1 - w_1)f_{2,d} + w_1(1 - w_1)K_2 - (1 - w_1)\frac{\lambda_1 K_2}{f_{1,d}}. \tag{11.31}$$

Taking the first derivative of the above equation with respect to w_1, we get

$$\frac{\partial \lambda_2}{\partial w_1} = -f_{2,d} + K_2 - 2w_1 K_2 + \frac{\lambda_1 K_2}{f_{1,d}}. \tag{11.32}$$

Since K_2 is non-negative, the second derivative is negative, hence, the cost function in (11.31) is concave in w_1 and the necessary conditions for optimality KKT [15] are also sufficient. Equating the first derivative in (11.32) to zero, the solution w_1^* is given by

$$w_1^* = \frac{1}{2K_2}\left(K_2 - f_{2,d} + \frac{\lambda_1 K_2}{f_{1,d}}\right). \tag{11.33}$$

From the first constraint in (11.28), the minimum value for w_1 that guarantees the stability of queue 1 in system S_1 is given by $w_{1,\min} = \lambda_1/f_{1,d}$. Hence, if $w_1^* > w_{1,\min}$, and given concavity of the cost function, the optimal solution is just w_1^*, otherwise it is given by $w_{1,\min}$. Characterizing this in terms of the channel parameters, the optimal solution for the optimization problem in (11.28) is given by

$$
w_1^* = \begin{cases} \frac{1}{2K_2}\left(K_2 - f_{2,d} + \frac{\lambda_1 K_2}{f_{1,d}}\right), & \lambda_1 \le f_{1,d} - \frac{f_{1,d}f_{2,d}}{K_2} \\ \frac{\lambda_1}{f_{1,2}}, & f_{1,d} - \frac{f_{1,d}f_{2,d}}{K_2} < \lambda_1 < f_{1,d}. \end{cases}
\tag{11.34}
$$

If $\lambda_1 > f_{1,d}$, then the first queue can never be stable by construction of S_1.

Now we solve for the other branch in the stability region given by the dominant system S_2. The equations for this branch are given by (11.20). Similar to the first stability region branch, solving for the Lagrangian of this branch gives the necessary condition $w_1 + w_2 = 1$. First, we find the maximum achievable stable rate for the first queue, which is achieved when $\lambda_2 = 0$. Substituting in (11.20), the arrival rate λ_1 can be written as

$$
\lambda_1 = w_1 f_{1,d} + w_1(1 - w_1)K_1.
\tag{11.35}
$$

The above equation is obviously concave. Taking the first derivative and equating to zero we get

$$
w_1^*|_{\lambda_2=0} = \frac{1}{2K_1}(f_{1,d} + K_1).
\tag{11.36}
$$

Since $w_1 \le 1$, then if $w_1^*|_{\lambda_2=0} > 1$, i.e., $f_{1,d} > K_1$, and given the concavity of the cost function, we let $w_1^*|_{\lambda_2=0} = 1$. Substituting $w_1^*|_{\lambda_2=0}$ in (11.35), the maximum achievable rate λ_1 can be given by

$$
\lambda_1^* = \begin{cases} f_{1,d}, & f_{1,d} > K_1, \\ \frac{1}{4K_1}(f_{1,d} + K_1)^2, & 0 \le f_{1,d} \le K_1. \end{cases}
\tag{11.37}
$$

Next for a fixed λ_1, we solve for the optimal λ_2 that can be achieved from the second branch. We can write (11.20) in terms of λ_2 as follows:

$$
\lambda_2 = \frac{f_{1,d}f_{2,d}}{K_1} + (1 - w_1)f_{2,d} - \lambda_1\frac{f_{2,d}}{w_1 K_1}.
\tag{11.38}
$$

Taking the first derivative with respect to w_1 we get

$$
\frac{\partial \lambda_2}{\partial w_1} = -f_{2,d} + \lambda_1\frac{f_{2,d}}{K_1 w_1^2}.
\tag{11.39}
$$

The second derivative is negative, which renders the whole function concave. Equating the first derivative to zero we get

$$
w_1^* = \sqrt{\frac{\lambda_1}{(1 - f_{1,d})f_{1,l}f_{1,d}}}.
\tag{11.40}
$$

From (11.20), we have the following constraint for the stability of the second queue, $\lambda_2 \le w_2 f_{2,d}$, which can be written in terms of w_1 as follows:

$$
w_1 \le 1 - \frac{\lambda_2}{f_{2,d}}.
\tag{11.41}
$$

Substituting λ_2 from (11.41) into (11.38), we find that the maximum value for w_1 in terms of λ_1 is given by $w_1 \leq \lambda_1/f_{1,d}$. Therefore, the optimal value for w_1 can be written as

$$w_1^* = \min\left(\sqrt{\frac{\lambda_1}{(1-f_{1,d})f_{1,l}f_{1,d}}}, \frac{\lambda_1}{f_{1,d}}\right). \tag{11.42}$$

The value of the expression in (11.42) can be shown to never exceed 1 by substituting the maximum value of λ_1 given by (11.37) in the above equation. After some manipulations, the optimum w_1^* in (11.42) can be further simplified as follows:

$$w_1* = \begin{cases} \frac{\lambda_1}{f_{1,d}}, & \text{if } \lambda_1 \leq \frac{f_{1,d}^2}{(1-f_{1,d})f_{1,l}f_{1,d}}, \\ \sqrt{\frac{\lambda_1}{(1-f_{1,d})f_{1,l}f_{1,d}}}, & \text{otherwise.} \end{cases} \tag{11.43}$$

We summarize equations (11.31), (11.34), (11.38), and (11.43) describing the envelopes of the first and second branches as follows. For the first branch given by equations (11.31) and (11.34), substituting (11.34) in (11.31) we get

$$g_1(\lambda_1) = \begin{cases} K_2\left(\frac{\lambda_1}{2f_{1,d}} + 0.5 - \frac{f_{2,d}}{2K_2}\right)^2 - \frac{K_2\lambda_1}{f_{1,d}} + f_{2,d}, & 0 \leq \lambda_1 \leq f_{1,d} - \frac{f_{1,d}f_{2,d}}{K_2}, \\ f_{2,d} - \frac{f_{2,d}}{f_{1,d}}\lambda_1, & f_{1,d} - \frac{f_{1,d}f_{2,d}}{K_2} < \lambda_1 \leq f_{1,d}. \end{cases} \tag{11.44}$$

Similarly, by substituting (11.43) in (11.38), we get for the second branch

$$g_2(\lambda_1) = \begin{cases} f_{2,d} - \frac{f_{2,d}}{f_{1,d}}, & \lambda_1 < \frac{f_{1,d}^2}{(1-f_{1,d})f_{1,l}f_{1,d}}, \\ \frac{f_{1,d}f_{2,d}}{K_1} + f_{2,d} - 2f_{2,d}\sqrt{\frac{\lambda_1}{K_1}}, & \frac{f_{1,d}^2}{(1-f_{1,d})f_{1,l}f_{1,d}} \leq \lambda_1 \leq \lambda_1^*. \end{cases} \tag{11.45}$$

The value for λ_1^* is given by (11.37). Since $g_1(\lambda_1)$ and $g_2(\lambda_2)$ are achieved by the dominant systems S_1 and S_2 respectively, then they are both achieved by CCMA-S. This proves the theorem. ∎

An interesting observation that we make from the above theorem is that both functions $g_1(\lambda_1)$ and $g_2(\lambda_1)$ are linear over some part of their domain and strictly convex over the other part. It is not obvious, however, whether both functions are strictly convex over their domain of definition, and hence, whether the boundary of the stability region of CCMA-S given by $\max\{g_1, g_2\}$ is convex or not. In the following lemma, we prove this property.

Lemma 11.3.2   *The boundary of the stability region of system CCMA-S given by* $\max\{g_1, g_2\}$ *is convex.*

Proof   We prove the convexity of the envelope of region $\mathcal{R}(\text{CCMA-S})$. First we consider the envelope $g_1(\lambda_1)$ for the first branch. As shown in Figure 11.3, $g_1(\lambda_1)$ is defined over two regions: for $\lambda_1 \in [0, f_{1,d} - f_{1,d}f_{2,d}/K2]$, $g_1(\lambda_1) \triangleq g_{11}(\lambda_1) = K_2\left(\lambda_1/2f_{1,d} + 0.5 - f_{2,d}/2K_2\right)^2 - K_2\lambda_1/f_{1,d} + f_{2,d}$; while for $\lambda_1 \in (f_{1,d} - f_{1,d}f_{2,d}/K2, f_{1,d}]$, $g_1(\lambda_1) \triangleq g_{12}(\lambda_1)$ is a straight line given by $f_{2,d} - f_{2,d}/f_{1,d}$. It can be readily seen that $g_1(\lambda_1)$ is convex over both regions because its second derivative

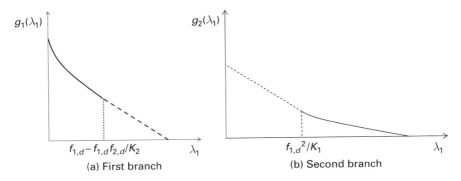

Fig. 11.3   Envelopes for the stability region of CCMA-S.

is non-negative. Thus, to prove that $g_1(\lambda_1)$ is convex over the whole region, we check the continuity and the first derivative at the intersection point $f_{1,d} - f_{1,d}f_{2,d}/K2$. It is simple to show that $g_1(\lambda_1)$ is continuous at this point. Now, it remains to check the slope of the tangent at the intersection point. Taking the first derivative of g_{11} and g_{12}, we can show that

$$\frac{\partial g_{11}(\lambda_1)}{\partial \lambda_1} = \frac{\partial g_{12}(\lambda_1)}{\partial \lambda_1}\Big|_{\lambda_1 = f_{1,d} - \frac{f_{1,d}f_{2,d}}{K2}}. \tag{11.46}$$

Therefore, $g_1(\lambda_1)$ is also differentiable, which proves that $g_1(\lambda_1)$ is convex over its domain of definition. Similar arguments apply for $g_2(\lambda_1)$ depicted in Figure 11.3 to prove its convexity.

The envelope of the stability region of $\mathcal{R}(\text{CCMA-S})$ is given by $\max[g_1(\lambda_1), g_2(\lambda_2)]$. The maximum of two convex functions can be shown to be convex as follows. Let $0 < a < 1$, and λ_{11}, λ_{12} belong to the feasible region of λ_1, then we have

$$\begin{aligned}
&\max[g_1(a\lambda_{11} + (1-a)\lambda_{12}), g_2(a\lambda_{11} + (1-a)\lambda_{12})] \\
&\leq \max[ag_1(\lambda_{11}) + (1-a)g_1(\lambda_{12}), ag_2(\lambda_{11}) + (1-a)g_2(\lambda_{12})], \quad (1) \\
&\leq \max[ag_1(\lambda_{11}), ag_2(\lambda_{11})] + \max[(1-a)g_1(\lambda_{12}), (1-a)g_2(\lambda_{12})] \,(2)
\end{aligned} \tag{11.47}$$

where (1) follows by the convexity of $g_1(\cdot)$ and $g_2(\cdot)$, and (2) follows by the properties of the max function. This proves that the envelope of the stability region for system CCMA-S is convex.   ∎

The above Lemma will prove useful in characterizing the relation among the stability regions of the different multiple access protocols considered in this paper. The first relation that we state is that between the stability regions of TDMA and CCMA-S.

Lemma 11.3.3   *The stability region of TDMA is contained inside that of CCMA-S. In other words*

$$\mathcal{R}(\text{TDMA}) \subseteq \mathcal{R}(\text{CCMA} - \text{S}). \tag{11.48}$$

The two regions are identical if the following two conditions are satisfied simultaneously

$$(1 - f_{2,d}) f_{2,l} f_{l,d} < f_{2,d}, \qquad (11.49)$$

$$(1 - f_{1,d}) f_{1,l} f_{l,d} < f_{1,d}. \qquad (11.50)$$

Proof We use Theorem 11.3.1 and Lemma 11.3.2 in the proof of the this lemma. The stability region of TDMA is determined according to the following parametric inequalities:

$$\lambda_1 \leq \omega_1 f_{1,d}, \qquad \lambda_2 \leq \omega_2 f_{2,d} \qquad (11.51)$$

or, equivalently,

$$\frac{\lambda_1}{f_{1,d}} + \frac{\lambda_2}{f_{2,d}} \leq 1. \qquad (11.52)$$

From Lemma 11.3.2, both functions g_1 and g_2 that determine the boundary of the stability region of CCMA-S are convex. From the proof of the convexity, we note that the straight line $\lambda_2 = f_{2,d} - f_{2,d}/f_{1,d}\lambda_1$ is a tangent for both functions, hence, it lies below both functions. Since this straight line is itself the boundary for the TDMA stability region, then the stability region of TDMA is a subset of that of CCMA-S. To prove the second part of the Lemma we use the definitions of the functions g_1, g_2 in Theorem 11.3.1. From (11.37), observe that if $(1 - f_{1,d}) f_{1,l} f_{l,d} < f_{1,d}$ then the maximum λ_1 is determined by $f_{1,d}$. If simultaneously $(1 - f_{2,d}) f_{2,l} f_{2,d} < f_{2,d}$, then by substituting both conditions in the domain definitions of the functions g_1, g_2, it can be seen that both functions reduce to

$$g_1(\lambda_1) = g_2(\lambda_1) = f_{2,d} - \frac{f_{2,d}}{f_{1,d}}\lambda_1, \quad 0 \leq \lambda_1 \leq f_{1,d}, \qquad (11.53)$$

which is the boundary for the stability region of TDMA. Hence if both conditions in (11.50) are satisfied, CCMA-S and TDMA have the same stable throughput regions. This completes the proof of the lemma. ∎

11.3.1.2 The symmetric *M*-terminals case

Stability analysis for the general M-terminals case is very complicated. For the ALOHA case, only bounds on the stability region have been derived before. In this section, we only focus on the symmetric scenario. We define the dominant system for M-terminals CCMA-S as follows. For $1 \leq j \leq N$, define S_j^M as follows:

- arrivals at queue i in S_j^M are the same as in CCMA;
- the channel realizations h_{kl}, where $k \in \mathcal{T}$ and $l \in \mathcal{R}$, are identical for both S_j^M and CCMA;
- user $i \in \mathcal{M}$ is assigned the same time slots in both systems;
- the noise generated at the receiving ends of both systems is identical;
- the packets served by the relay for the first j terminals are not removed from these terminals' queues.

The last rule implies that the first j queues act as in a TDMA system without a relay. The relay, however, can help the other users $j + 1 \leq k \leq N$ in the empty slots of the TDMA frame.

Now consider system S_M^M in which the relay does not help any of the users. It is clear that the queue sizes in this system are never smaller than those in the original system CCMA-S. For S_M^M, the success probability of transmitting a packet is equal for all terminals and is given by

$$P_s(S_M^M) = \Pr[\text{SNR} \geq \beta] = f_{1,d}. \tag{11.54}$$

The service rate per terminal is thus given by $\mu(S_M^M) = f_{1,d}/M$, due to the symmetry of the problem. Since system S_M^M acts as a TDMA system without a relay, the queues are decoupled and hence the arrival process and departure process of each of them is strictly stationary. Applying Loynes' theorem, the stability condition for S_M^M is given by

$$\lambda < f_{1,d}, \tag{11.55}$$

where λ is the aggregate arrival rate for the M terminals.

Next, let us consider the stability of symmetric CCMA-S. Since S_M^M as described before dominates CCMA-S, if S_M^M is stable then CCMA-S is also stable. Therefore, for $\lambda < f_{1,d}$, system CCMA-S is stable. On the other hand, if all the queues in S_M^M are unstable, then none of these queues ever empty, hence, the relay loses its role and both systems CCMA-S and S_M^M are *indistinguishable* if both started with the same initial conditions. Therefore, if we have $\lambda > f_{1,d}$ then all the queues in S_M^M are unstable and accordingly system CCMA-S is unstable as well. Therefore, the maximum stable throughput for system CCMA-S can be summarized in the following theorem.

THEOREM 11.3.2   *The maximum stable throughput $\lambda_{\text{MST}}(\text{CCMA} - \text{S})$ for system CCMA-S is equal to that of a TDMA system without a relay and is given by*

$$\lambda_{\text{MST}}(\text{CCMA-S}) = f_{1,d}. \tag{11.56}$$

However, we conjecture that for the general asymmetric M-terminals scenario, the whole stability region of TDMA will be contained inside that of CCMA-S. Another important issue to point out is that although CCMA-S and TDMA have the same maximum stable throughput for the symmetrical case, the two systems do not have the same delay performance as will be discussed later.

11.3.2   Stability analysis of CCMA-M

In CCMA-M the relay's queue can possibly grow and hence should be taken into account when studying the system stability. This means that for stability we require both the M terminals' queues and the relay's queue to be stable. The stability region of the whole system is the intersection of the stability regions of the M terminals and that of the relay. First, we consider the $M = 2$-terminals case. According to the operation of system CCMA-M, a terminal succeeds in transmitting a packet if either the destination or the relay receives this packet correctly. The success probability of terminal i in CCMA-M can thus be calculated as

$$\Pr[\text{Success of terminal } i] = \Pr\left[\overline{O_{i,l}} \cup \overline{O_{i,d}}\right], \tag{11.57}$$

where $\overline{O_{i,l}}$ denotes the event that the relay received the packet successfully, and $\overline{O_{i,d}}$ denotes the event that the destination received the packet successfully. The success probability of terminal $i \in \{1, 2\}$ in CCMA-M can thus be specified as follows

$$P_i = f_{i,d} + f_{i,l} - f_{i,d} f_{i,l}. \tag{11.58}$$

We first consider the stability region for the system determined just by the terminals' queues. Since for each queue $i \in \mathcal{M}$, the queue behaves exactly as in a TDMA system with the success probability determined by (11.58), the stability region $\mathcal{R}_{\mathcal{M}}(\text{CCMA-M})$ for the set of queues in $\mathcal{M}$ is given by

$$\mathcal{R}_{\mathcal{M}}(\text{CCMA-M}) = \left\{ [\lambda_1, \lambda_2, \dots, \lambda_M] \in R_+^M : \lambda_i < \omega_i P_i, \forall i \in \mathcal{M}, \right.$$
$$\left. \text{and } [\omega_1, \omega_2, \dots, \omega_M] \in F \right\}. \tag{11.59}$$

Next we study the stability of the relay's queue l. The evolution of the relay's queue can be modeled as

$$Q_l^t = \left(Q_l^t - Y_l^t\right)^+ + X_l^t, \tag{11.60}$$

where X_l^t denotes the number of arrivals at time slot t and Y_l^t denotes the possibility of serving a packet at this time slot from the relay's queue ($Y_l^t(G)$ takes values in $\{0, 1\}$). Now we establish the stationarity of the arrival and service processes of the relay. If the terminals' queues are stable, then by definition the departure processes from both terminals are stationary. A packet departing from a terminal queue is stored in the relay's queue (i.e., counted as an arrival) if simultaneously the following two events happen: the terminal–destination channel is in outage and the terminal–relay channel is not in outage. Hence, the arrival process to the queue can be modeled as follows:

$$X_l^t = \sum_{i \in \mathcal{M}} \mathbf{1} \left[A_i^t \bigcap \{Q_i^t \neq 0\} \bigcap O_{i,d}^t \bigcap \overline{O_{i,l}^t} \right]. \tag{11.61}$$

In (11.61), $\{Q_i^t \neq 0\}$ denotes the event that terminal i's queue is not empty, i.e., the terminal has a packet to transmit, and according to Little's theorem it has probability $\lambda_i/(\omega_i P_i)$, where P_i is terminal i's success probability and is defined in (11.58). The random processes involved in the above expressions are all stationary, hence the arrival process to the relay is stationary. The expected value of the arrival process can be computed as follows:

$$\lambda_l = \sum_{i \in \mathcal{M}} \lambda_i \frac{(1 - f_{i,d}) f_{i,l}}{P_i}. \tag{11.62}$$

Similarly, we establish the stationarity of the service process from the relay's queue. The service process of the relay's queue depends by definition on the empty slots available from the terminals, and on the channel from the relay to the destination not being in outage. By assuming the terminals' queues to be stable, they offer stationary empty slots (stationary service process) to the relay. Also the channel statistics is stationary, hence, the relay's service process is stationary. The service process of the relay's queue can be modeled as

$$Y_l^t = \sum_{i \in \mathcal{M}} \mathbf{1} \left[A_i^t \bigcap \{ Q_i^t = 0 \} \bigcap \overline{O_{1,d}^t} \right], \qquad (11.63)$$

and the average service rate of the relay can be determined from the following equation:

$$E[Y_l^t] = \sum_{i \in \mathcal{M}} \omega_i \left(1 - \frac{\lambda_i}{\omega_i P_i} \right) f_{1,d}. \qquad (11.64)$$

Using Loynes' theorem and (11.62) and (11.64), the stability region for the relay $\mathcal{R}_l$(CCMA-M) is determined by the condition $E[X_l^t] < E[Y_l^t]$. The total stability region for system CCMA-M is given by the intersection of two regions $\mathcal{R}_\mathcal{M}$(CCMA-M) $\bigcap \mathcal{R}_l$(CCMA-M) which is easily shown to be equal to $\mathcal{R}_l$(CCMA-M). The stability region for CCMA-M with two terminals is thus characterized as follows:

$$\mathcal{R}(S^2)$$
$$= \left\{ [\lambda_1, \lambda_2] \in R^{+2} : \frac{\lambda_1}{P_1} \left((1 - f_{1,d}) f_{1,l} + f_{1,d} \right) + \frac{\lambda_2}{P_2} \left((1 - f_{2,d}) f_{2,l} + f_{1,d} \right) < f_{1,d} \right\}$$
$$(11.65)$$

For $M = 2$, this reveals that the stability region of CCMA-M is bounded by a straight line. Since the stability region for TDMA is also determined by a straight line, when comparing both stability regions it is enough to compare the intersection of these lines with the axes. These intersections for CCMA-M are equal to

$$\lambda_1^*(\text{CCMA-M}) = \frac{f_{1,d} P_1}{f_{1,d} + (1 - f_{1,d}) f_{1,l}},$$
$$\lambda_2^*(\text{CCMA-M}) = \frac{f_{1,d} P_2}{f_{1,d} + (1 - f_{2,d}) f_{2,l}}, \qquad (11.66)$$

while the corresponding values for TDMA are given by

$$\lambda_1^*(\text{TDMA}) = f_{1,d},$$
$$\lambda_2^*(\text{TDMA}) = f_{2,d}. \qquad (11.67)$$

It is clear that the stability region for TDMA is completely contained inside the stability region of CCMA-M if $\lambda_1^*(\text{CCMA-M}) > \lambda_1^*(\text{TDMA})$ and $\lambda_2^*(\text{CCMA-M}) > \lambda_2^*(\text{TDMA})$. Using (11.66) and (11.67), these two conditions are equivalent to

$$f_{1,d} > f_{1,d}, \qquad f_{1,d} > f_{2,d}. \qquad (11.68)$$

These conditions have the following intuitive explanation. If the channel between the relay and destination has higher success probability than the channel between the terminal and destination, then it is better to have the relay help the terminal transmit its packets. Note that (11.68) implies that TDMA can offer better performance for the terminal whose success probability does not satisfy (11.68). This possible degradation in the performance does not appear in CCMA-S because the design of CCMA-S does not allow the relay to store the packets it received for ever. Hence, the performance of protocol CCMA-S can not be less than that of TDMA. In protocol CCMA-M, however, the relay becomes responsible for all of the packets it receives, and if the relay–destination channel has a higher probability of outage than the terminal–destination channel then the system encounters a loss in the performance. This calls for the development of an enhanced version of protocol CCMA-M to take this into account.

11.3.2.1   Enhanced protocol CCMA-Me

The previous discussion motivates the design of an *enhanced* version of CCMA-M, which we refer to as CCMA-Me. In this enhanced strategy, the relay only helps the terminals that are in a worse channel condition than the relay itself. In other words, the relay helps the terminal whose outage probability to the destination satisfies $f_{1,d} > f_{i,d}$ for $i \in \mathcal{M}$. Other terminals that do not satisfy this inequality operate as in TDMA, i.e., the relay does not help them.

Next let us calculate the stability region for the enhanced system and consider $M = 2$ terminals for illustration. Assume that the relay only helps terminal 1. Similar to our calculations for the arrival and service processes for the relay in CCMA-M, we can show that the average arrival rate to the relay in CCMA-Me is given by

$$E\,[X_l^t(\text{CCMA-Me})] = \frac{\lambda_1}{P_1}(1 - f_{1,d})f_{1,l}, \tag{11.69}$$

and the average service rate to the relay is given by

$$E\,[Y_l^t(\text{CCMA-Me})] = \left(\omega_1\left(1 - \frac{\lambda_1}{\omega_1 P_1}\right) + \omega_2\left(1 - \frac{\lambda_2}{\omega_2 f_{2,d}}\right)\right)f_{1,d}. \tag{11.70}$$

Using Loynes' theorem [124] and (11.69) and (11.70), the stability region $\mathcal{R}(\text{CCMA-Me})$ is given by

$$\mathcal{R}(\text{CCMA-Me}) = \left\{[\lambda_1, \lambda_2] \in R_+^2 : \frac{\lambda_1}{P_1}\left((1 - f_{1,d})f_{1,l} + f_{1,d})\right) + \lambda_2\frac{f_{1,d}}{f_{2,d}} < f_{1,d}\right\}. \tag{11.71}$$

The stability region for the enhanced protocol CCMA-Me is no less than the stability region of TDMA $\mathcal{R}(\text{TDMA}) \subseteq \mathcal{R}^{2,e}$, and the proof simply follows from the construction of the enhanced protocol CCMA-Me.

For a general M-terminal case, the analysis is the same and the stability region for CCMA-Me can be fully characterized as follows.

THEOREM 11.3.3   *The stability region for M-terminals CCMA-Me is specified as*

$$\mathcal{R}(\text{CCMA} - \text{Me}) = \left\{[\lambda_1, \lambda_2, \dots, \lambda_M] \in R_+^M : \sum_{i \in \mathcal{M}_1}\frac{\lambda_i}{P_i}\left((1 - f_{i,d})f_{i,l} + f_{1,d})\right)\right.$$

$$\left. + \sum_{j \in \mathcal{M}_2}\lambda_j\frac{f_{1,d}}{f_{jd}} < f_{1,d}\right\}.$$

$$\tag{11.72}$$

where $\mathcal{M}_1 = \{i \in \mathcal{M} : f_{1,d} > f_{i,d}\}$, or the set of terminals that the relay helps, and $\mathcal{M}_2 = \{i \in \mathcal{M} : f_{1,d} < f_{i,d}\}$ is the complement set.

We observe that the stability region of CCMA-Me is still bounded by a straight line. This follows because the stability of the system of queues in CCMA-Me is determined by the stability of a single queue, which is the relay's queue. It remains to specify the relation between the stability regions of CCMA-S and CCMA-Me, which is characterized in the following theorem.

THEOREM 11.3.4 *The stability region of CCMA-Me contains that of CCMA-S. In other words,*

$$\mathcal{R}(TDMA) \subseteq \mathcal{R}(CCMA - S) \subseteq \mathcal{R}(CCMA - Me). \tag{11.73}$$

Proof We now prove that the stability region of CCMA-S is a subset of that of CCMA-Me. In the proof of this theorem, we use two facts: the envelope of $\mathcal{R}(CCMA-S)$ is convex as proved in Lemma 11.3.2, and the envelope of $\mathcal{R}(CCMA-Me)$ is a straight line. Hence, to prove Theorem 11.3.4, it suffices to show that the intersections of the envelope of region $\mathcal{R}(CCMA-S)$ with the λ axes are not greater than those for region $\mathcal{R}(CCMA-Me)$. We consider the scenario when both CCMA-S and CCMA-Me have larger stability regions that TDMA, or equivalently $f_{i,d}f_{i,l}(1 - f_{i,d}) \ge f_{i,d}, i = 1, 2$, because if this condition is not satisfied the stability region of CCMA-S becomes identical to that of TDMA, and CCMA-Me was shown to outperform TDMA, and hence our theorem is true.

We will consider only the intersections with the λ_1 axis; similar arguments follow for the λ_2 axis. The intersection of $\mathcal{R}(CCMA-S)$ with the λ_1 axis, or the maximum stable arrival rate for the first queue, is given by (11.37), while that for $\mathcal{R}(CCMA-Me)$ is given by (11.66). Denote the difference between these two quantities by δ, given as follows:

$$\delta = \frac{f_{1,d}(f_{1,d} + f_{1,l} - f_{1,d}f_{1,l})}{f_{1,d} + (1 - f_{1,d})f_{1,l}} - \frac{[f_{1,d} + (1 - f_{1,d})f_{1,l}f_{1,d}]^2}{4(1 - f_{1,d})f_{1,l}f_{1,d}}. \tag{11.74}$$

If we prove that δ is nonnegative, then we are done. For an arbitrary fixed value for the pair $(f_{1,d}, f_{1,d})$, we consider the range of $f_{1,l}$ that satisfies the constraint $f_{1,d}f_{1,l}(1 - f_{1,d}) \ge f_{1,d}$. The first derivative of δ with respect to $f_{1,l}$ is given by

$$\frac{\partial \delta}{\partial f_{1,l}} = \frac{f_{1,d}(1 - f_{1,d})(f_{1,d} - f_{1,d})}{(f_{1,d} + (1 - f_{1,d})f_{1,l})^2} - \frac{1}{4}(1 - f_{1,d})f_{1,d} + \frac{f_{1,d}^2}{4(1 - f_{1,d})f_{1,l}^2 f_{1,d}}, \tag{11.75}$$

Since $f_{1,d} > f_{1,d}$, the second derivative of the above function is easily seen to be negative, hence δ is concave in $f_{1,l}$ in the region of interest.

$f_{1,l}$ takes values in the range $\left[\frac{f_{1,d}}{(1-f_{1,d})f_{1,d}}, 1\right]$. The function δ evaluated at the minimum value of $f_{1,l}$ is

$$\delta|_{f_{1,l,\min}} = \frac{f_{1,d}f_{1,d} + f_{1,d}}{f_{1,d} + (1 - f_{1,d})f_{1,l}} - f_{1,d} \tag{11.76}$$

$$= \frac{f_{1,d} - f_{1,d}(1 - f_{1,d})f_{1,l}}{f_{1,d} + (1 - f_{1,d})f_{1,l}} > 0. \tag{11.77}$$

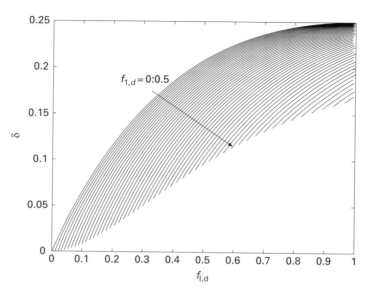

Fig. 11.4   Numerical evaluation of δ for $f_{1,l} = 1$.

Hence, δ is positive at the leftmost boundary of the feasible region of $f_{1,l}$. For the maximum feasible value of $f_{1,l}$, which is equal to 1, the function δ is given by

$$\delta = \frac{f_{1,d}}{1 - f_{1,d} + f_{1,d}} - \frac{(f_{1,d} + (1 - f_{1,d})f_{1,d})^2}{4(1 - f_{1,d})f_{1,d}}. \tag{11.78}$$

Figure 11.4 depicts δ for $f_{1,l} = 1$ over the feasible region of the pair $(f_{1,d}, f_{1,d})$. As shown in the figure, the value of δ is always positive.

Hence, for an arbitrary feasible value of the pair $(f_{1,d}, f_{1,d})$, δ is concave in $f_{1,l}$ and positive at the extreme end points of the feasible region of $f_{1,l}$; therefore, δ is positive for the whole range of $f_{1,l}$ for an arbitrary value of the pair $(f_{1,d}, f_{1,d})$. This proves that δ is always positive over the feasible region of success probabilities, and, therefore, that $\mathcal{R}(\text{CCMA-S})$ is a subset of $\mathcal{R}(\text{CCMA-Me})$. This proves Theorem 11.3.3.[3]   ∎

11.3.3   Existing cooperation protocols: stability analysis

In this section we discuss stability results for some existing decode-and-forward cooperation protocols. In particular, we consider the family of adaptive relaying discussed in previous chapters of the book, which comprises selection and incremental relaying. In the following, we discuss stability results for these two protocols and compare them to our proposed CCMA protocol.

---

[3] The above theorem has another proof that does not need numerical evaluation; however, it is complicated and will not add new insights to the results.

11.3.3.1 Stability region for selection decode-and-forward

In this subsection, we characterize the stability region of selection decode-and-forward (SDF) described before in the single relay chapter. For this chapter to be self-contained, we describe the protocol again here. The cooperation is done in two phases. In the first phase, the source transmits and both the relay and the destination listen. In the second phase, if the relay is able to decode the signal correctly, then it is going to forward the received packet to the destination, otherwise the source retransmits the packet. Accordingly, there is always a specified channel resource dedicated for the relay to help the source, which is different from the opportunistic nature of cooperation in our proposed algorithms. Note that we do not allow the destination to store analog signals in order to do maximum ratio combining (MRC). This is to enable a fair comparison with the other protocols presented in this chapter that do not utilize MRC. Note that all of the results described here can be extended to the scenario where the destination saves copies of received signals and applies MRC; however, it would not add new insights to the results. For the sake of the analysis of SDF, we also assume that the channel fade changes independently from one time slot to another. Note also that this assumption is in favor of SDF, and in general if the channel is correlated from one time slot to another, the performance of SDF will degrade because of diversity loss.

In the following we analyze the outage probability for SDF under the following two scenarios:

- In the first scenario, the structure of the packets arrivals at the terminal is not allowed to be altered. Thus, each packet is transmitted in two consecutive time slots with the original spectral efficiency (for example using the same modulation scheme). This, however, results in SDF having half the bandwidth efficiency of TDMA and CCMA because each packet requires two time slots for transmission.
- In the second scenario, the bandwidth efficiency is preserved among all protocols. This can be done by allowing the terminal and the relay to change the structure of the incoming packets so that each of them transmit at twice the incoming rate (twice the spectral efficiency). Hence each packet is now transmitted in one time slot again, and the average spectral efficiency for SDF under this scenario is equal to that of TDMA and CCMA.

For the first scenario, an outage occurs if both the source–destination link and the source–relay–destination link are in outage. This can be specified as follows:

$$
\text{Pr}_{\text{SDF}}(O) = \Pr\Big[\big(\{\text{SNR}_{i,d} < 2\beta\} \bigcap \{\text{SNR}_{i,l} < 2\beta\} \bigcap\big) \\
\bigcup \big(\{\text{SNR}_{i,d} < 2\beta\} \bigcap \{\text{SNR}_{i,l} > 2\beta\} \bigcap \{\text{SNR}_{l,d} < 2\beta\}\big)\Big],
\tag{11.79}
$$

for $i \in \mathcal{M}$. The factor 2 in front of the SNR threshold β is to account for the fact that the transmitted power is divided by 2 in the first scenario to have the same energy per bit. (Note that the bandwidth efficiency of SDF in the first scenario is half of that of CCMA, hence we need to reduce the transmit power by half.) The first term in the right-hand side of (11.79) corresponds to the event that both the source–destination and the source–relay links were in outage in the first time slot, and the source–destination link remained in outage in the second time slot. The second term in (11.79) corresponds to the event that the source–destination link was in outage and the source–relay link was not in outage in the first time slot, but the relay–destination link was in outage in the second time slot. The probability in (11.79) can be expressed as

$$\mathrm{Pr}_{i,\mathrm{SDF}}(O) = \left(1 - f_{i,d}^2\right)^2 \left(1 - f_{i,l}^2\right) + \left(1 - f_{i,d}^2\right) f_{i,l}^2 \left(1 - f_{1,d}^2\right), \tag{11.80}$$

where $f_{i,j}$ is defined in (11.9). Since a single packet is transmitted in two time slots, one can think of this protocol as a modified TDMA system with the cooperation time slot having twice the length of the time slot in TDMA. The average arrival rate per cooperation time slot is $2\lambda_i$ for $i \in \mathcal{M}$. Loynes' condition for stability is given by

$$\frac{\lambda_1}{1 - \mathrm{Pr}_{1,\mathrm{SDF}}(O)} + \frac{\lambda_2}{1 - \mathrm{Pr}_{2,\mathrm{SDF}}(O)} < \frac{1}{2}, \tag{11.81}$$

where $1 - \mathrm{Pr}_{i,\mathrm{SDF}}(O)$ is the success probability for terminal i.

For the second scenario, we need to calculate the SNR threshold corresponding to transmitting at twice the rate; denote this threshold by β'. The resulting SNR threshold β' should generally be larger than β required for transmission at the original rate. It is in general very difficult to find an explicit relation between the SNR threshold β and the transmission rate, and thus we turn to a special case to capture the insights of this scenario. Let the outage be defined as the event that the mutual information I between two terminals is less than some specific rate R [141]. If the transmitted signals are Gaussian, then according to our channel model, the mutual information between terminal $i \in \mathcal{T}$ and terminal $j \in \mathcal{D}$ is given by $I = \log(1 + \mathrm{SNR}_{i,j})$. The outage event for this case is defined as

$$O_I \triangleq \left\{h_{i,j} : I < R\right\}. \tag{11.82}$$

The above equation implies that if the outage is defined in terms of the mutual information and the transmitted signals are Gaussian, then the SNR threshold β and the spectral efficiency R are related as $\beta = 2^R - 1$, i.e., they exhibit an exponential relation. Hence, for protocol SDF when transmitting at twice the rate, the corresponding SNR threshold β' is given by $\beta' = 2^{2R} - 1$, and given β one can find β' through the previous equation. Note that we do not reduce the power in this second scenario because both SDF and CCMA have the same spectral efficiency. Intuitively, under a fixed modulation scheme and fixed average power constraint, one can think of the SNR threshold as being proportional to the minimum distance between the constellation points, which in turn depends on the number of constellation points for fixed average power, and the later has an exponential relation to the number of bits per symbol, which determines the spectral efficiency.

11.3.3.2   Stability region for incremental decode-and-forward

In such a strategy, feedback from the destination in the form of ACK or NACK is utilized at the relay node to decide whether to transmit or not. In amplify-and-forward incremental relaying (discussed in the relay chapter), the source transmits in the first phase, and if the destination is not able to receive correctly it sends a NACK that can be received by the relay. The relay then amplifies and forwards the signal it received from the source in the first phase. It can be readily seen that such a strategy is more bandwidth efficient than SDF because the relay only transmits if necessary.

We consider a modified version of the incremental relaying strategy. In particular, we consider a decode-and-forward incremental relaying scheme with selection capability at the relay (SIDF). In SIDF, the first phase is exactly as amplify-and-forward incremental relaying. In the second phase, if the destination does not receive correctly then the relay, if it was able to decode the source signal correctly, forwards the re-encoded signal to the destination, otherwise the source retransmits again. One can think of this protocol as combining the benefits of selection and incremental relaying.

Next we analyze the outage probability of SIDF. As we did when studying SDF, we are also going to consider two scenarios for SIDF, namely, when the packet structure is not allowed to be changed and the scenario of equal spectral efficiency. First we consider the first scenario where the packet structure is not allowed to be changed. The spectral efficiency of SIDF in this case is less than TDMA or CCMA because the relay is occasionally allocated some channel resources for transmission with positive probability. Since both SDF and SIDF have the same mechanism for the outage event, it is readily seen that the outage event for SIDF is also given by (11.79) with the difference that we only use β in this case without the term 2 because SIDF will use the same transmit power.[4] The outage event is thus given by

$$\mathrm{Pr}_{\mathrm{SIDF}}(O) = \left(1 - f_{i,d}\right)^2 \left(1 - f_{i,l}\right) + \left(1 - f_{i,d}\right) f_{i,l} \left(1 - f_{l,d}\right). \quad (11.83)$$

The above expression represents the success probability of transmitting a packet in one or two consecutive time slots. Terminal i uses one time slot with probability $f_{i,d}$ and two time slots with probability $1 - f_{i,d}$. The average number of time slots used by terminal i during a frame in SIDF is thus given by $2 - f_{i,d}$. The set of queues are not interacting in this case and the stability region is simply given by

$$\sum_{i \in M} \frac{\lambda_i \left(2 - f_{i,d}\right)}{1 - \mathrm{Pr}_{\mathrm{SIDF}}(O)} < 1. \quad (11.84)$$

Next, we consider the second scenario of SIDF where the spectral efficiency is preserved for SIDF as for TDMA or CCMA. In this scenario, both the terminals and the relay will be transmitting at a higher rate $\tilde{R}$ such that the average spectral efficiency is equal to the spectral efficiency R of TDMA or CCMA. The average spectral efficiency $R(\mathrm{SIDF})$ of SIDF when transmitting at a spectral efficiency $\tilde{R}$ is given by

$$R(\mathrm{SIDF}) = \tilde{R} \, \tilde{f}_{i,d} + \frac{\tilde{R}}{2} \left(1 - \tilde{f}_{i,d}\right) \triangleq v(\tilde{R}), \quad (11.85)$$

where $\tilde{f}_{i,d}$ is the success probability for the link between terminal i and the destination when operating at spectral efficiency $\tilde{R}$, and we denote the whole function in the above expression by $v(\cdot)$. For the sake of comparison $R(\mathrm{SIDF}) = v(\tilde{R})$ should be equal to R.

---

[4] This is in favor of SIDF because the transmit power should be reduced to account for the reduction in the average spectral efficiency.

Thus for a given R one should solve for $\tilde{R} = v^{-1}(R)$. This function can lead to many solutions for $\tilde{R}$, and we are going to choose the minimum $\tilde{R}$. The stability region is thus given by

$$\sum_{i \in \mathcal{M}} \frac{\lambda_i}{1 - \text{Pr}_{\text{SIDF}}(\tilde{O})} < 1. \qquad (11.86)$$

where $\text{Pr}_{\text{SIDF}}(\tilde{O})$ has the same form as $\text{Pr}_{\text{SIDF}}(O)$ but evaluated at spectral efficiency $\tilde{R}$.

11.3.4   Numerical examples

In the following examples, we compare the stability regions of $M = 2$-users TDMA, CCMA-S, CCMA-Me, the two forms of adaptive relaying (selection and incremental relaying), and ALOHA as an example of random access. In ALOHA, a terminal transmits a packet with some positive probability p if it has a packet to transmit. This means that there can be collisions among different terminals due to simultaneous transmissions in a time slot. The stability region for a general multipacket reception (MPR) ALOHA system was characterized in [NMT05].

Example 11.5   (Stability region)   In Figures 11.5 and 11.6 we plot the stability regions for TDMA, CCMA-S, CCMA-Me, selection decode-and-forward (SDF), incremental decode-and-forward (SIDF), and ALOHA for a SNR threshold of $\beta = 35$, and $\beta = 64$, respectively. For SDF and SIDF we use the first scenario in which the packet structure is not changed. The parameters used to depict these results are as follows. The distances

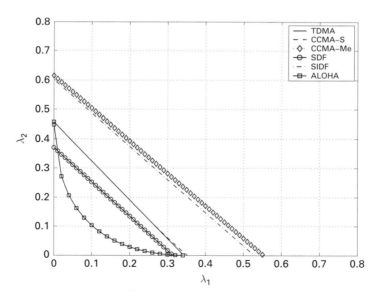

Fig. 11.5   Stability regions for the different considered protocols at a SNR threshold of $\beta = 35$. For this value of β CCMA-S is equivalent to TDMA as depicted. CCMA-Me has the largest throughput region.

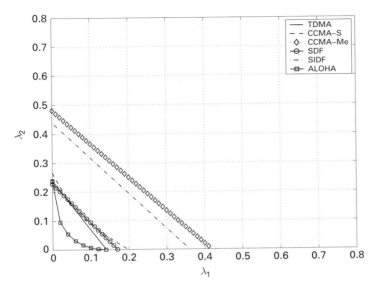

Fig. 11.6   Stability regions for the different considered protocols at a SNR threshold of $\beta = 64$. TDMA is contained in CCMA-S, and the gap between SIDF and CCMA-Me increases in this case.

in meters between different terminals are given by $r_{1,d} = 120$, $r_{2,d} = 110$, $r_{1,d} = 40$, $r_{1,l} = 85$, $r_{2,l} = 80$. The propagation path loss is given by $\gamma = 3.6$, the transmit power $G = 0.01$ W, and $N_0 = 10^{-11}$. In both Figures 11.5 and 11.6, CCMA-Me has the largest stable throughput region. In Figure 11.5, CCMA-S and TDMA have identical stable throughput regions, and it can be checked that the conditions in (11.50) are satisfied for $\beta = 35$. In Figure 11.6, TDMA is contained inside CCMA-S. Both CCMA-S and CCMA-Me provide a larger stable throughput region over SDF and ALOHA. This is because of the lost bandwidth efficiency in SDF and the interference in ALOHA due to collisions. SIDF is very close to CCMA-Me for smaller values of β, and the gap between them increases with increasing β as depicted in Figure 11.6. This is because the bandwidth efficiency of SIDF reduces with increasing β, which increases the probability of using a second time slot by the relay.   ▲

Example 11.2   Next, we demonstrate the tradeoff between the maximum stable throughput (MST) versus the SNR threshold β and the transmission rate R. For SDF and SIDF, we consider the two scenarios described in Section 11.3.3.1, where in the first scenario the incoming packet structure is not changed and hence the two protocols have less bandwidth efficiency compared to TDMA and CCMA. While in the second strategy, the packet structure is changed to preserve the bandwidth efficiency. The maximum stable throughput results for the two scenarios are depicted in Figures 11.7 and 11.8, respectively. In both figures, the relative distance between terminals are, $r_{1,d} = r_{2,d} = 130$,

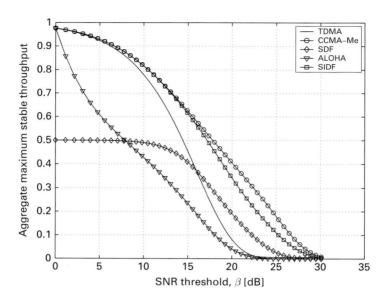

Fig. 11.7   Aggregate maximum stable throughput versus SNR threshold β in dB. The propagation path loss is set to $\gamma = 3.5$. First scenario is used for SDF and SIDF. CCMA-Me has the best tradeoff curve among all the other protocols.

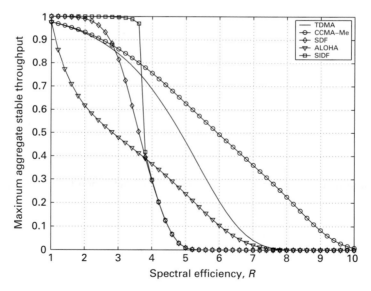

Fig. 11.8   Aggregate maximum stable throughput versus spectral efficiency R in bits/s/Hz. The propagation path loss is set to $\gamma = 3.5$. Second scenario is used for SDF and SIDF. SIDF has the best performance for low spectral efficiency but it suffers from a catastrophic degradation with increasing spectral efficiency R. CCMA-Me has a graceful degradation due to its bandwidth efficiency, and it has the best tradeoff for medium to high spectral efficiency.

$r_{1,d} = 50$, and $r_{1,l} = r_{2,l} = 80$. In Figure 11.7, the MST is plotted against the SNR threshold β and the propagation path loss is set to $\gamma = 3.5$. For SDF and SIDF, we use the first scenario. CCMA-Me has the best tradeoff for the whole range. TDMA and CCMA-S have identical performance as proven before for the symmetric case. The maximum attained MST for SDF is 0.5 as expected because of the time slot repetition. ALOHA has better performance over SDF for a low SNR threshold, but for medium and high values of β, SDF has better performance. SIDF has a close performance to CCMA-Me for low values of β, and CCMA-Me outperforms it for the rest of the SNR threshold range.

In Figure 11.8 the MST is plotted against the transmission rate R, and we use the second scenario for SDF and SIDF. For this case the MST of SDF starts from 1 for low rates R but decays exponentially after that. SIDF and SDF have the best performance for low spectral efficiency regimes where sacrificing the bandwidth by transmitting at higher rate is less significant than the gains achieved by diversity. SIDF performs better than SDF because it is more bandwidth efficient. For higher spectral efficiency regimes, the proposed CCMA-Me provides significantly higher stable throughput compared to SDF or SIDF. An important point to observe from Figure 11.8 is the graceful degradation in the performance of CCMA-Me, but the sudden catastrophic performance loss in SDF and SIDF. The rationale here is that the cognitive feature of the proposed CCMA-Me results in no bandwidth loss because cooperation occurs in the idle time slots, while both SDF and SIDF suffer from a bandwidth loss that increases for SIDF with increasing R. Our results show that utilizing empty time slots to increase system reliability via cooperation is a very promising technique in designing cooperative relaying strategies for wireless networks. ▲

11.4 Throughput region

In the characterization of the stable throughput region in the previous section, the source burstiness is taken into consideration. Consider now the scenario under which the queues of all terminals are saturated, i.e., each terminal has an infinite number of packets waiting transmission. The maximum throughput supported by any terminal can be defined under such a scenario by the average maximum number of packets that can be transmitted successfully by that terminal. The set of all such saturated throughput for different resource-sharing vectors defines the throughput region.

Since the relay role in both CCMA-S and CCMA-Me depends on having empty time slots to enable cooperation, there is no surprise that under the saturated queues scenario the relay loses its role and CCMA-S and CCMA-Me reduce to TDMA without relaying. We state this in the following corollary.

Corollary 11.4.1 *The throughput regions of TMDA, CCMA-S, and CCMA-Me are equivalent.*

$$\mathcal{C}\,(\text{TDMA}) \equiv \mathcal{C}\,(\text{CCMA} - \text{S}) \equiv \mathcal{C}\,(\text{CCMA} - \text{Me})\,. \tag{11.87}$$

From the above corollary, we conclude that the saturated throughput region is a subset of the stability region for both CCMA-S and CCMA-Me. This is an important observation, because for ALOHA systems it is conjectured in [129] that the maximum stable throughput region is identical to the throughput region. It is of interest then to point out that CCMA-S and CCMA-Me are examples of multiple access protocols where the stable throughput region is different from the throughput region.

11.5   Delay analysis

In this section we characterize the delay performance of the cognitive cooperative multiple access protocols, CCMA-S and CCMA-Me.

11.5.1   Delay performance for CCMA-S

In CCMA-S, a packet does not depart a terminal's queue until it is successfully transmitted to the destination. Therefore, the delay encountered by a a packet is the one encountered in the terminal's queue. Delay analysis for interacting queues in ALOHA has been studied before, and it turns out to be a notoriously hard problem. Most of the known results are only for the two-user ALOHA case. In this section we consider a symmetric two-user CCMA-S scenario and characterize its delay performance.

Define the moment-generating function of the joint queues' sizes processes (Q_1^t, Q_2^t) as follows:

$$G(u, v) = \lim_{t \to \infty} E\left[u^{Q_1^t} v^{Q_2^t}\right]. \tag{11.88}$$

From the queue evolution equations in (11.13), we have

$$E\left[u^{Q_1^{t+1}} v^{Q_2^{t+1}}\right] = E\left[u^{X_1^t} v^{X_2^t}\right] E\left[u^{(Q_1^t - Y_1^t)^+} v^{(Q_2^t - Y_2^t)^+}\right], \tag{11.89}$$

where the above equation follows from the independence assumption of the future arrival processes from the past departure and arrival processes. Since the arrival processes are assumed to follow a Bernoulli random process, the moment-generating function of the joint arrival process is given by

$$A(u, v) \triangleq \lim_{t \to \infty} E\left[u^{X_1^t} v^{X_2^t}\right] = (u\lambda + 1 - \lambda)(v\lambda + 1 - \lambda), \tag{11.90}$$

where, due to the symmetry of the two terminals, each has an arrival rate λ. From the definition of the service process of CCMA-S (11.17), it follows that

$$E\left[u^{(Q_1^t - Y_1^t)^+} v^{(Q_2^t - Y_2^t)^+}\right] = E\left[\mathbf{1}(Q_1^t = 0, Q_2^t = 0)\right] + B(u)E\left[\mathbf{1}(Q_1^t > 0, Q_2^t = 0)u^{Q_1^t}\right]$$
$$+ B(v)E\left[\mathbf{1}(Q_1^t = 0, Q_2^t > 0)v^{Q_2^t}\right] + D(u, v)E\left[\mathbf{1}(Q_1^t > 0, Q_2^t > 0)u^{Q_1^t} v^{Q_2^t}\right],$$

where

$$B(z) = \frac{wf_{1,d} + w^2 f_{1,l} f_{1,d}(1 - f_{1,d})}{z} + 1 - \left(wf_{1,d} + w^2 f_{1,l} f_{1,d}(1 - f_{1,d})\right), \tag{11.91}$$

$$D(u, v) = wf_{1,d}(\tfrac{1}{u} + \tfrac{1}{v}) + 2w(1 - f_{1,d}), \tag{11.92}$$

in which we use w to denote the symmetric resource-sharing portion of each terminal.

Substituting (11.90) and (11.91) into (11.89) and taking the limits we get

$$G(u, v) = A(u, v) (G(0, 0) + B(u) [G(u, 0) - G(0, 0)] + B(v) [G(0, v)$$
$$- G(0, 0)] + D(u, v) [G(u, v) + G(0, 0) - G(u, 0) - G(0, v)]) .$$
(11.93)

We can rewrite the above equation as $G(u, v) = H(u, v)/F(u, v)$, where

$$H(u, v) = G(0, 0) + B(u) [G(u, 0) - G(0, 0)] + B(v) \times [G(0, v) - G(0, 0)]$$
$$+ D(u, v)(G(0, 0) - G(u, 0) - G(0, v)),$$

and $F(u, v) = 1 - A(u, v)D(u, v)$.

Define

$$G_1(u, v) \triangleq \frac{\partial G(u, v)}{\partial u}.$$
(11.94)

Due to symmetry, the average queue size is given by $G_1(1, 1)$. Hence, to find the average queuing delay, we need to compute $G_1(1, 1)$.

First we find a relation between $G(0, 0)$ and $G(1, 0)$ using the following two properties which follow from the symmetry of the problem: $G(1, 1) = 1$ and $G(1, 0) = G(0, 1)$. Applying these two properties to (11.93) along with a simple application of the L'Hopital limit theorem we get

$$\left(w^2 f_{1,l} f_{1,d}(1 - f_{1,d})\right) G(0, 0) + \left(w f_{1,d} - w^2 f_{1,l} f_{1,d}(1 - f_{1,d})\right) G(1, 0) = w f_{1,d} - \lambda.$$
(11.95)

Taking the derivative of (11.93) with respect to u, applying L'Hopital twice, and using the relation in (11.95) we get

$$G_1(1, 1) = \frac{\lambda(1 - \lambda)}{w f_{1,d} - \lambda} - \frac{w^2 f_{1,l} f_{1,d}(1 - f_{1,d})}{w f_{1,d} - \lambda} G_1(1, 0).$$
(11.96)

To find another equation relating $G_1(1, 1)$ and $G_1(1, 0)$, we compute $\partial G(u, u)/\partial u$ at $u = 1$. After some tedious but straightforward calculations, we get

$$\frac{G(u, u)}{u}\bigg|_{u=1} = 2\lambda - 1 + \frac{w f_{1,d} - w^2 f_{1,l} f_{1,d}(1 - f_{1,d})}{w f_{1,d} - \lambda}$$
$$G_1(1, 0) - \frac{4w f_{1,d}\lambda - 2w f_{1,d} - \lambda^2}{2(w f_{1,d} - \lambda)}.$$
(11.97)

Due to the symmetry of the problem, we have the following property:

$$\frac{\partial G(u, u)}{\partial u}\bigg|_{u=1} = 2G_1(1, 1).$$
(11.98)

Using the above equation, and solving (11.96) and (11.97), we get

$$G_1(1, 1) = \frac{-\left(2w f_{1,d} + w^2 f_{1,l} f_{1,d}(1 - f_{1,d})\right)\lambda^2 + 2w f_{1,d}\lambda}{2(w f_{1,d} + w^2 f_{1,l} f_{1,d}(1 - f_{1,d}))(w f_{1,d} - \lambda)}.$$
(11.99)

The queueing delay for system CCMA-S can thus be determined, as in the following theorem.

THEOREM 11.5.1   *The average queueing delay for a symmetrical two-terminals CCMA-S system is given by*

$$D(\text{CCMA} - \text{S}) = \frac{G_1(1,1)}{\lambda} = \frac{-\left(2wf_{1,d} + w^2 f_{1,l} f_{1,d}(1 - f_{1,d})\right)\lambda + 2wf_{1,d}}{2(wf_{1,d} + w^2 f_{1,l} f_{1,d}(1 - f_{1,d}))(wf_{1,d} - \lambda)}.$$

(11.100)

From Theorem 11.5.1, it can be observed that at $\lambda = wf_{1,d}$ the delay of the system becomes unbounded, i.e., the system becomes saturated. This confirms our previous results in Corollary 11.4.1 that, for a symmetrical system, both TDMA and CCMA-S have the same maximum stable throughput of $\lambda = wf_{1,d}$.

11.5.2   Delay performance of CCMA-Me

Due to the symmetrical scenario considered in analyzing the delay performance, if the relay helps one terminal then it helps all terminals, in which case both CCMA-M and CCMA-Me become equivalent. In CCMA-M, a packet can encounter two queuing delays: the first in the terminal's queue and the second in the relay's queue. If a packet successfully transmitted by a terminal goes directly to the destination, then this packet is not stored in the relay's buffer. Denote this event by ξ. The total delay encountered by a packet in CCMA-M can thus be modeled as

$$T(\text{CCMA-M}) = \begin{cases} T_t, & \xi, \\ T_t + T_l, & \bar{\xi}, \end{cases}$$

(11.101)

where T_t is the queuing delay in the terminal's queue, and T_l is the queuing delay at the relay's queue. We can elaborate more on (11.101) as follows. For a given packet in the terminal's queue, if the first successful transmission for this packet is to the destination, then the delay encountered by this packet is only the queuing delay in the terminal's queue. On the other hand, if the first successful transmission for this packet is not to the destination, then the packet will encounter a queuing delay in the terminal's queue in addition to the queuing delay in the relay's queue.

First, we find the queuing delay in either the terminal's or the relay's queue, as both queues have similar evolution equations, with the difference being in the average arrival and departure rates. Using the same machinery utilized in the analysis of the queuing delay in CCMA-S to analyze the delay performance of CCMA-M, the average queue size can be found as

$$E[N] = \frac{\lambda(1 - \lambda)}{\mu - \lambda},$$

(11.102)

where λ denotes the average arrival rate and μ denotes the average departure rate. We now compute the average delay in (11.101). The probability that, for any packet, the first successful transmission from the terminal's queue is to the destination is given by

$$\Pr[\xi] = \frac{f_{1,d}}{f_{1,d} + f_{1,l} - f_{1,d} f_{1,l}} = \frac{f_{1,d}}{P_1}. \tag{11.103}$$

From (11.101), (11.102), and (11.103), the average delay for system CCMA-M is thus given by

$$D(\text{CCMA} - \text{M}) = \frac{f_{1,d}}{P_1} \frac{1-\lambda}{w P_1 - \lambda} + \frac{f_{1,l}(1 - f_{1,d})}{P_1} \left(\frac{1-\lambda}{w P_1 - \lambda} + \frac{1-\lambda_l}{\mu_l - \lambda_l} \right), \tag{11.104}$$

where λ_l and μ_l are the average arrival and departure rates, respectively, for the relay's queue defined in (11.61) and (11.63). After simplifying the above equation, the average queuing delay for system CCMA-M can be summarized in the following theorem.

THEOREM 11.5.2   *The average queuing delay for a packet in a symmetrical two-terminal CCMA-M system is given by*

$$D(\text{CCMA} - \text{M}) = \frac{1-\lambda}{w P_1 - \lambda} + \frac{f_{1,l}(1 - f_{1,d})}{P_1} \left(\frac{1-\lambda_l}{\mu_l - \lambda_l} \right). \tag{11.105}$$

11.5.3   Numerical examples for delay performance

We illustrate the delay performance of the presented multiple access schemes with varying SNR threshold β through some numerical examples. As in the case for the stability region, we include the delay performance of ALOHA, TDMA without relaying, SDF, and SIDF in our results.

Example 11.3   To compare the delay performance of the different multiple access protocols considered in this chapter, we plotted the analytical expressions obtained for the queueing delay. The system parameters are the same used to generate the MST plots in Figures 11.7 and 11.8. Figures 11.9 and 11.10 depict the delay results for SNR thresholds $\beta = 15$ and $\beta = 64$, respectively. From Figure 11.9, at very low arrival rates, ALOHA has the best delay performance. Increasing the arrival rate λ, both CCMA-S and SDF outperforms other strategies. For higher values of λ, CCMA-Me has the best performance

The situation changes in Figure 11.10 for $\beta = 64$ as both CCMA-S and SIDF outperform ALOHA even for very small arrival rates. The intuition behind this is the more stringent system requirements reflected by the higher SNR threshold $\beta = 64$, which makes the interference in ALOHA more severe. This makes our cognitive multiple access protocol CCMA-S and CCMA-M perform better than ALOHA because of its high bandwidth efficiency and the gains of cooperation. Another important

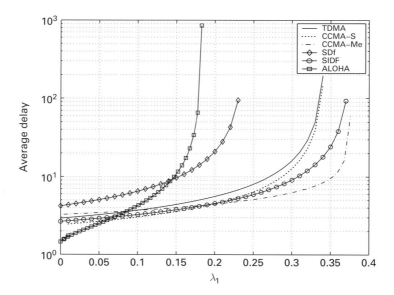

Fig. 11.9   Average queuing delay per terminal versus the arrival rate for a SNR threshold of $\beta = 15$.

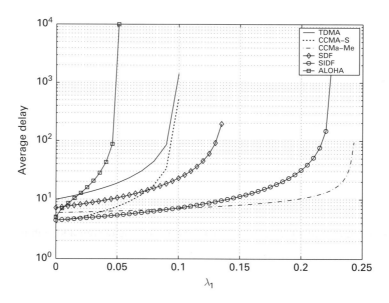

Fig. 11.10   Average queuing delay per terminal versus the arrival rate for a SNR threshold of $\beta = 64$.

remark is that, although CCMA-S and TDMA have the same MST for the symmetric case, as proven before and as clear from Figures 11.9 and 11.10 where both proto-cols saturate at the same arrival rate, CCMA-S always has a better delay performance than TDMA.   ▲

11.6   Chapter summary and bibliographical notes

In this chapter, we have studied the impact of cooperative communications at the multiple access layer. We described a cognitive multiple access protocol in the presence of a relay in the network. The relay senses the channel for idle channel resources and exploits them to cooperate with the terminals in forwarding their packets. Two protocols to implement the proposed multiple access strategy, namely, CCMA-S and CCMA-Me, were described. We characterized the maximum stable throughput region of the proposed protocols and compared them to some existing adaptive relaying strategies: non-cooperative TDMA and random-access ALOHA.

We studied the delay performance of the described protocols. The analysis reveals significant performance gains of the cognitive protocols over their non-cognitive counterparts. This is because the described cognitive multiple access strategies do not result in any bandwidth loss, as cooperation is enabled only in idle "*unused*" channel resources, which results in a graceful degradation of the maximum stable throughput when increasing the communication rate. On the other hand, the maximum stable throughput of non-cognitive relaying strategies as selection and incremental relaying suffer from catastrophic degradation with increasing communication rate, because of their inherent bandwidth inefficiency.

Analyzing the stability of interacting queues is a difficult problem that has been addressed for ALOHA systems initially in [220]. Later in [149], the dominant system approach was explicitly introduced and employed to find bounds on the stable throughput region of ALOHA with collision channel model. Many other works followed that to study the stability of ALOHA. In [209], necessary and sufficient conditions for the stability of a finite number of queues were provided; however, the stable throughput region was only explicitly characterized for a three-terminals system. In [127], the authors provided tighter bounds on the stable throughput region for the ALOHA system using the concept of *stability ranks*, which was also introduced in the same paper. The stability of ALOHA systems under a multi-packet reception model (MPR) was considered in [138, 129]. Characterizing the stable throughput region for interacting queues with $M > 3$ terminals is still an open problem.

Delay analysis for interacting queues is a notoriously hard problem that has been investigated in [136, 185] for ALOHA. Studies on the design and stability analysis of the cognitive collaborative multiple access protocol can be found in [159, 160, 162].

Exercises

11.1   In Section 11.3.3, the stability regions for selective and incremental relaying were derived.

(a) Derive the outage probability for both decode-and-forward relaying and amplify-and-forward relaying and use the results to characterize the stable throughput regions for both relaying techniques using Loynes' theorem.

(b) Find the maximum stable throughput for these two protocols and plot the results against the SNR threshold for a fixed spectral efficiency, and against the spectral efficiency when fixing the SNR. Compare the performance to CCMA-S and CCMA-Me.

11.2 The stability results for CCMA-S and CCMA-Me were derived under the Rayleigh flat fading channel model.

(a) Derive the stability regions for the two protocols under an AWGN channel model including the propagation path loss.

(b) Characterize the relation between the stability regions of the proposed protocols and TDMA under the AWGN channel.

11.3 Use the moment-generating function approach to analyze the delay performance of decode-and-forward and amplify-and-forward relaying under a Rayleigh flat fading channel model for a two-user system. Plot the delay performance versus the arrival rate for a symmetric case. Find the arrival rate at which the system saturates. Perform the previous calculations for the following three scenarios:

(a) Relay close to the source.

(b) Relay close to the destination.

(c) Relay in the middle.

Compare the delay performance results of decode-and-forward and amplify-and-forward to the CCMA-S and CCMA-Me protocols presented in this chapter.

11.4 Find another approach to prove the results of Theorem 11.3.3 without using numerical techniques.

[Hint: Find an auxiliary system whose stability region contains that of CCMA-S and at the same time is a subset from the stability region of CCMA-Me.]

11.5 Characterize the maximum sum stable throughput that can be achieved by of two-user TDMA, CCMA-S, and CCMA-Me for a general asymmetric case.

What is the impact of a scheme that maximizes the sum stable throughput on the system fairness? In other words, do both users transmit at the same rate?

11.6 In Exercise 11.5 the goal was to maximize the system sum-stable throughput. Consider now that the objective is to achieve an equal grade of service, i.e., all users in the system achieve the same throughput.

Find the maximum common throughput that all users in the system can achieve. [An equal grade of service corresponds to absolute fairness because all users get to transmit at the same throughput.]

[Hint: An equal grade of service corresponds to the point on the stable throughput region that intersects with a line that makes a 45° with the x-axis.]

11.7 Write down a Matlab code that draws the stability region of CCMA-S using the expression in (11.22), and without resorting to the closed-form expression in Theorem 11.3.1. Next compare your numerical results to this closed-form expression.

11.8 In (11.42), the optimal amount of resources allocated for the first user was specified as

$$w_1* = \begin{cases} \frac{\lambda_1}{f_{1,d}}, & \text{if } \lambda_1 \leq \frac{f_{1,d}^2}{(1-f_{1,d})f_{1,l}f_{1,d}}, \\ \sqrt{\frac{\lambda_1}{(1-f_{1,d})f_{1,l}f_{1,d}}}, & \text{otherwise.} \end{cases} \qquad \text{(E11.1)}$$

Prove that this expression never exceeds one.

11.9   A moment-generating function approach was used to analyze the delay of CCMA-S. The moment-generating function expression was given as (11.93)

$$G(u, v) = A(u, v) (G(0, 0) + B(u) [G(u, 0) - G(0, 0)] + B(v) [G(0, v)$$
(E11.2)

$$-G(0, 0)] + D(u, v) [G(u, v) + G(0, 0) - G(u, 0) - G(0, v)]) .$$
(E11.3)

Prove this expression.

11.10 Applying the L' Hopital theorem twice prove the expression in (11.99), i.e., prove

$$G_1(1, 1) = \frac{- \left(2wf_{1,d} + w^2 f_{1,l} f_{1,d}(1 - f_{1,d})\right) \lambda^2 + 2wf_{1,d}\lambda}{2(wf_{1,d} + w^2 f_{1,l} f_{1,d}(1 - f_{1,d}))(wf_{1,d} - \lambda)} .$$
(E11.4)

12 Content-aware cooperative multiple access

The previous chapter studied the effects and use of cooperation on the multiple access channel. In this chapter we look further into using the properties of the source traffic to improve the efficiency of cooperative multiple access. Because the presentation in this chapter is highly dependent on the characteristics of the source, we will focus on the communication of packet speech. Nevertheless, the main underlying ideas can be extended to other types of sources.

Speech communication has a distinctive characteristic that differentiates it from data communication, which was the main focus of the previous chapter. Speech sources are characterized by periods of silence in between talk spurts. The speech talk–silence patterns could be exploited in statistical multiplexing-like schemes where silent users release their reserved channel resources, which can then be utilized to admit more users to the network. This comes at the cost of requiring a more sophisticated multiple access protocol. One well-known protocol that uses this approach for performance improvement is the packet reservation multiple access (PRMA) protocol [49], which can be viewed as a combination of TDMA and slotted ALOHA protocols. In PRMA, terminals in talk spurts contend for the channel in empty time slots. If a user contends succesfully, then the slot used for contention is reserved for the user. Users with reservations transmit their voice packets in their reserved slots. If a user fails to transmit its packet due to channel errors, the user looses the reservation and the reserved slot becomes free for contention again. Although the effects of this operation is practically unnoticeable in channels where errors are very infrequent, in channels with frequent errors (such as the wireless channel) the loss of renovation due to an error has the potential to adversely affect the efficiency of the protocol. This is because channel errors not only force the discarding of the damaged packets, but also increase the network traffic and access delay, as users with lost packets have to repeat the contention process again.

This chapter will focus on a cooperative multiple access protocol that uses the properties of speech to increase the network capacity and the user cooperation efficiency. As is the case with established multiple access protocols, the network capacity is increased by reserving the network resources only to the users in a call spurt. Those users finishing a silence period need to contend for channel access over a shared resource. At the same time, the use of cooperation increases the system performance by helping users in talk spurts to reduce the probability of dropping packets and having to contend again. Cooperation is achieved through the deployment of a relay node. This relay node exploits the silence periods typical of speech communications in a new way, since it cognitively forwards speech packets for active calls using part of the free time slots left available by users that are silent. Most importantly, because the resources allocated to the relay were previously available for other users' contention for channel access, no new exclusive channel resources are needed for cooperation and the system encounters no bandwidth losses. On the other hand, the use of cooperation imposes a tradeoff between the amount of help offered to active calls and the probability of a successful contention for channel access. This kind of tradeoff is, in some sense, similar to the diversity–multiplexing tradeoff in MIMO and other user-cooperative systems. By judicious control of this tradeoff it is possible to achieve significant performance improvements through cooperation.

12.1 System model

12.1.1 speech source model

Speech sources are characterized by periods of silence in between talk spurts that account for roughly 60% of the conversation time. This key property could be exploited to significantly improve the utilization of channel resources but with the cost of requiring a more sophisticated multiple access protocol.

Example 12.1 Figure 12.1 shows the voice signal amplitude for a speech sequence and illustrates the alternation of speech between talking and silence periods. In the figure, the dashed line indicates the state of the speech sequence, "on" or "off," over the time. In order to make the figure sufficiently clear we have chosen arbitrarily a value of 0.3 when

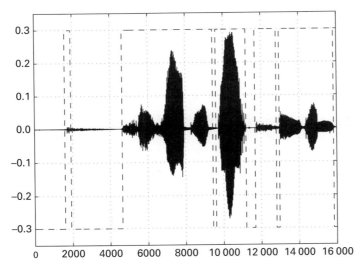

Fig. 12.1 A typical speech segment illustrating the on/off characteristic of speech. The dashed lines take a value of 0.3 (chosen arbitrarily so the figure is sufficiently clear) when speech is detected "on" and a value of −0.3 when speech is detected "off".

speech is detected "on"(talking state) and a value of -0.3 when speech is detected "off" (silence state). In practice, the detection of the speech state is performed in the source encoder through an algorithm named VAD (voice activity detector).   ▲

To model the alternations between periods of silence and talk spurts, each speech source in a conversation is modeled as a Markov chain as shown in Figure 12.2 with two states: talk (TLK) and silence (SIL). In the figure, γ represents the transition probability from the talking state to the silence state and σ is the transition probability from the silence state to the talking state. The value of these two probabilities depend, of course, on the speech model but also on the time unit used to model state transitions in the Markov chain. In packet speech communications scenarios, as in the one considered in this chapter, it is convenient for the purpose of mathematical analysis, to choose the same basic time unit for the Markov chain as the one used for channel access. As will be discussed in the next section, the channel is divided into TDMA time frames, each of duration T seconds. Hence, it is suitable to also choose the basic time unit for the Markov chain to be equal to T seconds, which means that state transitions are only allowed at the frame boundaries.

It is also customary in the Markov chain modeling of speech sources to assume that the waiting time in any state has an exponential distribution. Then, with these assumptions in mind, the transition probability from the talking state to the silence state is the probability that a talk spurt with mean duration t_1 ends in a frame of duration T, which can be calculated as

$$\gamma = 1 - e^{-T/t_1}. \tag{12.1}$$

Similarly, the transition probability from the silence state to the talking state is the probability that a silence gap of mean duration t_2 ends during a frame of duration T, and can be calculated as

$$\sigma = 1 - e^{-T/t_2}. \tag{12.2}$$

12.1.2   Network model

We consider a network with three types of nodes as illustrated in Figure 12.3: source nodes, associated with each user of the network; a relay node; and a base station. We will focus exclusively on the uplink channel and assume a packet network carrying speech traffic. Medium access in the network is based on the packet reservation multiple

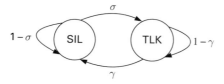

Fig. 12.2   Speech source model using a Markov chain. The two talk states TLK and SIL correspond to a speaker being in a talk spurt or a period of silence, respectively.

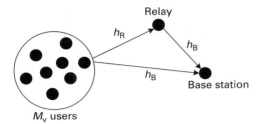

Fig. 12.3 Network and channel model.

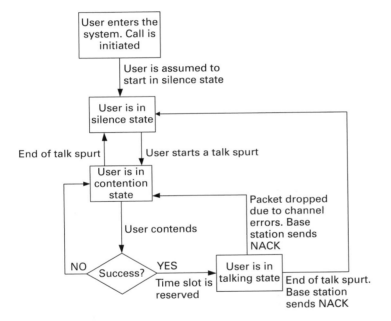

Fig. 12.4 Flowchart of the PRMA protocol.

access (PRMA) protocol [49]. The PRMA protocol can be viewed as a combination of the TDMA and slotted ALOHA protocols, where the channel is subdivided into time frames and each frame is in turn subdivided into N time slots. Figure 12.4 shows a simple diagram of how PRMA works. Those users in the process of starting a talk spurt contend for the channel over empty time slots, independently of each other and with a fixed access probability that we will denote as p_V. If a user is successful in the contention process, then a slot is reserved for that user; otherwise, the base station feeds back a NULL message to make the slot available for contention in the next time frame.

Users with reserved slots keep the reservation, in principle for the duration of the talk spurt, and use them to transmit their corresponding speech packets. Upon ending a talk spurt, a user enters a silence state where it is not generating or transmitting any packets. In this case, the base station feeds back a NULL message to declare that the previously reserved time slot is free again for other users to use. Note that the PRMA protocol exploits the on–off nature of speech to improve the utilization of the channel by reserving slots only for calls in a talk spurt. Nevertheless, note also that because users contend for an empty time slot with certain probability, some slots may be left unused even in situations of access congestion.

Note that the state of every time slot (free or reserved) in the current frame is determined by the base station feedback at the end of each time slot in the previous frame. It is assumed that the feedback channel is error free, thus there is no uncertainty in the state of any time slot. Moreover, it is assumed that the base station will also feed back a NULL message in response to errors due to contention and due to wireless channel impairments. This means that a user will loose its reservation if it faces a channel error while holding a reserved slot and transmitting its packet. Here, we ideally assume that the feedback message is immediate.

Because speech communication is very sensitive to delay, speech packets require prompt delivery. In PRMA, the voice packets from calls that fail the contention to access the channel are placed in a waiting queue. If a packet remains undelivered for a pre-specified maximum delay of $D_{\max}$ frames, the packet is dropped from the user's queue.

12.1.3   Channel model

The received signal at the base station can be written as

$$y_{\mathrm{B}} = \sqrt{P_1 r_{\mathrm{B}}^{-\alpha}} h_{\mathrm{B}} x + \eta_{\mathrm{B}};$$ (12.3)

similarly, the received signal at the relay can be written as

$$y_{\mathrm{R}} = \sqrt{P_1 r_{\mathrm{R}}^{-\alpha}} h_{\mathrm{R}} x + \eta_{\mathrm{R}};$$ (12.4)

where x is the transmitted signal, P_1 the transmit power, assumed to be the same for all users, r_{B} and r_{R} denote the distance from any user to the base station and to the relay, respectively, α is the path loss exponent, and h_{B} and h_{R} are the channel fading coefficients for the user–base station and user–relay links, respectively, which are modeled as zero-mean, complex Gaussian random variables with unit variance. The additive noise terms η_{B} and η_{R} are modeled as zero-mean, complex Gaussian random variables with variance N_0. We assume that the channel coefficients are constant for the transmission duration of one packet and that the network can be considered as symmetric, i.e., all the inter-user channels are assumed to be statistically identical (see Figure 12.3).

As in earlier chapters, the success and failure of packet reception is characterized by outage events and outage probabilities. In this case we consider a slightly more general definition of the outage probability, not necessarily tied to the notion of capacity, and we define the outage probability as the probability that the signal-to-noise ratio (SNR) at the receiver is less than a given SNR threshold β, called the *outage SNR*. For the channel model in (12.3) and (12.4), the received SNR of a signal transmitted between any user and the base station can be specified as follows:

$$\mathrm{SNR}_{\mathrm{B}} = \frac{|h_{\mathrm{B}}|^2 r_{\mathrm{B}}^{-\alpha} P_1}{N_0},$$ (12.5)

where $\mid h_B \mid^2$ is the random channel gain magnitude squared, which has an exponential distribution with unit mean. Since the SNR in (12.5) is a monotone function of $\mid h_B \mid^2$, the outage event for an outage SNR β is equivalent to

$$\{h_B : \text{SNR}_B < \beta\} = \left\{ h_B : \mid h_B \mid^2 < \frac{\beta N_0 r_B^\alpha}{P_1} \right\}. \tag{12.6}$$

Accordingly, and from the exponential distribution of the received SNR, the outage probability is

$$P_{\text{OB}} = \Pr\left\{ \mid h_B \mid^2 < \frac{\beta N_0 r_B^\alpha}{P_1} \right\} = 1 - \exp\left(-\frac{\beta N_0 r_B^\alpha}{P_1} \right). \tag{12.7}$$

Similar relations hold for the outage probability between any user and the relay.

12.2 Content-aware cooperative multiple access protocol

Transmission errors, which are inherent to wireless communication channels, have a significant impact on the PRMA network performance. On one side, if a user experiences an error while contending for access to a time slot, it would fail on the try and would have to contend again in another free slot. Moreover, if a user that already holds a reserved slot experiences an error while sending a speech packet, the user would have to give it up and go through the contention process again because the base station would send a NULL feedback upon receiving a packet with errors, which would also indicate that the reserved slot is free. These effects translate into an increase in the number of contending users and, thus, a significant increase in network traffic and in delay to gain a slot reservation. These effects ultimately severely degrade the speech quality. In fact, the congestion may reach a level where all users experience reduced speech quality due to packets dropped due to excessive delay.

By enabling cooperation in the voice network, one can benefit from the spatial diversity offered by cooperation to mitigate the wireless channel impairments. In the cooperative multiple access protocol we will consider here, a single relay node is deployed into the network. This node will have the task of helping users holding slot reservations forward their packets by operating in an incremental decode-and-forward mode. As explained in Section 4.2.2.2, in order for the relay to decide whether to forward the packet or not, it utilizes limited feedback from the base station in the form of an automatic repeat request. This means that the relay will only forward the packets that were not successfully received by the base station. The rationale in introducing a relay is that it would result in a more reliable end-to-end link and, hence, a reduction in the number of users losing their reserved time slots. This leads to a further reduction in the average number of contending users, and therefore, a much lower access delay and packet dropping probability, which ultimately improves speech quality.

The incorporation of the relay operation into the network involves modifying the frame structure as illustrated in Figure 12.5. The first N_T slots create a variable size (from frame to frame) compartment of slots reserved for the talking users. Of the

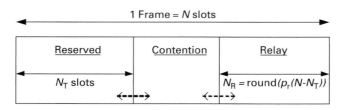

Fig. 12.5   The organization of time slots in a frame.

remaining $(N - N_T)$ free slots, a fraction p_R is assigned to the relay and the remaining free slots are made available for contention. The ordering of slots in a frame is first the N_T slots reserved for the talking users, followed by the slots used in the contention process and the last N_R slots are those assigned to the relay. When a user gives up its reservation or gains a new reservation, the slots are rearranged in order to maintain this frame structure. In any specific frame, the number of slots assigned to the relay will be

$$N_R = \text{round}(p_r(N - N_T)).$$

It is clear that the value of p_r determines how much help the relay will offer to talking users, also it determines the reduction in the number of free time slots available for contention. Therefore, the introduction of cooperation poses a tradeoff between the amount of help the relay offers to existing users and the ability of the network to admit new users because of the reduction in the number of contention slots. Since such tradeoff is governed by p_r, the choice of the value of this parameter is crucial for the optimal performance of the system. This issue will be addressed later in the chapter.

12.3   Dynamic state model

In this section, we develop an analytical model to study and measure the network performance. Based on the models discussed above, a user can be in one of three states: "SIL" when in a silence period, "CON" when contending for channel access, and "TLK" when holding a reserved slot. The dynamics of user transitions between these three states can be described by the Markov chain of Figure 12.6. A user in SIL state moves to CON state when a new talk spurt begins. When there is an available slot, with probability p_v, a user in CON state will send the packet at the head of its queue. If contention succeeds, a user in CON state transits to TLK state, where it will have the slot reserved in subsequent frames. A user moves from CON state to SIL state if its talk spurt ends before gaining access to the channel. A user in TLK state transits to SIL state when its talk spurt ends, and transits to CON state if its packet is not received correctly by the base station. This later transition could be avoided if the relay is able to help the user.

Again, we will consider one complete frame as the time step for the Markov chain. Although the actions of different users are independent, the transition probabilities between different states for a given user are in general dependent on the number of users in CON and TLK states. These numbers will affect the probability with which a

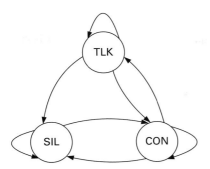

Fig. 12.6 User's terminal model

user succeeds in contention. Moreover, the number of users in TLK state will determine the number of slots assigned to the relay, and hence the relay's ability to help users.

In order to take these dependencies into consideration, the whole network will be modeled as the two-dimensional Markov chain (M_C, M_T), where M_C and M_T are random variables denoting the number of users in CON and TLK states, respectively. Assuming there is M_v users in the network, then the number of users in the SIL state is $M_S = M_v - M_C - M_T$. In what follows, we will analyze this Markov chain and calculate its stationary distribution which will allow for the derivation of different performance measures.

Let $S_1 = (M_{C_1}, M_{T_1})$, and $S_2 = (M_{C_2}, M_{T_2})$ be the system states at two consecutive frames. Then,

$$M_{C_2} = M_{C_1} + m_{SC} + m_{TC} - m_{CS} - m_{CT}, \tag{12.8}$$

$$M_{T_2} = M_{T_1} + m_{CT} - m_{TS} - m_{TC}, \tag{12.9}$$

where m_{ij} denotes the number of users departing from state $i \in \{S, C, T\}$ to state $j \in \{S, C, T\}$, for example, m_{SC} is the number of users departing from SIL state to CON state. This implies that the transition probability between any two states can be determined in terms of the distributions of $m_{SC}, m_{CS}, m_{CT}, m_{TS}$, and m_{TC}. Next we will calculate these distributions.

(i) **Distribution of m_{SC}:** From Section 12.1.1, and since all users are independent, the number of users making a transition from the SIL state to the CON state, m_{SC}, follows a binomial distribution with parameter σ, where σ is defined in (12.2). Then,

$$\Pr(m_{SC} = i) = \binom{M_S}{i} \sigma^i (1 - \sigma)^{M_S - i}, \quad i = 0, \ldots, M_S. \tag{12.10}$$

(ii) **Distribution of m_{CT}:** Upon a successful contention, a user transits from the CON to the TLK state. This transition occurs at the end of each free slot where contention can take place. Thus, the number of contending users will vary from slot to slot. Suppose there is M_T reserved slots and M_R relay slots in a given frame, then there is $(N - M_T - M_R)$ free slots for contention. We want to calculate the distribution of the number of users that moved from CON state to TLK state at the end of the last free slot. This distribution could be calculated using the following recurrence model. Let $q(M_C')$ be the probability that a user succeeds in contention when there is M_C' contending users, then

$$q(M_C') = M_C' p_v (1 - p_v)^{M_C'-1}(1 - P_{OB}),$$

which is the probability that only one user has permission to transmit and the channel was not in outage during packet transmission.

Define $R_k(M_C')$ as the probability that M_C' terminals remain in the CON state at the end of the k-th available slot, $(k = 0, 1, 2, \ldots, N - M_T - M_R)$. Conditioning on the outcome of the $(k - 1)st$ time slot, it follows that

$$R_k(M_C') = R_{k-1}(M_C')[1 - q(M_C')] +$$
$$+ R_{k-1}(M_C' + 1)q(M_C' + 1), \tag{12.11}$$

for $M_C' = 0, 1, \ldots, M_C$ and where M_C is the number of users in the CON state at the beginning of the frame. The initial condition for this recursion is

$$R_0(M_C') = \begin{cases} 1 & M_C' = M_C \\ 0 & M_C' \neq M_C \end{cases} \tag{12.12}$$

and the boundary condition $q(M_C + 1) = 0$, which follows from the fact that the total number of contending users is M_C. Finally, the distribution of m_{CT} is

$$\Pr(m_{CT} = i) = R_{N - M_T - M_R}(M_C - i), \quad i = 0, \ldots, M_C.$$

(iii) **Distribution of m_{TS}:** It follows from Section 12.1.1, and from users independence, that the number of users making a transition from the TLK state to the SIL state, m_{TS}, is binomially distributed with parameter γ, where γ is defined in (12.1). Then,

$$\Pr(m_{TS} = i) = \binom{M_T}{i} \gamma^i (1 - \gamma)^{M_T-i}, \quad i = 0, \ldots, M_T.$$

(iv) **Distribution of m_{CS}:** A user makes a transition from the CON state to the SIL state if its talk spurt ends before gaining access to the channel. Conditioning on the number of users that successfully accessed the channel, m_{CT}, and through the same argument as in (iii) above, we have

$$\Pr(m_{CS} = i | m_{CT}) = \binom{M_C - m_{CT}}{i} \gamma^i (1 - \gamma)^{M_C - m_{CT}-i}, \tag{12.13}$$

for $i = 0, \ldots, M_C - m_{CT}$. In what follows, we will not seek to remove the conditioning on m_{CT} from this distribution because this is the form we will be interested in when calculating the state transition matrix later in this chapter.

(v) **Distribution of m_{TC}:** A user leaves the TLK state to the CON state if its transmitted packet fails to reach the base station successfully, and if the relay did not help that user. Also, a user in the TLK state will leave to SIL state if its talk spurt ends in the current frame irrespective of the reception state of its last transmitted packet. This means that this user will not attempt to retransmit its last packet in the talk spurt and the relay will not try to help this user.

Given the number of users making transitions from TLK state to SIL state, m_{TS}, the number of transmission errors from the remaining users in the TLK state, ε, follows a binomial distribution with parameter P_{OB}, the outage probability of the link between any user and the base station as defined in (12.7). Therefore,

$$\Pr\left(\varepsilon = i \mid m_{TS}\right) = \binom{M_T'}{i} P_{O^i B} \cdot (1 - P_{OB})^{M_T' - i}, \tag{12.14}$$

for $i = 0, \ldots, M_T'$ and where $M_T' = M_T - m_{TS}$, the number of remaining users in the TLK state. Assume that the relay can successfully receive ε_R packets of the ε erroneous packets. Then, conditioned on ε, the number of successfully received packets by the relay, ε_R, is also binomially distributed but with parameter P_{OR}, the outage probability of the link from any user to the relay,

$$\Pr\left(\varepsilon_R = i \mid \varepsilon\right) = \binom{\varepsilon}{i}(1 - P_{OR})^i P_{O^{\varepsilon - i} R}, \quad i = 0, \ldots, \varepsilon.$$

For each of the slots assigned to the relay, a packet among the ε_R packets in the relay's queue is selected at random and forwarded. It follows that the number of successfully forwarded packets ε_F is binomially distributed with parameter P_{OB},

$$\Pr\left(\varepsilon_F = i \mid \varepsilon_R\right) = \binom{\min(M_R, \varepsilon_R)}{i}(1 - P_{OB})^i P_{OB}^{\min(M_R, \varepsilon_R) - i},$$

for $i = 0, \ldots, \min(M_R, \varepsilon_R)$ and where M_R is the number of time slots assigned for the relay. The minimum of M_R and ε_R is taken, since the number of forwarded packets cannot exceed the number of slots assigned to the relay or the number of packets in the relay's queue. Now, the probability that i users make the transition from TLK state to CON state is the probability that from the ε erroneous packets, the relay successfully forwards ($\varepsilon_F = \varepsilon_R - i$) packets. Then the distribution of m_{TC} is given by

$$\Pr\left(m_{TC} = i \mid m_{TS}\right) = \sum_{k=0}^{M_T - m_{TS}} \sum_{l=0}^{k} \Pr\left(\varepsilon_F = \varepsilon_R - i \mid \varepsilon_R = l\right)$$
$$\times \Pr\left(\varepsilon_R = l \mid \varepsilon = k\right) \Pr\left(\varepsilon = k \mid m_{TS}\right). \tag{12.15}$$

At this point, it is important to remark that all the distributions calculated above are state dependent because they generally depend on M_C and M_T. This means that it is necessary to calculate a different set of distributions for each possible state of the system.

12.3.1 State transition probabilities

Having derived the probability distributions for the state transitions, the next step in mathematically characterizing a Markov chain is to write the state transition matrix **P**. For the present case, this matrix is square, and it can be shown that its dimension M is

$$M = \begin{cases} \frac{(N+1)(N+2)}{2} & \text{if } M_\text{v} \leq N \\ (N+1)(M_v - \frac{N}{2} + 1) & \text{if } M_\text{v} > N \end{cases}$$

An element $P(S_1, S_2)$ of this matrix is the transition probability from state $S_1 = (M_{C_1}, M_{T_1})$ to state $S_2 = (M_{C_2}, M_{T_2})$. We also have $P(S_1, S_2) = 0$ when

$$M_{T_2} > \min(M_{T_1} + M_{C_1}, N),$$

because the number of terminals in TLK state in the next frame cannot exceed the total number of time slots in a frame or the number of terminals in TLK and CON states in the current frame. From (12.8) and (12.9), and the distributions developed above, the transition probability $P(S_1, S_2)$ is given by

$$P(S_1, S_2) = \sum_{x=0}^{M_{C_1}} \sum_{y=0}^{M'} \sum_{z=0}^{M_{T_1}} \Pr(m_{CS} = x | m_{CT} = y, S_1) \Pr(m_{CT} = y | S_1)$$
$$\times \Pr(m_{TC} = M_{T_1} - M_{T_2} + y - z | m_{TS} = z, S_1)$$
$$\times \Pr(m_{TS} = z | S_1)$$
$$\times \Pr(m_{SC} = M_{C_2} - M_{C_1} + x + y - z | S_1), \qquad (12.16)$$

where

$$M' = \min(M_{C_1} - x, N - M_{T_1} - M_{R_1}).$$

It should be noted that

$$\Pr(m_{TC} = M_{T_1} - M_{T_2} + y - z | S_1) = 0$$

if $M_{T_1} - M_{T_2} + y - z > M_{T_1} - z$, because the number of users transiting from TLK state to CON state cannot exceed the difference between the number of users initially in the TLK state and the number of users leaving the TLK state to the SIL state. Also,

$$\Pr(m_{SC} = M_{C_2} - M_{C_1} + x + y - z | S_1) = 0$$

if $M_{C_2} - M_{C_1} + x + y - z > M_{S_1}$, because the number of users leaving the SIL state cannot be larger than the number of users initially in this state.

Finally, the stationary distribution vector π (indicating the probability of each state) can be calculated as the left eigenvector of the minimum eigenvalue of the matrix $\mathbf{P}$ or, in other words, can be obtained by solving $\pi = \pi \mathbf{P}$.

12.4 Performance analysis

To assess the performance of the voice network under the content-aware cooperative MAC protocol, four measures will be considered: network throughput, multiple access delay, packet dropping probability, and subjective speech quality.

12.4.1   Network throughput

The throughput can be defined as the aggregate average amount of data transported through the channel in a unit time. In the present case, the number of packets successfully transmitted in a given frame can be decomposed into two components linked by the tradeoff between the use of cooperation and the reduction in the number of contention slots. The first one comes from the contending users who succeed in gaining access to the channel. And the second component comes from the talking users who succeed in transmitting their packets to the base station, either by themselves or through the help of the relay. Thus, the throughput can be expressed as

$$\text{Th} = \frac{\text{E}\{\text{E}\{m_{\text{CT}}|S_1\} + M_{\text{T}} - \text{E}\{m_{\text{TC}}|S_1\}\}}{N}, \tag{12.17}$$

where $\text{E}\{\cdot\}$ is the expectation operator. The outer expectation is with respect to the stationary distribution of the system's Markov chain. In (12.17), the first component resulting in the number of packets successfully transmitted in a given frame is given by the contending users who succeed in gaining access to the channel $\text{E}\{m_{\text{CT}}|S_1\}$. The second component is given by the number of successfully transmitted packets, which is expressed as $M_{\text{T}} - \text{E}\{m_{\text{TC}}|S_1\}$, the number of users in TLK state minus the expected number of users leaving the TLK state to the CON state, who are the users with failures in their transmissions.

Example 12.2   Consider a system with $N = 10$ time slots in a frame and where the speech source model has a mean talk spurt and a mean silence period duration of $t_1 = 1$ and $t_2 = 1.35$ s, respectively, with a maximum delay of $D_{\max} = 2$ frames. Also, assume that the contention permission probability is $p_v = 0.3$, the SNR threshold is $\beta = 15$ dB and the path loss exponent is $\alpha = 3.7$. The distance between any user and the base station is 100 m, and between any user and the relay is 50 m.

Figures 12.7 shows the throughput as a function of transmit power for a fixed number of users $M_v = 25$. To get insight on how the performance changes based on the tradeoff around the amount of free resources allocated to the relay, the throughput is plotted for different values of p_r, namely for $p_r = 0.1, 0.3$ and 0.5. In order to understand how the introduction of the relay affects performance, the results are compared to a traditional, non-cooperative PRMA scheme. The figure shows that the cooperative protocol outperforms the non-cooperative protocol in terms of throughput for all values of p_r. For example, at a power level of 100 mW, the non-cooperative throughput is around 0.25 while the cooperative throughput with $p_r = 0.3$ is around 0.57, which amounts to a 128% increase. Note that increasing the amount of resources allocated to the relay increases the throughput gain, which is expected since with more resources the relay is able to help more users, hence the average number of successfully transmitted packets per frame increases.   ▲

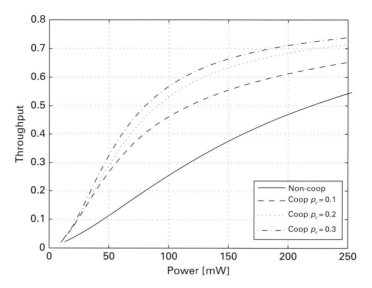

Fig. 12.7   Throughput for a network with 25 users and transmission power varying from 10 mW to 250 mW.

12.4.2   Multiple access delay

The delay is the number of frames a user remains in the CON state before gaining access to the channel. This delay is a function of the probability with which a user succeeds in contending during a given frame. This success probability depends on the network state at the instant the user enters the CON state, and will differ from frame to frame according to the path the network follows in the state space. Therefore, for exact evaluation of the multiple access delay, one should condition on the state at which our user of interest enters the CON state for the first time. Starting from this state, the delay is obtained from the calculation of the statistics of all possible paths the network follows in the state space till the user succeeds in the contention process. It is possible to show that for a network with N time slots per frame and M_v users, the total number of states is given by $(M_v - N/2 + 1)(N + 1)$ for $M_v \geq N$. For a network with $M_v = N = 10$, the number of states is 66. With such large number of states, finding an exact expression of the multiple access delay becomes prohibitively complex.

To get an approximate expression for the delay, let us assume that when the user enters the contention state the system state will not change until that user succeeds in contention. Thus, the success probability will be constant throughout the whole contention process, and the delay at any given state will follow a geometric distribution with parameter $p_s(i)$, the success probability at any state i. The approximate average delay is given by

$$D_{\text{avg}} = \sum_{i \in \Omega} \frac{\pi(i)}{p_s(i)},$$ (12.18)

where Ω is the set of states where $M_C \neq 0$, and $\pi(i)$ is the i-th element of the stationary distribution vector π.

The last step is to calculate the success probability $p_s(i)$. Given the assumption that all users are statistically identical, the probability that a user succeeds during contention in a given frame is equal to the probability that at least one user succeeds during contention in that frame, which can be shown to be equal to

$$p_s(i) = \left[1 - M_C p_v (1 - p_v)^{M_C - 1} (1 - P_{OB})\right]^{N - M_T - M_R}. \tag{12.19}$$

Example 12.3   Consider the same setup as in Example 12.2 but we are now interested in looking at the approximate average delay. As was also the case in Example 12.2, the results are compared to a traditional, non-cooperative PRMA scheme.

The delay performance is shown in Figure 12.8. It can be seen that, as the power increases, the delay starts to decrease and then, after some point, it increases again. This is because at low power levels the outage probability will be high and, even if a user succeeds in sending its packet with no collisions, its packet will be lost, with high probability due to channel outage. The delay is reduced by increasing the power because doing so also reduces the outage probability. It is in this region of reduced delay that the relay has a positive effect on performance, as seen for power levels less than 100 mW. Naturally, a relay is not necessary in any network where terminals are not constrained in transmit power. As power continues to increase, the outage probability will be almost negligible and the main cause of delay will be failed contentions. This justifies the fact that for high power levels the cooperative protocol exhibits a larger

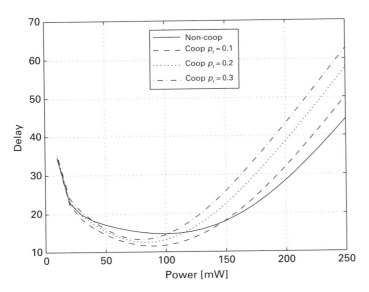

Fig. 12.8   Approximate delay as in (12.18) for 25 users and transmission power varying from 10 mW to 250 mW.

delay, as the presence of the relay decreases the number of slots available for contention and, hence, increases the probability of packet collisions. This is also why the delay performance is better for lower values of p_r.   ▲

12.4.3   Packet dropping probability

Speech communication is delay sensitive and requires prompt delivery of speech packets. In the PRMA protocol, packets start to be dropped if they are delayed in the network for more than a maximum allowable delay of D_{max} frames. Based on the assumption that the speech coder generates exactly one speech packet per frame, every user will maintain a buffer of length D_{max}. Whenever the buffer is full at the start of a frame, the oldest packet is dropped until the user succeeds in reserving a time slot. After gaining a reservation, in each frame the oldest packet in the queue will be transmitted and the new incoming packet is added at the end of the queue. If the talk spurt ends before getting a slot reservation, all the packets in the buffer are dropped. Because of channel errors, a user with a reserved time slot may lose its reservation and return to the group of contending users, thus risking further packet dropping.

First, we focus on the case when a user is trying to access the channel for the first time. Given that the system is at state i with N_C contending user and N_T users holding slot reservations, consider a contending user whose talk spurt started at the current frame. The talk spurt consists of L packets, where L is a random variable. The user will start to contend for a time slot in the current frame and continue in subsequent frames until it succeeds or the talk spurt ends. The user waits in the CON state for D frames to obtain a reservation. Using the assumption developed above that delay D is geometrically distributed, the probability that a user waits for d frames is

$$P_D(d) = (1 - p_s(i))(p_s(i))^d, \quad d = 0, 1, \ldots. \tag{12.20}$$

We need to distinguish between two different cases relating the length of the talk spurt L and the maximum allowable delay D_{max}:

(i) $L \leq D_{max}$: In this case, the buffer is long enough to store the whole talk spurt. If reservation is obtained before the talk spurt ends, no packets are lost. Otherwise, all the talk spurt packets are discarded. As a function of the waiting time d, the number of dropped packets is

$$n_d(d) = \begin{cases} 0, & 0 \leq d < L \\ L, & d \geq L \end{cases}$$

and the distribution of the number of dropped packets is given by

$$\Pr\{n_d | L \leq D_{max}, i\} = \begin{cases} \sum_{d=0}^{L} P_D(d), & n_d = 0 \\ \sum_{d=L+1}^{\infty} P_D(d), & n_d = L \end{cases}$$

(ii) $L > D_{max}$: In this case, after waiting D_{max} frames, one packet is dropped per frame until being able to reserve a slot. The dropped packet is the oldest in the queue with an associated delay of $D_{\max}$. The number of dropped packets as a function of the delay is given by

$$n_{\mathrm{d}}(d) = \begin{cases} 0, & 0 \le d \le D_{\max} - 1 \\ k, & d = D_{\max} + k - 1, \quad k = 1, 2, \ldots, (L - D_{\max}) \\ L, & d \ge L \end{cases}$$

and its distribution

$$\Pr\{n_{\mathrm{d}}|L > D_{\max}, i\} = \begin{cases} \sum_{d=0}^{D_{\max}-1} P_D(d), & n_{\mathrm{d}} = 0 \\ (1 - p_s(i))(P_s(i))^{n_{\mathrm{d}}}, & n_{\mathrm{d}} = 1, 2, \ldots, (L - D_{\max}) \\ \sum_{d=L}^{\infty} P_D(d), & n_{\mathrm{d}} = L \end{cases}$$

We note here that although all the summations mentioned above have closed-form expressions, they tend to become complex and lengthy. Therefore, we avoid writing them here, so as to keep the presentation compact. This will also apply to the next section.

The expected number of dropped packets for the above two cases, namely $\mathrm{E}\{n_{\mathrm{d}}|L \le D_{\max}, i\}$ and $\mathrm{E}\{n_{\mathrm{d}}|L > D_{\max}, i\}$, can be easily calculated using the corresponding distributions and, then, combined to get the total expected number of dropped packets as

$$\mathrm{E}\{n_{\mathrm{d}}|i\} = \sum_{l=1}^{D_{\max}} \mathrm{E}\{n_{\mathrm{d}}|L \le D_{\max}, i\} P_L(l)$$

$$+ \sum_{l=D_{\max}+1}^{\infty} \mathrm{E}\{n_{\mathrm{d}}|L > D_{\max}, i\} P_L(l), \tag{12.21}$$

where $P_L(l)$ is the probability mass function of the length of the talk spurt. From the speech source model of Figure 12.2, the talk spurt duration, L, is geometrically distributed with parameter γ, i.e.,

$$P_{\mathrm{r}}\{L\} = \gamma(1 - \gamma)^{l-1}, \quad l = 1, 2, \ldots. \tag{12.22}$$

Finally, the packet dropping probability is the ratio between the average number of dropped packets per talk spurt to the average number of packets generated per talk spurt, i.e,

$$P_{d_0} = \frac{1}{\gamma} \sum_{i \in \Omega} \mathrm{E}\{n_{\mathrm{d}}|i\} \pi(i), \tag{12.23}$$

where the sum is over Ω, the set of states with $N_C \ne 0$ (because packets are dropped only when the user is in the CON state).

Next we consider the packet dropping probability due to the first transition from the TLK state to the CON state caused by channel errors. First, we need to make the following assumptions:

- Any user in a TLK state has obtained its reservation with the first packet in the talk spurt. This means no packets were dropped in the first contention process. Furthermore, this packet is delayed by $D_0 = D_{\mathrm{avg}}$ frames, i.e., this packet is delayed by the average multiple access delay calculated in the last section.

- The first channel error occurs while transmitting the j-th packet of the talk spurt. Since the first packet was delayed by D_0 frames, the remaining maximum delay for the subsequent packets in the talk spurt is $D_1 = D_{\max} - D_0$ frames.
- There are L packets in the talk spurt, and L_1 packets following and including the j-th packet which encountered a channel error.

Based on the time instant when the user left TLK state to CON state, we need to analyze three cases:

- **Case 1:** Transmission instant of the j-th packet is after the end of the talk spurt. This means, $D_0 + (j - 1) \geq L$, or $L_1 \leq D_0$. In this case all the remaining L_1 packets are discarded without any contentions and
$$E\{n_d|L_1 \leq D_0, i\} = L_1.$$

- **Case 2:** Transmission instant of the j-th packet is before the end of the talk spurt and the remaining time till the end of the talk spurt is less than the maximum remaining delay D_1. That is, $0 < L - D_0 - (j - 1) \leq D_1$, or $D_0 < L_1 \leq D_{\max}$. In this case, no packets are dropped if the user gets a reserved slot before the end of the talk spurt. Otherwise, all L_1 are discarded. The number of dropped packets as a function of the waiting time is
$$n_d(d) = \begin{cases} 0, & 0 \leq d \leq L_1 - D_0 \\ L_1, & d > L_1 - D_0 \end{cases} \tag{12.24}$$

The waiting time is distributed according to (12.20), and
$$\Pr\{n_d|D_0 < L_1 \leq D_{\max}, i\} = \begin{cases} \displaystyle\sum_{d=0}^{L_1-D_0} P_D(d), & n_d = 0 \\ \displaystyle\sum_{d=L_1-D_0+1}^{\infty} P_D(d), & n_d = L_1 \end{cases} \tag{12.25}$$

- **Case 3:** Here $L - D_0 - (j - 1) > D_1$, or $L_1 > D_{\max}$. In this case, the j-th packet is dropped after waiting for D_1 frames and a packet will be dropped every frame till the user gets access to the channel. If the talk spurt ends before accessing the channel, all the packets in the buffer are discarded. We have
$$n_d(d) = \begin{cases} 0, & 0 \leq d \leq D_1 - 1 \\ k, & d = D_1 + k - 1, \quad k = 1, 2, \ldots, (L_1 - D_{\max}) \\ L_1, & d > L - D_0 \end{cases} \tag{12.26}$$

and
$$\Pr\{n_d|L_1 > D_{\max}, i\} = \begin{cases} \displaystyle\sum_{d=0}^{D_1-1} P_D(d), & n_d = 0 \\ (1 - P_s(i))(P_s(i))^{n_d}, & n_d = 1, \ldots, (L_1 - D_{\max}) \\ \displaystyle\sum_{d=L_1-D_0}^{\infty} P_D(d), & n_d = L_1. \end{cases}$$

Having the distributions of the number of dropped packets for each case, one can calculate the corresponding expected number of dropped packets, $E\{n_d|L1 \leq D_0, i\}$, $E\{n_d|D_0 < L_1 \leq D_{\max}, i\}$, and $E\{n_d|L_1 > D_{\max}, i\}$.

The next step is to average with respect to L_1, the number of remaining packets in the talk spurt after the first error. From the earlier assumptions, we have

$$\Pr\{L_1 = l_1 | L = l\} = u(1-u)^{(l-l_1-1)}, \quad l_1 = 1, 2, \ldots, l-1,$$

where u is the user's transition probability from the TLK state to the CON state. A user leaves the TLK state to the CON state if: (i) the packet transmission failed, (ii) a relay did not help that user, and (iii) the talk spurt did not end during current frame (had the talk spurt ended, the transition would have been to the SIL state). Now, we need to consider the talk spurt length L. If $(L-1) \leq D_{\max}$, cases one and two above would occur, otherwise all three cases would occur. Therefore,

$$E\{n_d|L, i\} = E\{n_d|L-1 \leq D_{\max}, i\} + E\{n_d|L-1 > D_{\max}, i\},$$

where

$$E\{n_d|L-1 \leq D_{\max}, i\} = \sum_{L_1=1}^{D_0} E\{n_d|L_1 \leq D_0, i\} \Pr\{L_1|L\}$$

$$+ \sum_{L_1=D_0+1}^{D_{\max}} E\{n_d|D_0 < L_1 \leq D_{\max}, i\} \Pr\{L_1|L\},$$

$$E\{n_d|L-1 > D_{\max}, i\} = \sum_{L_1=1}^{D_0} E\{n_d|L_1 \leq D_0, i\} \Pr\{L_1|L\}$$

$$+ \sum_{L_1=D_0+1}^{D_{\max}} E\{n_d|D_0 < L_1 \leq D_{\max}, i\} \Pr\{L_1|L\}$$

$$+ \sum_{L_1=D_{\max}+1}^{L-1} E\{n_d|L_1 > D_{\max}, i\} \Pr\{L_1|L\}.$$

Finally,

$$E\{n_d|i\} = \sum_{L=2}^{D_{\max}+1} E\{n_d|L-1 \leq D_{\max}, i\} \Pr\{L\}$$

$$+ \sum_{L=D_{\max}+2}^{\infty} E\{n_d|L-1 > D_{\max}, i\} \Pr\{L\},$$

where $\Pr\{L\}$ is defined in (12.22). As in (12.23), the packet dropping probability due to the first error in the talk spurt is

$$P_{d_1} = \frac{1}{\gamma} \sum_{i \in \Omega} E\{n_d|i\} \pi(i). \tag{12.27}$$

After regaining access to the channel, a second transmission error may occur and the user has to go through contention again and may lose some packets. This process may be repeated several times till the end of the talk spurt. The number of such cycles is a random variable because of the random nature of channel errors. Furthermore, calculation of the packet dropping probability due to the second error and above becomes intractable due to the more complex scenarios to be considered. To deal with this issue, the following approximation will be used. Let k be the average number of transitions from TLK to CON states during a talk spurt, $k = u/\gamma$, where u is the user's transition probability from the TLK state to the CON state. If u is small, $k \leq 1$, and P_d can be approximated by

$$P_d = P_{d_0} + P_{d_1},$$

and if u is large, then $k > 1$ and P_d can be approximated by

$$P_d = P_{d_0} + k P_{d_1}.$$

Example 12.4   In this example, we continue working with a system with the same setup as in Examples 12.2 and 12.3, but we are now interested in looking at the packet dropping probability. As was also the case with the previous examples, the results are compared to a traditional, non-cooperative PRMA scheme so as to learn how the use of a relay affects performance.

The packet dropping probability is shown in Figure 12.9 for $D_{max} = 2$ frames, where it can be seen that it is a decreasing function of the power level. The effect of the relay is also apparent, e.g., at 100 mW power level there is about 38 % decrease in packet dropping probability. Note that the gain from cooperation decreases as the power increases

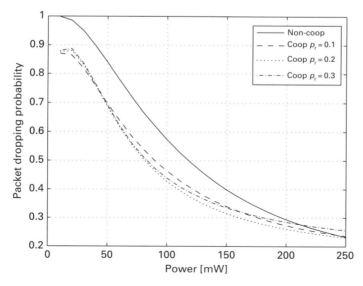

Fig. 12.9   Packet dropping probability for 25 users and transmission power varying from 10 mW to 250 mW.

and at some point (250 mW in Figure 12.9) the non-cooperative protocol will be as good as the cooperative protocol and start to have lower packet dropping probability. Again, this is because at high power, there is little need to use a relay that decreases the number of contention slots to reduce an already low outage probability. ▲

12.4.4 Subjective voice quality

Since we are considering speech traffic, an important metric to study end-to-end performance is some measure of how the speech is subjectively perceived at the receiver. For many years, the established method to obtain such measure was by conducting a mean opinion score (MOS) test. In this test, a group of listeners are polled to grade a number of speech sequences. More recently, and in many ways driven by the need of voice over IP systems for an automated subjective speed quality test, a number of algorithms have been developed that output a result highly correlated to those obtained with MOS, but this time without human intervention. Here, we will base the voice quality assessment on one such algorithm that is based on a predictive model. This model uses parameters from the source codec, the end-to-end delay and the packet dropping probability to predict the value of the conversational mean opinion score (MOS_c). This variant of the MOS is a perceptual voice quality measure based on the ITU-T PESQ quality measure standard that takes values in the range from 1 (bad quality) to 5 (excellent quality). The algorithm estimated the (MOS_c) as a polynomial of two variables, packet dropping probability and average delay, with coefficients tailored to the voice codec and the source encoding rate. This polynomial is of the form

$$MOS_c = c_0 + c_1 P_d + c_2 D + c_3 P_d^2 + c_4 P_d^3 + c_5 D^3$$
$$+ c_6 P_d D^2 + c_7 D^2 + c_8 P_d D + c_9 P_d^2 D,$$

where P_d is the packet dropping probability and D is the average delay we calculated earlier.

Example 12.5 In this example we look at the network with the same setup as in Examples 12.2–12.4, but we compare performance for the schemes with and without cooperation in terms of conversational Mean Opinion Score. To encode speech we assume the use of the GSM AMR voice codec operating at a coding rate of 12.2 kbps. For this case, the polynomial to estimate the (MOS_c) is,

$$MOS_c = 3.91 - 0.17 P_d + 1.57 \cdot 10^{-3} D + 6.51 \cdot 10^{-3} P_d^2$$
$$- 10^{-4} P_d^3 + 2.62 \cdot 10^{-8} D^3 + 1.38 \cdot 10^{-7} P_d D^2$$
$$- 2.40 \cdot 10^{-5} D^2 - 7.53 \cdot 10^{-6} P_d D - 5.51 \cdot 10^{-8} P_d^2 D,$$

Using this relation, Figure 12.10 shows the resulting conversational mean opinion score. Since this quality measure is a function of both the delay and packet dropping

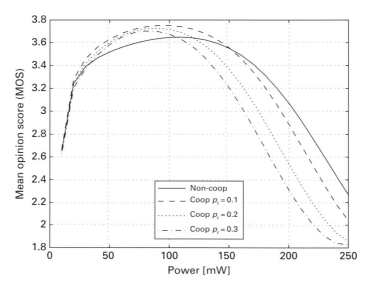

Fig. 12.10　Subjective speech quality, (MOS$_c$), for 25 users and transmission power varying from 10 mW to 250 mW.

probability, it can be seen that the cooperative protocol has better performance at low power level, while the non-cooperative one is better at high power levels. ▲

12.5　Access contention–cooperation tradeoff

As was discussed earlier, and illustrated in the previous examples, the content-aware cooperative MAC protocol is efficient in that it does not use resources reserved for users in a talk spurt. At the same time, the relay uses slots that would be otherwise available for contention. This established a tradeoff between how many packets the relay could potentially help to deliver and the delay in successfully contending for channel access. This tradeoff is controlled through the parameter p_r. We have observed that, at low power levels, the performance of the network is limited by the channel outage events, this is the case when the cooperative protocol outperforms the non-cooperative protocol because of the spatial diversity introduced by the relay.

On the other hand, for high power levels, the performance is limited by packet collisions. Since the relay uses part of the free time slots available for contentions, the probability of collision increases, and the performance of the cooperative protocol degrades in this region. One can conclude that, the more the assigned resources to the relay the better the throughput is, because of the increased relay ability to help more users. On the other hand, the converse is true for the delay, because the relay reduces the number of contention slots.

This idea is confirmed in Figures 12.11 and 12.12 depicting the variation of through-put and delay, respectively, as a function of p_r compared with the non-cooperative case

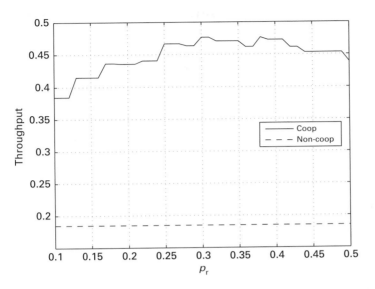

Fig. 12.11   Network performance measures as a function of p_r for 75 mW transmission power level and 25 users: throughput.

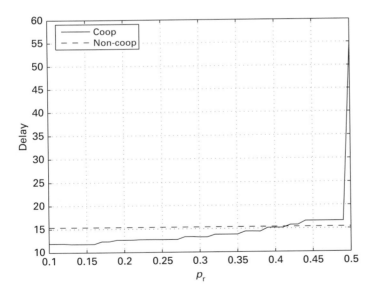

Fig. 12.12   Network performance measures as a function of p_r for 75 mW transmission power level and 25 users: delay.

for a network with 25 users and transmission power $P = 75$ mW. Figures 12.11 and 12.12 show that p_r can be chosen so as to obtain best overall performance by assigning about 25% to 40% of the free resources to the relay. With this assignment, the throughput is almost maximized while the delay is kept strictly less than the non-cooperative

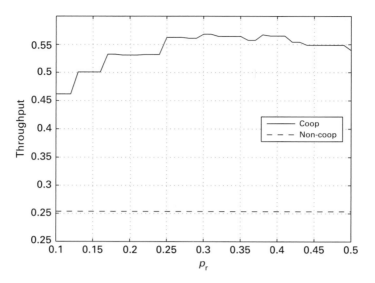

Fig. 12.13 Network performance measures as a function of p_r for 100 mW transmission power level and 25 users: throughput.

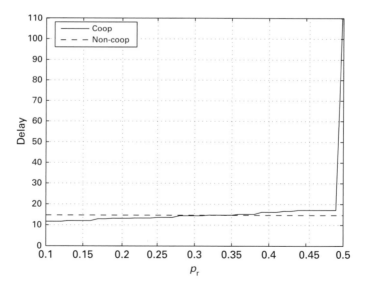

Fig. 12.14   Network performance measures as a function of p_r for 100 mW transmission power level and 25 users: delay.

protocol. If the transmission power is increased to 100 mW, Figures 12.13 and 12.14 show that in order to maximizes the throughput while constraining the delay to be less than the non-cooperative case, the optimum choice for p_r is between 25% and 30%. In these results we can see again that for higher power levels, less help from the relay is needed.

12.6   Chapter summary and bibliographical notes

In this chapter we have shown how user cooperation can be used to improve the performance of multiple access protocols. In particular, we considered protocols that take advantage of the statistical properties of the traffic, which in this chapter is real-time speech. From the point of view of traditional approaches that do not consider cooperation, perhaps the best known multiple access protocols that take advantage of the traffic characteristics of packet speech is the packet reservation multiple access (PRMA) protocol, which was studied in [49] and other derived literature [229, 95]. More specifically, the impact of channel losses on performance is studied in [95]. Indeed, much of the packet dropping probability analysis in this chapter is based on the work in [137] and [147] for the PRMA protocol. The talk spurt–silence speech model that is used by this protocol have been studied in different contexts but one of the most frequently cited references for the Markov chain modeling of speech sources and the exponential distribution model for the waiting time in any state is [16]. Also, we note here that the subjective speech quality measure used in Section 12.4.4 is based on the algorithm developed in [208], which is in turn based on the ITU-T PESQ quality measure standard [88]. More information on the MOS test can be found in [86].

In this chapter we showed that the use of cooperation reduces, for those users holding a time slot, the number of packets that are lost due to channel errors. Since time slot reservations are lost by those users experiencing dropped packets, the use of cooperation reduces the traffic load from users contending for channel access. Also, in this chapter we have shown how cooperation can be implemented efficiently by taking advantage of idle times in the sources and using resources that are not exclusively reserved for active calls. The approach explained in this chapter entails the implementation of cooperation through the use of time slots left unused by those users that are in a period of silence during the conversation. These free time slots can also be used for random access contention by those users transitioning from a silence period into a talk spurt. The dual use establishes a tradeoff between how many packets the relay could potentially help to deliver and the delay in successfully contending for channel access. We named this tradeoff the access contention–cooperation tradeoff. This tradeoff is controlled through the parameter p_r. By judiciously choosing this parameter, the protocol that uses cooperation shows a clear performance advantage compared with a protocol with no cooperation.

As we have noted at different places in this book, previously published literature on cooperative communications has focused for a large part on the physical layer. The study in this chapter is a clear example of the impact and implementation of cooperation at higher network layers, for which there are, nevertheless, some previous published works. Such is the case in [120], where the authors proposed a distributed version of the network diversity multiple access (NDMA) protocol [218]. Also, in [167], the notion of utilizing the spatial separation between users to assign cooperating pairs was presented. Further study of the content-aware multiple access protocol studied in this chapter can be found in [35, 34].

Exercises

12.1   Consider the derivation of the dynamic state model:
    (a) Derive (12.10) and discuss, also, the conditions under which the derivation is valid.
    (b) Derive (12.13) and discuss, also, the conditions under which the derivation is valid

12.2   Consider the derivation of the multiple access delay and derive (12.19).

12.3   In packet wireless communications there are two causes for packet dropping: packets that are dropped due to channel errors and packets that are dropped due to excessive delay. As we have discussed in this chapter, these two causes not only interact in the usual way, but also through the access contention–cooperation tradeoff. Assume a system with $N = 10$ time slots in a frame, where the speech source model has a mean talk spurt and a mean silence period duration of $t_1 = 1$ and $t_2 = 1.35$ s, respectively, with a maximum delay of $D_{max} = 2$ frames, the contention permission probability $p_v = 0.3$, SNR threshold $\beta = 15$ dB, and path loss exponent $\alpha = 3.7$. Also assume that the transmit power level is 100 mW and that distance between any user and the base station is 100 m, and between any user and the relay is 50 m. Plot the packet dropping probability due to excessive delay and the probability of a call transitioning to a contention state from the talk state (the probability of a packet loss due to channel errors) assuming $M_v = 25$ and $p_r = 0.1$, 0.3, and 0.5. Comment on the results observed on both curves from the point of view of the involved tradeoffs.

12.4   (Simulation project)
    (a) Build and run a Monte Carlo simulation to find the throughput and delay performance of the content-aware cooperative multiple access protocol. Compare the results with the analytical results presented in this chapter.
    (b) In this chapter, the results were focused on a network setup where all channel statistics for the sources were assumed identical. For a more general case, modify your Monte Carlo simulator so this condition on the network setup is relaxed. Within the new setup, examine the tradeoff between relaying and contention.
    (c) Note that in this chapter the focus was on evaluating performance through the use of average magnitudes, such as average throughput or average delay. In this problem we take a look at other statistics, which are also helpful in studying the overall performance.
    (d) Use Monte Carlo simulation to find the complete CDF curves of throughput and delay. Use these results to calculate the tenth percentile of the throughput (which is important to ensure fairness for users with bad channel conditions). Also calculate the 90th percentile of the delay (which is important when evaluating quality-of-service). Comment on the results.
    (e) Study the effects of changing p_r on the results.

13 Distributed cooperative routing

Routing is the process of transferring data packets from one terminal to another. Routing aims to find the optimal path according to some criterion. Shortest-path routing is a common scheme used for routing in data networks. It depends on assigning a length to each link in the network. A path made up of a series of links will have a path length equal to the sum of the lengths of the links in the route. Then, it chooses the path between source and destination that has the shortest route. The shortest-path route can be implemented using one of two well-known techniques, namely, the Bellman–Ford algorithm or the Dijkstra's algorithm [11].

In mobile ad hoc networks (MANETs), data packet transmissions between source and destination nodes are done through relaying the data packets by intermediate nodes. Hence, the source needs to locate the destination and set up a path to reach it. There are two types of routing algorithms in MANETs, namely, table-based and on-demand algorithms. In table-based routing algorithms, each node in the network stores a routing table, which indicates the geographic locations of each node in the network. These routing tables are updated periodically, through a special HELLO message sent by every node. Table-based routing protocols for MANETs include the destination sequence distance vector routing protocol (DSDV), wireless routing protocol (WRP), and cluster-head gateway switch routing (CGSR). The periodical updating of the routing tables makes table-based routing algorithms inefficient.

On the other hand, in on-demand routing protocols, only when a terminal needs to send data packets to a destination does it discover and maintain a route to that destination. A source initiates a route discovery request which goes from one node to another until it reaches the destination or a node which has a route to the destination. Hence, each node does not have to store information about the other nodes in the whole network or to store a route to every node. On-demand protocols in use include Ad-hoc on-demand distance vector routing (AODV), dynamic source routing (DSR), and temporary ordered routing algorithm (TORA).

In Part II, the merits of cooperative communications in the physical layer were explored; however, the impact of the cooperative communications on the design of the higher layers was not covered. Routing algorithms, which are based on cooperative communications, are known in the literature as *cooperative routing* algorithms. Designing cooperative routing algorithms can lead to significant power savings. Cooperative routing can make use of two facts: the wireless broadcast advantage in the broadcast mode and the wireless cooperative advantage in the cooperative mode. In the broadcast mode each node sends its data to more than one node, while in the cooperative mode many nodes send the same data to the same destination.

A straightforward approach for cooperation-based routing algorithms is implemented by finding a shortest-path route first and then building the cooperative route based on the shortest-path one. But such routing algorithms do not fully exploit the merits of cooperative communications at the physical layer, since the optimal cooperative route might be completely different from the shortest-path route. In addition, most of these cooperation-based routing algorithms require a central node, which has global information about all the nodes in the network, in order to calculate the best route given a certain source–destination pair. Having such a central node may not be possible in some wireless networks. Particularly, in infrastructureless networks (e.g., mobile ad hoc networks) routes should be constructed in a *distributed* manner, i.e., each node is responsible for choosing the next node towards the destination. Thus, the main goal of this chapter is to consider a distributed cooperation-based routing algorithm that takes into consideration the effect of the cooperative communications while constructing the minimum-power route.

In this chapter, we consider the minimum-power routing problem with cooperation in wireless networks. The optimum route is defined as the route that requires the minimum transmitted power while guaranteeing a certain end-to-end throughput. First, we derive a cooperation-based link cost formula, which represents the minimum transmitted power over a particular link, required to guarantee the desired QoS: a cooperation-based routing algorithm, namely the minimum-power cooperative routing (MPCR) algorithm, which can choose the minimum-power route while guaranteeing the desired QoS will be presented in this chapter.

13.1   Network model and transmission modes

In this section, we describe the network model and formulate the minimum-power routing problem. Then, we present the direct transmission and cooperative transmission modes.

13.1.1   Network model

We consider a graph $G(N, E)$ with N nodes and E edges. Given any source–destination pair (S, D), the goal is to find the route $S-D$ that minimizes the total transmitted power, while satisfying a specific throughput. For a given source–destination pair, denote Ω as the set of all possible routes, where each route is defined as a set consisting of its hops. For a route $\omega \in \Omega$, denote ω_i as the i-th hop of this route. Thus, the problem can be formulated as

$$\min_{\omega \in \Omega} \sum_{\omega_i \in \omega} P_{\omega_i} \quad \text{s.t.} \quad \eta_\omega \geq \eta_0 , \tag{13.1}$$

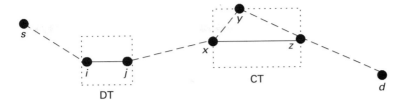

Fig. 13.1 Cooperative transmission (CT) and direct transmission (DT) modes as building blocks for any route.

where P_{ω_i} denotes the transmitted power over the i-th hop, η_ω is the end-to-end throughput, and η_o represents the desired value of the end-to-end throughput. Let η_{ω_i} denote the throughput of the i-th hop, which is defined as the number of successfully transmitted bits per second per hertz (bits/s/Hz) of a given hop. Furthermore, the end-to-end throughput of a certain route ω is defined as the minimum of the throughput values of the hops constituting this route, i.e.,

$$\eta_\omega = \min_{\omega_i \in \omega} \eta_{\omega_i} \,. \tag{13.2}$$

One can readily see that the minimum-energy cooperative path (MECP) routing problem, i.e., find the minimum energy route using cooperative radio transmission, is *NP-complete*. This is due to the fact that the optimal path could be a combination of cooperative transmissions and broadcast transmissions. Therefore, we consider two types of building block:

- direct transmission (DT);
- cooperative transmission (CT).

In Figure 13.1 the DT block is represented by the link (i, j), where node i is the sender and node j is the receiver. In addition, the CT block is represented by the links (x, y), (x, z), and (y, z), where node x is the sender, node y is a relay, and node z is the receiver. The route can be considered as a cascade of any number of these two building blocks, and the total power of the route is the summation of the transmitted powers along the route. Thus, the minimization problem in (13.1) can be solved by applying any distributed shortest-path routing algorithm such as the Bellman–Ford algorithm [11].

13.1.2 Direct and cooperative transmission modes

Let $h_{u,v}$, $d_{u,v}$, and $n_{u,v}$ represent the channel coefficient, length, and additive noise of the link (u, v), respectively. For the direct transmission between node i and node j, the received symbol can be modeled as

$$r_{i,j}^{\mathrm{D}} = \sqrt{P_D \, d_{i,j}^{-\alpha}} \, h_{i,j} \, s + n_{i,j} \,, \tag{13.3}$$

where P_D is the transmitted power in the direct transmission mode, α is the path loss exponent, and s is the transmitted symbol.

For the cooperative transmission, we consider a modified version of the decode-and-forward incremental relaying cooperative scheme previously described in Section 4.2.2.1. The transmission scheme for a sender x, a relay y, and a receiver z, can be described as follows. The sender sends its symbol in the current time slot. Due to the broadcast nature of the wireless medium, both the receiver and the relay receive noisy versions of the transmitted symbol. The received symbols at the receiver and the relay can be modeled as

$$r_{x,z}^C = \sqrt{P_C \, d_{x,z}^{-\alpha}} \, h_{x,z} \, s + n_{x,z} \tag{13.4}$$

and

$$r_{x,y}^C = \sqrt{P_C \, d_{x,y}^{-\alpha}} \, h_{x,y} \, s + n_{x,y} , \tag{13.5}$$

respectively, where P_C is the source transmitted power in the cooperative transmission mode.

Once the symbol is received, the receiver and the relay decode it. We assume that the relay and the receiver decide that the received symbol is correctly received if the received signal-to-noise ratio (SNR) is greater than a certain threshold, which depends on the transmitter and the receiver structures. Such system suffers from error propagation but its effect can be neglectable. The rationale behind this is that when the relays operate in a high SNR regime, the dominant source of error is the channel being in outage, i.e., deep fade, which corresponds to the SNR falling below some threshold. This issue had been illustrated at length in Example 6.1.

If the receiver decodes the symbol correctly, then it sends an acknowledgment (ACK) to the sender and the relay to confirm a correct reception. Otherwise, it sends a negative acknowledgment (NACK) that allows the relay, if it received the symbol correctly, to transmit this symbol to the receiver in the next time slot. This model represents a modified form of the automatic repeat request (ARQ), where the relay retransmits the data instead of the sender, if necessary. The received symbol at the receiver can be written as

$$r_{y,z}^C = \sqrt{P_C \, d_{y,z}^{-\alpha}} \, h_{y,z} \, s + n_{y,z} . \tag{13.6}$$

In general, the relay can transmit with a power that is different from the sender power P_C. However, this complicates the problem of finding the minimum-power formula, as will be derived later. For simplicity, we consider that both the sender and the relay send their data using the same power P_C.

Flat quasi-static fading channels are considered, hence, the channel coefficients are assumed to be constant during a complete frame, and may vary from a frame to another. We assume that all the channel terms are independent complex Gaussian random variables with zero mean and unit variance. Finally, the noise terms are modeled as zero-mean, complex Gaussian random variables with equal variance N_0. In this section, we have formulated the minimum-power routing problem and we have defined the two main transmission modes. In the next section, we derive the closed-form expressions for the transmitted power in both direct and cooperative transmission modes required to achieve the desired throughput.

13.2   Link analysis

Since the throughput is a continuous monotonously increasing function of the transmission power, the optimization problem in (13.1) has the minimum when $\eta_\omega = \eta_0$, $\forall \omega \in \Omega$. Since the end-to-end throughput $\eta_\omega = \min_{\omega_i \in \omega} \eta_{\omega_i}$, then the optimum power allocation, which achieves a desired throughput η_0 along the route ω, forces the throughput at all the hops η_{ω_i} to be equal to the desired one, i.e.,

$$\eta_{\omega_i} = \eta_0 , \quad \forall \ \omega_i \in \omega. \tag{13.7}$$

This result can be explained as follows. Let $P_{\omega_1}^*, P_{\omega_2}^*, \dots, P_{\omega_n}^*$ represent the required powers on a route consisting of n hops, where P_i^* results in $\eta_{\omega_i} = \eta_0$ for $i = 1, \dots, n$. If we increase the power of the i-th block to $P_{\omega_i} > P_{\omega_i}^*$ then the resulting throughput of the i-th block increases, i.e., $\eta_{\omega_i} > \eta_0$, while the end-to-end throughput does not change as $\min_{\omega_i \in \omega} \eta_{\omega_i} = \eta_0$. Therefore, there is no need to increase the throughput of any hop over η_0, which is indicated in (13.7).

Since the throughput of a given link ω_i is defined as the number of successfully transmitted bits per second per hertz, it can thus be calculated as

$$\eta_{\omega_i} = p_{\omega_i}^S \times R_{\omega_i} , \tag{13.8}$$

where $p_{\omega_i}^S$ and R_{ω_i} denote the per-link probability of success and transmission rate, respectively. We assume that the desired throughput can be factorized as

$$\eta_0 = p_0^S \times R_0 , \tag{13.9}$$

where p_0^S and R_0 denote the desired per-link probability of success and transmission rate, respectively. Next, we calculate the required transmitted power in order to achieve the desired per-link probability of success and transmission rate for both the direct and cooperative transmission modes. As previously noted (see (1.13) in Section 1.1.5), the channel gain $|h_{u,v}|^2$ between any two nodes u and v, is exponentially distributed with parameter one [146].

For the direct transmission mode in (13.3), the mutual information between sender i and receiver j is

$$I_{i,j} = \log \left(1 + \frac{P_D \, d_{i,j}^{-\alpha} \, |h_{i,j}|^2}{N_0} \right). \tag{13.10}$$

Without loss of generality, we have assumed unit bandwidth in (13.10). The outage probability is defined as the probability that the mutual information is less than the required transmission rate R_0. Thus, the outage probability of the link (i, j) is calculated as

$$p_{i,j}^O = \Pr \left(I_{i,j} \leq R_0 \right) = 1 - \exp \left(-\frac{(2^{R_0} - 1) \, N_0 \, d_{i,j}^\alpha}{P_D} \right). \tag{13.11}$$

If an outage occurs, the data is considered lost. The probability of success is calculated as $p_{i,j}^S = 1 - p_{i,j}^O$. Thus, to achieve the desired p_0^S and R_0 for direct transmission mode, the required transmitted power is

$$P_D(d_{i,j}) = \frac{(2^{R_0} - 1) \, N_0 \, d_{i,j}^\alpha}{-\log(p_0^S)} . \qquad (13.12)$$

For the cooperative transmission mode, the total outage probability is given by

$$p_{x,y,z}^O = \Pr\,(I_{x,z} \le R_C) \cdot \left(1 - \Pr\,(I_{x,y} \le R_C)\right) \cdot \Pr\,(I_{y,z} \le R_C)$$
$$+ \Pr\,(I_{x,z} \le R_C) \cdot \Pr\,(I_{x,y} \le R_C) , \qquad (13.13)$$

where R_C denotes the transmission rate for each time slot. In (13.13), the first term corresponds to the event when both the sender–receiver and relay–receiver channels are in outage but the sender–relay is not, and the second term corresponds to the event when both the sender–receiver and the sender–relay channels are in outage. Consequently, the probability of success of the cooperative transmission mode can be calculated as

$$p_s = \exp\left(-g \, d_{x,z}^\alpha\right) + \exp\left(-g \, (d_{x,y}^\alpha + d_{y,z}^\alpha)\right) - \exp\left(-g \, (d_{x,y}^\alpha + d_{y,z}^\alpha + d_{x,z}^\alpha)\right), \qquad (13.14)$$

where

$$g = \frac{(2^{R_C} - 1) \, N_0}{P_C} . \qquad (13.15)$$

In (13.13) and (13.14), we assume that the receiver decodes the signals received from the relay either at the first time slot or at the second time slot, instead of combining the received signals together. In general, maximum ratio combining (MRC) at the receiver gives a better result. However, it requires the receiver to store an analog version of the received data from the sender, which requires huge storage capacity. The probability that the source transmits only, denoted by $\Pr(\phi)$, is calculated as

$$\Pr(\phi) = 1 - \Pr(I_{x,z} \le R_C) + \Pr(I_{x,z} \le R_C) \Pr(I_{x,y} \le R_C)$$
$$= 1 - \exp\left(-g \, d_{x,y}^\alpha\right) + \exp\left(-g \, (d_{x,y}^\alpha + d_{x,z}^\alpha)\right), \qquad (13.16)$$

where the term $\left(1 - \Pr(I_{x,z} \le R_C)\right)$ corresponds to the event when the sender–receiver channel is not in outage, while the other term corresponds to the event when both the sender–receiver and sender–relay channels are in outage. The probability that the relay cooperates with the source is calculated as

$$\overline{\Pr(\phi)} = 1 - \Pr(\phi). \qquad (13.17)$$

Thus, the average transmission rate of the cooperative transmission mode can be calculated as

$$R = R_C \cdot \Pr(\phi) + \frac{R_C}{2} \cdot \overline{\Pr(\phi)} = \frac{R_C}{2}\left(1 + \Pr(\phi)\right), \qquad (13.18)$$

where R_C corresponds to the transmission rate if the sender is sending alone in one time slot and $R_C/2$ corresponds to the transmission rate if the relay cooperates with the sender in the consecutive time slot.

We set the probability of success in (13.14) as $p_S = p_0^S$ and the average transmission rate in (13.18) as $R = R_0$. By approximating the exponential functions in (13.14) as $\exp(-x) \approx 1 - x + x^2/2$, we obtain

$$g \approx \sqrt{\frac{1 - p_{\mathrm{o}}^{\mathrm{S}}}{d_{\mathrm{eq}}}} \,, \tag{13.19}$$

where $d_{\mathrm{eq}} \triangleq d_{x,z}^{\alpha}(d_{x,y}^{\alpha} + d_{y,z}^{\alpha})$. Thus, R_C can be obtained using (13.18) as

$$
\begin{aligned}
R_C &= \frac{2\,R_{\mathrm{o}}}{1 + \Pr(\phi)} \\
&\approx \frac{2\,R_{\mathrm{o}}}{2 - \exp\left(-\sqrt{\frac{1 - p_{\mathrm{o}}^{\mathrm{S}}}{d_{\mathrm{eq}}}}\, d_{x,y}^{\alpha}\right) + \exp\left(-\sqrt{\frac{1 - p_{\mathrm{o}}^{\mathrm{S}}}{d_{\mathrm{eq}}}}\,(d_{x,y}^{\alpha} + d_{x,z}^{\alpha})\right)} \,,
\end{aligned}
\tag{13.20}
$$

where we substituted (13.19) in (13.16). In addition, the required power per link can be calculated using (13.15) and (13.19) as

$$P_C \approx (2^{R_C} - 1)\, N_0 \sqrt{\frac{d_{\mathrm{eq}}}{1 - p_{\mathrm{o}}^{\mathrm{S}}}} \,. \tag{13.21}$$

Finally, the total transmitted power of the cooperative transmission mode can be calculated as

$$P_{\mathrm{tot}}^C(d_{x,z}, d_{x,y}, d_{y,z}) = P_C \cdot \Pr(\phi) + 2\,P_C \cdot \overline{\Pr(\phi)} = P_C\big(2 - \Pr(\phi)\big) \,, \tag{13.22}$$

where $\Pr(\phi)$ and P_C are given in (13.16) and (13.21), respectively.

13.3 Cooperation-based routing algorithms

In this section, we present two cooperation-based routing algorithms, which require polynomial complexity to find the minimum-power route. Then, we discuss the impact of cooperation on the routing in specific regular wireless networks, which are the regular linear and grid networks. We assume that each node periodically broadcasts a HELLO packet to its neighbors to update the topology information. In addition, we consider a simple medium access control (MAC) protocol, which is the conventional time division multiple access (TDMA) scheme with equal time slots.

First, let us describe a cooperation-based routing algorithm, namely, the minimum-power cooperative routing (MPCR) algorithm. The MPCR algorithm takes into consideration the cooperative communications while constructing the minimum-power route. The derived power formulas for direct transmission and cooperative transmission are utilized to construct the minimum-power route. It can be distributively implemented by the Bellman–Ford shortest path algorithm [11]. In the conventional Bellman–Ford shortest path algorithm, each node $i \in \{1, \ldots, N\}$ executes the iteration

$$D_i = \min_{j \in N(i)} (d_{i,j}^{\alpha} + D_j) \,, \tag{13.23}$$

where $N(i)$ denotes the set of neighboring nodes of node i, $d_{i,j}^{\alpha}$ denotes the effective distance between node i and j, and D_j represents the latest estimate of the shortest path from node j to the destination that is included in the HELLO packet.

The MPCR algorithm is implemented as follows:

- First, each node calculates the costs (required powers) of its outgoing links, and then applies the shortest-path Bellman–Ford algorithm using these newly calculated costs. The required transmission power between two nodes is the minimum power obtained by searching over all the possible nodes in the neighborhood to act as a relay. If there is no available relay in the neighborhood, a direct transmission mode is considered.
- Second, the distributed Bellman–Ford shortest-path routing algorithm is implemented at each node. Each node updates its cost toward the destination as

$$P_i = \min_{j \in N(i)} (P_{i,j} + P_j), \qquad (13.24)$$

where P_i denotes the required transmitted from node i to the destination and $P_{i,j}$ denotes the minimum transmission power between node i and node j. $P_{i,j}$ is equal to either P_D in (13.12) if direct transmission is considered or P_{tot}^C in (13.22) if cooperative transmission is considered employing one of the nodes in the neighborhood as a relay.

Table 13.1 describes the MPCR algorithm in details. The worst-case computational complexity of calculating the costs at each node is $O(N^2)$ since it requires two nested loops, and each has the maximum length of N to calculate all the possible cooperative transmission blocks.

Now, let us present a cooperation-based routing algorithm, namely, the cooperation along the shortest non-cooperative path (CASNCP) algorithm. The CASNCP algorithm is a heuristic algorithm that applies cooperative communications upon the shortest-path route. However, it is implemented in a different way using the proposed cooperation-based link cost formula. First, it chooses the shortest-path route then it applies the cooperative transmission mode upon each three consecutive nodes in the chosen route; first node as the sender, second node as the relay, and third node as the receiver. Table 13.2 describes the CASNCP algorithm.

Table 13.1 MPCR algorithm.

Step 1: Each node $x \in \{1, \ldots, N\}$ behaving as a sender calculates the cost of the its outgoing link (x, z), where $z \in N(x)$ is the receiver as follows. For each other node $y \in N(x)$, $y \neq z$, node x calculates the cost of the cooperative transmission in (13.22) employing node y as a relay.

Step 2: The cost of the (x, z)-th link is the minimum cost among all the costs obtained in *Step 1*.

Step 3: If the minimum cost corresponds to a certain relay $y*$, node x employs this relay to help the transmission over that hop. Otherwise, it uses the direct transmission over this hop.

Step 4: Distributed Bellman–Ford shortest-path algorithm is applied using the calculated cooperation-based link costs. Each node $i \in \{1, \ldots, N\}$ executes the iteration $P_i = \min_{j \in N(i)} (P_{i,j} + P_j)$, where $N(i)$ denotes the set of neighboring nodes of node i and P_j represents the latest estimate of the shortest path from node j to the destination.

Table 13.2 CASNCP algorithm.

Step 1: Implement the shortest non-cooperative path (SNCP) algorithm using the distributed
Bellman–Ford algorithm to choose the conventional shortest-path route ω_S as follows. Each
node $i \in \{1, \ldots, N\}$ executes the iteration $D_i = \min_{j \in N(i)} (d_{i,j}^\alpha + D_j)$, where $N(i)$ denotes
the set of neighboring nodes of node i and D_j represents the latest estimate of the shortest
path from node j to the destination.

Step 2: For each three consecutive nodes on ω_S, the first, second, and third nodes behave as the
sender, relay, and receiver, respectively, i.e., the first node sends its data using to the third
node with the help of the second node as discussed in the cooperative transmission mode.

13.3.1 Performance analysis: regular linear networks

The regular linear network, shown in Figure 13.2, is a one-dimensional chain of nodes
placed at equal intervals d_0. Without taking into consideration the interference effect,
nodes are placed at equal intervals to achieve the best performance in terms of the
throughput and the energy consumption.

We explain the route chosen by each algorithm when the source is node 0 and the destination is node $N - 1$. The SNCP routing algorithm, described in Step 1 of Table 13.2,
constructs the shortest route as a sequence of all the nodes between the source and destination, i.e., $\omega_{\text{SNCP}} = \{(0, 1), (1, 2), \ldots, (N - 2, N - 1)\}$, where (i, j) denotes the
direct transmission building block between sender i and receiver j. The CASNCP routing algorithm applies cooperative transmission mode on each three consecutive nodes
in the SNCP route, i.e., $\omega_{\text{CASNCP}} = \{(0, 1, 2), (2, 3, 4), \ldots, (N - 3, N - 2, N - 1)\}$,
where (x, y, z) denotes a cooperative transmission building block with x, y, and z
denoting the sender, relay, and receiver, respectively. We note that the direct transmission mode can be used in case there is no available relay. Finally, the MPCR
routing algorithm, applied on this linear network, chooses a different route, which is
$\omega_{\text{MPCR}} = \{(0, 1), (1, 0, 2), (2, 1, 3), \ldots, (N - 2, N - 3, N - 1)\}$. In other words, each
node sends its data to the adjacent node towards the destination utilizing its other adjacent node towards the source as a relay. In the following, we calculate the average
required transmitted power by each algorithm in a linear network.

For any routing scheme, the average end-to-end transmitted power can be calculated as

$$P(\text{route}) = \sum_{l=1}^{N-1} P(\text{route}|l) \times \Pr(l) , \qquad (13.25)$$

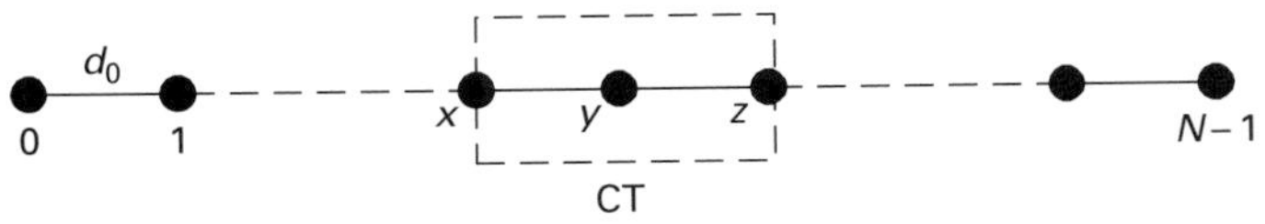

Fig. 13.2 Linear wireless network, d_0 denotes the distance between two adjacent nodes.

where $P(\text{route}|l)$ is the end-to-end transmitted power when the destination is l hops away from the source and $\Pr(l)$ denotes the probability mass function (PMF) of having l hops between any source–destination pair. The PMF $\Pr(l)$ can be calculated as

$$\Pr(l) = \begin{cases} \frac{1}{N}, & l = 0 \\ \frac{2\,(N-l)}{N^2}, & l = 1, 2, \ldots, N-1 \end{cases}. \tag{13.26}$$

Next, we illustrate how (13.26) is derived. The probability of choosing a certain node is $1/N$. Thus, the probability of having the source and destination at certain locations is given by $1/N \times 1/N = 1/N^2$. At $l = 0$ hops there is N possible combinations of this event, where the source and destination are the same. Therefore, $\Pr(0) = N/N^2 = 1/N$. Considering one direction only (e.g., from left to right in Figure 13.2), at $l = 1$ there is $N - 1$ distinct source–destination pairs: the first is the 0-to-1 pair and the last is the $(N - 1)$-to-N pair. By considering the other direction, the number of different source–destination pairs is $2 \times (N - 1)$. Therefore, the probability of having a source–destination pair with $l = 1$ hop in between is $\Pr(1) = 2(N - 1)/N^2$. In general, there is $2(N - l)$ different source–destination pairs with l hops in between, hence, the PMF of having source–destination pairs with l hops in between is given by (13.26).

For a route of l hops, the MPCR end-to-end transmitted power can be calculated as

$$P_{\text{MPCR}}(\text{route}|l) = P_D(d_0) + P_{\text{tot}}^C(d_0, d_0, 2\,d_0) \times (l - 1), \tag{13.27}$$

where the term $P_D(d_0)$ accounts for the first transmission from the source to its adjacent node towards the destination and $P_{\text{tot}}^C(d_0, d_0, 2\,d_0)$ is the required cooperative transmission power over one hop, which is given in (13.22) with $d_{x,z} = d_0$, $d_{x,y} = d_0$, and $d_{y,z} = 2d_0$. The CASNCP end-to-end transmitted power can be given as

$$P_{\text{CASNCP}}(\text{route}|l) = \begin{cases} P_{\text{tot}}^C(2\,d_0, d_0, d_0) \times \frac{l}{2} & l \text{ is even} \\ P_{\text{tot}}^C(2\,d_0, d_0, d_0) \times \frac{l-1}{2} + P_D(d_0) & l \text{ is odd}. \end{cases} \tag{13.28}$$

If l is even, there exist $l/2$ cooperative transmission blocks and each block requires a total power of $P_{\text{tot}}^C(2\,d_0, d_0, d_0)$. If l is odd, then a direct transmission mode is done over the last hop. Finally, the SNCP end-to-end transmitted power is calculated as

$$P_{\text{SNCP}}(\text{route}|l) = P_D(d_0) \times l. \tag{13.29}$$

The average end-to-end transmitted power for any routing scheme can be calculated by substituting the corresponding power formulas, which are (13.27), (13.28), and (13.29) for the MPCR, CASNCP, and SNCP, respectively in (13.25).

13.3.2   Performance analysis: regular grid networks

Figure 13.3 shows a regular 4×4 grid topology and d_0 denotes the distance between each two nodes in the vertical or horizontal directions. To illustrate the routes selected by different routing schemes, we assume that the source is node 0 and the destination is node 11. The SNCP routing algorithm chooses one of the possible shortest routes. For instance, the chosen shortest-route is $w_{\text{SNCP}} = \{(0, 1), (1, 5), (5, 6),$

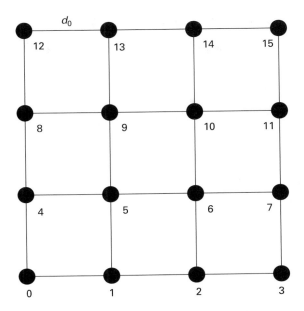

Fig. 13.3 Grid wireless network; d_0 denotes the distance between two adjacent nodes.

(6, 10), (10, 11)}. The CASNCP routing algorithm applies cooperation among each three consecutive nodes on the shortest-route, and the resulting route is $w_{\text{CASNCP}} = \{(0, 1, 5), (5, 6, 10), (10, 11)\}$. By applying the MPCR algorithm on this example, we find that MPCR chooses the route given by $w_{\text{MPCR}} = \{(0, 1, 5), (5, 6, 10), (10, 15, 11)\}$. Therefore, the MPCR algorithm shares part of the route with the CASNCP routing algorithm, when MPCR is routing the data across the diagonal walk. If the MPCR is routing the data in the horizontal (vertical) direction only, MPCR considers the receiver to be the sender's nearest node towards the destination and the relay to be the node nearest to the receiver along the vertical (horizontal) direction.

The average required transmitted power by each algorithm can be calculated as

$$P(\text{route}) = \sum_{i=1}^{N-1} \sum_{j=0}^{i} P(\text{route}|\sqrt{i^2 + j^2}) \times \Pr(\sqrt{i^2 + j^2}), \tag{13.30}$$

where i and j denote the number of hops between the source and destination in the horizontal and vertical directions, respectively. In addition, $\sqrt{i^2 + j^2}$ denotes the distance between the source and the destination. The PMF $\Pr(\sqrt{i^2 + j^2})$, which depends on the number of hops between the source and destination as well as their relative locations, is given by

$$\Pr(\sqrt{i^2 + j^2}) = \begin{cases} \frac{1}{N^2}, & i = j = 0; \\ \frac{4\,(N-i)\,(N-j)}{N^4}, & i = j \text{ or } j = 0; \quad \text{for } j \leq i \text{ and } 0 \leq i \leq (N-1). \\ \frac{8\,(N-i)\,(N-j)}{N^4}, & \text{otherwise} \end{cases}$$

$$\tag{13.31}$$

We explain (13.31) in a similar way to (13.26) as follows. The probability of choosing a certain node to be the source or the destination is $1/N^2$. Thus, the probability of choosing any source–destination pair is given by $(1/N^2) \times (1/N^2) = 1/N^4$. There are N^2 possible combinations, in which the source and the destination are the same. Hence at $i = j = 0$, $\Pr(0) = N^2/N^4 = 1/N^2$. In the following, we consider only the lower triangular part, i.e., $j \le i$. At $j = 0$, the grid network reduces to the linear case with $N - i$ possible source–destination pairs. For source–destination pair separated by $i = j$ hops in the horizontal and vertical directions, the number of possible source–destination pairs in one direction (e.g. left to right) is $(N - i) \times (N - j)$. This result is very similar to the one in (13.26) with considering the nodes on two dimensions instead of one dimension only in the linear case. At $i = j$ or $j = 0$, and considering the upper triangular part ($\times 2$) and reversing the source–destination pairs ($\times 2$), then the probability of having such source–destination pairs is $4(N - i)(N - j)/N^4$. For the third component in (13.31) i.e., at $j < i$, we additionally multiply this number by 2 to compensate the other combinations when i and j can be interchanged while giving the same distance of $\sqrt{i^2 + j^2}$, which results in a total of 8.

The MPCR end-to-end transmitted power can be calculated as

$$P_{\text{MPCR}}(\text{route}|\sqrt{i^2 + j^2}) = P_{\text{tot}}^C(\sqrt{2}\,d_0, d_0, d_0) \times j + P_{\text{tot}}^C(d_0, \sqrt{2}\,d_0, d_0) \times |i - j|, \tag{13.32}$$

where the first term represents the diagonal walk for j steps and the second term represents the horizontal $|i - j|$ steps. The CASNCP end-to-end transmitted power is calculated by

$$P_{\text{CASNCP}}(\text{route}|\sqrt{i^2 + j^2})$$
$$= \begin{cases} P_{\text{tot}}^C(\sqrt{2}\,d_0, d_0, d_0) \times j + P_{\text{tot}}^C(2\,d_0, d_0, d_0) \times \frac{|i-j|}{2} & (|i - j|) \text{ is even;} \\ \\ P_{\text{tot}}^C(\sqrt{2}\,d_0, d_0, d_0) \times j + P_{\text{tot}}^C(2\,d_0, d_0, d_0) \times \frac{|i-j-1|}{2} & (|i - j|) \text{ is odd.} \end{cases} \tag{13.33}$$

Finally, the SNCP end-to-end transmitted power is given by

$$P_{\text{SNCP}}(\text{route}|\sqrt{i^2 + j^2}) = P_D(d_0) \times (i + j), \tag{13.34}$$

which represents a direct transmission over $i + j$ hops, each of length d_0. The average end-to-end transmitted power for any routing scheme can be calculated by substituting the power formulas for the MPCR, CASNCP, and SNCP (given by (13.32), (13.33), and (13.34), respectively) in (13.30).

13.4   Simulation examples

Example 13.1   In this example we take a look at transmitted power required by the
MPCR, CASNCP and SNCP algorithms for different number of nodes when considering both linear and grid network topologies. Figure 13.4 depicts the end-to-end transmitted power for throughput $\eta_o = 1.96$ bits/s/Hz, transmission rate $R_o = 2$ bits/s/Hz,
noise variance $N_0 = -70$ dBm, and path loss $\alpha = 4$. The Figure also highlights the correspondence between the number of nodes and the distance d_0 as defined in Figure 13.2
for the linear network and Figure 13.3 for the grid network.

   The figure shows a monotone behavior in that the MPCR algorithm requires the minimum end-to-end transmitted power compared to both CASNCP and SNCP routing
algorithms for both the linear and grid networks.   ▲

Example 13.2   For this example, let us define the power saving ratio of scheme 2 with
respect to scheme 1 as

$$\text{Power saving} = \frac{P_{\text{Scheme1}} - P_{\text{Scheme2}}}{P_{\text{Scheme1}}}. \tag{13.35}$$

For the linear network, Figure 13.5 depicts the power saving versus the network size
for the network setup defined above. It is shown that at $N = 100$ nodes, the power

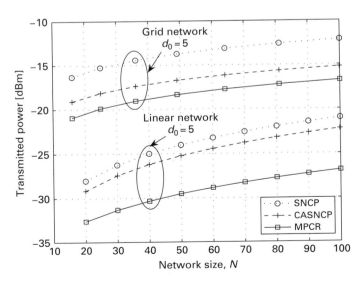

Fig. 13.4   Required transmitted power per route versus the network size for $N_0 = -70$ dBm, $\alpha = 4$,
$\eta_o = 1.96$ bits/s/Hz, and $R_o = 2$ bits/s/Hz in regular linear and grid networks.

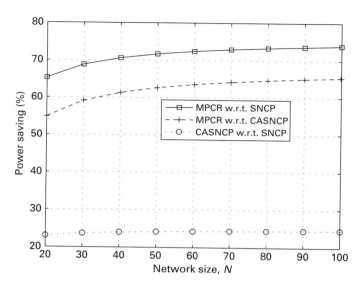

Fig. 13.5   Power saving due to cooperation versus the network size for $N_0 = -70\,\text{dBm}$, $\alpha = 4$, $\eta_0 = 1.96\,\text{bits/s/Hz}$, and $R_0 = 2$ in regular linear networks.

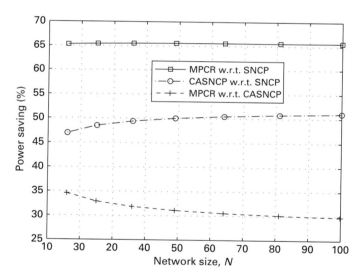

Fig. 13.6   Power saving due to cooperation versus the network size for $N_0 = -70\,\text{dBm}$, $\alpha = 4$, $\eta_0 = 1.96\,\text{bits/s/Hz}$, and $R_0 = 2$ in regular grid networks.

savings of the MPCR with respect to SNCP and CASNCP algorithms are 73.91% and 65.61%, respectively. On the other hand, applying cooperation over the shortest-path route results in a power saving of only 24.57%, as illustrated in the the CASNCP with respect to the SNCP curve. Similarly, Figure 13.6 depicts the power savings for the grid network. At $N = 100$ nodes, the power savings of the MPCR with respect to SNCP and

CASNCP algorithms are 65.63% and 29.8%, respectively. Applying cooperation over the shortest-path route results in a power saving of 51.04%. ▲

Example 13.3 In this example we look at the effect of varying the desired throughput on the required transmitted power per route. We now consider a network located within a 200 m × 200 m square area, where N nodes are uniformly distributed. The additive white Gaussian noise has variance $N_0 = -70$ dBm and the path loss exponent is $\alpha = 4$. Given a certain network topology, we randomly choose a source–destination pair and apply the various routing algorithms discussed in Section 13.3 to choose the corresponding route. For each algorithm, we calculate the total transmitted power per route. Finally, these quantities are averaged over 1000 different network topologies.

Figure 13.7 depicts the transmitted power per route, as required by the different routing algorithms. It is shown that the SNCP algorithm, which applies the Bellman–Ford shortest-path algorithm, requires the most transmitted power per route. Applying the cooperative communication mode to each of three consecutive nodes in the SNCP route results in a reduction in the required transmitted power as shown in the CASNCP routing algorithm's curve. Moreover, the MPCR algorithm requires the least transmitted power among the other routing algorithms. One of the major observations is that the MPCR algorithm requires less transmitted power than the CASNCP algorithm. Intuitively, this result is because the MPCR applies the cooperation-based link cost formula to construct the minimum-power route. On the other hand, the CASNCP algorithm first constructs the shortest-path route then applies the cooperative communication protocol to the established route. Therefore, the CASNCP algorithm is limited to applying the

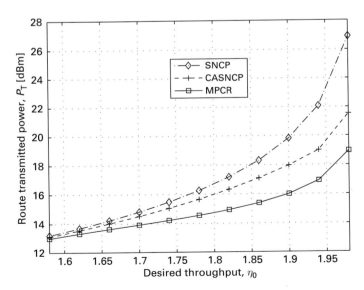

Fig. 13.7 Required transmitted power per route versus the desired throughput for $N = 20$ nodes, $\alpha = 4$, $N_0 = -70$ dBm, and $R_0 = 2$ bits/s/Hz in a 200 m × 200 m square.

cooperative-communication protocol to a certain number of nodes, while the MPCR algorithm can consider any node in the network to be in the CT blocks which constitute the route. Thus, the MPCR algorithm reduces the required transmitted power more than the CASNCP algorithm.   ▲

Example 13.4   In this example we consider the effect of varying the number of nodes on the required transmitted power per route. We consider the same network setup as in Example 13.3, i.e, the network is located within a 200 m × 200 m square area, with the N nodes uniformly distributed in space, additive white Gaussian noise with variance $N_0 = -70$ dBm, and path loss exponent $\alpha = 4$. As in Example 13.3, given a certain network topology, we randomly choose a source–destination pair and apply the various routing algorithms to choose the corresponding route. For each algorithm, we calculate the total transmitted power per route. Finally, these quantities are averaged over 1000 different network topologies.

Figure 13.8 depicts the required transmitted power per route by the different routing algorithms for different numbers of nodes at $p_0^S = 0.95$ and $\eta_o = 1.9$ bits/s/Hz. As shown, the required transmitted power by any routing algorithm decreases with the number of nodes. Intuitively, the higher the number of nodes in a fixed area, the closer the nodes are to each other, and the lower the required transmitted power between these nodes, which results in a lower required end-to-end transmitted power. We also calculate the power saving ratio, introduced in Example 13.2, as a measure of the improvement of the MPCR algorithm. At $N = 100$ nodes, $p_0^S = 0.95$, and $\eta_o = 1.9$ bits/s/Hz, the power savings of the MPCR algorithm with respect to the SNCP and CASNCP algorithms

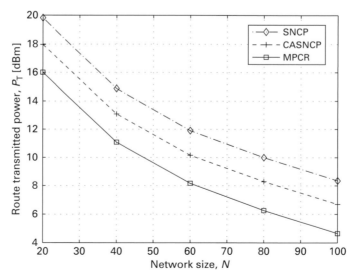

Fig. 13.8   Required transmitted power per route versus the number of nodes for $\eta_o = 1.9$ bits/s/Hz and $\alpha = 4$ in a 200 m × 200 m square.

are 57.36% and 37.64%, respectively. In addition, the power saving of the CASNCP algorithm with respect to the SNCP algorithm is 31.62%. ▲

Example 13.5 We now consider the effect of the routing algorithms on the average number of hops per route considering the same network setup as in Examples 13.3 and 13.4.

In Figure 13.9 the average number of hops in each route, constructed by the different routing algorithms, is shown versus the number of nodes in the network. For the cooperative transmission mode, the average number of hops is defined as

$$h^C = 1 \cdot \Pr(\phi) + 2 \cdot \overline{\Pr(\phi)} = 2 - \Pr(\phi) , \qquad (13.36)$$

and the average number of hops for the direct transmission mode is one. As shown, the routes constructed by either the CASNCP or the MPCR algorithms consist of a number of hops that is less than the routes constructed by the SNCP algorithm. Moreover, the average number of hops increases with N as there are more available nodes in the network, which can be employed to reduce the transmitted power. Although the MPCR scheme requires less power than the CASNCP routing algorithm, it also requires a longer delay. Intuitively, this is because the minimum-power routes may involve more nodes. This shows the tradeoff between the required power and the delay in the routes chosen by the MPCR and CASNCP routing schemes. ▲

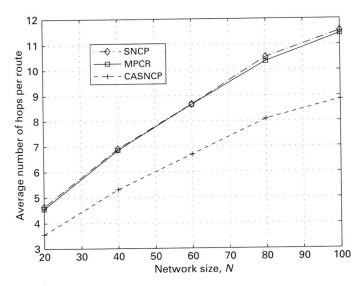

Fig. 13.9   Average number of hops per route versus the number of nodes for $\eta_o = 1.9$ bits/s/Hz and $\alpha = 4$ in a 200 m × 200 m square.

13.5   Chapter summary and bibliographical notes

In this chapter, we have considered the impact of cooperative communication on the minimum-power routing problem in wireless networks. For a given source–destination pair, the optimum route requires the minimum end-to-end transmitted power while guaranteeing certain throughput. This chapter first focused on the MPCR algorithm, which applies the cooperative communication while constructing the route. The MPCR algorithm constructs the minimum-power route using any number of the proposed cooperation-based building blocks, which require the least possible transmitted power. This chapter also presented the CASNCP algorithm, which is similar to most of the existing cooperative routing algorithms. The CASNCP algorithm first constructs the conventional shortest-path route then applies a cooperative-communication protocol to the established route. As we have seen, for random networks of $N = 100$ nodes, the power savings of the MPCR algorithm with respect to the conventional shortest-path and CASNCP routing algorithms are 57.36% and 37.64%, respectively. In addition to random networks, we have considered regular linear and grid networks, and have derived the analytical results for the power savings due to cooperation in these cases. In the case of a regular linear network with $N = 100$ nodes, the power savings of the MPCR algorithm with respect to the shortest-path and CASNCP routing algorithms are 73.91% and 65.61%, respectively. Similarly, the power savings of the MPCR algorithm with respect to the shortest-path and CASNCP routing algorithms in a grid network of 100 nodes are 65.63% and 29.8%, respectively.

The topic of routing in MANETs has resulted in a number of publications addressing different problems. A useful reference to learn about different routing protocols is [119]. One of the main concerns when designing routing algorithms for different wireless networks, such as mobile ad hoc networks, is to ensure that the result provides energy savings to extend the device's battery life. The goal of energy savings is one of the main objectives in [234] (for sensor networks). Also, it was shown in [38] that in some wireless networks, such as ad hoc networks, nodes spend most of their power in communication, either sending their own data or relaying other nodes' data. An analysis of the complexity associated with finding the solution of the minimum-energy cooperative path routing problem can be found in [115], where it was shown that finding the minimum-energy route using cooperative radio transmission requires solving a NP-complete problem. In addition to saving more energy, the routes selected by the routing protocol may be required to guarantee a certain quality of Service (qoS), as it is of great importance for some wireless applications (e.g., multimedia applications) [236]. The use of cross-layer techniques in routing protocols, such as providing the physical information about the wireless medium to upper layers, is a relative new approach aimed at achieving efficient scheduling, routing, resource allocation, and flow control algorithms. One work following this approach is [55], where it was shown that for high end-to-end delivery probabilities and given a certain delay constraint, long-hop schemes save more energy than that of the traditional nearest-neighbor routing algorithm.

As in other areas of cooperative communications, the cooperative routing problem has received recent attention from the research community, resulting in some noteworthy published research. In [96], two centralized heuristic algorithms (cooperation along the minimum energy non-cooperative path (CAN-L) and progressive cooperation (PC-L)) are proposed. In these algorithms, the optimum route is found through a dynamic programming mechanism which is NP hard. Also, [115] proposed the cooperative shortest path (CSP) algorithm, which chooses the next node in the route that minimizes the power transmitted by the last L nodes added to the route, while [186] presented an information-theoretic viewpoint of the cooperative routing in linear wireless network for both the power-limited and bandwidth-limited regimes. In addition, this paper presents an analysis of the transmission power required to achieve a desired end-to-end rate. For the case of a wireless sensor network, [143] studied the impact of cooperative communication on maximizing the lifetime of the network. Finally, [128] considered three cooperative routing algorithms, namely, relay-by-flooding, relay-assisted routing, and relay-enhanced routing. In the relay-by-flooding, the message is propagated by flooding and multiple hops. The relay-assisted routing uses cooperative nodes of an existing route and the relay-enhanced routing adds cooperative nodes to an existing route. Both of these routing schemes start with a route determined without cooperation. The authors also assume that more than one relay can send the data in the same time period, while not considering the synchronization problem. More discussion on the topic of distributed energy-efficient cooperative routing can be found in [81] and [85].

Exercises

13.1 In this exercise, we consider minimum-power cooperative routing problems with QoS, which are determined by the end-to-end probability of success and bandwidth efficiency. We aim at designing centralized cooperation-based routing algorithms. For a route $\omega \in \Omega$, denote ω_i as the i-th block of this route. Thus, the problem can be formulated as

$$\min_{\omega \in \Omega} \sum_{\omega_i \in \omega} P_{\omega_i} \quad \text{s.t.} \quad p_S \geq p_d^S \quad \text{and} \quad R \geq R_d \,,$$

where P_{ω_i} denotes the transmitted power over the i-th block and p_S and R are the end-to-end probability of success and bandwidth efficiency, respectively. Similarly, p_d^S and R_d represent the minimum desired values of the probability of success and the bandwidth efficiency, respectively.

(a) Given that the bandwidth efficiency $R = \min_{\omega_i \in \omega} R_{\omega_i}$, what is the desired bandwidth efficiency at all the hops R_{ω_i} that require minimum transmitted power and achieve the desired bandwidth efficiency?

(b) The p_S of the route ω can be calculated as

$$p_S = \prod_{\omega_i \in \omega} p_{\omega_i}^S \,, \quad R = \min_{\omega_i \in \omega} R_{\omega_i} \,,$$

where $p_{\omega_i}^S$ and R_{ω_i} denote the probability of success and the bandwidth efficiency of the i-th block, respectively. In order to get the per-link probability of success, consider a solution for a route of n hops that assumes that the probabilities of success over all the links constituting this route are equal to the n-th root of the desired probability of success. i.e.,

$$p_{\omega_i}^S = p_h^S = \sqrt[n]{p_d^S}, \text{ for } i = 1, \ldots, n,$$

where p_h^S is the probability of success per hop. Compare the results obtained by this solution with the optimal results obtained numerically for a two-stage route. The central node approximates the number of hops in a given route as the one given by the shortest-path routing algorithm.

(c) Based on the obtained probability of success and bandwidth efficiency per link, calculate the link costs for both the direct transmission and cooperative transmission building blocks.

(d) Implement the centralized MPCR and CASNCP routing algorithms based on the obtained link-level QoS.

(e) (Computer simulations) Consider a 200 m × 200 m grid, where N nodes are uniformly distributed. The additive white Gaussian noise has variance $N_0 = -70$ dBm.

(i) Assume that the desired bandwidth efficiency is $R_d = 2$ bits/s/Hz. Show the transmitted power per route, required by the different routing algorithms for $\alpha = 2$ and $\alpha = 4$.

(ii) Show the resulting probability of success for the three routing algorithms at $N = 20$, $R_d = 2$ bits/s/Hz, and $\alpha = 4$.

(iii) Show the average number of hops per route versus the number of nodes for $p_d^S = 0.95$, $R_d = 2$ bits/s/Hz, and $\alpha = 4$.

(iv) Plot the transmitted power per route versus the desired transmission rate for $p_d^S = 0.95$ bits/s/Hz and $\alpha = 4$.

(v) Finally, plot the transmitted power per route and the power saving versus the number of nodes for $p_d^S = 0.95$, $R_d = 2$ bits/s/Hz, and $\alpha = 4$.

13.2   Consider the three consecutive nodes x, y, and z in Figure 13.2, where node x needs to transmit its data to node z. Illustrate the route chosen by each of the routing algorithms in this case.

(a) The SNCP routing algorithm transmits the data directly from node x to node y then from node y to node z. Calculate the required power for the SNCP routing algorithm.

(b) The CASNCP routing algorithm applies cooperative communication transmission to the shortest-path route. Calculate the transmitted power for the CASNCP routing algorithm.

(c) By applying the MPCR algorithm, we find that the route is chosen in two consecutive phases as follows. First, node x transmits its data directly to node y utilizing direct transmission mode. Second, node y transmits its data to node z in a cooperative transmission mode utilizing node x as a relay. In other words, if node z does not receive the data correctly from node y, then node x will retransmit the data to node z. Calculate the transmitted power required by the MPCR algorithm.

(d) Plot the required transmitted power per block (x, y, z) as a function of the distance d_0 at throughput $\eta_0 = 1.96$ bits/s/Hz, transmission rate $R_0 = 2$ bits/s/Hz, noise variance $N_0 = -70$ dBm, and path loss exponent $\alpha = 4$.

13.3 In a 5 m × 5 m grid network, consider the lower left corner node to be the source and the upper right corner to be the destination. Apply the MPCR and CASNCP routing algorithms and show the route chosen by each algorithm. Calculate the energy saving due to cooperation in this case?

13.4 (Computer simulations) Verify by numerical results that the probability density function (PDF) of the distance between any two nodes in linear or grid networks, are given by (13.26) and (13.31), respectively.

13.5 Consider a linear network of $N = 5$ nodes. Let the source be the first node 0 and the destination be node 9. Apply the MPCR and CASNCP routing algorithms and show the route chosen by each algorithm. Calculate the energy saving due to cooperation in this case.

14 Source–channel coding with cooperation

The design of wireless devices and networks present unique design challenges due to the combined effects of physical constraints, such as bandwidth and power, and communication errors introduced through channel fading and noise. The limited bandwidth has to be addressed through a compression operation that removes redundant information from the source. Unfortunately, the removal of redundancy makes the transmitted data not only more important but also more sensitive to channel errors. Therefore, it is necessary to complement the source compression operation with the application of an error control code to do channel coding. The goal of the channel coding operation is to add redundancy back to the source coded data that has been designed to efficiently detect and correct errors introduced in the channel. In this chapter we will study the use of cooperation to transmit multimedia traffic (such as voice or video conferencing). This involves studying the performance of schemes with strict delay constraint where user cooperation is combined with source and channel coding.

Although the source and channel codecs complement each other, historically their design had been approached independently of each other. The basic assumptions in this design method are that the source encoder will present to the channel encoder a bit stream where all source redundancy have been removed and that the channel decoder will be able to present to the source decoder a bit stream where all channel errors have been corrected. This approach is justified in Shannon's separation theorem [181, 182], which states that for certain systems there is no loss of optimality in designing source and channel coding each independently of the other. Unfortunately, this theorem does not apply in many practical systems where its assumptions of arbitrary large complexity and infinite delay do not hold. Networks carrying delay sensitive traffic, such as multimedia, are one example of systems for which the assumptions in Shannon's separation theorem are not practical. For these systems, it is better to jointly design the source and channel units, an approach that is named joint source-channel coding (JSCC).

14.1 Joint source–channel coding bit rate allocation

In this chapter we will study the use of cooperation as part of a network where performance is best studied from an end-to-end viewpoint. Specifically, we will now study the application of user cooperation to transmit an interactive media. We will assume that the transmitted source may be speech, video, or some other synthetic source that is useful

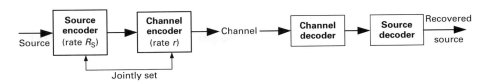

Fig. 14.1 Block diagram of a tandem source and channel coding scheme.

for analytical purposes and can also represent any multimedia source. The multimedia component in the traffic setup basically means that there is a strict delay constraint for the transmission of a source sample (which in practice takes values around 100 ms). This condition implies constraints in both delay and complexity that do not match the assumptions in Shannon's separation theorem. Because of this, the design of the source and channel codecs as two independent components would translate in performance loss and inefficiencies. In these cases a JSCC approach results in a more efficient design.

In practice, there are many approaches to the joint design of source and channel coding. In this chapter we will focus on a technique known as joint source–channel coding bit rate allocation because it is arguably the most straightforward in terms of practical implementation. In this technique the source and channel codecs are two separate units working in tandem and designed jointly (Figure 14.1). Because the source and channel codecs are two separated units, the benefit with this approach is that it retains the advantage in Shannon's separation theorem where much of the knowledge on individual source or channel codecs can be applied in the design of the constituent source and channel codecs. Yet, in this technique the two constituent codecs are jointly optimized to realize the advantages of JSCC design.

The technique of JSCC bit rate allocation consists of, given the channel signal-to-noise ratio (SNR), to allocate a fixed bit budget (in principle, the transmission capacity made available to a call) between source encoding rate and channel coding rate. In essence, this implies picking the source compression rate and an amount of error protection that yields the best end-to-end performance.

The goal of the JSCC bit rate allocation is to obtain the best end-to-end performance. It still remains unspecified how performance is defined and measured. For multimedia data, it is best to measure performance through the end-to-end distortion. The end-to-end distortion comprises of the source encoder distortion (which depends on the source encoding rate) and the channel-induced distortion (which depends on the channel SNR, error protection scheme, channel coding rate and error concealment operations). In general, the end-to-end distortion, D, associated with the transmission of a single frame of coded source data can be written as

$$D = D_{\mathrm{F}}P(f) + D_{\mathrm{S}}(1 - P(f)), \qquad (14.1)$$

where $P(f)$ is the probability that the transmitted frame has channel errors after channel decoding and has to be discarded, D_{F} is the distortion when the source frame is received with errors (channel-induced distortion) and D_{S} is the distortion due to quantization and coding (may be compression) of the source (source encoder distortion).

Furthermore, when including the channel state, the performance is measured in terms of the end-to-end distortion as a function of the channel SNR, a function we name the *D-SNR curve*. One of the problems when studying JSCC bit rate allocation is that it is not possible to obtain closed form solutions for the D-SNR curves, especially for real multimedia data. This is because generally the design has to be based on iterative methods or on an exhaustive search over reduced spaces. A simple solution to obtain a bound on the D-SNR curve is to ignore delay and complexity constraints and assume that the source is encoded optimally and transmitted at a rate equal to the channel capacity. This results in the OPTA (optimum performance theoretically attainable) curve. Another approach aimed at obtaining performance bounds is based on resorting to high and low SNR approximations, error exponents, asymptotically large source code dimension, and infinite complexity and delay. This approach presents a valuable step forward in understanding the D-SNR curve for jointly source–channel coded sources because it suggests the D-SNR relation of the type that will be discussed later in this chapter. Nevertheless, by the own nature of this approach, it cannot answer questions arising in more practical designs for multimedia communications, such as if performance can still be characterized in the same way, and how it will change depending on the source and channel codes being used.

The discussion so far points to the need for a new approach to studying the D-SNR curve that is suitable for system with strict delay constraints and featuring practical (non-capacity achieving) source and channel codes. In an effort to model the D-SNR curve of video signals with practical channel codes, a model can be selected from several candidates by choosing the best fit to measured data. As it turns out, it is possible to exploit the properties of how the D-SNR curve is obtained from solving the JSCC bit rate allocation problem and arrive at a good closed-form characterization of the D-SNR curve without the need for proposing a model beforehand. This close-form approximation will be derived later in this chapter as our main tool to study the interaction between cooperation and JSCC.

14.2   Joint source–channel coding with user cooperation

Another limitation associated with the delay constraint we will be dealing with in this chapter is the type of error correcting techniques that can be used. This is because after accounting for the many sources of delay (such as encoding, decoding, propagation, routing, transcoding, etc.), there is little time left for error correcting techniques that introduce significant delay, such as ARQ. Because of this, techniques that improve the link quality, such as user cooperation, become a necessity.

Nevertheless, the application of user cooperation to multimedia traffic brings up the important design consideration that for user cooperation there is a tradeoff between received signal quality and bandwidth efficiency. As we have studied in previous chapters, typical implementations of user cooperation improve signal quality through the combination of signals received from multiple paths, but at the same time, it can reduce the bandwidth efficiency. This is important for schemes that adapt source and channel

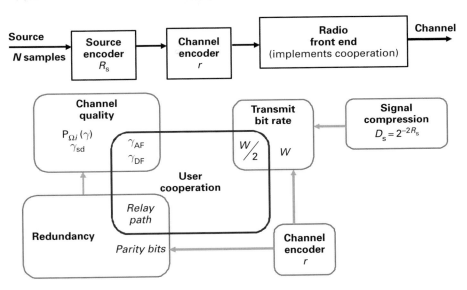

Fig. 14.2   A conceptual view at the different interdependences studied in this chapter.

coding rate, because while the presence of multiple paths can be seen as extra redundancy and error protection, the ensuing sacrifice in communication capacity forces a higher compression of the source or a choice of a weaker channel code.

This chapter will address these issues by studying the interaction between three main components: the improvement in signal quality from using cooperation, the level of compression of the source and the amount of redundancy introduced though different techniques to protect against channel errors. These interactions are schematized in Figure 14.2. The concepts represented in this figure will be developed and studied throughout this chapter.

It should be clear from the discussion so far that, while the combination of user cooperation with the JSCC design adds a new dimension to the problem, the correct approach is still a study from an end-to-end viewpoint. With this in mind, this chapter will center around studying the tradeoffs involved in combining user cooperation with JSCC bit rate allocation. We call this problem, the source–channel-cooperation tradeoff problem. One of the challenges in studying the source–channel–cooperation tradeoff problem is that it becomes necessary to find a unified framework that is able to jointly include the effects on end-to-end performance of user cooperation and the JSCC bit rate allocation stage.

Our approach to this problem is to use an expanded view of the concept of diversity as the unifying tool in our analysis. We will approach the concept of diversity with a more general view than the one limited to the presence of multiple channels or fading. In essence, the concept of diversity, as that of transmitting multiple copies of a message, can be extended to include the notion of redundant information. In fact, error correcting codes exhibit an inherent diversity since the copies of the message (which may be partial in this case) are coded as parity data of a channel code. Also, residual redundancy after source encoding can be seen as diversity in the form of partial information that can be used to improve the received message quality. Then, since redundancy can be seen as a form of diversity, we explore both in a unified way and we apply the concept to the study and characterization of end-to-end performance resulting from source–channel–cooperation tradeoffs. We will do so by using a unified framework for the modeling, analysis and performance measurement of multimedia systems that use two magnitudes: the coding gain and the diversity gain.

In conjunction with the need to study the source–channel–cooperation tradeoffs in a unified way, it is also important not to loose track of considerations typical of multimedia systems, such as delay constraints precluding the use of capacity achieving codes or typical operational envelopes that are not necessarily restricted to high SNR channels. Even when including the extra dimension of user cooperation, these are qualities of the approach to characterize the D-SNR curve that will be developed in this chapter. Because of this, the characterization of the D-SNR curves will be suitable not only to study the combination of cooperation with practical source and channel codes, but also to analyze from an end-to-end viewpoint the practical design issues associated with this setup such as how performance is affected by the use of a stronger channel code or a less efficient source encoder.

14.3   The Source–channel–cooperation tradeoff problem

We now formalize the description of the source–channel–cooperation tradeoff problem. Let us consider a wireless network carrying conversational multimedia traffic between a source node and a destination node, i.e., each ongoing call carries the information from a source, which has been coded for transmission over noisy and bandwidth-limited channels, while also being subject to strict delay and jitter constraints. Bandwidth is shared between users by allocating to each call an orthogonal channel that delivers a constant number of bits, W, per transmission period. Without loss of generality, we will assume that these bits are transmitted using BPSK modulation with coherent detection and maximum-likelihood decoding in the receiver. As shown in Figures 14.1 and 14.2, for each transmission period, the transmission of a message is divided in three main stages: a source encoder, which compresses a source, a channel encoder, which adds protection against channel errors, and a radio link. In order to minimize end-to-end distortion, both the source coding and channel coding rates can be adapted to channel conditions through jointly allocating the W available bits. Also, cooperative communication may or may not be used in the radio link

We assume a low mobility scenario, where the communication is carried over a quasi-static fading channel; i.e., fading may be considered constant during the transmission of a frame. Furthermore, we will consider the source–channel–cooperation tradeoffs within each transmission period so as to obtain best performance when having perfect knowledge of the channels. For this, let us implicitly assume that the network offers a mechanism that allows the transmitters to estimate the channel states (similarly to that of MIMO communications). This can be achieved in a time division duplexed network by using techniques such as channel sounding. In this case, after assuming reciprocity of the channels, the source can estimate the source–relay channel from the signal emitted by the relay (likely containing pilots) and the source–destination and relay–destination channels from estimates fed back from the destination. As long as the quasi-static fading assumption holds, the difference between the actual channels states and their estimate (due to actual errors in channel estimation and due to the time evolution of the channel) should have little effect on the final result. This is because, under these conditions, there would be no difference in the parameters chosen for source and channel coding as the range of possible choices is discrete and the selection of parameters remains the same over a range of channel SNRs.

In general terms, in most scenarios involving node movement at speeds consistent with pedestrian setups, the channel is expected to stay constant over the duration of several transmit frames (for example, at a carrier frequency of 2.5 GHz and a node velocity of 5 mph the channel coherence time is approximately 50 ms). As the channel starts to change more rapidly, there may be differences between the actual channels states and their estimates that would translate into a mismatched choice of source and channel coding parameters, which would result in performance loss. The magnitude of this loss depends on several variables, such as the rate at which the channel is changing in relation with the duration of a transmission period, the coarseness in the range of choices for source and channel coding, the performance of both the source and the channel codecs, etc. Nevertheless, the effect of a rapidly changing channel affects the considered scheme whether user cooperation is used or not.

Since the main goal of this chapter is to study the interaction between the received signal quality-bandwidth efficiency tradeoff (stemming from user cooperation) and the source compression-error protection tradeoff, the effects of channel mismatch can be decouple from the results by assuming that channel states are ideally estimated at the transmitter. This will lead to results that reveal the pure source–channel–cooperation tradeoffs and that reflect the best performance that can be achieved in each of the considered schemes. Finally, to follow these assumptions, we will consider that performance measurement and setting design for the three transmission stages is done on a frame-by-frame basis.

Note that the source–channel–cooperation tradeoff problem encompasses many layers of the communication stack. This affects the way in which we measure performance. Measuring performance using relations between some measure involving channel errors, such as bit error rate (BER), and channel SNR is applicable for studies focused on lower layers of the communication stack, such as modulation schemes, user cooperation or channel coding, but is not useful to represent all the effect of the higher layers. At the application layer and for the case of source codecs, performance is best measured through the distortion-rate (D-R) function; which characterizes the distortion observed in a reconstructed source after it had been compressed at a given rate.

However, for the source–channel–cooperation tradeoff problem, it is best to use a performance measure that reflects tradeoff effects at all layers of the source–channel–cooperation problem and that is consistent with multimedia traffic. Thus, it is best to measure performance through the relation between end-to-end distortion and channel

SNR, a function we call the *"D-SNR curve"*. Even more, in order to study the source–channel–cooperation problem it is necessary to have a clear understanding of the main components of the system, namely the source codec, the error control coding block and the transmission with and without cooperation. In the next two sections, we momentarily pause in the study of user cooperative systems to present the main concepts of source and channel coding that are pertinent to the source–channel–cooperation problem.

14.4   Source codec

During each sample period, a block of N input signal samples are presented to the source encoder at the transmitter node (see Figure 14.2). The function of the source encoder is to quantize and compress these samples. This operation results in a coded representation of a sample using a number of bits that is called the source encoding rate. On the receiver side, a decoder performs the reverse operations to those done by the encoder so as to obtain a reconstruction of the source. Since the operations performed at the encoder and the decoder add some distortion to the source samples, the decoder actually obtains an approximate reconstruction of the original input signal samples. Then, a key figure of merit for a source codec is the measurement input-output distortion when using a certain number of bits to represent the source coded signal. This relation is the achievable D-R (distortion-rate) function, where distortion is measured in the majority of cases as the mean-squared error between the original and the reconstructed source samples.

Let's denote by R_S the source encoding rate (measured in number of bits per source sample) and by D_S the distortion associated with the source encoding operation. The D-R function is frequently modeled as

$$D_S(R_S) = c_1 2^{-c_2 R_S}, \tag{14.2}$$

because this function can be used to approximate or bound many practical, well-designed codecs. Examples of this are video coding using an MPEG codec [60, 103], speech coding using a CELP-type codec [101], or coding when the high rate approximation holds. In the study of problems that involves the use of D-R functions without assuming a particular type of source, it is a common practice to assume that the input samples are memoryless, with a zero-mean, unit-variance Gaussian distribution and are coded with a long block source code. In this case, the D-R function was shown to be specified by $c_1 = 1$ and $c_2 = 2$, i.e.,

$$D_S(R_S) = 2^{-2R_S}. \tag{14.3}$$

For illustrative purposes, this function is plotted in Figure 14.3.

In this chapter, we will use in most of the formulation the D-R function as given by (14.3). Note that the exponential relation between distortion and rate makes the D-R function to be convex and decreasing. All these properties are not exclusive to the particular D-R function in (14.3), but, in fact, can be also observed in practical, well-designed encoders for any type of multimedia source. Because the overall behavior of

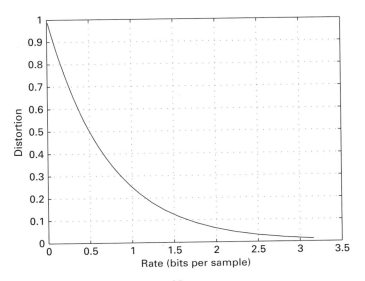

Fig. 14.3   The D-R function $D_S(R_S) = 2^{-2R_S}$.

D-R functions play a role in understanding the source–channel–cooperation tradeoff, we will take next a closer look at the D-R function for some practical sources.

14.4.1   Practical source codecs

In the previous section, we explained that (14.3) is a quite general form of D-R function. We also highlighted that the exponential relation between distortion and rate makes the D-R function to be convex and decreasing and that these properties can be also observed in practical, well-designed encoders. We now illustrate this concept when the source being encoded is speech or video.

The vast majority of today's speech codecs are based on the principle of analysis-by-synthesis. In this approach, the human speech production mechanism is modeled as a time-varying filter (representing the actions of the glottal flow, the vocal tract, and the lips) that is excited by a signal representing the vocal cords, which is modeled as an impulse train (for voiced speech) or random noise (for unvoiced speech). During each encoding period the encoder inputs a block of samples corresponding to a few tenths of milliseconds, or less, of speech. The encoder applies to each block of samples the speech filter model so as to calculate a set of parameter characterizing the synthetic filter and excitation signal. These parameters are calculated with the goal of minimizing the distortion between the input and reconstructed speech samples, which is estimated by running an instance of the decoder at the encoder. The output of the encoding process is a quantized version of the voice model parameters, as well as a compressed version of the input-output error signal (obtained using the instance of the decoder in the encoder).

As a distortion measure, the mean squared error has the advantage of being simple to calculate and of frequently leading to mathematically tractable analysis. Nevertheless, when dealing with sources such as speech, this distortion measure needs to be used judiciously. This is because an objective distortion measure as the mean squared error does not take into consideration how the properties of human sound perception affect the perceived quality of a speech sequence. In reality, the best way to obtain an assessment of speech quality that accurately incorporates subjective human perception is to assemble a group of human testers that listen and grade the target speech sequences. This procedure has proven to be effective and has been standardized under the name of the mean opinion score (MOS) [86]. For all the value offered by this method in providing a perceptual quality measure, the clear main drawback is the need to assemble a panel of listeners, which may be impractical in many situations.

The solution to this problem is to develop distortion measures that still result in perceptual-based measurements and, also, that can be obtained in a computer by following some algorithm. Among these measures, the International Telecommunications Union (ITU) has standardized in 2001 a quality measure algorithm called "perceptual evaluation of speech quality"–PESQ (ITU-T standard P.862 [88]). We have already used a variation of this speech quality evaluation algorithm in Chapter 12. Recall that the most important property of PESQ is that its results accurately predict those that would be obtained in a typical subjective test using the method of polling a panel of human listeners (the MOS test). In the variant of the algorithm we will use in this chapter, PESQ will reduce measurements of quality as numbers between 4.5 and 1 [89]. The higher the quality measurement, the better the perceptual quality is. This means that a quality measure above 4 corresponds to excellent quality, around 4 corresponds to good quality, around 3 is associated with fair but still acceptable quality, around 2 means poor quality and approximately 1 corresponds to bad quality.

Figure 14.4 shows the D-R function for 18 speech sequences (half males and half females), all from the NIST corpus [29] encoded using the widely-used speech vocoder GSM AMR codec [36]. This codec, based on the analysis-by-synthesis principle, encodes speech at eight possible rates: 12.2, 10.2, 7.95, 7.4, 6.7, 5.9, 5.1, and 4.75 Kbps. In this encoder, source encoding rate adaptation is achieved by changing the encoding parameters and internal quantization tables. In essence, the encoder operates as if it were eight different encoders that perform the same encoding operations but only one is selected at a time. In the figure, the distortion is not measured using the mean squared error but through a measure derived from the PESQ quality measure. To turn the quality measure into a distortion measure, the figure simply shows the values $4.5 - Q$, where Q is the quality measure obtained by using the PESQ algorithm. By doing this transformation the distortion will be equal to zero when quality is best, and will increase as the quality degrades.

The applicability of the properties of the D-R function to practical sources and source encoder, as illustrated in Figure 14.4, can also be extended to other sources. Specifically, Figure 14.5 illustrates the same concept when the source is video. The majority of video codecs perform compression through a combination of motion prediction and transform coding. To do video compression, the encoder divides the frames in a video sequence

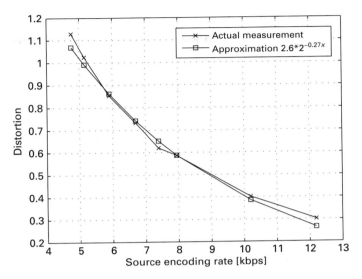

Fig. 14.4   The D-R function, and an exponential approximation, for speech encoded with the GSM AMR codec and distortion measured using the PESQ algorithm.

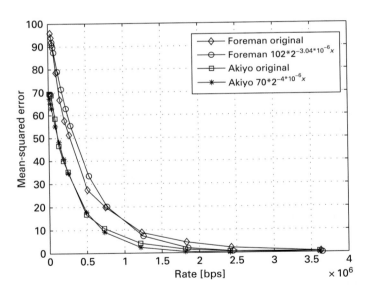

Fig. 14.5   The D-R function, and an exponential approximation, for two video sequences encoded with the MPEG4 FGS codec.

into groups, called group of frames (GOF). The first frame in a GOF, named an "I" frame, is encoded essentially as if it were a single picture, i.e., it is divided in 8×8 blocks, each of which is transformed using the Discrete Cosine Transform followed by some sort of entropy coding to achieve compression. For the rest of the frames, an estimate is calculated by using the changes due to movement in past (for which the frames are named "P" frames) and, sometimes, future frames. Subsequently, the difference between the actual and estimated frame is calculated. It is this rrsidual error that is quantized and compressed as if it was an individual picture. The coded error is finally transmitted along with compressed information about the movement in the video frame being encoded.

Figure 14.5 shows the D-R function of one representative frame (number 182) from two video sequences, named "Foreman" (which is considered to have high motion) and "Akiyo" (which is considered to have low motion). Distortion in this figure is measured using the objective mean squared error between each pixel in the original and reconstructed frame. The video encoder used is an extension of the MPEG4 video codec, which is designed for easy coding rate adaptation. This version, known as the MPEG4 FGS (fine granularity scalability) codec [114], generates a two-layer coded representation of the video sequence. The two layers are known as base and enhancement layers. The base layer provides a basic, coarse representation of the video sequence of low quality. The enhancement layer contains the coded representation of the difference between the base layer and the frame encoded into a representation with good quality. To be useful, the enhancement layer needs to be combined with the base layer so as to generate a reconstruction of the video sequence with good quality.

The easy adaptation in the coding rate is achieved by making the enhancement layer be encoded into an embedded bit stream. An embedded bit stream is created in such a way that it can be truncated at any point and still be decodable using the bits remaining from the origin of the stream up to the truncation point. Note that this illustrates another possible method, different from the one used for speech, to change the source encoding rate. We briefly remark here that there are other methods to control the bit rate of compressed video sequences but most of them are based on buffering several frames, which add annoying delay for conversational applications.

Finally, we highlight here that Figure 14.5 illustrates how the approximately exponential relation for the D-R function, as well as its associated properties of being convex and decreasing, can still be observed in practical, well-designed encoders, such as video in this case.

14.5   Channel codec

In the scheme of the source–channel–cooperation problem, the source-encoded bits are organized into a source frame and protected against transmission errors through a channel code. The ratio between the number of bits at the input and output of the channel encoder is called the rate of the channel code, which we will denote by r. The channel code rate is an important parameter describing a channel code, since it specifies the amount of redundancy bits added to the stream, which is the cost associated with the error protection scheme. As is the common practice in communications, we assume that the receiver discards source frames containing errors after channel decoding. Although error concealment schemes are typically used to replace the lost information, this is not our focus here, so we will assume that missing data is simply concealed with its expected value. For input samples that are memoryless, with a zero-mean, unit-variance Gaussian distribution, the mean squared distortion associated with this simple error concealment scheme is equal to 1, the variance of the input samples.

Due to the strict delay constraints associated with multimedia data, our interest here is to consider practical channel codes. We cannot consider the use of capacity-achieving codes or even practical ARQ schemes. Therefore, we will carry on the following study assuming the use of a family of convolutional codes. It is possible to consider other types of codes, such as some block codes, but this will lead into similar results that would only divert the presentation from the focus on user cooperation. Because their channel coding rate can easily be changed, we will implement the family of variable rate convolutional codes with rate compatible punctured convolutional (RCPC) codes. For these codes, the channel coding rate is controlled by eliminating bits from the channel encoded stream. We assume that decoding is performed with a soft input Viterbi decoder.

Convolutional encoders, as illustrated in Figure 14.6, operate by processing a sequence of input bits through a tandem of one-bit memory elements. Each output bit of the encoder is generated as the sum in modulo 2 of the values at the output of certain memories. The collection of all the output bits, generated by feeding the input sequence into the encoder, form a codeword. At the receiver side, the decoding operation often employs a Viterbi decoder, which starts by associating each combination of the encoder memory states with a code state. In this way, the encoding operation can be thought of as a series of transitions through the different states where each transition is driven by an input data bit and results in a number of coded bits. All the possible transitions over time are represented in a trellis diagram. The decoder, then, uses the bits of the received codeword (which are the product of each transition through the state diagram with errors introduced in the noisy channel) to estimate the most likely path through the trellis. A decoding error occurs when the decoder picks as the most likely path, one that does not corresponds to the one followed by the encoder during encoding. This occurs when one or more incorrect state transitions are calculated as having a larger probability (a larger path metric, strictly speaking) than the correct one.

Rate compatible punctured convolutional (RCPC) codes are convolutional codes with an output rate that is controlled by deleting bits from the convolutional code codeword. This operation is called *puncturing* and the convolutional code from which codewords are punctured is called the *mother code*. The more bits that are punctured from the codeword, the largest the channel coding rate is and the less capable of correcting errors the punctured codeword is. With RCPC, a family of punctured codewords of increasing

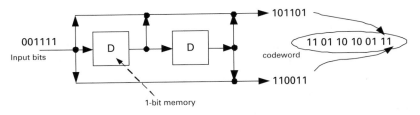

Fig. 14.6   An example of a memory 2 convolutional encoder.

channel coding rate is built by incrementally removing bits from the lower coding rate codewords in such a way that once a bit has been removed, it remains punctured from all the codewords of higher coding rate.

To consider in the analysis the contribution of channel errors to end-to-end distortion, we are interested in finding the probability of having a source frame with errors after channel decoding. Since RCPC codes are convolutional codes, which in turn are linear codes, the probability that the decoder will be unable to correct channel errors can be obtained by assuming that a sequence with all its bits equal to zero is transmitted, and determining the probability that the decoder decides in error for a different sequence, i.e chooses a different path in the decoding trellis. This is essentially the probability that the all-zero sequence is sent, but a different sequence is decoded. With this assumption, the probability of a decoding error can be obtained by first calculating the probability that a path with d input bits has a larger probability (a larger path metric, strictly speaking) than the correct one, and thus is chosen. Given the channel SNR γ, this can be shown to equal

$$P_e(d|\gamma) = 0.5\,\mathrm{erfc}\sqrt{d\gamma}, \tag{14.4}$$

for BPSK modulation [56], with $\mathrm{erfc}(\gamma)$ being the complementary error function defined as $\mathrm{erfc}(\gamma) = 2/\pi \int_\gamma^\infty e^{-u^2}\,du$. Since we are assuming that the all-zero sequence is sent, considering sequences that differ in d bits is equivalent to consider that the sequences in error have Hamming weight d.

Next, to obtain the probability of a decoding error, it is necessary to compute the probability of decoding error associated with all paths with Hamming weight d. This can be upper-bounded as $a(d)P_e(d|\gamma)$, where $a(d)$ is the number of valid codewords with Hamming weight d (i.e., d bits differing from the all-zero codeword). Finally, the probability of having a source frame with errors after channel decoding results from combining the effects of all possible paths with nonzero Hamming weight and can be upper-bounded using the union bound as [130]

$$P(\gamma) \leq 1 - \left[1 - \sum_{d=d_f}^{W} a(d)P_e(d|\gamma)\right]^{NR_s}, \tag{14.5}$$

where d_f is the free distance of the code, which is the minimum number of bits equal to 1 in a valid codeword.

14.6   Analysis of source–channel–cooperation performance

We approach the study of the source–channel–cooperation tradeoff by first characterizing the end-to-end distortion as a function of SNR, and then using this characterization to compare the performance of different schemes. Recall that we call this function the *D-SNR curve*. The end-to-end distortion is formed by the source encoder distortion (which depends on the source encoding rate) and the channel-induced distortion (which depends on the channel SNR, use of cooperation, channel coding rate and error con-cealment operations). In the system setup we are considering, the best adaptation to the known channels SNRs is achieved by jointly setting both the source and channel coding rates so as to minimize the end-to-end distortion. At the same time, the choice for source and channel coding rate cannot add to a total communication capacity exceeding the maximum available W.

Typically both the source and the channel coding rates are chosen from a finite set, resulting in a finite number of possible useful combinations. We call an operating mode Ω_i, the triplet that describes a choice of source encoding rate, a choice of channel coding rate and the indication of whether user cooperation is to be used or not. Let $\Omega = \{\Omega_i, \forall i\}$, be the set of all operating modes. Note that since we are assuming fixed transmit bit rate, without loss of generality we have $W/2 = N R_S/r$ if using coopera-tion and $W = N R_S/r$ if not, i.e., any Ω_i could be specified through either the source or the channel coding rate. In what follows, we will make explicit the dependence of the probability of a frame error on Ω_i by using the notation $P_{\Omega_i}(\gamma)$.

14.6.1 Amplify-and-forward cooperation

We consider a scheme as in Figure 14.7 where an encoded source is transmitted using AF user cooperation. Let R_{SC} be the number of bits used by the source encoder to represent each of the N source samples. Let this source encoded data be protected with a rate r RCPC code. As discussed, we have that $R_{SC} = Wr/2N$. Let $\Omega^{(c)}$ be the subset of Ω such that user cooperation is used. Based on (14.1), the D-SNR function for AF cooperation is the solutions to

$$D_{CAF} = \min_{\Omega_i \in \Omega^{(c)}} \left\{ D_F P_{AF\Omega_i}(\gamma_{AF}) + 2^{-2R_{SC}} \left(1 - P_{AF\Omega_i}(\gamma_{AF}) \right) \right\}, \tag{14.6}$$

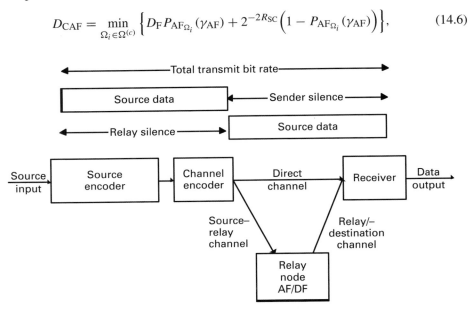

Fig. 14.7   Configuration of the scheme featuring a source encoder with user cooperation.

where D_F is the distortion when the source frame is received with errors. When using mean-squared error distortion measure, $D_F = E[(s - \hat{s})^2]$, where s is the random variable representing the input sample, $\hat{s}$ is the random variable representing its reconstructed value and E is the expectation operator. Since when the source frame is received with errors, the coded source is estimated by its mean, $D_F = E[(s - E[s])^2]$. For our source model (zero mean and unit variance), $D_F = 1$. In (14.6), $P_{AF_{\Omega_i}}(\gamma_{AF})$ is the probability of having a source frame with errors after channel decoding when using AF user cooperation. Its expression can be simply approximated from (14.5) as

$$P_{AF_{\Omega_i}}(\gamma_{AF}) \approx 1 - \left[1 - \sum_{d=d_f}^{W/2} a(d)P(d|\gamma_{AF}) \right]^{NR_{SC}}, \tag{14.7}$$

where γ_{AF} is the SNR at the receiver after the MRC. From (5.49), it is straightforward to see that γ_{AF}, is equal to

$$\gamma_{AF} = \gamma_{s,d} + \frac{\gamma_{s,r}\gamma_{r,d}}{1 + \gamma_{s,r} + \gamma_{r,d}}, \tag{14.8}$$

where $\gamma_{s,d}$ is the SNR in the source–destination link, $\gamma_{s,r}$ is the SNR in the source–relay link and $\gamma_{r,d}$ is the SNR in the relay–destination link.

Finding a close form expression for the D-SNR curve is a challenging problem. Nevertheless, a good approximation can be found by considering the following. Let

$$f_{\Omega_i}(\gamma_{AF}) = P_{AF_{\Omega_i}}(\gamma_{AF}) + 2^{-2R_{SC}}\left(1 - P_{AF_{\Omega_i}}(\gamma_{AF})\right),$$

as per (14.3) and (14.6), be the D-SNR function that results from the choice of one operating mode $\Omega_i \in \Omega^{(c)}$. We will call $f_{\Omega_i}(\gamma_{AF})$ the *single-mode D-SNR curve*. Figure 14.8 shows all the possible single-mode D-SNR curves for an AF system with

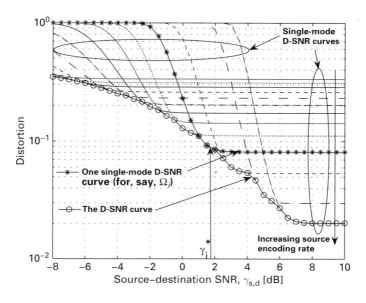

Fig. 14.8   The overall D-SNR curve as an interleaving of single-mode D-SNR curves.

$\gamma_{s,r} = \gamma_{r,d} = 1$ dB and using a memory 4, mother code rate $1/4$, RCPC code family from [56]. In addition, the Figure shows that the envelope formed by the single-mode D-SNR curves is the solution to (14.6), i.e, the D-SNR curve. As implied in (14.6) and seen in Figure 14.8, the D-SNR curve is made from the interleaving of sections of D-SNR curves.

The comparison of two single-mode D-SNR curves shows the basic tradeoff associated with bit rate allocation process in (14.6). Let's consider that the indexes of the operating modes are sorted in increasing source encoding rate order; i.e., $i > j$ (from Ω_i and Ω_j) if $R_{SCi} > R_{SCj}$. Note that, due to the relation $W = 2NR_{SC}/r$, $i > j$ means that $r_i > r_j$ also. For each single-mode D-SNR curve the distortion only increases due to channel errors because source and channel coding are fixed. Figure 14.8 illustrates the behavior of single-mode D-SNR curves. By examining any of the of single-mode D-SNR curves it is possible to identify three distinct parts:

- At high SNR the channel errors are few and of negligible impact. This is because the channel code is strong enough to correct most of channel errors, making the end-to-end distortion be practically constant and equal to the source encoder distortion $D_S(\Omega_i)$ for operating mode Ω_i.
- At low SNR, the distortion increases rapidly because the channel code of the corresponding operating mode is too weak to correct the errors introduced by the channel.
- Between the section where the single-mode D-SNR curve is practically constant and the section where distortion increases rapidly, there is a section that acts as a transition since it corresponds to the range of channel SNR for which the distortion starts to increase due to channel errors. Let γ_i^* be the SNR value at which the distortion of the mode-Ω_i single-mode D-SNR curve starts to increase.

Next, we have that for $i > j$, $D_S(\Omega_i) < D_S(\Omega_j)$. This implies that, over the section where the single-mode D-SNR curve is approximately constant, a single-mode D-SNR curve associated with a larger source encoding rate will present a lower distortion. Also, we have that for $i > j$, $\gamma_i^* > \gamma_j^*$ because $r_i > r_j$, i.e., channel errors start to become significant for weaker channel codes at higher SNR values. The overall effect is that for two single-mode D-SNR curves, $f_{\Omega_i}(\gamma)$ and $f_{\Omega_{i+1}}(\gamma)$, we have $f_{\Omega_{i+1}}(\gamma) < f_{\Omega_i}(\gamma)$ for $\gamma > \gamma_{i+1}^*$ and $f_{\Omega_{i+1}}(\gamma) > f_{\Omega_i}(\gamma)$ for $\gamma < \gamma_i^*$ Operating modes for which these conditions do not hold are not useful because either one or both are not active in the solution to (14.6) (they are a degenerate mode for which, at any SNR, there is always another operating mode with lower distortion) or they are over a negligible range of SNRs. As Figure 14.8 illustrates, this means that the D-SNR performance resulting from solving (14.6) is the sequential interlacing of sections of single-mode D-SNR curves. These sections are the portion of single-mode D-SNR curves where channel-induced distortion is such that $D_S(\Omega_{i+1}) \leq f_{\Omega_{i+1}}(\gamma) \leq D_S(\Omega_i)$, approximately.

Note that these observations follow from the mechanics associated with problem (14.6), they can be found as well in applications of this problem with setups very different from the one we are considering. Also note in Figure 14.8, that at very low SNR (and large distortion) there is an operating region where the smallest of the channel codes had already been chosen and there is no more use of JSCC bit rate allocation. Performance in this region follows that of a system with fixed channel coding, thus it is of no interest to us.

The challenge in characterizing the D-SNR curve is that, even when there is no use of cooperation, the expressions that can be derived do not lead to a closed-form solution to (14.6). Because of this, we study the D-SNR curve by approximating its representation through a curve determined by a subset of its points. Each of these points is selected from a different single-mode D-SNR curve. The criterion used to select each point is based on the observation that the segment of a single-mode D-SNR curve that belongs to the solution of (14.6) follows $D_S(\Omega_{i+1}) \leq f_{\Omega_{i+1}}(\gamma) \leq D_S(\Omega_i)$, approximately; which means that a point where the relative contribution of channel errors to the end-to-end distortion is small is likely to belong to the overall D-SNR curve. Points with a relative contribution that is practically negligible or is large may or may not belong to the D-SNR curve and, thus, are not good choices to form the approximation. Then, for each single-mode D-SNR curve, the point chosen to represent the D-SNR curve is such that $D = (1 + \Delta)D_S(\Omega_i)$, $\Omega_i \in \Omega^{(c)}$, where Δ is a small number. Formally, we are considering those points where

$$D = (1 + \Delta)D_S(\Omega_i)$$
$$= D_F P_{\mathrm{AF}_{\Omega_i}}(\gamma_{\mathrm{AF}}) + D_S(\Omega_i)\left(1 - P_{\mathrm{AF}_{\Omega_i}}(\gamma_{\mathrm{AF}})\right), \tag{14.9}$$

Equivalently, from (14.9), the D-SNR curve is formed by those points where

$$P_{\mathrm{AF}_{\Omega_i}}(\gamma_{\mathrm{AF}}) = \frac{\Delta}{\frac{D_F}{D_S(\Omega_i)} - 1}. \tag{14.10}$$

This expression offers the advantage that it simplifies the D-SNR curve characterization problem by translating it into a problem essentially in error control coding. By examining Figure 14.8 it is possible to see that a choice of Δ as a small number means that the points used in the characterization of the D-SNR function falls into the "elbow" region of typical single-mode D-SNR curves. Therefore, the choice of Δ as a small number, besides being justified by the above discussion, have the advantage that the results will show little changes to different values of Δ. This is because in the "elbow" region of typical single-mode D-SNR curves, relatively large changes in distortion translate in relatively small changes in SNR. This idea will be further explored in the problems at the end of this chapter. Even more, to solve the characterization problem, it would be possible to use the related fact that channel-induced errors are relatively few and apply approximations that are accurate at low BER regime. Following these ideas and combining (14.7) and (14.10), it follows that

$$\frac{\Delta}{\frac{D_F}{D_S(\Omega_i)} - 1} \approx 1 - \left(1 - \frac{a(d_f)}{2} \operatorname{erfc}\sqrt{d_f \gamma_{\mathrm{AF}}}\right)^{N R_{\mathrm{SC}}}, \tag{14.11}$$

where we have used (14.4).

Equation (14.11) shows an implicit dependence, through d_f and $a(d_f)$, between $P_{\mathrm{AF}_{\Omega_i}}(\gamma_{\mathrm{AF}})$ and the rate of the particular code chosen from the RCPC family, i.e. the

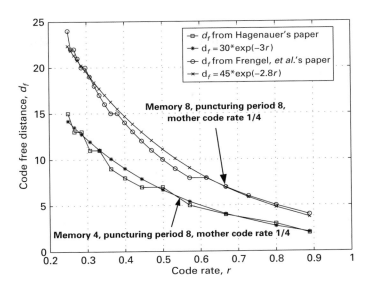

Fig. 14.9   Code rate versus free distance for two families of RCPC codes.

particular operating mode. To make this dependence explicit, we study next the relation between d_f and $a(d_f)$ with r. Figure 14.9 shows how d_f changes as a function of
the code rate r for the best known RCPC codes with mother code rate 1/4, puncturing
period 8 and memory 4 or 8 (as specified in [56] and [41], respectively). Figure 14.9
also shows that the free distance of a family of RCPC codes can be approximated by a
function of the form

$$d_{\mathrm{f}} \approx \kappa \, \exp(-cr) \approx \kappa \exp\left(-\frac{2cN R_{\mathrm{SC}}}{W}\right),$$

where κ and c are two constants. A similar study for $a(d_f)$ shows that it is not possible
to find a practical functional approximation. Therefore, we will roughly approximate
$a(d_f)$ by its average value, $\bar{a}$, an approximation that will prove to yield good results.
Using the approximations for d_f and $a(d_f)$, (14.11) becomes

$$\frac{\Delta}{\frac{D_{\mathrm{F}}}{D_{\mathrm{S}}(\Omega_i)} - 1} \approx 1 - \left(1 - \frac{\bar{a}}{2} \, \mathrm{erfc}\left(\sqrt{\kappa e^{-2cN R_{\mathrm{SC}}/W}} \gamma_{\mathrm{AF}}\right)\right)^{N R_{\mathrm{SC}}}. \tag{14.12}$$

Using (14.10), (14.12) can be solved for γ_{AF} to show the explicit relation between γ_{AF}
and source coding rate R_{SC},

$$\gamma_{\mathrm{AF}} \approx \frac{\exp\left(\frac{2cN R_{\mathrm{SC}}}{W}\right)}{\kappa} \left(\mathrm{erfc}^{-1}\left(\frac{2}{\bar{a}}\left[1 - \left(1 - \frac{\Delta}{\frac{D_{\mathrm{F}}}{D_{\mathrm{S}}(\Omega_i)} - 1}\right)^{\frac{1}{N R_{\mathrm{SC}}}}\right]\right)\right)^2. \tag{14.13}$$

Equation (14.13) shows that two factors contribute to γ_{AF}: one that is the inverse of the
free distance and another that depends on the inverse complementary error function. If
we take the logarithm of γ_{AF}, these two factors become the two terms of a sum where

the first term is linear in R_{SC}. Let us work next with this second term, which we denote as $\log\left(g(R_{SC})\right)$, where

$$g(R_{SC}) = \left(\operatorname{erfc}^{-1}\left(\frac{2}{\bar{a}}\left[1 - \left(1 - \frac{\Delta}{\frac{D_F}{D_S(\Omega_i)} - 1}\right)^{1/(N R_{SC})}\right]\right)\right)^2.$$

Using the approximation $\operatorname{erfc}^{-1}(y) \approx \sqrt{-\ln(y)}$, [24], we can write

$$g(R_{SC}) \approx \ln\left(\frac{\bar{a}}{2}\right) - \ln\left(1 - \left(1 - \frac{\Delta}{\frac{D_F}{D_S(R_{SC})} - 1}\right)^{1/(N R_{SC})}\right). \tag{14.14}$$

Before continuing, we note that the approximation $\operatorname{erfc}^{-1}(y) \approx \sqrt{-\ln(y)}$ is not tight, yet it is still useful for us here because it is very good in representing the functional behavior of the inverse error function. In essence, this means that the probability of error associated with the modulation scheme approximately behaves as $\exp(-x^2)$ with respect to SNR. Working with the second term of the right-hand side of (14.14),

$$\theta = -\ln\left(1 - \left(1 - \frac{\Delta}{\frac{D_F}{D_S(R_{SC})} - 1}\right)^{1/(Nx)}\right) \tag{14.15}$$

$$\Rightarrow \ln\left(1 - e^{-\theta}\right) = \frac{1}{N R_{SC}} \ln\left(1 - \frac{\Delta}{\frac{D_F}{D_S(R_{SC})} - 1}\right) \tag{14.16}$$

$$\Rightarrow -e^{-\theta} \approx \frac{1}{N R_{SC}}\left(-\frac{\Delta}{\frac{D_F}{D_S(R_{SC})} - 1}\right) \tag{14.17}$$

$$\Rightarrow \theta \approx \ln\left(\frac{N}{\Delta} R_{SC}\left(\frac{D_F}{D_S(R_{SC})} - 1\right)\right) \tag{14.18}$$

$$\approx \ln\left(\frac{N}{\Delta} R_{SC}\right) + \ln\left(\frac{D_F}{D_S(R_{SC})}\right), \tag{14.19}$$

where (14.17) follows from the fact that both logarithms in (14.16) are of the form $\ln(1 - y)$ with $y \ll 1$ and the approximation in (14.19) follows from recognizing that in general $D_F/D_S(x) - 1 \approx D_F/D_S(x)$, i.e., the reconstructed source error in the case of transmission errors is typically much larger than the quantization distortion. Following (14.3), θ in (14.18) is the sum of the logarithm of $N R_{SC}/\Delta$ and a term that is approximately linear in R_{SC}, where generally $N/\Delta \gg R_{SC}$. Then, we can approximate

$$\ln\left(\frac{N R_{SC}}{\Delta}\right) \approx \ln\left(\frac{N \bar{R}_{SC}}{\Delta}\right),$$

with $\bar{R}_{SC}$ being the average value of R_{SC}, and consider that the approximately linear term would determine the overall behavior of $\theta(R_{SC})$. Using (14.3), it follows that

$$g(R_{SC}) \approx \ln\left(\frac{\bar{a} N D_F \bar{R}_{SC}}{2\Delta}\right) + 2 R_{SC} \ln(2).$$

Recall that we are interested in finding a more useful expression for $\log(g(R_{SC}))$. Through algebraic operations it can be shown that the coefficients of Taylor's expansion of $\log(g(R_{SC}))$ around $\bar{R}_{SC}$ are of the form $t^i/i!$, with

$$t = \ln(4) \left(\ln \left(\frac{\bar{a} N D_F \bar{R}_{SC}}{2\Delta D_S(\bar{R}_{SC})} \right) \right)^{-1}$$

being much smaller than 1 for typical system parameters. This means that the weight of coefficients in Taylor's expansion fall quite rapidly and $\log(g(R_{SC}))$ can be accurately approximated through the first order expansion

$$\log(g(R_{SC})) \approx \log(\Psi + 2\bar{R}_{SC}\ln(2)) + \frac{(R_{SC} - \bar{R}_{SC})\log(4)}{\Psi + 2\bar{R}_{SC}\ln(2)} \tag{14.20}$$

where

$$\Psi = \ln \left(\frac{\bar{a} N D_F \bar{R}_{SC}}{2\Delta} \right).$$

Recall that we noticed that the logarithm of γ_{AF} can be decomposed into two terms: one that is linear in R_{SC} and the other being $\log(g(R_{SC}))$, which we have just shown is approximately linear also. Therefore, we can see that, when measured in decibels, γ_{AF} is approximately a linear function of R_{SC}. Specifically, using (14.13) and (14.20), we can write γ_{AF} in decibels as

$$\gamma_{AF_{dB}} \approx \frac{20cN R_{SC}\log(e)}{W} - 10\log(\kappa) + 10\log(\Psi + 2\bar{R}_{SC}\ln(2))$$
$$+ 10\frac{\log(4)}{\Psi + 2\bar{R}_{SC}\ln(2)}(R_{SC} - \bar{R}_{SC}). \tag{14.21}$$

Then, we have $\gamma_{AF_{dB}}(R_{SC}) \approx A_1 x + B_1$, or, equivalently, $R_{SC} \approx A_2 \gamma_{AF_{dB}} + B_2$. Since $D = (1 + \Delta)D_S = (1 + \Delta)2^{-2R_{SC}}$, we have

$$D = (1 + \Delta)10^{-(\log 4)R_{SC}}$$
$$\approx (1 + \Delta)10^{-(\log 4)\left(A_2 \gamma_{AF_{dB}} + B_2\right)}$$
$$= (1 + \Delta)10^{-B_2(\log 4)}\gamma_{AF}^{-10A_2\log 4}. \tag{14.22}$$

This is a relation of the form $D \approx (G_c \gamma)^{-10m}$ with, from (14.21),

$$D_{CAF} \approx (G_c \gamma_{AF})^{-10m_c}, \tag{14.23}$$

$$m_c = \frac{\log(4)}{A_1} = \left[20 \left(\frac{cN}{W\ln(4)} + \frac{1}{2\Psi + 2\bar{R}_{SC}\ln(4)} \right) \right]^{-1}, \tag{14.24}$$

$$G_c = \left[(1 + \Delta) \right]^{\frac{-1}{10m}} 10^{\frac{-B_1}{10}}$$
$$= \frac{\kappa(1 + \Delta)^{\frac{-1}{10m}}}{\Psi + \bar{R}_{SC}\ln(4)} 10^{\frac{\log(4)\bar{R}_{SC}}{\Psi + \bar{R}_{SC}\ln(4)}}. \tag{14.25}$$

Equation (14.23) has the appeal of being of the same form as the error rate-SNR function found in the study of communication systems, as is illustrated in Section 1.4. The main differences are that we are considering distortion instead of average error probability, there is no assumption of asymptotically large SNR, and that (14.23) models performance of systems with practical components instead of being a bound at high

SNR. Following this similarity, we can derive the analogy that G_c is the coding gain and m_c is the diversity gain.

The coding gain G_c represents the SNR for a reference distortion value of 1. The diversity gain m_c determines the slope (i.e., rate of change) of the logarithm of the distortion when SNR is measured in decibels (the factor of 10 multiplying m_c in (14.23) is to set these units). Here, we are using the concept of diversity beyond its natural realm of transmitting multiple copies of a message over fading channels. This expanded use of the concept of diversity beyond its natural realm of systems with fading is based on considering that there is a significant overlap between the concept of diversity and redundancy.

The general concept of diversity, as that of transmitting multiple copies of a message, can be extended to other forms of redundancy. Such is the case for error correcting codes when considering that they exhibit an inherent diversity because the (maybe partial) copies of the message are coded in the form of parity data. In this sense, such an expanded view of diversity aims at measuring more general forms of redundant information in cross-layer designs, which allows us to consider in a unified way the use of cooperation to deliver multiple copies of the message and the tradeoffs in redundancy at the source–channel bit rate allocation stages. Also, it is worth highlighting that the basic premise on which we base the D-SNR characterization, i.e (14.9), follows from the problem setup and the general dynamics of the interaction of source encoding and channel-induced distortion. This is not exclusive to a particular choice of source, channel code family, or even a particular distortion measure. This implies that the analysis could be extended to other types of channel code, yielding similar qualitative results.

14.6.2   Decode-and-forward cooperation

We now consider the case when transmission is carried on using DF user cooperation. In this case, the system setup can still be schematized as in Figure 14.7 and we still assume a source coding rate R_{SC} and channel coding rate r. For DF cooperation the relation $R_{SC} = Wr/2N$ still holds. Thus, following (14.30), the D-SNR performance function for DF cooperation is the solutions to

$$D_{CDF} = \min_{\Omega_i \in \Omega^{(c)}} \left\{ D_F P_{DF_{\Omega_i}}(\gamma_{DF}) + 2^{-2R_{SC}} \left(1 - P_{DF_{\Omega_i}}(\gamma_{DF}) \right) \right\}.$$

Here, $P_{DF_{\Omega_i}}(\gamma_{DF})$ is the probability of having a source frame with errors after channel decoding when using DF user cooperation. If we denote by Ψ the event "correct decoding at relay" and by $\bar{\Psi}$ the event "incorrect decoding at relay," this probability is

$$P_{DF_{\Omega_i}}(\gamma_{DF}) = P_{\Omega_i}(\gamma_{DF}|\Psi) P_{\Omega_i}(\Psi) + P_{\Omega_i}(\gamma_{DF}|\bar{\Psi}) P_{\Omega_i}(\bar{\Psi})$$

$$= P_{\Omega_i}(\gamma_{s,d} + \gamma_{r,d}) \left(1 - P_{\Omega_i}(\gamma_{s,r}) \right) + P_{\Omega_i}(\gamma_{s,d}) P_{\Omega_i}(\gamma_{s,r}).$$

$$(14.26)$$

Note that this relation is derived by considering that the first term in the right-hand side of (14.26) represents the probability of successful transmission from the source to the relay and then to the destination. The second term represents the case when the relay fails decoding. Here, γ_{DF} represents the received SNR at the output of the MRC when using DF cooperation.

Therefore, to derive a close form approximation to the D-SNR curve, it is necessary to appropriately consider the structure of $P_{DF_{\Omega_i}}(\gamma_{DF})$ as in (14.26). Consequently, we consider three cases: the source–relay channel is "good", the source–relay channel is "bad", and the channel states are such that there is no solution to (14.10) when considering DF cooperation. One element introduced by the adaptability of channel coding rate is that the classification of "good" or 'bad" source–relay channel does not depend only on $\gamma_{s,r}$, but also on the operating mode and the strength of its associated channel code. This is because for a given channel state the strongest channel codes may be operating at the low BER regime but the weakest codes may be operating in the high BER regime. This is why the source–relay channel needs to be qualified in terms of the frame error probability $P_{\Omega_i}(\gamma_{s,r})$.

In the case of a "good" source–relay channel the source frame error probability (14.26) can be approximated as $P_{DF_{\Omega_i}}(\gamma_{DF}) \approx P_{\Omega_i}(\gamma_{s,d} + \gamma_{r,d})$ for most operating modes. In this case (14.9) becomes

$$D = (1 + \Delta)D_S(\Omega_i) = D_F P_{\Omega_i}(\gamma_{s,d} + \gamma_{r,d}) + D_S(\Omega_i)\Big(1 - P_{\Omega_i}(\gamma_{s,d} + \gamma_{r,d})\Big),$$

and we have

$$D_{CDF} \approx \big(G_c(\gamma_{s,d} + \gamma_{r,d})\big)^{-10m_c}. \tag{14.27}$$

In the case of a "bad" source–relay channel, (14.26) can be approximated as $P_{DF_{\Omega_i}}(\gamma_{DF}) \approx P_{\Omega_i}(\gamma_{s,d})$ for most operating modes. In this case we have

$$D_{CDF} \approx \big(G_c\gamma_{s,d}\big)^{-10m_c}. \tag{14.28}$$

The differentiation between "good" and "bad" source–relay channels is done by setting a threshold on $P_{\Omega_i}(\gamma_{s,r})$ such that the operating modes are separated into two regimes. These regimes are those where $P_{DF}(\gamma_{DF})$ in (14.26) can approximate either $P_{DF_{\Omega_i}}(\gamma_{DF}) \approx P_{\Omega_i}(\gamma_{s,d} + \gamma_{r,d})$ (applicable to operating modes with strong error protection such that $P_{\Omega_i}(\gamma_{s,r})$ is less than the threshold) or $P_{DF_{\Omega_i}}(\gamma_{DF}) \approx P_{\Omega_i}(\gamma_{s,d})$ (applicable to operating modes with weak error protection such that $P_{\Omega_i}(\gamma_{s,r})$ is more than the threshold). Although there may be one or two operating modes in borderline cases with $P_{\Omega_i}(\gamma_{s,r})$ near the threshold, the rest of the modes will be clearly in a good or bad channel regime. Because of this, the composition of the two sets of modes in good or bad channel regime show little sensitivity to the value chosen as a threshold. Following this, we considered the threshold to be $P_{\Omega_i}(\gamma_{s,r}) \approx 0.1$. Also, we note that because the bandwidth efficiency of DF cooperation is the same as in AF cooperation, and because both schemes share the same value for the variables that affect m_c and G_c, the value of these two magnitudes in (14.27) and (14.28) are given by (14.24) and (14.25).

The third modeling case corresponds to situations where the channel states are such that there is no solution to (14.10). This may occur when $\gamma_{s,d}$ is very low, both $\gamma_{s,r}$ and $\gamma_{r,d}$ are such that

$$P_{\mathrm{DF}_{\Omega_i}}(\gamma_{\mathrm{DF}}) < \Delta\left(\frac{D_{\mathrm{F}}}{D_{\mathrm{S}}(\Omega_i)} - 1\right)^{-1}$$

for some Ω_i. In this case, the distortion in our analysis becomes an asymptotic value. Because Δ is chosen to be small, we approximate this value by solving for the operating mode for which

$$\lim_{\gamma_{s,d}\to 0} P_{\mathrm{DF}_{\Omega_i}}(\gamma_{\mathrm{DF}}) = \Delta\left(\frac{D_{\mathrm{F}}}{D_{\mathrm{S}}(\Omega_i)} - 1\right)^{-1}.$$

Using this fact and (14.3) into (14.12), the asymptotic distortion value can be approximated as $D_{\mathrm{a}} \approx 2^{-2R_{\mathrm{SC}}}$ after solving for the source encoding rate R_{SC} in

$$R_{\mathrm{SC}} = \frac{1}{2}\log_2\left(1 + \frac{\Delta}{1 - \left[1 - \frac{\bar{a}}{2}\mathrm{erfc}\left(\sqrt{\kappa\exp\left(-\frac{2cNR_{\mathrm{SC}}}{W}\right)\gamma_{r,d}}\right)\right]^{NR_{\mathrm{SC}}}}\right). \qquad (14.29)$$

14.6.3   No use of cooperation

We consider also a scheme with no user cooperation. This case, schematized in Figure 14.10, corresponds to a direct communication between source and destination. Let R_{SN} be the number of bits used by the source encoder to represent each of the N source samples and $\Omega^{(nc)}$ be the subset of Ω such that there is no use of cooperation. Then, based on (14.1), the D-SNR curve is the solution to

$$D_{\mathrm{SN}} = \min_{\Omega_i \in \Omega^{(nc)}}\left\{D_{\mathrm{F}}P_{\Omega_i}(\gamma_{s,d}) + 2^{-2R_{\mathrm{SN}}}\left(1 - P_{\Omega_i}(\gamma_{s,d})\right)\right\}. \qquad (14.30)$$

Equation (14.30) is of the same form as (14.6). Therefore, a closed-form approximation for the D-SNR curve could be derived as in Section 14.6.1. The most important factor to consider now is that because transmission is done in only one phase, we will have twice the bandwidth efficiency as in cases using user cooperation. This implies that $\bar{R}_{\mathrm{SN}} = 2\bar{R}_{\mathrm{SC}}$. Considering this, the approximate expression for the D-SNR curve can be derived from algebraic manipulations of (14.23)–(14.25) and shown to be

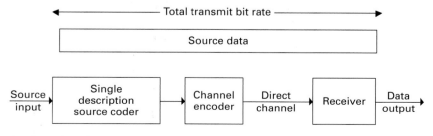

Fig. 14.10   Configuration of the scheme featuring a source encoder without user cooperation.

$$D_N \approx (G_N \gamma_{s,d})^{-10m_N}, \tag{14.31}$$

$$m_N = \left[10 \left(\frac{cN}{W \ln(4)} + \frac{1}{\Psi + \ln(2)(1 + 4\bar{R}_{SC})} \right) \right]^{-1}, \tag{14.32}$$

$$G_N = \frac{\kappa (1 + \Delta)^{-1/(10m_N)}}{\Psi + \ln(2)(1 + 4\bar{R}_{SC})} 10^{\frac{2\log(4)\bar{R}_{SC}}{\Psi + \ln(2)(1 + 4\bar{R}_{SC})}}. \tag{14.33}$$

We briefly pause now our study to illustrate our results so far with a series of examples.

Example 14.1 Consider a system where the source encoder has a D-R performance as in (14.3) and where variable-rate error protection is implemented with the family of the best known RCPC codes with mother code rate 1/4, puncturing period 8 and memory 4 (as specified in [56]). Also, assume that $N = 150$ samples, $W = 950$ bits per transmission period and $\Delta = 0.1$. Figure 14.11 shows the D-SNR curves obtained from simulations along with those derived using the characterization developed in the previous sections. In the figure, $\gamma_{s,d}$ is treated as the main variable, while $\gamma_{s,r}$ and $\gamma_{r,d}$ are treated as parameters. In doing so, it is possible to consider different scenarios that can be thought of as associated with the distance between the relay and the other nodes. For the family of RCPC codes that is being used we have $\bar{a} = 6.1$ (obtained from the table of $a(d)$ in [56]), and to approximate d_f, we have $\kappa = 30$ and $c = 3$ (see Figure 14.9). It can be seen in Figure 14.11 that the characterization developed in the previous sections is able to accurately represent the behavior of the schemes under study.

Also, the figure shows that the largest difference between simulation and analytical results occurs at a region of relative low SNR or large distortion (this is more clear

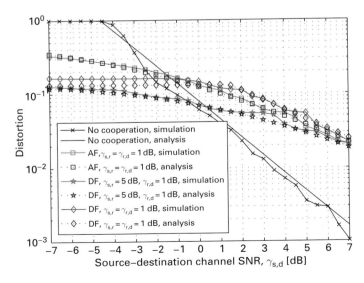

Fig. 14.11 D-SNR curves obtained from simulations and applying analysis in Section 14.6.

for the results of the scheme without cooperation for $\gamma_{s,d}$ in the range between -3 and -1 dB). This difference is because in this range there is no adaptation of the source or channel code because the strongest channel code has already been chosen at $\gamma_{s,d} \approx -1$ dB. In other words, the D-SNR curve has effectively become the single-mode D-SNR curve with the strongest channel code (see Figure 14.8 for an illustration), thus, the system is operating outside the range where the above analysis applies.
▲

Example 14.2   The D-SNR characterization can also be applied to practical schemes. This is because, as noted earlier, D-R functions of the form we have considered can approximate or bound the performance of many practical well-designed source codecs. Figure 14.12 illustrated this by showing the D-SNR performance of the GSM AMR speech codec [36], coupled to the memory 4 RCPC codes just used in Example 14.1. The figure shows seven single mode D-SNR curves, obtained through Monte Carlo method, corresponding to source encoding rates 12.2, 10.2, 7.4, 6.7, 5.9, 5.15, and 4.75 Kbps; where the last three modes are used in AF cooperation and the first four without cooperation. Figure 14.12 shows how the D-SNR curve is made from the interleaving of single-mode D-SNR curves. To emphasize the wide applicability of the D-SNR characterization, distortion is measured using the perceptually-based measure related to the ITU-T PESQ quality measure standard P.862 [88], where distortion is measured as $4.5 - Q$ with Q being the output from the PESQ algorithm. The figure also includes the performance approximation as in equations (14.23) and (14.31), with parameters obtained by measuring two performance points (as will be discussed in detail in the next section). The figure shows that, despite the much more coarse set

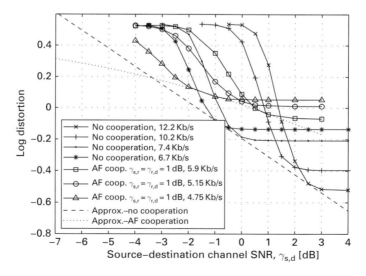

Fig. 14.12   Application of schemes to GSM AMR speech codec.

of available operating modes, the above characterization of the D-SNR still applies in these cases. Also, the figure illustrates how the points following (14.10) are used in the characterization. ▲

Example 14.3 Figure 14.13 shows another example of a practical application, this time video. The results shown in the figure correspond to a frame from the QCIF video sequence "Akiyo" playing at 30 frames per second. The video sequence was encoded using the MPEG-4 video codec and protected with a memory 8, puncturing period 8, mother code rate 1/4 family of RCPC codes [41]. In correspondence with the system model, the source encoding rate is controlled on a frame-by-frame basis by changing the "Q factor" of the encoder. Roughly, the Q factor controls the level of quantization noise of the DCT coefficients, obtained during compression of a frame. As a representative example, Figure 14.13 shows results for frame number 190. This frame is the tenth P frame[1] after an I frame.[2] Here, again, it is possible to observe how the D-SNR curve is made from the interleaving of single-mode D-SNR curves and that the D-SNR characterization still applies even though there are only a few operating modes available. Interestingly, this is the case even when the video source is not memoryless, as previously assumed. This is because, as shown in (14.9), the D-SNR characterization focuses on points where the channel-induced distortion starts to become important, which is an effect that is somewhat decoupled from the source memory (i.e., the channel states for

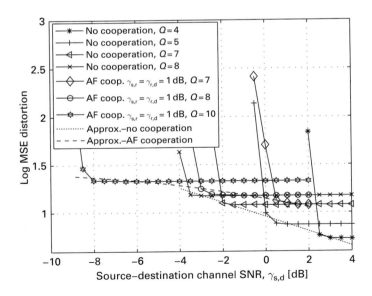

Fig. 14.13   Application of schemes to MPEG-4 video codec.

---

[1] Obtained through linear prediction from the preceding frame.
[2] Encoded as an independent image.

which the channel-induced distortion becomes important is such that it significantly
affects all memory interdependent frames).   ▲

14.7   Validation of D-SNR characterization

Through the series of steps shown in (14.9)–(14.22) it was possible to find an accurate
characterization for the D-SNR function. Nevertheless, the series of steps involved a
number of approximations that are likely to accumulate error into the final distortion
approximation. Therefore, it is important, as a way to evaluate the approximation, to
look at the error between the distortion and its approximation. Due to the approach
followed to characterize the D-SNR curve, the error should be measured as the relative
error between the logarithm of the actual and the approximated distortion values at those
specific points used to approximate the D-SNR curve.

Recall from the discussion that led to (14.10), that the actual distortion value at these
points is, by definition, $(1 + \Delta)D_S(\Omega_i)$, where Ω_i may or may not involve the use of
cooperation. For each point used to approximate the D-SNR curve (one per operating
mode) it is possible to find the actual SNR value by solving for γ in

$$P_{\Omega_i}(\gamma) = \frac{\Delta}{\frac{D_F}{D_S(\Omega_i)} - 1}. \qquad (14.34)$$

When the operating modes Ω_i in (14.34) are those with no cooperation, the resulting γ
will correspond to the $\gamma_{s,d}$ of the points used to approximate the D-SNR curve. Because
(14.34) does not depend on the specific type of cooperation and we are solving for γ
as the unknown variable, when the operating modes are those with cooperation, the
resulting γ at the points used to approximate the D-SNR curve, will correspond to γ_{AF}
if using AF cooperation or either $\gamma_{s,d} + \gamma_{r,d}$ or $\gamma_{s,d}$ if using DF cooperation. From the
knowledge of γ for each scenario, it is possible to find the approximate values of the
distortion at the points used to approximate the D-SNR curve through (14.24), (14.27),
(14.28), or (14.31).

Figure 14.14 shows, for the memory 4 RCPC code already used before, the relative
error between the logarithm of the actual and the approximated distortion values at the
points used to approximate the D-SNR curve for schemes with and without cooperation.
Only one curve is needed for all the schemes with cooperation because the abscissa rep-
resents γ_{AF}, $\gamma_{s,d} + \gamma_{r,d}$ or $\gamma_{s,d}$, depending on the case and because G_c and m_c are the
same for all schemes with cooperation. In the figure, it can be seen that the approxima-
tion is sufficiently good with most points having relative errors of less than 5% (many,
in fact, less than 2.5%) and only one slightly exceeding 10%. The source for most of
the error can be traced to the tightness of the approximation $d_f \approx \kappa \exp(-cr)$ and the
approximation of $a(d_f)$ by its average value.

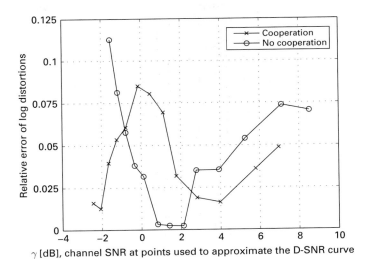

Fig. 14.14   Relative error between the logarithm of the actual and the approximated distortion values
at the points used to approximate the D-SNR curve.

14.8   Effects of source–channel–cooperation tradeoffs

From the previous results, it follows that to study the tradeoff between source and channel coding and use of cooperation, it is important to study the relation between m_N and
m_c, and G_N and G_c, since they characterize the D-SNR performance.

We first consider the relation between m_c and m_N, as given by (14.24) and (14.32).
Because in practice $\Psi > \ln(2)$, we have that $2\Psi + 2\bar{R}_{SC}\ln(4) > \Psi + \ln(2)(1 + 4\bar{R}_{SC})$,
which implies that $m_N/m_c < 2$. Also, it can be shown through algebraic operation that
the condition $m_N/m_c < 1$ requires $cN/(W\ln(4)) < 0$, which is not possible. Therefore,
we conclude that

$$1 < \frac{m_N}{m_c} < 2. \tag{14.35}$$

Furthermore, for typical system setups the term $cN/(W\ln 4)$ is the one that contributes most to the value of m_c or m_N because it typically differs from $(2\Psi +
2\bar{R}_{SC}\ln(4))^{-1}$ or $(\Psi + \ln(2)(1 + 4\bar{R}_{SC}))^{-1}$ by approximately an order of magnitude.
This implies that we can approximate

$$\frac{m_N}{m_c} \approx 2. \tag{14.36}$$

Physically, (14.35) means that the distortion will always decrease at a faster rate for systems without user cooperation and (14.36) means that, in fact, distortion will decrease
at a rate approximately twice as fast in systems without user cooperation (as can be seen
in Figure 14.11, for example). Interestingly, note that this later relation follows mostly
from the fact that when using cooperation, the communication capacity is halved.

Secondly, to study the relation between G_N and G_c, we recall that (14.23), (14.27), (14.28), and (14.31) show that the functional relation between end-to-end distortion and received SNR (be it $\gamma_{s,d}$, γ_{AF}, or γ_{DF}) is linear in log-log scales. Therefore, m_N, G_N, m_c and G_c can be calculated from the knowledge of only two points of the D-SNR curve. Let (D_{N1}, γ_{N1}) and (D_{N2}, γ_{N2}) be the coordinates of these points for schemes without cooperation and (D_{C1}, γ_{C1}) and (D_{C2}, γ_{C2}) be those for schemes using cooperation. The coordinates of these points could be approximated using again the observation that each single-mode D-SNR curve contributes to the overall D-SNR curve over a section where the contribution of channel-induced error is relatively small. Then, from (14.9) and (14.10), and for i equalt to 1 and 2,

$$D_i = \left(1 + \Delta\right)2^{-2R_i}, \quad \gamma_{dBi} = 10\log\left[P_{\Omega_i}^{-1}\left(\frac{\Delta}{2^{2R_i} - 1}\right)\right], \tag{14.37}$$

where the subscript dB denotes a magnitude expressed in decibels. Knowing the points coordinates, it is then possible to find

$$m_N = \frac{\log\left(D_{N1}/D_{N2}\right)}{\gamma_{N2,dB} - \gamma_{N1,dB}}, \quad G_{NdB} = -\left(\gamma_{N1,dB} + \frac{\log D_{N1}}{m_N}\right),$$

$$m_c = \frac{\log\left(D_{C1}/D_{C2}\right)}{\gamma_{C2,dB} - \gamma_{C1,dB}}, \quad G_{CdB} = -\left(\gamma_{C1,dB} + \frac{\log D_{C1}}{m_c}\right).$$

Focusing on G_N and G_c, note that $\log D_{C1} = \log(1 + \Delta) - 2R_{SC1}\log 2$. Since Δ is small, $\log D_{C1} \approx -2R_{SC1}\log 2 = -R_{SN1}\log 2 \approx (\log D_{N1})/2$. Then, since $m_N/m_c \approx 2$, we have $\log D_{C1}/m_c \approx \log D_{N1}/m_N$, which implies that the relation between G_N and G_c, approximately only depends on the relation between $\gamma_{N1,dB}$ and $\gamma_{C1,dB}$. Furthermore, when the channel-induced errors are few, (14.5) can be approximated using a first-order Taylor approximation as

$$P(\gamma) \lesssim N R_s \left(\sum_{d=d_f}^{W} a(d)P(d|\gamma)\right).$$

Differentiating explicitly between the use or not of cooperation by using the subscripts C and N, respectively, it follows that $P_C(\gamma) \approx P_N(\gamma)/2$ for a given operating mode Ω_i. Considering these observations along with (14.9) and (14.10), leads to (14.37) being re-written as

$$\gamma_{Ci,dB} = 10\log\left[P_{N\Omega_i}^{-1}\left(\frac{2\Delta}{2^{R_{SN_i}} - 1}\right)\right], \quad i = 1, 2, \tag{14.38}$$

which implies that $\gamma_{C1,dB} < \gamma_{Ni,dB}$. Therefore, we conclude that

$$G_{CdB} > G_{NdB}. \tag{14.39}$$

Consider next the value of $\gamma_{s,d}$ for which $D_{CAF} = D_N$.

$$\left(G_c \left(\gamma_{s,d} + \frac{\gamma_{s,r}\gamma_{r,d}}{1 + \gamma_{s,r} + \gamma_{r,d}} \right) \right)^{-10m_c} = (G_N\gamma_{s,d})^{-10m_N}$$

$$\Rightarrow \quad \frac{G_N^{m_N}}{G_c^{m_c}} = \frac{\left(\gamma_{s,d} + \frac{\gamma_{s,r}\gamma_{r,d}}{1+\gamma_{s,r}+\gamma_{r,d}} \right)^{m_c}}{\gamma_{s,d}^{m_N}}$$

$$\Rightarrow \quad \frac{G_N^2}{G_c} \approx \frac{\left(\gamma_{s,d} + \frac{\gamma_{s,r}\gamma_{r,d}}{1+\gamma_{s,r}+\gamma_{r,d}} \right)}{\gamma_{s,d}^2} \qquad (14.40)$$

$$\Rightarrow \quad \gamma_{s,d} \approx \frac{G_c^2}{G_N} \left(1 + \sqrt{1 + 4\left(\frac{\gamma_{s,r}\gamma_{r,d}}{1 + \gamma_{s,r} + \gamma_{r,d}} \right) \frac{G_N^2}{G_c}} \right), \qquad (14.41)$$

where in (14.40), we have used the approximation (14.36). Equation (14.41) shows that there is only one value of $\gamma_{s,d}$ for which the distortion using cooperation equals the one without cooperation. The combined effect of (14.35), (14.39) and (14.41) is that there are a range of values of $\gamma_{s,d}$ (those that correspond to relative high SNR) for which it is better not to use cooperation and a range of values for which it is better to use cooperation. The same analysis that leads to (14.41) could be carried out for DF cooperation and reach the same conclusion.

We explore further this issue by studing the ratio D_C/D_N between distortion with and without cooperation. Figures 14.15 and 14.16 show results for source–relay channels that are classified as "good" or "bad," respectively, for most operating modes in DF cooperation. These figures also explore the effects of the relative strength of the channel

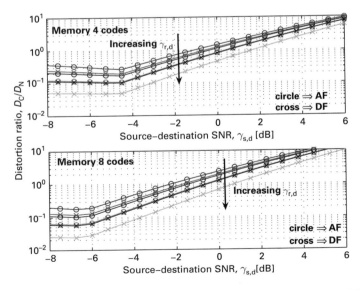

Fig. 14.15 **Fig. 14.15**   Ratio between distortions with and without cooperation when $\gamma_{s,r} = 5$ dB and $\gamma_{r,d} = -2$ dB, $\gamma_{r,d} = 1$ dB and $\gamma_{r,d} = 4$ dB.

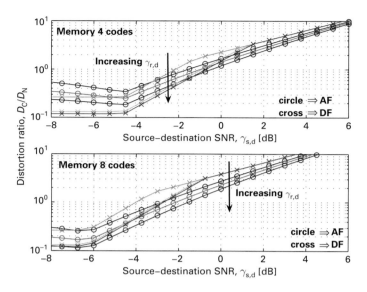

Fig. 14.16 Ratio between distortions with and without cooperation when $\gamma_{s,r} = 1$ dB and $\gamma_{r,d} = -2$ dB, $\gamma_{r,d} = 1$ dB and $\gamma_{r,d} = 4$ dB.

code family by comparing results for the memory 4 family of RCPC codes, to those obtained for a memory 8 RCPC codes from [41] (for these codes we have $\bar{a} = 7.1$, $\kappa = 45$ and $c = 2.8$). The results confirm that non-cooperation is better than cooperation at large $\gamma_{s,d}$, i.e. D_C/D_N. This relation is inverted at low $\gamma_{s,d}$. As it is to expect, the value of $\gamma_{s,d}$ for which cooperation is better increases with $\gamma_{s,r}$. Also, in the case of AF cooperation we note that there is a value of $\gamma_{s,d}$ for which the ratio is minimum (cooperation yields best performance gain). This value does not depend on $\gamma_{s,r}$, $\gamma_{r,d}$ or the type of cooperation, it only depends on the channel code family used and is close to the value of $\gamma_{s,d}$ for which the non-cooperative distortion reaches 1. Comparison between AF and DF cooperation shows that the later is better for all values of source–destination and relay–destination channel SNRs when the source–relay channel is "good." In the case of a "bad" source–relay channel, DF cooperation outperforms AF only at low $\gamma_{s,d}$, and this advantage can be overcome by choosing a family of channel codes that is strong enough.

Among other results, Figures 14.15 and 14.16 also show the effects on performance due to a change in the channel codec. We now briefly consider the influence that a change in source codec efficiency has on the results. In practice, different source codecs exhibit different compression efficiency, i.e. different source codecs achieve the same distortion values at different encoding rates. Thus, the efficiency of the source codec can be incorporated into the formulation by writing the D-R function as

$$D_S(R_S) = 2^{-2(1-\lambda)R_S} = 2^{-\hat{\lambda}R_S},$$

where λ determines the coding inefficiency and $\hat{\lambda} \triangleq 2(1-\lambda)$. By modifying the previous results accordingly, we have that (14.24), (14.25), (14.32), and (14.33) become

$$m_{\mathrm{c}} = \left[20 \left(\frac{cN}{W\hat{\lambda}\ln(2)} + \frac{1}{2\Psi + 2\hat{\lambda}\ln(2)\bar{R}_{\mathrm{SN}}} \right) \right]^{-1},$$

$$G_{\mathrm{c}} = \frac{\kappa(1+\Delta)^{-1/(10m_{\mathrm{c}})}}{\Psi + \hat{\lambda}\bar{R}_{\mathrm{SN}}\ln(2)} 10^{\frac{\hat{\lambda}\log(2)\bar{R}_{\mathrm{SN}}}{\Psi + \hat{\lambda}\bar{R}_{\mathrm{SN}}\ln(2)}},$$

$$m_{\mathrm{N}} = \left[10 \left(\frac{cN}{W\hat{\lambda}\ln(2)} + \frac{1}{\Psi + \ln(2)(1+2\hat{\lambda}\bar{R}_{\mathrm{SN}})} \right) \right]^{-1},$$

$$G_{\mathrm{N}} = \frac{\kappa(1+\Delta)^{-1/(10m_{\mathrm{N}})}}{\Psi + \ln(2)(2\hat{\lambda}\bar{R}_{\mathrm{SN}}+1)} 10^{\frac{\hat{\lambda}\log(2)\bar{R}_{\mathrm{SN}}}{\Psi + \ln(2)(2\hat{\lambda}\bar{R}_{\mathrm{SN}}+1)}}.$$

These equations show that m_{c} and m_{N} change with $\hat{\lambda}$ approximately in a linear fashion. Thus, the relation $m_{\mathrm{N}}/m_{\mathrm{c}} \approx 2$ is maintained. Figures 14.17 and 14.18 show the ratio $D_{\mathrm{C}}/D_{\mathrm{N}}$ when $\lambda = 0.2$. It can be seen that in the case of a "bad" source–relay channel, the increased inefficiency of the source codec translates into an increase in the range of values of $\gamma_{\mathrm{s,d}}$ for which no cooperation outperforms AF cooperation. This effect is compensated by using stronger channel coding. Also, it can be seen that DF cooperation now outperforms AF cooperation for a larger range of values of $\gamma_{\mathrm{s,d}}$. In the case of a "good" source–relay channel, DF cooperation is the cooperative scheme that is outperformed by no cooperation over a larger range of $\gamma_{\mathrm{s,d}}$. Also, we can see that the results show a reduced sensitivity to the value of $\gamma_{\mathrm{r,d}}$.

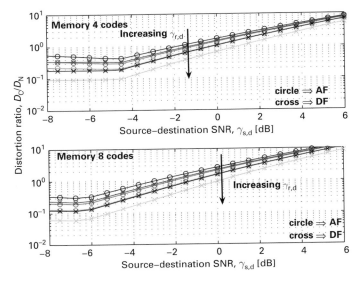

Fig. 14.17   Effect of source codec efficiency on the ratio between distortions with and without cooperation when $\lambda = 0.2$, $\gamma_{\mathrm{s,r}} = 5\,\mathrm{dB}$ and $\gamma_{\mathrm{r,d}} = -2\,\mathrm{dB}$, $\gamma_{\mathrm{r,d}} = 1\,\mathrm{dB}$ and $\gamma_{\mathrm{r,d}} = 4\,\mathrm{dB}$.

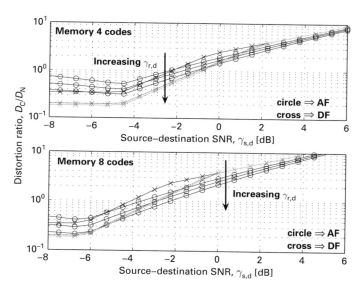

Fig. 14.18   Effect of source codec efficiency on the ratio between distortions with and without cooperation when $\lambda = 0.2$, $\gamma_{s,r} = 1\,dB$ and $\gamma_{r,d} = -2\,dB$, $\gamma_{r,d} = 1\,dB$ and $\gamma_{r,d} = 4\,dB$.

14.9   Chapter summary and bibliographical notes

In this chapter, we have studied the effects that the tradeoffs among source coding, channel coding, and use of cooperation have on the performance of multimedia communication systems. We considered practical source and channel codecs and for user cooperation we considered amplify-and-forward and decode-and-forward schemes. As such, the topics treated in this chapter are very much based on the theory of source and channel coding. Being source and channel coding two of the most important parts in the process of communication, their treatment could encompass several books. A classic book covering many topics in source coding, including the OPTA (optimum performance theoretically attainable) curve, is [10]. In addition, source coding is a topic covered in most information theory text books, of which, the book by Cover and Thomas [26] is a good source. On the channel coding side, general theory can be found in information theory text books and in specialized error correcting codes books such as [228]. The specifics details of RCPC codes can be found in [56].

As was discussed at the start of this chapter, the problem of whether to jointly design the source and channel coding stages considering end-to-end performance was first studied in Shannon's separation theorem [181, 182]. This theorem states the conditions under which it is optimal to design the source and channel codecs as two independent components. Since these conditions do not hold in many practical designs, JSCC solutions for these cases had been the subject of much research. In terms of studying the achievable performance, in [68, 69] bounds were developed by resorting to high and low SNR approximations, error exponents, asymptotically large source code dimension, and infinite complexity and delay. In practice, there are many approaches to the joint design of source and channel coding. Hence, there are many JSCC techniques, encompassing digital, analog and hybrid digital-analog (HDA) implementations. Among the not all-digital techniques, many are built on the idea of designing mappings from the source space into the channel space. While for the analog JSCC techniques the mapping is direct [61], in HDA techniques there is a quantization step that is applied to the source samples [42, 148, 113, 27].

With respect to all the digital techniques, many are reviewed in [51, 30]. Integrated source–channel coders (those where the source and channel codecs are two separate units working in tandem and designed jointly) are studied in [31], and specific variations of this technique are studied in [37] for channel optimized quantization, [238] for index assignment, [194] for unequal error protection and [139] for joint bit rate allocation. In this chapter we have focused on the joint source–channel coding bit rate allocation technique because is one of the most practical and straightforward to implement. There are many examples in the research literature where JSCC bit rate allocation has been applied to transmission of multimedia sources. For illustrative purposes, we highlight here only the representative works [48, 221] for speech transmission in mobile channels, [134] for image transmission and [19, 197, 98] for video communication. Also, there are many works presenting different design methods for different problem setups, as is the case in [227, 183, 118, 102], which deals with iterative solutions, or [20], which presents a design based on exhaustive search over reduced spaces.

The study in this chapter is based on measuring performance using the D-SNR curve. This curve represents the relation between end-to-end distortion and channel SNR. We saw that the D-SNR curve can be accurately approximated as a linear function in a log–log scale. To show this, we used an approach that considers practical source and channel codes and their settings, while it avoids resorting to high SNR asymptotic analysis. While further study of the D-SNR curve can be found in [106] and [100], the extension of the theory to consider cooperation can be found in [107].

We have also seen that the schemes using cooperation have better coding gain (due to the better error performance when using cooperation) but they show a decrease of distortion at approximately half the rate of schemes not using cooperation (due to the sacrifice in bandwidth efficiency). The overall effect is that non-cooperative schemes have better performance at high source–destination SNR, while AF or DF coopera-tive schemes have better performance in the rest of cases. Further analysis showed that the best performance among cooperative schemes is achieved with DF coopera-tion in most of the cases but, when the source–relay channel is bad the performance advantage of DF cooperation is reduced by choosing a channel code family sufficiently strong.

In addition, in this chapter we have studied the effects that the efficiency of the source codec has on system performance and we showed that it reduces the diversity gain proportionally to the codec loss of efficiency. Also, we saw that for a "bad" source–relay channel, AF cooperation is outperformed by DF cooperation and no cooperation over a larger range of channel SNRs but this effect could be compensated by choosing a stronger family of channel codes. Similar observations apply to DF cooperation in the case of a "good" source–relay channel. Further study of the interaction between practical source and channel coding with cooperation for multimedia communications can be found in [104] and [105].

Exercises

14.1 In this problem we explore the dependence of the diversity and coding gain on the choice of the parameter Δ. We do so in two steps.

(a) We first study how a small charge in Δ affects the calculation of m_c and G_{cdB}. Formally, this is measured by calculating the sensitivity of m_c and G_{cdB} with respect to Δ, which is respectively defined as

$$S_\Delta^{m_c} = \frac{dm_c/m_c}{d\Delta/\Delta}$$

$$S_\Delta^{G_{cdB}} = \frac{dG_{cdB}/G_{cdB}}{d\Delta/\Delta}$$

Find the expression for the sensitivity of m_c and G_{cdB} with respect to Δ and plot these results for Δ between 0.05 and 0.2. What conclusions can be derived? To plot the results use the same setup as in Example 14.1 on page 501, i.e., the source encoder has a D-R performance as in (14.3), the variable-rate error protection is implemented with the family of memory 4, puncturing period 8, mother code rate $1/4$, RCPC codes from [56] ($\bar{a} = 6.1$, $\kappa = 30$, and $c = 3$), $N = 150$ samples, and $W = 950$ bits per transmission period.

(b) We now study how relatively large charges in Δ (while still remaining a small number) affects the calculation of m_c and G_{cdB}. Taking the setup from example 14.1, in page 501, plot the characterization of the D-SNR curve for AF cooperation, as in 14.23, for $\Delta = 0.05$, $\Delta = 0.1$, and $\Delta = 0.2$. What conclusions can be derived?

14.2 Derive the expressions for m_N and G_N in (14.32) and (14.33).

14.3 Derive the expression for the source encoding rate associated with the asymptotic distortion in decode-and-forward given by (14.29).

14.4 In this problem we study how channel encoding affects the performance of a variation of a decode-and-forward scheme. Consider a decode-and-forward where two transmitting nodes are paired-up. Phase 1 of the scheme remains uncharged from the scheme used in this chapter. In phase 2 a node will forward the message from the partner only if its reception was correct, otherwise it will send a copy of its own transmitted message. Both paired nodes will symmetrically operate with the same protocol. Consider that variable source and channel coding is still used but now the performance measure of interest is the block error rate after channel decoding.

(a) What would be the advantage of this DF scheme instead of the one considered in this chapter?

(b) Derive the expression for the probability of having a source frame with errors after channel decoding.

(c) Assume that the variable-rate error protection is implemented with the family of memory 4, puncturing period 8, mother code rate $1/4$, RCPC codes from [56] ($\bar{a} = 6.1$, $\kappa = 30$, and $c = 3$), $N = 150$ samples, and $W = 950$ bits per transmission period. Applying the modeling equations used during the analysis in this chapter, plot the probability of having a source frame with errors after channel decoding as a function of the channel SNR for $-4\,\text{dB} \le \gamma_{s,d} \le 4\,\text{dB}$ and

  (i) $\gamma_{s,r} = 6\,\text{dB}$, $\gamma_{r,d} = 0\,\text{dB}$.
  (ii) $\gamma_{s,r} = 3\,\text{dB}$, $\gamma_{r,d} = 0\,\text{dB}$.
  (iii) $\gamma_{s,r} = 0\,\text{dB}$, $\gamma_{r,d} = 0\,\text{dB}$.

  Explain the differences in the results and draw conclusions.

15 Asymptotic performance of distortion exponents

In Chapter 13 we argued that cooperation can benefit the layers of the communication stack that are above the physical layer, where the idea was approached through the study of distributed cooperative routing. In Chapter 14 cooperation was studied as combined with source and channel coding, which extends the use of cooperation with the higher application layer. In fact, when considering that the benefits of cooperation come from the degree of diversity it provides, it is important to also recognize that diversity is not exclusive to implementations at the physical layer. Diversity can also be formed when multiple channels are provided to the application layer, where they are exploited through the use of multiple description source encoders. In *multiple description coding* different descriptions of the source are generated with the property that they can each be individually decoded or, if possible, be jointly decoded to obtain a reconstruction of the source with lower distortion. More importantly, a multiple description stream provides diversity that can be exploited by sending each description through an independent channel. This form of diversity has been called source coding diversity. Similarly, if we consider channel coding instead of source coding as the originator of diversity, we would be generating channel coding diversity. This chapter focuses on studying systems that exhibit three forms of diversity: source coding diversity, channel coding diversity, and cooperation. More specifically, in this chapter we consider how cooperative diversity interacts with source and channel coding diversity in its traditional form (when compared with Chapter 14)

We will be considering source coding diversity through the particular form of multiple description coding where the number of descriptions is two. This form of multiple description coding is known as dual description coding and, because it is the simplest form of multiple description coding, it is by far the best understood. On the side of user-cooperation diversity, we will consider both relay channels and multi-hop channels, with a single or M relays helping the source by repeating its information either using the amplify-and-forward or the decode-and-forward protocols. As understood in this chapter, the differentiation between relay channels and multi-hop channels means whether there is or not a direct communication path between the source and the destination, a system design variable that is worth considering in this study.

Given the presence of fading channels, we will study the system behavior at large SNR where system performances can be compared in terms of the distortion exponent, which measures the rate of decay of the end-to-end distortion at high SNR. Based on this, the goal will be to study the tradeoff between the diversity gain (which will be directly related to the number of relays) and the quality of the source encoder.

15.1   Systems setup for source–channel diversity

Let us focus on systems that communicate a source signal over a wireless multi-hop or relay channel. Let the input to the system be a memoryless sourcer. Let us also assume that communication is performed over a complex, additive white Gaussian noise (AWGN) fading channel. Denoting by I the maximum average mutual information between the channel input and output for a given channel realization, from Section 1.2

$$I = \log(1 + |h|^2 \mathrm{SNR}), \tag{15.1}$$

where h is the fading value. Because of the random nature of the fading, I and the ability of the channel to support transmission at some rate are themselves random. Recall that the probability of the channel not being able to support a rate R is called the outage probability and was given in (1.29), or in short form by $P_0 = \Pr[I < R]$.

It will be convenient for us to work with the random function e^I. At large SNR, e^I from (15.1) can be simplified to

$$e^I \approx |h|^2 \mathrm{SNR}. \tag{15.2}$$

This implies that the random function e^I will have at high SNR approximately the same statistics as $|h|^2$ (although the SNR scaling will trivially affect them). The study in this chapter will frequently make use of the cumulative distribution function (cdf) for e^I. In particular, at high SNR e^I will have approximately the same cumulative distribution function (cdf) as $|h|^2$, e.g., if the channel is Rayleigh, e^I will have the cdf of an exponential random variable. Furthermore, with the goal of using a sufficiently general model for the cdf of e^I (or $|h|^2$) beyond the Rayleigh case, we will assume that the cdf of $|h|^2$ is smooth enough that it can be approximated through the first term of its Taylor series around the origin. In other words, we will assume that the cdf, F_{e^I}, of e^I can be approximated at high SNR as

$$F_{e^I}(t) \approx c \left(\frac{t}{\mathrm{SNR}} \right)^p. \tag{15.3}$$

Both c and p are model-dependent parameters. The parameter c can be seen as a normalizing constant, which will play a rather minor role in the study in this chapter. The parameter p is the order of the first nonzero derivative in the Taylor series. For the case of Rayleigh fading we have $p = 1$. Other values of p may be used to consider other distributions suitable for modeling $|h|^2$, such as the Nakagami distribution.

Let us consider a communication system, as shown in Figure 15.1, consisting of a source, a source encoder and a channel encoder. Let the input to the system be a memoryless sourcer. The source samples are fed into the source encoder for quantization and compression. The output of the source encoder are fed into a channel encoder which

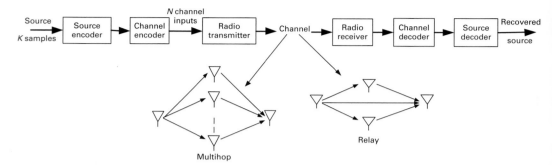

Fig. 15.1   Simplified block diagram of the system studied in this chapter.

outputs N channel inputs. For K source samples and N channel inputs, we denote by $\beta \triangleq N/K$, the bandwidth expansion factor or processing gain. We assume that K is large enough to average over the statistics of the source but N is not sufficiently large to average over the statistics of the channel, i.e., we assume a block fading wireless channel.

In this chapter, we are specifically interested in systems where the source signal average end-to-end distortion is the figure of merit. Thus, performance will be measured in terms of the expected distortion

$$E[D] = E[d(\mathbf{s}, \hat{\mathbf{s}})],$$

where

$$d(\mathbf{s}, \hat{\mathbf{s}}) = \frac{1}{K} \sum_{k=1}^{K} d(s_k, \hat{s}_k)$$

is the average distortion between a sequence $\mathbf{s}$ of K samples and its corresponding reconstruction $\hat{\mathbf{s}}$ and $d(s_k, \hat{s}_k)$ is the distortion between a single sample s_k and its reconstruction $\hat{s}_k$. We will assume $d(s_k, \hat{s}_k)$ to be the mean-squared distortion measure.

Following the fading channels assumption, we will be interested in studying the system behavior at large channel SNRs where system performances can be compared in terms of the rate of decay of the end-to-end distortion. This figure of merit called the *distortion exponent* is defined as

$$\Delta \triangleq - \lim_{\text{SNR} \to \infty} \frac{\log E[D]}{\log \text{SNR}}. \tag{15.4}$$

We will consider two types of source encoders: a *single description* (SD) and a *dual description* source encoder. The SD codec corresponds to the cases of source codecs found in earlier chapters. In an SD encoder, a source is encoded into one stream or description. A dual description code is a special case of a *multiple description* (MD) source codec, where the encoder generates two coded descriptions of the source (Figure 15.2). Multiple description codecs have the property that each of the source descriptions can be decoded independently of each other, thus obtaining separate representations of the source. Even more, at the MD decoder, the descriptions can also be combined together to obtain a representation of the source of better quality than the

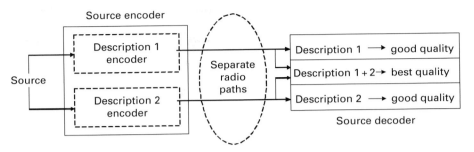

Fig. 15.2 Simplified diagram of a dual description source codec as a special case of a multiple description source codec.

representations provided by each separate description. In a communications application, MD codecs can be used in such a way that as long as any one description can be received error-free, it is possible to obtain a representation of the source. Now, if more than one description can be received error-free then these descriptions can be combined and decoded together to obtain a representation of the source with better quality. As a simple example of a MD codec, consider a video codec that generates two coded streams: one from the odd numbered frames and one from the even numbered frames. At the receiver, it is possible to play either one of the stream but when the two streams are available with no errors, playing them interleaved will result in a better quality video.

The performance of source encoders can be measured through its achievable rate-distortion (R-D) function, which characterizes the tradeoff between source encoding rate and distortion. As discussed in Section 14.4, the R-D function for SD source encoders is frequently considered to be of the form

$$R = \frac{1}{c_2} \log \left(\frac{c_1}{D} \right),$$

where, for convenience, we are taking now the logarithm with base e and hence, R, the source encoding rate, is measured in nats per channel use. This form of R-D function can approximate or bound a wide range of practical systems such as video coding with an MPEG codec, speech using a CELP-type codec, or when the high rate approximation holds. Assuming that high resolution approximation can be applied to the source encoding operation, each of the input samples can be modeled as a memoryless Gaussian source, showing a zero-mean, unit-variance Gaussian distribution. In this case, the R-D function can be approximated without loss of generality, as

$$R = \frac{1}{2\beta} \log \left(\frac{1}{D} \right). \tag{15.5}$$

For MD source encoders, the R-D region is only known for the dual description source encoders. Let R_1 and R_2 be the source encoding rates of descriptions 1 and 2, respectively, and $R_{md} = R_1 + R_2$. Let D_1 and D_2 be the reconstructed distortions associated with descriptions 1 and 2, respectively, when each is decoded alone. Let D_0 be the source distortion when both description are combined and jointly decoded. For the same source model and assumptions as in the single description case, R_1 and D_1, and R_2 and D_2 are related through,

$$R_1 = \frac{1}{2\beta} \log\left(\frac{1}{D_1}\right), \quad R_2 = \frac{1}{2\beta} \log\left(\frac{1}{D_2}\right). \tag{15.6}$$

The R-D function when both descriptions can be combined at the source decoder differs depending on whether distortions can be considered low or high [32]. The low distortion scenario corresponds to $D_1 + D_2 - D_0 < 1$, in which case we have,

$$R_{md} = \frac{1}{2\beta} \log\left(\frac{1}{D_0}\right)$$
$$+ \frac{1}{2\beta} \log\left(\frac{(1-D_0)^2}{(1-D_0)^2 - \left[\sqrt{(1-D_1)(1-D_2)} - \sqrt{(D_1-D_0)(D_2-D_0)}\right]^2}\right). \tag{15.7}$$

All the schemes we will consider in this work present the same communication conditions to each description. Therefore, it will be reasonable to assume $R_1 = R_2 = R_{md}/2$. Under this condition, it is possible to derive from (15.7) the bounds

$$(4D_0 D_1)^{-1/(2\beta)} \lesssim e^{R_{md}} \lesssim (2D_0 D_1)^{-1/(2\beta)}, \tag{15.8}$$

where the lower bound requires $D_0 \to 0$ and the upper bound requires also $D_1 \to 0$.

In the case of the high distortion scenario, $D_1 + D_2 - D_0 > 1$, the R-D function equals

$$R_{md} = \frac{1}{2\beta} \log\left(\frac{1}{D_0}\right), \tag{15.9}$$

because in this scenario there is no penalty due to representing the source into two streams with dual-description properties.

The channel-encoded message is then sent from the source node to a destination node with or without user cooperation. In a setup with user cooperation, the relay nodes are associated with the source node to achieve user-cooperation diversity as has been explained earlier. Communication in a cooperative setup with one relay node takes place in two phases. In phase 1, the source node sends information to its destination node. In phase 2, the relay node cooperates by forwarding to the destination the information received from its associated source node. At the destination node, both signals received from the source and the relay are combined and detected.

For each additional relay used during transmission, a new phase, similar to phase 2, is added to allow transmission of the new relay. Because of this multi-phase transmission and for fair comparison of the different schemes that will be consider, we need to fix the total number of channel uses for a source block of size K and change the bandwidth expansion factor accordingly to each scheme, as will be seen in detail later in this chapter. We will consider two techniques that implement user cooperation, amplify-and-forward and decode-and-forward, each differing in the processing done at the relay. Recall that in amplify-and-forward, the relay retransmits the source's signal without further processing other than power amplification. In decode-and-forward, the relay first decodes the message from the source. If the decoded message has no error, the relay re-encodes it and transmits a copy. If the relay fails to decode the message, it idles until the next is received.

15.2 Multi-hop channels

In this section, we consider the distortion exponents of multi-hop networks using amplify-and-forward and decode-and-forward user-cooperation protocols. By multi-hop channel we mean it is a channel where there is no direct path between the source and destination; i.e., the information path between source and destination contains one or more relaying nodes. Without loss of generality we consider the two-hop case. The analysis can be easily extended to scenarios with larger number of hops.

15.2.1 Multi-hop amplify-and-forward protocol

In this section, we will consider the analysis for multi-hop amplify-and-forward schemes with different channel and source coding diversity achieving schemes. We derive the distortion exponent for the two-hop single relay channel with an SD source encoder and extend the result to the case of M relays with repetition channel coding diversity.

15.2.1.1 Single relay

The system under consideration consists of a source, a relay and a destination as shown in Figure 15.3. Transmission of a message is done in two phases. In phase 1, the source sends its information to the relay node. The received signal at the relay node is given by

$$y_{r_1} = h_{s,r_1}\sqrt{P}x_s + n_{s,r_1},\qquad(15.10)$$

where h_{s,r_1} is the channel gain between the source and the relay node, x_s is the transmitted source symbol, P is the source transmit power with $E[\|x_s\|^2] = 1$, and n_{s,r_1} is the

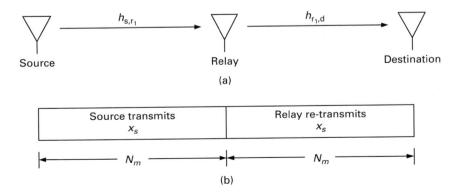

Fig. 15.3 Two-hop single relay system: (a) system model, (b) time frame structure.

noise at the relay node modeled, as zero-mean, circularly symmetric complex Gaussian noise with variance $N_0/2$ per dimension. In phase 2, the relay normalizes the received signal by the factor

$$\alpha_1 \leq \sqrt{\frac{P}{P|h_{s,r_1}|^2 + N_0}},$$

as discussed in previous chapters, and retransmits to the destination. The received signal at the destination is given by

$$y_d = h_{r_1,d}\alpha_1 y_{r_1} + n_{r_1,d}$$
$$= h_{r_1,d}\alpha_1 h_{s,r_1}\sqrt{P}x_s + h_{r_1,d}\alpha_1 n_{s,r_1} + n_{r_1,d}, \qquad (15.11)$$

where $n_{r_1,d}$ is the noise at the destination node and is modeled as zero-mean, circularly symmetric complex Gaussian noise with variance $N_0/2$ per dimension. Following (4.17), the mutual information in this case is given by

$$I(x_s, y_d) = \log\left(1 + \frac{|h_{s,r_1}|^2\mathrm{SNR}|h_{r_1,d}|^2\mathrm{SNR}}{|h_{s,r_1}|^2\mathrm{SNR} + |h_{r_1,d}|^2\mathrm{SNR} + 1}\right), \qquad (15.12)$$

where $\mathrm{SNR} = P/N_0$. At high SNR, we have

$$I(x_s, y_d) \approx \log\left(1 + \frac{|h_{s,r_1}|^2\mathrm{SNR}|h_{r_1,d}|^2\mathrm{SNR}}{|h_{s,r_1}|^2\mathrm{SNR} + |h_{r_1,d}|^2\mathrm{SNR}}\right)$$

$$\approx \log\left(\frac{|h_{s,r_1}|^2\mathrm{SNR}|h_{r_1,d}|^2\mathrm{SNR}}{|h_{s,r_1}|^2\mathrm{SNR} + |h_{r_1,d}|^2\mathrm{SNR}}\right). \qquad (15.13)$$

Equation (15.13) indicates that the two-hop amplify-and-forward channel appears as a link with signal-to-noise ratio that is the harmonic mean of the source–relay and relay–destination channels signal-to-noise ratios. To calculate the distortion exponent, consider the source–relay and relay–destination channels for which the maximum average mutual information is, respectively,

$$I_1 = \log(1 + |h_{s,r_1}|^2\mathrm{SNR}),$$
$$I_2 = \log(1 + |h_{r_1,d}|^2\mathrm{SNR}).$$

At high SNR we can write

$$\exp(I_1) \approx |h_{s,r_1}|^2\mathrm{SNR},$$
$$\exp(I_2) \approx |h_{r_1,d}|^2\mathrm{SNR}.$$

Denoting

$$Z_1 = |h_{s,r_1}|^2\mathrm{SNR},$$
$$Z_2 = |h_{r_1,d}|^2\mathrm{SNR},$$

and assuming statistical symmetry between the source–relay and relay–destination channels, we have

$$F_{Z_1}(t) \approx c \left(\frac{t}{\text{SNR}} \right)^p$$
$$F_{Z_2}(t) \approx c \left(\frac{t}{\text{SNR}} \right)^p, \tag{15.14}$$

where $F_{Z_1}(.)$ and $F_{Z_2}(.)$ are the cdf of Z_1 and Z_2, respectively.

The harmonic mean of two nonnegative random variables can be upper and lower bounded as

$$\frac{1}{2} \min(Z_1, Z_2) \leq \frac{Z_1 Z_2}{Z_1 + Z_2} \leq \min(Z_1, Z_2), \tag{15.15}$$

where the lower bound is achieved if and only if $Z_1 = Z_2$, $Z_1 = 0$ or $Z_2 = 0$, and the upper bound is achieved if and only if $Z_1 = 0$ or $Z_2 = 0$.

Define the random variable

$$Z = \frac{Z_1 Z_2}{Z_1 + Z_2},$$

from (15.15) we have

$$\Pr[\min(Z_1, Z_2) < t] < \Pr[Z < t] \leq \Pr[\min(Z_1, Z_2) < 2t]. \tag{15.16}$$

Then we have

$$\Pr[\min(Z_1, Z_2) < t] = 2F_{Z_1}(t) - \left(F_{Z_1}(t) \right)^2$$
$$\approx 2c \left(\frac{t}{\text{SNR}} \right)^p - c^2 \left(\frac{t}{\text{SNR}} \right)^{2p} \tag{15.17}$$
$$\approx c_1 \left(\frac{t}{\text{SNR}} \right)^p,$$

where $c_1 = 2c$. Similarly, we have

$$\Pr[\min(Z_1, Z_2) < 2t] \approx c_2 \left(\frac{t}{\text{SNR}} \right)^p, \tag{15.18}$$

where $c_2 = 2^{p+1} c$. From (15.17) and (15.18) we get

$$c_1 \left(\frac{t}{\text{SNR}} \right)^p \lessgtr F_Z(t) \lessgtr c_2 \left(\frac{t}{\text{SNR}} \right)^p, \tag{15.19}$$

where $F_Z(t)$ is the cdf of the random variable Z. The minimum expected end-to-end distortion can now be computed as

$$E[D] = \min_D \left\{ \Pr[I(x_s, y_d) < R(D)] + D \Pr[I(x_s, y_d) \geq R(D)] \right\}, \tag{15.20}$$

where D is the source encoder distortion and R is the source encoding rate. Note that (15.20) implicitly assumes that in case of an outage the missing source data is concealed by replacing the missing source samples with their expected value (equal to zero). Using the bounds in (15.19) the minimum expected distortion can be upper and lower bounded as

$$
\min_D \left\{ c_1 \left(\frac{\exp(R(D))}{\text{SNR}} \right)^p + \left[1 - c_2 \left(\frac{\exp(R(D))}{\text{SNR}} \right)^p \right] D \right\}
$$

$$
\lesssim E[D] \lesssim \min_D \left\{ c_2 \left(\frac{\exp(R(D))}{\text{SNR}} \right)^p + \left[1 - c_1 \left(\frac{\exp(R(D))}{\text{SNR}} \right)^p \right] D \right\}.
$$
(15.21)

For sufficiently large SNRs, we have

$$
\min_D \left\{ c_1 \left(\frac{\exp(R(D))}{\text{SNR}} \right)^p + D \right\} \lesssim E[D] \lesssim \min_D \left\{ c_2 \left(\frac{\exp(R(D))}{\text{SNR}} \right)^p + D \right\}. \quad (15.22)
$$

From (15.5), $\exp(R(D)) = D^{-1/2\beta_m}$, where $\beta_m = N_m/K$ as illustrated in Figure 15.3, which leads to

$$
\min_D c_1 \frac{D^{\frac{-p}{2\beta_m}}}{\text{SNR}^p} + D \lesssim E[D] \lesssim \min_D c_2 \frac{D^{\frac{-p}{2\beta_m}}}{\text{SNR}^p} + D. \quad (15.23)
$$

Differentiating the lower bound and setting equal to zero we get the optimizing distortion

$$
D^* = \left(\frac{2\beta_m}{c_1 p} \right)^{\frac{-2\beta_m}{2\beta_m + p}} \text{SNR}^{\frac{-2\beta_m p}{2\beta_m + p}}. \quad (15.24)
$$

Substituting from (15.24) into (15.23) we get

$$
C_{\text{LB}} \, \text{SNR}^{\frac{-2\beta_m p}{2\beta_m + p}} \lesssim E[D] \lesssim C_{\text{UB}} \, \text{SNR}^{\frac{-2\beta_m p}{2\beta_m + p}}, \quad (15.25)
$$

where C_{LB} and C_{UB} are terms that are independent of the SNR.

The distortion exponent is now given by the following theorem.

THEOREM 15.2.1   *The distortion exponent of the two-hop single-relay amplify-and-forward protocol is*

$$
\Delta_{\text{SH-1R-AMP}} = \frac{2p\beta_m}{p + 2\beta_m}, \quad (15.26)
$$

where $\beta_m = N_m/K$, and N_m is the number of the source channel uses and K is the number of source samples.

Before continuing, we note here that in the sequel, for simplicity of presentation we will use

$$
F_Z(t) \approx \acute{c} \left(\frac{t}{\text{SNR}} \right)^p, \quad (15.27)
$$

where Z is the harmonic mean of the source–relay and relay–destination signal-to-noise ratios and $\acute{c}$ is a constant. The basis for the simplification is that the analysis will not be affected by this substitution as we can always apply the analysis presented here by forming upper and lower bounds on the expected distortion and this will yield the same distortion exponent.

15.2.1.2 Multiple relays

We consider now a system consisting of a source, M relay nodes and a destination as shown in Figure 15.4. The M relay nodes amplify the received signals from the source and then retransmit to the destination. The destination selects the signal of the highest quality (highest SNR) to recover the source signal. The distortion exponent of this system is given by the following theorem.

THEOREM 15.2.2 *The distortion exponent of the two-hop M-relay selection channel coding diversity amplify-and-forward protocol is*

$$\Delta_{\text{SH-MR-AMP}} = \frac{4M p \beta_m}{M(M+1)p + 4\beta_m}. \qquad (15.28)$$

Proof Let y_{d_i} be the signal received at the destination due to the i-th relays transmission. At sufficiently high SNR, the mutual information between x_s and y_{d_i} is given by

$$I(x_s, y_{d_i}) \approx \log\left(\frac{|h_{s,r_i}|^2 \text{SNR} |h_{r_i,d}|^2 \text{SNR}}{|h_{s,r_i}|^2 \text{SNR} + |h_{r_i,d}|^2 \text{SNR}} \right), \quad i = 1, 2, \ldots, M.$$

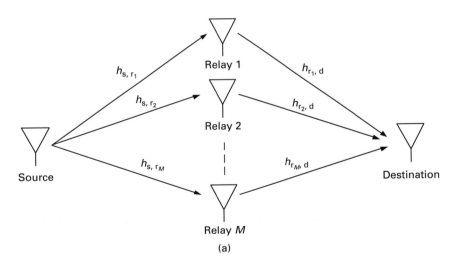

(a)

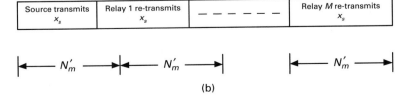

(b)

Fig. 15.4 Two-hop M-relay system: (a) system model, (b) time frame structure.

Define the random variable

$$W_i = \frac{|h_{s,r_i}|^2 \text{SNR} |h_{r_i,d}|^2 \text{SNR}}{|h_{s,r_i}|^2 \text{SNR} + |h_{r_i,d}|^2 \text{SNR}}, \quad i = 1, 2, \ldots, M.$$

The cdf of W_i can be approximated at high SNR as

$$F_{W_i}(t) \approx \acute{c} \left(\frac{t}{\text{SNR}} \right)^p. \tag{15.29}$$

The minimum end-to-end expected distortion can be computed as

$$
\begin{aligned}
\text{E}[D] = \min_D \Bigg\{ & \Pr\left[\max(I(x_s, y_{d_1}), I(x_s, y_{d_2}), \ldots, I(x_s, y_{d_M})) < R(D) \right] \\
& + \Pr\left[\max(I(x_s, y_{d_1}), I(x_s, y_{d_2}), \ldots, I(x_s, y_{d_M})) \geq R(D) \right] D \Bigg\} \\
= \min_D & \Bigg\{ \prod_{i=1}^{M} F_{W_i}(\exp(R(D))) + \left[1 - \prod_{i=1}^{M} F_{W_i}(\exp(R(D))) \right] D \Bigg\} \\
\approx \min_D & \Bigg\{ \acute{c}^M \frac{D^{\frac{-Mp}{2\beta'_m}}}{\text{SNR}^{Mp}} + \left[1 - \acute{c}^M \frac{D^{\frac{-Mp}{2\beta'_m}}}{\text{SNR}^{Mp}} \right] D \Bigg\} \\
\approx \min_D & \Bigg\{ \acute{c}^M \frac{D^{\frac{-Mp}{2\beta'_m}}}{\text{SNR}^{Mp}} + D \Bigg\}, \tag{15.30}
\end{aligned}
$$

where D is the source encoding distortion, $\beta'_m = N'_m / K$, and N'_m is the number of the source channel uses, as illustrated in Figure 15.4. Differentiating and setting equal to zero we get the optimizing distortion

$$D^* = \left(\acute{c}^M \frac{Mp}{2\beta'_m} \right)^{\frac{2\beta'_m}{Mp + 2\beta'_m}} \text{SNR}^{\frac{-2M\beta'_m p}{2\beta'_m + Mp}}. \tag{15.31}$$

Substituting we get

$$\text{E}[D] \approx C_{\text{MR}} \, \text{SNR}^{\frac{-2M\beta'_m p}{2\beta'_m + Mp}}, \tag{15.32}$$

where C_{MR} is a term that does not depend on the SNR. Hence, the distortion exponent is given as

$$\Delta_{\text{SH-MR-AMP}} = \frac{2M\beta'_m p}{2\beta'_m + Mp}. \tag{15.33}$$

For fair comparison with the single relay case, we should compare the different systems under the same number of channel uses. So that we have $2N_m = (M+1)N'_m$, from which we have $\beta'_m = (2/M + 1)\beta_m$. Substituting in (15.33) we get

$$\Delta_{\text{SH-MR-AMP}} = \frac{4Mp\beta_m}{M(M+1)p + 4\beta_m}. \tag{15.34}$$

∎

The distortion exponent shows a tradeoff between the diversity and the source encoder performance. Increasing the number of relay nodes increases the diversity of the system at the expense of using lower rate source encoder, e.g., higher distortion under no outage.

Now let us calculate the optimal number of relays nodes to maximize the distortion exponent in (15.28). To get the optimal number of relays, M_{opt}, note that the distortion exponent in (15.28) can be easily shown to be concave in the number of relays (if we think of M as a continuous variable). Differentiating and setting equal to zero, we get

$$\frac{\partial}{\partial M}\Delta_{SH-MR-AMP} = 0 \longrightarrow M_{opt} = 2\sqrt{\frac{\beta_m}{p}}. \tag{15.35}$$

If M_{opt} in (15.35) is an integer number then it is the optimal number of relays. If M_{opt} in (15.35) is not an integer, substitute in (15.28) with the largest integer that is less than M_{opt} and the smallest integer that is greater than M_{opt} and choose the one that yields the higher distortion exponent as the optimum number of relay nodes. From the result in (15.35) it is clear that the number of relays decreases, for a fixed β_m, as p increases.

For higher channel quality (higher p) the system performance is limited by the distortion introduced by the source encoder in the absence of outage. Then, as p increases, the optimum number of relays decreases to allow for the use of a better source encoder with lower source encoding distortion. In this scenario, the system is said to be a quality limited system because the dominant phenomena in the end-to-end distortion is source encoding distortion and not outage.

Similarly as β_m increases (higher bandwidth), for a fixed p, the performance will be limited by the outage event rather than the source encoding distortion. As β_m increases, the optimum number of relays increases to achieve better outage performance. In this case, the system is said to be an outage limited system.

15.2.1.3 Optimal channel coding diversity with two relays

We now expand the study of performance in terms of distortion exponent by exploiting the cooperative diversity at higher layers of the communication stack. In this section, this idea translates into splitting the output of the channel coding stage into parts that are transmitted through a different relay node, i.e., a separate path. Let us consider now a system as shown in Figure 15.5 comprising a source, two relays and a destination. After channel encoding, the resulting block is split into two blocks: x_{s_1} and x_{s_2}, which are transmitted to the relay nodes. The first relay will only forward the block x_{s_1} and the second relay will only forward x_{s_2} as shown in Figure 15.5. From (15.13), it can be shown that the mutual information is

$$I \approx \log\left(1 + \frac{|h_{s,r_1}|^2 SNR |h_{r_1,d}|^2 SNR}{|h_{s,r_1}|^2 SNR + |h_{r_1,d}|^2 SNR}\right)$$

$$+ \log\left(1 + \frac{|h_{s,r_2}|^2 SNR |h_{r_2,d}|^2 SNR}{|h_{s,r_2}|^2 SNR + |h_{r_2,d}|^2 SNR}\right), \tag{15.36}$$

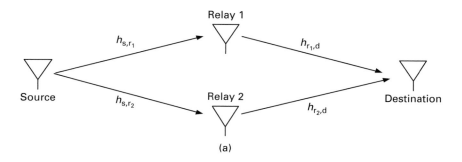

Fig. 15.5   Two-hop two-relay optimal channel coding diversity (source coding diversity) system: (a) system model, (b) time frame structure.

where x_{s_1} and x_{s_2} are independent zero-mean, circularly symmetric complex Gaussian random variables with variance $1/2$ per dimension. We can show that the distortion exponent of this system is given by the following theorem.

THEOREM 15.2.3   *The distortion exponent of the two-hop two-relay optimal channel coding diversity amplify-and-forward system is*

$$\Delta_{\text{SH}-2\text{R}-\text{OPTCH}-\text{AMP}} = \frac{2p\beta_m}{p + \beta_m}. \qquad (15.37)$$

Proof   Consider a setup where the physical layer presents to the upper layers two parallel and independent channels with the same statistical characteristics. The distortion exponent for the optimal channel coding diversity over two parallel channels can be written as

$$\Delta_{\text{SH}-2\text{R}-\text{OPTCH}-\text{AMP}} = \frac{4p\beta_m''}{p + 2\beta_m''}, \qquad (15.38)$$

Using (15.27) and (15.36) and considering $\beta_m'' = N_m''/K$ where N_m'' is the number of source channel uses for the x_{s_1} (x_{s_2}) block (refer to Figure 15.5) we get for our system the same distortion exponent as (15.38). For fair comparison with the previous schemes we should have $2N_m = 4N_m''$, which means that $\beta_m'' = 1/2\beta_m$. Finally, substituting this relation in (15.38) yields (15.37). ∎

In previous studies of multiplexed channel coding diversity the setting assumes that at the physical layer there exist a number of parallel channels, which can be exploited at higher layers to obtain diversity. The existence of the parallel channels is not related with the use of cooperation; which translates into the observation of a gain in the distortion exponent for the multiplexed channel coding diversity scheme (compared to the direct transmission) due to the increase in the bandwidth obtained from the simultaneous use of the parallel channels. When introducing cooperative diversity into the system setup, there is no gain in using multiplexed channel coding diversity because the overall bandwidth of the channel is kept fixed and needs to be distributed among the transmissions in the different cooperative phases, i.e., using either one relay or two relays does not increase the bandwidth of the channel. This is because only one node, either the source or a relay, is transmitting at a given time slot. The multiplexed channel coding diversity in this case is equivalent to allowing one relay helping the source to forward an SD source-coded message during one block and using the other relay for the next block. Hence, when using cooperative diversity, the multiplexed channel coding diversity is equivalent to the two-hop single-relay system with the same distortion exponent.

15.2.1.4 Source coding diversity with two relays

In the previous section we exploited cooperative diversity to transmit different parts of the output of the channel coding stage through different relay nodes. The same general idea can be extended to the source encoder, with the difference that now the source encoder is a dual description encoder, which implies that its output will have some properties that are not present in SD codecs and that can be used at the decoder to provide source coding diversity. Let us consider again a system with one source, two relays, and one destination nodes as shown in Figure 15.5. The source transmits two blocks x_{s_1} and x_{s_2} to the relay nodes. Each block represents one of the two descriptions generated by the dual descriptions source encoder. In this case, the two blocks are broken up before the channel encoder, that is, each description is fed to a different channel encoder. The first relay will only forward the block x_{s_1} and the second relay will only forward x_{s_2} as shown in Figure 15.5. The distortion exponent of this system is given by the following theorem.

THEOREM 15.2.4 *The distortion exponent of the two-hop two-relay source coding diversity amplify-and-forward protocol is*

$$\Delta_{\text{SH}-2\text{R}-\text{SRC}-\text{AMP}} = \max\left[\frac{4p\beta_m}{3p + 2\beta_m}, \frac{2p\beta_m}{p + 2\beta_m}\right]. \tag{15.39}$$

Proof The distortion exponent for the source coding diversity over two parallel channels can be written as

$$\Delta_{\text{SH}-2\text{R}-\text{SRC}-\text{AMP}} = \max\left[\frac{8p\beta_m''}{3p + 4\beta_m''}, \frac{4p\beta_m''}{p + 4\beta_m''}\right], \tag{15.40}$$

Using (15.27) and (15.36) and considering $\beta_m'' = N_m''/K$ (refer to Figure 15.5) we get for our system the same distortion exponent as (15.40). For fair comparison with the previous schemes, $2N_m = 4N_m''$; which leads to $\beta_m'' = 1/2\beta_m$. Substituting this equality in (15.40) completes the proof. ∎

15.2.2   Multi-hop decode-and-forward protocol

In this section, we will analyze schemes using multi-hop decode-and-forward user coop-eration under different channel and source coding diversity schemes. In these cases, the relay nodes decode the received source symbols. Only those relay nodes that had cor-rectly decoded the source symbols will proceed to retransmit them to the destination node. When a relay fails in decoding the source symbols we say that an outage has occurred. Furthermore, an outage occurs when either the source–relay or the relay–destination channel are in outage, as discussed in Sections 4.2.1.2 and 15.1. That is, the quality of the source–relay–destination link is limited by the minimum of the source–relay and relay–destination channels. Therefore, for the single–relay case we can formulate the outage as

$$P_{\text{outage}} = \Pr\left[\min(I(x_s, y_{r_1}), I(x_{r_1}, y_d)) < R(D)\right], \tag{15.41}$$

where x_{r_1} is the transmitted signal from the relay node.

Note that in those schemes using decode-and-forward the mutual information of any source–relay–destination link is limited by the minimum of the source–relay and relay–destination links SNRs. On the other hand, for two-hop amplify-and-forward schemes, the performance is limited by the harmonic mean of the source–relay and the relay–destination links SNRs which is strictly less than the minimum of the two links SNRs. Hence, the multi-hop amplify-and-forward protocol has a higher outage probability (lower quality) than the multi-hop decode-and-forward protocol. That is, in terms of out-age probability, the multi-hop decode-and-forward protocol outperforms the multi-hop amplify-and-forward protocol.

The above argument is also applicable under different performance measures (for example, if the performance measure was symbol error rate). From our presentation so far it is clear that the distortion exponents for multi-hop decode-and-forward schemes are the same as their corresponding multi-hop amplify-and-forward schemes for the repetition channel coding diversity and source coding diversity cases. For example, for the two-hop single–relay decode-and-forward scheme, the minimum expected distortion is given by the lower bound in (15.25), which has the same distortion exponent as the two-hop single–relay amplify-and-forward scheme. The following theorem summarizes the results (see Exercise 15.5).

THEOREM 15.2.5   *The distortion exponent of the multi-hop decode-and-forward schemes are:*

- *for the two-hop single relay*

$$\Delta_{\text{SH}-1\text{R}-\text{DEC}} = \frac{2p\beta_m}{p + 2\beta_m}, \tag{15.42}$$

- *for the two-hop M-relay selection channel coding diversity*

$$\Delta_{\text{SH}-\text{MR}-\text{DEC}} = \frac{4Mp\beta_m}{M(M+1)p + 4\beta_m}, \tag{15.43}$$

- *for the two-hop two-relay source coding diversity*

$$\Delta_{SH-2R-SRCDEC} = \max \left[\frac{4p\beta_m}{3p + 2\beta_m}, \frac{2p\beta_m}{p + 2\beta_m} \right]. \tag{15.44}$$

15.2.2.1 Optimal channel coding diversity with two relays

We consider now the use of optimal channel coding with two-relay decode-and-forward protocols. In this case, each relay will treat the two blocks x_{s_1} and x_{s_2} together, combining them for joint decoding. This means that when any relay decodes correctly, it will have useful copies of both x_{s_1} and x_{s_2} and could forward both blocks, as illustrated in Figure 15.6. Restricting the first relay to forward only x_{s_1} (if it has decoded correctly) will cause a performance degradation in those cases when the second relay had decoded erroneously. Hence, if the first relay decoded correctly and the second did not, it is better (in terms of outage probability) for the first relay to forward both x_{s_1} and x_{s_2}.

Clearly, a similar argument could be applied to the operation of the second relay. Also, when both relays decode correctly, allowing the second relay to transmit also x_{s_1} and x_{s_2} will cause a loss of diversity. To gain both advantages (lower outage probability when only one relay decodes correctly and diversity when both correctly decode) we propose to use a space–time transmission scheme. Without loss of generality, we choose the Alamouti scheme introduced in Example 1.5 and also discussed in Section 2.2.2, with the time frame structure as shown in Figure 15.6. Then, the distortion exponent of this system is given by the following theorem.

THEOREM 15.2.6 *The distortion exponent of the two-hop two-relay optimal channel coding diversity decode-and-forward protocol is*

$$\Delta_{SH-2R-OPTCH-DEC} = \frac{2p\beta_m}{p + \beta_m}. \tag{15.45}$$

Proof We start by deriving the expression for the outage probability. Let $S \longrightarrow R_i$ and $R_i \longrightarrow D$ denote the channel between the source and the i-th relay and the channel between the i-th relay and the destination, respectively. Let $R_1, R_2 \longrightarrow D$ denote the channel between the two relays and the destination when both relays decode correctly.

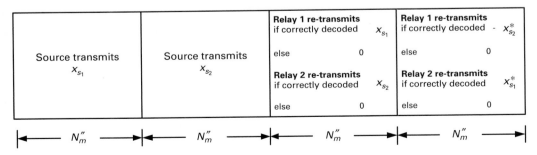

Fig. 15.6 Two-hop two-relay decode-and-forward optimal channel coding diversity system time frame structure.

Also, we denote by $R_i \xrightarrow{\text{o}} D$ as the event that a channel is in outage and by $R_i \xrightarrow{\hat{\text{o}}} D$ the event that it is not in outage. We calculate the outage probability by splitting the outage event into disjoint events, i.e., $P_{\text{outage}} = P_{o_1} + P_{o_2} + P_{o_3} + P_{o_4}$, where

$$
\begin{aligned}
P_{o_1} &= \Pr\left[S \xrightarrow{\text{o}} R_1, \ S \xrightarrow{\text{o}} R_2\right] \\
&= \Pr\left[S \xrightarrow{\text{o}} R_1\right] . \Pr\left[S \xrightarrow{\text{o}} R_2\right] \\
&= \Pr\left[2\log(1 + |h_{s,r_1}|^2 \text{SNR}) < R(D)\right] \\
&\quad \Pr\left[2\log(1 + |h_{s,r_2}|^2 \text{SNR}) < R(D)\right] \\
&\approx c_{o_1}\left(\frac{\exp(pR(D))}{\text{SNR}^{2p}}\right).
\end{aligned}
\tag{15.46}
$$

$$
\begin{aligned}
P_{o_2} &= \Pr\left[S \xrightarrow{\text{o}} R_1, \ S \xrightarrow{\hat{\text{o}}} R_2, \ R_2 \xrightarrow{\text{o}} D\right] \\
&= \Pr\left[S \xrightarrow{\text{o}} R_1\right] . \Pr\left[S \xrightarrow{\hat{\text{o}}} R_2\right] . \Pr\left[R_2 \xrightarrow{\text{o}} D\right] \\
&\approx c_{o_2}\left(\frac{\exp(pR(D))}{\text{SNR}^{2p}}\right).
\end{aligned}
\tag{15.47}
$$

$$
\begin{aligned}
P_{o_3} &= \Pr\left[S \xrightarrow{\text{o}} R_2, \ S \xrightarrow{\hat{\text{o}}} R_1, \ R_1 \xrightarrow{\text{o}} D\right] \\
&\approx c_{o_3}\left(\frac{\exp(pR(D))}{\text{SNR}^{2p}}\right).
\end{aligned}
\tag{15.48}
$$

$$
\begin{aligned}
P_{o_4} &= \Pr\left[S \xrightarrow{\hat{\text{o}}} R_1, \ S \xrightarrow{\hat{\text{o}}} R_2, \ R_1, R_2 \xrightarrow{\text{o}} D\right] \\
&= \Pr\left[S \xrightarrow{\hat{\text{o}}} R_1\right] . \Pr\left[S \xrightarrow{\hat{\text{o}}} R_2\right] . \Pr\left[R_1, R_2 \xrightarrow{\text{o}} D\right] \\
&\approx \Pr\left[2\log(1 + \frac{1}{2}(|h_{r_1,d}|^2 \text{SNR} + |h_{r_2,d}|^2 \text{SNR})) < R(D)\right],
\end{aligned}
\tag{15.49}
$$

where the factor $1/2$ in (15.49) is due to the loss in SNR because of the use of transmit diversity. To calculate P_{o_4} in (15.49) we need to calculate the cdf of the random variable $|h_{r_1,d}|^2 \text{SNR} + |h_{r_2,d}|^2 \text{SNR}$. Let $W_1 = |h_{r_1,d}|^2 \text{SNR}$ and $W_2 = |h_{r_2,d}|^2 \text{SNR}$. The pdf of $W_1 + W_2$ can be computed as

$$
\begin{aligned}
f_{W_1+W_2}(w) &= \int_0^w f_{W_1}(\tau) f_{W_2}(w - \tau) d\tau \\
&\approx \frac{c_{11} c_{22} p^2}{\text{SNR}^{2p}} \int_0^w \tau^{p-1} (w - \tau)^{p-1} d\tau \\
&= c_{11} c_{22} p^2 \frac{w^{2p-1}}{\text{SNR}^{2p}} B(p, p),
\end{aligned}
\tag{15.50}
$$

where $B(.,.)$ is the Beta function [50]. The cdf of $W_1 + W_2$ can be computed as

$$F_{W_1+W_2}(w) = \int_0^w f_{W_1+W_2}(\tau)d\tau = c_{33} \left(\frac{w}{\text{SNR}}\right)^{2p}, \tag{15.51}$$

from which we have

$$P_{o_4} \approx c_{o_4} \left(\frac{\exp(pR(D))}{\text{SNR}^{2p}}\right). \tag{15.52}$$

Then, the outage probability is

$$P_{\text{outage}} = P_{o_1} + P_{o_2} + P_{o_3} + P_{o_4}$$

$$= c_o \left(\frac{\exp(pR(D))}{\text{SNR}^{2p}}\right).$$

In the proof, we have assumed that x_{s_1} and x_{s_2} are independent zero-mean, complex Gaussian with variance $1/2$ per dimension. We can easily show that this choice of x_{s_1} and x_{s_2} is the optimal choice for maximizing the mutual information (minimizing the outage probability) by inspection of the individual outage events in (15.53).

Following this result, the minimum expected distortion can now be computed as

$$E[D] = \min_D \left\{ P_{\text{outage}} + D(1 - P_{\text{outage}}) \right\}$$

$$\approx \min_D \left\{ c_o \left(\frac{\exp(pR(D))}{\text{SNR}^{2p}}\right) + D\left[1 - c_o\left(\frac{\exp(pR(D))}{\text{SNR}^{2p}}\right)\right] \right\}$$

$$\approx \min_D \left\{ c_o \left(\frac{\exp(pR(D))}{\text{SNR}^{2p}}\right) + D \right\} \tag{15.53}$$

$$\approx \min_D \left\{ c_o \frac{D^{-\frac{p}{2\beta_m''}}}{\text{SNR}^{2p}} + D \right\}, \tag{15.54}$$

where D is the source encoder distortion, (15.53) follows from high SNR approximation and (15.54) follows from (15.5). Differentiating and setting equal to zero we get the optimizing distortion

$$D^* = \left(\frac{2\beta_m''}{c_o p}\right)^{\frac{-2\beta_m''}{2\beta_m''+p}} \text{SNR}^{\frac{-4\beta_m'' p}{2\beta_m''+p}}. \tag{15.55}$$

Hence, the distortion exponent is given as

$$\Delta_{\text{RC}-1\text{R}-\text{AMP}} = \frac{4\beta_m'' p}{2\beta_m'' + p}. \tag{15.56}$$

For fair comparison, the total number of channel uses should be kept fixed for all schemes. Thus, we have $N_m = 2N_m''$, from which we have $\beta_m'' = (1/2)\beta_m$. Substituting in (15.56) we get

$$\Delta_{\text{SH}-2\text{R}-\text{OPTCH}-\text{DEC}} = \frac{2p\beta_m}{p + \beta_m}. \tag{15.57}$$

∎

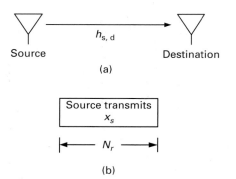

Source   Destination

(a)

Source transmits
x_s

$\mid\!\!\longleftarrow\quad N_r\quad\longrightarrow\!\!\mid$

(b)

Fig. 15.7   No diversity (direct transmission) system: (a) system model, (b) time frame structure.

15.3   Relay channels

We now extend the analysis on distortion exponents to the case of a relay channel when using either amplify-and-forward or decode-and-forward user cooperation. By a relay channel, we now consider that in addition to the source–relay–destination channels there is also a direct communication channel between the source and destination nodes. For comparison purposes, we consider the case when the source transmits a single description source coded message over the source–destination channel without the help of any relay node. The system is shown in Figure 15.7. In this case, the distortion exponent can be derived in a similar way as Theorem 15.2.1 and is given by

$$\Delta_{\text{NO–DIV}} = \frac{2p\beta_r}{p + 2\beta_r},\tag{15.58}$$

where $\beta_r = N_r/K$ and N_r is the number of channel uses for the source block.

15.3.1   Amplify-and-forward protocol

In this section, we analyze the same schemes presented for the multi-hop channels when now they are used over the amplify-and-forward relay channel. We will consider both single and M-relay repetition channel coding diversity and two-relay optimal channel coding diversity and source coding diversity.

15.3.1.1   Single relay
Consider a system comprising a source, a relay, and a destination as shown in Figure 15.8. We consider that the relay operates following an amplify-and-forward user-cooperation scheme. We also assume that the destination applies a maximum ratio combiner (MRC) to detect the transmitted signal from those received in each phase. The mutual information of this system is given by

$$I(x_s, y_d) = \log\left(1 + |h_{\text{s,d}}|^2\text{SNR} + \frac{|h_{\text{s,r}_1}|^2\text{SNR}|h_{\text{r}_1,\text{d}}|^2\text{SNR}}{|h_{\text{s,r}_1}|^2\text{SNR} + |h_{\text{r}_1,\text{d}}|^2\text{SNR} + 1}\right),\tag{15.59}$$

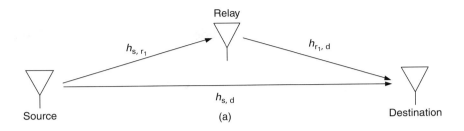

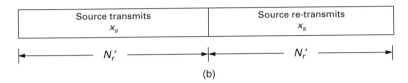

(b)

Fig. 15.8   Single-relay system: (a) system model, (b) time frame structure.

where $\mathrm{SNR} = P/N_0$ and $h_{\mathrm{s,d}}$ is the channel between the source and the destination. At high SNR, we have

$$I(x_s, y_d) \approx \log\left(1 + |h_{\mathrm{s,d}}|^2\mathrm{SNR} + \frac{|h_{\mathrm{s,r_1}}|^2\mathrm{SNR}|h_{\mathrm{r_1,d}}|^2\mathrm{SNR}}{|h_{\mathrm{s,r_1}}|^2\mathrm{SNR} + |h_{\mathrm{r_1,d}}|^2\mathrm{SNR}}\right)$$

(15.60)

$$\approx \log\left(|h_{\mathrm{s,d}}|^2\mathrm{SNR} + \frac{|h_{\mathrm{s,r_1}}|^2\mathrm{SNR}|h_{\mathrm{r_1,d}}|^2\mathrm{SNR}}{|h_{\mathrm{s,r_1}}|^2\mathrm{SNR} + |h_{\mathrm{r_1,d}}|^2\mathrm{SNR}}\right).$$

The distortion exponent of this system is given by the following theorem.

THEOREM 15.3.1   *The distortion exponent of the single-relay amplify-and-forward scheme is*

$$\Delta_{\mathrm{RC-1R-AMP}} = \frac{2p\beta_r}{2p + \beta_r}.$$

(15.61)

Proof   Let

$$W_1 = |h_{\mathrm{s,d}}|^2\mathrm{SNR},$$

and

$$W_2 = \frac{|h_{\mathrm{s,r_1}}|^2\mathrm{SNR}|h_{\mathrm{r_1,d}}|^2\mathrm{SNR}}{|h_{\mathrm{s,r_1}}|^2\mathrm{SNR} + |h_{\mathrm{r_1,d}}|^2\mathrm{SNR}}.$$

The outage probability can be calculated as

$$P_{\mathrm{outage}} = \Pr\left[\log(1 + W_1 + W_2) < R(D)\right]$$
$$\approx \Pr\left[W_1 + W_2 < \exp(R(D))\right].$$

(15.62)

From the proof of Theorem 15.2.6, the cdf of $W_1 + W_2$ is given by

$$F_{W_1+W_2}(w) \approx c_{33} \left(\frac{w}{\text{SNR}} \right)^{2p}. \tag{15.63}$$

The minimum expected distortion can now be computed as

$$
\begin{aligned}
E[D] &\approx \min_D \left\{ \Pr\left[W_1 + W_2 < \exp(R(D))\right] + \Pr\left[W_1 + W_2 \geq \exp(R(D))\right] D \right\} \\
&= \min_D \left\{ F_{W_1+W_2}(\exp(R(D))) + \Pr[1 - F_{W_1+W_2}(\exp(R(D)))] D \right\} \\
&\approx \min_D \left\{ c_{33} \left(\frac{\exp(2pR(D))}{\text{SNR}^{2p}} \right) + \left[1 - c_{33} \left(\frac{\exp(2pR(D))}{\text{SNR}^{2p}} \right) \right] D \right\} \\
&\approx \min_D \left\{ c_{33} \left(\frac{D^{\frac{-p}{\beta_r'}}}{\text{SNR}^{2p}} \right) + D \right\}, \tag{15.64}
\end{aligned}
$$

where $\beta_r' = N_r'/K$ and N_r' is the number of source channel uses (refer to Figure 15.8). Differentiating and setting equal to zero we get the optimizing distortion

$$D^* = \left(\frac{\beta_r}{c_{33} p} \right)^{\frac{-\beta_r'}{\beta_r'+p}} \text{SNR}^{\frac{-2\beta_r' p}{\beta_r'+p}}. \tag{15.65}$$

Hence, the distortion exponent is given as

$$\Delta_{\text{RC–1R–AMP}} = \frac{2\beta_r' p}{\beta_r' + p}. \tag{15.66}$$

For fair comparison we should have $N_r = 2N_r'$ from which we have $\beta_r' = (1/2)\beta_r$. Substituting into (15.66) we get

$$\Delta_{\text{RC–1R–AMP}} = \frac{2\beta_r p}{\beta_r + 2p}. \tag{15.67}$$

∎

If we asymptotically compare the distortion exponents for the cases of no diversity and a single relay we have

$$
\begin{aligned}
\lim_{\beta_r/p \to \infty} \frac{\Delta_{\text{RC–1R–AMP}}}{\Delta_{\text{NO–DIV}}} &= 2, \\
\lim_{\beta_r/p \to 0} \frac{\Delta_{\text{RC–1R–AMP}}}{\Delta_{\text{NO–DIV}}} &= \frac{1}{2}.
\end{aligned}
\tag{15.68}
$$

Note that as β_r/p increases (bandwidth increases) the system becomes outage limited because the performance is limited by the outage event. In this case, the single-relay amplify-and-forward system will achieve a higher distortion exponent since it achieves diversity. Conversely, as β_r/p tends to zero (higher channel quality) the performance is not limited by the outage event, but is limited by the source encoder quality performance.

The ongoing analysis can be easily extended to the case of M amplify-and-forward relay nodes. The distortion exponent in this case is given by the following theorem.

THEOREM 15.3.2 *The distortion exponent of the M-relay-nodes amplify-and-forward protocol is*

$$\Delta_{\text{RC}-\text{MR}-\text{AMP}} = \frac{2(M+1)p\beta_r}{2\beta_r + (M+1)^2 p}. \tag{15.69}$$

Again we can think of selecting the optimum number of relays to maximize the distortion exponent. This is again a tradeoff between the diversity and the quality of the source encoder.

15.3.1.2 Optimal channel coding diversity with two relays

We consider the same problem as in Section 15.2.1.3, but new for the relay channel. Once again, we assume a system consisting of a source, two relays and a destination as shown in Figure 15.9. The source transmits two channel-coded blocks x_{s_1} and x_{s_2} to the destination and the relay nodes. The first relay will only forward the block x_{s_1} and the second relay will only forward x_{s_2} as shown in Figure 15.9. First, we calculate the mutual information for optimal channel coding. In phase 1, the source broadcasts its information to the destination and two relay nodes. The received signals are

$$y_{s,d_m} = \sqrt{P} h_{s,d} x_{s_m} + n_{s,d}, \tag{15.70}$$

$$y_{s,r_i} = \sqrt{P} h_{s,r_i} x_{s_m} + n_{s,r_i}, \quad i = 1, 2, \ m = 1, 2. \tag{15.71}$$

Relay 1 will only forward x_{s_1} and relay 2 will only forward x_{s_2}. The received signals at the destination due to relay 1 and relay 2 transmissions are given by

$$y_{r_i,d} = h_{r_i,d} \zeta_i y_{s,r_i} + n_{r_i,d}, \ i = 1, 2, \tag{15.72}$$

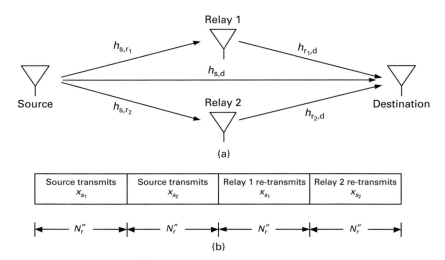

(a)

Source transmits x_{s_1}	Source transmits x_{s_2}	Relay 1 re-transmits x_{s_1}	Relay 2 re-transmits x_{s_2}
$\xleftarrow{\quad N_r'' \quad}$	$\xleftarrow{\quad N_r'' \quad}$	$\xleftarrow{\quad N_r'' \quad}$	$\xleftarrow{\quad N_r'' \quad}$

(b)

Fig. 15.9   Two-relay system: (a) system model, (b) time frame structure.

where ζ_i is the signal amplification performed at the relay which satisfies the power constraint with equality, that is,

$$\zeta_i = \sqrt{\frac{P}{P|h_{s,r_i}|^2 + N_0}}, \tag{15.73}$$

and where all the noise components are modeled as independent zero-mean, complex Gaussian random variables with variance $N_0/2$ per dimension.

Define the 4×1 vector, $\mathbf{y} = [y_{s,d_1}, y_{s,d_2}, y_{r_1,d}, y_{r_2,d}]^T$. To get the mutual information between $\mathbf{x} = [x_{s_1}, x_{s_2}]$ and $\mathbf{y}$, we consider that the receiver has to detect the two blocks x_{s_1} and x_{s_2} by operating separately on the two amplify-and-forward channels that are established through each of the relays. In this way, an MRC detector is applied on $y_{s,d_1}, y_{r_1,d}$ and another MRC detector is applied on $y_{s,d_2}, y_{r_2,d}$. The output of the first MRC detector is given by

$$r_1 = \alpha_s y_{s,d_1} + \alpha_1 y_{r_1,d}, \tag{15.74}$$

where $\alpha_s = \sqrt{P} h^*_{s,d}/N_0$ and

$$\alpha_1 = \frac{\sqrt{P}\zeta_1 h^*_{r_1,d} h^*_{s,r_1}}{(\zeta_1^2 |h_{r_1,d}|^2 + 1)N_0}.$$

We can write r_1 in terms of x_{s_1} as

$$r_1 = \left(|h_{s,d}|^2 \text{SNR} + \frac{|h_{r_1,d}|^2 \text{SNR}|h_{s,r_1}|^2 \text{SNR}}{|h_{s,r_1}|^2 \text{SNR} + |h_{r_1,d}|^2 \text{SNR} + 1}\right) x_{s_1} + n_1, \tag{15.75}$$

where n_1 is a zero-mean, complex Gaussian noise of variance

$$|h_{s,d}|^2 \text{SNR} + \frac{|h_{r_1,d}|^2 \text{SNR}|h_{s,r_1}|^2 \text{SNR}}{|h_{s,r_1}|^2 \text{SNR} + |h_{r_1,d}|^2 \text{SNR} + 1}.$$

Similarly we can have r_2, representing the output of the second MRC detector, given by

$$r_2 = \left(|h_{s,d}|^2 \text{SNR} + \frac{|h_{r_2,d}|^2 \text{SNR}|h_{s,r_2}|^2 \text{SNR}}{|h_{s,r_2}|^2 \text{SNR} + |h_{r_2,d}|^2 \text{SNR} + 1}\right) x_{s_2} + n_2, \tag{15.76}$$

where n_2 is a zero-mean, complex Gaussian noise of variance

$$|h_{s,d}|^2 \text{SNR} + \frac{|h_{r_2,d}|^2 \text{SNR}|h_{s,r_2}|^2 \text{SNR}}{|h_{s,r_2}|^2 \text{SNR} + |h_{r_2,d}|^2 \text{SNR} + 1}.$$

This means that, given the channel coefficients, r_1 and r_2 act as the received x_{s_1} and x_{s_2} when transmitted over an additive white Gaussian channel. Because of this, it follows the well known fact that r_1 and r_2, given the channel coefficients, are sufficient statistics for $\mathbf{x}$ [223]. This implies that the mutual information between $\mathbf{x}$ and $\mathbf{y}$ equals the mutual information between $\mathbf{x}$ and $\mathbf{r} = [r_1, r_2]$, that is

$$I(\mathbf{x}; \mathbf{r}) = I(\mathbf{x}; \mathbf{y}). \tag{15.77}$$

As was discussed in the proof of Theorem 15.2.6, for any covariance matrix of $\mathbf{x}$ the mutual information is maximized when $\mathbf{x}$ is a zero-mean, complex Gaussian random vector. Define

$$\gamma_1 = |h_{s,d}|^2 \text{SNR} + \frac{|h_{r_1,d}|^2 \text{SNR} |h_{s,r_1}|^2 \text{SNR}}{|h_{s,r_1}|^2 \text{SNR} + |h_{r_1,d}|^2 \text{SNR} + 1}$$

$$\gamma_2 = |h_{s,d}|^2 \text{SNR} + \frac{|h_{r_2,d}|^2 \text{SNR} |h_{s,r_2}|^2 \text{SNR}}{|h_{s,r_2}|^2 \text{SNR} + |h_{r_2,d}|^2 \text{SNR} + 1}.$$

The mutual information can be computed as

$$I(\mathbf{x}, \mathbf{y}) = I(\mathbf{x}, \mathbf{r}) = \log\left(\det\left[\mathbf{I}_2 + \begin{bmatrix} \gamma_1 & \alpha\gamma' \\ \alpha^*\gamma'' & \gamma_2 \end{bmatrix}\right]\right), \tag{15.78}$$

where $\mathbf{I}_2$ is the 2×2 identity matrix, γ' and γ'' are functions of the channel coefficients and the noise variance and $\alpha = \mathrm{E}\left[x_{s_1} x_{s_2}^*\right]$. It is next necessary to maximize (15.78) in order to find the maximum mutual information. From (15.78) it is clear that both α and $-\alpha$ will give the same mutual information. From the concavity of the log-function we can see that the mutual information maximizing α is $\alpha = 0$, that is x_{s_1} and x_{s_2} are independent (since both x_{s_1} and x_{s_2} are Gaussian). The maximum mutual information can now be given as

$$I \approx \log\left(1 + |h_{s,d}|^2 \text{SNR} + \frac{|h_{s,r_1}|^2 \text{SNR} |h_{r_1,d}|^2 \text{SNR}}{|h_{s,r_1}|^2 \text{SNR} + |h_{r_1,d}|^2 \text{SNR}}\right)$$

$$+ \log\left(1 + |h_{s,d}|^2 \text{SNR} + \frac{|h_{s,r_2}|^2 \text{SNR} |h_{r_2,d}|^2 \text{SNR}}{|h_{s,r_2}|^2 \text{SNR} + |h_{r_2,d}|^2 \text{SNR}}\right). \tag{15.79}$$

The distortion exponent of this system is given by the following theorem.

THEOREM 15.3.3 *The distortion exponent of the two-relay optimal channel coding diversity amplify-and-forward protocol is*

$$\Delta_{\text{RC-2R-OPTCH-AMP}} = \frac{3p\beta_r}{3p + \beta_r}. \tag{15.80}$$

Proof   To compute the distortion exponent we start with the analysis of a suboptimal system at the destination node. In the suboptimal system, the detector (suboptimal detector) selects the paths with the highest SNR and does not apply an MRC detector (the optimal detector is the one that applies MRC on the received signals). For example, for x_{s_1}, it either selects the source–destination link or the source–relay–destination link based on which one has higher SNR. The mutual information for the suboptimal system can be easily proved to be

$$I_{\text{sub}} \approx \log\left(1 + \max\left(|h_{s,d}|^2 \text{SNR}, \frac{|h_{s,r_1}|^2 \text{SNR} |h_{r_1,d}|^2 \text{SNR}}{|h_{s,r_1}|^2 \text{SNR} + |h_{r_1,d}|^2 \text{SNR}}\right)\right)$$

$$+ \log\left(1 + \max\left(|h_{s,d}|^2 \text{SNR}, \frac{|h_{s,r_2}|^2 \text{SNR} |h_{r_2,d}|^2 \text{SNR}}{|h_{s,r_2}|^2 \text{SNR} + |h_{r_2,d}|^2 \text{SNR}}\right)\right). \tag{15.81}$$

Let

$$W_1 = |h_{s,d}|^2 \text{SNR},$$

$$W_2 = \frac{|h_{s,r_1}|^2 \text{SNR}|h_{r_1,d}|^2 \text{SNR}}{|h_{s,r_1}|^2 \text{SNR} + |h_{r_1,d}|^2 \text{SNR}},$$

$$W_3 = \frac{|h_{s,r_2}|^2 \text{SNR}|h_{r_2,d}|^2 \text{SNR}}{|h_{s,r_2}|^2 \text{SNR} + |h_{r_2,d}|^2 \text{SNR}}.$$

The outage probability of the suboptimal system is given by

$$
\begin{aligned}
P_{\text{outage}} &= \Pr[I_{\text{sub}} < R] \\
&= \Pr\left[\log(1 + \max(W_1, W_2)) + \log(1 + \max(W_1, W_3)) < R\right] \\
&= \Pr\Big[\{2\log(1 + W_1) < R, W_1 > W_2, W_1 > W_3\} \\
&\quad \bigcup\{\log(1 + W_1) + \log(1 + W_3) < R, W_1 > W_2, W_3 > W_1\} \\
&\quad \bigcup\{\log(1 + W_2) + \log(1 + W_1) < R, W_2 > W_1, W_1 > W_3\} \\
&\quad \bigcup\{\log(1 + W_2) + \log(1 + W_3) < R, W_2 > W_1, W_3 > W_1\}\Big] \\
&= \Pr[2\log(1 + W_1) < R, W_1 > W_2, W_1 > W_3] \\
&\quad + \Pr[\log(1 + W_1) + \log(1 + W_3) < R, W_1 > W_2, W_3 > W_1] \\
&\quad + \Pr[\log(1 + W_2) + \log(1 + W_1) < R, W_2 > W_1, W_1 > W_3] \\
&\quad + \Pr[\log(1 + W_2) + \log(1 + W_3) < R, W_2 > W_1, W_3 > W_1],
\end{aligned}
$$

$$(15.82)$$

where the last equality follows from the events being disjoint. In the last equation we used R instead of $R(D)$ for simplicity of presentation. The joint pdf of W_1, W_2, and W_3, which are independent random variables, is given by

$$f(w_1, w_2, w_3) \approx c_j p^3 \left(\frac{w_1^{p-1} w_2^{p-1} w_3^{p-1}}{\text{SNR}^{3p}}\right), \qquad (15.83)$$

where c_j is a constant. To find the outage probability, we calculate the probability of the individual outage events in (15.82),

$$
\begin{aligned}
P_1 &= \Pr[2\log(1 + W_1) < R, W_1 > W_2, W_1 > W_3] \\
&= \int_{w_1=0}^{\exp(R/2)} \int_{w_3=0}^{w_1} \int_{w_2=0}^{w_1} f(w_1, w_2, w_3) dw_2 dw_3 dw_1 \\
&\approx \int_{w_1=0}^{\exp(R/2)} \int_{w_3=0}^{w_1} \int_{w_2=0}^{w_1} c_j p^3 \left(\frac{w_1^{p-1} w_2^{p-1} w_3^{p-1}}{\text{SNR}^{3p}}\right) dw_2 dw_3 dw_1 \\
&= \frac{c_j \exp\left(\frac{3pR}{2}\right)}{3\text{SNR}^{3p}}.
\end{aligned}
$$

$$P_2 = \Pr[\log(1 + W_1) + \log(1 + W_3) < R, W_1 > W_2, W_3 > W_1]$$
$$\approx \Pr[\log(W_1) + \log(W_3) < R, W_1 > W_2, W_3 > W_1]$$

$$= \int_{w_1=0}^{\exp(R/2)} \int_{w_3=w_1}^{\frac{\exp(R)}{w_1}} \int_{w_2=0}^{w_1} f(w_1, w_2, w_3) dw_2 dw_3 dw_1$$

$$\approx \int_{w_1=0}^{\exp(R/2)} \int_{w_3=w_1}^{\frac{\exp(R)}{w_1}} \int_{w_2=0}^{w_1} c_j p^3 \left(\frac{w_1^{p-1} w_2^{p-1} w_3^{p-1}}{\mathrm{SNR}^{3p}} \right) dw_2 dw_3 dw_1$$

$$= \frac{c_j p^2}{\mathrm{SNR}^{3p}} \int_{w_1=0}^{\exp(R/2)} \int_{w_3=w_1}^{\frac{\exp(R)}{w_1}} w_1^{2p-1} w_3^{p-1} dw_3 dw_1$$

$$= \frac{2c_j \exp\left(\frac{3pR}{2}\right)}{3\mathrm{SNR}^{3p}}.$$

$$P_3 = \Pr[\log(1 + W_2) + \log(1 + W_1) < R, W_2 > W_1, W_1 > W_3]$$
$$\approx \Pr[\log(W_1) + \log(W_3) < R, W_1 > W_2, W_3 > W_1]$$

$$\approx \frac{2c_j \exp\left(\frac{3pR}{2}\right)}{3\mathrm{SNR}^{3p}}.$$

$$P_4 = \Pr[\log(1 + W_2) + \log(1 + W_3) < R, W_2 > W_1, W_3 > W_1]$$
$$\approx \Pr[\log(W_2) + \log(W_3) < R, W_2 > W_1, W_3 > W_1]$$

$$= \int_{w_1=0}^{\exp(R/2)} \int_{w_2=w_1}^{\frac{\exp(R)}{w_1}} \int_{w_3=w_1}^{\frac{\exp(R)}{w_2}} f(w_1, w_2, w_3) dw_3 dw_2 dw_1$$

$$\approx \int_{w_1=0}^{\exp(R/2)} \int_{w_2=w_1}^{\frac{\exp(R)}{w_1}} \int_{w_3=w_1}^{\frac{\exp(R)}{w_2}} c_j p^3 \left(\frac{w_1^{p-1} w_2^{p-1} w_3^{p1}}{\mathrm{SNR}^{3p}} \right) dw_3 dw_2 dw_1$$

$$= \frac{4c_j \exp\left(\frac{3pR}{2}\right)}{3\mathrm{SNR}^{3p}},$$

where we have $\lim_{w_1 \to 0} w_1^p \log w_1 = 0$ for $p \geq 1$. The outage probability for the suboptimal system is

$$P_{\mathrm{outage}} = P_1 + P_2 + P_3 + P_4 \approx \frac{c_m \exp\left(\frac{3pR}{2}\right)}{\mathrm{SNR}^{3p}}, \tag{15.84}$$

where c_m is a constant. The minimum expected end-to-end distortion can now be computed as

$$E[D] = \min_D \left\{ P_{\mathrm{outage}} + \left(1 - P_{\mathrm{outage}}\right) D \right\}$$

$$\approx \min_D \left\{ \frac{c_m \exp\left(\frac{3pR}{2}\right)}{\mathrm{SNR}^{3p}} + \left(1 - \frac{c_m \exp\left(\frac{3pR}{2}\right)}{\mathrm{SNR}^{3p}}\right) D \right\}$$

$$\approx \min_{D} \left\{ \frac{c_m D^{\frac{-3p}{4\beta_r''}}}{\mathrm{SNR}^{3p}} + \left(1 - \frac{c_m D^{\frac{-3p}{4\beta_r''}}}{\mathrm{SNR}^{3p}}\right) D \right\}$$

$$\approx \min_{D} \left\{ \frac{c_m D^{\frac{-3p}{4\beta_r''}}}{\mathrm{SNR}^{3p}} + D \right\}, \tag{15.85}$$

where $\beta_r'' = N_r''/K$ (refer to Figure 15.9), D is the source encoder distortion and we have used both high SNR approximations and (15.5). Differentiating and setting equal to zero, we get the optimizing distortion

$$D^* = \left(\frac{4\beta_r''}{3c_m p}\right)^{\frac{-4\beta_r''}{4\beta_r''+3p}} \mathrm{SNR}^{\frac{-12\beta_r'' p}{4\beta_r''+3p}}.$$

Substituting we get the distortion exponent for this suboptimal system as

$$\Delta_{\mathrm{SUBOPTIMAL}} = \frac{12\beta_r'' p}{4\beta_r'' + 3p}.$$

For fair comparison the total number of channel uses is fixed and, thus, $\beta_r'' = (1/4)\beta_r$, which leads to the distortion exponent of the suboptimal system be given by

$$\Delta_{\mathrm{SUBOPTIMAL}} = \frac{3\beta_r p}{\beta_r + 3p}.$$

For the optimal detector (the one using an MRC detector), the distortion exponent satisfies

$$\Delta_{\mathrm{RC-2R-OPTCH-AMP}} \geq \Delta_{\mathrm{SUBOPTIMAL}} = \frac{3\beta_r p}{\beta_r + 3p}. \tag{15.86}$$

Next, we find an upper bound on the distortion exponent for the optimal system. In this case, the mutual information in (15.79) can be upper and lower bounded as

$$\log(1 + 2W_1 + W_2 + W_3) \leq \log(1 + W_1 + W_2) + \log(1 + W_1 + W_3)$$

$$\leq 2\log(1 + W_1 + \frac{1}{2}W_2 + \frac{1}{2}W_3),$$

where W_1, W_2, and W_3 are non-negative numbers. The upper bound follows from the concavity of the log-function. Therefore, the outage probability P_o of the optimal system can be upper and lower bounded as

$$\Pr[2\log(1 + W_1 + \frac{1}{2}W_2 + \frac{1}{2}W_3) < R] \leq P_o \leq \Pr[\log(1 + 2W_1 + W_2 + W_3) < R]. \tag{15.87}$$

From (15.87) we can easily show that

$$C_{\mathrm{L}} \frac{\exp\left(\frac{3pR}{2}\right)}{\mathrm{SNR}^{3p}} \lesssim P_o \lesssim C_{\mathrm{U}} \frac{\exp(3pR)}{\mathrm{SNR}^{3p}}, \tag{15.88}$$

where C_L and C_U are two constants that do not depend on the SNR. Similar to the suboptimal system, and using (15.88), the minimum expected end-to-end distortion for the optimal system can be lower bounded as

$$
\begin{aligned}
E[D] &\gtrsim \min_D \left\{ C_L \frac{\exp\left(\frac{3pR}{2}\right)}{\text{SNR}^{3p}} + \left(1 - C_U \frac{\exp(3pR)}{\text{SNR}^{3p}}\right) D \right\} \\
&\approx \min_D \left\{ \frac{C_L D^{\frac{-3p}{4\beta_r''}}}{\text{SNR}^{3p}} + \left(1 - \frac{C_U D^{\frac{-3p}{2\beta_r''}}}{\text{SNR}^{3p}}\right) D \right\} \\
&\approx \min_D \left\{ \frac{C_L D^{\frac{-3p}{4\beta_r''}}}{\text{SNR}^{3p}} + D \right\}.
\end{aligned}
\tag{15.89}
$$

Differentiating the lower bound and setting equal to zero we get the optimizing distortion as

$$
D^* = \left(\frac{4\beta_r''}{3pC_L}\right)^{\frac{-4\beta_r''}{4\beta_r''+3p}} \text{SNR}^{\frac{-12\beta_r'' p}{4\beta_r''+3p}}.
\tag{15.90}
$$

Substituting, we get

$$
E[D] \gtrsim C_{LO} \text{SNR}^{\frac{-12\beta_r'' p}{4\beta_r''+3p}},
\tag{15.91}
$$

from which we can upper bound the distortion exponent of the optimal system as

$$
\Delta_{RC-2R-OPTCH-AMP} \leq \frac{12\beta_r'' p}{4\beta_r'' + 3p} = \frac{3\beta_r p}{\beta_r + 3p}.
\tag{15.92}
$$

Finally, from (15.86) and (15.92) we get

$$
\Delta_{RC-2R-OPTCH-AMP} = \frac{3\beta_r p}{\beta_r + 3p}.
\tag{15.93}
$$

∎

15.3.1.3 Source coding diversity with two relays

We continue analyzing a system as in Figure 15.9 but now we assume that each of the two blocks sent from the source, x_{s_1} and x_{s_2}, represents one description generated from a dual descriptions source encoder. The first relay will only forward the block x_{s_1} and the second relay will only forward x_{s_2} as shown in Figure 15.9. The distortion exponent of this system is given by the following theorem.

THEOREM 15.3.4 *The distortion exponent of the two-relay source coding diversity amplify-and-forward protocol is*

$$
\Delta_{RC-2R-SRC-AMP} = \max\left[\frac{2p\beta_r}{2p + \beta_r}, \frac{3p\beta_r}{4p + \beta_r}\right],
\tag{15.94}
$$

Proof The receiver applies an MRC detector on the received data to detect x_{s_1} and x_{s_2}. Let

$$W_1 = |h_{s,d}|^2 SNR,$$

$$W_2 = \frac{|h_{s,r_1}|^2 SNR |h_{r_1,d}|^2 SNR}{|h_{s,r_1}|^2 SNR + |h_{r_1,d}|^2 SNR},$$

$$W_3 = \frac{|h_{s,r_2}|^2 SNR |h_{r_2,d}|^2 SNR}{|h_{s,r_2}|^2 SNR + |h_{r_2,d}|^2 SNR}.$$

The minimum expected end-to-end distortion is given by

$$
\begin{aligned}
E[D] \approx \min_{D_0,D_1} \Pr\Big[&\log(1 + W_1 + W_2) < R_{md}(D_0, D_1)/2, \\
&\log(1 + W_1 + W_3) < R_{md}(D_0, D_1)/2\Big] \\
+ \Pr\Big[&\log(1 + W_1 + W_2) < R_{md}(D_0, D_1)/2, \\
&\log(1 + W_1 + W_3) > R_{md}(D_0, D_1)/2\Big] \\
+ \Pr\Big[&\log(1 + W_1 + W_2) > R_{md}(D_0, D_1)/2, \\
&\log(1 + W_1 + W_3) < R_{md}(D_0, D_1)/2\Big]D_1 \\
+ \Pr\Big[&\log(1 + W_1 + W_2) > R_{md}(D_0, D_1)/2, \\
&\log(1 + W_1 + W_3) > R_{md}(D_0, D_1)/2\Big]D_0,
\end{aligned}
\tag{15.95}
$$

where R_{md}, D_0, and D_1 are as introduced in Section 15.1. To calculate the minimum expected distortion we need to calculate the following probabilities in (15.95):

$$
\begin{aligned}
P_1' &= \Pr\Big[\log(1 + W_1 + W_2) < R_{md}(D_0, D_1)/2, \\
&\qquad \log(1 + W_1 + W_3) < R_{md}(D_0, D_1)/2\Big] \\
&= \Pr\Big[\log(1 + W_1 + \max(W_2, W_3)) < R_{md}(D_0, D_1)/2\Big] \\
&\approx c_{s1} \frac{1}{SNR^{3p}} \exp\left(\frac{3p}{2} R_{md}(D_0, D_1)\right).
\end{aligned}
$$

$$
\begin{aligned}
P_2' &= \Pr\Big[\log(1 + W_1 + W_2) > R_{md}(D_0, D_1)/2, \\
&\qquad \log(1 + W_1 + W_3) > R_{md}(D_0, D_1)/2\Big] \\
&= \Pr\Big[\log(1 + W_1 + \min(W_2, W_3)) > R_{md}(D_0, D_1)/2\Big] \\
&= 1 - \Pr\Big[\log(1 + W_1 + \min(W_2, W_3)) < R_{md}(D_0, D_1)/2\Big] \\
&\approx 1 - c_{s2} \frac{1}{SNR^{2p}} \exp\left(p R_{md}(D_0, D_1)\right).
\end{aligned}
$$

$$P_3' = \Pr\left[\log(1 + W_1 + W_2) < R_{md}(D_0, D_1)/2, \right.$$
$$\left. \log(1 + W_1 + W_3) > R_{md}(D_0, D_1)/2 \right]$$
$$+ \Pr\left[\log(1 + W_1 + W_2) > R_{md}(D_0, D_1)/2, \right.$$
$$\left. \log(1 + W_1 + W_3) < R_{md}(D_0, D_1)/2 \right]$$
$$= 1 - P_1' - P_2'$$
$$\approx c_{s2} \frac{1}{\mathrm{SNR}^{2p}} \exp\left(p R_{md}(D_0, D_1)\right) - c_{s1} \frac{1}{\mathrm{SNR}^{3p}} \exp\left(\frac{3p}{2} R_{md}(D_0, D_1)\right)$$
$$\approx c_{s2} \frac{1}{\mathrm{SNR}^{2p}} \exp\left(p R_{md}(D_0, D_1)\right).$$

The minimum expected distortion in (15.95) can now be calculated as

$$E[D] \approx \min_{D_0, D_1} \left\{ c_{s1} \frac{1}{\mathrm{SNR}^{3p}} \exp\left(\frac{3p}{2} R_{md}(D_0, D_1)\right) \right.$$
$$+ c_{s2} \frac{1}{\mathrm{SNR}^{2p}} \exp\left(p R_{md}(D_0, D_1)\right) D_1$$
$$+ \left(1 - c_{s2} \frac{1}{\mathrm{SNR}^{2p}} \exp\left(p R_{md}(D_0, D_1)\right)\right) D_0 \right\}$$
$$\approx \min_{D_0, D_1} \left\{ c_{s1} \frac{1}{\mathrm{SNR}^{3p}} \exp\left(\frac{3p}{2} R_{md}(D_0, D_1)\right) \right.$$
$$\left. + c_{s2} \frac{1}{\mathrm{SNR}^{2p}} \exp\left(p R_{md}(D_0, D_1)\right) D_1 + D_0 \right\}.$$

Substituting from (15.8) yields upper and lower bound for the minimum expected end-to-end distortion as

$$E[D] \gtrsim \min_{D_0, D_1} \frac{c_{s1}}{\mathrm{SNR}^{3p}} \left(\frac{1}{4D_0 D_1}\right)^{\frac{3p}{4\beta_r''}} + \frac{c_{s2}}{\mathrm{SNR}^{2p}} \left(\frac{1}{4D_0 D_1}\right)^{\frac{p}{2\beta_r''}} . D_1 + D_0$$
$$\tag{15.96}$$
$$E[D] \lesssim \min_{D_0, D_1} \frac{c_{s1}}{\mathrm{SNR}^{3p}} \left(\frac{1}{2D_0 D_1}\right)^{\frac{3p}{4\beta_r''}} + \frac{c_{s2}}{\mathrm{SNR}^{2p}} \left(\frac{1}{2D_0 D_1}\right)^{\frac{p}{2\beta_r''}} . D_1 + D_0.$$

Note that for $p \geq 2\beta_r''$ the minimum expected distortion increases as D_1 decreases. Hence, the optimal choice of D_1 approaches a constant that is bounded away from zero. For $D_1 \geq 1/2$ the source coding rate is given by (15.9) and not (15.8). The optimal system in this case degenerates to the single relay system (the argument is the same as for the multi-hop channel). Thus, the distortion exponent is given by

$$\Delta_{RC-2R-SRC-AMP} = \frac{2p\beta_r}{2p + \beta_r}, \qquad p \geq \frac{1}{2}\beta_r = 2\beta_r''. \tag{15.97}$$

For $p < 2\beta_r''$, we can find the optimal value of D_1 by differentiating the lower bound in (15.96) and setting it equal to zero. We get

$$D_1^* = \left(\frac{c_{s_1}}{c_{s_2}} \left(\frac{3p}{(\beta_r - 2p)} \right) \right)^{\frac{\beta_r}{p+\beta_r}} \text{SNR}^{-\frac{p\beta_r}{p+\beta_r}} (4D_0)^{-\frac{p}{p+\beta_r}}, \quad p < \frac{1}{2}\beta_r, \qquad (15.98)$$

where, for fair comparison, we fix the total number of channel uses and get $\beta_r'' = (1/4)\beta_r$. For the case when $p < (1/2)\beta_r$, substituting (15.98) in the lower bound in (15.96) we get

$$E[D] \gtrsim \min_{D_0} C.D_0^{-\frac{3p}{p+\beta_r}}.\text{SNR}^{-\frac{3p\beta_r}{p+\beta_r}} + D_0, \quad p < \frac{1}{2}\beta_r, \qquad (15.99)$$

where C is a constant that does not depend on D_0 and the SNR. Differentiating and setting equal to zero we can get the expression for the optimizing D_0 as

$$D_0^* = C'.\text{SNR}^{-\frac{3p\beta_r}{4p+\beta_r}}, \quad p < \frac{1}{2}\beta_r. \qquad (15.100)$$

Hence, from (15.100) we have

$$C_{LB}'\text{SNR}^{-\frac{3p\beta_r}{4p+\beta_r}} \lesssim E[D] \lesssim C_{UB}'\text{SNR}^{-\frac{3p\beta_r}{4p+\beta_r}}, \quad p < \frac{1}{2}\beta_r. \qquad (15.101)$$

From (15.97) and (15.101) we conclude that the distortion exponent for the source diversity system is given by

$$\Delta_{RC-2R-SRC-AMP} = \max \left[\frac{2p\beta_r}{2p + \beta_r}, \frac{3p\beta_r}{4p + \beta_r} \right], \qquad (15.102)$$

where the second term in (15.102) is the maximum for the case $p < (1/2)\beta_r$. ∎

15.3.2 Decode-and-forward relay channel

We now analyze the decode-and-forward relay channel. The distortion exponents for the different schemes can be easily derived from the analysis presented in the previous sections. We collect the corresponding results in the following theorem.

THEOREM 15.3.5 *The distortion exponents of the decode-and-forward relay channel are:*

- *For the single-relay channel*

$$\Delta_{RC-1R-DEC} = \frac{2p\beta_r}{2p + \beta_r}. \qquad (15.103)$$

- *For the M-relay selection channel coding diversity*

$$\Delta_{RC-MR-DEC} = \frac{2(M + 1)p\beta_r}{2\beta_r + (M + 1)^2 p}. \qquad (15.104)$$

- *For the optimal channel coding with two relays, with the same time frame structure as in Figure 15.6,*

$$\Delta_{RC-2R-OPTCH-AMP} = \frac{3p\beta_r}{3p + \beta_r}. \qquad (15.105)$$

- *For the source coding diversity with two relays*

$$\Delta_{RC-2R-SRC-DEC} = \max\left[\frac{2p\beta_r}{2p+\beta_r}, \frac{3p\beta_r}{4p+\beta_r}\right]. \tag{15.106}$$

In summary, the distortion exponents for the decode-and-forward relay channel are the same as the amplify-and-forward relay channel.

15.4 Discussion

The distortion exponent for the various schemes analyzed in this chapter are summarized in Table 15.1. From the results in Table 15.1 we can see that the optimal channel coding diversity scheme always results in a higher distortion exponent than the source coding diversity scheme at any bandwidth expansion factor (the result is valid over both the multi-hop and relay channels). This means that, between source and channel coding, it is better to exploit diversity at the channel encoder level.

Comparing the expressions for the distortion exponents for the single-relay and M-relay nodes we can see that increasing the number of relays does not always result in an increase in the distortion exponent, showing that there is a tradeoff between the quality (resolution) of the source encoder and the amount of cooperation (number of relays).

Figure 15.10 compares the distortion exponent for the various systems as a function of β_m for the multi-hop channel. The results in Figure 15.10 confirms that the optimal channel coding diversity gives better distortion exponent than the source coding diversity. Note that as β_m increases, the factor that limits the distortion exponent performance is the diversity (number of relays nodes). In this case (high β_m), the system is said to be

Table 15.1 Distortion exponents for the amplify-and-forward (decode-and-forward) multi-hop and relay channels.

	Multi-hop channels	Relay channel
Single relay	$\dfrac{2p\beta_m}{p+2\beta_m}$	$\dfrac{2p\beta_r}{2p+\beta_r}$
Selective channel coding diversity with M relays	$\dfrac{4Mp\beta_m}{M(M+1)p+4\beta_m}$	$\dfrac{2(M+1)p\beta_r}{2\beta_r+(M+1)^2p}$
Optimal channel coding diversity with two relays	$\dfrac{2p\beta_m}{p+\beta_m}$	$\dfrac{3p\beta_r}{3p+\beta_r}$
Source coding diversity with two relays	$\max\left[\dfrac{4p\beta_m}{3p+2\beta_m}, \dfrac{2p\beta_m}{p+2\beta_m}\right]$	$\max\left[\dfrac{2p\beta_r}{2p+\beta_r}, \dfrac{3p\beta_r}{4p+\beta_r}\right]$

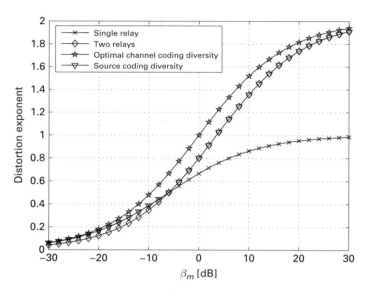

Fig. 15.10   Distortion exponents for two-hop amplify-and-forward (decode-and-forward) protocol. $P = 1$ (Rayleigh fading channel).

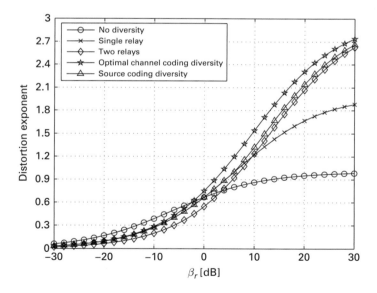

Fig. 15.11   Distortion exponents for the amplify-and-forward (decode-and-forward) relay channel. $P = 1$ (Rayleigh fading channel).

an outage limited system as the outage probability, rather than the quality of the source encoder, is the main limiting factor in the end-to-end distortion. Figure 15.10 shows that in this scenario, the distortion exponent performance is improved by increasing the number of relays so as to increase diversity.

At low β_m the system is said to be quality limited as the quality of the source encoder (distortion under no outage), rather than the outage probability, is the main limiting factor in the end-to-end distortion. In this case, the gain from using a better source encoder, that has a higher resolution, is more significant than the gain from increasing the number of relay nodes. Figure 15.10 shows that in this scenario, the distortion exponent performance is improved by using only a single relay node, which allows for the use of a higher resolution source encoder.

Figure 15.11 shows the distortion exponent versus β_r for the various relay channel schemes. Figure 15.11 confirms that the scheme with optimal channel coding diversity yields a better distortion exponent than the one with source coding diversity. As was the case for multi-hop schemes, as β_m increases, diversity becomes the limiting factor for the distortion exponent, in which case, Figure 15.11 also shows that increasing the number of relays improves the distortion exponent results. Again, at low β_m, direct transmission (no-diversity) results in a lower end-to-end distortion which can be interpreted in the same way as for the multi-hop channel.

15.5   Chapter summary and bibliographical notes

In this chapter, we have studied the performance of systems that combine user cooperation diversity with diversity generated at higher layers of the communication stack in the form of source coding and channel coding diversity. In the case of source coding, diversity is introduced through the use of dual-description source encoders. Channel coding diversity is obtained from joint decoding of channel coded blocks sent through different channels. From the viewpoint of user cooperation, diversity is provided to upper layers through configurations with a direct path between the source and the destination node (a relay channel), and without such direct path (a multi-hop channel). We have considered user cooperation using either the amplify-and-forward or the decode-and-forward techniques to study the achievable performance limits, which was measured in terms of the distortion exponent.

The study in this chapter shows that channel coding diversity provides better performance, followed by source coding diversity. For the case of having multiple relays that can help the source node to forward its information, and depending on the operating bandwidth expansion factor, we have determined the optimal number of relay nodes to cooperate with the source node to maximize the distortion exponent. The study shows a tradeoff between the source coding resolution and the number of relay nodes assigned for helping the source node. We note that at low bandwidth it is not the channel outage event, but the distortion introduced at the source coding stage the dominant factor limiting the distortion exponent performance. Therefore, in these cases it is better not to cooperate and use a lower distortion source encoder. Similarly, we showed that as the bandwidth expansion factor increases, the distortion exponent improves by allowing user cooperation. In these cases, the system is said to be an outage limited system and it is better to cooperate so as to minimize the outage probability and, consequently, minimize the end-to-end distortion.

The presentation in this chapter can be framed first within the idea that diversity can also be formed when multiple channels are provided to the application layer. A good study of this idea, from the perspective of the achievable performance of source–channel diversity over general parallel channels, can be found in [110]. The study in this paper is also focused on distortion exponent but without considering cooperation. An example of a work that uses distortion exponent to study single description source coded schemes with cooperation is [53]. In terms of multiple description coding, the work in [140] presents an early and useful study of the most important topics, such as rate-distortion performance. A further study on the achievable rates of multiple description coding can be found in [32] and further information on source–channel diversity in cooperative networks can be found in [173, 177, 169]

Exercises

15.1 For the dual description description source encoder operating in the region of low distortion ($D_1 + D_2 - D_0 < 1$), use (15.7) and the fact that for any well-designed MD codec $D_0 \leq D_1/2$, to derive the inequalities (15.8),

$$(4D_0 D_1)^{-1/(2\beta)} \lessgtr e^{R_{md}} \lessgtr (2D_0 D_1)^{-1/(2\beta)}.$$

[Hint: Recall that the lower bound requires $D_0 \to 0$ and the upper bound requires also $D_1 \to 0$.]

15.2 Using the techniques presented in this chapter prove that for the case when the source transmits a single description source coded message without the help of any relay node the distortion exponent is given by

$$\Delta_{NO-DIV} = \frac{2p\beta_r}{p + 2\beta_r},$$

where $\beta_r = N_r/K$ and N_r is the number of channel uses for the source block (refer to Figure 15.7).

15.3 Using the techniques presented in this chapter prove that Theorem 15.2.3 is true, i.e., show that for the two-hop two-relay optimal channel coding diversity amplify-and-forward system the distortion exponent is

$$\Delta_{SH-2R-OPTCH-AMP} = \frac{2p\beta_m}{p + \beta_m}.$$

15.4 Following the techniques used in this chapter show that the distortion exponent for the source coding diversity over two parallel channels can be written as

$$\Delta_{SH-2R-SRC-AMP} = \max \left[\frac{8p\beta_m''}{3p + 4\beta_m''}, \frac{4p\beta_m''}{p + 4\beta_m''} \right].$$

15.5 For the multi-hop channels with the decode-and-forward protocol, we have the following expressions for the distortion exponents, as stated in Theorem 15.2.5:

- for the two-hop single relay

$$\Delta_{\text{SH}-1\text{R}-\text{DEC}} = \frac{2p\beta_m}{p + 2\beta_m},$$

- for the two-hop M-relay selection channel coding diversity

$$\Delta_{\text{SH}-\text{MR}-\text{DEC}} = \frac{4Mp\beta_m}{M(M+1)p + 4\beta_m},$$

- for the two-hop two relay source coding diversity

$$\Delta_{\text{SH}-2\text{R}-\text{SRC}-\text{DEC}} = \max\left[\frac{4p\beta_m}{3p + 2\beta_m}, \frac{2p\beta_m}{p + 2\beta_m}\right].$$

Prove Theorem 15.2.5 using the techniques presented in this chapter.

15.6   For the relay channels with the decode-and-forward protocol, we have the following expressions for the distortion exponents, as stated in Theorem 15.3.5:

- For the single-relay channel

$$\Delta_{\text{RC}-1\text{R}-\text{DEC}} = \frac{2p\beta_r}{2p + \beta_r}.$$

- For the M-relay selection channel coding diversity

$$\Delta_{\text{RC}-\text{MR}-\text{DEC}} = \frac{2(M+1)p\beta_r}{2\beta_r + (M+1)^2 p}.$$

- For the optimal channel coding with two relays, with the same time frame structure as in Figure 15.6

$$\Delta_{\text{RC}-2\text{R}-\text{OPTCH}-\text{AMP}} = \frac{3p\beta_r}{3p + \beta_r}.$$

- For the source coding diversity with two relays

$$\Delta_{\text{RC}-2\text{R}-\text{SRC}-\text{DEC}} = \max\left[\frac{2p\beta_r}{2p + \beta_r}, \frac{3p\beta_r}{4p + \beta_r}\right].$$

Prove Theorem 15.3.5 using the techniques presented in this chapter.

15.7   In Section 15.1 we highlighted the fact that a high SNR approximation of the CDF of e^I as

$$F_{e^I}(t) \approx c\left(\frac{t}{\text{SNR}}\right)^p,$$

allows considering other distributions suitable for modeling $|h|^2$ besides the Rayleigh distribution. Assume now that the channel can be modeled with a Rice distribution and derive the value of the parameter p for this case. Use this result to redo Figures 15.10 and 15.11, and analyze the differences.

16 Coverage expansion with cooperation

For practical implementation of cooperative communications in wireless networks, we need to develop protocols by which nodes are assigned to cooperate with each other. In most of the previous chapters on cooperation, the cooperating relays are assumed to exist and are already paired with the source nodes in the network. A deterministic network topology, i.e., deterministic channel gain variances between different nodes in the network, was also assumed. If the random users' spatial distribution, and the associated propagation path losses between different nodes in the network, are taken into consideration, then these assumptions, in general, are no longer valid.

Moreover, it is of great importance for service providers to improve the coverage area in wireless networks without the cost of more infrastructure and under the same quality of service requirements. This poses challenges for deployment of wireless networks because of the difficult and unpredictable nature of wireless channels.

In this chapter, we address the relay assignment problem for implementing cooperative diversity protocols to extend the coverage area in wireless networks. We study the problem under the knowledge of the users' spatial distribution which determines the channel statistics, as the variance of the channel gain between any two nodes is a function of the distance between these two nodes. We consider an uplink scenario where a set of users are trying to communicate to a base station (BS) or access point (AP) and describe practical algorithms for relay assignment.

Another way to look at employing cooperation for coverage extension is to think of the relays as additional cheap access points that the service provider can deploy in the network. These relays communicate to the base station through the wireless channels, and can help those users away from the base station. Next, we elaborate more on this concept.

16.1 System model

We consider a wireless network with a circular cell of radius ρ. The BS/AP is located at the center of the cell, and N users are uniformly distributed within the cell. The probability density function of the user's distance r from the BS/AP is thus given by

$$q(r) = \frac{2r}{\rho^2}, \qquad 0 \le r \le \rho, \tag{16.1}$$

and the user's angle is uniformly distributed between $[0, 2\pi)$. Two communications schemes are going to be examined in the sequel. Non-cooperative transmission, or direct transmission, where users transmit their information directly to the BS/AP, and cooperative communications where users can employ a relay to forward their data.

In the direct transmission scheme, which is employed in current wireless networks, the signal received at the destination d (BS/AP) from source user s, can be modeled as

$$y_{s,d} = \sqrt{PKr_{s,d}^{-\alpha}} h_{s,d} x + n_{s,d}; \qquad (16.2)$$

where P is the transmitted signal power, x is the transmitted data with unit power, $h_{s,d}$ is the channel fading gain between the two terminals. The channel fade of any link is modeled as a zero-mean, circularly symmetric complex Gaussian random variable with unit variance. In (16.2), K is a constant that depends on the antennas design, α is the path loss exponent, and $r_{s,d}$ is the distance between the two terminals. K, α, and P are assumed to be the same for all users. The term $n_{s,d}$ in (16.2) denotes additive noise. All the noise components are modeled as white Gaussian noise (AWGN) with variance N_0. From (16.2), the received signal-to-noise ratio is

$$\text{SNR}(r_{s,d}) = \frac{|h_{s,d}|^2 Kr_{s,d}^{-\alpha} P}{N_0}. \qquad (16.3)$$

We characterize the system performance in terms of outage probability. Outage is defined as the event that the received SNR falls below a certain threshold γ_{nc}, where the subscript nc denotes non-cooperative transmission. The probability of outage $\mathcal{P}_{nc}$ for non-cooperative transmission is defined as,

$$\mathcal{P}_{nc} = P(\text{SNR}(r) \leq \gamma_{nc}). \qquad (16.4)$$

The SNR threshold γ_{nc} is determined according to the application and the transmitter/receiver structure. If the received SNR is higher than the threshold γ_{nc}, the receiver is assumed to be able to decode the received message with negligible probability of error. If an outage occurs, the packet is considered lost.

For the cooperation protocol, a hybrid version of the incremental and selection relaying is employed. In this hybrid protocol, if a user's packet is lost, the BS/AP broadcasts negative acknowledgement (NACK) so that the relay assigned to this user can re-transmit this packet again. This introduces spatial diversity because the source message can be transmitted via two independent channels as depicted in Figure 16.1. The relay will only transmit the packet if it is capable of capturing the packet, i.e., if the received SNR at the relay is above the threshold. In practice, this can be implemented by utilizing a cyclic redundancy check (CRC) code in the transmitted packet. The signal received from the source to the destination d and the relay l in the first stage can be modeled as

$$y_{s,d} = \sqrt{PKr_{s,d}^{-\alpha}} h_{s,d} x + n_{s,d},$$
$$y_{s,l} = \sqrt{PKr_{s,l}^{-\alpha}} h_{s,l} x + n_{s,l}. \qquad (16.5)$$

If the SNR of the signal received at the destination from the source falls below the cooperation SNR threshold γ_c, the destination requests a second copy from the relay.

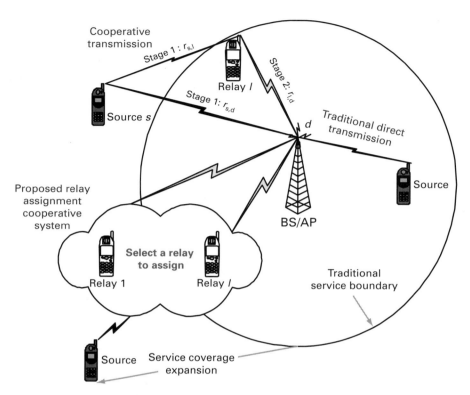

Fig. 16.1   Illustrating the difference between the direct and cooperative transmission schemes, and the coverage extension prospected by cooperative transmission.

Then if the relay was able to receive the packet from the source correctly, it forwards it to the destination:

$$y_{l,d} = \sqrt{PKr_{l,d}^{-\alpha}} h_{l,d}x + n_{l,d}. \qquad (16.6)$$

The destination will then combine the two copies of the message x as follows:

$$y_d = a_{s,d}y_{s,d} + a_{l,d}y_{l,d}; \qquad (16.7)$$

where $a_{s,d} = I\sqrt{PKl_{s,d}^{-\alpha}} h_{s,d}^*$, and $a_{r,d} = \sqrt{PKl_{r,d}^{-\alpha}} h_{r,d}^*$. The formulation in (16.7) allows us to consider two scenarios at the destination: if $I = 1$ then the combining at the destination is a maximal ratio combiner (MRC); on the other hand, if $I = 0$, then the destination only uses the relay message for decoding. The later scenario might be useful in the case where the destination cannot store an analogue copy of the source's message from the previous transmission.

16.2   Relay assignment: protocols and analysis

In this section, we start with driving the average outage for direct transmission. Then we calculate the conditional outage probability for cooperative transmission and try to use the formulas to deduce the best relay location.

16.2.1   Direct transmission

As discussed before, the outage is defined as the event that the received SNR is lower than a predefined threshold which we denote by γ_{nc}. The outage probability for the direct transmission mode $\mathcal{P}_{OD}$ conditioned on the user's distance can be calculated as

$$
\mathcal{P}_{OD}(r_{s,d}) = \mathcal{P}\left(\text{SNR}(r_{s,d}) \leq \gamma_{nc}\right)
$$
$$
= 1 - \exp\left(-\frac{N_0 \gamma_{nc} r_{s,d}^{\alpha}}{KP}\right)
$$
$$
\simeq \frac{N_0 \gamma_{nc} r_{s,d}^{\alpha}}{KP}, \tag{16.8}
$$

where the above follows because $\mid h_{s,d}\mid^2$, the magnitude of the channel fade squared, has an exponential distribution with unit mean. The approximation in (16.8) is at high SNR.

To find the average outage probability over the cell, we need to average over the user distribution in (16.1). The average outage probability is thus given by

$$
\mathcal{P}_{OD} = \int_0^{\rho} \mathcal{P}_{OD}(r_{s,d}) q(r_{s,d}) \mathrm{d}r_{s,d}
$$
$$
= \int_0^{\rho} \frac{2r_{s,d}}{\rho^2}\left(1 - \exp\left(-\frac{N_0 \gamma_{nc} r_{s,d}^{\alpha}}{KP}\right)\right) \mathrm{d}r_{s,d}
$$
$$
= 1 - \frac{2}{\alpha \rho^2}\left(\frac{KP}{N_0 \gamma_{nc}}\right)^{\frac{2}{\alpha}} \Gamma\left(\frac{2}{\alpha}, \frac{N_0 \gamma_{nc} \rho^{\alpha}}{KP}\right)
$$
$$
\simeq \frac{2\gamma_{nc}\rho^{\alpha} N_0}{KP(\alpha + 2)}, \tag{16.9}
$$

where $\Gamma(.,.)$ is the incomplete Gamma function, and it is defined as

$$
\Gamma(a, x) = \int_0^x \exp^{-t} t^{a-1} \mathrm{d}t. \tag{16.10}
$$

16.2.2   Cooperative transmission: conditional outage probability

Consider a source–destination pair that are $r_{s,d}$ units distance apart. Let us compute the conditional outage probability for given locations of the user and the helping relay. Using (16.5), the SNR received at the BS/AP d and the relay l from the source s is given by

$$\text{SNR}(r_{\text{s,d}}) = \frac{|\,h_{\text{s,d}}\,|^2\, K r_{\text{s,d}}^{-\alpha}\, P}{N_0},$$

$$\text{SNR}(r_{\text{s,l}}) = \frac{|\,h_{\text{s,l}}\,|^2\, K r_{\text{s,l}}^{-\alpha}\, P}{N_0}. \tag{16.11}$$

While from (16.7), the SNR of the combined signal received at the BS/AP is given by

$$\text{SNR}_d = \mathrm{I}\frac{|\,h_{\text{s,d}}\,|^2\, K r_{\text{s,d}}^{-\alpha}\, P}{N_0} + \frac{|\,h_{\text{l,d}}\,|^2\, K r_{\text{l,d}}^{-\alpha}\, P}{N_0}. \tag{16.12}$$

The terms $|\,h_{\text{s,d}}\,|^2$, $|\,h_{\text{s,l}}\,|^2$, and $|\,h_{\text{l,d}}\,|^2$ are mutually independent exponential random variables with unit mean. The outage probability of the cooperative transmission $\mathcal{P}_{\text{OC}}$ conditioned on the fixed topology of the user s and the relay l can be calculated as follows. Using the law of total probability we have

$$\mathcal{P}_{\text{OC}} = \Pr(\text{Outage}|\text{SNR}_{\text{s,d}} \le \gamma_{\text{c}})\Pr(\text{SNR}_{\text{s,d}} \le \text{SNR}_{\text{s,d}}) \tag{16.13}$$

where the probability of outage is zero if $\text{SNR}_{\text{s,d}} > \gamma_{\text{c}}$. The outage probability conditioned on the event that the source–destination link is in outage is given by

$$\Pr(\text{Outage}|\text{SNR}_{\text{s,d}} \le \gamma_{\text{c}}) = \Pr(\text{SNR}_{\text{s,l}} \le \gamma_{\text{c}}) + \Pr(\text{SNR}_{\text{s,l}} > \gamma_{\text{c}})$$
$$\times \Pr(\text{SNR}_d \le \gamma_{\text{c}}|\text{SNR}_{\text{s,d}} \le \gamma_{\text{c}}), \tag{16.14}$$

where the addition of the above probabilities is because they are disjoint events, and the multiplication is because the source–relay link is assumed to fade independently from the other links. The conditioning was removed for the same reason.

For the case where MRC is allowed at the destination, then the conditional outage probability at the destination is given by

$$\Pr(\text{SNR}_d \le \gamma_{\text{c}}|\text{SNR}_{\text{s,d}} \le \gamma_{\text{c}}) = \frac{\Pr(\text{SNR}_d \le \gamma_{\text{c}})}{\Pr(\text{SNR}_{\text{s,d}} \le \gamma_{\text{c}}}. \tag{16.15}$$

Using (16.15) and (16.14) in (16.13), the conditional outage probability for cooperative communications with MRC can be calculated as

$$\mathcal{P}_{\text{OC}}(r_{\text{s,d}}, r_{\text{s,l}}, r_{\text{l,d}}) = \left(1 - f(\gamma_{\text{c}}, r_{\text{s,d}})\right)\left(1 - f(\gamma_{\text{c}}, r_{\text{s,l}})\right) + f(\gamma_{\text{c}}, r_{\text{s,l}})$$
$$\times \left[1 - \frac{r_{\text{s,d}}^{-\alpha}}{r_{\text{s,d}}^{-\alpha} - r_{\text{l,d}}^{-\alpha}} f(\gamma_{\text{c}}, r_{\text{s,d}}) - \frac{r_{\text{l,d}}^{-\alpha}}{r_{\text{l,d}}^{-\alpha} - r_{\text{s,d}}^{-\alpha}} f(\gamma_{\text{c}}, r_{\text{l,d}})\right]$$

where $f(x, y) = \exp(-N_0 x y^{\alpha}/KP)$. The above expression can be simplified as follows:

$$\mathcal{P}_{\text{OC}}(r_{\text{s,d}}, r_{\text{s,l}}, r_{\text{l,d}}) = \left(1 - f(\gamma_{\text{c}}, r_{\text{s,d}})\right) - \frac{r_{\text{l,d}}^{-\alpha}}{r_{\text{l,d}}^{-\alpha} - r_{\text{s,d}}^{-\alpha}}$$
$$\times f(\gamma_{\text{c}}, r_{\text{s,l}})\left(f(\gamma_{\text{c}}, r_{\text{l,d}}) - f(\gamma_{\text{c}}, r_{\text{s,d}})\right).$$

For the I $= 1$ case, or when MRC is used at the destination, then using the approximation $\exp(-x) \simeq 1 - x + x^2/2$ for small x, the above outage expression can be approximated at high SNR to

$$\mathcal{P}_{OC}(r_{s,d}, r_{s,l}, r_{l,d}) \simeq \frac{N_0}{KP} r_{s,d}^{\alpha} - \frac{N_0^2}{2K^2P^2} r_{s,d}^{2\alpha} - \frac{r_{s,d}^{\alpha}}{r_{s,d}^{\alpha} - r_{l,d}^{\alpha}}$$

$$\times \left[\frac{N_0}{KP} (r_{s,d}^{\alpha} - r_{l,d}^{\alpha}) + \frac{N_0^2}{2K^2P^2} \left((r_{l,d}^{\alpha} - r_{s,d}^{\alpha})(2r_{s,l}^{\alpha} + r_{l,d}^{\alpha} + r_{s,d}^{\alpha}) \right) \right]$$

Simplifying the above expression, we get

$$\mathcal{P}_{OC}(r_{s,d}, r_{s,l}, r_{l,d}) \simeq \frac{N_0^2}{2K^2P^2} r_{s,d}^{2\alpha} \left[2\frac{r_{s,l}^{\alpha}}{r_{s,d}^{\alpha}} + \frac{r_{l,d}^{\alpha}}{r_{s,d}^{\alpha}} \right] \tag{16.16}$$

For the $I = 0$ case, or when no MRC is used at the destination, then the conditional outage expression in (16.15) simplifies to

$$\Pr(SNR_d \leq \gamma_c | SNR_{s,d} \leq \gamma_c) = \Pr(SNR_d \leq \gamma_c). \tag{16.17}$$

This is because the SNR received at the destination in this case is just due to the signal received from the relay–destination path. The conditional outage expression in this case can be shown to be given by

$$\mathcal{P}_{OC}(r_{s,d}, r_{s,l}, r_{l,d}) = \left(1 - f(\gamma_c, r_{s,d}) \right) \left[1 - f(\gamma_c, r_{l,d}) f(\gamma_c, r_{s,l}) \right]. \tag{16.18}$$

16.2.3 Optimal relay position

To find the optimal relay position, we need to find the pair $(r_{s,l}, r_{l,d})$ that minimizes the conditional outage probability expression in (16.16). First we consider the $I = 1$ scenario, where MRC is utilized at the receiver.

16.2.3.1 MRC case

In the following, we will prove that the optimal relay position, under fairly general conditions, is towards the source and on the line connecting the source and destination. Examining the conditional outage expression in (16.16), it is clear that, for any value of $r_{l,d}$, the optimal value for $r_{s,l}$ that minimizes the outage expression is the minimum value for $r_{s,l}$. And since for any value of $r_{l,d}$ the minimum $r_{s,l}$ lies on the straight line connecting the source and destination, we get the first intuitive result that the optimal relay position is on this straight line.

Now, we prove that the optimal relay position is towards the source. Normalizing with respect to $r_{s,d}$ by substituting $x = r_{l,d}/r_{s,d}$ in (16.16) and $1 - x = r_{s,l}/r_{s,d}$, we have

$$\mathcal{P}_{OC}(x) = \frac{N_0^2}{2K^2P^2} r_{s,d}^{2\alpha} \left[2(1-x)^{\alpha} + x^{\alpha} \right]. \tag{16.19}$$

Taking the derivative with respect to x we get

$$\frac{\partial \mathcal{P}_{OC}(x)}{\partial x} = \frac{N_0^2}{2K^2P^2} r_{s,d}^{2\alpha} \left[-2\alpha(1-x)^{\alpha-1} + \alpha x^{\alpha-1} \right]. \tag{16.20}$$

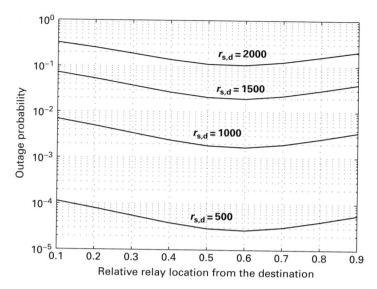

Fig. 16.2   The effect of the relay location on the outage probability.

Equating the above derivative to zero we get the unique solution

$$x^* = \frac{1}{1 + \left(\frac{1}{2}\right)^{\frac{1}{\alpha-1}}}. \tag{16.21}$$

Checking for the second-order conditions, we get, that $\mathcal{P}''_{OC}(x) \geq 0$, which shows that the problem is convex, and x^* specified in (16.21) is indeed the optimal relay position. Note from the optimal relay position in (16.21), that for propagation path loss $\alpha \geq 2$, we have that $x^* > 0.5$, which means that the optimal relay position is closer to the source node. In Figure 16.2, we plot the conditional outage probability expression for different source–destination separation distances, and different values for the relay location x. It is clear from the figure that the optimal relay position is, for a lot of cases, around $x = 0.6$ to $x = 0.75$. This is the motivation for considering the nearest-neighbor relay selection protocol in the next section.

16.2.3.2 No-MRC case

Next, we determine the optimal relay location for the $I = 0$ case. From the conditional outage expression in (16.18), it can be seen that if we have the freedom to put the relay anywhere in the two-dimensional plane of the source–destination pair, then the optimal relay position should be on the line joining the source and the destination – this is because if the relay is located at any position in the two-dimensional plane, then its distances to both the source and the destination are always larger than their corresponding projections on the straight line joining the source–destination pair.

In this case, we can substitute for $r_{\mathrm{l,d}}$ by $r_{\mathrm{s,d}} - r_{\mathrm{s,l}}$. The optimal relay position can be found via solving the following optimization problem:

$$r_{\mathrm{s,l}}^* = \arg\min_{r_{\mathrm{s,l}}} \mathcal{P}_{\mathrm{OC}}(r_{\mathrm{s,d}}, r_{\mathrm{s,l}}),$$

$$\text{s.t.} \quad 0 \le r_{\mathrm{s,l}} \le r_{\mathrm{s,d}}. \tag{16.22}$$

Since the minimization of the expression in (16.16) with respect to $r_{\mathrm{s,l}}$ is equivalent to minimizing the exponent in the second bracket, solving the optimization problem in (16.22) is equivalent to solving

$$r_{\mathrm{s,l}}^* = \arg\min_{r_{\mathrm{s,l}}} r_{\mathrm{s,l}}^{\eta} + \left(r_{\mathrm{s,d}} - r_{\mathrm{s,l}}\right)^{\eta},$$

$$\text{s.t.} \quad 0 \le r_{\mathrm{s,l}} \le r_{\mathrm{s,d}}. \tag{16.23}$$

The above optimization problem can be simply analytically solved, and the optimal relay position can be shown to be equal to $r_{\mathrm{s,l}}^* = r_{\mathrm{s,d}}/2$ for $\eta > 1$. Therefore, the optimal relay position is exactly in the middle between the source and destination when no MRC is used at the destination. For this case, we are able to drive a lower bound on the performance of any relay assignment protocol as will be discussed later.

16.3 Relay assignment algorithms

In this section, we describe two distributed relay assignment algorithms. The first is a user cooperation protocol in which the nearest neighbor is assigned as a relay. The second considers the scenario where fixed relays are deployed in the network to help the users.

16.3.1 Nearest-neighbor protocol

In this subsection, we describe the nearest-neighbor protocol for relay assignment, which is both distributed and simple to implement. In this protocol, the relay assigned to help is the nearest neighbor to the source as demonstrated in Figure 16.3. The source sends a "Hello" message to its neighbors and selects the signal received with the largest SNR, or the shortest arrival time, to be its closest neighbor.

The outage probability expression, which we refer to as $\mathcal{P}_{\mathrm{ONN}}$, for given source–relay–destination locations is still given by (16.16). To find the total probability, we need to average over all possible locations of the user and the relay. The user's location distribution with respect to the BS/AP is still given as in the direct transmission case (16.1). The relay's location distribution, however, is not uniform. In the sequel we calculate the probability density function of the relay's location. According to our protocol, the relay is chosen to be the nearest neighbor to the user. The probability that the nearest neighbor is at distance $r_{\mathrm{s,l}}$ from the source is equivalent to calculating the probability that the shaded area in Figure 16.3 is empty.

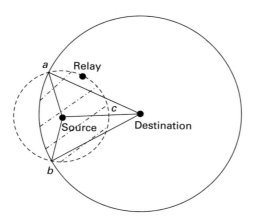

Fig. 16.3   Illustrating cooperation under the nearest-neighbor protocol: the nearest neighbor is at a distance $r_{s,1}$ from the source. Therefore, the shaded area should be empty from any users.

Denote this area, which is the intersection of the two circles with centers s and d, by $A(r_{s,d}, r_{s,1})$. For $0 < r_{s,1} \leq \rho - r_{s,d}$, the area of intersection is a circle with radius $r_{s,1}$ and center s. The probability density function of $r_{s,1}$, $p_{r_{s,1}}(x)$, can be calculated as

$$p_{r_{s,1}}(x) = \frac{\partial}{\partial x}\left(1 - \mathcal{P}(r_{s,1} > x)\right) \tag{16.24}$$

$$= \frac{\partial}{\partial x}\left(1 - \left(1 - \frac{x^2}{\rho^2}\right)^{N-1}\right) \tag{16.25}$$

$$= \frac{2(N-1)x}{\rho^2}\left(1 - \frac{x^2}{\rho^2}\right)^{N-2}, \quad 0 < r_{s,1} \leq \rho - r_{s,d}. \tag{16.26}$$

For $\rho - r_{s,d} < r_{s,1} \leq \rho + r_{s,d}$, the intersection between the two circles can be divided into three areas: (1) the area of the sector acb in circle s; (2) area of the triangle asb; (3) area enclosed by the chord ab in circle d. Hence, the intersection area, denoted by $A(r_{s,d}, r_{s,1})$ can be written as

$$A(r_{s,d}, r_{s,1}) = r_{s,1}^2\theta + \frac{1}{2}r_{s,1}^2\sin(2\theta) + \left(\rho^2\phi - \frac{1}{2}\rho^2\sin(2\phi)\right) \tag{16.27}$$

where $\theta = \cos^{-1}\left(\frac{\rho^2 - r_{s,1}^2 - r_{s,d}^2}{2r_{s,1}r_{s,d}}\right)$, and $\phi = \cos^{-1}\left(\frac{r_{s,1}^2 - \rho^2 - r_{s,d}^2}{2\rho r_{s,d}}\right)$. The probability density function for $r_{s,1}$ for this range is given by

$$p_{r_{s,1}}(x) = \frac{\partial}{\partial x}\left(1 - \left(1 - \frac{A(r_{s,d}, r_{s,1})}{\pi\rho^2}\right)^{N-1}\right), \quad \rho - r_{s,d} < r_{s,1} \leq \rho + r_{s,d}. \tag{16.28}$$

This completely defines the probability density function for the nearest neighbor and the average can be found numerically as the integrations are extremely complex. In the sequel, we derive an approximate expression for the outage probability for both the MRC and no-MRC scenarios under the following two assumptions. Since the relay is chosen to be the nearest neighbor to the source, the SNR received at the relay from the source is rarely below the threshold γ_c, hence, we assume that the event of the relay being in outage is negligible. The second assumption is that the nearest neighbor always lies on the intersection of the two circles, as points a or b in Figure 16.3. This second assumption is a kind of worst-case scenario, because a relay at distance $r_{s,1}$ from the source can be anywhere on the arc $\overparen{acb}$, and a worst-case scenario is to be at points a or b. This simplifies the outage calculation as the conditional outage probability (16.16) for the MRC case is now only a function of the source distance $r_{s,1}$ as follows:

$$\mathcal{P}_{ONN}(r_{s,d}) \simeq 1 - f(\gamma_c, r_{s,d}) - \frac{N_0 \gamma_c}{KP} r_{s,d}^\gamma f(\gamma_c, r_{s,d}). \qquad (16.29)$$

Averaging (16.29) over the user distribution (16.1) and using the definition of the incomplete Gamma function in (16.10), we get

$$\mathcal{P}_{ONN} \simeq 1 - \frac{2}{\alpha \rho^2} \left(\frac{KP}{N_0 \gamma_c} \right)^{\frac{2}{\alpha}} \Gamma \left(\frac{2}{\alpha}, \frac{N_0 \gamma_c \rho^\alpha}{KP} \right)$$
$$- \frac{2}{\alpha \rho^2} \left(\frac{KP}{N_0 \gamma_c} \right)^{\frac{2}{\alpha}} \Gamma \left(\frac{2}{\alpha} + 1, \frac{N_0 \gamma_c \rho^\alpha}{KP} \right).$$

Using the same approximation as above, the conditional outage probability for the no-MRC case is given by

$$\mathcal{P}_{ONN}(r_{s,d}) = \left(1 - f(\gamma_c, r_{s,d}) \right)^2. \qquad (16.30)$$

Averaging the above expression over $r_{s,d}$ we have

$$\mathcal{P}_{ONN} \simeq 1 - \frac{4}{\alpha \rho^2} \left(\frac{KP}{N_0 \gamma_c} \right)^{\frac{2}{\alpha}} \Gamma \left(\frac{2}{\alpha}, \frac{N_0 \gamma_c \rho^\alpha}{KP} \right)$$
$$+ \frac{2}{\alpha \rho^2} \left(\frac{KP}{2 N_0 \gamma_c} \right)^{\frac{2}{\alpha}} \Gamma \left(\frac{2}{\alpha}, \frac{2 N_0 \gamma_c \rho^\alpha}{KP} \right).$$

16.3.2   Fixed relays strategy

In some networks, it might be easier to deploy fixed nodes in the cell to act as relays. This will reduce the overhead of communications between users to pair for cooperation. Furthermore, in wireless networks users who belong to different authorities might act selfishly to maximize their own gains, i.e., selfish nodes. For such scenarios protocols for enforcing cooperation or to introduce incentives for the users to cooperate need to be implemented. In this subsection, we consider deploying nodes in the network that act as relays and do not have their own information. Each user will be associated with one relay to help in forwarding the dropped packets. The user can select the closets relay, which can be implemented using the exchange of "Hello" messages and selecting the signal with shortest arrival time, for example.

Continuing with our circular model for the cell, with uniform users distribution, the relays are deployed uniformly by dividing the cell into a finite number m of equal sectors, equal to the number of fixed relays to be deployed. Figure 16.4 depicts a network

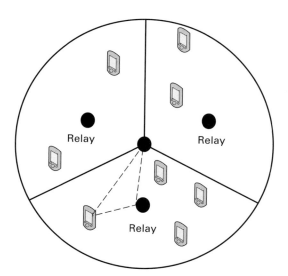

Fig. 16.4   Illustrating cooperation under the nearest-neighbor protocol: the nearest neighbor is at a distance $r_{s,1}$ from the source. Therefore, the shaded area should be empty from any users.

example for $m = 3$. The relays are deployed at a distance $r_{1,d}$ from the destination. This distance should be designed to minimize the average outage probability as follows:

$$r_{1,d}^* = \arg \min \mathcal{P}_{OC}(r_{1,d}),$$
$$\text{s.t.} \; 0 < r_{1,d} < \rho \tag{16.31}$$

where the average outage probability $\mathcal{P}_{OC}(r_{1,d})$ is defined as

$$\mathcal{P}_{OC} = \int_0^\rho \frac{2l_{s,d}}{\rho^2} \int_{-\frac{\pi}{m}}^{\frac{\pi}{m}} \mathcal{P}_{OC}(r_{s,d}, r_{s,1}(\theta), r_{1,d}) \frac{m}{2\pi} d\theta \, dl_{s,d} \tag{16.32}$$

where $\mathcal{P}_{OC}(l_{s,d}, l_{s,r}, l_{r,d})$ is defined in (16.16), and the distance from the source to the fixed relay is given by

$$r_{s,1}(\theta) = \sqrt{r_{s,d}^2 + r_{1,d}^2 - 2r_{s,d}r_{1,d}\cos(\theta)}, \tag{16.33}$$

where θ is uniformly distributed between $[-\pi/m, \pi/m]$. Solving the above optimization problem is very difficult, hence, let us consider the following heuristic. In the nearest-neighbor protocol, the relay was selected to be the nearest neighbor to the user. Here, we can calculate the relay position that minimizes the mean square distance between the users in the sector and the relay. Without loss of generality, assuming the line dividing the sector to be the x-axis, the mean square distance between a user at distance r and angle θ from the center of the cell and the relay is given by the following function:

$$q(r_{1,d}) = E\left(\|re^{j\theta} - r_{1,d}\|^2\right), \tag{16.34}$$

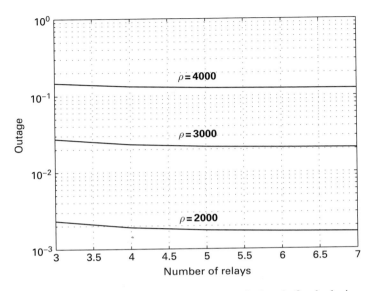

Fig. 16.5 Average outage probability versus the number of relays in fixed relaying.

where $j = \sqrt{-1}$, and E denotes the joint statistical expectation over the random variables r and θ. Solving for the optimal $r_{1,d}$ that minimizes $q(r_{1,d})$,

$$r_{1,d}^* = \arg \min E\left(\|re^{j\theta} - r_{1,d}\|^2\right), \tag{16.35}$$

we get

$$r_{1,d}^* = \frac{2m}{3\pi} \sin\left(\frac{\pi}{m}\right)\rho. \tag{16.36}$$

Figure 16.5 depicts the average outage probability versus the number of relays deployed in the network for different cell sizes. The numerical results are for the following parameters: $K = 1$, $\alpha = 3$, $P = 0.05$, $R = 1$, and $N_0 = 10^{-12}$. We can see from the results that the performance saturates at approximately $m = 6$ relays, which suggests that dividing the cell into six sectors with a relay deployed in each sector can provide good enough performance.

16.3.3 Lower bound: the Genie-aided algorithm

For both the MRC and no-MRC cases, we determined the optimal relay location. For the MRC case, the optimal relay position is towards the source. For the no-MRC case, we showed that the optimal relay position is in the mid-point between the source and the destination. We will derive a lower bound on the outage probability for any relay assignment protocol based on a Genie-aided approach. This bound serves as a benchmark for the performance of the nearest-neighbor protocol, and the fixed relaying scheme described in the chapter.

The Genie-aided protocol works as follows. For any source node in the network, a Genie is going to put a relay at the optimal position on the line joining this source node and the destination (BS/AP).

Next we analyze the average outage performance of the Genie-aided protocol. For the MRC case, substituting the optimal relay position in (16.21) in the conditional outage expression in (16.16) we get

$$
\mathcal{P}_{\mathrm{OC}}(r_{\mathrm{s,d}}) = 1 - f(\gamma_c, r_{\mathrm{s,d}}) - \frac{1}{1-x^*} f(\gamma_c, (1-x^*)r_{\mathrm{s,d}}) \left(f(\gamma_c, x^* r_{\mathrm{s,d}}) - f(\gamma_c, r_{\mathrm{s,d}}) \right).
$$
(16.37)

Averaging the above expression over the user distribution, the average outage probability for the Genie-aided lower bound for the MRC case $I = 1$ is given by

$$
\mathcal{P}_{\mathrm{OG},1} = 1 - \frac{2}{\alpha\rho^2} \left(\frac{KP}{N_0\gamma_c} \right)^{\frac{2}{\alpha}} \Gamma\left(\frac{2}{\alpha}, \frac{N_0\gamma_c\rho^\alpha}{KP} \right)
$$

$$
- \frac{1}{1-(x^*)^\alpha} \frac{2}{\alpha\rho^2} \left(\frac{KP}{N_0\gamma_c((1-x^*)^\alpha + (x^*)^\alpha)} \right)^{\frac{2}{\alpha}}
$$
$$
\Gamma\left(\frac{2}{\alpha}, \frac{((1-x^*)^\alpha + (x^*)^\alpha) N_0\gamma_c\rho^\alpha}{KP} \right)
$$

$$
+ \frac{1}{1-(x^*)^\alpha} \frac{2}{\alpha\rho^2} \left(\frac{KP}{N_0\gamma_c((1-x^*)^\alpha + 1)} \right)^{\frac{2}{\alpha}}
$$
$$
\Gamma\left(\frac{2}{\alpha}, \frac{((1-x^*)^\alpha + 1) N_0\gamma_c\rho^\alpha}{KP} \right)
$$
(16.38)

We will denote the average probability of outage for the no-MRC case by $\mathcal{P}_{\mathrm{OG},2}$. Substituting the optimal relay position $r_{\mathrm{s,1}}^*$ in the conditional outage expression (16.16), we get

$$
\mathcal{P}_{\mathrm{OG}}(r_{\mathrm{s,d}}) = \left(1 - \exp\left(-\frac{N_0\gamma_c r_{\mathrm{s,d}}^\alpha}{KP} \right) \right) \left(1 - \exp\left(-\frac{2N_0\gamma_c \left(\frac{r_{\mathrm{s,d}}}{2} \right)^\alpha}{KP} \right) \right).
$$
(16.39)

Averaging the above expression over all possible users' locations,

$$
\mathcal{P}_{\mathrm{OG},2} = 1 + \frac{2}{\alpha\rho^2} \left(\frac{kP}{N_0\gamma_c(1+2^{1-\alpha})} \right)^{\frac{2}{\alpha}} \Gamma\left(\frac{2}{\alpha}, \frac{N_0\gamma_c(1+2^{1-\alpha})\rho^\alpha}{kP} \right)
$$

$$
- \frac{2}{\alpha\rho^2} \left(\frac{kP}{N_0\gamma_c} \right)^{\frac{2}{\alpha}} \Gamma\left(\frac{2}{\alpha}, \frac{N_0\gamma_c\rho^\alpha}{kP} \right)
$$

$$
- \frac{2}{\alpha\rho^2} \left(\frac{kP}{N_0\gamma_c 2^{1-\alpha}} \right)^{\frac{2}{\alpha}} \Gamma\left(\frac{2}{\alpha}, \frac{N_0\gamma_c 2^{1-\alpha}\rho^\alpha}{kP} \right).
$$
(16.40)

16.4 Numerical results

We performed some computer simulations to compare the performance of the above discussed relay assignment protocols and validate the theoretical results we derived in this chapter. In all of our simulations, we compared the outage performance of three different transmission schemes: direct transmission, nearest-neighbor protocol, and fixed relaying. In all of the simulations, the channel fading between any two nodes (either a user and the BS/AP or two users) is modeled as a random Rayleigh fading channel with unit variance.

For fairness in comparison between the described cooperative schemes and the direct transmission scheme, the spectral efficiency is kept fixed in both cases and this is done as follows. Since a packet is either transmitted once or twice in the cooperative protocol, the average rate in the cooperative case can be calculated as

$$E(R_c) = R_c \mathcal{P}_{OD,\gamma_c}(r_{s,d}) + \frac{R_c}{2}\mathcal{P}_{OD,\gamma_c}(r_s d). \tag{16.41}$$

where R_c is the spectral efficiency in bits/s/Hz for cooperative transmission, and $\mathcal{P}_{OD,\gamma_c}(r_{s,d})$ denotes the outage probability for the direct link at rate R_c. In (16.41), note that one time slot is utilized if the direct link is not in outage, and two time slots are utilized if it is in outage. Note that the later scenario is true even if the relay does not transmit because the time slot is wasted anyway. Averaging over the source–destination separation, the average rate is given by

$$\bar{R}_c = \frac{R_c}{2}\left(1 + \frac{2}{\alpha\rho^2}\left(\frac{KP}{\gamma_c N_0}\right)^{\frac{2}{\alpha}}\Gamma\left(\frac{2}{\alpha}, \frac{N_0\gamma_c\rho^\alpha}{KP}\right)\right) \tag{16.42}$$

We need to calculate the SNR threshold γ_c corresponding to transmitting at rate R_c. The resulting SNR threshold γ_c should generally be larger than γ_{nc} required for non-cooperative transmission. It is in general very difficult to find an explicit relation between the SNR threshold γ_c and the transmission rate R_c, and thus we render to a special case to capture the insights of this scenario. Let the outage be defined as the event that the mutual information I between two terminals is less than some specific rate R. If the transmitted signals are Gaussian, then according to our channel model, the mutual information is given by $I = \log(1 + SNR_{s,d})$. The outage event for this case is defined as

$$O_I \triangleq \{h_{s,d} : I < R\} = \{h_{s,d} : SNR_{s,d} < 2^R - 1\}. \tag{16.43}$$

The above equation implies that if the outage is defined in terms of the mutual information and the transmitted signals are Gaussian, then the SNR threshold γ_c and the spectral efficiency R are related as $\gamma_c = 2^R_c - 1$, i.e., they exhibit an exponential relation. Intuitively, under a fixed modulation scheme and fixed average power constraint, one can think of the SNR threshold as being proportional to the minimum distance between the constellation points, which in turn depends on the number of constellation points for fixed average power, and the later has an exponential relation to the number of bits per symbol that determines the spectral efficiency R. For the sake of comparison $\bar{R}_c$ should be equal to R, the spectral efficiency of direct transmission. Thus for a given R one should solve for R_c. This can lead to many solutions for R_c, and we are going to choose the minimum R_c as we have discussed before in Chapter 11.

Example 16.1   In the following simulation comparisons, we study the outage probability performance when varying three basic quantities in our communication setup: the transmission rate, the transmit power, and the cell radius. In all the scenarios, we consider direct transmission, nearest-neighbor, fixed relaying with six relays deployed in the network, and the Genie-aided lower bound. For all the cooperative transmission cases, both MRC and no-MRC is examined.

Figure 16.6 depicts the outage probability versus the transmit power in dBW. It is clear from the slopes of the curves that cooperation yields more steeper curves due to the diversity gain. Fixed relaying with MRC has the best performance, and it is very close to the Genie lower bound with no MRC. Fixed relay has a generally better performance than the nearest-neighbor protocols. Cooperation yields around 7 dbW savings in the transmit power with respect to direct transmission

Figure 16.7 depicts the outage probability curves versus the cell radius. Fixed relaying also has the best performance. There is a 200% increase in the cell radius at an 0.01 outage. We can see that the gap between direct transmission and cooperation decrease with increasing the cell size. The rationale here is that with increasing the cell size, the probability of packets in outage increases, and hence, the probability that the relay will forward the source's packet increases. This reduces the bandwidth efficiency of the system, and hence increases the overall outage probability. This tradeoff between the

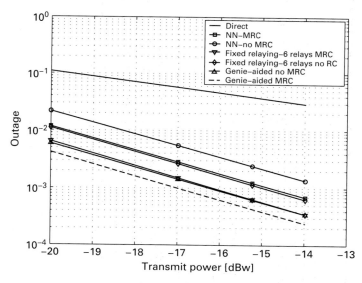

Fig. 16.6   Average outage probability versus the transmit power.

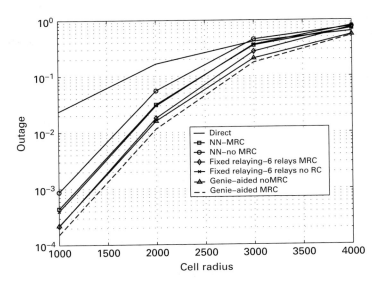

Fig. 16.7   Average outage probability versus the cell radius.

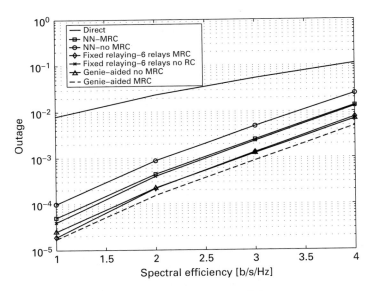

Fig. 16.8   Average outage probability versus the spectral efficiency.

spectral efficiency and the diversity gain of cooperation makes direct transmission good enough for larger cell sizes.

Similar conclusions can be drawn from Figure 16.8, which plots the outage probability versus the spectral efficiency.   ▲

16.5   Chapter summary and bibliographical notes

Most of the literature on relay assignment assumes the availability of a list of candidate relays and develop relay-selection algorithms from among the list [123]. In [128], two approaches for selecting a best relay are provided: best-select in the neighbor set and best-select in the decoded set. The best-select in the neighbor set algorithm is based on the average received SNRs, or equivalently the distance, while the latter is based on the instantaneous channel fading realization.

In this chapter, we addressed the relay assignment problem for coverage extension in cooperative transmission over wireless networks based on the knowledge of the channel statistics governed by the users' spatial distribution. The subject has been treated in the literature in [155, 158, 157, 153]. We described in this chapter two distributed relay assignment protocols. The nearest-neighbor protocol is a simple algorithm in which the relay is selected to be the nearest neighbor to the user. We also considered the scenario where fixed relays are deployed in the network to help the existing users. Outage performance of the protocols was analyzed. We further developed lower bounds on the performance of any relay assignment protocol via a Genie-aided method. Our numerical results indicate significant gains in the system performance. In particular, fixing the average transmit power, significant increase in the coverage area (more than 200%) of the network can be achieved by our simple distributed protocols. Similarly, for fixed cell radius, the average power required to achieve a certain outage probability is significantly reduced by more than 7 dbW in our numerical examples. We have also shown that cooperation can allow, at the same quality of service, transmitting at higher rates compared to direct transmission; more than 2 bits/s/Hz can be gained at the same transmit power. We can also see that for larger cell sizes, the performance gap between direct and cooperative transmission diminishes.

Exercises

16.1   In this chapter, the distribution of the nearest neighbor was derived to find the outage performance of the nearest-neighbor protocol. Consider the scenario where instead of assigning the first neighbor to act as a relay, the second neighbor is assigned this role.
   (a) Find the distribution of the second nearest neighbor.
   (b) Find the outage probability when utilizing the second nearest neighbor as a relay. Compare the performance to the scenario described in this chapter in which the nearest neighbor acts as the relay.

16.2   Numerical examples for the fixed relays scheme with six relays were developed in the chapter. Simulate the performance of the fixed relay scenarios with $N > 6$ relays. Find the value of N at which the performance starts to saturate.

16.3 Assume a cell with a fixed user density of 20 users/km^2.

   (a) Write a Matlab code to simulate the outage probability performance of the nearest neighbor protocol with increasing cell radius from 1 km to 5 km.

   (b) Assuming six relays are deployed in the cell, simulate the outage performance as a function of the cell radius.

16.4 Consider the scenario when you can at most deploy three relays in the cell in the fixed-relaying technique. Compare the performance to the nearest-neighbor protocol assuming a fixed user density as in the previous problem.

16.5 Prove that the optimal relay location in the fixed relays strategy is given by (16.36), i.e., prove that

$$r_{1,d}^* = \frac{2m}{3\pi} \sin\left(\frac{\pi}{m}\right) \rho.$$ (E16.1)

16.6 Prove the outage probability expression when maximal ratio combiner (MRC) is used at the destination. Prove the following simplified form of the outage expression:

$$\mathcal{P}_{OC}(r_{s,d}, r_{s,1}, r_{1,d}) = \left(1 - f(\gamma_c, r_{s,d})\right) - \frac{r_{1,d}^{-\alpha}}{r_{1,d}^{-\alpha} - r_{s,d}^{-\alpha}}$$
$$\times f(\gamma_c, r_{s,1}) \left(f(\gamma_c, r_{1,d}) - f(\gamma_c, r_{s,d})\right).$$

16.7 Prove that the average outage probability for the Genie-aided lower bound for the MRC case $I = 1$ is given by (16.38):

$$\mathcal{P}_{OG,1} = 1 - \frac{2}{\alpha\rho^2} \left(\frac{KP}{N_0\gamma_c}\right)^{\frac{2}{\alpha}} \Gamma\left(\frac{2}{\alpha}, \frac{N_0\gamma_c\rho^\alpha}{KP}\right)$$

$$- \frac{1}{1 - (x^*)^\alpha} \frac{2}{\alpha\rho^2} \left(\frac{KP}{N_0\gamma_c \left((1 - x^*)^\alpha + (x^*)^\alpha\right)}\right)^{\frac{2}{\alpha}}$$
$$\times \Gamma\left(\frac{2}{\alpha}, \frac{((1-x^*)^\alpha + (x^*)^\alpha) N_0\gamma_c\rho^\alpha}{KP}\right)$$

$$+ \frac{1}{1 - (x^*)^\alpha} \frac{2}{\alpha\rho^2} \left(\frac{KP}{N_0\gamma_c \left((1 - x^*)^\alpha + 1\right)}\right)^{\frac{2}{\alpha}}$$
$$\times \Gamma\left(\frac{2}{\alpha}, \frac{((1-x^*)^\alpha + 1) N_0\gamma_c\rho^\alpha}{KP}\right).$$

16.8   Prove that the average outage probability for the Genie-aided lower bound for the no-MRC case is given by (16.40):

$$
\mathcal{P}_{\text{OG},2} = 1 + \frac{2}{\alpha\rho^2} \left(\frac{kP}{N_0\gamma_c \left(1 + 2^{1-\alpha}\right)} \right)^{\frac{2}{\alpha}} \Gamma \left(\frac{2}{\alpha}, \frac{N_0\gamma_c \left(1 + 2^{1-\alpha}\right) \rho^\alpha}{kP} \right)
$$
$$
- \frac{2}{\alpha\rho^2} \left(\frac{kP}{N_0\gamma_c} \right)^{\frac{2}{\alpha}} \Gamma \left(\frac{2}{\alpha}, \frac{N_0\gamma_c\rho^\alpha}{kP} \right) - \frac{2}{\alpha\rho^2} \left(\frac{kP}{N_0\gamma_c 2^{1-\alpha}} \right)^{\frac{2}{\alpha}}
$$
$$
\times \Gamma \left(\frac{2}{\alpha}, \frac{N_0\gamma_c 2^{1-\alpha}\rho^\alpha}{kP} \right).
$$

17 Broadband cooperative communications

In broadband communications, OFDM is an effective means to capture multipath energy, mitigate the intersymbol interference, and offer high spectral efficiency. OFDM is used in many communications systems, e.g., wireless local area networks (WLANs) and wireless personal area networks (WPANs). Recently, OFDM together with time–frequency interleaving across subbands, the so-called multiband OFDM [9], has been adopted in the ultra-wideband (UWB) standard for wireless personal area networks (WPANs).

To improve the performance of OFDM systems, the fundamental concept of cooperative diversity can be applied. Nevertheless, special modulations/cooperation strategies are needed to efficiently exploit the available multiple carriers.

In this chapter, we study an OFDM cooperative protocol that improves spectral efficiency over those based on fixed relaying protocols while achieving the same performance of full diversity. By exploiting limited feedback from the destination node, the described protocol allows each relay to help forward information of multiple sources in one OFDM symbol. We also describe a practical relay assignment scheme for implementing this cooperative protocol in OFDM networks.

17.1   System model

In this section, we describe the system model of a wireless network, in which we take into consideration the random users' spatial distribution. The channel model, the signal model, and the performance measure in term of outage probability are discussed.

We consider an OFDM wireless network such as a WLAN or a WPAN with a circular cell of radius ρ. The cell contains one central node and multiple users, each communicating with the central node. The central node can be a base station or an access point in case of the WLAN, and it can be a piconet coordinator in case of the WPAN. Suppose the central node is located at the center of the cell, and K users are uniformly located within the cell. Then, the user's distance $D_{s,d}$ from the central node has the following probability density function (pdf):

$$p_{D_{s,d}}(D) = 2D/\rho^2, \quad 0 \le D \le \rho, \tag{17.1}$$

and the user's angle is uniformly distributed over $[0, 2\pi)$. We assume that each node is equipped with single antenna, and its transmission is constrained to half-duplex mode.

We consider an uplink scenario where all users transmit their information to the central node. Channel access within the cell is based on orthogonal multiple access mechanism as used in many current OFDM wireless networks.

The channel fades for different propagation links are assumed to be statistically independent. This assumption is reasonable since nodes in the network are usually spatially well separated. The data packet of each user consists of preamble, header, and frame payload which carries several OFDM data symbols. The header includes the pilot symbols which allow channel estimation to be performed at the central node.

For subsequent analysis, we consider a stochastic tapped-delay-line channel model as in (1.6), which allows to take into consideration the frequency selectivity of wireless channels. If the channel between each pair of transmit-receive nodes have L independent delay paths, the channel impulse response from node i to node j can be modeled as

$$h_{i,j}(t) = \sum_{l=0}^{L-1} \alpha_{i,j}(l)\delta(t - \tau_{i,j}(l)),$$ (17.2)

where the subscript $\{i, j\}$ indicates the channel link from node i to node j, $\tau_{i,j}(l)$ is the delay of the l-th path, and $\alpha_{i,j}(l)$ is the complex amplitude of the l-th path. The path amplitude $|\alpha_{i,j}(l)|$ is Rayleigh distributed, whereas the phase $\angle\alpha_{i,j}(l)$ is uniformly distributed over $[0, 2\pi)$. Specifically, The fade $\alpha_{i,j}(l)$ is modeled as zero-mean, complex Gaussian random variables with variances $E\left[|\alpha_{i,j}(l)|^2\right] = \Omega_{i,j}(l)$,

The received signal at subcarrier n of destination d from source user s can be modeled as

$$y_{s,d}(n) = \sqrt{P_{nc}\kappa D_{s,d}^{-\nu}} H_{s,d}(n)x_s(n) + z_{s,d}(n),$$ (17.3)

where P_{nc} is the transmitted power in each subcarrier at the source in non-cooperative mode, $x_s(n)$ denotes an information symbol to be transmitted from the source s at subcarrier n, $z_{s,d}(n)$ is an additive white Gaussian noise with zero mean and variance N_0, and $H_{s,d}(n)$ is the frequency response of the channel at the n-th subcarrier. If we consider OFDM with N subcarriers and assume that the length of cyclic prefix in the OFDM symbol is longer than the channel delay spread, then we can express the channel frequency response as

$$H_{s,d}(n) = \sum_{l=0}^{L-1} \alpha_{s,d}(l) \exp\left(-j2\pi n\tau_{s,d}(l)/N\right).$$ (17.4)

In (17.3), κ is a constant whose value depends on the propagation environment and antenna design, ν is the propagation loss factor, and $D_{s,d}$ represents the distance between source node s and destination node d. The parameters P_{nc}, κ, and ν are assumed to be the same for all users.

Let each information symbol have unit energy. For a given distance $D_{s,d}$ between the source and the destination, the received signal-to-noise ratio (SNR) for the n-th subcarrier can be given by

$$\zeta_{s,d}^{(n)} = P_{nc}\kappa D_{s,d}^{-\nu}|H_{s,d}(n)|^2/N_0.$$ (17.5)

In case of direct transmission between two nodes that are $D_{s,d}$ apart, the maximum average mutual information in a subcarrier n, which is achieved for the independent and identically distributed zero-mean, circularly symmetric complex Gaussian inputs, is given by

$$I^{(n)}(D_{s,d}) = \log(1 + \zeta_{s,d}^{(n)}). \qquad (17.6)$$

Let R denote a target rate for each subcarrier, then the probability that a subcarrier is in outage can be given by

$$P_{out}^{D}(D_{s,d}) = \Pr\left(I^{(n)}(D_{s,d}) \leq R\right) = \Pr\left(\zeta_{s,d}^{(n)} \leq 2^R - 1\right). \qquad (17.7)$$

If an outage occurs in a subcarrier, then the transmitted information in that subcarrier is considered loss. Otherwise, the receiver is assumed to be able to decode the received information with negligible probability of error.

17.2 Cooperative protocol and relay-assignment scheme

In this section, we first describe the cooperative protocol for OFDM wireless networks, and then present the relay assignment scheme. Consider a cooperation scenario where each source can employ a relay to forward its information to the destination.

In phase 1, each user transmits its packet to the destination (central node) and the packets are also received at the relay. After receiving the user's packet, the destination performs channel estimation using the OFDM pilot symbols in the packet header. Based on the estimated channel coefficients, the destination is able to specify which subcarrier symbols are not received successfully (i.e., those in the subcarriers of which the combined SNRs fall below the required SNR threshold), and then broadcasts the indices of the subcarriers carrying those symbols. Such feedback enables the assigned relay to help forward the source information only when necessary.

In phase 2, the relay decodes the source symbols that are unsuccessfully received at the destination via direct transmission, and then forwards the decoded information to the destination. The relay will only decode the source symbol that the relay is capable of capturing, i.e., when the received SNR at the relay is above the target threshold. Since it is unlikely that all subcarrier symbols are sent unsuccessfully by the source, the protocol makes efficient use of the available bandwidth by allowing the relay to help forward the information of multiple source users in one OFDM block.

Specifically, suppose a relay is assigned to help k users, then after all k users transmit their packets in phase 1, the relay sends in phase 2 an additional packet containing these users' symbols that are not captured at the destination in phase 1. In order for the destination to know which relay subcarriers carry information of which users, the relay can send an additional overhead together with the source symbols. Or the relay can use a fixed subcarrier assignment scheme such that that the destination can specify the relay subcarrier assignment without the use of additional overhead.

For instance, if n_i subcarriers of user i are in outage, then in phase 2, the relay can use the first n_1 subcarriers to transmit the data of user 1, the next n_2 subcarriers to transmit the data of user 2, and so on. The fixed subcarrier assignment scheme requires no additional overhead, but it does not fully utilize the resources when both relay and destination cannot decode the source symbols; nevertheless, this event occurs with small probability.

Figure 17.1 illustrates as example of the signal transmissions of the described protocol with fixed subcarrier assignment scheme for an OFDM system with two source users and one relay. Figure 17.1(a) and (b) depict transmission in phase 1 and phase 2, respectively.

In phase 1, the received signals at the destination and the relay are

$$y_{s,d}(n) = \sqrt{P_c \kappa D_{s,d}^{-\nu}} H_{s,d}(n) x_s(n) + z_{s,d}(n) \tag{17.8}$$

and

$$y_{s,r}(n) = \sqrt{P_c \kappa D_{s,r}^{-\nu}} H_{s,r}(n) x_s(n) + z_{s,r}(n), \tag{17.9}$$

respectively, where P_c is the transmitted power in the cooperative mode. In phase 2, the signal received at the destination from the relay is

$$y_{r,d}(n) = \sqrt{P_c \kappa D_{r,d}^{-\nu}} H_{r,d}(n) \tilde{x}_s(n) + z_{r,d}(n), \tag{17.10}$$

where $\tilde{x}_s(n)$ denotes the source symbols that are not captured by the destination in phase 1.

To decide the pairing of sources and relays, we describe in this subsection a practical relay assignment scheme for cooperative OFDM networks. We focus on a fixed-relay-location scenario in which a certain number of relays are installed in fixed locations

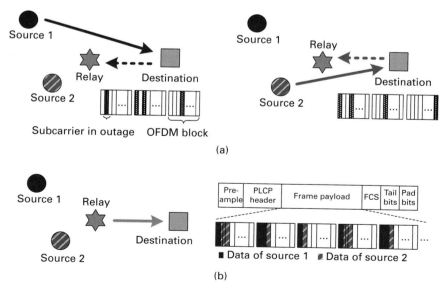

(a)

(b)

Fig. 17.1   Illustrations of the described cooperative protocol for OFDM system with two users and one relay. (a) Phase 1: Source transmits information to the destination. (b) Phase 2: Relay forwards source information to the destination.

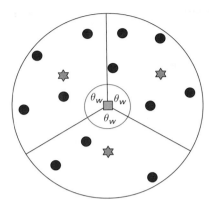

Fig. 17.2   An example of relay assignment for a multi-user OFDM system. A cell with three relays.

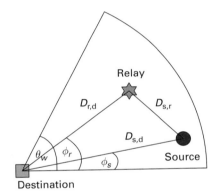

Fig. 17.3   An example of relay assignment for a multi-user OFDM system. One sector.

in the network. Let W denote the number of relays in a circular cell with radius ρ, as illustrated in Figure 17.2. Then, the relay assignment scheme is as follows:

- The cell is equally divided into W sectors, each with central angle $2\pi/W$. One relay is assigned to help users within each sector.
- In each sector, the relay is placed at an optimum relay location which minimizes the outage probability for all possible source–destination pairs within the sector (Figure 17.3).

The optimum relay location will be determined in Section 17.5.

17.3   Performance analysis

In this section, we first analyze the outage probability of the described cooperative protocol based on fixed subcarrier assignment scheme, then provide its closed-form lower bound, and finally determine the optimum relay location.

17.3.1   Outage probability

In the non-cooperative protocol, the channel resources (time and frequency) are divided among K users. In the cooperative protocol, on the other hand, W relays are also utilized, so the same channel resources are divided among $K + W$ users. Thus, for the cooperative protocol, the mutual information in each subcarrier is given by $K/(K + W)I^{(n)}(D_{i,j})$, and the event that a subcarrier in the i–j transmit–receive link is in outage is

$$\frac{K}{K + W}I^{(n)}(D_{i,j}) \leq R, \tag{17.11}$$

which corresponds to the event

$$|H_{i,j}(n)|^2 \leq \frac{(2^{(K+W)R/K} - 1)N_0}{\kappa P_c}D_{i,j}^{-\nu} \triangleq \beta D_{i,j}^{-\nu}. \tag{17.12}$$

where β is a constant.

Under the described cooperative protocol, the average probability that a subcarrier is in outage can be expressed as

$$P_{\text{out}} = \sum_{k=0}^{K} P_{\text{out}|k}\Pr(k \text{ users in the sector}), \tag{17.13}$$

where $P_{\text{out}|k}$ denotes the outage probability per subcarrier given that k users are in the sector. With W sectors in the cell and the assumption that the users are uniformly located in the cell, the chance that a user is located in a specific sector is given by $1/W$. Accordingly, the probability that k users are located in one sector follows the binomial distribution as

$$\Pr(k \text{ users in the sector}) = \binom{K}{k}\left(\frac{1}{W}\right)^k\left(1 - \frac{1}{W}\right)^{K-k}$$

$$\triangleq c(k), \quad k \in \{0, 1, \ldots, K\}. \tag{17.14}$$

The conditional outage probability per subcarrier, $P_{\text{out}|k}$, can be determined as follows. In the case that the relay has available resources to help forward the source information, the outage event in subcarrier n is equivalent to the event

$$\Phi = \left(\{I^{(n)}(D_{\text{s,d}}) \leq \tilde{R}\} \cap \{I^{(n)}(D_{\text{s,r}}) \leq \tilde{R}\}\right) \cup \left(\{I^{(n)}(D_{\text{s,d}}) \leq \tilde{R}\}\right.$$

$$\left.\cap \{I^{(n)}(D_{\text{r,d}}) \leq \tilde{R}\} \cap \{I^{(n)}(D_{\text{s,r}}) > \tilde{R}\}\right), \tag{17.15}$$

where $\tilde{R} = (K+W)R/K$. The first term in the union of (17.15) corresponds to the event that both the source–destination and source–relay links are in outage, while the second term corresponds to the event that both the source–destination and relay–destination links are in outage while the source–relay link is not. According to (17.12) and Rayleigh distribution, we have

$$\Pr(\{I^{(n)}(D_{i,j}) \leq \tilde{R}\}) = \Pr(|H_{i,j}(n)|^2 \leq \beta D_{i,j}^{\nu}) = 1 - \exp(-\beta D_{i,j}^{\nu}), \tag{17.16}$$

where β is defined in (17.12). Note that the events in the union in (17.15) are mutually exclusive. Hence, the conditional outage probability under the cooperative mode can be determined as

$$P_{\text{out}}^{C} = \left[1 - \exp\left(-\frac{(2^{\tilde{R}} - 1)N_0 D_{s,d}^{\nu}}{\kappa P_c} \right) \right] \left[1 - \exp\left(-\frac{(2^{\tilde{R}} - 1)N_0}{\kappa P_c}(D_{s,r}^{\nu} + D_{r,d}^{\nu}) \right) \right].$$

(17.17)

In case that the relay does not have resources to help forward the source information, the outage event in subcarrier n is equivalent to that of direct transmission between the source and destination, which is given by

$$P_{\text{out}}^{D} = \Pr\left(I^{(n)}(D_{s,d}) \leq \tilde{R} \right) = 1 - \exp\left(-\frac{(2^{(K+W)R/K} - 1)N_0 D_{s,d}^{\nu}}{\kappa P_c} \right). \qquad (17.18)$$

Given k users in the sector, the conditional outage probability per subcarrier can be obtained from (17.17) and (17.18) as

$$P_{\text{out}}|_k = P_{\text{out}}^{C}(1 - Q|_k) + P_{\text{out}}^{D} Q|_k. \qquad (17.19)$$

In (17.19), $Q|_k$ is defined as the probability that a source subcarrier that is in outage is not helped by the relay, under the condition that there are k users in the sector. In other words, the relay uses all of its resources to help forward the k users' information carried in other subcarriers. The physical meaning of (17.19) is that a diversity gain of two is achieved when the relay has available resource to help forward the source information, which occurs with probability $Q|_k$, whereas the diversity gain reduces to one when all the resources (allocated subcarriers) at the relay are fully used.

The probability $Q|_k$ can be determined as follows. Consider a sector with k users, each transmitting information in N subcarriers during one OFDM symbol period. Equivalently, the total number of kN subcarriers (in k OFDM symbol periods) are used to transmit the information from all k users in the sector. Assume that the subcarriers in outage are helped by the relay with equal probability. Then, given that x out of kN subcarriers are in outage, the probability that a subcarrier in outage is not helped by the relay is $(x - N)/x$ for $N + 1 \leq x \leq kN$. Thus, the probability $Q|_k$ can be obtained as

$$Q|_k = \sum_{x=N+1}^{kN} \frac{x - N}{x} \Pr(x \text{ out of } kN \text{ subcarriers in outage}). \qquad (17.20)$$

Since each subcarrier is in outage with probability P_{out}^{D}, the probability that x out of kN subcarriers are in outage can be approximated by

$$\Pr(x \text{ out of } kN \text{ subcarriers in outage}) = \binom{kN}{x} \left(P_{\text{out}}^{D} \right)^{x} \left(1 - P_{\text{out}}^{D} \right)^{kN-x}. \qquad (17.21)$$

In (17.21), it is difficult, if not impossible, to derive a closed form formulation that includes all possible combinations of the users' locations. For analytical tractability of the analysis, we resort to an approximate probability which is based on an assumption that all the x subcarriers are in outage with the same probability P_{out}^{D}. This approximation would be effective for small cell sizes where path loss is not the dominant fading effect. As we will show later, the analytical outage probability obtained from this approximation is very close to the simulation results.

Substituting (17.18) and (17.21) into (17.20), we obtain the probability $Q|_k$ as

$$Q|_k = \sum_{x=N+1}^{kN} \frac{x-N}{x} \binom{kN}{x} \left[1 - \exp(-\beta D_{s,d}^v)\right]^x \left[\exp(-\beta D_{s,d}^v)\right]^{kN-x}. \quad (17.22)$$

From (17.22) and (17.19), we can determine the conditional outage probability of the described scheme as

$$P_{\text{out}}|_k = \left(1 - \exp(-\beta D_{s,d}^v)\right)\left(1 - \exp(-\beta (D_{s,r}^v + D_{r,d}^v))\right)(1 - Q|_k)$$
$$+ \left(1 - \exp(-\beta D_{s,d}^v)\right)Q|_k. \quad (17.23)$$

Substituting (17.14) and (17.23) into (17.13), the outage probability of the described scheme can be obtained as

$$P_{\text{out}} = \left(1 - \exp(-\beta D_{s,d}^v)\right)\left[1 - \exp(-\beta (D_{s,r}^v + D_{r,d}^v))(1 - g(D_{s,d}))\right]; \quad (17.24)$$

where

$$g(D_{s,d}) = \sum_{k=1}^{K} c(k) \sum_{x=N+1}^{kN} \frac{x-N}{x} \binom{kN}{x} \left[1 - \exp(-\beta D_{s,d}^v)\right]^x \left[\exp(-\beta D_{s,d}^v)\right]^{kN-x}. \quad (17.25)$$

It is worth noting that when spectral efficiency goes to infinity, direct transmission outperforms cooperative communications. As the number of users increases, for a fixed number of relays, the performance of the cooperative scheme approaches that of direct transmission.

Finally, we determine the average outage probability by averaging (17.24) over the distribution of the user's distance as follows. Without loss of generality, we consider a sector as shown in Figure 17.2 where the relay is located at $D_{r,d}e^{j\phi_r}$ and a source user is located at $D_{s,d}e^{j\phi_s}$ ($0 \leq \phi_r, \phi_s \leq \theta_w$). The distance between the source and the relay can be expressed as

$$D_{s,r} = \left[D_{s,d}^2 + D_{r,d}^2 - 2D_{s,d}D_{r,d}\cos(\phi_r - \phi_s)\right]^{\frac{1}{2}} \triangleq f(D_{s,d}, \phi_s). \quad (17.26)$$

Assuming that users are uniformly distributed within the cell, the pdf of the user's distance D from the destination conditioned that the user is located in the sector can be given by

$$p_D(D \mid 0 \leq \phi_s \leq \theta_W) = \frac{2D}{W\rho^2}, \quad 0 \leq D \leq \rho. \quad (17.27)$$

Therefore, given specific relay locations, the average outage probability of the cooperative protocol can be obtained as

$$\bar{P}_{\text{out}} = \frac{1}{\pi\rho^2} \int_0^{\rho} \int_0^{\frac{2\pi}{W}} \left[1 - e^{-\beta(f^v(D_{s,d},\phi_s)+D_{r,d}^v)}(1 - g(D_{s,d}))\right]$$
$$\times D_{s,d}\left(1 - e^{-\beta D_{s,d}^v}\right)d\phi_s dD_{s,d}. \quad (17.28)$$

From (17.28), we can clearly see that the performance of the cooperative protocol depends on the number of relays in the cell and how each relay are assigned to help the source users. To get more insights of the cooperation systems, we provide the lower bound on the outage probability of the cooperative protocol and the performance of the relay-assignment scheme in the following subsections.

17.4 Performance lower bound

We exploit the hypothetical Genie-aided relay assignment scheme to find a lower bound on the outage probability of any practical relay assignment schemes as follows. The Genie-aided relay assignment scheme assumes that the assigned relay for any source in the network is located in the optimum location that minimize the outage probability for the fixed source–destination pair.

Observe that if the relay can be placed anywhere in the cell, the optimum relay location for a source–destination pair is on the line joining the source and destination. Accordingly, the optimum distance between the relay and destination can be written as

$$D_{s,r} = D_{s,d} - D_{r,d}. \tag{17.29}$$

Thus, from (17.17), the optimum relay location for a source–destination pair can be obtained by solving

$$D_{r,d}^* = \arg\min_{D_{r,d}} \left\{ 1 - \exp\left(-\beta((D_{s,d} - D_{r,d})^\nu + D_{r,d}^\nu) \right)(1 - g(D_{s,d))) \right\}$$

$$\text{s.t. } 0 \le D_{r,d} \le D_{s,d}. \tag{17.30}$$

The optimization problem in (17.30) is equivalent to find $D_{r,d}$ that minimizes $(D_{s,d} - D_{r,d})^\nu + D_{r,d}^\nu$. Thus, (17.30) can be analytically solved simply, and the optimum solution can be shown to be

$$D_{r,d}^* = D_{s,d}/2, \tag{17.31}$$

i.e., the optimum relay location for a specific source–destination pair is at the mid point of the line joining the source and the destination.

Using the Genie-aided relay assignment scheme and the optimum relay location $D_{r,d}^*$, we can determine the lower bound on the outage probability as follows. Substitute $D_{r,d}^*$ and $D_{s,r}^* = D_{s,d} - D_{r,d}^* = D_{s,d}/2$ into the outage probability formulation in (17.28), then the outage probability of the cooperative protocol can be lower bounded by

$$P_{out} \ge \left(1 - \exp(-\beta D_{s,d}^\nu) \right)\left[1 - \exp(-\beta D_{s,d}^\nu/(2^{\nu-1}))(1 - g(D_{s,d})) \right]. \tag{17.32}$$

Averaging (17.32) over all users' possible locations, we have

$$\bar{P}_{out} \ge \frac{2}{\rho^2} \int_0^\rho D_{s,d}\left(1 - e^{-\beta D_{s,d}^\nu}\right)\left[1 - e^{-\beta D_{s,d}^\nu/(2^{\nu-1})}\right] dD_{s,d}$$

$$+ \frac{2}{\rho^2} \sum_{k=0}^K c(k) \sum_{x=N+1}^{kN} \frac{x-N}{x}\binom{kN}{x} \int_0^\rho D_{s,d}\left(1 - e^{-\beta D_{s,d}^\nu}\right)^{x+1}$$

$$\times \left(e^{-\beta D_{s,d}^\nu}\right)^{kN-x+\frac{1}{2^{\nu-1}}} dD_{s,d}. \tag{17.33}$$

At high SNR, the effect of the second term in (17.33) on the outage probability is insignificant. Thus, neglecting the second term in (17.33), the average outage probability of the protocol can be lower bounded by

$$
\begin{aligned}
\bar{P}_{out} \geq 1 + \frac{2}{v\rho^2} & \left(\frac{1}{\beta(1+2^{1+v})} \right)^{\frac{2}{v}} \Upsilon \left(\frac{2}{v}, \beta(1+2^{1+v})\rho^v \right) \\
& - \frac{2}{v\rho^2} \left(\frac{1}{\beta} \right)^{\frac{2}{v}} \Upsilon \left(\frac{2}{v}, \beta\rho^v \right) - \frac{2}{v\rho^2} \left(\frac{1}{\beta(1+2^{1+v})} \right)^{\frac{2}{v}} \Upsilon \left(\frac{2}{v}, 2^{1-v}\beta\rho^v \right),
\end{aligned}
$$

(17.34)

where $\Upsilon(a, x)$ is the incomplete Gamma function, defined as

$$
\Upsilon(a, x) = \int_0^x e^{-t} t^{a-1} dt.
$$

17.5   Optimum relay location

Based on the average outage probability (17.28), we determine in this subsection the optimum relay location in each sector for a cell with W sectors. Since the users are uniformly located in the cell, one can show that the optimum relay location is on the line that divides the central angle θ_w into two equal parts, i.e., the optimum relay angle is

$$
\phi_r^* = \theta_w/2.
$$

(17.35)

Now, the remaining problem is to determine the optimum relay distance $\hat{D}_{r,d}$. Substitute ϕ_r^* into (17.28) and take the first derivative of $\bar{P}_{out}$ with respect to $D_{r,d}$, then the optimum relay distance $D_{r,d}^*$ can be obtained by solving

$$
\begin{aligned}
\int_0^\rho D_{s,d} & \left(1 - \exp(-\beta D_{s,d}^v) \right) (1 - g(D_{s,d})) \int_0^{\frac{2\pi}{W}} \eta(D_{s,d}, \phi_s) \\
& \times \exp(-\beta(f^v(D_{s,d}, \phi_s) + D_{r,d}^v)) d\phi_s dD_{s,d} = 0,
\end{aligned}
$$

(17.36)

in which

$$
\begin{aligned}
\eta(D_{s,d}, \phi_s) = [D_{s,d}^2 + D_{r,d}^2 & - 2D_{s,d}D_{r,d}\cos(\pi/W - \phi_s)]^{\frac{v}{2}-1} \\
& (D_{r,d} - D_{s,d}\cos(\pi/W - \phi_s)) + D_{r,d}^{v-1}.
\end{aligned}
$$

(17.37)

To gain further insight, we also provide here an explicit relay location that achieves close performance to that of optimum relay location. First, we find the average value of all possible users' locations as

$$
\bar{D}_{s,d} = \int_0^\rho D_{s,d} p_{D_{s,d}}(D_{s,d}) dD_{s,d} = 2\rho/3.
$$

(17.38)

From (17.19) and (17.38), an approximate relay location can be obtained by solving

$$
\bar{D}_{r,d} = \arg \min_{0 \leq D_{r,d} \leq \bar{D}_{s,d}} 1 - e^{-\beta((\bar{D}_{s,d} - D_{r,d})^v + D_{r,d}^v)} (1 - g(\bar{D}_{s,d})),
$$

(17.39)

which results in

$$
\bar{D}_{r,d} = \bar{D}_{s,d}/2 = \rho/3.
$$

(17.40)

The approximate relay location can be shown to achieve very close performance to the optimum relay location.

Example 17.1 We performed simulation scenarios under WLAN scenario. Each OFDM symbol has 64 subcarriers and the total bandwidth is 20 MHz. The target rate is

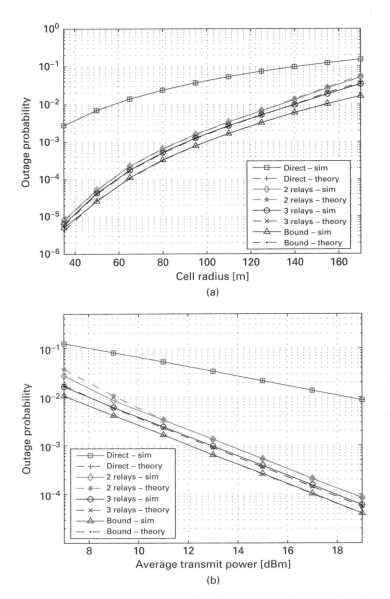

Fig. 17.4 Outage probability performance in WLANscenario. (a) Outage probability versus cell radius. (b) Outage probability versus average transmitted power.

fixed at $R = 1$ bit/s/Hz per user per subcarrier. The propagation loss factor is $v = 2.6$, and the number of users in the cell is set at 10 users.

Figure 17.4(a) and (b) compare the performance of the cooperative protocol with that of direct transmission scheme and the lower bound. The theoretical performance is plotted along with the simulation curves.

Figure 17.4(a) depicts the outage probability versus the cell radius. The average transmitted power is fixed at 12 dBm. If the outage performance is required to be at most 0.01, then the maximum cell size achieved by the direct transmission scheme is about 60 m. With the cooperative protocol, the maximum cell size of 132 m and 140 m can be achieved when 2 and 3 relays are deployed, respectively. The cooperative protocol can increase the cell size by about 130% in this case. Moreover, the described scheme achieves close performance to the lower bound, especially when the cell size is small.

In Figure 17.4(b), we study the performance gain that can be achieved by the cooperation protocol in terms of energy efficiency. The cell radius is fixed at 100 m, and the average transmitted power varies from 7 dBm to 19 dBm. If the outage is required to be at most 0.01, then the direct transmission scheme requires the transmitted power of 18 dBm, whereas the cooperation scheme requires only about 8 dBm; in other words, 10 dB power saving is achieved. It is worth pointing out that diversity is a high SNR concept and the cooperative scheme is able to achieve the diversity gain of two at practical transmit powers.

In both figures, the theoretical curves closely match to the simulation results, which validates the analysis. Note that the simulation and theoretical results completely match in the direct transmission case. ▲

17.6 Chapter summary and bibliographical notes

For broadband networks, we described a bandwidth-efficient cooperative protocol for multiuser OFDM systems. The destination broadcasts subcarrier indices of which the received SNR falls below a specific SNR threshold, and the relay forwards only the source symbols carried in those subcarriers. In this way, the relay can help forward the data of multiple sources in one OFDM symbol, and the described protocol greatly improves the spectral efficiency, while still achieving full diversity at high SNR. Relay assignment algorithms have been described. On the topic of cooperative communications protocol for multiuser OFDM networks, more information can be found in [192, 191].

An application of space–time cooperation in OFDM systems was investigated in [233]. In [57], pairing of users and level of cooperation are jointly determined to minimize overall transmitted power of OFDM system. Most of the existing works are based on fixed relaying protocols, in which the relays always repeat the source information. Moreover, these works rely on an assumption of fixed channel variances which implies a fixed network topology and fixed source–relay pairs.

Exercises

17.1 Prove the conditional outage probability expression under the cooperative mode in 17.17, i.e., prove

$$P_{\text{out}}^C = \left[1 - \exp\left(-\frac{(2^{\tilde{R}} - 1)N_0 D_{\text{s,d}}^v}{\kappa P_c}\right)\right]$$
$$\times \left[1 - \exp\left(-\frac{(2^{\tilde{R}} - 1)N_0}{\kappa P_c}(D_{\text{s,r}}^v + D_{\text{r,d}}^v)\right)\right].$$

17.2 Prove the expression in (17.23):

$$P_{\text{out}}|_k = \left(1 - \exp(-\beta D_{\text{s,d}}^v)\right)\left(1 - \exp(-\beta(D_{\text{s,r}}^v + D_{\text{r,d}}^v))\right)(1 - Q|_k)$$
$$+ \left(1 - \exp(-\beta D_{\text{s,d}}^v)\right)Q|_k.$$

17.3 Conditioned on the user's location, prove that the outage probability of the cooperative broadband scheme described in this chapter is given by

$$P_{\text{out}} = \left(1 - \exp(-\beta D_{\text{s,d}}^v)\right)\left[1 - \exp(-\beta(D_{\text{s,r}}^v + D_{\text{r,d}}^v))(1 - g(D_{\text{s,d}}))\right];$$

where

$$g(D_{\text{s,d}}) = \sum_{k=1}^{K} c(k) \sum_{x=N+1}^{kN} \frac{x - N}{x}\binom{kN}{x}$$
$$\times \left[1 - \exp(-\beta D_{\text{s,d}}^v)\right]^x \left[\exp(-\beta D_{\text{s,d}}^v)\right]^{kN-x}.$$

17.4 Consider a WLAN system. Each OFDM symbol has 128 subcarriers and the total bandwidth is 20 MHz. The target rate is fixed at $R = 1$ bit/s/Hz per user per subcarrier. The propagation loss factor is $v = 2.6$, and the number of users in the cell is set at 10 users.
   (a) Write Matlab code to simulate the outage performance of the broadband cooperative protocol with direct transmission versus cell radius when the power is fixed at 12 dBm.
   (b) Repeat part (a) by fixing the cell radius at 100 m and varying the transmit power.
   (c) Consider now an OFDM system that uses 512 subcarriers. Repeat parts (a) and (b). What is the effect of increasing the number of subcarriers on the performance of the cooperation protocol.

17.5 In this chapter, the relay is assumed to use an incremental relaying scheme with decode-and-forward. Assume instead that the relay uses only selective decode-and-forward. In such scenario the relay will only forward the packets from users that where decoded correctly irrespective of whether the destination was able to decode the packet correctly or not.
   (a) Find the outage probability for a single user.
   (b) Find the outage probability assuming N users in each sector.

(c) Compare the performance of such relaying scheme with that described in the chapter.

17.6 The uplink case is studied in this chapter. Consider now the downlink where the base station transmits data to different users. Assume there are N users in each sector, and that the base station schedules the users in N consecutive time slots. In each time slot the base station transmits an OFDM symbol to a user. A single relay is deployed in each sector to help the users using a similar approach as described in the chapter.

(a) Find the probability that a user has n subcarriers in outage.

(b) Find the outage probability of m subcarriers at the relay node for a given base station transmission.

(c) Find the total average outage probability taking into account the finite resources at the relay.

18 Network lifetime maximization via cooperation

Extending the lifetime of battery-operated devices is a key design issue that allows uninterrupted information exchange among distributed nodes in wireless networks. Cooperative communications enables and leverages effective resource sharing among cooperative nodes. This chapter provides a general framework for lifetime extension of battery-operated devices by exploiting cooperative diversity. The framework efficiently takes advantage of different locations and energy levels among distributed nodes. First, a lifetime maximization problem via cooperative nodes is considered and performance analysis for M-ary PSK modulation is provided. With an objective to maximize the minimum device lifetime under a constraint on bit error rate performance, the optimization problem determines which nodes should cooperate and how much power should be allocated for cooperation. Moreover, the device lifetime is further improved by a deployment of cooperative relays in order to help forward information of the distributed nodes in the network. Optimum location and power allocation for each cooperative relay are determined with an aim to maximize the minimum device lifetime. A suboptimal algorithm is presented to solve the problem with multiple cooperative relays and cooperative nodes.

18.1 Introduction

In many applications of wireless networks, extending the lifetime of battery-operated devices is a key design issue that ensures uninterrupted information exchange and alleviates burden of replenishing batteries. Lifetime extension of battery-limited devices has become an important issue due to the need in sensor and ad-hoc networks. However, there is not a broadly accepted definition of the concept of network lifetime because it often depends on the application of the networks under consideration. For example, a network lifetime can be defined as the time until the first node in the network dies or as the time until a certain percentage of nodes die. Alternatively, concerning the quality of communication, network lifetime can be defined in terms of the packet delivery rate or in terms of the number of alive flows.

Consider contemporary wireless networks which comprise heterogeneous devices such as mobile phones, laptop computers, personal digital assistants (PDAs), etc. These devices have limited lifetime; nevertheless, some of them may have longer lifetimes due to their location or energy advantages. For instance, a devices in some ideal location may have location advantage, while another devices may have energy advantage if they are equipped with high initial energy. By introducing a cooperation protocol among distributed nodes, a portion of energy from these devices can be allocated to help forward information of other energy depleting devices in the network. In this way, the lifetime of the energy depleting devices can be greatly improved, and hence, the minimum device lifetime of the network is increased.

In this chapter, we show a framework to increase the device lifetime by exploiting cooperative diversity and leveraging both location and energy advantages in wireless networks. The framework is based on the decode-and-forward (DF) cooperation protocol, which is well-suited for wireless LAN or cellular settings; nevertheless, other cooperation protocols such as amplify-and-forward can be used as well. We first describe a signal model for a non-cooperative network. Then, we present a signal model for a cooperative networks employing the DF protocol. After that, we consider an optimization problem with the goal of maximizing the minimum device lifetime under a bit error rate (BER) constraint. Following this, we analyze the problem based on the use of an M-ary phase shift keying (M-PSK) modulation scheme. From this, it becomes possible to derive an analytical solution for a two-node cooperative network, which provides some insights into the optimization problem. The challenge in this stage is that the problem is NP hard. Based on the two-node solution, we consider a fast suboptimal algorithm to reduce the complexity of the formulated problem. Furthermore, we consider device lifetime improvement by deploying additional relays over a network with energy depleting nodes. This will lead into determining where to place the relays and how much power should these relays use for cooperation in order to maximize the device lifetime. In addition, to reduce complexity of the optimization problem, we restrict the number of relays per nodes to at most one, in which case we study an efficient suboptimal algorithm.

18.2   System models

Consider a wireless network with N randomly deployed nodes as shown in Figure 18.1. Each node knows its next node in a predetermined route that connects to the destination to deliver information. The destination node can be a base station or an access point in wireless local area networks, a piconet coordinator in wireless personal area network, or a data-gathering unit in wireless sensor networks. Next, we describe the system model for a non-cooperative network and for a cooperative network in two separate subsections, starting with the non-cooperative scenario.

18.2.1   Non-cooperative wireless networks

In a non-cooperative wireless network, each source node transmits only its own information to the destination node through a predetermined route. Figure 18.1 shows an example of a wireless network with several randomly deployed nodes. Suppose there are N nodes in the network, and let x_j denotes a symbol to be transmitted from node j to

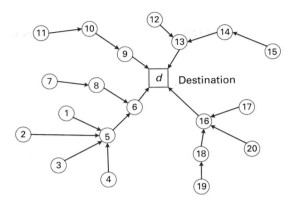

Fig. 18.1 An example of a wireless network with a destination (d) and N distributed nodes ($N = 20$).

its next node, defined as n_j, in its predetermined route. The symbol x_j can be the information of node j itself, or it can be the information of other nodes that node j routes through to the destination. The received signal at n_j due to the transmitted information from node j can be expressed as

$$y_{j,n_j} = \sqrt{P_j} h_{j,n_j} x_j + w_{j,n_j}, \tag{18.1}$$

where P_j is the transmit power of node j, h_{j,n_j} is the fading coefficient from node j to n_j, and w_{j,n_j} is an additive noise. The channel coefficient h_{j,n_j} is modeled as a complex Gaussian random variable with zero mean and variance σ_{j,n_j}^2, i.e., $\mathcal{CN}(0, \sigma_{j,n_j}^2)$, and w_{j,n_j} is $\mathcal{CN}(0, N_0)$ distributed. The channel variance σ_{j,n_j}^2 is modeled as

$$\sigma_{j,n_j}^2 = \eta D_{j,n_j}^{-\alpha}, \tag{18.2}$$

where D_{j,n_j} denotes the distance between node j and n_j, α is the propagation loss factor, and η is a constant whose value depends on the propagation environment. Considering an uncoded system and from Section 5.2.4, the average BER performance for a non-cooperative node with M-PSK modulation is upper bounded by

$$\text{BER}_j \leq \frac{N_0}{4b P_j \sigma_{j,n_j}^2 \log_2 M}, \tag{18.3}$$

where $b = \sin^2(\pi/M)$.

Let the performance requirement of node j be $\text{BER}_j \leq \varepsilon$ in which ε represents the maximum allowable BER. We assume that ε is the same for every node. Accordingly, the optimum transmit power of a non-cooperative node is given by

$$P_j = \frac{N_0}{4b\varepsilon \sigma_{j,n_j}^2 \log_2 M}. \tag{18.4}$$

We denote E_j as the initial battery energy for node j, and denote P_s as an amount of processing power (i.e. power used for encoding information, collecting data, and etc.) at the source node. Let $\lambda_{l,j}$ ($l = 1, 2, \ldots, N$ and $l \neq j$) be the data rate (symbol rate) that node l sends information to node j, and λ_j be a data rate that node j sends information to its next node n_j. Assume that the symbol length is fixed and equal for all nodes in the network and that the symbol length is normalized such that the energy required to transmit a symbol is given by the transmit power. Furthermore assume that the symbol rate generated at any node is low which is typical for most sensor networks applications. Also assume in the analysis that the cumulative data rate that need to be transmitted at any node (node's own generated data plus the data being routed for other nodes) is lower than the node's service rate. This guarantees that any node's buffer in the network does not overflow.

The average power that node j uses to send information to n_j is $\lambda_j P_s + \sum_{l=1}^{N} \lambda_{l,j} P_j$, where $\lambda_j P_s$ is the average processing power at node j, $\lambda_j P_j$ represents the average power that node j sends its own formation, and $\sum_{l=1,l\neq j}^{N} \lambda_{l,j} P_j$ corresponds to the average power that node j routes information of other nodes. Accordingly, the lifetime of node j can be determined as

$$T_j = \frac{E_j}{\lambda_j P_s + P_j \sum_{l=1}^{N} \lambda_{l,j}} = \frac{4b\varepsilon\sigma_{j,n_j}^2 E_j \log_2 M}{4b\varepsilon\sigma_{j,n_j}^2 \lambda_j P_s \log_2 M + N_0 \sum_{l=1}^{N} \lambda_{l,j}}.$$

From this equation, we can see that the lifetime of each node relies on both the initial energy and the geographical location of the node. Clearly, the node whose energy is small and location is far away from its next node tends to have small lifetime. Such node can result in small network lifetime. In the following subsection, we introduce the use of cooperative communications to prolong the network lifetime.

18.2.2   Cooperative DF protocol

We consider a cooperative wireless network where all nodes can transmit information cooperatively. Each node can be a source node that transmits its information or it can be a relay node that helps forward information of other nodes. The cooperation strategy is based on the DF protocol. We assume that each signal transmission is constrained to half-duplex mode, the system is uncoded, and the source and the relay transmit signals through orthogonal channels by using existing TDMA, FDMA, or CDMA schemes. For subsequent derivations, we define a power allocation matrix $\mathbf{P}$ as an $N \times N$ matrix with the following properties:

(i) Each element $P_{i,j} \geq 0$, for $i, j = 1, 2, \ldots, N$.
(ii) P_j represents a power that node j uses to transmit its own information to its next node n_j and the relays.
(iii) $P_{i,j}$ represents a power that node i helps forward information of node j (information of other nodes) to the next node n_j.

Let us assume that all nodes have their information to be transmitted, then $P_j > 0$ for all j. Figure 18.2(a) illustrates a cooperative network with $N = 4$ nodes. Each solid line represents a transmission link from a source node to its next node, and each dash line represents a link from a source to a relay. In addition, Figure 18.2(b) shows a power allocation matrix $\mathbf{P}$ which corresponds to the cooperative network in Figure

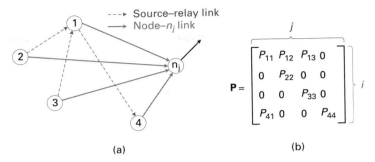

Fig. 18.2 Illustration of a cooperative wireless network: (a) a network with four nodes; (b) the corresponding power allocation matrix.

18.2(a). Each nonzero diagonal element of **P** represents a transmit power of a source node. In Figure 18.2(a), node 1 helps relay information of node 2 and 3 to their intended destination. Therefore, P_{12} and P_{13} in the first row of **P** contains nonzero elements, they represent power that node 1 helps node 2 and node 3, respectively. Similarly, P_{41} is a nonzero element because node 4 helps forward information of node 1.

Suppose node j acts as a source (or a helped node) and node i acts as a relay (or a helping node). When node j sends information to n_j in phase 1, the received signal at n_j is given in (18.1). However, the received signal at the helping node i is given by

$$y_{j,i} = \sqrt{P_j} h_{j,i} x_j + w_{j,i}, \qquad (18.5)$$

where $h_{j,i}$ denotes a channel coefficient from node j to node i, and $w_{j,i}$ represents an additive noise. In phase 2, the relay (node i in this case) forwards the information of node j to n_j only if the symbol is correctly decoded. The received signal at n_j can be expressed as

$$y_{i,n_j} = \sqrt{\tilde{P}_{i,j}} h_{i,n_j} x_j + w_{i,n_j}, \qquad (18.6)$$

where $\tilde{P}_{i,j} = P_{i,j}$ if the relay correctly decodes the symbol, and $\tilde{P}_{i,j} = 0$ otherwise. In (18.6), h_{i,n_j} and w_{i,n_j} are modeled as $\mathcal{CN}(0, \sigma_{i,n_j}^2)$ and $\mathcal{CN}(0, N_0)$, respectively. After that, the destination (n_j in this case)uses a maximum ratio combiner (MRC) to combine the signal received directly from the source in Phase 1 and that from the relay in Phase 2. Assuming that x_j has unit energy, the instantaneous SNR at the MRC output of n_j is from (5.9)

$$\gamma_{n_j} = \frac{P_j |h_{j,n_j}|^2 + \tilde{P}_{i,j} |h_{i,n_j}|^2}{N_0}. \qquad (18.7)$$

By taking into account the decoding result at the relay and averaging the conditional BER over the Rayleigh distributed random variables, according to (5.15), the average BER in case of M-PSK modulation can be expressed as

$$\mathrm{BER}_j = \frac{1}{\log_2 M} F\left(1 + \frac{bP_j\sigma_{j,n_j}^2}{N_0 \sin^2\theta}\right) \cdot F\left(1 + \frac{bP_j\sigma_{j,i}^2}{N_0 \sin^2\theta}\right)$$
$$+ \frac{1}{\log_2 M} F\left(\left(1 + \frac{bP_j\sigma_{j,n_j}^2}{N_0 \sin^2\theta}\right)\left(1 + \frac{bP_{i,j}\sigma_{i,n_j}^2}{N_0 \sin^2\theta}\right)\right)$$
$$\times \left[1 - F\left(1 + \frac{bP_j\sigma_{j,i}^2}{N_0 \sin^2\theta}\right)\right], \tag{18.8}$$

where

$$F(x(\theta)) = \frac{1}{\pi}\int_0^{(M-1)\pi/M} [x(\theta)]^{-1} d\theta \tag{18.9}$$

and b is defined in (18.3). The first term on the right-hand-side of (18.8) corresponds to an incorrect decoding at the relay whereas the second term corresponds to a correct decoding at the relay. By assuming that all channel links are available, i.e., $\sigma_{j,n_j}^2 \neq 0$ and $\sigma_{j,i}^2 \neq 0$, the BER upper bound of (18.8) can be obtained by removing the negative term and all one's in (18.8), from Theorem 5.2.2, we have

$$\mathrm{BER}_j \leq \frac{N_0^2}{b^2 \log_2 M} \cdot \frac{A^2 P_{i,j}\sigma_{i,n_j}^2 + B P_j\sigma_{j,i}^2}{P_j^2 P_{i,j}\sigma_{j,n_j}^2 \sigma_{j,i}^2 \sigma_{i,n_j}^2}, \tag{18.10}$$

where

$$A \triangleq \frac{M-1}{2M} + \frac{\sin(2\pi/M)}{4\pi}, \tag{18.11}$$

and

$$B \triangleq 3\frac{M-1}{8M} + \frac{\sin(2\pi/M)}{4\pi} - \frac{\sin(4\pi/M)}{32\pi}. \tag{18.12}$$

We can see from (18.10) that cooperative transmission obtains a diversity order of two as indicated in the power of N_0. Hence, with cooperative diversity, the total power required at the source and the relay is less than that required for non-cooperative transmission in order to obtain the same BER performance. Therefore, by properly allocating the transmit power at the source (P_j) and the transmit power at the relay ($P_{i,j}$), the lifetime of the source can be significantly increased whereas the lifetime of the relay is slightly decreased.

18.3　Lifetime maximization by employing a cooperative node

In this section, we aim to maximize the minimum device lifetime among all cooperative nodes in the network. First, we formulate the lifetime maximization problem. Then, an analytical solution is provided for a network with two cooperative nodes. After that, based on the solution for the two-node network, a fast suboptimal algorithm is developed to solve a problem for a network with multiple cooperative nodes.

18.3.1 Problem formulation

As shown in the previous section, the cooperative scheme requires less power to achieve the same performance as the non-cooperative scheme, thus it can be used to improve the minimum device lifetime. Note that different nodes may have different remaining energy, and they may contribute to different performance improvement due to their different locations. So the nodes with energy or location advantages can help forward information of other energy depleting nodes. In what follows, we formulate an optimization problem to determine which node should be a helping node and how much power should be allocated in order to efficiently increase the minimum device lifetime.

Let us first determine the device lifetime in a non-cooperative network. From Section 18.2.1, the non-cooperative device lifetime of node j is given by

$$T_j = \frac{\kappa \varepsilon \sigma_{j,n_j}^2 E_j}{(\kappa \varepsilon \sigma_{j,n_j}^2 \lambda_j P_s + N_0 \sum_{l=1}^{N} \lambda_{l,j})}, \tag{18.13}$$

where $\kappa \triangleq 4b \log_2 M$, and P_s represents the processing power. From (18.13), we can see that the lifetime of each node depends on its initial energy and its geographical location. Intuitively, the node whose energy is small and location is far away from its next node tends to have small device lifetime.

In the case of a cooperative network, the overall transmit power of each node is a summation of the power used by the node to transmit its own information and the power used by the node to cooperatively help forward information of other nodes. Let P_r be the processing power at each relay node, i.e., the power that the relay uses for decoding and forwarding information. From the power allocation matrix $\mathbf{P}$ in Section 18.2.2, the overall average transmit power of the cooperating node i is

$$P_i \sum_{l=1}^{N} \lambda_{li} + \sum_{\substack{j=1 \\ j \neq i}}^{N} P_{i,j} \left(\sum_{l=1}^{N} \lambda_{l,j} \right), \tag{18.14}$$

and the overall average processing power of node i is

$$\lambda_i P_s + \sum_{\substack{j=1 \\ j \neq i}}^{N} P_r \, \text{sgn}(P_{i,j}) \left(\sum_{l=1}^{N} \lambda_{l,j} \right) \tag{18.15}$$

where $\text{sgn}(P_{i,j})$ represents the sign function that returns 1 if $P_{i,j} > 0$, and 0 otherwise. Therefore, the lifetime of the cooperative node i can be written as

$$T_i(\mathbf{P}) = \frac{E_i}{\lambda_i P_s + P_i \sum_{l=1}^{N} \lambda_{li} + \Lambda(P_r, P_{i,j}, \lambda_{l,j})}, \tag{18.16}$$

where

$$\Lambda(P_r, P_{i,j}, \lambda_{l,j}) \triangleq \sum_{\substack{j=1 \\ j \neq i}}^{N} \left(P_r \, \text{sgn}(P_{i,j}) + P_{i,j} \right) \left(\sum_{l=1}^{N} \lambda_{l,j} \right), \tag{18.17}$$

and E_i is an initial energy of node i. Obviously, the lifetime of node i reduces if node i helps transmit information of other nodes. However, the more the power $P_{i,j}$ that node i uses in helping forward the information of node j, the longer the lifetime of node j. Therefore, it is crucial to properly design the power allocation matrix $\mathbf{P}$ such that the minimum device lifetime is maximized.

With an objective to maximize the minimum device lifetime under the BER constraint on each node, the optimization problem can be formulated as

$$\max_{\mathbf{P}} \min_{i} T_i(\mathbf{P}) \tag{18.18}$$

$$\text{s.t.} \begin{cases} \text{Performance: } \text{BER}_i \leq \varepsilon, \; \forall i; \\ \text{Power: } 0 < P_i \leq P_{\max}, \; \forall i; \\ \text{Power: } 0 \leq P_{i,j} \leq P_{\max}, \; \forall j \neq i, \end{cases}$$

where ε denotes a BER requirement. In (18.18), the first constraint is to satisfy the BER requirement as specified in (18.8), the second constraint guarantees that each node has information to be transmitted and the transmit power is no greater than $P_{\max}$ and the third constraint ensures that all the allocated power is nonnegative and no greater than $P_{\max}$. Due to its assignment and combinatorial nature, the formulated problem is NP hard. Even though each source–destination route is already known, this framework needs to optimize the pairing between each source and its relay. This problem of choosing relay is an assignment problem.

18.3.2   Analytical solution for a two-node wireless network

To get some insightful understanding on the formulated problem, we provide in this section a closed-form analytical solution at high SNR scenario for a network with two cooperative nodes ($N = 2$). Each node transmits its information directly to the destination d. In this two-node network, there are three possible transmission strategies, namely:

(i) each node transmits non-cooperatively;
(ii) one node helps forward information of the other;
(iii) both nodes help forward information of each other.

In the sequel, we will maximize the minimum device lifetime for each strategy. Without loss of generality, we assume that the transmit power required for a non-cooperative transmission is less than $P_{\max}$.

18.3.2.1   Non-cooperative transmission among nodes

Based on the discussion in Section 18.2.1, the optimum power allocation for non-cooperative case is $P_j = N_0/(\kappa \varepsilon \sigma_{jd}^2)$ for $j = 1, 2$, and $P_{i,j} = 0$ for $i \neq j$. Using (18.13), the optimum device lifetime for this transmission strategy is given by

$$T^*_{\text{non-coop}} = \min \left[\frac{\kappa \varepsilon \sigma_{1d}^2 E_1}{\lambda_{11}(\kappa \varepsilon \sigma_{1d}^2 P_s + N_0)}, \frac{\kappa \varepsilon \sigma_{2d}^2 E_2}{\lambda_{22}(\kappa \varepsilon \sigma_{2d}^2 P_s + N_0)} \right]. \tag{18.19}$$

18.3.2.2 Cooperative transmission when one node helps the other node

Without loss of generality, we will provide a solution for a case that node i helps relay information of node j to the destination. In this case, the lifetimes of node i and node j are given by

$$T_i = \frac{E_i}{\lambda_i(P_s + P_i) + \lambda_j(P_r + P_{i,j})} \tag{18.20}$$

and

$$T_j = \frac{E_j}{\lambda_j(P_s + P_j)}, \tag{18.21}$$

respectively. Hence, appropriately choosing P_i, $P_{i,j}$, and P_j can improve the minimum device lifetime while maintaining a specified BER requirement.

In order for node i to satisfy the BER requirement ε, the optimum transmit power of node i is

$$P_i = \frac{N_0}{\kappa \varepsilon \sigma_{id}^2}.$$

To determine P_j and $P_{i,j}$, we first note that, according to the BER upper bound in (18.10), P_j and $P_{i,j}$ must satisfy

$$\frac{N_0^2 A^2 P_{i,j} \sigma_{id}^2 + N_0^2 B P_j \sigma_{j,i}^2}{b^2 \log_2(M) P_j^2 P_{i,j} \sigma_{jd}^2 \sigma_{j,i}^2 \sigma_i^2} = \varepsilon. \tag{18.22}$$

Then, we can express $P_{i,j}$ in term of P_j as

$$P_{i,j} = \frac{P_j}{C_{i,j} P_j^2 - D_{i,j}} \triangleq f(P_j), \tag{18.23}$$

where

$$C_{i,j} = \frac{\varepsilon \sigma_{id}^2 \sigma_{jd}^2 b^2 \log_2 M}{B N_0^2},$$

and

$$D_{i,j} = \frac{A^2 \sigma_{id}^2}{B \sigma_{j,i}^2}.$$

From the expressions for P_i and $P_{i,j}$, we have

$$T_i = \frac{E_i}{\lambda_i(P_s + \frac{N_0}{\kappa \varepsilon \sigma_{id}^2}) + \lambda_j(P_r + f(P_j))}, \tag{18.24}$$

which is a function of P_j. Therefore, the optimization problem (18.18) is simplified to

$$T_{i\text{-helps-}j}^* = \max_{P_j} \left[\min \left(\frac{E_i}{\lambda_i(P_s + \frac{N_0}{\kappa \varepsilon \sigma_{id}^2}) + \lambda_j(P_r + f(P_j))}, \right. \right.$$
$$\left. \left. \frac{E_j}{\lambda_j(P_s + P_j)} \right) \right]. \tag{18.25}$$

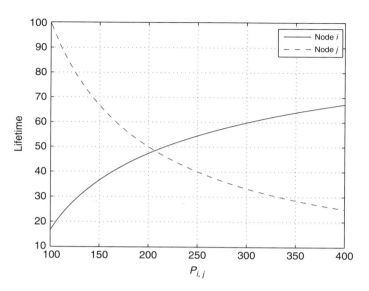

Fig. 18.3   Lifetimes of the two cooperative nodes as functions of the transmit power of the helped node (P_{22}).

As an illustrative example, Figure 18.3 plots the lifetime T_i and T_j as functions of P_j for a specific set of parameters. For unconstrained optimization of (18.25), Figure 18.3 shows that the optimum power P_j in (18.25) is the one that results when $T_i = T_j$. Therefore, the optimum device lifetime in the case when node i helps node j is

$$T^*_{i\text{-helps-}j} = \frac{E_i}{\lambda_i\left(P_s + \frac{N_0}{\kappa\varepsilon\sigma_{id}^2}\right) + \lambda_j(P_r + f(P_j^*))} = \frac{E_j}{\lambda_j(P_s + P_j^*)}, \qquad (18.26)$$

where P_j^* is the solution to

$$C_{i,j}E_i\lambda_j P_j^3 + KC_{i,j}P_j^2 - (D_{i,j}\lambda_j E_i + \lambda_j E_j)P_j - \Upsilon D_{i,j} = 0$$

in which

$$\Upsilon = E_i\lambda_j P_s - E_j\lambda_i P_s - \lambda_j E_j P_r + (\lambda_i N_0 E_j)/(\kappa\varepsilon\sigma_{id}^2).$$

Accordingly, we can find $P_{i,j}^* = f(P_j^*)$ from (18.23).

If the resulting $P_{i,j}^*$ is not larger than $P_{\max}$, then (18.26) is the optimum device lifetime for this scenario. Otherwise, let $P_{i,j}^* = P_{\max}$ and find P_j^* that satisfies the BER requirement (18.22). After some manipulations, we have

$$P_j^* = \frac{-Q_1 + \sqrt{Q_1^2 + Q_2Q_3 P_{\max}^2}}{Q_2 P_{\max}}, \qquad (18.27)$$

where

$$Q_1 = \frac{B\sigma_{j,i}^2 N_0^2}{b^2 \log_2 M},$$

$$Q_2 = 2\varepsilon\sigma_{id}^2\sigma_{j,i}^2\sigma_{jd}^2,$$

and

$$Q_3 = \frac{2A^2\sigma_{id}^2 N_0^2}{b^2 \log_2 M}.$$

Therefore, the lifetimes of nodes i and node j are $T_i^* = E_i/(\lambda_i(P_s + P_i) + \lambda_j(P_r + P_{\max}))$ and $T_j^* = E_j/(\lambda_j(P_s + P_j^*))$, respectively. Hence, the optimum device lifetime when $P_{i,j}^* > P_{\max}$ is the minimum among T_i^* and T_j^*. As a result, the optimum device lifetime when node i helps node j can be summarized as follows:

$$T_{i\text{-helps-}j}^* = \begin{cases} \frac{E_j}{\lambda_j(P_s+P_j^*)}, & P_{i,j}^* \le P_{\max}; \\ \min\{T_i^*, T_j^*\}, & P_{i,j}^* > P_{\max}. \end{cases} \tag{18.28}$$

18.3.2.3 Cooperative transmission when both nodes help each other

When both nodes help each other, $P_{i,j}$ and T_i are given in (18.23) and (18.24), respectively. The optimum device lifetime in this case can be obtained by finding P_i^* and P_j^* that maximizes T_i (or T_j) under the condition $T_i = T_j$, we have

$$T_{\text{both-help}}^* = \frac{E_i}{\lambda_i(P_s + P_i^*) + \lambda_j\left(P_r + \frac{P_j^*}{C_{i,j}(P_j^*)^2 - D_{i,j}}\right)}$$

$$= \frac{E_j}{\lambda_j(P_s + P_j^*) + \lambda_i\left(P_r + \frac{P_i^*}{C_{j,i}(P_i^*)^2 - D_{j,i}}\right)}, \tag{18.29}$$

where P_i^* and P_j^* are the solutions to:

$$\underset{P_i,P_j}{\arg\max} \; \frac{E_i}{\lambda_{id}(P_s + P_i) + \lambda_{j,i}\left(P_r + \frac{P_j}{C_{i,j}P_j^2 - D_{i,j}}\right)} \tag{18.30}$$

$$\text{s.t.} \begin{cases} \dfrac{[(\lambda_i P_s + \lambda_j P_r + \lambda_i P_i)(C_{i,j}P_j^2 - D_{i,j}) + \lambda_j P_j](C_{j,i}P_i^2 - D_{j,i})}{[(\lambda_j P_s + \lambda_i P_r + \lambda_j P_j)(C_{j,i}P_i^2 - D_{j,i}) + \lambda_i P_i](C_{i,j}P_j^2 - D_{i,j})} = \dfrac{E_i}{E_j}; \\[12pt] P_j > \sqrt{\dfrac{D_{i,j}}{C_{i,j}}}, \; \forall j \ne i, \end{cases}$$

in which, the first constraint ensures that $T_i = T_j$, and the second constraint guarantees that $P_{i,j} = P_j/(C_{i,j}P_j^2 - D_{i,j}) > 0$.

If $P_{i,j}^* \le P_{\max}$ and $P_{j,i}^* \le P_{\max}$, then the solution to (18.29) is the optimum device lifetime for this transmission strategy. Otherwise, the optimization problem is separated into two subproblems:

- In the first subproblem, we let $P_{i,j}^* = P_{\max}$ and find P_j^* from (18.27). This leads to both T_i and T_j being functions of P_i. Therefore, the optimum device lifetime for this subproblem is to maximize $\min\{T_i, T_j\}$ over P_i.
- In the second subproblem, we let $P_{j,i}^* = P_{\max}$ and find P_i^* from (18.27). In this case, T_i and T_j are functions of P_j. The optimum device lifetime results from maximizing $\min\{T_i, T_j\}$ over P_j.

After solving the two subproblems, the optimum device lifetime when $P_{i,j}^* > P_{\max}$ or $P_{j,i}^* > P_{\max}$ is the maximum among the two solutions. Consequently, the optimum device lifetime when both nodes help each other can be summarized as follows:

$$T_{\text{both-help}}^* = \begin{cases} T_i, & P_{i,j}^* \text{ and } P_{j,i}^* \le P_{\max}, \\ \max\left(\mathcal{T}(ii), \mathcal{T}(jj)\right), & P_{i,j}^* \text{ or } P_{j,i}^* > P_{\max}, \end{cases} \tag{18.31}$$

where

$$T_i \triangleq \frac{E_i}{\lambda_i\left(P_s + P_i^*\right) + \lambda_j\left(P_r + \dfrac{P_j^*}{C_{i,j}(P_j^*)^2 - D_{i,j}}\right)}.$$

In (18.31), we denote

$$\mathcal{T}(ii) \triangleq \max_{P_{ii}} \min\{T_i, T_j\},$$

and

$$\mathcal{T}(jj) \triangleq \max_{P_{jj}} \min\{T_i, T_j\}.$$

Finally the optimum device lifetime for the two-node cooperative network is

$$T_D^* = \max\left\{T_{\text{non-coop}}^*, T_{1\text{-helps-2}}^*, T_{2\text{-helps-1}}^*, T_{\text{both-help}}^*\right\}, \tag{18.32}$$

where T_D^* is the maximum among lifetime of these four possible transmission strategies. Although the optimum solution can be obtained through a full search, it is computationally expensive for a large cooperative network. To reduce complexity of the problem, we will use next a greedy suboptimal algorithm to determine the power allocation and the corresponding device lifetime.

18.3.3   Multi-node wireless network

The basic idea of the greedy suboptimal algorithm is to find with a step-by-step approach a node to be helped and a helping node. In each step, the algorithm selects a node to be helped as the one with minimum lifetime and that has never been helped by others. Then, the algorithm chooses a helping node as the one that maximizes the minimum device lifetime after the helped node has been served. In this way, the minimum device lifetime can be increased step by step. The iteration stops when the device lifetime cannot be significantly improved or all cooperative nodes have been helped. A flowchart that summarizes the algorithm is shown in Figure 18.4. Note that the greedy suboptimal approach can be applied to any multi-node cooperation strategy.

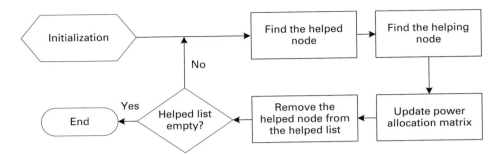

Fig. 18.4 A flowchart illustrates the suboptimal algorithm.

In what follows, we first maximize the minimum device lifetime for a given pair of helped and helping nodes, and then we describe the algorithm in details. For a given pair of helped and helping nodes, their transmit power and the corresponding lifetime can be determined in a similar way as those for the two-node network in the previous subsection. Specifically, consider a two-node cooperation strategy, then the optimum device lifetime when node i helps node j can be obtained by solving

$$
T^*_{i\text{-helps-}j} = \max_{P_j} \left[\min \left(\frac{E_i}{\Psi_i + (P_r + f(P_j)) \sum_{l=1}^{N} \lambda_{l,j}}, \ \frac{E_j}{\Psi_j + P_j \sum_{l=1}^{N} \lambda_{l,j}} \right) \right] \tag{18.33}
$$

where

$$
\Psi_i = \lambda_i P_s + P_i \sum_{l=1}^{N} \lambda_{li} + \sum_{\substack{k=1 \\ k \neq i,j}}^{N} \left(P_r \operatorname{sgn}(P_{ik}) + P_{ik} \right) \left(\sum_{l=1}^{N} \lambda_{lk} \right), \tag{18.34}
$$

and

$$
\Psi_j = \lambda_j P_s + \sum_{\substack{k=1 \\ k \neq j}}^{N} \left(P_r \operatorname{sgn}(P_{jk}) + P_{jk} \right) \left(\sum_{l=1}^{N} \lambda_{lk} \right), \tag{18.35}
$$

in which Ψ_i and Ψ_j are constants that do not depend on P_j. Using the equality $T_i = T_j$, and after some manipulations, we can find that

$$
T^*_{i\text{-helps-}j} = \frac{E_j}{\Psi_j + P_j^* \sum_{l=1}^{N} \lambda_{l,j}}, \tag{18.36}
$$

where P_j^* is the solution to

$$
C_{i,j} E_i \Sigma_{l=1}^{N} \lambda_{l,j} P_j^3 + G C_{i,j} P_j^2 - (D_{i,j} E_i + E_j)\left(\sum_{l=1}^{N} \lambda_{l,j}\right) P_j - G D_{i,j} = 0, \tag{18.37}
$$

in which $G \triangleq E_i \Psi_j - E_j \Psi_i - E_j P_r \sum_{l=1}^{N} \lambda_{l,j}$. If the resulting $P_{i,j}^* = f(P_j^*)$ is larger than $P_{\max}$ then the same calculation steps as in the previous subsection can be used to determine $T^*_{i\text{-helps-}j}$. This formulation is used to find the device lifetime at each step in the algorithm.

Initially, the power allocation matrix **P** is assigned as a diagonal matrix with its diagonal component

$$P_j = \frac{N_0}{\kappa \varepsilon \sigma_{jd}^2},$$

that is, the initial scheme is the non-cooperative transmission scheme. The corresponding lifetime of node j is $T_j = E_j/(\lambda_j P_s + P_j \Sigma_{l=1}^{N} \lambda_{l,j})$. Next, construct a *helped list*, which is a list of all possible nodes to be helped, $H_{\text{list}} = \{1, 2, \ldots, N\}$, by:

- First, the algorithm finds a helped node from the helped list by choosing the node who has minimum lifetime, i.e., the helped node $\hat{j}$ is given by

$$\hat{j} = \arg \min_{j \in H_{\text{list}}} T_j. \tag{18.38}$$

- Second, the algorithm finds a node to help node $\hat{j}$ from all nodes i, $i = 1, 2, \ldots, N$ and $i \neq \hat{j}$. For each possible helping node i, the algorithm uses (18.33) to find a power allocation for the helping node i and the helped node $\hat{j}$.
- Thirdly, the algorithm determines $T^*_D(i)$ as the minimum lifetime among cooperative nodes after node i finishes helping node $\hat{j}$. The obtained $T^*_D(i)$ from all possible helping nodes are compared, and then the algorithm selects node $\hat{i} = \arg\max_i T^*_D(i)$ as the one who helps node $\hat{j}$.
- Next, the algorithm updates **P** and the helped list by removing node $\hat{j}$ from the helped list.
- Then, the algorithm goes back to the first step. The iteration continues until all nodes have been helped, i.e., the helped list is empty, or the device lifetime cannot be significantly increased.

The resulting **P** is the optimum power allocation which gives answer to the questions: which node should help which node, and how much power should be used for cooperation? The detailed algorithm is shown in Table 18.1.

Note that the algorithm is suboptimal. Even though it is based on a cooperation strategy with only one relay ($K = 1$), the algorithm significantly improves the device lifetime as will be confirmed by simulation examples in Section 18.5. In terms of complexity of the algorithm, it increases quadratically with the number of cooperative nodes. In addition, the minimum device lifetime can be further improved by a cooperation with more than one relay; nevertheless, such lifetime improvement trades off with higher complexity. Note also that all necessary computations can be performed offline. Once the algorithm is executed, each cooperative node follows the determined power allocation and cooperation strategy.

Since the algorithm allocates power based on the average channel realizations, it is updated only when the network topology changes considerably. Furthermore, additional overhead for the cooperation assignment is required only at the beginning of the transmission. In (18.33), it is obvious that the helped node and the helping node should be

Table 18.1 Suboptimal algorithm for maximizing the minimum device lifetime of a wireless network with multiple cooperative nodes.

Initialization: $P_j = N_0/(\kappa \varepsilon \sigma_{jd}^2), T_j = E_j/\lambda_j(P_s + P_j),$
$\quad\quad T_D^* = \min T_j,$ and $H_{\text{list}} = \{1, 2, \dots, N\}.$

Iteration:
(1) Select the helped node with the minimum lifetime from the helped list:
$\quad \hat{j} = \arg\min_{j \in H_{\text{list}}} T_j,$
$\quad$ where $T_j = E_j/(\lambda_j P_s + P_j \sum_{l=1}^N \lambda_{l,j}).$
(2) Select the helping node from $\phi_{\hat{j}} = \{1, 2, \dots, N\} - \{\hat{j}\}.$
$\quad$ • For each $i \in \phi_{\hat{j}}$, solve (18.25) for T_i and $T_{\hat{j}}$, and then find the corresponding minimum device lifetime $T_D^*(i).$
$\quad$ • Select $\hat{i}$ that results in maximum of minimum device lifetime, $\hat{i} = \arg\max_{i \in \phi_{\hat{j}}} T_D^*(i),$ as the helping node.
(3) Update power allocation matrix **P** and helped list $H_{\text{list}}.$ Go to (1).

End: If the helped list is empty: $H_{\text{list}} = \emptyset,$ or the device lifetime cannot be significantly increased. return **P**.

close to each other. According to this observation, we can further reduce the complexity of the algorithm by searching for a helping node among cooperative nodes that are in the vicinity of the helped node. In this way, only local information is needed to compute the power allocation matrix. Although this may lead to some performance degradations, we will see in the computer simulations in Section 18.5 that such performance loss is insignificant.

18.4 Deploying relays to improve device lifetime

In this section, we improve the device lifetime by exploiting cooperative diversity through a deployment of cooperative relays in an energy depleting network. Each of these relays does not have information to be transmitted; however, they help forward information of all energy depleting nodes. The relay deployment reduces the need of frequent battery changing for each node which in turn helps reduce maintenance cost. In addition, the relay deployment does not require any modification in the cooperative nodes. An additional implementation cost is installation cost of the relays. By using a proper number of cooperative relays and placing these relays in appropriate locations, the device lifetime can be greatly increased while the overall cost is minimized. In the sequel, we determine location of each cooperative relay in the network with an objective to maximize the minimum device lifetime.

We consider a wireless network with N randomly located nodes, K cooperative relays, and a destination. The cooperative nodes are denoted as nodes $1, 2, \dots, N,$ and the cooperative relays are represented by $R_1, R_2, \dots, R_K.$ Since there is no cooperation among the cooperative nodes, the power allocation matrix $\mathbf{P}$ as defined in Section 18.2.2 is an $N \times N$ diagonal matrix whose diagonal element, P_j, represents a power that node j transmits information to its next node n_j. We assume that all cooperative nodes have information to be transmitted, i.e., $P_j > 0$ for all j. Hence, the lifetime of node j is given by

$$T_j(\mathbf{P}) = \frac{E_j}{\lambda_j P_s + P_j \sum_{l=1}^{N} \lambda_{l,j}}. \tag{18.39}$$

In addition, we also define a $K \times N$ relay power allocation matrix $\hat{\mathbf{P}}$ whose $(i, j)^{\text{th}}$ element, $\hat{P}_{i,j}$, represents a power that the relay R_i helps the node j. We assume that each relay does not have its own information to transmit; it only helps transmit information of other cooperative nodes. By denoting E_{R_i} as an initial energy of a relay R_i, the lifetime of R_i is

$$T_{R_i}(\hat{\mathbf{P}}) = \frac{E_{R_i}}{\sum_{j=1}^{N} (P_r \text{sgn}(\hat{P}_{i,j}) + \hat{P}_{i,j})(\sum_{l=1}^{N} \lambda_{l,j})}. \tag{18.40}$$

Example 18.1   As an illustrative example, consider a wireless network with four cooperative nodes and two cooperative relays as depicted in Figure 18.5(a). In the figure, the solid line represents a link from a node (source j or relay) to the next node n_j, and the dashed line represents a link from a source to a relay. Figure 18.5(b) shows the power allocation matrix $\mathbf{P}$ and the relay power allocation matrix $\hat{\mathbf{P}}$ which correspond to the wireless network in Figure 18.5(a). Since all four cooperative nodes transmit their information to n_j, then all diagonal elements of $\mathbf{P}$ are nonzeros. As shown in Figure 18.5(a), solid lines with square ("□") and circle ("o") represent the case when relay R_1 helps transmit information of node 1 and node 2, respectively. Accordingly, $\hat{P}_{11}$ and $\hat{P}_{12}$

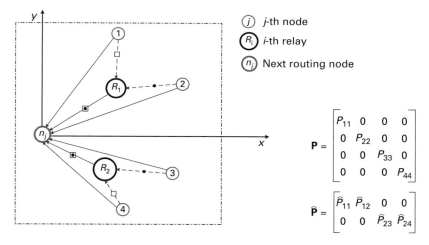

Fig. 18.5   Cooperative wireless network with relay deployment: (a) one cluster with four nodes, two relays, and one destination; (b) the corresponding power allocation matrices ($\mathbf{P}$) and ($\hat{\mathbf{P}}$) for the nodes and the relays, respectively.

are nonzero elements in the first row of $\hat{P}$. Similarly, relay R_2 helps transmit information of node 3 and node 4 to the next node n_j; $\hat{P}_{23}$ and $\hat{P}_{24}$ are nonzero elements in the second row of $\hat{P}$. ▲

Denote x_j and y_j as a location of node j on the x-axis and the y-axis, respectively. Then we represent a location of node j in a vector form as $\bar{\mathbf{D}}_j = [x_j \ y_j]^{\mathrm{T}}$. Accordingly, the channel variance between node j and its next node n_j is given by

$$\sigma^2_{j,n_j} = \eta \|\bar{\mathbf{D}}_j - \bar{\mathbf{D}}_{n_j}\|^{-\alpha},$$

where $\|\cdot\|$ denotes the Frobenius norm. The locations of the cooperative relays are specified by a $2 \times K$ matrix $\mathbf{D}_R = [\bar{\mathbf{D}}_{R_1} \ \bar{\mathbf{D}}_{R_2} \ \cdots \ \bar{\mathbf{D}}_{R_K}]$ in which the ith column indicates the location of relay R_i, i.e., $\bar{\mathbf{D}}_{R_i} = [x_{R_i} \ y_{R_i}]^{\mathrm{T}}$ is the location vector of the relay R_i. Then, the channel variance between R_i and node n_j is

$$\sigma^2_{R_i,n_j} = \eta \|\bar{\mathbf{D}}_{R_i} - \bar{\mathbf{D}}_{n_j}\|^{-\alpha},$$

and the channel variance between node j and R_i is

$$\sigma^2_{j,R_i} = \eta \|\bar{\mathbf{D}}_j - \bar{\mathbf{D}}_{R_i}\|^{-\alpha}.$$

If node j is helped by R_i, then the BER of node j has a similar form as (18.8) with $P_{i,j}$ and $\sigma^2_{j,i}$ replaced by $\hat{P}_{i,j}$ and σ^2_{j,R_i}, respectively. Our objective is to determine $\mathbf{D}_R$, $\mathbf{P}$, and $\hat{P}$ such that the minimum device lifetime is maximized. The optimization problem can formulated as

$$\max_{\mathbf{D}_R, \mathbf{P}, \hat{\mathbf{P}}} \min_{i,j} \{T_j(\mathbf{P}), T_{R_i}(\hat{\mathbf{P}})\} \tag{18.41}$$

$$\text{s.t.} \begin{cases} \text{Performance: } \mathrm{BER}_j \leq \varepsilon, \ \forall j; \\ \text{Power: } 0 < P_i \leq P_{\max}, \ P_{i,j} = 0 \ \forall i, j \neq i; \\ \text{Power: } 0 \leq \hat{P}_{i,j} \leq P_{\max}, \ \forall i, j. \end{cases}$$

In (18.41), the first constraint is to satisfy the BER requirement. The second constraint guarantees that all nodes transmit their information with power no greater than $P_{\max}$ and there is no cooperation among nodes. The third constraint ensures that the power that each cooperative relay helps a node is nonnegative and not greater than $P_{\max}$. Due to the assignment and combinatorial nature of the formulated problem, the problem in (18.41) is NP hard. Since it is computationally expensive to obtain the optimum solution to (18.41), a fast suboptimal algorithm will be considered to solve the formulated problem.

The basic idea of the fast suboptimal algorithm is to add one cooperative relay at a time into the network. Each time the optimum location of the added relay is chosen as the one, among all possible locations, that maximizes the minimum device lifetime. The algorithm stops when the device lifetime improvement is insignificant after adding another cooperative relay or when the maximum number of relays is reached. In the sequel, we first describe the algorithm to determine the device lifetime in each step, and then we describe the algorithm in details.

To maximize the minimum device lifetime when the number of relays and their locations are given, a step algorithm, similar to the one described in Figure 18.4 in Section 18.3.3, can be used:

- Initially, all nodes are sorted in ascending order according to their non-cooperative lifetimes, as specified in (18.13), and then register them in a helped list H_{list}.
- In each iteration, first, select the first node in the helped list as the one to be helped.
- Second, determine the minimum device lifetime after all of the cooperative relay R_i's ($i = 1, 2, \ldots, K$) finish helping the selected node,
- Choose the relay $R_{\hat{i}}$ where $\hat{i}$ is the relay that maximizes the minimum device lifetime to help the selected node.
- Update the power allocation matrices $\mathbf{P}$ and $\hat{\mathbf{P}}$, and remove the selected node in the first step from the helped list.
- Continue the iteration until all nodes have been helped and the helped list is empty or until the device lifetime improvement is insignificant.

The algorithm to find an optimum location of each cooperative relay is described as follows:

- Denote K_{max} as the maximum number of cooperative relays and denote Φ_D as a set of all possible relay locations.
- Initially, the number of relays is set to zero.
- In each iteration, the number of relay is increased by one, and the optimum relay location $\hat{\mathbf{D}}$ is determined using one of the heuristic search methods (e.g., local search or simulated annealing) together with the algorithm in Table 18.2.
- The location $\hat{\mathbf{D}}$ that results in the maximum of the minimum device lifetime is selected as the optimum relay location and, then, the device lifetime is updated.

Table 18.2 Suboptimal algorithm to determine device lifetime when relay locations are fixed.

Initialization: $P_j = \dfrac{N_0}{\kappa \varepsilon \sigma_{jd}^2}$, $T_j = \dfrac{E_j}{\lambda_j(P_s + P_j)}$, $T_D^* = \min T_j$,

Sort N nodes by their lifetimes in ascending order and list in H_{list}.

Iteration:
(1) Select the first node in the H_{list} as the helped node.
(2) Select the helping relay $R_{\hat{i}}$ from the set of K relays.
- For each i, use the heuristic algorithm to maximize the minimum device lifetime, $T_D^*(i)$.
- Select $R_{\hat{i}}$ that results in maximum of minimum device lifetime to help the node $\hat{j}$.
(3) Update $P_{\hat{j}\hat{j}}$ in $\mathbf{P}$ and update $\hat{P}_{\hat{i}\hat{j}}$ in $\hat{\mathbf{P}}$.

Set $\hat{P}_{i\hat{j}} = 0$ for all $i \neq \hat{i}$ and set $T_D^* = T_D(\hat{i})$.

Remove node $\hat{j}$ from the helped list H_{list}. Go to (1).

End: If the helped list is empty: $H_{\text{list}} = \emptyset$, or the device lifetime cannot be significantly increased. Return $\mathbf{P}$, $\hat{\mathbf{P}}$, T_D^*.

Table 18.3 Algorithm to determine relay locations.

Initialization: $q = 0$

Iteration:
(1) Increase number of relays: $q = q + 1$
(2) For each location $\mathbf{D}_l \in \Phi_D$. Set $\mathbf{D}_{R_q} = \mathbf{D}_l$.

    Find T_D^*, $\mathbf{P}$ and $\hat{\mathbf{P}}$ using the algorithm in Table 18.2

    Denote the obtained results by $T_D^*(l)$, $\mathbf{P}(l)$ and $\hat{\mathbf{P}}(l)$
(3) Find the relay location $R_q: \mathbf{D}_{R_q} = \mathbf{D}_{l^*}$, where $l^* = \arg\max_l T_D^*(l)$
(4) Update $\mathbf{P}$, $\hat{\mathbf{P}}$, and T_D^*. Go to (1).

End: If the device lifetime improvement is insignificant, or $q = K_{\max}$. Return $\mathbf{P}$, $\hat{\mathbf{P}}$, T_D^*.

- Finally, the algorithm goes back to the first step.
- The algorithm stops if the device lifetime improvement is insignificant or the number of relays reaches $K_{\max}$.

The detailed algorithm is presented in Table 18.3.

    Note that the fast suboptimal algorithm allows at most one relay to help each node. Although the algorithm is suboptimal, simulation example in Section 18.5 shows that the algorithm significantly improves the device lifetime. In addition, all of the required computations can be performed offline. In addition, the problems and algorithms in Section 18.3 and Section 18.4 are closely related. Specifically, both sections aim to extend the device lifetime by exploiting cooperative diversity. In Section 18.3, the cooperative diversity is exploited by cooperation among devices. In Section 18.4, however, cooperative relays are deployed, and the cooperative diversity is exploited by cooperation between each device and one of these additional cooperative relays. Therefore, the basic algorithm of finding $\mathbf{P}$ and $\hat{\mathbf{P}}$ in Section 18.4 is similar to that in Section 18.3. The search process of $\mathbf{P}$ and $\hat{\mathbf{P}}$ is done under the given locations of relays (i.e., for a fixed $\bar{\mathbf{D}}$); the process is not affected by the method of finding $\bar{\mathbf{D}}$.

18.5   Simulation examples

In all examples, BPSK modulation is used in the system, the propagation loss factor is $\alpha = 3$, $\eta = 1$, and $\varepsilon = 10^{-3}$ (unless stated otherwise). The processing power of each node (P_s) is set at 25% of transmit power of the node whose location is at $(10\text{m}, 0)$. The processing power of each relay (P_r) is set at 50% of P_s. All nodes are equipped with equal initial energy of $E_j = 10^5$. The noise variance is set at $N_0 = 10^{-2}$. The nodes are randomly distributed based on uniform distribution and the destination is located in the center of the area. Each node sends information to the destination via a route that is determined by the Dijkstra's algorithm.

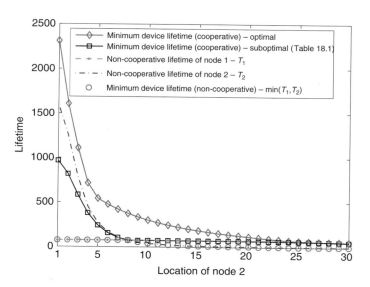

Fig. 18.6   Device lifetime in a two-node wireless network.

Example 18.2   Figure 18.6 considers a two-node wireless network where the destination is located at coordinate $(0, 0)$. Node 1 is fixed at coordinate $(0, 8m)$. The location of node 2 varies from $(0, 1)$ to $(0, 30\,m)$. We can see that the minimum device lifetime of the non-cooperative scheme is determined by the lifetime of the node who is located farther from the destination as shown by the curve marked with circles ("o"). Under cooperative transmission, the minimum device lifetime is significantly increased, especially when node 2 is located close to the destination.

The reason is that node 2 requires small transmit power to reach the destination, after node 2 helps node 1, the transmit power of node 2 slightly increases, while the transmit power of node 1 greatly reduces due to the cooperative diversity. With the suboptimal algorithm (Table 18.1), the minimum device lifetime is improved to almost the same as the lifetime of the node who is closer to the destination (see the curve with rectangular "□" markers). By using the optimum power allocation obtained from Section 18.3.2, the minimum device lifetime can be further increased (see the curve with diamond "◇" markers) since both nodes take advantage of the cooperative diversity while using smaller amount of their transmit power.   ▲

Example 18.3   Figure 18.7 depicts the minimum device lifetime according to the density of cooperative nodes in a square area. The number of randomly located nodes vary from 20 to 50 over an area of size $100\,m \times 100\,m$. In the simulation, we normalize the transmission rate to be the same for all network sizes. For local search, the helping node is chosen among the nodes whose distances from the source node are less than $20\,m$. From the figure, we can see that the minimum device lifetime of the cooperative network is higher than that of the non-cooperative network for all network sizes. For

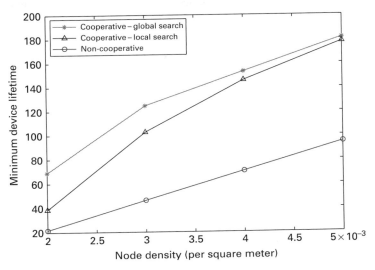

Fig. 18.7   Minimum device lifetime with different numbers of randomly located nodes.

example, in the cooperative network the minimum device lifetime is twice that of the non-cooperative network when there are 50 nodes in the network. Note that the performance gain is calculated as $T_{\text{coop}}/T_{\text{non-coop}}$, where T_{coop} and $T_{\text{non-coop}}$ represent the lifetime of cooperative and non-cooperative networks, respectively.

From Figure 18.7, the cooperative scheme with local search yields similar performance to the one with global search, especially when the node density is high. This confirms our expectation that the helping node is chosen as the one that is located close to the helped node. Note that the minimum device lifetimes for non-cooperative and cooperative networks increase with the number of nodes because the chance of being helped by a node with good location and high energy increases.   ▲

Example 18.4   Figure 18.8 illustrates an improvement of minimum device lifetime according to different BER requirements. We assume in the simulation that there are 30 randomly located cooperative nodes in an area of size 100 m × 100 m. We can see from the figure that the minimum device lifetime is small at a BER requirement of 10^{-6} under both non-cooperative and cooperative networks. This is because each node requires large transmit power to satisfy such small BER requirement. As the BER constraint increases, the minimum device lifetime also increases since the transmit power required to satisfy the BER constraint decreases.

Note that the cooperative network achieves longer device lifetime than that for the non-cooperative network over the entire range of BER requirement. For example, the cooperative network achieves $125/49 = 2.6$ times longer lifetime than the non-cooperative network at a BER requirement of 10^{-3}. However, both cooperative and

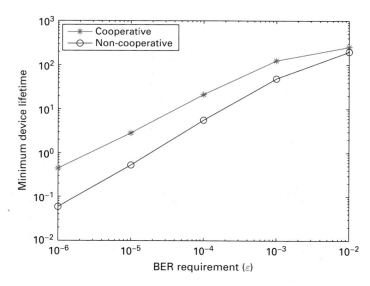

Fig. 18.8   Minimum device lifetime improvement according to different BER targets.

non-cooperative networks yield almost the same device lifetime at a BER constraint of 10^{-2}. The reason is that the transmit power required to satisfy the BER of 10^{-2} is much smaller than the processing power. The effect of processing power on the device lifetime dominates that of transmit power in this case.   ▲

Example 18.5   Figure 18.9 shows the minimum device lifetime for different relay locations. We consider a case where there are 20 randomly located nodes and a relay with initial energy of $E_{R_i} = 10^6$ in area of $100\,\text{m} \times 100\,\text{m}$. In the figure, the node with a circular ("o") marker represents a randomly-located node, and the node with a rectangular ("□") marker shows the location of the destination. We vary the relay location in a grid area of $100\,\text{m} \times 100\,\text{m}$. From the figure, the minimum device lifetime of the non-cooperative network is the same for all possible relay locations (as indicated by a point with "◇"). However, the minimum device lifetime of the cooperative network gradually increases when the relay moves closer to the destination.

Specifically, the minimum device lifetime is the same as that of non-cooperative network when the relay is far away from the destination. But the minimum device lifetime further improves to $216/18 = 12$ times longer than that of the non-cooperative network when the relay is close to the center of the area. This is because the node that is nearest to the destination tends to drain out its battery first, and its lifetime can be greatly improved by placing the relay near the destination.   ▲

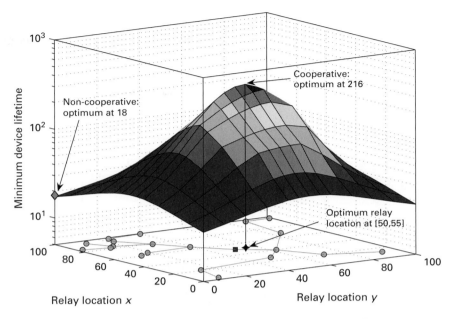

Fig. 18.9 Minimum device lifetime for a cooperative network with a cooperative relay.

Example 18.6 Figure 18.10 shows the minimum devicelifetime according to the density of cooperative relays (i.e., the number of relays per square meter). We consider a cooperative network with 20 randomly located nodes in an area of 100 m × 100 m. The initial energy of each relay is 10^5. The minimum device lifetime of the cooperative network with one randomly-added cooperative relay is about $28/11 = 2.55$ times longer than that of the non-cooperative network (as shown by the curve with the circular "∘" marker). If the relay is placed at its optimum location, the minimum device lifetime (the curve with the star "∗" marker) can be improved to $42/11 = 3.83$ times longer than that of the non-cooperative network. Furthermore, when two to four relays are added into the network, the minimum device lifetime can be further increased under a case with optimally-placed relays as well as a case with randomly-placed relays. However, the minimum device lifetime is almost saturated when more than two relays are deployed in the network. ▲

18.6 Chapter summary and bibliographical notes

This chapter considers lifetime maximization by employment of a cooperative-node and the deployment of a relay. By introducing cooperation protocols among nodes, both energy advantage and location advantage can be exploited such that the device lifetime is improved. First, the decode-and-forward cooperation protocol is employed among

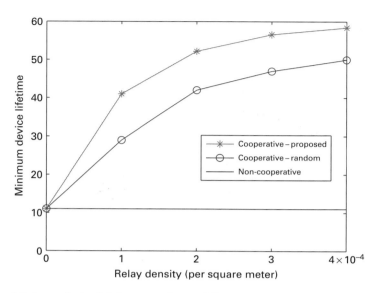

Fig. 18.10   Minimum device lifetime according to different numbers of deployed cooperative relays.

nodes. We determined which nodes should cooperate and how much power should be allocated for cooperation. An optimization problem is formulated with an aim to maximize the minimum device lifetime under a BER constraint. An analytical solution for a two-node cooperative network is provided.

In the case of a multiple-node scenario, it turns out that the formulated problem is NP hard. A suboptimal algorithm is developed to reduce the complexity of the formulated problem. By using the suboptimal algorithm, simulation results show that the minimum device lifetime of the two-node cooperative network can be increased to almost the same as the lifetime of the node that is closer to the destination.

In the case of the multiple cooperative nodes, the minimum device lifetime of the cooperative network is twice that of the non-cooperative network. Furthermore, we considered device lifetime improvement by adding cooperative relays into an energy depleting cooperative network. An optimization problem is formulated to determine the power allocation as well as the relay locations. By optimally placing a cooperative relay with energy 10 times higher than energy of the nodes, the device lifetime increases 12 times over that for the non-cooperative network. Furthermore, when the energy of each cooperative relay is equal to the energy of each cooperating node, the algorithm shows that only a few cooperative relays are required in order to improve the device lifetime.

As mentioned at the beginning of the chapter, there is not a broadly accepted definition for network lifetime. A network lifetime is defined as the time until the first node die in [190]. In [231] an alternative definition is considered where network lifetime is the time until a certain percentage of nodes die. Also, in [23] network lifetime is defined in terms of the packet delivery rate and in [18] in terms of the number of flows alive.

The problem of network algorithm designs that are aware of network lifetime have been the subject of some research interest. A data routing algorithm that aims at maximizing the minimum lifetime among nodes in wireless sensor networks was proposed in [22]. Later, considerable research efforts have been devoted to maximize such minimum lifetime. For example, upper bounds on lifetime of various wireless networks with energy-constrained nodes are derived in [12, 235, 76] and references therein. The problem of finding an energy-efficient tree for network lifetime maximization has been considered in [12, 132] for a broadcasting scenario, and in [39] for a multicasting scenario. The works in [4, 21, 117] considered the problem of minimum-energy broadcasting, which is proved to be NP-complete. In [133], a technique for network lifetime maximization by employing a ccumulative broadcast strategy was considered. The proposed work relies on the assumption that nodes cooperatively accumulate energy of unreliable receptions over the relay channels. The work in [75] considered provisioning additional energy on existing nodes and deploying relays to extend the network lifetime.

In terms of previous approaches to the problem tackled in this chapter, it is worth noticing that the works in [109, 179, 180, 222, 79, 204, 57, 128] have proved the significant potential of using cooperative diversity in wireless networks, but most focus on improving physical layer performance or minimizing energy consumption. On the other hand, note that there are many works on extending lifetime [22, 12, 235, 76, 132, 39, 4, 21, 117, 133, 75] but they all concentrate on settings with non-cooperative transmissions. Cases of works that focus on lifetime maximization via cooperative nodes and relay deployment are [64, 65].

Exercises

18.1 The device lifetime was derived in this chapter assuming MPSK modulation. Assume that the nodes in the network use MQAM instead.
   (a) Find the bit error rate performance for non-cooperative and cooperative transmission for MQAM.
   (b) Find the optimal power allocation for the two-node network scenario for the three cases discussed in the chapter: no-cooperation; one node only helps; both node help each other.
   (c) Derive device lifetime in case of MQAM modulation and find optimum lifetime for two-node wireless networks. Compare and discuss lifetime improvement for cooperative nodes with MPSK or MQAM.

18.2 Show that $T^*_{i\text{-helps-}j}$ can be expressed as in equation (18.28).

18.3 Derive the expression of $T^*_{\text{both-helps}}$ as shown in equation (18.31).

18.4 An analytical solution for the two-node network scenario was discussed in this chapter. Consider the three-node network scenario and denote the nodes by $\{1, 2, 3\}$. Find the network lifetime for each one of the following scenarios:
   (a) No cooperation.
   (b) Node 1 helps both nodes 2 and 3.

(c) Nodes 1 and 2 only help each other.

(d) Node 1 helps node 2; node 2 helps node 3; and node 3 helps node 1.

18.5 In the simulation examples in the chapter, all nodes were assumed to have the same initial battery energy of $E = 10^5$. In practice, different nodes can have different initial batteries.

Assume that the initial battery energy is a uniformly distributed random variable that takes value between 10^3 and 10^5.

(a) Repeat Example 18.3 and write a program to find the minimum device life time according to the density of the cooperative nodes. Compare to the results in the chapter where all nodes have the same initial battery life.

(b) Repeat Example 18.6 and find the effect of the relay density. Assume that the relays' initial energy is fixed and equal to 10^5.

18.6 The modulation scheme used in the chapter was assumed to be always fixed to MPSK whether the nodes cooperate or not. Assume that if a node cooperates it has to transmit at twice the rate. Consider the two-node network model. Assume that a non-cooperative node uses BPSK and a cooperative node uses QPSK. Find the minimum device life time for the following cases.

(a) Both nodes do not cooperate.

(b) Node 1 helps node 2.

(c) Both nodes cooperate.

18.7 Prove the device lifetime expression in (18.36).

References

1. M. Abramowitz and I. A. Stegun. *Handbook of Mathematical Functions with Formulas, Graphs, and Mathematical Tables*. Dover Publications, 1970.

2. S. Aeron and V. Saligrama. Wireless ad hoc networks: strategies and scaling laws for the fixed SNR regime. *IEEE Transactions on Information Theory*, 53(6), June 2007.

3. D. Agrawal, V. Tarokh, A. Naguib, and N. Seshadri. Space-time coded OFDM for high data-rate wireless communication over wideband channels. In *Proceedings of the IEEE Vehicular Technology Conference*, pp. 2232–2236, 1998.

4. A. Ahluwalia, E. Modiano, and L. Shu. On the complexity and distributed construction of energy-efficient broadcast trees in static ad hoc networks. In *Proceedings of the 36th Annual Conference on Information Sciences and Systems*, pp. 807–813, March 2002.

5. S. M. Alamouti. A simple transmit diversity technique for wireless communications. *IEEE Journal on Selected Areas in Communications*, 16(8):1451–1458, 1998.

6. P. A. Anghel and M. Kaveh. Exact symbol error probability of a cooperative network in a Rayleigh-fading environment. *IEEE Transactions on Wireless Communications*, 3(5):1416–1421, September 2004.

7. P. A. Anghel, G. Leus, and M. Kaveh. Multi-user space-time coding in cooperative networks. In *Proceedings of the IEEE International Conference on Acoustics, Speech and Signal Processing (ICASSP)*, volume 4, pp. 73–76, April 6–10, 2003.

8. S. Barbarossa, L. Pescosolido, D. Ludovici, L. Barbetta, and G. Scutari. Cooperative wireless networks based on distributed space-time coding. In *Proceedings of the IEEE International Workshop on Wireless Ad-hoc Networks (IWWAN)*, pp. 30–34, May 31–June 3, 2004.

9. A. Batra, J. Balakrishnan, G. R. Aiello, J. R. Foerster, and A. Dabak. Design of a multiband ofdm system for realistic UWB channel environments. *IEEE Transactions on Microwave Theory and Techniques*, 52(9):2123–2138, September 2004.

10. T. Berger. *Rate Distortion Theory*. Prentice-Hall, 1971.

11. D. Bertsekas and R. Gallager. *Data Networks*. Prentice Hall, 1992.

12. M. Bhardwaj, T. Garnett, and A. P. Chandrakasan. Upper bounds on the lifetime of sensor networks. In *Proceedings of the IEEE International Conference on Communications (ICC)*, volume 3, pp. 785–790, 2001.

13. R. Blum and Y. Li. Improved space-time coding for MIMO-OFDM wireless communications. *IEEE Transactions on Communications*, 49(11):1873–1878, November 2001.

14. J. Boutros and E. Viterbo. Signal space diversity: a power- and bandwidth-efficient diversity technique for the Rayleigh fading channel. *IEEE Transactions on Information Theory*, 44(4):1453–1467, July 1998.

15. S. Boyd and L. Vandenberghe. *Convex Optimization*. Cambridge University Press.

16. P. T. Brady. A model for on-off speech patterns in two-way conversation. *Bell System Technical Journal*, 48:2445–2472, 1969.

17. D. G. Brennan. Linear diversity combining techniques. *Proceedings of the IEEE*, 91(2):331–356, February 2003.

18. M. T. Brown, H. Gabow, and Q. Zhang. Maximum flow-life curve for a wireless ad hoc network. In *Proceedings of ACM MobiHoc*, pp. 128–136, 2001.

19. M. Bystrom and J. W. Modestino. Recent advances in joint source-channel coding of video. In *Proceedings of the 1998 URSI International Symposium on Signals, Systems, and Electronics*, pp. 332–337, September 1998.

20. M. Bystrom and T. Stockhammer. Dependent source and channel rate allocation for video transmission. *IEEE Transactions on Wireless Communications*, 3(1):258–268, January 2004.

21. M. Cagalj, J. Hubaux, and C. Enz. Energy-efficient broadcast in all-wireless networks. In *Proceedings of ACM/Kluwer Mobile Networks and Applications (MONET)*, pp. 177–188, June 2003.

22. J.-H. Chang and L. Tassiulas. Maximum lifetime routing in wireless sensor networks. *IEEE/ACM Transactions on Networking*, 12(4):609–619, August 2004.

23. B. Chen, K. Jamieson, H. Balakrishnan, and R. Morris. Span: an energy-efficient coordination algorithm for topology maintainance in ad hoc wireless networks. In *Proceedings of the 33rd Hawaii International Conference on System Sciences*, pp. 1–10, 2000.

24. M. Chiani, D. Dardari, and M. K. Simon. New exponential bounds and approximations for the computation of error probability in fading channels. *IEEE Transactions on Wireless Communications*, 2(4):840–845, July 2003.

25. T. Cover and A. E. Gamal. Capacity theorems for the relay channel. *IEEE Transactions on Information Theory*, 25(5):572–584, September 1979.

26. T. Cover and J. Thomas. *Elements of Information Theory*. John Wiley Inc., 1991.

27. H. Coward and T. A. Ramstad. Robust image communication using bandwidth reducing and expanding mappings. In *Proceedings of the ASILOMAR Conference on Signals, Systems, and Computers*, Monterey, California, pp. 1384–1388 October 2000.

28. M. O. Damen, K. Abed-Meraim, and J. C. Belfiore. Diagonal algebraic space-time block codes. *IEEE Transactions on Information Theory*, 48(2):628–636, March 2002.

29. DARPA TIMIT. Acoustic-phonetic continuous speech corpus CD-ROM. In *Document NISTIR 4930, NIST Speech Disk 1-1.1*.

30. N. Demir and K. Sayood. Joint source/channel coding for variable length codes. In *Proceedings of the Data Compression Conference (DCC)*, Snowbird, Utah, pp. 139–148, April 1998.

31. J. G. Dunham and R. M. Gray. Joint source and noisy channel trellis encoding. *IEEE Journal on Selected Areas in Communications*, 27(4):516–519, July 1981.

32. A. A. El Gamal and T. M. Cover. Achievable rates for multiple descriptions. *IEEE Transactions on Information Theory*, 28(6):851–857, November 1982.

33. A. A. El Gamal and S. Zahedi. Capacity of a class of relay channels with orthogonal components. *IEEE Transactions on Information Theory*, 51(5):1815–1817, May 2005.

34. A. El-Sherif, A. Kwasinski, A. Sadek, and K. J. R. Liu. Content-aware cognitive multiple access protocol for cooperative packet speech communications. To appear in *IEEE Transactions on Wireless Communications*.

35. A. El-Sherif, A. Sadek, A. Kwasinski, and K. J. R. Liu. Content-aware cooperative multiple access protocol for packet speech communications. In *Proceedings of IEEE GLOBECOM*, pp. 2952–2956, November 2007.

36. ETSI/GSM. Digital cellular telecommunications system (phase 2+); adaptive multi-rate (AMR) speech transcoding (GSM 06.90 version 7.2.1). In *Document ETSI EN 301 704 V7.2.1 (2000–04)*, 1998.

37. N. Farvardin and V. Vaishampayan. Optimal quantizer design for noisy channels: an approach to combined source-channel coding. *IEEE Journal on Selected Areas in Communications*, 33(6):827–838, November 1987.

38. L. M. Feeney and M. Nilsson. Investigating the energy consumption of a wireless network interface in an ad hoc networking environment. In *Proceedings of the IEEE Conference on Computer Communications (INFOCOM)*, volume 3, pp. 1458–1557, 2001.

39. P. Floreen, P. Kaski, J. Kohonen, and P. Orponen. Lifetime maximization for multicasting in energy-constrained wireless networks. *IEEE Journal on Selected Areas in Communications*, 23(1):117–127, January 2005.

40. G. J. Foschini and M. Gans. On the limits of wireless communication in a fading environment when using multiple antennas. *Wireless Personal Communication*, 6:311–335, March 1998.

41. P. Frenger, P. Orten, T. Ottosson, and A. B. Svensson. Rate-compatible convolutional codes for multirate DS-CDMA systems. *IEEE Transactions on Communications*, 47:828–836, June 1999.

42. A. Fuldseth and T. A. Ramstad. Bandwidth compression for continuous amplitude channels based on vector approximation to a continuous subset of the source signal space. In *Proceedings of the International Conference on Acoustics, Speech and Signal Processing ICASSP*, volume 4, pp. 3093–3096, 1997.

43. G. Ganesan and P. Stoica. Space-time block codes: a maximum SNR approach. *IEEE Transactions on Information Theory*, 47:1650–1656, May 2001.

44. X. Giraud, E. Boutillon, and J. C. Belfiore. Algebraic tools to build modulation schemes for fading channels. *IEEE Transactions on Information Theory*, 43(3):938–952, May 1997.

45. A. Goldsmith. *Wireless Communications*. Cambridge University Press, 2005.

46. Y. Gong and K. B. Letaief. Space-frequency-time coded OFDM for broadband wireless communications. In *Proceedings of the IEEE Global Telecommunications Conference, GLOBECOM*, 2001.

47. Y. Gong and K. B. Letaief. An efficient space-frequency coded wideband OFDM system for wireless communications. In *Proceedings of the IEEE International Conference on Communications (ICC)*, volume 1, pp. 475–479, 2002.

48. D. J. Goodman and C. E Stundberg. Combined source and channel coding for variable-bit-rate speech transmission. *Bell System Technical Journal*, 62:2017–2036, September 1983.

49. D. Goodman and S. X. Wei. Efficiency of packet reservation multiple access. *IEEE Transactions on Vehicular Technology*, 40(1):170–176, February 1991.

50. I. S. Gradshteyn and I. M. Ryshik. *Table of Integrals, Series and Products*, 6th edn. Academic Press, 2000.

51. R. M. Gray and D. L. Neuhoff. Quantization. *IEEE Transactions on Information Theory*, 44(6):2325–2383, October 1998.

52. J.-C. Guey, M. P. Fitz, M. R. Bell, and W.-Y. Kuo. Signal design for transmitter diversity wireless communication systems over Rayleigh fading channels. *IEEE Transactions on Communications*, 47:527–537, April 1999.

53. D. Gunduz and E. Erkip. Source and channel coding for cooperative relaying. *IEEE Transactions on Information Theory*, 53(10):3454–3475, October 2007.

54. P. Gupta and P. R. Kumar. The capacity of wireless networks. *IEEE Transactions on Information Theory*, pp. 388–404, March 2000.

55. M. Haenggi. On routing in random Rayleigh fading networks. *IEEE Transactions on Wireless Communications*, 4:1553—1562, July 2005.

56. J. Hagenauer. Rate compatible punctured convolutional (RCPC) codes and their applications. *IEEE Transactions on Communications*, 36(4):389–399, April 1988.

57. Z. Han, T. Himsoon, W. Siriwongpairat, and K. J. R. Liu. Energy efficient cooperative transmission over multiuser OFDM networks: who helps whom and how to cooperate. In *Proceedings of the IEEE Wireless Communications and Networking Conference (WCNC)*, New Orleans, Lousiana, volume 2, pp. 1030–1035, March 2005.

58. M. O. Hasna and M.-S. Alouini. Performance analysis of two-hop relayed transmissions over rayleigh fading channels. In *Proceedings of the IEEE Vehicular Technology Conference*, volume 4, pp. 1992–1996, 2002.

59. B. Hassibi and B. M. Hochwald. High-rate codes that are linear in space and time. *IEEE Transactions on Information Theory*, 48(7):1804–1824, July 2002.

60. Z. He, J. Cai, and C. W. Chen. Joint source channel rate-distortion analysis for adaptive mode selection and rate control in wireless video coding. *IEEE Transactions on Circuits and Systems for Video Technology*, 12(6):511–523, June 2002.

61. F. Hekland, G. E. Oien, and T. A. Ramstad. Using 2:1 Shannon mapping for joint source-channel coding. In *Proceedings of the Data Compression Conference (DCC)*, Snowbird, Utah, pp. 223–232, March 2005.

62. T. Himsoon, W. Su, and K. J. R. Liu. Differential transmission for amplify-and-forward cooperative communications. *IEEE Signal Processing Letters*, 12:597—600, September 2005.

63. T. Himsoon, W. Su, and K. J. R. Liu. Differential unitary space-time signal design using matrix rotation structure. *IEEE Signal Processing Letters*, 12:45—48, January 2005.

64. T. Himsoon, W. Siriwongpairat, Z. Han, and K. J. R. Liu. Lifetime maximization by cooperative sensor and relay deployment in wireless sensor networks. In *Proceedings of the IEEE Wireless Communications and Networking Conference (WCNC)*, Las Vegas, 1:439–444, April 2006.

65. T. Himsoon, W. P. Siriwongpairat, Z. Han, and K. J. R. Liu. Lifetime maximization via cooperative nodes and relay deployment in wireless networks. *IEEE Journal on Selected Areas in Communications*, 25(2):306–317, February 2007.

66. T. Himsoon, W. P. Siriwongpairat, W. Su, and K. J. R. Liu. Differential modulation with threshold-based decision combining for cooperative communications. *IEEE Transactions on Signal Processing*, 55(7):3905–3923, July 2007.

67. T. Himsoon, W. P. Siriwongpairat, W. Su, and K. J. R. Liu. Differential modulation for multi-node cooperative communications. *IEEE Transactions on Signal Processing*, 56(7):2941–2956, July 2008.

68. B. Hochwald and K. Zeger. Tradeoff between source and channel coding. *IEEE Transactions on Information Theory*, 43(5):1412–1424, September 1997.

69. B. Hochwald. Tradeoff between source and channel coding on a Gaussian channel. *IEEE Transactions on Information Theory*, 44(7):3044–3055, October 1998.

70. B. M. Hochwald and W. Swelden. Differential unitary space-time modulation. *IEEE Transactions on Communications*, 48(6):2041–2052, December 2000.

71. B. M. Hochwald and T. L. Marzetta. Unitary space-time modulation for multiple-antenna communication in Rayleigh flat fading. *IEEE Transactions on Information Theory*, 46(2):543–564, February 2000.

72. B. M. Hochwald, T. L. Marzetta, T. J. Richardson, W. Swelden, and R. Urbanke. System-atic design of unitary space-time constellations. *IEEE Transactions on Information Theory*, 46(6):1962–1973, June 2000.

73. R. A. Horn and C. R. Johnson. *Matrix Analysis*. Cambridge University Press, 1985.

74. R. A. Horn and C. R. Johnson. *Topics in Matrix Analysis*. Cambridge University Press, 1991.

75. Y. T. Hou, Y. Shi, H. D. Sherali, and S. F. Midkiff. On energy provisioning and relay node placement for wireless sensor networks. *IEEE Transactions on Communications*, 53:2579–2590, September 2005.

76. Z. Hu and B. Li. On the fundamental capacity and lifetime limits of energy-constrained wireless sensor networks. In *Proceedings of the 10th IEEE Proceedings of Real-Time and Embedded Technology and Applications Symposium*, pp. 2–9, 2004.

77. B. L. Hughes. Differential space-time modulation. *IEEE Transactions on Information Theory*, 46:2567–2578, November 2000.

78. T. E. Hunter and A. Nosratinia. Cooperation diversity through coding. In *Proceedings of the IEEE International Symposium on Information Theory*, p. 220, July 2002.

79. T. E. Hunter, S. Sanayei, and A. Nosratinia. Outage analysis of coded cooperation. *IEEE Transactions on Information Theory*, 52:375–391, February 2006.

80. J. Hwang, R. R. Consulta, and H. Yoon. 4G mobile networks - technology beyond 2.5G and 3G. In *PTC*, January 2007.

81. A. S. Ibrahim, Z. Han, and K. J. R. Liu. Distributed energy-efficient cooperative routing in wireless networks. To appear in *IEEE Transactions on Wireless Communications*.

82. A. S. Ibrahim, A. K. Sadek, W. Su, and K. J. R. Liu. Cooperative communications with relay selection: when to cooperate and whom to cooperate with? *IEEE Transactions on Wireless Communications*, 7(7):2814–2817, July 2008.

83. A. Ibrahim, A. Sadek, W. Su, and K. J. R. Liu. Cooperative communications with par-tial channel state information: when to cooperate? In *Proceedings of the IEEE Global Telecommunications Conference, GLOBECOM*, 5:3068–3072, November 2005.

84. A. Ibrahim, A. Sadek, W. Su, and K. J. R. Liu. Relay selection in multi-node cooperative com-munications: when to cooperate and whom to cooperate with? In *Proceedings of the IEEE Global Telecommunications Conference, GLOBECOM*, November 2006.

85. A. Ibrahim, Z. Han, and K. J. R. Liu. Distributed energy-efficient cooperative routing in wireless networks. In *Proceedings of the IEEE Global Telecommunications Conference, GLOBECOM*, pp. 4413–4418, November 2007.

86. ITU-T. Recommendation p.800: Methods for subjective determination of transmission qual-ity, 1996.

87. ITU-T. Guidelines for evaluation of radio transmission technologies for IMT-2000. *ITU-R M.1225*, 1997.

88. ITU-T. Recommendation p.862: Perceptual evaluation of speech quality (pesq): an objective method for end-to-end speech quality assessment of narrow-band telephone networks and speech codecs, 2001.

89. ITU-T. Recommendation p.862.1: Mapping function for transforming p.862 raw result scores to MOS-LQO. 2003.

90. H. Jafarkhani. A quasi-orthogonal space-time block code. *IEEE Transactions on Information Theory*, 49(1):1–4, January 2001.

91. M. Jankiraman. *Space-time Codes and MIMO Systems*. Artech House Publishers, 2004.

92. M. Janani, A. Hedayat, T. E. Hunter, and A. Nosratinia. Coded cooperation in wireless communications: space-time transmission and iterative decoding. *IEEE Transactions on Signal Processing*, 52:362–370, February 2004.

93. Y. Jing and B. Hassibi. Distributed space-time coding in wireless relay networks. In *Proceedings of the 3rd IEEE Sensor Array and Multi-Channel Signal Processing Workshop*, July 18–21, 2004.

94. Y. Jing and H. Jafarkhani. Using orthogonal and quasi-orthogonal designs in wireless relay networks. In *Proceedings of the IEEE Global Telecommunications Conference, GLOBECOM*, November 2006.

95. M. M. Khan and D. Goodman. Effects of channel impairments on packet reservation multiple access. In *Proceedings of the 44th IEEE Vehicle Technology Conference*, Stockholm, Sweden, pp. 1218–1222, June 1994.

96. A. E. Khandani, E. Modiano, L. Zheng, and J. Abounadi. Cooperative routing in wireless networks. In *Advances in Pervasive Computing and Networking*, Kluwer Academic Publishers, 2004.

97. T. Kiran and B. S. Rajan. Distributed space-time codes with reduced decoding complexity. In *Proceedings of the IEEE International Symposium on Information Theory (ISIT)*, July 2006.

98. L. M. Kondi, F. Ishtiaq, and A. K. Katsaggelos. Joint source-channel coding for motion-compensated DCT-based SNR scalable video. *IEEE Transactions on Image Processing*, 11(9):1043–1052, September 2002.

99. G. Kramer, M. Gastpar, and P. Gupta. Cooperative strategies and capacity theorems for relay networks. *IEEE Transactions on Information Theory*, 51(9):3037–306, September 2005.

100. A. Kwasinski and K. J. R. Liu. Source-channel-cooperation tradeoffs for adaptive coded communications. To appear in *IEEE Transactions on Wireless Communications*.

101. A. Kwasinski, Z. Han, K. J. R. Liu, and N. Farvardin. Power minimization under real-time source distortion constraints in wireless networks. In *Proceedings of the IEEE Wireless Communications and Networking Conference (WCNC)*, volume 1, New Orleans, Lousiana, pp. 532–536, March 2003.

102. A. Kwasinski and N. Farvardin. Optimal resource allocation for CDMA networks based on arbitrary real-time source coders adaptation with application to MPEG4 FGS. In *Proceedings of the IEEE Wireless Communications and Networking Conference (WCNC)*, Atlanta, Georgia, volume 4, pp. 2010–2015, March 2004.

103. A. Kwasinski. *Cross-Layer Resource Allocation Protocols for Multimedia CDMA Networks*. PhD thesis, University of Maryland, College Park, 2004.

104. A. Kwasinski, Z. Han, and K. J. R. Liu. Cooperative multimedia communications: Joint source coding and collaboration. In *Proceedings of the IEEE Global Telecommunications Conference, GLOBECOM*, pp. 374–378, 2005.

105. A. Kwasinski, Z. Han, and K. J. R. Liu. Joint source coding and cooperation diversity for multimedia communications. In *Proceedings of the IEEE Workshop on Signal Processing Advances in Wireless Communications, SPAWC*, pp. 129–133, 2005.

106. A. Kwasinski and K. J. R. Liu. Characterization of distortion-snr performance for practical joint source-channel coding systems. In *Proceedings of the IEEE Global Telecommunications Conference, GLOBECOM*, pp. 3019–3023, 2007.

107. A. Kwasinski and K. J. R. Liu. Towards a unified framework for modeling and analysis of diversity in joint source-channel coding. *IEEE Transactions on Communications*, 56:90–101, January 2008.

108. J. N. Laneman and G. W. Wornell. Distributed space-time coded protocols for exploiting cooperative diversity in wireless networks. *IEEE Transactions Information Theory*, 49(10):2415–2425, October 2003.

109. J. N. Laneman, D. N. C. Tse, and G. W. Wornell. Cooperative diversity in wireless networks: efficient protocols and outage behavior. *IEEE Transactions on Information Theory*, 50(12):3062–3080, December 2004.

110. J. N. Laneman, E. Martinian, G. W. Wornell, and J. G. Apostolopoulos. Source-channel diversity for parallel channels. *IEEE Transactions on Information Theory*, 51(10):3518–3539, October 2005.

111. K. Lee and D. Williams. A space-frequency transmitter diversity technique for ofdm systems. In *Proceedings of the IEEE Global Telecommunications Conference, GLOBECOM*, volume 3, pp. 1473–1477, 2000.

112. A. Leon-Garcia. *Probability and Random Processes for Electrical Engineering*, 2nd edn. Addison-Wesley, 1993.

113. J. M. Lervik and T. A. Ramstad. Robust image communication using subband coding and multilevel modulation. In *Proceedings of the SPIE Symposium on Visual Communications and Image Processing VCIP*, pp. 524–535, Orlando, Florida, March 1996.

114. W. Li. Overview of fine granularity scalability in MPEG-4 video standard. *IEEE Transactions on Circuits and Systems for Video Technology*, 11(3):301–317, March 2001.

115. F. Li, K. Wu, and A. Lippman. Energy-efficient cooperative routing in multi-hop wireless ad hoc networks. In *Proceedings of the IEEE International Performance, Computing, and Communications Conference*, pp. 215–222, 2006.

116. Y. Li, W. Zhang, and X.-G. Xia. Distributive high-rate full-diversity space-frequency codes achieving full cooperative and multipath diversity for asynchronous cooperative communications. In *Proceedings of the IEEE Global Telecommunications Conference, GLOBECOM*, December 2006.

117. W. Liang. Constructing minimum-energy broadcast trees in wireless ad hoc networks. In *Proceedings of the International Symposium on Mobile and Ad Hoc Networking and Computing*, pp. 112–122, June 2002.

118. P. Ligdas and N. Farvadin. Optimizing the transmit power for slow fading channels. *IEEE Transactions on Information Theory*, 46(2):565–576, March 2000.

119. X. Lin, Y. Kwok, and V. Lau. A quantitative comparison of ad hoc routing protocols with and without channel adaptation. *IEEE Transactions on Mobile Computing*, 4(2):111–128, March 2005.

120. R. Lin and A. P. Petropulu. Cooperative transmission for random access wireless networks. In *Proceedings of the International Conference on Acoustics, Speech and Signal Processing*, pp. 19–23, March 2005.

121. Z. Liu, Y. Xin, and G. Giannakis. Space-time-frequency coded OFDM over frequency selective fading channels. *IEEE Transactions on Signal Processing*, 50:2465–2476, October 2002.

122. Z. Liu, Y. Xin, and G. B. Giannakis. Linear constellation precoding for ofdm with maximum multipath diversity and coding gains. *IEEE Transactions on Communications*, 51(3):416–427, March 2003.

123. B. Liu, B. Chen, and R. S. Blum. Exploiting the finite-alphabet property for cooperative relay. In *Proceedings of the IEEE International Conference on Acoustics, Speech and Signal Processing ICASSP*, 3:357–360, 2005.

124. R. M. Loynes. The stability of a queue with non-independent inter-arrival and service times. *Proceedings of the Cambridge Philosophical Society*, pp. 497–520, 1962.

125. B. Lu and X. Wang. Space-time code design in OFDM systems. In *Proceedings of the IEEE Global Telecommunications Conference, GLOBECOM*, pp. 1000–1004, 2000.

126. B. Lu, X. Wang, and K. R. Narayanan. Ldpc-based space-time coded OFDM systems over correlated fading channels: performance analysis and receiver design. *IEEE Transactions on Communications*, 50(1):74–88, January 2002.

127. W. Luo and A. Ephremides. Stability of n interacting queues in random-access systems. *IEEE Transactions on Information Theory*, 45(5):1579–1587, July 1999.

128. J. Luo, R. S. Blum, L. J. Greenstein, L. J. Cimini, and A. M. Haimovich. New approaches for cooperative use of multiple antennas in ad hoc wireless networks. In *Proceedings of the IEEE Vehicular Technology Conference*, volume 4, pp. 2769–2773, September 2004.

129. J. Luo and A. Ephremides. On the throughput, capacity and stability regions of random multiple access. *IEEE Transactions on Information Theory*, 52(6):2593–2607, June 2006.

130. E. Malkamaki and H. Leib. Evaluating the performance of convolutional codes over block fading channels. *IEEE Transactions on Information Theory*, 45(5):1643–1646, July 1999.

131. R. K. Mallik. On multivariate Rayleigh and exponential distributions. *IEEE Transactions on Information Theory*, 49(6):1499–1515, June 2003.

132. R. J. Marks, A. K. Das, and M. El-Sharkawi. Maximizing lifetime in an energy constrained wireless sensor array using team optimization of cooperating systems. In *Proceedings of the International Joint Conference on Neural Networks*, pp. 371–376, 2002.

133. I. Maric and R. D. Yates. Cooperative multicast for maximum network lifetime. *IEEE Journal on Selected Areas in Communications*, 23:127–135, January 2005.

134. J. W. Modestino and D. G. Daut. Combined source-channel coding of images. *IEEE Transactions on Communications*, 27(11):1644–1659, November 1979.

135. A. F. Molisch, M. Z. Win, and J. H. Winters. Space-time-frequency (STF) coding for MIMO-OFDM systems. *IEEE Communications Letters*, 9:370—372, September 2002.

136. P. Nain. Analysis of a two node aloha network with infinite capacity buffers. In *Proceedings of the International Seminar on Computer Networking, Performance Evaluation*, p. 2.2.1-2.2.16, 1985.

137. S. Nanda, D. J. Goodman, and U. Timor. Performance of PRMA: a packet voice protocol for cellular systems. *IEEE Transactions on Vehicular Technology*, 40:585–598, August 1991.

138. V. Naware, G. Mergen, and L. Tong. Stability and delay of finite-user slotted aloha with multipacket reception. *IEEE Transactions on Information Theory*, 51(7):2636–2656, July 2005.

139. A. Ortega and K. Ramchandran. Rate-distortion methods for image and video compression. *IEEE Signal Processing Magazine*, pp. 23–50, November 1998.

140. L. Ozarov. On a source coding problem with two channels and three receivers. *Bell Systems Technical Journal*, 59(10):1909–1921, December 1980.

141. L. H. Ozarow, S. Shamai, and A. D. Wyner. Information theoretic considerations for cellular mobile radio. *IEEE Transactions on Information Theory*, 43(2):359–378, May 1994.

142. A. Ozgur, O. Leveque, and D. N. C. Tse. Hierarchical cooperation achieves optimal capacity scaling in ad hoc networks. *IEEE Transactions on Information Theory*, 53(10):3549–3572, October 2007.

143. C. Pandana, W. P. Siriwongpairat, T. Himsoon, and K. J. R. Liu. Distributed cooperative routing algorithms for maximizing network lifetime. In *Proceedings of the IEEE Wireless Communications and Networking Conference (WCNC)*, volume 1, pp. 451–456, 2006.

144. A. J. Paulraj, D. A. Gore, R. U. Nabar, and H. Bolcskei. An overview of MIMO communications: a key to gigabit wireless. *Proceedings of the IEEE*, 92(2):198–218, February 2004.

145. R. F. Pawula. Generic error probabilities. *IEEE Transactions on Communications*, 47(5):697–702, May 1999.

146. J. G. Proakis. *Digital Communications*. McGraw-Hill Inc., 2001.

147. H. Qi and R. Wyrwas. Performance analysis of joint voice-data PRMA over random packet error channels. *IEEE Transactions on Vehicular Technology*, 45(2):332–345, May 1996.

148. T. A. Ramstad. Signal processing for multimedia – robust image and video communication for mobile multimedia. In *NATO Science Series: Computers & Systems Sciences*, volume 174, pp. 71–90. IOS Press, 1999.

149. R. Rao and A. Ephremides. On the stability of interacting queues in a multi-access system. *IEEE Transactions on Information Theory*, 34:918–930, September 1988.

150. T. S. Rappaport. *Wireless Communications, Principles and Practice*. Prentice Hall PTR, 2002.

151. A. Reznik, S. R. Kulkarni, and S. Verdu. Degraded Gaussian multirelay channel: capacity and optimal power allocation. *IEEE Transactions on Information Theory*, 50(12):3037–3046, December 2004.

152. A. Robeiro, X. Cai, and G. B. Giannakis. Symbol error probabilities for general cooperative links. *IEEE Transactions on Wireless Communications*, 52(10):1820–1830, October 2004.

153. A. K. Sadek, Z. Han, and K. J. R. Liu. Distributed relay-assignment protocols for coverage extension in cooperative wireless networks. Submitted to *IEEE Transactions on Mobile Computing Communications*.

154. A. K. Sadek, W. Yu, and K. J. R. Liu. On the energy efficiency of cooperative communications in wireless sensor networks. Submitted to *ACM Transactions on Sensor Networks*.

155. A. K. Sadek, Z. Han, and K. J. R. Liu. Multi-node cooperative resource allocation to improve coverage area in wireless networks. In *Proceedings of the IEEE Global Telecommunications Conference, GLOBECOM*, pp. 3058–3062, 2005.

156. A. K. Sadek, W. Su, and K. J. R. Liu. A class of cooperative communication protocols for multi-node wireless networks. In *Proceedings of the IEEE International Workshop on Signal Processing Advances in Wireless Communications SPAWC*, pp. 560–564, June 2005.

157. A. K. Sadek, Z. Han, and K. J. R. Liu. A distributed relay-assignment algorithm for cooperative communications in wireless networks. In *Proceedings of the IEEE International Conference on Communications (ICC)*, Istanbul, 4:1592–1597, June 2006.

158. A. K. Sadek, Z. Han, and K. J. R. Liu. An efficient cooperation protocol to extend coverage area in cellular networks. In *Proceedings of 2006 Wireless Communications and Networking Conference (WCNC)*, pp. 1687–1692, March 2006.

159. A. K. Sadek, K. J. R. Liu, and A. Ephremides. Collaborative multiple access for wireless networks: Protocols design and stability analysis. In *Proceedings of IEEE CISS*, Princeton, New Jersey, pp. 1224–1229, March 2006.

160. A. K. Sadek, K. J. R. Liu, and A. Ephremides. Collaborative multiple access protocols for wireless networks. In *Proceedings of the IEEE International Conference on Communications (ICC)*, Istanbul, pp. 4495–4500, June 2006.

161. A. K. Sadek, W. Yu, and K. J. R. Liu. When does cooperationn have better performance in sensor networks? In *Proceedings of the IEEE International Conference on Sensor, Mesh, and Ad Hoc Communications and Networks (SECON)*, Reston, Virginia pp. 188–197, September 2006.

162. A. K. Sadek, K. J. R. Liu, and A. Ephremides. Cognitive multiple access via cooperation: protocol design and stability analysis. *IEEE Transactions on Information Theory*, 53(10):3677–3696, October 2007.

163. A. K. Sadek, W. Su, and K. J. R. Liu. Multinode cooperative communications in wireless networks. *IEEE Transactions on Signal Processing*, 55(1):341–355, January 2007.

164. Z. Safar and K. J. R. Liu. Systematic design of space-time trellis codes for diversity and coding advantages. *EURASIP Journal on Applied Signal Processing, Systematic Design of Space-Time Trellis Codes for Diversity and Coding Advantages*, 2002(3):221–235, 2002.

165. Z. Safar and K. J. R. Liu. Systematic space-time trellis code construction for correlated Rayleigh fading channels. *IEEE Transactions on Information Theory*, 50(11):2855–2865, November 2004.

166. Z. Safar, W. Su, and K. J. R. Liu. A fast sphere decoding framework for space-frequency block codes. In *Proceedings of the 2004 International Conference on Communications, ICC'04*, volume 5, pp. 2591–2595, 2004.

167. G. Scutari, S. Barbar, and D. Ludovici. Cooperation diversity in multihop wireless networks using opportunistic driven multiple access. In *Proceedings of the IEEE Workshop on Signal Processing Advances in Wireless Communications (SPAWC '03)*, pp. 170–174, June 2003.

168. K. G. Seddik, A. S. Ibrahim, and K. J. R. Liu. Trans-modulation in wireless relay networks. *IEEE Communications Letters*, 121(3):170–172, March 2008.

169. K. G. Seddik, A. Kwasinski, and K. J. R. Liu. Source-channel diversity for multi-hop and relay channels. Submitted to *IEEE Transactions on Wireless Communications*.

170. K. G. Seddik and K. J. R. Liu. Distributed space-frequency coding over broadband relay channels. To appear in *IEEE Transactions on Wireless Communications*.

171. K. G. Seddik, A. K. Sadek, A. S. Ibrahim, and K. J. R. Liu. Design criteria and performance analysis for distributed space-time coding. *IEEE Transactions on Vehicular Technology*, 57(4):2280–2292, July 2008.

172. K. G. Seddik, A. K. Sadek, and K. J. R. Liu. Protocol-aware design criteria and performance analysis for distributed space-time coding. In *Proceedings of the IEEE Global Telecommunications Conference, GLOBECOM*, December 2006.

173. K. G. Seddik, A. Kwasinski, and K. J. R. Liu. Distortion exponents for different source-channel diversity achieving schemes over multi-hop channels. In *Proceedings of the IEEE International Conference on Communications (ICC)*, Glasgow, Scotland, June 2007.

174. K. G. Seddik and K. J. R. Liu. Distributed space-frequency coding over relay channels. In *Proceedings of the IEEE Global Telecommunications Conference, GLOBECOM*, pp. 3524–3528, November 2007.

175. K. G. Seddik, A. K. Sadek, A. Ibrahim, and K. J. R. Liu. Synchronization-aware distributed space-time codes in wireless relay networks. In *Proceedings of the IEEE Global Telecommunications Conference, GLOBECOM*, pp. 3452–3456, November 2007.

176. K. G. Seddik, A. K. Sadek, W. Su, and K. J. R. Liu. Outage analysis and optimal power allocation for multi-node relay networks. *IEEE Signal Processing Letters*, 14:377—380, June 2007.

177. K. Seddik, A. Kwasinski, and K. J. R. Liu. Asymptotic distortion performance of source-channel diversity schemes over relay channels. In *Proceedings of the IEEE Wireless Communications and Networking Conference (WCNC)*, Las Vegas, Nevada, 2008.

178. K. G. Seddik and K. J. R. Liu. Distributed space-frequency coding over amplify-and-forward relay channels. In *Proceedings of the 2008 IEEE Wireless Communications and Networking Conference (WCNC 08)*, 2008.

179. A. Sendonaris, E. Erkip, and B. Aazhang. User cooperation diversity, part I: System description. *IEEE Transactions on Communications*, 51(11):1927–1938, November 2003.

180. A. Sendonaris, E. Erkip, and B. Aazhang. User cooperation diversity, part II: implementation aspects and performance analysis. *IEEE Transactions on Communications*, 51(11):1939–1948, November 2003.

181. C. E. Shannon. A mathematical theory of communication. *Bell Systems Technical Journal*, 27:379–423, 1948.

182. C. E. Shannon. Coding theorems for a discrete source with a fidelity criterion. *IRE International Convertion Record*, pages 142–163, March 1959.

183. Y. Shoham and A. Gersho. Efficient bit allocation for an arbitrary set of quantizers. *IEEE Transactions on Acoustics, Speech, Signal Processing*, 36(9):1445–1453, September 1988.

184. A. Shokrollahi, B. Hassibi, B. M. Hochwald, and W. Sweldens. Representation theory for high-rate multiple-antenna code design. *IEEE Transactions on Information Theory*, 47:2335–2367, September 2001.

185. M. Sidi and A. Segall. Two interfering queues in packet-radio networks. *IEEE Transactions of Communications*, 31(1):123–129, January 1983.

186. M. Sikora, J. N. Laneman, M. Haenggi, D. J. Costello, and T. E. Fuja. Bandwidth- and power-efficient routing in linear wireless networks. *IEEE Transactions on Information Theory*, 52:2624–2633, June 2006.

187. M. K. Simon and M.-S. Alouini. A unified approach to the performance analysis of digital communication over generalized fading channels. *Proceedings of the IEEE*, 86(9):1860–1877, September 1998.

188. M. K. Simon and M.-S. Alouini. A unified approach to the probability of error for noncoherent and differentially coherent modulations over generalized fading channels. *IEEE Transactions on Communications*, 46(12):1625–1638, December 1998.

189. M. K. Simon and M.-S. Alouini. *Digital Communication over Fading Channels: A Unified Approach to Performance Analysis*. John Wiley & Sons, 2000.

190. S. Singh, M. Woo, and C. Raghavendra. Power-aware routing in mobile as hoc networks. In *ACM Mobicom*, pp. 181–190, 1998.

191. W. P. Siriwongpairat, A. K. Sadek, and K. J. R. Liu. Cooperative communications protocol for multiuser OFDM networks. *IEEE Transactions on Wireless Communications*, 7(7):2430–2435, July, 2008.

192. W. P. Siriwongpairat, A. K. Sadek, and K. J. R. Liu. Bandwidth-efficient cooperative protocol for OFDM wireless networks. In *Proceedings of the IEEE Global Telecommunications Conference, GLOBECOM*, December 2006.

193. B. Sirkeci-Mergen and A. Scaglione. Randomized distributed space-time coding for cooperative communication in self-organized networks. In *Proceedings of the IEEE Workshop on Signal Processing Advances for Wireless Communications (SPAWC)*, pp. 500–504, June 2005.

194. M. Srinivasan and R. Chellappa. Adaptive source-channel subband video coding for wireless channels. *IEEE Transactions on Information Theory*, 16(9):1830–1839, December 1998.

195. H. Stark and J. W. Woods. *Probability and Random Processes with Applications to Signal Processing*. Prentice Hall, 2002.

196. G. Stuber. *Principles of Mobile Communication*. Kluwer, 2001.

197. K. Stuhlmüller, N. Färber, M. Link, and B. Girod. Analysis of video transmission over lossy channels. *IEEE Journal on Selected Areas in Communications*, 18(6):1012–1032, June 2000.

198. W. Su, Z. Safar, and K. J. R. Liu. Space-time signal design for time-correlated Rayleigh fading channels. In *Proceedings of the 2003 International Conference on Communications, ICC'03*, volume 5, pp. 3175–3179, 2003.

199. W. Su, Z. Safar, M. Olfat, and K. J. R. Liu. Obtaining full-diversity space-frequency codes from space-time codes via mapping. *IEEE Transactions on Signal Processing*, 51:2905–2916, November 2003.

200. W. Su and X.-G. Xia. On space-time block codes from complex orthogonal designs. *Wireless Personal Communications*, 25(1):1–26, April 2003.

201. W. Su, Z. Safar, and K. J. R. Liu. Diversity analysis of space-time modulation for time-correlated Rayleigh fading channels. *IEEE Transactions on Information Theory*, 50(8):1832–1839, August 2004.

202. W. Su and X.-G. Xia. Signal constellations for quasi-orthogonal space-time block codes with full diversity. *IEEE Transactions on Information Theory*, 50(10):2331–2347, October 2004.

203. W. Su, X.-G. Xia, and K. J. R. Liu. A systematic design of high-rate complex orthogonal space-time block codes. *IEEE Communications Letters*, 8(6):380–382, June 2004.

204. W. Su, A. K. Sadek, and K. J. R. Liu. SER performance analysis and optimum power allocation for decode-and-forward cooperation protocol in wireless networks. In *Proceedings of 2005 Wireless Communications and Networking Conference*, volume 2, 984–989, March 2005.

205. W. Su, Z. Safar, and K. J. R. Liu. Full-rate full-diversity space-frequency codes with optimum coding advantage. *IEEE Transactions on Information Theory*, 51(1):229–249, January 2005.

206. W. Su, Z. Safar, and K. J. R. Liu. Towards maximum achievable diversity in space, time and frequency: performance analysis and code design. *IEEE Transactions on Wireless Communications*, 4:1847—1857, July 2005.

207. W. Su, A. K. Sadek, and K. J. R. Liu. Cooperative communications in wireless networks: performance analysis and optimum power allocation. *Wireless Personal Communications*, 44:181–217, January 2008.

208. L. Sun and E. C. Ifeachor. Voice quality prediction models and their application in VoIP networks. *IEEE Transactions on Multimedia*, 8(4):809–820, August 2006.

209. W. Szpankowski. Stability conditions for some multiqueue distributed systems: buffered random access systems. *Advanced Applied Probability*, 26:498–515, 1994.

210. O. Y. Takeshita and D. J. Constello Jr. New classes of algebraic interleavers for turbo-codes. In *Proceedings of the IEEE International Symposium on Information Theory*, p. 419, 1998.

211. V. Tarokh, H. Jafarkhani, and A. R. Calderbank. Space-time block coding for wireless communications: performance results. *IEEE Journal on Select Areas in Communications*, 17(3):451–460, March 1999.

212. V. Tarokh, N. Seshadri, and A. R. Calderbank. Space-time codes for high data rate wireless communication: performance criterion and code construction. *IEEE Transactions on Information Theory*, 44(2):744–765, March 1998.

213. V. Tarokh, N. Seshadri, and A. R. Calderbank. Space-time block codes from orthogonal designs. *IEEE Transactions on Information Theory*, 45(5):1456–1467, July 1999.

214. E. Telatar. Capacity of multi-antenna Gaussian channels. *AT&T Bell Labs, Technical Report*, June 1995.

215. E. Telatar. Capacity of multi-antenna Gaussian channels. *European Transactions on Telecommunications*, 10(6):585–595, November/December 1999.

216. O. Tirkkonen, A. Boariu, and A. Hottinen. Minimal non-orthogonality rate 1 space-time block code for 3+ tx antennas. In *Proceedings of the IEEE 6th International Symposium on Spread-Spectrum Technology & Appl. (ISSSTA 2000)*, pp. 429–432, September 2000.

217. O. Tirkkonen and A. Hottinen. Square-matrix embeddable space-time block codes for complex signal constellations. *IEEE Transactions on Information Theory*, 48(2):384–395, February 2002.

218. M. K. Tsatsanis, R. Zhang, and S. Banerjee. Network-assisted diversity for random access wireless networks. *IEEE Transactions on Signal Processing*, 48:702–711, March 2000.

219. D. Tse and P. Viswanath. *Fundamentals of Wireless Communications*. Cambridge University Press, 2005.

220. B. Tsybakov and W. Mikhailov. Ergodicity of slotted aloha system. *Problemy Peredachi Informatsii*, 15:73–87, 1997.

221. A. Uvliden, S. Bruhn, and R. Hagen. Adaptive multi-rate: a speech service adapted to cellular radio network quality. In *Proceedings of the 32nd Asilomar Conference on Signals, Systems & Computers*, volume 1, pp. 343–347, 1998.

222. M. C. Valenti and B. Zhao. Distributed turbo codes: towards the capacity of the relay channel. In *Proceedings of the IEEE Vehicular Technology Conference, Fall 2003*, pp. 322–326, October 2003.

223. H. L. Van Trees. *Detection, Estimation, and Modulation Theory-Part (I)*. Wiley, 1968.

224. E. C. van der Meulen. Three-terminal communication channels. *Advances in Applied Probability*, 3:120–154, 1971.

225. E. C. van der Meulen. A survey of multi-way channels in information theory: 1961–1976. *IEEE Transactions on Information Theory*, 23(1):1–37, January 1977.

226. H. S. Wang and P.-C. Chang. On verifying the first-order Markovian assumption for a Rayleigh fading channel model. *IEEE Transactions on Vehicular Technology*, 45(2):353–357, 1996.

227. P. H. Westerink, J. Biemond, and D. E. Boekee. An optimal bit allocation algorithm for sub-band coding. In *Proceedings of the IEEE International Conference on Acoustics, Speech and Signal Processing, ICASSP'88*, pp. 757–760, 1988.

228. S. Wicker. *Error Control Systems for Digital Communication and Storage*. Prentice Hall, 1995.

229. G. Wu, K. Mukumoto, and A. Fukuda. Analysis of an integrated voice and data transmission system using packet reservation multiple access. *IEEE Transactions on Vehicular Technology*, 43(2):289–297, May 1994.

230. Z. Xiong, A. D. Liveris, and S. Cheng. Distributed source coding for sensor networks. *IEEE Signal Processing Magazine*, 21:80–94, September 2004.

231. Y. Xu, J. Heidemann, and D. Estrin. Geography-informed energy conservation for ad hoc routing. In *Proceedings of ACM Mobicom*, pp. 70–84, 2001.

232. Q. Yan and R. Blum. Optimum space-time convolutional codes. In *Proceedings of the IEEE Wireless Communications and Networking Conference (WCNC)*, volume 3, pp. 1351–1355, 2000.

233. S. Yatawatta and A. P. Petropulu. A multiuser OFDM system with user cooperation. In *Proceedings of the 38th Asilomar Conference on Signals, Systems & Computers*, volume 1, pp. 319–323, 2004.

234. M. Younis, M. Youssef, and K. Arisha. Energy-aware management for cluster-based sensor networks. *Journal of Computer Networks*, 45:539–694, December 2003.

235. H. Zhang and J. Hou. On deriving the upper-bound of lifetime for large sensor networks. In *Proceedings of ACM MobiHoc*, pp. 121–132, 2004.

236. B. Zhang and H. T. Mouftah. Qos routing for wireless ad hoc networks: problems, algorithms, and protocols. *IEEE Communications Magazine*, 43:110–117, October 2005.

237. B. Zhao and M. C. Valenti. Practical relay networks: a generalization of hybrid-arq. *IEEE Journal on Selected Areas in Communications*, 23(1):7–18, January 2005.

238. K. Zeger and A. Gersho. Pseudo-gray coding. *IEEE Transactions on Communications*, 38(12):2147–2158, December 1990.

239. L. Zheng and D. N. C. Tse. Diversity and multiplexing: a fundamental tradeoff in multiple antenna channels. *IEEE Transactions on Information Theory*, 49:1073–1096, May 2003.

240. M. Zorzi and R. R. Rao. Geographic random forwarding (geraf) for ad hoc and sensor networks: multihop performance. *IEEE Transactions on Mobile Computing*, 2(4):337–348, Oct.-Dec. 2003.

Index